Lecture Notes in Mathematics

Edited by A. Dold and B. Eckmann

732

Algebraic Geometry

Summer Meeting, Copenhagen,
August 7–12, 1978

Edited by K. Lønsted

Springer-Verlag
Berlin Heidelberg New York 1979

Editor

Knud Lønsted
Københavns Universitets
Matematiske Institut
Universitetsparken 5
DK-2100 København Ø

AMS Subject Classifications (1970): 14-XX

ISBN 3-540-09527-6 Springer-Verlag Berlin Heidelberg New York
ISBN 0-387-09527-6 Springer-Verlag New York Heidelberg Berlin

Library of Congress Cataloging in Publication Data
Copenhagen Summer Meeting in Algebraic Geometry, 1978.
Algebraic geometry.
(Lecture notes in mathematics ; 732)
Bibliography: p.
Includes index.
1. Geometry, Algebraic--Congresses. I. Lønsted, Knud, 1942- II. Title. III. Series:
Lecture notes in mathematics (Berlin) ; 732.
QA3.L28 no. 732 [QA564] 510'.8s [512'.33] 79-17367

Printing and binding: Beltz Offsetdruck, Hemsbach/Bergstr.
2141/3140-543210

PREFACE

These Proceedings contain the papers contributed to the Copenhagen Summer Meeting in Algebraic Geometry 1978, held August 7-12, 1978, at the H. C. Ørsted Institut of the University of Copenhagen. In addition to the papers presented at lectures this volume also includs a few ones by other participants, as well as a couple by some who were prevented from participating.

It is the editors hope that the present volume will offer a representative coverage of the actual activities in algebraic GEOMETRY, and that it may be found useful to workers in this field.

The meeting was finanically supported by the Danish Natural Science Research Council under grant no. 511-10092. Matematisk Institut and the H. C. Ørsted Institut offered hospitality and practical help. On the behalf of the organizers, who also included H.-B. Foxby, A. Thorup and N. Yui, I should like to express our gratitude for this, and also to all the participants for their collaboration, which made the organization much easier than expected. Special thanks go to Mrs. A. Tarnov and to Ms. Ulla Jacobsen, whose assistance during critical moments of the meeting and at the final preparation of the manuscript proved to be invaluable.

Knud Lønsted

LIST OF PARTICIPANTS

M. ARTIN
H. BASS
J.F. BOUTOT
R.O. BUCHWEITZ
F. CATANESE
P. CHERENACK
N. CHIARLI
A. COLLINO
D. CORAY
D. COX
M. DESCHAMPS
G. ELLINGSRUD
F. ELZEIN
H. ESNAULT
D. FERRAND
R. FOSSUM
H.B. FOXBY
S. GRECO
M. HAZEWINKEL
A. HOLME
J. HUBBARD
C. HØSTMAEHLINGEN
B. IVERSEN
J.P. JOUANOLOU
J.M. KANTOR
M. KATO
T. KATSURA
Y. KAWAMATA
G. KEMPF
H. KLEPPE
J. KLEPPE
A. KURIBAYASHI
D. LAKSOV
O.A. LAUDAL
E. LLUIS
S. LUBKIN
A. LUBOTZKY
K. LØNSTED
A. MAGID
G. MALTSINIOTIS
Z. MEBKHOUT
J.Y. MERINDOL
A. MILHØJ
C. MITSCHI
Y. MIYAOKA
L. NESS
H.A. NIELSEN
N.O. NYGAARD
L.D. OLSON
F. OORT
S. PEDERSEN
U. PERSSON
R. PIENE
H. PINKHAM
H. POPP
M. vdPUT
A. RAMANATHAN
M. REID
F. SAKAI
M. SCHAPS
C.S. SESHADRI
T. SHIODA
J.R. STROOKER
U. STUHLER
S.A. STRØMME
M. TEICHER
A. THORUP
R. TREGER
E. VIEHWEG
S. USUI
VAJNRYB
L. VERMEULEN
C.H. WACHMANN
G.E. WELTERS
N. YUI
S. ZUCKER

CONTENTS

SURFACES WITH $K^2 = p_g = 1$ AND THEIR PERIOD MAPPING.

Fabrizio Catanese - Università di Pisa - Harvard University[+]

Introduction.

Recently a result of Kynef ([14]) drew attention on minimal surfaces S with $K^2 = p_g = 1$: he constructed a quotient of the Fermat sextic in $\mathbb{P}^3$ by a suitable action of $\mathbb{Z}/6$, with these invariants, such that the differential of the period mapping (see [7], [9]) is not injective at it, thus answering negatively a problem posed by Griffiths in [8].

One may remark however that the local Torelli theorem (injectivity of the infinitesimal period mapping) fails, for curves, exactly when one has an hyperelliptic curve ([7]), though the global Torelli theorem holds.

So one is motivated to study these surfaces and their period mapping.

They were first considered by Enriques in 1897, who proved their existence in [5] (see also [6] pag. 305) ; Bombieri ([1], pag. 201) proved rigorously that for these surfaces the tricanonical map is birational.

(+) The author was partly supported by a N.A.T.O.-C.N.R. fellowship during his stay at Harvard University.

Here we prove that the bicanonical map $\Phi = \Phi_{2K}$ is a morphism and that[++)] any such surface is a weighted complete intersection of type (6,6) in the weighted projective space $\mathbb{P}(1,2,2,3,3)$ (see [4], [15] about the theory of weighted complete intersections).

We also show that these surfaces have equations in canonical form: this is a first step towards an explicit description of their moduli space, that we hope to accomplish in the future. Then we describe a geometric construction giving all the "special" surfaces, i.e. those for which Φ is a Galois covering (and it turns out that the Galois group is $\mathbb{Z}/2 + \mathbb{Z}/2$).

Using this explicit description of our surfaces we prove that they are all diffeomorphic and simply connected, and that when K is ample the Kuranishi family is smooth of dimension 18 (as their local period space): our main result is that the differential of the period mapping is invertible outside an hypersurface, so that the period mapping is generally finite.

The 12 dimensional subfamily parametrizing "special" surfaces is strictly contained in the subvariety where the rank of the differential drops by 2 (the maximum possible amount) and we prove, by means of a more general result on deformations of cyclic coverings, that the restriction of the period mapping to this subfamily is locally 1-1 : this suggests that the period mapping might have no positive dimensional fibres, but we have not yet pursued such investigation.

One last remark is that our results on the failure of the local Torelli theorem for weighted complete intersections (w.c.i.) show that the

---

++) The proof which appears here of this result is due to collaboration with Miles Reid.

restrictions put by S. Usui in his work [20] cannot all be eliminated.

I would like here to thank P. Griffiths for suggesting this research and I. Dolgachev for useful conversations.

Notations throughout the paper:

S is a minimal smooth surface with $p_g = K^2 = 1$

$x_\circ \in H^\circ(S, \mathcal{O}(K))$ the unique (up to constants) non zero section

$C = \mathrm{div}(x_\circ)$ the canonical curve

R the graded ring $\mathbb{C}[X_\circ, Y_1, Y_2, Z_3, Z_4]$, where $\deg X_\circ = 1$, $\deg Y_i = 2$, $\deg Z_j = 3$ $(i = 1,2, \quad j = 3,4)$

$W = \mathbb{C}[Y_1, Y_2, Z_3, Z_4]$ as a graded subring of R

R_m, W_m the graded parts of degree m of R, resp. W

$Q = Q(1,2,2,3,3) = \mathrm{proj}(R)$

$R(S) = \sum_{m=o}^{\infty} H^\circ(S, \mathcal{O}(mK))$ the canonical ring of S

$h^i(S,L) = \dim H^i(S,L)$ if L is a coherent sheaf on S

§ 1. STRUCTURE OF SURFACES S WITH $K^2 = P_g = 1$.

LEMMA 1. $P_m = h^{\circ}(S, \mathcal{O}(mK)) = \frac{1}{2} m(m-1) + 2.$

Proof. $P_m = \frac{1}{2} m(m-1)K^2 + \chi(\mathcal{O}_S)$ (see [1] pag. 185, or [11]) and by Theorems 11,15 of [1] $q=0$ and S has no torsion, so $\chi = 2$.

One can choose therefore y_1, y_2, z_3, z_4 such that x_o^2, y_1, y_2 are a basis of $H^{\circ}(S, \mathcal{O}(2K))$, and $x_o^3, x_o y_1, x_o y_2, z_3, z_4$ are a basis of $H^{\circ}(S, \mathcal{O}(3K))$.

Write now $C = \mathrm{div}(x_o) = \Gamma + Z$, where $K \cdot \Gamma = 1$, $K \cdot Z = 0$.

LEMMA 2. If $D \in |2K|$ and $D \geq \Gamma$, $D = 2C$.

Proof. Write $D = D' + \Gamma$ and let D'' be the movable part of $|D'|$: $D'' \cdot K = 1$, so by the index theorem either D'' is homologous (hence linearly equivalent, as S has no torsion) to K, or $D''^2 \leq -1$, hence in both cases $h^{\circ}(S, \mathcal{O}(D'')) = 1$.

COROLLARY 3. $H^{\circ}(S, \mathcal{O}(4K)) = x_o \cdot H^{\circ}(S, \mathcal{O}(3K)) \oplus (\mathbb{C}y_1^2 \oplus \mathbb{C}y_1 y_2 \oplus \mathbb{C}y_2^2)$.

Proof. Because $P_4 = 8$ it is enough to prove that the two vector subspaces have no common line. Supposing the contrary, there would exist a section $s \in H^{\circ}(\mathcal{O}(3K))$, and costants $\lambda_1, \mu_1, \lambda_2, \mu_2$ such that $x_o \cdot s = (\lambda_1 y_1 + \mu_1 y_2)(\lambda_2 y_1 + \mu_2 y_2)$.

Taking the associated divisors $C+\mathrm{div}(s)=D_1+D_2$, where $D_i \in$ $\in |2K|$, and one of the D_i, say D_1, is $\geq \Gamma$ therefore. By lemma 2 $D_1=2C$, hence $\lambda_1 y_1+\mu_1 y_2=cx_o^2$ for a suitable $c \in \mathbb{C}$, contradicting the independence of x_o^2, y_1, y_2.

THEOREM 1. $|2K|$ has no base points, so that $\Phi = \Phi_{2K}: S \longrightarrow \mathbb{P}^2$ is a morphism of degree 4.

Proof. If b were a base point of $|2K|$, then x_o, y_1, y_2 would vanish at b ; by coroll. 3 b would be a base point of $|4K|$, contradicting Theorem 2 of [1] .

Define an homomorphism $\alpha^*: R \longrightarrow R(S)$ by sending X_o to $x_o, \ldots . Z_4$ to z_4: by theorem 1 α^* induces a morphism $\alpha: S \longrightarrow Q=Q(1,2,2,3,3)$.

Remark that Q is smooth outside the two $\mathbb{P}^1$'s $S_2=\{x_o=z_3=z_4=0\}$, $S_3=\{x_o=y_1=y_2=0\}$, and if $\mathbb{P}=\mathbb{P}(1,2,2,3,3)=$ $=Q-S_2-S_3$, on $\mathbb{P}$ $\mathcal{O}_Q(m)$ is an invertible sheaf for every integer m and $\forall\, a,b \in \mathbb{Z}$ one has an isomorphism $\mathcal{O}_Q(a) \otimes \mathcal{O}_Q(m)^{\otimes b} \longrightarrow \mathcal{O}_Q(a+bm)$ (compare [15] , exp. pages 619-624, and also cf. [4]).

PROPOSITION 4. $|3K|$ has no base points and $\alpha(S) \subset \mathbb{P}$.

Proof. If $\alpha(b) \in S_3$, then b is a base point of $|2K|$, if $\alpha(b) \in S_2$ b is a base point of $|3K|$: in view of theorem 1 we need to show that $|3K|$ has no base points.
If b is a base point of $|3K|$, we have that $b \in \Gamma - Z$: in fact $|3K|$ has no fixed part, but if $b \in E$, E irreducible with $K \cdot E=0$, any section of $\mathcal{O}(3K)$ vanishing on b vanishes

on E too.
Because of the exact sequence

$$0 \to H^{\circ}(\mathcal{O}(2K-\Gamma)) \to H^{\circ}(\mathcal{O}(2K)) \to H^{\circ}(\mathcal{O}(2K)\otimes \mathcal{O}_{\Gamma}) \to$$

of lemma 2, and the fact that $\Gamma \cdot 2K=2$, a general curve D of $|2K|$ passing through b is smooth at b, and one can suppose that the component of D to which b belongs is not rational (S being of general type).
It follows by prop. B of [2] that b is not a base point of $\omega_D = \mathcal{O}_D(3K)$: this is a contradiction by the exact sequence

$$0 \to H^{\circ}(\mathcal{O}(K)) \to H^{\circ}(\mathcal{O}(3K)) \to H^{\circ}(\mathcal{O}_D(3K)) \to 0 \quad .$$

Denote by I the ideal ker α^{*}: because dim $R_6=19$, $P_6=\dim R(S)_6=17$, there exist two independent elements $f,g \in I_6$.

PROPOSITION 5. f,g are irreducible and $\alpha(S)=Y=\{f=g=0\}$.

Proof. If f is reducible, by coroll. 3 $f=X_o \cdot f'$, $f' \in R_5$, and $\alpha(S) \subset \{f'=g=0\}$. Denote now by $p:Q \to \mathbb{P}^2$ the rational map given by (X_o^2, Y_1, Y_2) : clearly $\Phi = p \circ \alpha$. Now one gets a contradiction considering that

i) $p : \{f'=g=0\} \to \mathbb{P}^2$ is of degree ≤ 2, because $\{f'=0\}$ is irreducible and the variables Z_3, Z_4 appear at most quadratically in g, and linearly in f'(observe that $(-X_o, Y_1, Y_2, -Z_3, -Z_4) \cong (X_o, Y_1, Y_2, Z_3, Z_4)$).

ii) α is birational because the tricanonical map is such ([1], pag. 202).

iii) Φ is of degree four.

Finally $p : \{f=g=0\} \longrightarrow \mathbb{P}^2$ is of degree four by the same argument of i), hence $Y=\alpha(S)$ and is irreducible.

PROPOSITION 6. The subscheme of $\mathbb{P}$ weighted complete intersection of type (6,6), $Y=\{f=g=0\}$ is isomorphic to the canonical model of S. Therefore I is generated by f,g, and α^* induces an isomorphism $\alpha^*=R'=R/_I \longrightarrow R(S)$.

Proof. $\alpha : S \longrightarrow Y$ is a desingularization such that the pull back of the dualizing sheaf on Y is the canonical bundle K of S (as $\omega_Y \cong \mathcal{O}_Q(1)$ by [15], prop. 3.3.): therefore Y has only rational double points as singularities and is the canonical model of S (cf. [1], [16]).

THEOREM 2. The canonical models of minimal surfaces with $K^2=p_g=1$ correspond to weighted complete intersections Y of type (6,6) in $\mathbb{P}(1,2,2,3,3)$, with at most rational double points as singularities, and two surfaces are isomorphic iff their canonical models are projectively equivalent in $\mathbb{P}$.

Proof. If Y is as above, $\mathcal{O}_Y(1)$ is the canonical sheaf and by prop. 3.2. of [15] $\mathcal{O}_Y(1)^2=1$; again by prop. 3.3 of [15] R'_m is isomorphic to $H^\circ(Y,\mathcal{O}_Y(m))$, so our first assertion follows immediately.

Note that an isomorphism of two surfaces gives isomorphisms of the vector spaces $H^\circ(\mathcal{O}(mK))$, so the second statement is obvious after we describe the projective group of $\mathbb{P}$: it con sists of the invertible transformations of the following form

$$X_o \longrightarrow dX_o$$

$$Y_i \longrightarrow d_{i1}Y_1 + d_{i2}Y_2 + d_{i0}X_o^2 \qquad i=1,2$$

$$Z_j \longrightarrow C_{j3}Z_3+C_{j4}Z_4+C_{j0}X_o^3 + C_{j1}X_oY_1 + C_{j2}X_oY_2 .$$

PROPOSITION 6. There exists a projective change of coordinates such that Y is defined by 2 equations in canonical form

$$f = Z_3^2+X_oZ_4(a_oX_o^2+a_1Y_1+a_2Y_2) + F_3(X_o^2,Y_1,Y_2)$$

$$g = Z_4^2 + X_oZ_3(b_oX_o^2+b_1Y_1+b_2Y_2) + G_3(X_o^2,Y_1,Y_2)$$

where F_3, G_3 are cubic forms.

<u>Proof</u>. Write $f = Q_1(Z_3,Z_4)+\ldots\ldots$(terms of deg ≤ 1 in the Z_j)

$$g = Q_2(Z_3,Z_4)+\ldots .$$

I claim that the quadratic forms Q_1, Q_2 are not proportional: otherwise, by taking a linear combination of the 2 equations one would have $Q_2=0$, but then $p:Y \to \mathbb{P}^2$ would have degree 2 and not 4.

By a transformation $Z_j \to C_{j3}Z_3+C_{j4}Z_4$ one can suppose $Q_1=Z_3^2$, $Q_2=Z_4^2$: this is immediate if both Q_1,Q_2 have rank 1, while if, say, Q_1 has rank 2, one proceeds as follows.

First take coordinates such that $Q_1=Z_3 \cdot Z_4$, then, subtracting to g a multiple of f one can get $Q_2=m_3Z_3^2+m_4Z_4^2$.

If m_3 and m_4 are $\neq 0$, one takes first new variables

$\sqrt{m_3}\,Z_3$, $\sqrt{m_4}Z_4$, so that for $f/\sqrt{m_3m_4}$, g, Q_1, Q_2 have now the form Z_3Z_4 , $Z_3^2+Z_4^2$: then one takes variables Z_3', Z_4' with $Z_3=Z_3'-Z_4'$, $Z_4=Z_3'+Z_4'$ so $Q_1 = Z_3'^2-Z_4'^2$, $Q_2=2(Z_3'^2+Z_4'^2)$, and finally $\frac{2Q_1+Q_2}{4}$, $\frac{Q_2-2Q_1}{4}$ are in the desired form.

If, say m_4 is zero, one can suppose $Q_2=Z_3^2$: but we have a contradiction because then the point $(0,0,0,0,1)$ satisfies $f=g=0$, against the fact that $Y\subset \mathbb{P}$.

Finally, if now $f=Z_3^2+X_oZ_3(a_o'X_o^2+a_1'Y_1+a_2'Y_2)+\ldots$ one kills the a_j' by completing the square, i.e. by taking $Z_3+\frac{1}{2}X_o(a_o'X_o^2+a_1'Y_1+a_2'Y_2)$ as new Z_3 coordinate, and analogously one then does for g acting on the Z_4 variable.

REMARK 7. If Y and Y' are defined by two canonical forms, they are isomorphic iff the canonical equations are equivalent under the projective subgroup

$$\begin{pmatrix} d & d_{10} & d_{20} & 0 & 0 \\ 0 & d_{11} & d_{21} & 0 & 0 \\ 0 & d_{12} & d_{22} & 0 & 0 \\ 0 & 0 & 0 & c_{33} & 0 \\ 0 & 0 & 0 & 0 & c_{44} \end{pmatrix}$$

§ 2. A GEOMETRIC CONSTRUCTION OF THE "SPECIAL" SURFACES (Φ A GALOIS COVERING).

Consider $\mathbb{P}^2$ with coordinates (Y_o, Y_1, Y_2), denote by ℓ the line $\{Y_o=0\}$ and choose a reducible sextic curve F+G. In $\mathcal{O}_{\mathbb{P}^2}(3)$ take the double covering X of $\mathbb{P}^2$ branched along F+G, and let F+G have neither multiple components nor singular points of multiplicity ≥ 4, or of type (3,3): then (see e.g. [10], pag. 47-50) X has only rational double points as singularities, and its minimal resolution $\tilde{X}$ is a K^3 surface (in fact if $p:\tilde{X} \longrightarrow \mathbb{P}^2$ is the double cover, $K_{\tilde{X}} = p^*(\mathcal{O}_{\mathbb{P}^2}(3)+K_{\mathbb{P}^2})=0)$.

Suppose deg F = deg G = 3 and denote by $L=p^*(\ell)$, by $E_1,\ldots.E_p$ the rational curves with self-intersection -2 coming from the resolution, by F',G' the strict transforms of F,G.

One has $p^*(\mathcal{O}_{\mathbb{P}^2}(3)) \equiv 2F'+\sum r_iE_i$ (for some positive integers r_i)$\equiv$ $\equiv 2G'+\sum s_iE_i$. If we set $\mathcal{L} \doteq p^*(\mathcal{O}_{\mathbb{P}^2}(2))\otimes\mathcal{O}_{\tilde{X}}(-F'-\sum\left[\frac{r_i}{2}\right]E_i)$, and $J\subset\{1,\ldots p\}$ is the subset of indexes J for which r_j is odd, one can easily check that $L+\sum E_j\equiv 2\mathcal{L}$; therefore one takes a double cover $\tilde{S}$ of $\tilde{X}$ in $\mathcal{L}$, ramified over $L+\sum E_j$.

The E_j's become exceptional of the first kind in $\tilde{S}$, and after blowing them down I get a surface $S \xrightarrow{\Phi} \mathbb{P}^2$, for which $\Phi^*(\mathcal{O}_{\mathbb{P}^2}(1))\equiv 2K_s$; clearly then $K^2=1$, and $p_g=1$ because $H^{\circ}(\tilde{S},K_{\tilde{S}}) \cong H^{\circ}(\tilde{X},\mathcal{L}) + H^{\circ}(\tilde{X},\mathcal{O}_{\tilde{X}})$.

REMARK 8. If one chooses F of degree 5, G of degree 1, $H^{\circ}(\tilde{X},\mathcal{L})$ becomes 1-dimensional, so that one gets thus a surface S with $K^2=1$, $p_g=2$.

DEFINITION 9. A surface S is called "special" if it is obtained in the above described way.

PROPOSITION 10. A special S is simply connected and is a Galois covering of $\mathbb{P}^2$ with group $\mathbb{Z}/2 + \mathbb{Z}/2$. The canonical model of S in $\mathbb{P}(1,2,2,3,3)$ has equations

$$\begin{cases} z_3^2 = F(x_o^2, y_1, y_2) \\ z_4^2 = G(x_o^2, y_1, y_2) \end{cases}$$

F,G being the equations of the 2 cubics.

<u>Proof</u>. For any S denote by R the ramification divisor of Φ, and by $B=\Phi(R)$. $\pi_1(S)$ is a quotient of $\pi_1(S-R)$, a subgroup of index 4 in $\pi_1(\mathbb{P}^2-B)$.
If S is special $B=\ell+F+G$; if then F,G are smooth, the mutual intersections of ℓ,F,G are transversal, $\pi_1(\mathbb{P}^2-B)$ is abelian ([21]).
But S has no torsion, $q=0$ ([1] theorems 11 and 15), so $\pi_1(S)=0$. Take now $\zeta_i=(Y_j/Y_i)^3\zeta_j$ as a fibre coordinate in $\mathcal{O}_{\mathbb{P}^2}(3)$ (X is defined by $\zeta_i^2 = F\cdot G\cdot Y_i^{-6}$), and $\xi_i = (Y_j/Y_i)^2 \frac{\zeta_{j3}}{\zeta_{i3}}\xi_j$ as a fibre coordinate in $\mathcal{L}$, where

ζ_{i3} is a local coordinate for F', ζ_{i4} for G'. (So that $\zeta_i = \zeta_{i3} \cdot \zeta_{i4}$).

If X_o is the section corresponding to (ζ_i), Z_h to (ζ_{ih}), after easy manipulations one obtains the desired equations.

THEOREM 3. The special surfaces are exactly those for which Φ is a Galois covering, and they form a 12-dimensional family.

Proof. If $\Phi : S \rightarrow \mathbb{P}^2$ is Galois, let G be the Galois group (a priori it can be $\mathbb{Z}/_4$ or $\mathbb{Z}/_2 + \mathbb{Z}/_2$, but our proof will imply that the first case does not occur). Consider that if $C = \Gamma + Z$ is, as usual, the canonical curve, Φ makes Γ a double cover of a line; hence the image of G in $Aut(\Gamma)$ has order two, so exists an involution $\sigma \in G$ (i.e. σ^2=Identity) leaving Γ pointwise fixed, and σ is biregular, S being of general type.

We will first give a proof in the case when K is ample (so $\Gamma = C$). The proof of the following lemma is elementary and well known.

LEMMA 11. Let V be an n-dimensional manifold, σ a biregular involution, F the set of fixed points of σ.

Then $F = \bigcup_{r=0}^{n-1} F_r$, where each F_r is a closed submanifold (possibly empty or disconnected) of dimension r, and if $P \in F_r$ one can choose local coordinates $(z_1, \ldots z_r, z_{r+1}, \ldots, z_n)$ such that $\sigma(z_1, \ldots z_n) = (z_1, \ldots z_r, -z_{r+1}, \ldots -z_n)$.

In case K is ample $C \subset F_1$, but we have equality because if $D \subset F_1$, $C \cdot D \geq 1$, contradicting the smoothness of F_1. Therefore $S/_{\sigma} = X$ is a, possibly singular, K3 surface and $\Phi : S \rightarrow \mathbb{P}^2$ factors as $S \xrightarrow{\pi} X \xrightarrow{p} \mathbb{P}^2$. Because the ramification divisor R of Φ is $\equiv 7K$, by Porteous' formula or an explicit computation that we will see in prop. 12, X is a double cover of $\mathbb{P}^2$ ramified over a sextic curve B': now in $\tilde{X}$ the divisor $\pi(C) + \sum E_j$, the E_j's being the rational (-2) curves coming from the resolution of the isolated fixed points of σ, is divisible by 2 in $\mathrm{Pic}(\tilde{X})$ iff B' is reducible in components of odd degree, and by remark 8 one concludes that B' consists of two cubics.

When K is not ample this proof becomes more cumbersome, so we use a different idea in the general case, using the representation of σ on the vector spaces $H^0(\mathcal{O}(mK))$, hence on $\mathbb{P}(1,2,2,3,3)$.

Clearly $\sigma(x_0) = \pm x_0$, σ is the identity on $H^0(\mathcal{O}(2K))$, and one can choose z_3, z_4 so that they are eigenvectors for σ: therefore one can assume that σ acts on $\mathbb{P}$ by one of the two trasformations $(X_0,Y_1,Y_2,Z_3,Z_4) \rightarrow (-X_0,Y_1,Y_2,Z_3,Z_4)$ or $(-X_0,Y_1,Y_2,-Z_3,Z_4)$.

In the second case the fixed locus of σ on S is contained in $\{x_0=z_3=0\} \cup \{x_0=z_4=0\}$, i.e. a finite set of points ($|3K|$ has no base points): but the whole curve Γ is pointwise fixed, so this case cannot occur.

$I_6 \subset R_6$ being σ invariant, one can assume that f, g are eigenvectors for σ themselves: but the monomials in R_6

are eigenvectors of eigenvalue (-1) if X_o appears with an odd power, and $(+1)$ if X_o appears with even power. f,g not being divisible by X_o, the corresponding eigenvalues are both $+1$, and f,g are sum of monomials where X_o appears only with even power: proceeding as in proposition 6 one can find coordinates where f,g have the form

$$\begin{cases} Z_3^2 - F_3(X_o^2, Y_1, Y_2) = 0 \\ Z_4^2 - G_3(X_o^2, Y_1, Y_2) = 0 \end{cases}$$

The last statement follows from remark 7.

PROPOSITION 12. If B is the branch locus of $\Phi : S \to \mathbb{P}^2$, and the canonical model of S has equations as in prop. 6

$$f = Z_3^2 + X_o Z_4 \alpha_1(X_o^2, Y_1, Y_2) + F_3(X_o^2, Y_1, Y_2)$$

$$g = Z_4^2 + X_o Z_3 \beta_1(X_o^2, Y_1, Y_2) + G_3(X_o^2, Y_1, Y_2)$$

(α_1, β_1 being linear forms)

the equation of B is $Y_o^2(F^2G^2 + \alpha_1^2 Y_o G^3 + F^3\beta_1^2 Y_o - \frac{27}{256}\alpha_1^4 \beta_1^4 \cdot Y_o^4 + \frac{9}{8} Y_o^2 \beta^2 \alpha^2 FG)$. If S is not special $\pi_1(\mathbb{P}^2 - B)$ is not abelian.

<u>Proof</u>. We first write the equation of R, given by the vanishing of the jacobian of f, g, $Y_o = X_o^2, Y_1 Y_2$, easily computed to be $2X_o\left(\frac{\partial f}{\partial Z_3} \cdot \frac{\partial g}{\partial Z_4} - \frac{\partial f}{\partial Z_4} \cdot \frac{\partial g}{\partial Z_3}\right) = 2X_o(4Z_3 \cdot Z_4 - X_o^2 \alpha_1 \beta_1)$.

It is clear that $\Phi_*(X_o=0)$ gives twice the line $Y_o=0$, while, to compute the branch locus when $Y_o \neq 0$ we proceed as follows: given X_o, Y_1, Y_2, f and g can be considered as two conic

equations in the plane of coordinates (Z_3, Z_4) and one must write when two conics intersect in less than 4 points.
Now the pencil of conics $\lambda f+\mu g$ has ≤ 3 base points iff there are not 3 distinct degenerate conics in the pencil, i.e. when the discriminant of the cubic equation in (λ,μ), given by

$$\det \begin{vmatrix} \lambda & 0 & \mu X_o\beta_1/2 \\ 0 & \mu & \lambda X_o\alpha_1/2 \\ \frac{\mu\beta_1 X_o}{2} & \frac{\lambda X_o\alpha_1}{2} & \lambda F+\mu G \end{vmatrix} = 0$$

vanishes: using the expression for the discriminant of a cubic equation one obtains the above written equation for B.
For the second statement, consider that the group of covering transformations of $\Phi : S-R \longrightarrow \mathbb{P}^2-B$ is given by $N/_{\pi_1(S-R)}$, where N is the normalizer of $\pi_1(S-R) \hookrightarrow \pi_1(\mathbb{P}^2-B)$, so that if $\pi_1(S-B)$ is abelian Φ is a Galois covering (because a covering transformation can be extended to a biregular automorphism of S).
One can indeed check that the general B has 6 tacnodes on the line $Y_o=0$.

PROPOSITION 13. All the minimal surfaces with $K^2=p_g=1$ are diffeomorphic. In particular they are all simply connected.

Proof. By proposition 6 there exists a family with connected smooth base containing all the canonical models. By the results of [18] follows that all the nonsingular models are deformation of each other, hence they are all diffeomor-

phic.
We have already proven in prop. 10 that a general special S is simply connected.

REMARK 14. One can easily compute that the surface constructed by Kynef in [14] is a "special" surface corresponding to the following choice of the two cubics: $F=2Y_1^3-(Y_2^3-Y_0^3)$.

$$G=2Y_1^3+(Y_2^3-Y_0^3)$$

F and G have double contact on 3 points lying in the line $Y_1=0$, and these contribute 3 points of type A_5 on the singular K3 surface X, 3 points of type A_2 on the canonical model of S , which however are disjoint from the canonical curve. The (-2) rational curves, the smooth curve C, the two elliptic curves with self-intersection -1 , each covering twice via Φ the line $Y_1=0$, form an interesting configuration on the surface S .

REMARK 15. If S is special, the Galois group being $\mathbb{Z}/_2+\mathbb{Z}/_2$, it is easy to see that there are two more geometric constructions for S: take the double cover of $\mathbb{P}^2$ branched on $\ell + F$, then the double cover branched on the inverse image of G plus some rational (-2) curves (and the same with G in the place of F).

§ 3. THE INFINITESIMAL PERIOD MAPPING.

Consider the Kuranishi family of deformations of S: its tangent space at the point representing S is naturally identified with $H^1(S,T_S)$, T_S denoting the tangent sheaf of S. By [7], the differential of the period mapping $\mu: H^1(S,T_S) \longrightarrow \mathrm{Hom}(H^{\circ}(\Omega_S^2), H^1(\Omega_S^1))$ is obtained via the bilinear mapping $H^1(S,T_S) \times H^{\circ}(\Omega_S^2) \longrightarrow H^1(T_S \otimes \Omega_S^2)$, and the natural isomorphism $\Omega_S^1 \cong T_S \otimes \Omega_S^2$.

The injectivity of μ ("Local Torelli" problem) assumes an easy form when $p_g=1$: it means that if $x_{\circ}$ is the usual non zero section of $H^{\circ}(\Omega_S^2) = H^{\circ}(\mathcal{O}_S(K))$, $H^1(T_S) \xrightarrow{\text{mult by } x_\circ} H^1(\Omega_S^1)$ multiplication by $x_{\circ}$ is injective.

S being of general type $H^{\circ}(T_S)=0$, the morphism μ fits into the exact sequence of cohomology

$$H^{\circ}(\Omega_S^1) \longrightarrow H^{\circ}(\Omega_S^1 \otimes \mathcal{O}_C) \rightarrow H^1(T_S) \xrightarrow{\mu} H^1(\Omega_S^1) \longrightarrow$$
$$\begin{matrix} \| \\ 0 \end{matrix}$$

and here μ is injective iff $h^{\circ}(\Omega_S^1 \otimes \mathcal{O}_C)=0$.

In the rest of the paragraph we will assume that the canonical model of S is smooth, hence isomorphic to S.

PROPOSITION 16. If S is special $\ker \mu$ is 2 dimensional.

Proof. There is an involution σ on S leaving C pointwise fixed and C is smooth, so $\Omega_S^1 \otimes \mathcal{O}_C$ splits into the (+1) and (-1) eigenspaces for σ; then $\Omega_S^1 \otimes \mathcal{O}_C \cong \mathcal{O}_C(-C) \oplus \omega_C$, and C being of genus 2 $h^{\circ}(\Omega_S^1 \otimes \mathcal{O}_C)=h^{\circ}(\omega_C)=2$.

THEOREM 4. Suppose S is a smooth w.c.i. of type (6,6) in $\mathbb{P}$, with canonical equations (where $Y_o=X_o^2$):

$$\begin{cases} f=Z_3^2+X_oZ_4(\sum_{i=0}^{2} a_iY_i) + \sum_{0\le i\le j\le k} f_{ijk}Y_iY_jY_k=0 \\ g=Z_4^2+X_oZ_3(\sum_{i=0}^{2} b_iY_i)+ \sum_{0\le i\le j\le k} g_{ijk}Y_iY_jY_k=0 . \end{cases}$$

Then the Kuranishi family of S is smooth of dimension 18. Moreover there exists a non zero polynomial

$\Delta(a_1,a_2,b_1,b_2,f_{111},f_{112},f_{122},f_{222}, g_{111},g_{112},g_{122},g_{222})$

such that μ is injective iff $\Delta \neq 0$ for S.

Proof. One has the following exact sequences, where $T_{\mathbb{P}}$ is the tangent sheaf to $\mathbb{P}$, $e_o=1, e_1=e_2=2, e_3=e_4=3$ are the weights of $\mathbb{P}$:

$$\text{i) } 0 \to \mathcal{O}_S \xrightarrow{\alpha} \bigoplus_{i=0}^{4} \mathcal{O}_S(e_i) \xrightarrow{\beta} T_{\mathbb{P}} \otimes \mathcal{O}_S \to 0$$

$$\text{ii) } 0 \to T_S \xrightarrow{\gamma} T_{\mathbb{P}} \otimes {}_S \xrightarrow{\delta} \mathcal{O}_S(6) \oplus \mathcal{O}_S(6) \to 0$$

Here $\mathcal{O}_S(6) \oplus \mathcal{O}_S(6)$ is the normal bundle to S in $\mathbb{P}$, α is given by the transpose of $(X_o, 2Y_1, 2Y_2, 3Z_3, 3Z_4)$, $\beta({}^t(f_o,\dots f_4))=$

$= f_o \frac{\partial}{\partial X_o} + \dots f_4 \frac{\partial}{\partial Z_4}$.

We know from [15] prop. 3.3 that $h^1(S,\mathcal{O}_S(n))=0$ $\forall n$, and $H^o(S,\mathcal{O}_S(n))=R'_n$, where $R'=R/_I$ (I the ideal of S).

Therefore from the exact sequence of cohomology of i) we infer that $H^1(S,T_{\mathbb{P}}\otimes\mathcal{O}_S)=0$, provided $H^2(\mathcal{O}_S)\longrightarrow\bigoplus_{i=0}^{4}H^2(\mathcal{O}_S(e_i))$ is injective: this is equivalent to the surjectivity of the dual mapping $\bigoplus_{i=0}^{4}H^0(\mathcal{O}_S(1-e_i))\xrightarrow{t_\alpha}H^0(\mathcal{O}_S(1))$. This last can be written as $R'_0\xrightarrow{\text{mult by } x_0}R'_1$, hence α is an isomorphism ($R'_0=\mathbb{C}$, $R'_1=\mathbb{C}X_0$): thus we also obtain that $H^2(S,T_{\mathbb{P}}\otimes\mathcal{O}_S)=0$, which, together with the exact cohomology sequence of ii), and the above cited vanishing of $H^1(S,\mathcal{O}_S(6))$, gives $H^2(S,T_S)=0$.

So the Kuranishi family of S is smooth of $\dim = h^1(S,T_S)=$ $=-\chi(S,T_S)$ (cf. [12], [13]) $=-\frac{7}{6}K^2+\frac{5}{6}C_2=18$ (this formula, where $C_2=$ the Euler number of S, is obtained applying Hirzebruch's Riemann-Roch theorem to T_S).

As $H^1(\mathcal{O}_S)=0$ $\bigoplus_{i=0}^{4}R'_{e_i}\xrightarrow{\beta}H^0(T_{\mathbb{P}}\otimes\mathcal{O}_S)\longrightarrow 0$ is exact, as well as $H^0(T_{\mathbb{P}}\otimes\mathcal{O}_S)\xrightarrow{\delta}R'_6\oplus R'_6\xrightarrow{\partial}H^1(T_S)\longrightarrow 0$: it is important to notice that $\delta\circ\beta$ is given by the matrix

$$\begin{pmatrix}\partial_0 f & \cdots\cdots\cdots & \partial_4 f\\ \partial_0 g & \cdots\cdots\cdots & \partial_4 g\end{pmatrix}\quad \text{where } \partial_0 f=\frac{\partial f}{\partial X_0},\ \ldots\ldots .$$

We can tensor the two above sequences by $\mathcal{O}_S(1)$, and multiplication by X_0 gives a morphism of the former to the latter: again $\bigoplus_{i=0}^{4}R'_{e_i+1}\xrightarrow{\beta}H^0(T_{\mathbb{P}}\otimes\mathcal{O}_S(1))\longrightarrow 0$ is exact (because $H^1(\mathcal{O}_S(1))=0$) while $H^2(\mathcal{O}_S(1))$ is 1-dimensional,

$H^2(\mathcal{O}_S(e_i+1))=0$, being dual to $H^{\circ}(\mathcal{O}_S(-e_i))$, hence $h^1(T_{\mathbb{P}}\times\mathcal{O}_S(1))=1$, and one has the following commutative diagram, which is exact in the rows, and where the vertical arrows are given by multiplication by X_o

$$\begin{array}{ccccccccc}
\bigoplus_{i=0}^{4} R'_{e_i} & \xrightarrow{\delta\circ\beta} & R'_6\oplus R'_6 & \xrightarrow{\partial} & H^1(T_S) & \longrightarrow & 0 & & \\
\downarrow & & \downarrow & & \downarrow{\scriptstyle\mu} & & & \searrow & \\
\bigoplus_{i=0}^{4} R'_{e_i+1} & \xrightarrow{\delta\circ\beta} & R'_7\oplus R'_7 & \xrightarrow{\partial'} & H^1(T_S\otimes\mathcal{O}(1)) & \longrightarrow & H^1(T_{\mathbb{P}}\otimes\mathcal{O}_S(1)) & \longrightarrow & 0
\end{array}$$

We remark that $h^1(\Omega^1_S)=19$, so $\mathrm{Im}(\partial')$ has dimension 18: as $\mu(H^1(T_S))\subset\mathrm{Im}(\partial')$, μ is injective iff it is surjective on $\mathrm{Im}(\partial')$.

This last condition means that $R'_7\oplus R'_7=(X_oR'_6\oplus X_oR'_6)+ +\delta\circ\beta(\bigoplus_{i=0}^{4} R'_{e_i+1})$. Now $R'_m=R_m/I_m$, and $I_m=0$ for $m\leq 5$, $I_7=\mathbb{C}X_of+\mathbb{C}X_og$, and we observe that we have the splitting $R_m=W_m+X_oR_{m-1}$.

Given an element $\alpha\in R$, one can use the splitting $R=W\oplus X_oR$ to write $\alpha=\alpha_W+X_o\,\alpha_R$ in an unique way, and the same can be done for the matrix of $\delta\circ\beta$: $\bigoplus_{i=0}^{4} R_{e_i+1}\longrightarrow R_7\oplus R_7$.

$R'_7\oplus R'_7=(X_oR'_6\oplus X_oR'_6)+\delta\circ\beta\,(\bigoplus_{i=0}^{4} R'_{e_i+1})$ is equivalent to $R_7\oplus R_7=(X_oR_6+X_oR_6)+(\delta\circ\beta)_W(\bigoplus_{i=0}^{4}\cdot W_{e_i+1})\Longleftrightarrow(\delta\circ\beta)_W$: $:\bigoplus_{i=0}^{4} W_{e_i+1}\longrightarrow W_7\oplus W_7$ is an isomorphism.

Now we can write the matrix associated to $(\delta\circ\beta)_W$: it is

$$\begin{pmatrix} Z_4(a_1Y_1+a_2Y_2), & 3f_{111}Y_1^2+2f_{112}Y_1Y_2+f_{122}Y_2^2, \\ Z_3(b_1Y_1+b_2Y_2), & 3g_{111}Y_1^2+2g_{112}Y_1Y_2+g_{122}Y_2^2, \end{pmatrix.$$

$$\left. \begin{matrix} f_{112}Y_1^2+2f_{122}Y_1Y_2+3f_{222}Y_2^2, & 2Z_3, & 0 \\ g_{112}Y_1^2+2g_{122}Y_1Y_2+3g_{222}Y_2^2, & 0, & 2Z_4 \end{matrix} \right) .$$

To simplify the computations, observe that $W_7=Z_3W_4 \oplus Z_4W_4$, and one can write $(W_2 \oplus W_3 \oplus W_3) \oplus (W_4 \oplus W_4) = \bigoplus_{i=0}^{4} W_{e_i+1}$, $W_7 \oplus W_7 = (Z_4W_4 \oplus Z_3W_4) \oplus (Z_3W_4 \oplus Z_4W_4)$: $(\delta\circ\beta)_W$ is also a direct sum map, and its second summand, being given by

$$(W_4 \oplus W_4) \xrightarrow{\begin{pmatrix} 2Z_3 & 0 \\ 0 & 2Z_4 \end{pmatrix}} Z_3W_4 \oplus Z_4W_4,$$ is an obvious isomorphism.

So we conclude that $(\delta\circ\beta)_W$ is an isomorphism iff the first summand is such: to see this last map we pick bases of these 2 6-dimensional vector spaces over $\mathbb{C}$, and write the associa_ted matrix A.

We choose for $(W_2 \oplus W_3 \oplus W_3)$ the ordered basis

$$\begin{pmatrix} Y_1 \\ 0 \\ 0 \end{pmatrix} \begin{pmatrix} Y_2 \\ 0 \\ 0 \end{pmatrix} \begin{pmatrix} 0 \\ Z_3 \\ 0 \end{pmatrix} \begin{pmatrix} 0 \\ Z_4 \\ 0 \end{pmatrix} \begin{pmatrix} 0 \\ 0 \\ Z_3 \end{pmatrix} \begin{pmatrix} 0 \\ 0 \\ Z_4 \end{pmatrix},$$ and for $(Z_4W_4 \oplus Z_3W_4)$ the or-

dered basis

$$\begin{pmatrix} Y_1^2 Z_4 \\ 0 \end{pmatrix} \begin{pmatrix} Y_2^2 Z_4 \\ 0 \end{pmatrix} \begin{pmatrix} Y_1 Y_2 Z_4 \\ 0 \end{pmatrix} \begin{pmatrix} 0 \\ Y_1^2 Z_3 \end{pmatrix} \begin{pmatrix} 0 \\ Y_2^2 Z_3 \end{pmatrix} \begin{pmatrix} 0 \\ Y_1 Y_2 Z_3 \end{pmatrix} :$$

then A is the matrix (where an empty array means that the correspon_ding entry is zero)

a_1			$3f_{111}$		f_{112}
	a_2		f_{122}		$3f_{222}$
a_2	a_1		$2f_{112}$		$2f_{122}$
b_1		$3g_{111}$		g_{112}	
	b_2	g_{122}		$3g_{222}$	
b_2	b_1	$2g_{112}$		$2g_{122}$	

Call Δ the determinant of A: then μ is injective iff $\Delta(a_1,\ldots g_{222})\neq 0$ for S, and it remains only to show that Δ is not identically zero. But the monomial $a_2 b_1 f_{111} f_{222} g_{111} g_{222}$ appears in Δ with coefficient -36.

COROLLARY 17. The subfamily of special surfaces is contained in a 14 dimensional family where $\mathrm{Ker}\ \mu$ is 2 dimensional.

Proof. It is evident that the rank of μ drops by 2 iff the rank of A drops by 2: but 2 columns of A vanish in the 3 following cases:

$$\begin{cases} a_1 = a_2 = b_1 = b_2 = 0 \\ g_{111} = g_{112} = g_{122} = g_{222} = 0 \\ f_{111} = f_{112} = f_{122} = f_{222} = 0 \end{cases}$$

We have only to recall to our mind that S is special iff $a_o=a_1=a_2=b_o=b_1=b_2=0$.

REMARK 18. When the w.c.i. Y has rational double points, one considers as usual $T_Y = \underline{Hom}(\Omega_Y^1, \mathcal{O}_Y)$, and by [3] prop. 1.2. one has, if $S \xrightarrow{\pi} Y$ is the desingularization, $\pi_*(T_S)=T_Y$, $h^2(T_S)=h^2(T_Y)$.
The analogous of the exact sequence ii) is now

$$0 \longrightarrow T_Y \longrightarrow T_{\mathbb{P}} \otimes \mathcal{O}_Y \longrightarrow \underbrace{\mathcal{O}_Y(6) \oplus \mathcal{O}_Y(6)}_{= N_Y} \xrightarrow{\gamma} \underbrace{\underline{Ext}^1(\Omega_Y^1, \mathcal{O}_Y)}_{= M} \longrightarrow 0$$

where M is supported only at the singular points.
If $N_Y'=\ker \gamma$, we have 2 short exact sequences

$$0 \longrightarrow T_Y \longrightarrow T_{\mathbb{P}} \otimes \mathcal{O}_Y \longrightarrow N_Y' \longrightarrow 0$$

$$0 \longrightarrow N_Y' \longrightarrow N_Y \longrightarrow M \longrightarrow 0$$

As in the proof of theorem 4 $h^i(T_{\mathbb{P}} \otimes \mathcal{O}_Y)=0$ for $i=1,2$,

hence $H^2(T_Y) \cong H^1(N_Y')$, and (as $H^1(N_Y)=0$), one has

$$H^o(N_Y) \xrightarrow{r} H^o(M) \longrightarrow H^1(N_Y') \longrightarrow 0 .$$

So the Kuranishi family of S is obstructed iff r is not surjective: we have however not yet found Y's for which r is not onto.

§ 4. THE RESTRICTION OF THE LOCAL PERIOD MAPPING TO SPECIAL SURFACES.

We begin by fixing a rather general situation: X a smooth compact manifold of dimension k, L a line bundle on X $p:L \to X$ the projection map, s,σ two independent sections in $H^o(X,L^{\otimes n})$, S_t the nth cyclic covering of X in L branched over $D_t = div(s+t\sigma)$, which we assume smooth for $|t|<\varepsilon$. Let moreover M be another line bundle on X, m a non zero section of $H^o(S,p^*M)$, (where $S=S_o$), $\left(\frac{\partial S_t}{\partial t}\right)_{t=0} = \theta \in H^1(S,T_S)$ the infinitesimal deformation of S (see e.g. [12] pag. 36 and following).

<u>THEOREM 5</u>. $\theta \cdot m \in H^1(T_S \otimes M)$ is non zero if T_X does not split on $D=D_o$ as $T_D \oplus N_D$.

<u>Proof</u>. Take an acyclic covering $\{V_{o\alpha}, V_{1\alpha}, V_{2\alpha}\}_{\alpha\in A}$ of X satisfying the following properties (here we consider the Haussdorff topology of X)

i) $V_{2,\gamma} \cap V_{o,\alpha} = \emptyset \ \forall \alpha,\gamma$

ii) $X - \bigcup_{\alpha\in A}\{V_{1\alpha}, V_{2\alpha}\} = F$ is a closed neighbourhood of D

iii) $D \cap V_{o,\alpha} \neq \emptyset$, $V_{o\alpha} \cap V_{1\alpha} \neq \emptyset$

iv) $V_{o,\alpha} \cap V_{o\beta} \cap D \neq \emptyset \Leftrightarrow V_{1\alpha} \cap V_{1\beta} \neq \emptyset$

v) $\{V_{o\alpha} \cup V_{1\alpha}, V_{2\alpha}\}_{\alpha\in A}$ is a trivializing cover for L, M, T_X.

This can be achieved by taking a Stein covering (V'_α) of D, trivializing L, local coordinates $(z^1_\alpha,\dots z^{k-1}_\alpha, s_\alpha)$ near the points of V'_α, and choosing conveniently $\delta_{1\alpha} < \eta_\alpha < \delta_{2\alpha}$ so that $V_{o\alpha} = \{(z_\alpha, s_\alpha) z_\alpha \in V'_\alpha,\ |s_\alpha| < \eta_\alpha\}$, $V_{1\alpha} = \{(z_\alpha, s_\alpha) z_\alpha \in V'_\alpha\ \delta_{1\alpha} < |s_\alpha| < \delta_{2\alpha}\}$ satisfy iii), iv).

On $L|_{V_{o\alpha} \cup V_{1\alpha}}$ S_t is defined by $y_\alpha(t)^n = s_\alpha + t\sigma_\alpha$.

If $U_{i\alpha} = p^{-1}(V_{i\alpha})$, on $U_{o\alpha}$ take as local coordinates $(z^1_\alpha,\dots,z^{k-1}_\alpha, y_\alpha(t))$, on $U_{1\alpha}$ $(z^1_\alpha,\dots,y_\alpha(0))$ (because we assume ε so small that for $|t| < \varepsilon$ D_t is contained in the interior of F), on $U_{2\alpha}$ the lifting of any coordinates on X. Observe that the coordinate changes which depend on t are consequences of the coordinate change from $U_{o\alpha}$ to $U_{1\alpha}$: we get that $\theta_{1\alpha,o\alpha} = -\frac{1}{n}\frac{\sigma_\alpha}{y_\alpha^{n-1}}\frac{\partial}{\partial y_\alpha}$.

Now $(U_{i\alpha})$ is an acyclic covering, so if $\theta\cdot m$ were a coboundary, there would exist vector fields $\eta_{i\alpha}$ such that

$$\eta_{o\alpha} - \eta_{1\alpha} = m_\alpha \frac{\sigma_\alpha}{y_\alpha^{n-1}}\frac{\partial}{\partial y_\alpha}, \quad \eta_{j\alpha} = \frac{m_\alpha}{m_\beta}\eta_{j\beta} \ (j=0,1).$$

These equations, plus condition iv) imply that

$$\frac{\sigma_\alpha}{y_\alpha^{n-1}}\frac{\partial}{\partial y_\alpha} = \frac{\sigma_\beta}{y_\beta^{n-1}}\frac{\partial}{\partial y_\beta} \text{ : being } (y_\alpha)^n = s_\alpha, \frac{\partial}{\partial y_\alpha} = n y_\alpha^{n-1}\frac{\partial}{\partial s_\alpha}, \text{ then}$$

$$\frac{\sigma_\alpha}{\sigma_\beta}\cdot\frac{\partial}{\partial s_\alpha} = \frac{\partial}{\partial s_\beta}.$$

This equality says that T_X in a neighbourhood of D has a 1-dimensional subbundle, isomorphic to $\mathcal{O}(D)$, giving on D a direct summand for T_D.

We can apply this criterion to obtain the following :

THEOREM 6. If S is a general "special" surface, in the Kuranishi family of S the special surfaces form a submanifold P and the restriction of the local period mapping to P is an embedding .

Proof. Here we will adopt the notations of § 2, expecially prop. 10 . $\tilde{S}$ is a double cover of the smooth K3 surface $\tilde{X}$, branched on the smooth curve $L + \sum E_j$.
As the holomorphic 2-form on $\tilde{S}$ is the pull back of the 2-form on $\tilde{X}$, and the 2-dimensional homology of $\tilde{S}$ is the direct sum of the (+1) and (-1) eigenspaces for the involution, by the local Torelli theorem for K3 surfaces ([19] pag. 202 and foll.) we can only limit ourselves to consider infinitesimal deformations arising from moving L in $|p^*(\mathcal{O}_{\mathbb{P}^2}(1))|$, and apply theorem 5 with $m = x_o \in H^0(\mathcal{O}_S(K))$: we have then only to prove that on L the exact sequence

$$0 \longrightarrow \mathcal{O}_L(-L) \longrightarrow \Omega^1_X \otimes \mathcal{O}_L \longrightarrow \mathcal{O}_L(L) \longrightarrow 0$$

does not split, or that the extension class in $H^1(\mathcal{O}_L(-2L))$ is not zero .
To this purpose we first choose the following covering of L :
$V_1 = \{Y_1 FG \neq 0\}$, $V_2 = \{Y_2 FG \neq 0\}$, V_3 a neighbourhood of the six points of L where $FG = 0$.
Denote by $x = Y_1/Y_2$, $x' = Y_2/Y_1$, and write $F =$
$= F'(Y_1,Y_2) + Y_o F''$, $G = G'(Y_1,Y_2) + Y_o G''$; F , G being general one may take as a basis of Ω^1_X on V_1 $d(Y_o/Y_1)$, dx' , on V_2 $d(Y_o/Y_2)$, dx , on V_3 $d(Y_o/Y_1)$, $d\zeta_1$.

One easily computes the extension class : $\tau_{12} = 0$, $\tau_{13} = \frac{F'G'' + F''G'}{2\zeta_1 Y_1^5}$ (τ_{23} is deduced by the cocycle condition) .

By Serre duality we can interpret (τ_{ij}) as an element in the dual of $H^0(\omega_L(2L))$: to prove that it is non zero we check its value on the form $\frac{dx}{\zeta_1} \in H^0(\omega_L(2L))$.

We associate a repartition to (τ_{ij}) by setting $\tau_1 = \tau_{13}$, $\tau_3 = 0$, $\tau_2 = \tau_{13}$, and taking in V_i the repartition corresponding to τ_i .

Then $\left\langle (\tau_{ij}) , \frac{dx}{\zeta_1} \right\rangle = \sum_{P \in V_1 \cup V_2} \operatorname{Res}_P \left(\frac{dx}{\zeta_1} \frac{F'G'' + F''G'}{2\zeta_1 Y_1^5} \right)$.

Now on V_2 dx is holomorphic , and $\frac{F'G'' + F''G'}{\zeta_1^2 Y_1^5} = \frac{G''Y_1}{G'} + \frac{F''Y_1}{F'}$ is holomorphic too.

On V_1 $dx \left(\frac{G''Y_1}{G'} + \frac{F''Y_1}{F'} \right) = -\frac{dx'}{x'^2} \left(\frac{\frac{G''F' + F''G'}{Y_1^5}}{\frac{F'G'}{Y_1^6}} \right)$

so that the sum of the residues is non zero iff $\frac{G''F' + F''G'}{Y_1^5}$ has a non zero term of order 1 in x' , and this is clearly achieved for F,G general .

References

[1] Bombieri, E. "Canonical models of surfaces of general type", Publ. Math. IHES 42, pp. 171-219 .

[2] Bombieri, E.-Catanese,F. "The tricanonical map of a surface with $K^2 = 2$, $p_g = 0$, to appear in "Algebraic Geometry", Cambridge University Press.

[3] Burns,D. - Wahl,J. "Local contributions to global deformations of surfaces Inv. Math. 26, pp. 67-88 (1974).

[4] Dolgachev, I. "Weighted projective varieties" , to appear.

[5] Enriques, F. "Le superficie algebriche di genere lineare $p^{(1)} = 2$ " Rend. Accad. Lincei, s. 5ª , vol. VI (1897) pp. 139-144 .

[6] Enriques, F. "Le superfici algebriche", Zanichelli, Bologna (1949).

[7] Griffiths, P. "Periods of integrals on algebraic manifolds", I,II, Amer. J. Math. 90 (1968) pp. 568-626, 805-865 .

[8] Griffiths, P. "Periods of integrals on algebraic manifolds: summary of main results and discussion of open problems", Bull. Amer. Math. Soc. 76 (1970), pp. 228 - 296 .

[9] Griffiths, P. - Schmid, W. "Recent developments in Hodge theory: a discussion of techniques and results", Proc. Int. Coll. Bombay, (1973), Oxford Univ. Press.

[10] Horikawa, E. "On deformations of quintic surfaces", Inv. Math. 31 (1975) pp. 43-85 .

[11] Kodaira, K. "Pluricanonical systems an algebraic surfaces of general type", J. Math. Soc. Japan, 20 (1968) pp. 170-192 .

[12] Kodaira, K. - Morrow,J. "Complex Manifolds",Holt Rinehart and Winston,New-York.

[13] Kuranishi, M. "New proof for the existence of locally complete families of complex structures", Proc. Conf. Compl. Analysis, Minneapolis, pp. 142-154, Springer (1965).

[14] Kynef, V.I. "An example of a simply connected surface of general type for which the local Torelli theorem does not bold"(Russian), C.R. Acad. Bulgare Sc. 30, n. 3 (1977) pp. 323-325.

[15] Mori, S. "On a generalization of complete intersections", Jour. Math. Kyoto Univ. Vol. 15, N. 3 (1975) pp. 619-646.

[16] Mumford, D. "The canonical ring of an algebraic surface", Annals of Math. 76, (1962), pp. 612-615 .

[17] Peters,C. "The local Torelli theorem I. Complete intersections", Math. Ann. 217, (1975) pp. 1 - 16 .

[18] Tjurina , G.N. "Resolution of singularities of flat deformations of rational double points", Funk. Anal. i Prilozen 4, N. 1, pp.77-83.

[19] Shafarevich, I.R.-others "Algebraic Surfaces" , Moskva (1965)

[20] Usui, S. "Local Torelli for some non singular weighted complete intersections", to appear.

[21] Zariski,O. "On the problem of existence of algebraic functions of two variables possessing a given branch curve", Amer. J. Math., 51 (1929), pp. 305-328.

Added: Prop. 4 has a shorter proof.

In fact in prop. 5 is proven that x_o, y_1, y_2, z_3, z_4 generate $H^o(5K)$, so a base point for $|3K|$ would be a base point for $|5K|$, against Th. 2 of [1] .

Università di Pisa
Istituto di Matematica
Via Derna, 1
I-56100 Pisa
Italien

INTERNAL HOM-SETS IN AN EXTENSION OF AFFINE SCHEMES OVER A FIELD

PAUL CHERENACK

Let V be a vector space over k and $\dim_k V<\infty$. Then, the collection of linear maps $\mathrm{Hom}(V,V)$ is again a finite dimensional vector space. However if one considers polynomial maps, i.e. one considers affine schemes over k, then $\mathrm{Hom}(V,V)$ can be identified with a vector space such that $\dim_k \mathrm{Hom}(V,V)=\omega$. For example, if $V = k$, $\mathrm{Hom}(V,V) = \{(a_i) \mid a_0+a_1X^1+..+a_nX^n+.. \in k[X]\}$. Thus, if we wish hom-sets of affine schemes to be affine schemes, our category of affine schemes must at least be enlarged to contain ringed spaces whose k-valued points are vector spaces of countable dimension.

Notation <u>Aff</u> denotes the category of affine schemes of countable type over k. In order to always have the Hilbert Nullstellensatz at hand we assume k is an algebraically closed field and card $(k)>\omega$. See [5]. $k^{\mathbb{N}}$ denotes the element Spec $(k[X_1,X_2,..,X_n,..])$ in <u>Aff</u>. $\underset{\rightarrow}{k}^{\mathbb{N}}$ is the vector space $\{(a_i) \mid i\in\mathbb{N}\}$. If V is an object of <u>Aff</u> (or a subset of an object in <u>Aff</u>) then $V(k)$ is the collection of all k-valued points in V. <u>Rngsp</u> is the category of ringed spaces.

The language of scheme theory may be found in [1], [3], [8]; of category theory in [4], [7].

In §1 we define a category ind-<u>Aff</u> such that the inclusions

$$\underline{\mathrm{Aff}} \hookrightarrow \text{ind-}\underline{\mathrm{Aff}} \hookrightarrow \underline{\mathrm{Rngsp}}$$

are faithful and where every vector subspace V of $\underset{\rightarrow}{k}^{\mathbb{N}}$ is of the form $L(k)$ for a suitable element L of ind-<u>Aff</u>. The elements of ind-<u>Aff</u> will be called ind-affine schemes and the name arises from the fact that ind-<u>Aff</u> is formed by adding certain inductive limits of arrows in Aff, the inductive limits being taken in Rngsp.

In §2 we outline a proof of the following result, an assertion that internal hom-sets exists in ind-Aff.

Theorem ind-Aff is Cartesian closed. This means that ind-Aff has products and there is a bifunctor X^Y on ind-Aff with values in ind-Aff such that a natural equivalence

(*) $\mathrm{Hom}_{\text{ind-Aff}}(X\times Y, Z) \cong \mathrm{Hom}_{\text{ind-Aff}}(X, Z^Y)$ exists.

Our proof will be restricted to the case where X, Y and Z are in Aff in which case $X\times Y$ clearly exists. The extension to arbitrary elements of ind-Aff is relatively straight forward.

In §3 we show that the inclusion

$$i : \text{Aff} \hookrightarrow \text{ind-Aff}$$

has a left adjoint left inverse (lali) and that Aff is a bireflective subcategory of ind-Aff. One consequence is that the reflection map $\eta_A : A \to r(A)$ is both epi and mono but it is not an isomorphism. Define the bifunctor $X:Y$ on objects by setting $X:Y = r(Y^X)$ and extend to arrows in the obvious way. Then one can show (we will not) that there is a natural equivalence $(X\times Y):Z \cong X:(Y:Z)$ in Aff. Thus, although Aff is not Cartesian closed, it does have a type of internal hom-functor (perhaps even more natural than that of ind-Aff).

We now make two applications of the above theory. For these applications we assume char (k)=o.

In §4 we show that "almost every projective curve is non-singular", i.e. the homogeneous polynomials in $k^{(k^3)}(k)$ defining non-singular projective curves form an open and hence dense subset of all homogeneous polynomials. Then, we define the notion of local stability for polynomials F in $k^{(k^2)}(k)$ sending (o,o) to o, show that isomorphism of $F \in k^{(k^2)}(k)$ implies birational equivalance and thus deduce that no polynomial such as F above is locally stable.

is an affine closed subscheme of $k^{\mathbb{N}}$. A subset C of $H \cap X$ is closed if $C \cap X \cap H_i$ is closed for each $i \in I$ (one must show that this definition is independent of the choice of H_i).

One can turn I into a directed set by setting $i \leq j$ if $H_i \subset H_j$ and then regard I as a category. In this way the association of $H_i \cap X$ to i defines a functor

$$F_{X \cap H}: I \longrightarrow \underline{\text{Rngsp}}$$

Definition 1.4 A H-ind-affine scheme of countable type over k is the ringed space found by taking the inductive limit of $F_{X \cap H}$. ind-Aff is the faithful subcategory of Rngsp consisting of all H-ind-affine schemes for arbitrary H and maps induced from morphisms $k^{\mathbb{N}} \to k^{\mathbb{N}}$ in Aff.

§2 Internal hom-functors in ind-Aff

Let X, Y belong to Aff and $k[X], k[Y]$ be their affine rings. A morphism $f: X \to Y$ is given by a countable number of coordinates $f_i(x) \in k[X]$. Suppose $\{e_j\}_{j \in \mathbb{N}}$ is a basis for $k[X]$ and $I(Y)$ is the ideal defining Y. If $f(x) = (f_i(x)) = (\sum_j a_i^{\,j} e_j)$, then $o = F(f(x)) = \sum_p F_p(a_i^{\,j}) e_p^{\,Y}$ for $F \in I(Y)$ implies $F_p(a_i^{\,j}) = o$. We let Z be the affine closed subset of $k^{\mathbb{N}} \times k^{\mathbb{N}}$ defined by $F_p(x_i^{\,j}) = o$ for $F \in I(Y)$.

Now $\underset{\to}{k}^{\mathbb{N}} \times \underset{\to}{k}^{\mathbb{N}} = \{(a_i^{\,j}) \mid i, j \in \mathbb{N}\}$ can be (by diagonal counting) be identified with $\underset{\to}{k}^{\mathbb{N}}$. Let $V = \{(a_i^{\,j}) \mid$ for fixed i, $a_i^{\,j} = o$ except for a finite number of $j\}$. V is then a vector subspace of $\underset{\to}{k}^{\mathbb{N}}$. Let $T = \{(t_s) \mid t_s \in \mathbb{N}, s \in \mathbb{N}\}$. If $t = (t_s)$, consider the ideal A_t generated by X_s^j for $j \geq t_s$. Then $H = \bigcup_{t \in T} H_t$ where

$$H_t = \operatorname{Spec}\,(k[X_i^{\,j}]/A_t)$$

is a linear subscheme of $k^{\mathbb{N}}$ such that $H(k) = V$.

Definition 2.1 $Y^X = Z \cap H$.

One must show that this definition as well as all subsequent

In §5 we show how to identify the collection of irreducible algebraic curves in the projective plane $\mathbb{P}^2(k)$ of genus $\leq g$ with only nodes (ordinary double crossings) for singularities with a locally closed subset of an ind-affine scheme. We do not attempt to find any moduli spaces although this seems an appropriate next application.

Finally, the study of internal hom-functors was instigated in order to develop a purely scheme oriented homotopy theory but nothing is said about that here.

§1 ind-affine schemes

Definition 1.1 A subset H of $k^{\mathbb{N}}$ is called a linear subscheme of $k^{\mathbb{N}}$ if

i) $H = \bigcup_{i\in I} H_i$ for some indexing set I where H_i is a closed linear subscheme of $k^{\mathbb{N}}$ (defined by linear equations in $k[X_1, \ldots, X_n, \ldots]$).

ii) for each $i,j \in I$ there is a $m \in I$ such that

$$H_i \cup H_j \subset H_m \quad .$$

iii) $H(k)$ is a vector subspace of $k^{\mathbb{N}}$.

Proposition 1.2 For every vector subspace V of $k^{\mathbb{N}}$ there is a linear subscheme L (not necessarily unique) such that $L(k) = V$.

Proof Let $\{H_i\}_{i\in I}$ be the collection of all finite dimensional linear subschemes of $k^{\mathbb{N}}$ such that $H_i(k) \subset V$ and $L = \bigcup_{i\in I} H_i$. Then the conditions of definition 1.1 are easy to verify and clearly $L(k) = V$.

Let now H be a fixed linear subscheme of $k^{\mathbb{N}}$.

Definition 1.3 A H-ind-affine scheme of countable type over k (without sheaf) is a subset of $k^{\mathbb{N}}$ of the form $H \cap X$ where X

all (a_{ij}^h) $(k^{\mathbb{N}}\times k^{\mathbb{N}}\times k^{\mathbb{N}}$ is identified with $k^{\mathbb{N}})$ satisfying

A) $F_{pq}(a_{ij}^h)=o$ for all $F\in I(Z)$.

B) $a_{ij}^h=o$ except for a finite number of i,j.

Let $w_{jh}=\sum_i b_{ij}^h e_i^X$ be the jhth coordinate of an element in $\text{Hom}_{\text{ind-}\underline{\text{Aff}}}(X,Z^Y)$. Let $u_h=\sum_j w_{jh}e_j^Y$ be the hth coordinate of an element of $Z^Y(k)$ and $F\in I(Z)$. Then

$$o=F(u_h)=F(\sum_j(\sum_i b_{ij}^h e_i^X)e_j^Y)$$

$$=F(\sum_{i,j} b_{ij}^h e_i^X e_j^Y)$$

$$=\sum_{p,q} F_{pq}(b_{ij}^h)e_p^X e_q^Y$$

Since $u_h=\sum_{i,j} b_{ij}^h e_i^X e_j^Y$, the (b_{ij}^h) satisfy the same conditions as the (a_{ij}^h) above and the required identification can be made.

§3 The reflection of ind-Aff into ind-Aff

We use the language of Herrlich [4].

Let ι:Aff $\hookrightarrow$ ind-Aff be the inclusion functor.

Proposition 3.1 ι has a left adjoint left inverse.

Proof If $X\in$ind-Aff, $X\subset k^{\mathbb{N}}$. Define r:ind-Aff $\to$ Aff on objects by setting $r(X)=\overline{X}$, the Zariski closure of X in $k^{\mathbb{N}}$. If $f:X\to Y$ is an arrow in ind-Aff, f induces a map $r(f):\overline{X}\to\overline{Y}$ in Aff and thus a functor r:ind-Aff $\to$ Aff.

We need to show that there is a natural isomorphism

$$\text{Hom}_{\text{ind-}\underline{\text{Aff}}}(X,\iota(A))\simeq\text{Hom}_{\underline{\text{Aff}}}(r(X),A) .$$

Let $g\in\text{Hom}_{\text{ind-}\underline{\text{Aff}}}(X,\iota(A))$. Then, g defines a map

arguments is free of choice of the basis $\{e_j\}_{j\in\mathbb{N}}$ but this is straightforward.

We now show how to define Y^X as a functor from ind-Aff to ind-Aff in Y and X separately. It is then easy to deduce that Y^X is a bifunctor.

Let us define the map $g^X: Y^X \to W^X$ for a map $g: Y \to W$ in Aff. As

$$g(f(x)) = (g_m(f(x))) = (g_m(\sum_j a_i{}^j e_j))$$

$$= (\sum_p g_m^p(a_i{}^j)e_p),$$

one must set $g^X(a_i{}^j) = (g_m^p(a_i{}^j))$.

Clearly, Y^X is a functor in Y.

Next, let $g: W \to X$. g induces a map $g^*: k[X] \to k[W]$. Let $\{d_m\}_{m\in\mathbb{N}}$ be a basis for $k[W]$.

$$f(g(x)) = (f_i(g(x))) = (g^*(f_i(x)))$$

$$= (g^*(\sum_j a_i{}^j e_j)) = (\sum_j a_i{}^j g^*(e_j))$$

$$= (\sum_m L_i{}^m(a_i{}^j)d_m)$$

where $L_i{}^m(a_i{}^j)$ is a linear (!) function in $(a_i{}^j)$. Then, define a functor $Y^g: Y^X \to Y^W$ by setting $Y^g(a_i{}^j) = (L_i{}^m(a_i{}^j))$ and Y^X is in this way a functor in X.

Finally, let us show that the natural equivalence of Theorem 1 arises for X,Y,Z in Aff from an identification of mappings in $\mathrm{Hom}_{\text{ind-Aff}}(X\times Y,Z)$ and $\mathrm{Hom}_{\text{ind-Aff}}(X,Z^Y)$ viewed as points in $k_{\to}^{\mathbb{N}}$.

Let $\{e_i{}^X\}_{i\in\mathbb{N}}$, resp. $\{e_j{}^Y\}_{j\in\mathbb{N}}$, be a basis of $k[X]$, resp. $k[Y]$. Then, $\{e_i{}^X e_j{}^Y\}$ form a basis for $k[X]\otimes k[Y] = k[X\times Y]$. If $z_h = \sum_{i,j} a_{ij}^h\, e_i{}^X e_j{}^Y$ is the hth tuple of an element of $\mathrm{Hom}_{\text{ind-Aff}}(X\times Y,Z)$, write

$$F(z_h) = \sum_{p,q} F_{pq}(a_{ij}^h) e_p{}^X e_q{}^Y .$$

Then, the mappings in $\mathrm{Hom}_{\text{ind-Aff}}(X\times Y,Z)$ correspond to the set of

$\alpha(g):r(X)=\overline{X} \to \overline{A}=A$. Conversely, let $h\in \mathrm{Hom}_{\underline{\mathrm{Aff}}}(r(X),A)$. As X is a H-ind-affine scheme, for some linear subscheme H of k^N, $H_i \cap X \subset r(X)$ for the H_i associated to H. Thus there are maps $h_i:H_i \cap X \to A$ in ringed spaces. Taking inductive limits defines a map $\beta(h):X \to \iota(A)$ in ringed spaces. As β and α are inverses to each other, the natural isomorphism exists and thus Aff is a reflective subcategory of ind-Aff. As $r\circ\iota$ is the identity functor on Aff, r is a lali. q.e.d.

Proposition 3.2 Aff is a monoreflective subcategory of ind-Aff.

Proof We need to show that the unit of the adjunction and inclusion $\eta_X:X \to r(X)$ is a monomorphism. Suppose that $\alpha\circ f=\beta\circ g$ where $g,f:Y \to X$. Clearly one can assume that Y is an affine scheme of finite type over k . As $X=\bigcup_{i\in I} X_i$ where X_i are affine, the ascending chain condition in the affine ring of Y together with the directed nature of the X_i imply that $f(Y)\subset X_i$ and $g(Y)\subset X_i$ for some i. But then restricted to X_i α is the identity and thus $f=g$. q.e.d.

Lemma 3.3 A full monoreflective subcategory is also epireflective.

Proof See Herrlich[4] ,proposition 36.3.

From this lemma follows:

Proposition 3.4 Aff is a bireflective subcategory of ind-Aff.

For further applications of this proposition, see Herrlich [4],chapter X. As $\eta_X:X \to r(X)$ is an epimorphism and monomorphism, ind-Aff is not a balanced category.

§4 The generic character of non-singular projective curves and local stability

A homogenous non-zero $F \in k^{(k^3)}(k)$ can be identified (up to a multiple) with the curve $F=o$ in $\mathbb{P}^2(k)$. Suppose $\mathbb{P}^2(k)$ has homogeneous coordinates x_o, x_1, x_2. We now show that "almost all curves $F=o$ in $\mathbb{P}^2(k)$ are non-singular." More precisely, let Q be the closed subset of $k^{(k^3)}(k)$ consisting of homogeneous polynomials (including o). Then,

Proposition 4.1 There is an open subset U of Q such that $F=$ is non-singular if and only if $F \in U$ or $F=o$.

Proof Let $F \in Q$ be of degree n. Find the resultant system $D_1^n, D_2^n, \ldots, D_{h(n)}^n$ of the homogeneous polynomials $F, \frac{\partial F}{\partial x_o}, \frac{\partial F}{\partial x_1}, \frac{\partial F}{\partial x_2}$. As a consequence, see [9], page 8, $D_1^n, \ldots, D_{h(n)}^n$ are homogeneous polynomials in the coefficients of F and $F=o$ has a singularity if and only if $F \neq o$ and the coefficients of F satisfy

$$D_1^n = D_2^n = \ldots = D_{h(n)}^n = o.$$

Then, U is defined as the complement of the closed set of points satisfying

$$D_j^n = o$$

for $n=1,2,3,\ldots$ and $1 \leq j \leq h(n)$. q.e.d.

Note that Q is not an affine closed subscheme of $k^{\mathbb{N}}$.

Corollary 4.2 The collection of all $F \in k^{(k^2)}(k)$ such that $F=o$ defines a curve whose projective completion ($F^h=o$ where F^h is the homogenization of F) is non-singular form an open subset of $k^{(k^2)}(k)$.

To prove the corollary one considers the dehomogenization map from Q to $k^{(k^2)}(k)$. Then, the open set U in proposition 4.1 corresponds to the one needed in $k^{(k^2)}(k)$.

Next, we consider the $F \in k^{(k^2)}(k)$ as functions and examine the meaning of stability at a point.

Let k^2*k denote the closed subset of $k^{(k^2)}$ whose k-valued points are polynomials such that $F(o,o)=o$. Suppose $F,G\in(k^2*k)(k)$. We say that F and G are isomorphic at (o,o) if there are open neighborhoods N,N^1 of o and an isomorphism $\alpha:N\to N^1$ such that $F\circ\alpha=aG$ for some constant $a\in k-\{o\}$ on N. A function $F\in(k^2*k)(h)$ is locally stable at o if, in some neighborhood U of F in k^2*k, every $G\in U(k)$ is right equivalent to F. See Lu [6], page 15. F and G are birationally equivalent at o if there is a birational equivalence of F=o and G=o sending (o,o) to (o,o)

Proporition 4.2 Suppose F and G are irreducible curves in $(k^2*k)(k)$. Then, F is isomorphic to G if and only if the curves defined by F=o and G=o are birationally equivalent at (o,o).

Proof (⇒) Clearly the map α defining the isomorphism of F with G restricts to the required birational equivalence.

(⇐) Conversely, if F and G are birationally equivalent at (o,o) one has rational maps r,s defined on some neighborhoods N,N^1 of (o,o) respectively such that sr=1 when restricted to $N\cap\{(x,y)\,|\,G(x,y)=o\}$.

By choosing appropriate representatives for s and r, sr=1 on N. But r being injective by dimension theory must be dominant and hence epimorphic. Thus, rs=1 on N and F is isomorphic to G. q.e.d.

Corollary 4.3 No polynomial $F(x,y)\in k^2*k$ is locally stable.

Remarks i) If one lets N,N^1 above be neighborhoods of arbitrary points (not necessarily zero) birational equivalence implies isomorphism.

ii) Global stability for functions in k^2*k is rather

strong. It implies that the curves $F=0$ and $G=0$ are the same after a general linear change of coordinate of the domain space k^2.

§5 The identification of plane curves of genus g with a locally closed subset of an ind-affine scheme

Let Q be as in §4 the collection of homogeneous polynomials and ind-affine scheme.

Proposition 5.1 The collection of all singular projective curves in Q having d nodes (ordinary double crossings) is a locally closed subset N^d of Q.

Proof Using resultants as in proposition 4.1 the collection of all F having only nodes as singularities can be identified with the open subset W of Q consisting of all F such that the resultant system of

$$F, \frac{\partial F}{\partial x_0}, \frac{\partial F}{\partial x_1}, \frac{\partial F}{\partial x_2}, \left(\frac{\partial^2 F}{\partial x_0 \partial x_1}\right)^2 - 4\frac{\partial^2 F}{\partial x_0^2}\frac{\partial^2 F}{\partial x_1^2}, \left(\frac{\partial^2 F}{\partial x_1 \partial x_2}\right)^2 - 4\frac{\partial^2 F}{\partial x_1^2}\frac{\partial^2 F}{\partial x_2^2}$$

and $\left(\frac{\partial^2 F}{\partial x_0 \partial x_2}\right)^2 - 4\frac{\partial^2 F}{\partial x_0^2}\frac{\partial^2 F}{\partial x_2^2}$ (defined in a manner analagous to that of proposition 4.1) do not have a common zero. We will show that the collection of all $F \in Q$ having at least d nodes is a closed subset of W and this will complete the proof. Take the resultants of $F, \frac{\partial F}{\partial x_0}, \frac{\partial F}{\partial x_1}, \frac{\partial F}{\partial x_2}$ two by two with respect to x_0.

One obtains homogeneous polynomials $R_i(x_1, x_2), i=1,2,\ldots,6$. Because $F \in W$, the R_i have one linear factor for each node in common. We need then to determine a condition on the coefficients of F which guarantees that the R_i have at least d linear factors in common. There are algebraic conditions which guarantee that x_1 appears say m times. Thus, by dehomogenizing, one obtains polynomials $S_i(x)=R_i(1,x)$ for $i=1, \ldots, 6$ and we need algebraic conditions which guarantee that the $S_i(x)$ have a factor of least degree $p=d-m$ in common. These are found as follows. First, one proves (as in Walker [10] theorem 5.2):

Lemma 5.2 f and g (polynomials of degree n and m in x respectively) have a factor of degree p in common if and only if there exist polynomials Φ and Ψ of degrees less than $n-p+1$ and $m-p+1$ respectively such that $\Psi f=\Phi g$.

Writing out the condition $\Psi f=\Phi g$, one obtains (as in Walker [10], theorem 5.3)

Lemma 5.3 $f(x)=a_o+a_1+...+a_nx^n$ and $g(x)=b_o+b_1x+...+b_mx^m$ have a factor of degree p in common if and only if

$$R = \begin{vmatrix} a_o & a_1 & \cdots & a_n & & \\ & a_o & \cdots & a_{n-1} & a_n & \\ & & \cdots & & & \\ & & a_o & \cdots & & a_n \\ b_o & b_1 & \cdots & b_m & & \\ & b_o & \cdots & b_{m-1} & b_m & \\ & & \cdots & & & \\ & & b_o & \cdots & & b_m \end{vmatrix} = 0$$

with $n-p+1$ rows of a's and $m-p+1$ rows of b's.

Lemma 5.3 deals with only 2 polynomials. One can extend it to several polynomials as Van der Waerden [10], page 2, does. In particular, let

$$S_u(x)=u_1S_1+ \ldots +u_6S_6$$

$$S_v(x)=v_1S_1+ \ldots +v_6S_6$$

where $u_1, \ldots, u_6, v_1, \ldots, v_6$ are indeterminants. Calculate R (as in lemma 5.3) for $S_u(x)$ and $S_v(x)$. Then the coefficients of R as a polynomial in the u and v yield the required algebraic conditions. q.e.d.

Theorem 5.4 The collection G of curves in Q having only nodes (ordinary double crossings) as singularities and of genus $\leq g$ is a locally closed subset of Q.

Proof Let P_n be the collection of homogeneous polynomials of degree n. We must show that $G \cap P_n$ is locally closed. $F \in G \cap P_n$ if and only if F has more nodes than

$$d(n) = \frac{(n-1)(n-2)}{2} - g$$

nodes. Hence, $G \cap P_n = N^{d(n)} \cap P_n$.
Since by proposition 5.1, $N^{d(n)}$ is locally closed, so is $G \cap P_n$. q.e.d.

Corollary 5.5 The collection Γ of curves in Q having only nodes (ordinary double crossings) as singularities and of genus g is the difference of two locally closed subsets of Q (or the union of a finite number of locally closed subsets).

References

1. P. Cherenack, Basic objects for an algebraic homotopy theory, Can. J. Math. XXVI (1972), 155-166.
2. P. Cherenack, The topological nature of algebraic contractions, Comment. Math. Univ. Carolinae 15 (1974) 481-499.
3. A. Grothendieck, "Eléments de géométrie algébrique I," Springer, Berlin, 1970.

4. H. Herrlich and G.C. Strecker, "Category theory," Allyn and Bacon, Boston, 1973.

5. S. Lang, Hilbert's Nullstellensatz in infinite dimensional space, Proc. Amer. Math. Soc. 3 (1952), 407-410.

6. Y.-C. Lu, "Singularity theory and an Introduction to Catastrophe Theory," Springer-Verlag, Berlin, 1976.

7. S. Mac Lane, "Categories for the Working Mathematician," Springer-Verlag, Berlin, 1971.

8. D. Mumford, "Introduction to algebraic geometry," Harvard University Press, Cambridge, Mass.

9. B.L. van der Waerden, "Modern Algebra, Volume II", Ungar, New York, 1964.

10. R.J. Walker, "Algebraic Curves," Dover, New York, 1962.

Grants to the Topology Research Group from the University of Cape Town and from the South African Council for Scientific and Industrial Research are acknowledged.

University of Cape Town
7700 Rondebosch
South Africa

SOLUTIONS OF WEIERSTRASS EQUATIONS

DAVID A. COX

Introduction.

We will show how geometric techniques can be used to solve some arithmetic questions involving function fields.

Let K be a function field in one variable over $\mathbb{C}$, and let

(0.1) $y^2 = 4x^3 - g_2x - g_3 \qquad g_2, g_3 \in K$

be a Weierstrass equation over K (where $j = g_2^3/(g_2^3 - 27g_3^2) \notin \mathbb{C}$). We want to answer the following:

(0.2) Question. Given K-rational solutions $\sigma_i = (\alpha_i, \beta_i)$, $i = 1, \ldots, \ell$ of (0.1), do the σ_i form a basis (modulo torsion) for the group of K-rational solutions of (0.1)? [Since $\Delta \neq 0$, (0.1) defines an elliptic curve, so its K-rational solutions form a group $\mathfrak{S}$. Since $j \notin \mathbb{C}$, the Mordell-Weil theorem implies that $\mathfrak{S}$ is finitely generated.]

In general, this is a difficult question to answer. If, however, we put some restrictions on the Weierstrass equation (0.1), then (0.2) can be answered completely. The restrictions needed are described in terms of the geometric aspect of our situation: K is the function field of a compact Riemann surface S, and (0.1) defines the generic fiber of an elliptic fibration:

(0.3) $$f: X \to S .$$

In §4, we will prove the following:

(0.4) Theorem. If $p_g(X) = 0$ and $\#\mathfrak{S}_{tor}$ is known, then there is an effective algorithm (described in §4) for answering question (0.2).

The algorithm is complicated, but in practice it is straightforward to use. Here are two examples of what it can do:

(0.5) Example. The group of $\mathbb{C}(t)$-rational solutions of the equation:

$$y^2 = 4x^3-4t^2x+t^2$$

is a free abelian group with $(0,t)$ and (t,t) as a basis.

(0.6) Example. The group of $\mathbb{C}(t)$-rational solutions of the equation:

$$y^2 = 4x^3-4t^2x+4t^6$$

is a free abelian group with $(0,2t^3)$, $(t,2t^3)$, $(-t^2,2t^2)$ and $(\lambda t^2,2\lambda^2t^2)$ $(\lambda = e^{\pi i/3})$ as a basis.

Our methods use the map $f: X \to S$ in an essential way. $\mathfrak{S}$ is naturally isomorphic to the sections of f, and the solution $(0,1,0)$ (the identity of $\mathfrak{S}$) gives a section σ_0 of f (the zero section). We can assume that f is minimal (no exceptional curves of the first kind in the fibers), so that f (up to a fiber-preserving isomorphism) and the Weierstrass equation (0.1) (up to an isomorphism $(x,y) \to (u^2x,u^3y)$, $u \in K(S)^*$, which transforms (0.1) into:

(0.7) $$y^2 = 4x^3-g_2'x-g_3', \qquad g_2' = u^4g_2,\ g_3' = u^6g_3)$$

mutually determine each other.

Let $S' = \{s \in S:\ X_s = f^{-1}(s)$ is smooth$\}$. Then $\Sigma = S-S'$ is finite, and the bad fibers X_s, $s \in \Sigma$, are classified into types $I_b(b>0)$, $I_b^*(b\geq 0)$, II, II*, III, III*, IV and IV*--see Table I (and [3,§6] for more details). In Table I we label only components of multiplicity one, and types II and II* are omitted because they have only one component of multiplicity one. An important fact is

TABLE I

Structure of the bad fibers X_s

Type	Structure	Picture
I_b $(b>0)$	$C_0 + C_1 + \dots + C_{b-1}$	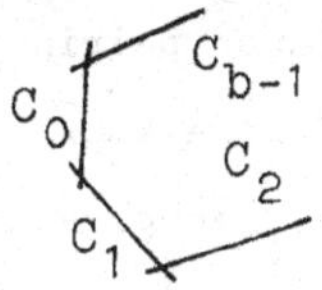
I_b^* $(b\geq 0)$	$C_0 + C_1 + 2C_2 + \dots + 2C_{b+2} + C_{b+3} + C_{b+4}$	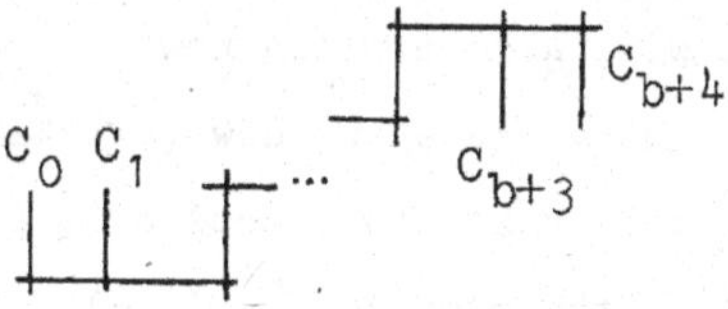
III	$C_0 + C_1$	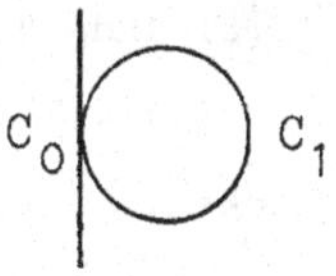
III*	$C_0 + C_1 + 2C_2 + 2C_3 + 3C_4 + 3C_5 + 4C_6 + 2C_7$	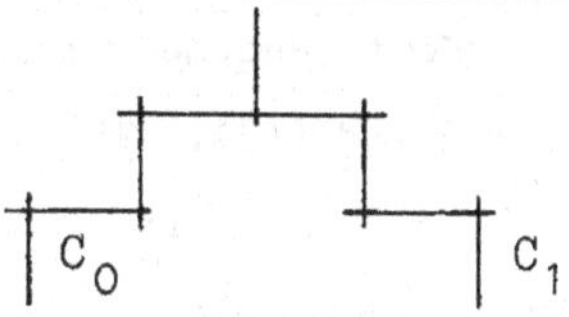
IV	$C_1 + C_2 + C_3$	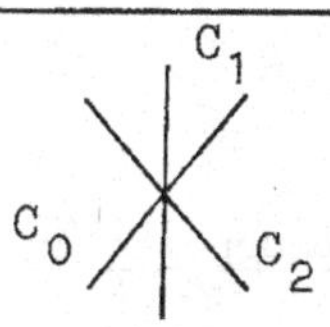
IV*	$C_0 + C_1 + C_2 + 3C_3 + 2C_4 + 2C_5 + 2C_6$	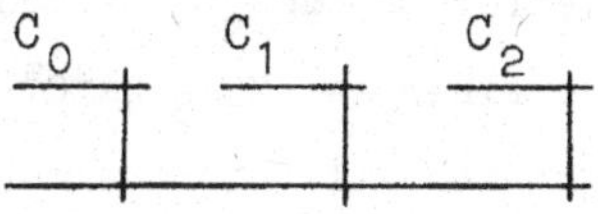

that the components of multiplicity one of X_s form a finite group G_s (see [4,III.17]), where addition of sections is compatible with addition in G_s.

We briefly outline the paper. In §1, we construct a homomorphism $\delta: \mathfrak{S} \to H^1(S,R^1f_*\mathbb{Q})$, and in §2 we use δ and cup product to construct a bilinear pairing $\langle\ ,\ \rangle$ on $\mathfrak{S}$. The crucial fact is that when $p_g = 0$, we know the discriminant of $\langle\ ,\ \rangle$ (see (2.8)). Computing $\langle\sigma,\sigma'\rangle$ involves some intersection products on X plus knowing which components of the bad fibers get hit by σ and σ'. We will also explain how $\langle\ ,\ \rangle$ and δ can be used to compute Tate heights. §3 shows how to take a solution of (0.1) and determine precisely how the section it gives hits the bad fibers. Then the algorithm promised in (0.4) is easy to state (§4), and we discuss the examples (0.5) and (0.6). The proofs given here are sometimes terse: full details will appear in [1]. Theorem (1.1) is joint work with S. Zucker.

The possibility of doing examples like these was first suggested by W. Hoyt, and he and C. Schwartz have worked out some examples (see [2] and [5]). The methods they use (automorphic forms, Eichler integrals, etc.) are quite cumbersome and so far have been applied to only the simplest examples.

§1. The homomorphism δ.

We let F denote a good fiber of f, and, for $s \in \Sigma$, we write the bad fiber as $X_s = \Sigma m_i^s C_i^s$. We label these so that σ_0 hits X_s in C_0^s (we call C_0^s the zero component).

The Leray spectral sequence $E_2^{p,q} = H^p(S,R^qf_*\mathbb{Q}) \Rightarrow H^{p+q}(X,\mathbb{Q})$ degenerates at E_2 (this is elementary), so that $H^2(X,\mathbb{Q})$ has a filtration $L^2 \subset L^1 \subset L^0 = H^2(X,\mathbb{Q})$ where:

$$L^1 = \mathrm{Ker}(H^2(X,\mathbb{Q}) \to H^0(S,R^2f_*,\mathbb{Q}))$$

$$L^1/L^2 \simeq H^1(S,R^1f_*\mathbb{Q})$$

$$L^2 = \mathrm{im}(H^2(S,\mathbb{Q}) \to H^2(X,\mathbb{Q})) = \mathbb{Q}\cdot[F].$$

(1.1) <u>Theorem</u> (Cox-Zucker). Let σ be in $\mathfrak{S}$. Then:

1. There is a rational linear combination $\sum_{s\in\Sigma} D_s$ of the components of bad fibers ($D_s = \sum_i a_i^s C_i^s$, $a_i^s \in \mathbb{Q}$) so that $[\sigma-\sigma_0+\sum_s D_s]$ lies in L^1.

2. The cohomology class $[\sigma-\sigma_0+\sum_s D_s]$ gives a well-defined element $\delta(\sigma)$ of $H^1(S,R^1f_*\mathbb{Q})$, and the map $\delta\colon \mathfrak{S} \to H^1(S,R^1f_*\mathbb{Q})$ is a homomorphism.

3. Each D_s, $s \in \Sigma$, is unique up to a rational multiple of X_s, and is computed as follows. Assume that σ hits C_k^s. If $k = 0$, we can choose $D_s = 0$; if $k \neq 0$, then D_s satisfies the equations:

$$(1.2) \qquad D_s\cdot C_i^s = \begin{cases} 1 & i = 0 \\ -1 & i = k \\ 0 & \text{otherwise} \end{cases}$$

<u>Proof</u>. L^1 consists of those classes which restrict to 0 in each fiber, so that $a \in L^1 \Leftrightarrow a\cdot C_i^s = 0$ for all s and i. Thus $[\sigma-\sigma_0+\sum_s D_s] \in L^1$ if and only if:

$$(1.3) \qquad (\sigma_0-\sigma)\cdot C_i^s = D_s\cdot C_i^s \qquad \text{for all } s \text{ and } i.$$

$D_s\cdot X_s = (\sigma_0-\sigma)\cdot X_s = 0$, so the equation for $i = 0$ is a consequence of the rest. Set $D_s' = D_s - a_0^s X_s$. Then (1.3) is equivalent to the equations:

$$(\sigma_0-\sigma)\cdot C_i^s = D_s'\cdot C_i^s \qquad \text{for all } s \text{ and } i > 0.$$

These have a unique solution since $(C_i^s\cdot C_j^s)_{i,j>0}$ is negative definite. Statements 1 and 3 follow immediately.

Since $[X_s] = [F]$ for $s \in \Sigma$, $\delta(\sigma) = [\sigma-\sigma_0+\sum_s D_s]$ gives us a well-defined element of $L^1/L^2 = H^1(S,R^1f_*\mathbb{Q})$. To show that δ is a homomorphism, let $X' = f^{-1}(S')$, so that we have a commutative diagram:

$$\begin{array}{ccc} X' & \hookrightarrow & X \\ {\scriptstyle f'}\downarrow & j & \downarrow{\scriptstyle f} \\ S' & \hookrightarrow & S \end{array}$$

where f' is proper and smooth. One easily checks that the map $R^1f_*\mathbb{Q} \to j_*R^1f'_*\mathbb{Q}$ is an isomorphism, so that we have an inclusion (from the Leray S.S. for j):

$$H^1(S,R^1f_*\mathbb{Q}) \cong H^1(S,j_*R^1f'_*\mathbb{Q}) \to H^1(S',R^1f'_*\mathbb{Q}).$$

Thus, we only have to show that the composed map $\mathfrak{S} \to H^1(S',R^1f'_*\mathbb{Q})$ is a homomorphism. This follows from the proof of Proposition 3.9 of [7].

§2. The pairing on sections.

For σ and σ' in $\mathfrak{S}$, we define:

(2.1) $\langle\sigma,\sigma'\rangle = -(\delta(\sigma)\cup\delta(\sigma'))$

where the cup product can be taken equivalently in $H^2(X,\mathbb{Q})$ or $H^1(S,R^1f_*\mathbb{Q})$ (the whole Leray S.S. has compatible cup products). There is a reason for the minus sign in (2.1):

(2.2) Lemma. $\langle\sigma,\sigma\rangle \geq 0$ and $\langle\sigma,\sigma\rangle = 0 \iff \sigma$ is torsion.

Proof. Choose an integer n so that $n\sigma$ hits the zero component of each X_s, $s \in \Sigma$. Then $\delta(n\sigma) = [n\sigma-\sigma_0]$ by (1.1). Thus

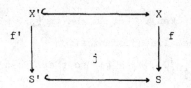

$n^2\langle\sigma,\sigma\rangle = \langle n\sigma,n\sigma\rangle = -(n\sigma)^2+2(n\sigma)\cdot\sigma_0-\sigma_0^2 = 2(n\sigma)\cdot\sigma_0-\sigma_0^2$. Since σ_0^2 is negative (see (4.2)), $\langle\sigma,\sigma\rangle > 0$ when $n\sigma$ and σ_0 are distinct divisors (i.e., when σ is not torsion).

From this, we see that $\delta\colon \mathfrak{S} \to H^1(S,R^1f_*Q)$ is injective modulo torsion. Also, if $\mathfrak{S}_0$ consists of those sections which always hit the zero component, the proof of (2.2) shows that $\mathfrak{S}_0$ is torsion free (this is also proved in [6]).

The first step in computing $\langle\sigma,\sigma'\rangle$ is:

(2.3) Lemma. For $\sigma,\sigma' \in \mathfrak{S}$, $\sigma\cdot D_s(\sigma') = \sigma'\cdot D_s(\sigma)$, and:

$$\langle\sigma,\sigma'\rangle = -(\sigma-\sigma_0)\cdot(\sigma'-\sigma_0)-\sum_{s\in\Sigma}\sigma\cdot D_s(\sigma').$$

Proof. Suppose that σ hits C_k^s, and write $D_s(\sigma') = \sum_{i>0} a_iC_i^s$. Then $\sigma\cdot D_s(\sigma') = a_k$, and from (1.2) we get $D_s(\sigma)\cdot D_s(\sigma') = -a_k$. Thus, the first equality follows by symmetry, and the second is now an easy computation.

Thus $\langle\sigma,\sigma'\rangle$ has a "geometric part", $-(\sigma-\sigma_0)\cdot(\sigma'-\sigma_0)$, and then "correction terms" $\sigma\cdot D_s(\sigma')$ coming from the behavior of σ and σ' at the bad fibers. These correction terms are easy to compute once we know which components of X_s σ and σ' hit. [For each type in the Kodaira classification, the picture in Table I (one needs to know the multiplicities of all the components) enables one to compute $D_s(\sigma')$ explicitly. From this, we read off $\sigma\cdot D_s(\sigma')$.] The results are listed in Table II.

Note that the greatest denominator that can occur is the exponent of the group G_s of components of multiplicity one of X_s. Thus $N\langle\sigma,\sigma'\rangle \in \mathbb{Z}$, where N is the l.c.m. of these exponents.

The final result we need relates the discriminant of $\langle\ ,\ \rangle$ on $\mathfrak{S}$ to the discriminant of the intersection form $(\ ,\ )$ on the

TABLE II. Local correction terms $\sigma\cdot D_s(\sigma')$.

Type of X_s	Criterion (see Table I)	Correction term
Arbitrary	σ or σ' hits the zero component	0
I_b	σ hits C_k, σ' hits $C_{k'}$, $0 < k \leq k'$	$k(b-k')/b$
I_b^*	σ,σ' hit C_1	1
	σ,σ' both hit C_{b+3} or C_{b+4}	$(b+4)/4$
	One hits C_{b+3}, the other C_{b+4}	$(b+2)/4$
	One hits C_1, the other C_{b+3} or C_{b+4}	1/2
III	σ,σ' hit C_1	1/2
III*	σ,σ' hit C_1	3/2
IV	σ,σ' both hit C_1 or C_2	2/3
	One hits C_1, the other hits C_2	1/3
IV*	σ,σ' both hit C_1 or C_2	4/3
	One hits C_1, the other hits C_2	2/3

Neron-Severi group NS(X). Recall that the discriminant of a $\mathbb{Q}$-valued form (,) on a finitely generated abelian group G is defined as follows: if $g_1,\ldots,g_r$ is a basis of G modulo torsion, then:

$$\operatorname{disc}(\ ,\ )_G = \begin{cases} \det(g_i,g_j)/(\#G_{tor})^2 & \text{if } r > 0 \\ 1/(\#G)^2 & \text{if } r = 0 \end{cases}$$

where G_{tor} is the torsion subgroup of G. Then we have:

(2.4) <u>Proposition</u>. Let m_s be the number of components of multiplicity one in the fiber X_s (so $m_s = \#G_s$). Then:

$$\operatorname{disc} \langle\ ,\ \rangle_{\mathfrak{S}} = |\operatorname{disc}(\ ,\ )_{NS(X)}| / \prod_s m_s .$$

<u>Proof</u>. If $\sigma = \sum_i b_i \sigma_i$ in $\mathfrak{S}$, in NS(X) we get a relation

(2.5) $[\sigma-\sigma_0] = \sum_i b_i[\sigma_i-\sigma_0] + a[F] + \sum_{i>0} b_i^s[C_i^s]$

where the F and C_i^s are as in §1. From this and Theorem 1.1 of [6] we see that the map sending σ to $\sigma-\sigma_0$ gives an isomorphism:

(2.6) $\mathfrak{S} \xrightarrow{\sim} NS(X)/(\mathbb{Z}[\sigma_0]+\mathbb{Z}[F]+\sum_{i>0} \mathbb{Z}[C_i^s])$.

Let H be the subgroup of $NS(X)$ generated by $[\sigma_0]$, $[F]$, $[C_i^s]$ (for $i>0$) and $[\sigma]$ (for $\sigma \in \mathfrak{S}_0$). If $\sigma_1,\ldots,\sigma_r$ is a basis of $_0$, then (2.5) shows that H is spanned by the elements $[\sigma_0]$, $[F]$, $[C_i^s]$ (for $i > 0$) and $a_i = [\sigma_i-\sigma_0-((\sigma_i-\sigma_0)\cdot\sigma_0)F]$ (note that $a_i\cdot C_i^s = a_i\cdot F = a_i\cdot\sigma_0 = 0$). Then one easily computes that:

(2.7) $|\mathrm{disc}(\ ,\ )_H| = \mathrm{disc}\langle\ ,\ \rangle_{\mathfrak{S}_0}\cdot\Pi m_s$

because $\det(a_i,a_j) = \det\langle\sigma_i,\sigma_j\rangle$ and $\det(C_i^s\cdot C_j^s) = m_s$ (see Lemma 1.3 in [6]--also $NS(X)$ is torsion-free by [1,(1.43)]).

The natural map $\mathfrak{S} \to \bigoplus_s G_s$ (evaluating which components a section hits--its kernel is $\mathfrak{S}_0$) gives us, via (2.6), a map $NS(X) \to \bigoplus_s G_s$. From (2.6) we see that its kernel is preciesly H, so that $[NS(X):H] = [\mathfrak{S}:\mathfrak{S}_0]$. Then we have:

$$\begin{aligned}\mathrm{disc}\langle\ ,\ \rangle_{\mathfrak{S}} &= \mathrm{disc}\langle\ ,\ \rangle_{\mathfrak{S}_0}/[\mathfrak{S}:\mathfrak{S}_0]^2\\ &= |\mathrm{disc}(\ ,\ )_H|/[\mathfrak{S}:\mathfrak{S}_0]^2\cdot\Pi_s m_s\\ &= |\mathrm{disc}(\ ,\ )_{NS(X)}|/\Pi_s m_s.\end{aligned}$$

A very restricted version of this is proved in [6].

If $p_g(X) = 0$, then $NS(X) = H^2(X,\mathbb{Z})$, and $(\ ,\ )$ has discriminant one by Poincare duality. Thus, we get:

(2.8) <u>Corollary</u>. Assume that $p_g(X) = 0$. Then:

1. If $\mathfrak{G}$ has rank 0, then $(\#\mathfrak{G})^2 = \prod_s m_s$.
2. If $\mathfrak{G}$ has rank $r > 0$, then $\sigma_1,\ldots,\sigma_r$ in $\mathfrak{G}$ are a basis modulo torsion if and only if:

$$\det\langle\sigma_i,\sigma_j\rangle = (\#\mathfrak{G}_{tor})^2/\prod_s m_s.$$

Question (0.2) is answered using Corollary (2.8). This means computing $\langle\sigma_i,\sigma_j\rangle$, which, by Lemma (2.3) and Table II, involves knowing which components of X_s get hit by σ_i and σ_j.

Let us also state (without proof) the relationship between Tate heights and $\langle\ ,\ \rangle$ and δ. Let D be a divisor on the generic fiber E of $f:X \to S$ (so that $\mathfrak{G}$ is just $E(K)$). Then we have a height function $h_D: \mathfrak{G} \to \mathbb{Z}$, and h_D gives a quadratic function $\hat{h}_D$ (see [8]). Then one can show that:

$$(2.9)\quad \hat{h}_D(\sigma) = 1/2\langle\sigma,\sigma\rangle \deg D + (\bar{D},\delta(\sigma))$$

where $\bar{D}$ is the divisor on X determined by D ($\bar{D}$ has no components in any bad fibers), and:

$$\delta: \mathfrak{G} \to H^2(X,Q)$$

defined by $\delta(\sigma) = \sigma-\sigma_0 + \sum_s D_s(\sigma)-((\sigma-\sigma_0)\cdot\sigma_0)\cdot F$. One can rewrite (2.9) as:

$$\hat{h}_D(\sigma) = (\bar{D},\sigma-\sigma_0 + \sum_s D_s(\sigma)-(1/2)(\sum_s \sigma\cdot D_s(\sigma))F)$$

and Theorem 4 of [8] can be strengthened to show that $\sigma \in \mathfrak{G}_0 \iff$ for every D, $\hat{h}_D(\sigma) = (\bar{D},\sigma-\sigma_0)$.

§3. The component of a bad fiber hit by a section.

This is a local problem, so we can assume that we have a solution $\sigma = (\alpha,\beta),\alpha,\beta \in \mathbb{C}((t))$, of a Weierstrass equation:

(3.1) $y^2 = 4x^3-g_2x-g_3$ $\qquad g_2,g_3 \in \mathbb{C}((t))$.

This equation has a local Neron model X over Spec ($\mathbb{C}[[t]]$), and we say that (3.1) has type I_b, I_b^*, etc., if the special fiber X_s of X is of that type. We want to know which component of X_s is hit by σ.

Since we are only interested in (3.1) up to an isomorphism as described in (0.7), we can transform (3.1) into a Weierstrass equation where ord $g_2 \geq 0$, ord $g_3 \geq 0$, and ord Δ = ord$(g_2^3-27g_3^2)$ is as small as possible. Such an equation is called minimal, and it defines a local surface Y over Spec($\mathbb{C}[[t]]$). Y is isomorphic to X after all of the non-zero components of X_s have been collapsed to a point. [Neron proves this for standard equations [4, III.7 and 16]. A minimal equation is standard except for types I_b and I_b^* $(b > 0)$, and in these two cases, a minimal equation is isomorphic (over $\mathbb{C}[[t]]$) to a standard one.]

The special fiber Y_s is defined by the cubic $y^2 = 4x^3-g_2(0)x-g_3(0)$, which has one singular point $(a,0)$ (and $a = 0$ except in case I_b). The solution $\sigma = (\alpha,\beta)$ gives compatible sections of X and Y over Spec ($\mathbb{C}[[t]]$), and determining where σ hits Y_s is easy: just evaluate (α,β) at $t = 0$. From this, one easily gets:

(3.2) Proposition. A solution (α,β) of a minimal Weierstrass equation hits a non-zero component of X_s iff ord $(\alpha-a) > 0$.

For a minimal equation of type III or III*, (3.2) is all we need: for these types, X_s has only one non-zero component of multiplicity one (see Table I).

For types IV, IV*, I_b and I_b^*, we will describe, case by case, which non-zero component of X_s (as labeled in Table I) is hit by a section σ satisfying (3.2). We will state the results without proof, though at the end of this section we give some brief indications of how the proofs go.

When we write an equation like:

$$\alpha = ut^k + \dots$$

we mean that the omitted terms have degree $> k$.

Minimal equations for these types can be written:

$$\text{(3.3)}\quad \begin{array}{lll} IV & : & y^2 = 4x^3 - rt^2x - st^2,\ s(0) \neq 0 \\ IV^* & : & y^2 = 4x^3 - rt^3x - st^4,\ s(0) \neq 0 \\ I_0^* & : & y^2 = 4x^3 - rt^2x - st^3,\ r(0)^2 - 27s(0)^3 \neq 0 \\ I_b\ (b > 0) & : & y^2 = 4x^3 - rx - s,\ r(0) \neq 0,\ s(0) \neq 0,\ \text{ord}\,\Delta = b \\ I_b^*\ (b > 0) & : & y^2 = 4x^3 - rt^2x - st^3,\ r(0) \neq 0,\ s(0) \neq 0,\ \text{ord}\,\Delta = b+6. \end{array}$$

(3.4) Proposotion. Suppose we have a minimal Weierstrass equation of type IV or IV*, as in (3.3). Pick a square root q of $-s(0)$. Then the non-zero components C_1 and C_2 of multiplicity one in X_s (see Table I) can be labeled so that a solution (α,β) of (3.3) hits C_1 (resp. C_2) if and only if:

1. (Type IV) $\beta = qt + \dots$ (resp. $\beta = -qt + \dots$)
2. (Type IV*) $\beta = qt^2 + \dots$ (resp. $\beta = -qt^2 + \dots$).

(3.5) Proposition. Suppose we have a minimal Weierstrass equation of type I_0^*, as in (3.3). Let u_1, u_2 and u_3 be the (distinct) roots of the cubic $4x^3 - r(0)x - s(0)$. Then the non-zero components C_1, C_2 and C_3 of multiplicity one in X_s (see Table I) can be labeled so that a solution (α,β) of (3.3) hits C_i if and only if $\alpha = u_it + \dots$.

(3.6) Proposition. Suppose we have a minimal Weierstrass equation of type I_b, $b > 0$, as in (3.3) (here Y_s has a singular point $(-3s(0)/2r(0),0)$). Pick a square root q of $-s(0)/2r(0)$. Then the non-zero components $C_1,\ldots,C_{b-1}$ of X_s (see Table I) can be labeled so that if a solution (α,β) of (3.3) misses the zero component, then:

1. (α,β) hits $C_{b/2}$ if and only if $2 \operatorname{ord} \beta \geq b$.
2. (α,β) hits C_k or C_{b-k} if and only if $\operatorname{ord} \beta = k$, $2k < b$. In this case, we can write $12\alpha^2-r = ut^k+\ldots$, $u \neq 0$, and then (α,β) hits C_k (resp. C_{b-k}) if and only if $\beta = (u/12q)t^k + \ldots$ (resp. $\beta = -(u/12q)t^k + \ldots$).

(3.7) Proposition. Suppose we have a minimal Weierstrass equation of type I_b^*, $b > 0$, as in (3.3). The cubic $4x^3-r(0)x-s(0)$ has a double root $a = -3s(0)/2r(0)$; the other root is $-2a$. Write $r^3-27s^2 = mt^b + \ldots$, and pick a square root q of $-m/3a$ (b odd) or of m (b even). Then the non-zero components C_1, C_{b+3}, C_{b+4} of multiplicity one of X_s (see Table I) can be labeled so that if (α,β) is a solution of (3.3), then:

1. (α,β) hits C_1 if and only if $\alpha = -2at + \ldots$
2. If b is odd, (α,β) hits C_{b+3} (resp. C_{b+4}) if and only if $\alpha = at + \ldots$, and $\beta = (q/12a)t^{(b+3)/2} + \ldots$ (resp. $\beta = -(q/12a)t^{(b+3)/2} + \ldots$).
3. If b is even, (α,β) hits C_{b+3} (resp. C_{b+4}) if and only if $\alpha = at + \ldots$, and $12\alpha^2-rt^2 = (q/3a)t^{(b+4)/2} + \ldots$ (resp. $12\alpha^2-rt^2 = -(q/3a)t^{(b+4)/2} + \ldots$).

All the labelings mentioned in these propositions preserve the incidence relations as pictured in Table I, and correspond to the labeling used in Table II.

The proofs of (3.4) and (3.5) are quite elementary. For example, consider type I_0^*. If a solution (α,β) of (3.3) misses the zero component, then $\alpha = ut + \ldots$ by (3.2). Substituting this in (3.3) gives $\beta^2 = (ru^3-r(0)u-s(0))t^3 + \ldots$, so that $u \in \{u_1,u_2,u_3\}$ and ord $\beta \geq 2$. Proving (3.5) then reduces to showing that:

(3.8) Solutions (α,β) and (α',β') hit the same non-zero component $\Longleftrightarrow$ α and α' have the same coefficient of t.

Since addition of solutions is compatible with addition of components of multiplicity one of X_s, (α,β) and (α',β') hit the same component $\Longleftrightarrow$ $(\alpha_1,\beta_1) = (\alpha,\beta)-(\alpha',\beta')$ hits the zero component. Since $\alpha_1 = -\alpha-\alpha' + (1/4)((\beta+\beta')/(\alpha-\alpha'))^2$, deducing (3.8) from (3.2) is an easy exercise.

The proofs of (3.6) and (3.7) make extensive use of Neron [4]. There, one has projections $\pi_i : X \to A_i$, where the A_i are certain very explicity plane curves over $\mathbb{C}[[t]]$ (see below). Each component of X_s maps onto a specific component of precisely one A_i^0 (the special fiber of A_i). This information is in the tables of generic pts on pages 104, 109 and 110 in [4]. For types I_b and I_b^* $(b > 0)$, the r of (3.3) can be written as $r = 12v^2$, where $v = a + \ldots$. Then the coordinate changes $(x,y) \to (x-v,y)$ (type I_b) and $(x,y) \to (x-vt,y)$ (type I_b^*) transform (3.3) into a standard equation A. The A_i are built out of A in a very explicit manner (see pages 103 and 108 in [4]). Thus, a solution of (3.3) gives explicit sections σ_i of the A_i. Since it is easy to tell where σ_i hits A_i^0 (substitute in $t = 0$), we can tell where σ hits X_s. The explicit coefficients ($u/12q$, $q/12a$, etc.) that occur in (3.6) and (3.7) reflect the equations defining the A_i^0.

§4. The algorithm and some examples.

Given any Weierstrass equation:

(4.1) $$y^2 = 4x^3-g_2x-g_3 \qquad g_2,g_3 \in K(S)$$

the first thing to compute is p_g. The formula is well-known [3,§12]:

(4.2) $\sigma_0^2 = -(p_g-q+1) = -(1/12)\,[\deg j + 6\sum_{b\geq 0} \nu(I_b^*)+2\nu(II)+$

$10\nu(II^*)+3\nu(III)+9\nu(III^*)+4\nu(IV)+8\nu(IV^*)]$

where $\nu(I_b^*)$, $\nu(II)$, etc. are the number of fibers of $f:X \to S$ of type I_b^*, II, etc. (and these are easy to determine from (4.1)--see [4,III.17]). Since $q = g$, the genus of S (see [1,1.43]), we can find p_g.

The next step is to compute $r = \operatorname{rank} \mathfrak{G}$. There are two ways to do this:

1. From [6] we have: $\operatorname{rank} NS(X) = r+2+\Sigma(m_s-1)$. When $p_g = 0$, $\operatorname{rank} NS(X) = 10$.
2. When $p_g = 0$, we have $r = -4+2\nu-\nu_1$, where ν is the number of bad fibers and $\nu_1 = \Sigma\nu(I_b)$ (see [1,§3]).

Now, the algorithm is clear: when $p_g = 0$ and $\#\mathfrak{G}_{tor}$ is known, (2.8) tells us precisely when $\sigma_1,\ldots,\sigma_r \in \mathfrak{G}$ are a basis modulo torsion, once we know the numbers $\langle\sigma_i,\sigma_j\rangle$. These are easy to compute: by (2.3), §3 and Table II, we need only compute $(\sigma_i-\sigma_0)\cdot(\sigma_j-\sigma_0)$. But for any sections σ and σ', $\sigma\cdot\sigma'$ is easy to compute.

If $\sigma = \sigma'$, then $\sigma\cdot\sigma' = \sigma^2 = \sigma_0^2$, which we know by (4.2). If $\sigma \neq \sigma'$, then $\sigma\cdot\sigma' = \sum_{P\in\sigma\cap\sigma'} (\sigma\cdot\sigma')_P$, a sum of local contributions. If $P \in \sigma\cap\sigma'$ lies on either a good fiber or the zero component of a bad fiber, then near P, X is defined by a minimal Weierstrass equation. This equation can be used to compute $(\sigma\cdot\sigma')_P$. If P is on a non-zero component, then $\sigma-\sigma'$ (the difference in $\mathfrak{G}$) and σ_0 meet at 0 (the identity) on the zero component, and $(\sigma\cdot\sigma')_P = (\sigma_0\cdot(\sigma-\sigma'))_0$.

We do not know an effective method for computing torsion. One fact, mentioned in [6], is that $\mathfrak{G}_{tor}$ injects into any fiber. If

X_s is a bad fiber not of type I_b, this gives us an injection:

(4.3) $$\mathfrak{S}_{tor} \to G_s.$$

Below we show another way to determine torsion in some examples.

The examples themselves are quite easy. Let us do the details of Example (0.5). Here we have a Weierstrass equation:

(4.4) $$y^2 = 4x^3-4t^2x+t^2$$

One computes that $\Delta = t^4(64t^2-27)$ and $j = 64t^2(64t^2-27)^{-1}$. There are 4 bad fibers of the following types: IV (over 0), I_0^* (over ∞), and I_1 (over $\pm 3\sqrt{3}/8$). From (4.2) we have $p_g = 0$ and $\sigma_0^2 = -1$. $\mathfrak{S}$ has rank 2 and no torsion (apply (4.3) to the fibers over 0 and ∞).

We will show that $(0,t)$ and (t,t) form a basis of $\mathfrak{S}$. It is easier to work with $\sigma_1 = (0,t)$ and $\sigma_2 = -(t,t) = (t,-t)$. By (2.8), we must show that $\det\langle\sigma_i,\sigma_j\rangle = 1^2/3\cdot4\cdot1\cdot1 = 1/12$.

It is clear that σ_1 and σ_2 cannot meet except possibly over 0 and ∞, and the same is true for σ_1 and σ_0, and σ_2 and σ_0. Let us see how σ_1 and σ_2 behave over 0 and ∞:

1. At $t = 0$, (4.4) is minimal, and (3.4) (with $q = 1$) says that σ_1 hits C_1 and σ_2 hits C_2 (this explains the minus sign).
2. At $t = \infty$, $T = 1/t$ is a local parameter, and we transform (4.4) to the minimal equation $y^2 = 4x^3-4T^2x+T^4$ Our solutions become $\sigma_1 = (0,T^2)$ and $\sigma_2 = (T,-T^2)$. Label the roots of $4x^3-4x$ as $u_1 = 0$, $u_2 = 1$ and $u_3 = -1$. Then, by (3.5), σ_1 hits C_1 and σ_2 hits C_2.

Thus, $\sigma_1\cdot\sigma_2 = \sigma_1\cdot\sigma_0 = \sigma_2\cdot\sigma_0 = 0$, so that (2.3) and Table II give us:

$$\langle\sigma_1,\sigma_1\rangle = -(-2+2/3+1) = 1/3$$
$$\langle\sigma_2,\sigma_2\rangle = -(-2+2/3+1) = 1/3$$
$$\langle\sigma_1,\sigma_2\rangle = -(-1+1/3+1/2) = 1/6,$$

and then $\det\langle\sigma_i,\sigma_j\rangle = 1/12$, as required.

Showing that $\mathfrak{S}$ is torsion-free in Example (0.6) is a little harder. Here, there are 7 bad fibers: one of type I_0^* (over ∞) and 6 of type I_1. If σ is torsion and $\sigma \neq \sigma_0$, then σ must hit a non-zero component over ∞ ($\mathfrak{S}_0$ is torsion-free). Thus $0 = \langle\sigma,\sigma\rangle = 2+2(\sigma\cdot\sigma_0)-1$, which is impossible. From here on, the verification of Example (0.6) is easy.

There are several more known examples--see [1,§2].

References.

1. D. Cox and S. Zucker: Intersection numbers of sections of elliptic surfaces. To appear.
2. W. Hoyt and C. Schwartz: Period relations for the Weierstrass equation $y^2 = 4x^3-3ux-u$. In preparation.
3. K. Kodaira: On compact analytic surfaces, II-III. Annals of Math. 77, 563-626; 78, 1-40 (1963).
4. A. Neron: Modèles minimaux des variétiés abéliennes sur les corps locaux et globaux. Pub. Math. I.H.E.S. 21 (1964).
5. C. Schwartz: On generators of the group of rational solutions of a certain Weierstrass equation. Trans. A.M.S., to appear.
6. T. Shioda: On elliptic modular surfaces. J. Math. Soc. Japan 24, 20-59 (1972).
7. S. Zucker: Generalized intermediate Jacobians and the theorem on normal functions. Inventiones math. 33, 185-222 (1976).
8. J. Manin: Tate heights of points of abelian varieties. Amer. Math. Soc. Transl., 59 (1966), p. 82-110.

Department of Mathematics
Rutgers University
New Brunswick, NJ 08903, USA

Instantons and sheaves on $\mathbb{C}\mathbb{P}^3$

V.G. Drinfel'd and Ju.I. Manin

Introduction.

0.1. Let S^4 be the four-dimensional sphere with the standard conformal structure, G one of the classical compact Lie groups, L a vector bundle over S^4 associated with some principal G-bundle and equipped with a G-invariant connection ∇ with selfdual curvature form. The pair (L, ∇) is said to be a selfdual Euclidean Yang-Mills field, or an instanton bundle. We shall replace the invariant G-structure by a fixed positive definite ∇-horizontal form on L (orthogonal, unitary or symplectic).

In the work [3] (see also [2]) it was shown how to reduce the problem of classification of instantons up to G-isomorphism (or "gauge equivalence") to the description of certain linear algebra data. In the work [5], these algebraic objects, their parameter spaces and the Yang-Mills fields corresponding to them were described in detail. The purpose of this paper is to show that the constructions [3],[5] give all instantons and that their gauge equivalence is faithfully described in the algebraic language given. A short exposition of the proof has been given in [2].

The main part of the proof consists in giving a direct construction of linear algebra data corresponding to all instanton bundles. This construction consists of several steps, some of which we describe in the introduction.

0.2. From instantons to holomorphic bundles on $\mathbb{C}\mathbb{P}^3$. This step has been described by Atiyah and Ward (see the more general formulation and proof in [1]). Let V be a four-dimensional complex space, $\mathbb{C}\mathbb{P}^3$ the space of hyperplanes in V (so that $V = H^0(\mathbb{C}\mathbb{P}^3, \mathcal{O}(1))$), $j : \mathbb{C}\mathbb{P}^3 \to \mathbb{C}\mathbb{P}^3$ the antiholomorphic mapping having in suitable homogenous coordinates the form : $j(z_1,z_2,z_3,z_4) = (-\bar{z}_2, \bar{z}_1, -\bar{z}_4, \bar{z}_3)$. There exists a projection $\pi : \mathbb{C}\mathbb{P}^3 \longrightarrow S^4$ whose fibers are all j-invariant lines (see the details in [5] and below, §1.6). Associated to an instanton (L, ∇) for the group $O(r)$ is a locally free holomorphic bundle over $\mathbb{C}\mathbb{P}^3$ with total space $\pi^*(L \otimes_{\mathbb{R}} \mathbb{C})$. Its local holomorphic sections ψ are characterised by the condition $(\pi^*\nabla)^{0,1} \psi = 0$. By Serre's theorem, this bundle has a unique algebraic structure.

Instantons for the groups $O(r)$ and $Sp(r)$ are interpreted as $O(2r)$- or $O(4r)$-instantons, equipped with additional structure: an orthogonal operator J with $J^2 = -1$, or a pair of orthogonal operators J_1, J_2 with $J_1^2 = J_2^2 = -1$, $J_1J_2 = -J_2J_1$. These operators must be ∇-horizontal.

<u>0.3. The vanishing theorem.</u> Let $\mathcal{L}$ be the sheaf of holomorphic (or algebraic) sections of $\pi^*(L \otimes_{\mathbb{R}} \mathbb{C})$, where L is an instanton bundle. Then $H^i(\mathbb{CP}^3, \mathcal{L}(\kappa)) = 0$ when $i = 0,1$; $i + \kappa \leq -1$ and when $i = 2,3$; $i + \kappa \geq 0$. This result is proven in §1 of the paper. Call locally free sheaves on $\mathbb{CP}^3$ with this property admissible. There is a simple way of constructing such sheaves, as follows. Let I be a finite-dimensional linear space over $\mathbb{C}$, $\tilde{I}$ the sheaf of sections of the trivial bundle $\mathbb{CP}^3 \times I$. Call a special monad $\mathcal{M}$ a sheaf complex of the form $0 \to \tilde{I}(-1) \overset{\alpha}{\to} \tilde{H} \overset{\beta}{\to} \tilde{I}'(1) \to 0$, where α is injective and β is surjective on geometric fibers. Associate with the special monad $\mathcal{M}$ the sheaf $\mathcal{F}(\mathcal{M}) = \operatorname{Ker}\beta/\operatorname{Im}\alpha$. It is not hard to see that $\mathcal{F}(\mathcal{M})$ is admissible (see Proposition 2.1).

<u>0.4. The classification theorem for admissible sheaves.</u> The functor $\mathcal{F}$ naturally defines an equivalence of the category of special monads and the category of admissible sheaves. This theorem is proven in §2. A basic moment of the proof consists in the explicit construction of the inverse functor $\mathcal{G}$. The spaces I, H, I' which the functor $\mathcal{G}$ associates with the sheaf $\mathcal{L}$ are, respectively, $H^1(\mathcal{L} \otimes \Omega^2(1))$, $H^1(\mathcal{L} \otimes \Omega^1)$ and $H^1(\mathcal{L}(-1))$; the morphisms α and β are described in a more complex and less explicit manner (see also [2]).

This construction was suggested in papers of G. Horrocks and W. Barth [8] and in turn was generalized significantly by A.A. Beylinson [6] and I.N. Bernstein, I.M. Gel'fand and S.I. Gel'fand [7]. Its application to our situation is made considerably easier because of the possibility of restricting to admissible sheaves.

Unfortunately, the proofs, especially in §2, need more homological techniques than expected. General facts on coherent analytic sheaves and their cohomologies, that suffice for understanding of §1 can be found in [12]. In [10], especially in lectures 6-7, the facts on higher direct images of sheaves needed in §2 are briefly described. Finally, in [11], one can find almost all results necessary for us with complete proofs.

Note that the theorem on the equivalence of categories of §2 is later used extensively : it is necessary not only for the proof of the whole construction, but for the correct description of gauge equivalence. Indeed, we show that each instanton bundle is canonically representable as a direct summand of a trivial bundle and the instanton connection is the projection of the trivial connection. The extension of the metric and connection on this trivial bundle are entirely based on the equivalence theorem, as is the construction itself.

0.5 From monads to linear algebra data. Giving a morphism $\beta : \tilde{H} \to \tilde{I}'(1)$ is equivalent to giving a linear mapping $H \to I' \times V$. The geometric fiber $\mathcal{O}(1)$ over the point $z \in \mathbb{CP}^3$ can be canonically identified with the factor space V/W_z, where W_z is the space of linear forms which are equal to zero at the point z. Hence, fiber surjectivity of β is equivalent to the surjectivity of all the induced mappings $H \to I' \otimes (V/W)$.

Similarly, giving a morphism $\alpha : \tilde{I}(-1) \to \tilde{H}$ is equivalent to giving a morphism $\alpha(1) : \tilde{I} \to \tilde{H}(1)$, or a mapping between the linear spaces $I \to V^* \otimes H$. For all hyperplanes $W \subset V$, the induced mappings $I \otimes (V/W)^* \to H$ have to be injective and their images have to be contained in the kernel of $H \to I' \otimes (V/W)$.

In this way, the classification of admissible sheaves reduces to the description of the diagram of the linear spaces $I \otimes V^* \to H \to I \otimes V$ with the required properties. But this is not yet a solution of the instanton problem : not all admissible sheaves come from instantons, and those that do are equipped with additional structure. A translation of this structure to the language of linear algebra is given in §3, and §4 contains a final theorem about the positive definiteness of the metric induced on the trivial bundle by the fiber H.

§1. Cohomology of instanton bundles.

1.1. In this section, we shall use the following notations : (L, ∇) a complex vector bundle over S^4 with selfdual connection and horizontal Hermitian metric $\langle\ ,\ \rangle$ (antilinear in the first variable); $\pi : \mathbb{CP}^3 \to S^4$ the mapping of Atiyah-Ward; $\mathcal{L}$ the sheaf of holomorphic (or algebraic) sections of $\pi^*(L)$ over $\mathbb{CP}^3$. The basic aim of this section is the proof of the following result :

1.2. Theorem. $H^i(\mathcal{L}(\kappa)) = 0$ when $i \le 1$, $i + \kappa \le -1$ and when $i \ge 2$, $i + \kappa \ge 0$.

The most technical part of the paper is the proof of the following proposition :

1.3. Proposition. $H^1(\mathcal{L}(-2)) = 0$.

1.4. Deduction of the theorem from the proposition. Let L^* be the dual bundle ; the dual connection ∇^* is selfdual and the dual metric is ∇^*-horizontal ; the sheaf $\mathcal{L}^*$ associated with L^* is dual to $\mathcal{L}$ as a holomorphic (algebraic) sheaf. Because of the duality theorem of Serre, $H^i(\mathcal{L}(\kappa))^* \cong H^{3-i}(\mathcal{L}^*(-4-\kappa))$, it is enough to show that $H^i(\mathcal{L}(\kappa)) = 0$ when $i \leq 1$, $i+\kappa \leq -1$. The fibers of π are projective lines in $\mathbb{CP}^3$ and the restriction of $\mathcal{L}$ to them is holomorphically trivial. Hence, (see [8]) the restriction of $\mathcal{L}$ to almost every line is trivial. Since $H^0(\mathbb{CP}^1, \mathcal{O}(\kappa)) = 0$ when $\kappa < 0$ each section $s \in H^0(\mathcal{L}(\kappa))$ has to reduce to zero on almost all lines and, consequently, is equal to zero.

Let, further, $E \subset \mathbb{CP}^3$ be a plane containing one of the fibers $\pi^{-1}(x)$, $x \in S^4$. Then the restriction of $\mathcal{L}$ to almost every line in E is trivial, so by the previous reasoning $H^0(E, \mathcal{L}|_E(\kappa)) = 0$ when $\kappa < 0$. The standard exact seqence

$$H^0(\mathcal{L}|_E(\kappa)) \longrightarrow H^1(\mathcal{L}(\kappa-1)) \longrightarrow H^1(\mathcal{L}(\kappa))$$

and an induction on κ starting with $\kappa = -2$ (Proposition 1.3) give that $H^1(\mathcal{L}(\kappa)) = 0$ when $\kappa \leq -2$.

1.5. The plan of the proof of the proposition. The group $H^1(\mathcal{L}(-2))$ can be computed as the cohomology group of the initial segment of the Dolbeault complex:

$$\Gamma(\pi^*(L)(-2)) \xrightarrow{\bar\partial} \Gamma(\pi^*(L)(-2) \otimes \Omega^{0,1}) \xrightarrow{\bar\partial} \Gamma(\pi^*(L)(-2) \otimes \Omega)^{0,1},$$

where Γ is the space of smooth sections corresponding to a bundle over $\mathbb{CP}^3$, and the operator $\bar\partial$ on $\Gamma(\pi^*(L))$ in the Atiyah-Ward construction coincides with the $(0,1)$ component of the lifted connection $\pi^*(\nabla)^{(0,1)}$.

Hence, we need to prove that if $\omega \in \Gamma(\pi^*(L)(-2) \otimes \Omega^{0,1})$ and $\bar\partial\omega = 0$, then $\omega = \bar\partial\nu$ for some $\nu \in \Gamma(\pi^*(L)(-2))$. Denote by ω_v the restriction of the form ω to the vertical vector fields (tangent to the fiber π). From local computations (see below) it will follow that if $\omega_v = 0$, then $\omega = \bar\partial\nu$. Hence we shall look for $\nu \in \Gamma(\pi^*(L)(-2))$ such that $(\omega - \bar\partial\nu)_v = 0$.

The last assertion is more or less equivalent to the condition that ω becomes $\bar\partial$-closed when restricted to the fiber π. Since $H^1(\mathbb{CP}^1, \mathcal{O}(-2)) \neq 0$, the verification of this is not automatic. But $H^1(\mathbb{CP}^1, \mathcal{O}(-1)) = 0$, so that if we fix an embedding $\mathcal{O}(-2) \hookrightarrow \mathcal{O}(-1)$, then on each fiber $\pi^{-1}(x)$ we can find a section ν_x of the sheaf $\mathcal{L}(-1)|_{\pi^{-1}(x)}$ such that $\omega|_{\pi^{-1}(x)} = \bar\partial\nu_x$. Moreover, $H^0(\mathbb{CP}^1, \mathcal{O}(-1)) = 0$, so that the forms ν_x are uniquely defined for a given embedding; we shall show that they glue together to a global section of the sheaf $\mathcal{L}(-1)$. In order to show that it comes from the sheaf

$\mathcal{L}(-2)$, we shall pass again to a more local consideration. Select a plane $E \subset \mathbb{CP}^3$ and represent $\mathcal{O}(-2)$ as a sheaf of local holomorphic equations of the union $E \cup jE$; correspondingly represent $\mathcal{L}(-2) \subset \mathcal{L}$. By the previous reasoning, $\omega_V = \bar\partial \nu_{jE,V}$, where ν_E is smoothly divisible by a local equation of E and ν_{jE} by an equation of jE. We shall show that $\nu_E = \nu_{jE}$; from this it would follow that $\nu = \nu_E = \nu_{jE}$ lies in $\Gamma(\mathcal{L}(-2))$. The equality $\nu_E = \nu_{jE}$ will follow from the fact that $\nu_E - \nu_{jE}$ proves to be a lifting of the smooth section L over S^4, lying in the kernel of the Laplace operator (of the connection ∇) and rapidly decreases near the point $\pi(E \cap jE)$. The first fact reflects the $\bar\partial$-closedness of ω (ν_E and ν_{jE} can be obtained from ω by additional "integration" so that $\nu_E - \nu_{jE}$ is anulled by an operator of the second order). For the estimate of $\nu_E - \nu_{jE}$ (in the Hermitian metric L) near $\pi(E \cap jE)$, choose again a pair of planes F, jF, perform similar constructions for them and compare the results. As for E,F, we choose the hyperplanes $z_3 = 0$, $z_1 = 0$ in the coordinate description of π, which was introduced in [5] and will be recalled below.

1.6. Computations in coordinates. Let $\mathbb{CP}^3 = (\mathbb{C}^4 \setminus \{0\})/\mathbb{C}^*$, let (z_1,z_2,z_3,z_4) be coordinates in $\mathbb{C}^4$, let S^4 be identified with $\mathbb{R}^4 \cup \{\infty\}$ by the stereographic projection, and let (x_1,x_2,x_3,x_4) be coordinates in $\mathbb{R}^4$. Let $\xi = x_2 + ix_1$, $\eta = x_4 + ix_3$. The mapping π is given in these coordinates by the formulas

$$z_1 = \bar\eta\, z_3 + \bar\xi\, z_4, \quad z_2 = -\xi\, z_3 + \eta\, z_4. \tag{1}$$

Its fibers are all possible j-invariant lines in $\mathbb{CP}^3$ (see [5], §1). Let $U_i \subset \mathbb{CP}^3$ be the set of points with $z_i \neq 0$. Set $\lambda = z_3/z_4$. Since $U_3 \cup U_4 = \pi^{-1}(\mathbb{R}^4)$, the functions $(\xi, \bar\xi, \eta, \bar\eta, \lambda, \bar\lambda)$ form a smooth system of coordinates in U_4 (we write ξ instead of $\pi^*(\xi)$ and so on). On the other hand, the functions $(z_1/z_4, z_2/z_4, \lambda)$ form a holomorphic coordinate system in U_4, and from (1), it is clear that

$$\frac{z_1}{z_4} = \bar\eta\,\lambda + \bar\xi, \quad \frac{z_2}{z_4} = -\,\xi\,\lambda + \eta\,. \tag{2}$$

Hence $\{d\bar\lambda, d''\bar\eta, d''\xi\}$ form a basis of the space of $(0,1)$-forms in each point of U_4.

Select once and for all a $\bar\partial$-closed form $\omega \in \Gamma(\pi^*(L)(-2) \otimes \Omega^{0,1})$. After identifying $\mathcal{O}(-2)$ with a sheaf of holomorphic equations of the pair of planes $(z_3 = 0)$ and $(z_4 = 0)$, ω can be represented by the form $\omega_{34} \in \Gamma(\pi^*(L) \otimes \Omega^{0,1})$, locally smoothly divisible by these equations. We represent ω_{34} on U_4 in the form

$$\omega_{34} = f\, d\bar{\lambda} + g\, d''\xi + h\, d''\bar{\eta} \tag{3}$$

and will consider f, g, h as smooth sections of the bundle L over $\mathbb{R}^4$, depending smoothly on λ as on the parameters. Covariant differentiation with respect to $\bar{\lambda}$ coincides with ordinary differentiation.

Holomorphic vanishing of ω_{34} on $z_3 = 0$ means that the sections $f\lambda^{-1}$, $g\lambda^{-1}$, $h\lambda^{-1}$ remain smooth for $\lambda = 0$. At each point of U_3, as follows from (2), the forms $\{d\bar{\lambda}^{-1}, d''\bar{\xi}, d''\eta\} = \{-\bar{\lambda}^{-2}d\bar{\lambda}, -\lambda d''\bar{\eta}, -\lambda d''\xi\}$ form a basis of the space of (0,1)-forms. Writing (3) in this basis, we find that holomorphic vanishing of ω_{34} on $z_4 = 0$ is equivalent to that the sections $f\bar{\lambda}\lambda^2$, g, h remain smooth at $\lambda = \infty$.

Denote $\nabla_\xi = \nabla(\frac{\partial}{\partial\xi})$ and so on. In the holomorphic structure of Atiyah-Ward, with $\bar{\partial} = (\pi^*\nabla)^{0,1}$, we have

$$\begin{aligned}\bar{\partial} f &= \frac{\partial f}{\partial\bar{\lambda}} d\bar{\lambda} + \nabla_\xi(f)\, d''\xi + \nabla_{\bar{\xi}}(f)\, d''\bar{\xi} + \nabla_\eta(f)\, d''\eta + \nabla_{\bar{\eta}}(f)\, d''\bar{\eta} \\ &= \frac{\partial f}{\partial\bar{\lambda}} + (\nabla_\xi + \lambda\nabla_\eta) f\, d''\xi + (\nabla_{\bar{\eta}} - \lambda\nabla_{\bar{\xi}}) f\, d''\bar{\eta}\,.\end{aligned} \tag{4}$$

Similar formulas holds for g and h. Hence the condition $\bar{\partial}\omega_{34} = 0$ means that

$$(\nabla_\xi + \lambda\nabla_\eta) f = \frac{\partial g}{\partial\bar{\lambda}}, \quad (\nabla_{\bar{\eta}} - \lambda\nabla_{\bar{\xi}}) f = \frac{\partial h}{\partial\bar{\lambda}}, \tag{5}$$

$$(\nabla_{\bar{\eta}} - \lambda\nabla_{\bar{\xi}}) g = (\nabla_\xi + \lambda\nabla_\eta) h. \tag{6}$$

Note that the operators $\nabla_\xi + \lambda\nabla_\eta$ and $\nabla_{\bar{\eta}} - \lambda\nabla_{\bar{\xi}}$ commute : this follows directly from the selfduality of ∇ as in the paper of Belavin-Zalharov [9], or from Lemma 2.2 in [5]. From this commutation, we directly obtain

<u>1.7. Lemma.</u> <u>If</u> $\omega_v = 0$, <u>then</u> $\omega = 0$.

Proof. In the representation (3), condition $\omega_v = 0$ means that $f = 0$. By the formula (5), the sections g and h depend holomorphically on λ. Consequently, they are constant along the fibers of π, because π^*L is holomorphically trivial along the fibers. On the other hand, they reduce to zero when $\lambda = 0$, hence they are zero.

<u>1.8. The sections</u> ν_3 <u>and</u> ν_4. By the argument of Paragraph 1.5, for each point $x \in \mathbb{R}^4$, there exists a unique smooth section ν_3 of the bundle π^*L along $\pi^{-1}(x)$, converging to zero at the point $\lambda = 0$ (i.e, at $\pi^{-1}(x) \cap \{z_3 = 0\}$) and such that $f = \partial\nu_3/\partial\bar{\lambda}$.

Similarly, we denote by ν_4 the section with a zero at $z_4 = 0$. The smooth dependence of ν_3 and ν_4 on x, can be deduced, on considering them as sections of π^*L over $U_3 \cup U_4$, from the integral formula

$$\nu_4(\lambda,x) = \frac{1}{2\pi i}\int \frac{f(\zeta,x)}{\zeta - \lambda}\, d\zeta \wedge d\bar{\zeta},$$

using the estimate $|f(\lambda,x)| = O((1+|\lambda|)^{-3})$, which follows from the smoothness of $f\bar{\lambda}^{-2}\lambda$ at $\lambda = \infty$, and similarly for ν_3. Clearly ν_3 and ν_4 are characterized by their zeros and by the condition $(\omega_{34} - \bar{\partial}\nu_3)_v = (\omega_{34} - \bar{\partial}\nu_4)_v = 0$.

Set $\gamma = \nu_3 - \nu_4$ and $\nabla_\eta = \nabla(\frac{\partial}{\partial x_\mu})$, $\mu = 1,\ldots,4$.

1.9. Lemma. The section γ is constant along the fibers of π and $(\sum_{\mu=1}^{4} \nabla_\mu^2)\gamma = 0$.

Proof. The first assertion follows from $\frac{\partial\gamma}{\partial\bar{\lambda}} = 0$ as in 1.7. For the proof of the second, note that

$$g = \nabla_\xi \nu_3 + \lambda\nabla_\eta \nu_4, \quad h = \nabla_{\bar{\eta}}\nu_3 - \lambda\nabla_{\bar{\xi}}\nu_4. \tag{7}$$

Indeed, $\partial/\partial\bar{\lambda}\,(g - \nabla_\xi\nu_3 - \lambda\nabla_\eta\nu_4) = 0$ by the formula (5); moreover, g, $\nabla_\xi\nu_3$ and $\lambda\nabla_\eta\nu_4$ are locally divisible by holomorphic equations of the plane $z_3 = 0$, so that not depending on λ, the section $g - \nabla_\xi\nu_3 - \lambda\nabla_\eta\nu_4$ must be zero. Similarly we can prove the formula for h.

Now substitute the right hand side of (7) in equation (6). Rewriting (7) in the form $g = (\nabla_\xi + \lambda\nabla_\eta)\nu_3 - \lambda\nabla_\eta\gamma$, $h = (\nabla_{\bar{\eta}} + \lambda\nabla_{\bar{\xi}})\nu_4 + \lambda\nabla_{\bar{\xi}}\gamma$ and using that $[\nabla_\xi + \lambda\nabla_\eta, \nabla_{\bar{\eta}} - \lambda\nabla_{\bar{\xi}}] = 0$, we get $(\nabla_\xi\nabla_{\bar{\xi}} + \nabla_{\bar{\eta}}\nabla_\eta)\gamma = 0$. To express the last operator in terms of ∇_μ (taking into account the selfduality of ∇: $[\nabla_1,\nabla_2] = [\nabla_3, \nabla_4]$ etc.), we get $\frac{1}{4}\sum_{\mu=1}^{4}\nabla_\mu^2$.

1.10. Computations on $U_1 \cup U_2$. Now we identify the sheaf $\mathcal{O}(-2)$ with the sheaf of the equation $\{z_1 = 0\} \cup \{z_2 = 0\}$ and denote by $\omega_{12} \in \Gamma(\pi^*(L) \otimes \Omega^{0,1})$ the corresponding image of the form ω. We may assume that $\omega_{12} = \frac{z_1 z_2}{z_3 z_4}\,\omega_{34}$ (with the standard choice of passage). As in point 1.8, construct sections ν_1 and ν_2 from $\Gamma(U_1 \cup U_2, \pi^*(L))$ with zeros at $z_1 = 0$ and $z_2 = 0$, respectively, and with the conditions $(\omega_{12} - \bar{\partial}\nu_1)_v = (\omega_{12} - \bar{\partial}\nu_2)_v = 0$. To write down the formulas connecting ν_1, ν_2 with ν_3, ν_4, we set $\hat{\xi} = \xi(|\xi|^2 + |\eta|^2)^{-1}$ and $\hat{\eta} = \eta(|\xi|^2 + |\eta|^2)^{-1}$. The functions $\{\hat{\xi}, \hat{\eta}, \bar{\hat{\xi}}, \bar{\hat{\eta}}\}$ are smooth coordinates on $(\mathbb{R}^4 \setminus \{0\}) \cup \{\infty\} = S^4 \setminus \{0\} = \pi(U_1 \cup U_2)$.

<u>1.11. Lemma</u>. $\nu_3 = z_3 z_2^{-1} \hat{\xi} \nu_2 + z_3 z_1^{-1} \bar{\hat{\eta}} \nu_1$, $\nu_4 = z_4 z_2^{-1} \hat{\eta} \nu_2 - z_4 z_1^{-1} \bar{\hat{\xi}} \nu_1$ <u>on</u> $(U_1 \cup U_2) \cap (U_3 \cup U_4)$.

Proof. The right hand sides clearly reduce to zero at $z_3 = 0$ and $z_4 = 0$, respectively. Hence it is sufficient to check that $[\omega_{34} - \bar{\partial}(z_3 z_2^{-1} \hat{\xi} \nu_2 + z_3 z_1^{-1} \bar{\hat{\eta}} \nu_1)]_v = 0$ and similarly for ν_4. We have

$$[\bar{\partial}(z_3 z_2^{-1} \hat{\xi} \nu_2 + z_3 z_1^{-1} \bar{\hat{\eta}} \nu_1)]_v = (z_3 z_2^{-1} \hat{\xi} + z_3 z_1^{-1} \bar{\hat{\eta}})\, \omega_{12,v}$$

$$= z_3 (z_1 z_2)^{-1} (|\xi|^2 + |\eta|^2)^{-1} (z_1 \xi + z_2 \bar{\eta})\, \omega_{12,v}\,.$$

But by (1), $z_1\xi + z_2\bar{\eta} = (\bar{\eta} z_3 + \bar{\xi} z_4)\xi + (-\xi z_3 + \eta z_4)\bar{\eta} = (|\xi|^2 + |\eta|^2) z_4$. Hence the last expression is equal to $z_3 z_4 (z_1 z_2)^{-1} \omega_{12,v} = \omega_{34,v}$. The second equality can be proved similarly.

<u>1.12. Corollary. Over</u> $(U_1 \cup U_2) \cap (U_3 \cup U_4)$, <u>we have</u>

$$\gamma = \nu_3 - \nu_4 = (|\xi|^2 + |\eta|^2)^{-1} (\nu_1 - \nu_2) = |x|^{-2} (\nu_1 - \nu_2).$$

<u>In particular, since</u> $\nu_1 - \nu_2$ <u>is a smooth section in the neighborhood of</u> ∞, <u>the section</u> γ <u>is also smooth near</u> ∞ <u>and</u> $|\gamma| = O(|x|^{-2})$ <u>in the Hermitian metric on fibers</u>.

Proof. By Lemma 1.11 and the formula (1),

$$\nu_3 - \nu_4 = \left(\frac{z_3}{z_2}\hat{\xi} - \frac{z_4}{z_2}\hat{\eta}\right)\nu_2 + \left(\frac{z_3}{z_1}\bar{\hat{\eta}} + \frac{z_4}{z_1}\bar{\hat{\xi}}\right)\nu_1$$

$$= (|\xi|^2 + |\eta|^2)^{-1} \left[\frac{z_3\xi - z_4\eta}{z_2}\nu_2 + \frac{z_3\bar{\eta} + z_4\bar{\xi}}{z_1}\nu_1\right]$$

$$= (|\xi|^2 + |\eta|^2)^{-1} (\nu_1 - \nu_2).$$

<u>1.13. Completion of the proof</u>. It remains to show that $\gamma = 0$. Set $I_R = \int_{|x| \le R} \sum_{\mu=1}^{4} \langle \nabla_\mu \gamma, \nabla_\mu \gamma \rangle d^4x$, where $d^4x = dx_1 \wedge \cdots \wedge dx_4$. From Lemma 1.9 and from Stokes' formula, we get

$$I_R = \int_{|x|\leq R} \sum_{\mu} \partial_{\mu} < \gamma, \nabla_{\mu}\gamma > d^4x = \int_{|x|\leq R} \sum_{\mu} < \gamma, \nabla_{\mu}\gamma > d^3_{\mu}x ,$$

where $d^3_{\mu}x = \bigwedge_{i\neq\mu} dx_i$. By Corollary 1.12, $|\gamma| = O(R^{-2})$, $|\nabla_{\mu}\gamma| = O(R^{-3})$ on a sphere of large radius R, so that $I_R = O(R^3 \cdot R^{-5}) = O(R^{-2})$. Since the metric $<\ ,\ >$ is positive definite, it follows that $\nabla_{\mu}\gamma = 0$ for all μ, so that $\partial_{\mu}|\gamma| = 0$, i.e, γ is constant. But $\gamma(\infty) = 0$, so $\gamma = 0$, which completes the proof.

§2. Admissible sheaves and monads.

In this paragraph, we shall prove the following

2.1. Theorem. *The functor* $\mathcal{F}$, *that associates to the special monad* $\mathcal{U} : 0 \to \tilde{I}(-1) \overset{\alpha}{\to} \tilde{H} \overset{\beta}{\to} \tilde{I}'(1) \to 0$ *the sheaf* $\mathcal{L} = \operatorname{Ker}\beta / \operatorname{Im}\alpha$ *defines an equivalence of the categories of special monads and admissible sheaves.*

Theorem 2.1 follows formally from three assertions :

2.2. Proposition. *For each special monad* $\mathcal{U}$, *the sheaf* $\mathcal{F}(\mathcal{U})$ *is admissible.*

2.3. Proposition. *For any two special monads* $\mathcal{U}_1, \mathcal{U}_2$, *the induced mapping* $\operatorname{Hom}(\mathcal{U}_1, \mathcal{U}_2) \to \operatorname{Hom}(\mathcal{F}(\mathcal{U}_1), \mathcal{F}(\mathcal{U}_2))$ *is an isomorphism.*

2.4. Proposition. *For each admissible sheaf* $\mathcal{L}$, *we can construct the following objects, functorially dependent on* $\mathcal{L}$:

(a) *The special monad* $\mathcal{G}(\mathcal{L}) = (I, H, I', \alpha, \beta)$ *with* $I = H^1(\mathcal{L} \otimes \Omega^2(1))$, $H = H^1(\mathcal{L} \otimes \Omega^1)$, $I' = H^1(\mathcal{L}(-1))$.

(b) The isomorphism $\mathcal{L} \to \operatorname{Ker}\beta / \operatorname{Im}\alpha$.

In other words, there exists a functor $\mathcal{G}$: *(sheaves)* $\to$ *(monads) and a morphism of functors* $\operatorname{Id} \to \mathcal{F}\mathcal{G}$, *inducing isomorphsims on the objects.*

2.5. Proof of Proposition 2.2. Set $\mathcal{L}' = \operatorname{Ker}\beta$. From the sequence $0 \to \mathcal{L}' \to \tilde{H} \to \tilde{I}'(1) \to 0$ we obtain $H^{i-1}(\tilde{I}(\kappa+1)) \to H^i(\mathcal{L}'(\kappa)) \to H^i(\tilde{H}(\kappa))$. In the domain of variation (i,κ) defining an admissible sheaf, the end terms $H^i(\tilde{H}(\kappa))$ are zero by the theorem of Serre, so that $H^i(\mathcal{L}'(\kappa)) = 0$. Moreover, from the exact sequence $0 \to \tilde{I}(-1) \to \mathcal{L}' \to \mathcal{L} \to 0$ we obtain $H^i(\mathcal{L}'(\kappa)) \to H^i(\mathcal{L}(\kappa)) \to H^{i+1}(\tilde{I}(\kappa-1))$, and in the same domain $H^{i+1}(\tilde{I}(\kappa-1)) = 0$.

2.6. Proof of Proposition 2.3. We shall use the fact that $\mathrm{Ext}^i(\mathcal{O}(\kappa), \mathcal{O}(\ell)) = 0$ when $i > 0$; $|\kappa|, |\ell| \le 1$ and $i = 0$, $\kappa > \ell$. The sheaves connected with the monads $\mathcal{M}_1$, $\mathcal{M}_2$ will be denoted as in the proceeding section by adding the indices 1, 2 : for example, $\mathcal{L}_2' = \mathrm{Ker}\ \beta_2$, etc. From the exact sequence $0 \to \tilde{I}_2(-1) \to \mathcal{L}_2' \to \mathcal{L}_2 \to 0$ we get

$$0 \to \mathrm{Hom}(\mathcal{L}_1', \tilde{I}_2(-1)) \to \mathrm{Hom}(\mathcal{L}_1', \mathcal{L}_2') \to \mathrm{Hom}(\mathcal{L}_1', \mathcal{L}_2) \to \mathrm{Ext}^1(\mathcal{L}_1', \tilde{I}_2(-1)).$$

Let us show that the two end sheaves become zero. From the exact sequence $0 \to \mathcal{L}_1' \to \tilde{H}_1 \to \tilde{I}_1'(1) \to 0$, we obtain

$$\mathrm{Hom}(\tilde{H}_1, \tilde{I}_2(-1)) \to \mathrm{Hom}(\mathcal{L}_1', \tilde{I}_2(-1)) \to \mathrm{Ext}^1(\tilde{I}_1'(1), \tilde{I}_2(-1)),$$

$$\mathrm{Ext}^1(\tilde{H}_1, \tilde{I}_2(-1)) \to \mathrm{Ext}^1(\mathcal{L}_1', \tilde{I}_2(-1)) \to \mathrm{Ext}^1(\tilde{I}_1'(1), \tilde{I}_2(-1)).$$

By the above remark, the end terms here reduce to zero, and hence so do the middle terms.

We have shown that $\mathrm{Hom}(\mathcal{L}_1', \mathcal{L}_2') \cong \mathrm{Hom}(\mathcal{L}_1', \mathcal{L}_2)$; in particular, each morphism of the sheaves $\mathcal{L}_1 \to \mathcal{L}_2$ is uniquely liftable to a morphism $\mathcal{L}_1' \to \mathcal{L}_2$ and then to a morphism $\mathcal{L}_1' \to \mathcal{L}_2'$, which clearly maps $\mathrm{Im}\ \alpha_1$ into $\mathrm{Im}\ \alpha_2$.

Now from the exact sequence $0 \to \mathcal{L}_1' \to \tilde{H}_1 \to \tilde{I}_1'(1) \to 0$, we obtain

$$0 \to \mathrm{Hom}(\tilde{I}_1'(1), \tilde{H}_2) \to \mathrm{Hom}(\tilde{H}_1, \tilde{H}_2) \to \mathrm{Hom}(\mathcal{L}_1', \tilde{H}_2) \to \mathrm{Ext}^1(\tilde{I}_1'(1), \tilde{H}_2).$$

As above, the two end terms are equal to zero, so that $\mathrm{Hom}(\tilde{H}_1, \tilde{H}_2) \cong \mathrm{Hom}(\mathcal{L}_1', \tilde{H}_2)$. In particular, each morphism $\mathcal{L}_1' \to \mathcal{L}_2'$ induces a morphism $\mathcal{L}_1' \to \tilde{H}_2$ which extends uniquely to a morphism $\tilde{H}_1 \to \tilde{H}_2$. If the morphism $\mathcal{L}_1' \to \mathcal{L}_2'$ lifts $\mathcal{L}_1 \to \mathcal{L}_2$, then the morphism $\tilde{H}_1 \to \tilde{H}_2$ constructed from it takes $\mathrm{Ker}\ \beta_1$ into $\mathrm{Ker}\ \beta_2$ and hence induces the morphism $\tilde{I}_1'(1) \to \tilde{I}_2'(1)$. Finally, we have established the fact that each morphism of sheaves $\mathcal{F}(\mathcal{M}_1) \to \mathcal{F}(\mathcal{M}_2)$ is induced uniquely by a monad morphism $\mathcal{M}_1 \to \mathcal{M}_2$.

2.7. Proof of Proposition 2.5. Let $\mathcal{O}_\Delta$ be the direct image of the structure sheaf of the diagonal on $\mathbb{CP}^3 \times \mathbb{CP}^3$; p_1, p_2 the projections of $\mathbb{CP}^3 \times \mathbb{CP}^3$ on the two components. Denote by $K_i \to \mathcal{O}_\Delta$ the Koszul resolution of the diagonal. Here $K_i = p_1^* \mathcal{O}(i+1) \otimes p_2^* \Omega^{-i}(-i-1)$, $i \le 0$, $K_i = 0$ when $i \le -4$. Each differential $d_i : K_i \to K_{i+1}$ is the multiplication (and the convolution in the second component) by a canonical element of

$H^0(p_1^* \mathcal{O}(1) \otimes p_2^* T(-1)) \cong V \otimes V^*$, coinciding with $\mathrm{id}: V \to V$ (where T is the tangent sheaf to $\mathbb{CP}^3$). In the coordinates (z_i), this element is equal to $\sum_{i=1}^{4} p_1^*(z_i) \otimes p_2^*(\partial/\partial z_i)$.

The complex $K. \otimes p_2^* \mathcal{L} \to \mathcal{O}_\Delta \otimes p_2^* \mathcal{L}$ is functorial in $\mathcal{L}$, the resolution of "the sheaf $\mathcal{L}$, concentrated in the diagonal".

Set $\mathcal{K} = \mathrm{Im}\, d_{-2} = \mathrm{Ker}\, d_{-1}$, $\mathcal{J} = \mathrm{Im}\, d_{-1}$ and write down a part of the exact sequence of higher direct images relative to the p_1 of the complex

$$0 \to \mathcal{K} \otimes p_2^* \mathcal{L} \to p_2^*(\mathcal{L} \otimes \Omega^1) \to \mathcal{J} \otimes p_2^* \mathcal{L} \to 0 \quad \text{(middle of resolution):}$$

$$p_{1*}(\mathcal{J} \otimes p_2^* \mathcal{L}) \to R^1 p_{1*}(\mathcal{K} \otimes p_2^* \mathcal{L}) \to R^1 p_{1*}(p_2^*(\mathcal{L} \otimes \Omega^1))$$

$$\to R^1 p_{1*}(\mathcal{J} \otimes p_2^* \mathcal{L}) \to R^2 p_{1*}(\mathcal{K} \otimes p_2^* \mathcal{L}) \qquad (8)$$

We shall compute in turn the terms of this sequence.

2.8. <u>The first term</u>. The embedding $0 \to \mathcal{J} \otimes p_2^* \mathcal{L} \to p_1^* \mathcal{O}(1) \otimes p_2^* \mathcal{L}(-1)$ induces the embedding

$$0 \to p_{1*}(\mathcal{J} \otimes p_2^* \mathcal{L}) \to p_{1*}(p_1^* \mathcal{O}(1) \otimes p_2^* \mathcal{L}(-1)).$$

In what follows, we shall frequently use the fact that for any locally free sheaves $\mathcal{M}, \mathcal{N}$ on $\mathbb{CP}^3$, there are isomorphisms $R^i p_{1*}(p_1^* \mathcal{M} \otimes p_2^* \mathcal{N}) \cong \mathcal{M} \otimes H^i(\mathcal{N})$ are functorial in $\mathcal{M}, \mathcal{N}$. In particular, $p_{1*}(p_1^* \mathcal{O}(1) \otimes p_2^* \mathcal{L}(-1)) = H^0(\mathcal{L}(-1))^\sim(-1) = 0$ by the admissibility of $\mathcal{L}$, so that

$$p_{1*}(\mathcal{J} \otimes p_2^* \mathcal{L}) = 0 . \qquad (9)$$

2.9. <u>The second term</u>. The beginning of the resolution $K. \otimes p_2^* \mathcal{L} \to \mathcal{O}_\Delta \otimes p_2^* \mathcal{L}$,

$$0 \to p_1^* \mathcal{O}(-2) \otimes p_2^*(\mathcal{L} \otimes \Omega^3(2)) \to p_1^* \mathcal{O}(-1) \otimes p_2^*(\mathcal{L} \otimes \Omega^2(1))$$

$$\to \mathcal{K} \otimes p_2^* \mathcal{L} \to 0,$$

gives the exact sequence of sheaves

$$R^1 p_{1*}(p_1^* \mathcal{O}(-2) \otimes p_2^*(\mathcal{L} \otimes \Omega^3(2))) \to R^1 p_{1*}(p_1^* \mathcal{O}(-1) \otimes p_2^*(\mathcal{L} \otimes \Omega^2(1)))$$

$$\to R^1 p_{1*}(\mathcal{K} \otimes p_2^* \mathcal{L}) \to R^2 p_{1*}(p_1^* \mathcal{O}(-2) \otimes p_2^*(\mathcal{L} \otimes \Omega^3(2))). \qquad (10)$$

The first term and the last term are equal to zero, since $H^i(\mathcal{L} \otimes \Omega^3(2)) = H^i(\mathcal{L}(-2)) = 0$ when $i = 1, 2$ by the admissibility of $\mathcal{L}$. Hence the middle terms in (10) define the isomorphism, functorial in $\mathcal{L}$,

$$R^1p_{1*}(\mathcal{K} \otimes p_2^*\mathcal{L}) = H^1(\mathcal{L} \otimes \Omega^2(1))^{\sim}(-1). \qquad (11)$$

<u>2.10. The third term.</u> By the remark in paragraph 2.8,

$$R^1p_{1*}(p_2^*\mathcal{L} \otimes \Omega^1) = H^1(\mathcal{L} \otimes \Omega^1)^{\sim}. \qquad (12)$$

<u>2.11. The fourth term.</u> The end of the resolution $K. \otimes p_2^*\mathcal{L} \to \mathcal{O}_\Delta \otimes p_2^*\mathcal{L}$:

$$0 \to \mathcal{J} \otimes p_2^*\mathcal{L} \to p_1^*\mathcal{O}(1) \otimes p_2^*(\mathcal{L}(-1)) \to \mathcal{O}_\Delta \otimes p_2^*\mathcal{L} \to 0$$

gives the sequence of sheaves

$$\begin{aligned} &p_{1*}(p_1^*\mathcal{O}(1) \otimes p_2^*\mathcal{L}(-1)) \to p_{1*}(\mathcal{O}_\Delta \otimes p_2^*\mathcal{L}) \to R^1p_{1*}(\mathcal{J} \otimes p_2^*\mathcal{L}) \\ &\to R^1p_{1*}(p_1^*\mathcal{O}(1) \otimes p_2^*\mathcal{L}(-1))) \to R^1p_{1*}(\mathcal{O}_\Delta \otimes p_2^*\mathcal{L}). \end{aligned} \qquad (13)$$

The first term reduces to zero, since $H^0(\mathcal{L}(-1)) = 0$. The second and fifth terms are computable by taking into account the fact that $p_{1*}|_\Delta : \Delta \to \mathbb{CP}^3$ is an isomorphism, so that $R^ip_{1*} = 0$ when $i > 0$ on $\mathcal{O}_\Delta$ sheaves. Hence the last term in (13) is zero, and the second term is functorially isomorphic to $\mathcal{L}$. Finally, the fourth term in (13) is functorially isomorphic to $H^1(\mathcal{L}(-1)^{\sim}(1)$. In conclusion, (13) gives the exact sequence functorial in $\mathcal{L}$

$$0 \to \mathcal{L} \to R^1p_{1*}(\mathcal{J} \otimes p_2^*\mathcal{L}) \to H^1(\mathcal{L}(-1))^{\sim}(1) \to 0. \qquad (14)$$

<u>2.12. The fifth term.</u> It appears in the extension of the exact sequence (10) :

$$\begin{aligned} &R^2p_{1*}(p_1^*\mathcal{O}(-1) \otimes p_2^*(\mathcal{L} \otimes \Omega^2(1))) \to R^2p_{1*}(\mathcal{K} \otimes p_2^*\mathcal{L}) \\ &\qquad \to R^3p_{1*}(p_1^*\mathcal{O}(-2) \otimes p_2^*(\mathcal{L} \otimes \Omega^3(2))). \end{aligned}$$

Here the third term is equal to zero, since $H^3(\mathcal{L} \otimes \Omega^3(2)) \cong H^3(\mathcal{L}(-2)) = 0$ by the admissibility of $\mathcal{L}$. To show that the first sheaf reduces to zero, we shall prove that $H^2(\mathcal{L} \otimes \Omega^2(1)) = 0$. This cohomology group can be computed as follows : multiply the standard exact sequence $0 \to \Omega^3(3) \to \widetilde{\Lambda^3 V} \to \Omega^2(3) \to 0$ by $\mathcal{L}(-2)$ and consider the exact sequence of cohomologies

$H^2(\widetilde{\Lambda^3 V \otimes \mathcal{L}(-2)}) \to H^2(\mathcal{L} \otimes \Omega^2(1)) \to H^3(\Omega^3(1))$ (note that $V = H^0(\mathcal{O}(1))$). Its end terms are equal to zero by the admissibility of $\mathcal{L}$ and by the theorem of Serre. Finally,

$$R^2 p_{1*}(\mathcal{K} \otimes p_2^* \mathcal{L}) = 0. \tag{15}$$

2.13. Put now in (8) the identities (9), (11) (12) and (15) to get the exact sequence of sheaves functorial in $\mathcal{L}$,

$$0 \to H^1(\mathcal{L} \otimes \Omega^2(1))^{\sim}(-1) \to H^1(\mathcal{L} \otimes \Omega^1)^{\sim} \to R^1 p_{1*}(\mathcal{J} \otimes p_2^* \mathcal{L}) \to 0$$

Exchange in it the third arrow by its composition with the third arrow in the sequence (14). We obtain the monad and isomorphisms associated with the sheaf $\mathcal{L}$, the existence of which was the assertion of Proposition 2.4. This completes the proof of Theorem 2.1.

The sheaves corresponding to instantons, we shall equip with a nondegenerate quadratic form in the next section. We shall show that this can be extended to the monads.

2.14. Sheaves with a bilinear form. Both categories, special monads and admissible sheaves, are equipped with the duality functor $*$, establishing their equivalence with the dual categories. Clearly the natural isomorphism $\mathcal{F}(\mathcal{M}^*) \cong \mathcal{F}(\mathcal{M})^*$ determines an isomorphism of the functors $\mathcal{F}*$ and $*\mathcal{F}$.

Hence the category of admissible sheaves $\mathcal{L}$, equipped with bilinear forms $\phi : \mathcal{L} \to \mathcal{L}^*$, and the isomorphisms that preserve the bilinear forms, is equivalent to the category of monads $\mathcal{M}$ with morphisms $\Phi : \mathcal{M} \to \mathcal{M}^*$, and the corresponding isomorphisms. The functorial isomorphism of $\mathcal{F}*$ with $*\mathcal{F}$ takes the morphism $\mathcal{F}(\Phi^*)$ into $\mathcal{F}(\Phi)^*$. Hence the form Φ is nondegenerate and symmetric (respectively, antisymmetric) on the middle object of the monad $\tilde{H}$, if and only if $\mathcal{F}(\Phi) \cong \phi$ is nondegenerate and symmetric (respectively, antisymmetric).

A monad, equipped with a symmetric isomorphism $\phi : \mathcal{M} \to \mathcal{M}^*$ can be given with a diagram of linear spaces

$$I \otimes V^* \overset{\alpha}{\to} H \cong H^* \overset{\alpha^*}{\to} I^* \otimes V,$$

where the middle isomorphism corresponds to the bilinear form Q induced on H by Φ. In the notation of Paragraph 0.5, for any hyperplane $W \subset V$ the kernel $H^* \overset{\alpha_W^*}{\to} I^* \otimes V/W$ coincides with the orthogonal (relative to Q) complement of the image $I \otimes (V/W)^* \overset{\alpha_W}{\to} H$. Hence in addition to the conditions of Paragraph 0.5, all these images need to be Q-isotropic, and the fiber $\mathcal{L}$ over the point $z \in \mathbb{CP}^3_-$, corresponding to W, is canonically isomorphic to $(\mathrm{Im}\, \alpha_W)^{\perp}/\mathrm{Im}\, \alpha_W$.

2.15. Invariants of symmetric sheaves. Let $\mathcal{L}$ correspond to a monad constructed from the given $(I \otimes V^* \to H, Q)$ as in the previous section. Then $\operatorname{rk} \mathcal{L} = \dim H - 2 \dim I$. The Chern class of $\mathcal{L}$ can be easily calculated by multiplying the Chern polynomials : $c_1(\mathcal{L}) = c_3(\mathcal{L}) = 0$, $c_2(\mathcal{L}) = \dim I$.

§3. Classification of instantons.

3.1. Linear algebra data. By orthogonal data, we shall mean a triple $(I, H_0 \underset{\mathbb{R}}{\otimes} V, Q)$ where V is a four-dimensional complex space, H_0 a finite-dimensional real space with the symmetric bilinear form Q, I a complex subspace of $H_0 \underset{\mathbb{R}}{\otimes} V$, such that the following conditions hold:

(a) Let $\sigma\colon V \to V$ be an antilinear mapping for which $\sigma : (z_1,z_2,z_3,z_4) \to (-z_2,z_1,-z_4,z_3)$, where $(z_\mu) \subset V$ is a basis of V as in Paragraphs 0.2, 1.6). Then $(\mathrm{id}_{H_0} \otimes \sigma)I = I$.

(b) For each complex subspace $D \subset V$, set $I_D = \sum_\ell (\mathrm{id}_{H_0} \otimes \ell)I = H$ where ℓ varies over the linear functions on V that are zero on D. Then for any hyperplane $W \subset V$ the space I_W is Q-isotropic and for any σ-invariant plane $E \subset V$ we have $\dim I_E^\perp = 2 \dim I$.

(c) The form Q is positive definite on all subspaces of $I_E^\perp \cap H_0$ (E as in (b)).

In §4, we shall show that (c) implies a stronger condition, which is actually the one used :

(c') The form Q is positive definite.

Unitary (respectively, symplectic) data are orthogonal data additionally equipped with the orthogonal operator $J' : H_0 \to H_0$ (respectively, two orthogonal operators $J_1', J_2' : H_0 \to H_0$) with the conditions :

(d) $J'^2 = -1$ (respectively, $J_1'^2 = J_2'^2 = -1$, $J_1'J_2' = -J_2'J_1'$).

(e) The subspace I is invariant relative to $J' \otimes \mathrm{id}_V$ (respectively, $J_1' \otimes \mathrm{id}_V$, $J_2' \otimes \mathrm{id}_V$).

3.2. The construction of instantons with respect to given linear algebra data. The data $(I, H_0 \otimes V, Q)$ correspond to the following instanton bundle (L, ∇) over the sphere S^4, realized as the space of σ-invariant planes $E \subset V$: the fiber L over E is $I_E^\perp \cap H_0$; the complex (respectively, quaternion) structure on the fiber is induced by the operator J (respectively, J_1', J_2') on H_0. The connection ∇ is the orthogonal projection of the trivial connection onto $S^4 \times H_0$.

The selfduality of the curvature form of ∇ has been shown directly in [5] (the slight difference in the constructions in [5] and in this paper is commented on in [5] and also below).

3.3. The basic theorem. All instanton bundles are obtainable from the prescribed construction ; they are isomorphic if and only if the corresponding given linear algebra data are isomorphic.

Proof. Basically, it consists of the construction of the linear data from the instanton (L, ∇) ; this data can be obtained by further work on the monad corresponding to the sheaf $\mathcal{L}$ of sections of $\pi^*(L \otimes_{\mathbb{R}} \mathbb{C})$.

3.4 Accuonting for the real structure. There exists a principal involution on the total space of the real bundle π^*L on $\mathbb{CP}^3$ that takes the point $z \in \mathbb{CP}^3$ into $j(z)$ and maps $\pi^*(L(z)) = L(\pi(z))$ onto $\pi^*L(j(z)) = L(\pi(j(z)) = L\pi(z))$. Extend it fiber-by-fiber antilinearly to the total space $\pi^*(L \otimes_{\mathbb{R}} \mathbb{C})$ and denote the extension by σ.

The involution σ acts on the smooth sections of $\pi^*(L \otimes_{\mathbb{R}} \mathbb{C})$ by the formula $(\sigma f)(jz) = \overline{f(z)}$, where the complex conjugation is naturally defined on the sections of $L \otimes_{\mathbb{R}} \mathbb{C}$. It is important that by this action, the subsheaf of holomorphic sections is antilinearly mapped into itself. It is sufficient to check this on a dense open set, say $U_3 \cap U_4$ in the notation of Paragraph 1.6. By the formula (4), the holomorphicity of f means that

$$\frac{\partial f}{\partial \bar{\lambda}} = (\nabla_\xi + \lambda \nabla_\eta)f = (\nabla_{\bar{\eta}} - \lambda \nabla_{\bar{\xi}})f = 0.$$

Let us prove, say, the equality $(\nabla_\xi + \lambda \nabla_\eta)(\sigma f) = 0$; the remaining proofs are similar. In view of $\lambda(jz) = -\bar{z}_4\bar{z}_3^{-1} = -\overline{\lambda(z)}^{-1}$, we have for $z \in U_3 \cap U_4$:

$$(\nabla_\xi + \lambda(j(z))\nabla_\eta)[\sigma f(jz)] = [\nabla_\xi - \overline{\lambda(z)}^{-1}\nabla_\eta](\overline{f(z)})$$

$$= \overline{-\lambda(z)^{-1}(\nabla_{\bar{\eta}} - \lambda(z)\nabla_{\bar{\xi}})[f(z)]} = 0.$$

We extend j equivariantly to the holomorphic sections of the sheaves $\mathcal{O}(i)$ and $\Omega^i(\kappa)$ by the formula $(\sigma f)(z) = \overline{f(j z)}$ where the complex conjugation at this step is as usual, in homogeneous holomorphic coordinates : $z_i \to \bar{z}_i$.

After computing all cohomologies, appearing in the construction of Proposition 2.4, on acyclic σ-invariant Čech coverings, we can define $\mathbb{C}$-antilinear automorphisms at all points in this construction of sheaves and spaces, that commute with all morphisms. (Here the σ-invariance of the Koszul complex and, in particular, its coboundary operators, are essential.)

In particular, we obtain an antilinear automorphism of monads : $\sigma : \mathcal{J}(\mathcal{L}) \to \mathcal{J}(\mathcal{L})$, induced by the previous σ on the fibers associated with its sheaf $\mathcal{L}$.

Such an automorphism is defined uniquely : if σ' is another such automorphism with the same property, then the composition $\sigma(\sigma - \sigma')$ is a linear endomorphism of the monad $\mathcal{J}(\mathcal{L})$, induced by the zero morphism of the sheaf. It is equal to zero by Proposition 2.4, so that $\sigma = \sigma'$. Moreover, $\sigma^2 = 1$, since this is so on the sheaf $\mathcal{L}$.

It follows from this that σ induces a natural structure on the space $H = H^1(\mathcal{L} \otimes \Omega^1)$ of the monad: setting $H_0 = \{ h \in H \mid \sigma h = h\}$, we have $H = H_0 \otimes_{\mathbb{R}} \mathbb{C}$. Hence $H \otimes_{\mathbb{C}} V = H_0 \otimes_{\mathbb{R}} V$ and the first arrow of the monad $0 \to \tilde{I}(-1) \to \tilde{H}$ defines an embedding $I \subset H_0 \otimes_{\mathbb{R}} V$.

3.5. <u>The form Q on H_0</u>. The orthogonal metric on the fibers of L, bilinearly extended to $\pi^*(L \otimes_{\mathbb{R}} \mathbb{C})$ and then to $\mathcal{L}$ induces a nondegenerate quadratic metric form on H by Paragraph 2.11. The action of σ on its arguments reduces to complex conjugation ; hence its restriction Q to H_0 defines a real nondegenerate quadratic form, and the form on H coincides with $Q \otimes \mathbb{C}$. We have constructed the data of Paragraph 3.1 in the orthogonal case.

3.6. <u>The operators</u> J', J_1', J_2'. The endomorphisms J, J_1, J_2 are transported from $\mathcal{L}$ onto H, functorially, and they then have well-defined restrictions to H_0, because they commute with σ. The orthogonality condition on $\mathcal{L}$ can be written in the form $J J^* = \mathrm{id}$, where $J^* : \mathcal{L}^* \to \mathcal{L}^*$ is transported to $\mathcal{L}$ by the isomorphism $\mathcal{L} \simeq \mathcal{L}^*$ corresponding to the form Q. In view of the functoriality, this condition is preserved on H_0 as are all needed relations : $J^2 = -1$ and so on. The uniqueness of these operators, as above, follows from the fact that their actions are known on fibers. These are the additional data of Paragraph 3.1 in the unitary and symplectic cases.

3.7. <u>Verification of the remaining assertions</u>. Properties 3.1 (a), (d) and (e) follow from the fact that the constructed operators σ , J, J_1', J_2' act on all of the monad, not only on its middle term. The fact that the space I_W is isotropic was shown in Paragraph 2.14.

We show now how to reconstruct L from the monad with the additional structure.

The fiber $L \otimes_{\mathbb{R}} \mathbb{C}$ over the point $x \in S^4$ is canonically isomorphic to the space of holomorphic sections of the restriction of $\mathcal{L}$ to $\pi^{-1}(x)$. Denote the restriction by $\mathcal{L}(x)$; similarly we shall denote the restrictions of other bundles on $\mathbb{CP}^3$. The monad $\mathcal{J}(\mathcal{L})$ restricted to $\pi^{-1}(x)$ gives two exact sequences $0 \to \tilde{I}(-1)(x) \to \mathcal{L}'(x) \to \mathcal{L}(x) \to 0$ and $0 \to \mathcal{L}'(x) \to \tilde{H}(x) \xrightarrow{\alpha^*(x)} \tilde{I}^*(1) \to 0$, which define the isomorphism $H^0(\mathcal{L}(x)) \cong H^0(\mathcal{L}'(x)) \cong \mathrm{Ker}(H \xrightarrow{\alpha^*(x)} I^* \otimes H^0(\pi^{-1}(x), \mathcal{O}(1)))$. If the point $x \in S^4$ corresponds to the j-invariant plane $E \subset V$, then $H^0(\pi^{-1}(x), \mathcal{O}(1))$ is canonically identifiable with V/E, and $\mathrm{Ker}\,\alpha^*(x)$ with the subspace $[\mathrm{Im}\, I \otimes (V/E)^*]^{\perp} = I_E^{\perp} \subset H$ by the selfduality of the monads. Since the complex dimension of the fibers $\mathcal{L}$ and $L \otimes \mathbb{C}$ is equal to $\dim H - 2 \dim I$, we must have $\dim I_E = \dim H - \dim I_E^{\perp} = 2 \dim I$. At the end, the fiber L over x consists of the σ-invariant elements of the fiber $L \otimes \mathbb{C}$, i.e, is identified with $I_E^{\perp} \cap H_0$. Hence the restriction of Ω to such subspaces needs to be positive definite.

More precisely, the previous reasoning shows that for any σ-invariant plane $E \subset V$ and hyperplane W that contains it the canonical mapping $I_E^{\perp} = (I_W + I_{\sigma W}) \to I_W^{\perp}/I_W$ is an isometry. Hence the direct summand of the trivial orthogonal bundle $S^4 \times H$, whose fiber over E is $I_E^{\perp}$, after lifting to $\mathbb{CP}^3$ is canonically identifiable with $\pi^*(L \otimes \mathbb{C})$ with the preservation of all additional structures (metric, σ , J, J_1, J_2). Consequently, this bundle is isomorphic to $L \otimes_{\mathbb{R}} \mathbb{C}$. The selfdual connection ∇ , reducing to the given holomorphic structure $\pi^*(L \otimes \mathbb{C})$ by the theorem of Atiyah-Ward is unique. Hence it has to coincide with the projection of the trivial connection (incidentally, this can be shown directly from its definition). The one-to-one correspondence between the instanton and the linear algebra data follows from the results of §2, and the verification of the uniqueness of the additional structure is obtained along the way.

§4. The theorem on positivity.

4.1. Theorem. Let H_0 be a finite-dimensional real space with the nondegenerate quadratic form Q, $I \subset H_0 \underset{\mathbb{R}}{\otimes} V$ a complex subspace, invariant relative to $id_{H_0} \otimes \sigma$. Then the conditions 3.1 (b) and (c) imply that $Q > 0$.

Proof. Extend Q to the Hermitian form $< \, , \, >$ on $H = H_0 \otimes \mathbb{C}$ by the formula $< h,h > = Q((id_{H_0} \otimes \sigma)h, h)$. Instead of the embedding $I \to H \otimes V$, we shall consider the mapping $\phi : I \otimes V^* \to H$ and denote the induced form on $I \otimes V^*$ by the same symbol $< \, , \, >$.

The space $\mathbb{CP}^3$ of hyperplanes $W \subset V$ is identifiable with the space of lines in V^*. Choose the basis $\varepsilon_1,\ldots,\varepsilon_4 \in V^*$, corresponding to the homogeneous coordinates $z_1,\ldots,z_4$ in $\mathbb{CP}^3$, that has been introduced in Paragraph 1.6 : to the point $z = (z_1,\ldots,z_4)$ associate the hyperplane $W_z : \sum_{i=1}^{4} z_\mu \varepsilon_\mu = 0$ in V.

For any vector $e \in I$, set $e_\mu = e \otimes \varepsilon_\mu$, $ez = \sum_{\mu=1}^{4} e_\mu z_\mu$, $I_z = \{ ez \mid e \in I \} \subset I \otimes V^*$. The last space depends only on $z \in \mathbb{CP}^3$. Since for each $\ell \in V^*$ we have $(id_{H_0} \otimes \ell)(e) = \phi(e \otimes \ell) \in H$, the space $I_{W_z} \subset H$ is the ϕ-image of the space $I_z \subset H \otimes V^*$. By the definition of $< \, , \, >$, the isotropy of I_{W_z} relative to Q is equivalent to the orthogonality of I_z and $I_{j(z)}$ in the metric $< \, , \, >$.

Any σ-invariant plane $E \subset V$ can be represented in the form $W_z \cap W_{j(z)}$ for suitable $z \in \mathbb{CP}^3$. By the previous reasoning, $I_E = \phi(I_z + I_{j(z)})$. Hence the condition $\dim I_E = 2 \dim I$ implies that ϕ is injective on the subspaces I_z, and $I_z \cap I_{j(z)} = \{ 0 \}$. Moreover, the metric is nondegenerate on $I_z + I_{j(z)}$.

The condition $Q \mid_{I_E^\perp \cap H_0} > 0$ for all E implies positive semi-definiteness of $< \, , \, >$ on all subspaces of $(I_z + I_{j(z)})^\perp$ (Hermitian orthogonal complement).

Let $\pi(z) = x \in \mathbb{R}^4 = S^4 \setminus \{ \infty \}$. To the point x, correspond the matrix $X_2 = X_2(x)$ defined by the formula

$$X_2^t = \sum_{a=1}^{3} i\, x_a \sigma_a + x_4 \begin{pmatrix} 1 & 0 \\ 0 & 1 \end{pmatrix} = \begin{pmatrix} x_4 + ix_3 & x_2 + ix_1 \\ -x_2 + ix_1 & x_4 - ix_3 \end{pmatrix}. \quad (16)$$

The formula (1) shows that the closure of the preimage of x in V^* relative to the mapping $V^* \setminus \{0\} \to \mathbb{CP}^3 \overset{\pi}{\to} S^4$ represents the plane P_x, defined by the equation

$$P_x = \{(z_1,\ldots,z_4) \mid \begin{pmatrix} z_1 \\ z_2 \end{pmatrix} = X^+ \begin{pmatrix} z_3 \\ z_4 \end{pmatrix} \}. \tag{17}$$

By the continuity, $P_\infty = \{(z_1,z_2,0,0)\}$. It follows easily that $I_z + I_{j(z)} = I \otimes P_x$ if $\pi(z) = x$.

Choose a basis $\{e_1,\ldots,e_n\}$ in I and set $e_{\kappa\mu} = e_\kappa \otimes e_\mu$. This is a basis in $I \otimes V^*$, which we order as $\{e_{11}, \ldots, e_{n1}; \ldots; e_{14},\ldots, e_{n4}\}$.

4.2. Lemma. The condition $\langle I_z, I_{j(z)} \rangle = 0$ is equivalent to the following properties of symmetry of the Gram matrix F of the basis $\{e_{\kappa\mu}\}$:

$$F = \begin{pmatrix} A & D^+ \\ D & R \end{pmatrix} \tag{18}$$

where

$$A = \begin{pmatrix} A_n & 0 \\ 0 & A_n \end{pmatrix}, \quad R = \begin{pmatrix} R_n & 0 \\ 0 & R_n \end{pmatrix}, \quad D = \begin{pmatrix} B & C \\ -C^+ & B^+ \end{pmatrix}, \quad A^+ = A, \quad R^+ = R$$

(+ denotes the Hermitian conjugation).

The proof follows from simple computations (compare [5], Proposition 5.3).

4.3. The block A in the formula (18) is the Gram matrix of the basis of the subspace $I \otimes \varepsilon_1 + I \otimes \varepsilon_2 = I \otimes P_\infty$. Since ϕ maps this into H and the metric on $\phi(I \otimes I_\infty)$ is non-degenerate, we have $\det A \neq 0$. We shall show below that $A > 0$. With the positive semi-definiteness of $\langle\ ,\ \rangle$ on $(I \otimes P_\infty)^\perp$ in mind, it will follows that $F \geq 0$, i.e, $\langle\ ,\ \rangle > 0$ on H (the metric is nondegenerate on H) and, finally, $Q > 0$ on H_0.

4.4. Set

$$X = X_2 \otimes E_n = \begin{pmatrix} (x_4-ix_3)E_n, & -(x_2+ix_1)E_n \\ (x_2-ix_1)E_n, & (x_4+ix_3)E_n \end{pmatrix}. \tag{19}$$

It follows from the formula (17) that the rows of the matrix $(\overline{X}, E_{2n})$ form a basis of the subspace $I \otimes P_x$, expressed in coordinates relative to the basis $(e_{\kappa\mu})$ of the space $I \otimes V^*$. Add to it the rows of the matrix $(E_{2n}, 0)$ and compute the new

Gram matrix:

$$F' = \begin{pmatrix} E_{2n} & 0 \\ X & E_{2n} \end{pmatrix} \begin{pmatrix} A & D^+ \\ D & R \end{pmatrix} \begin{pmatrix} E_{2n} & X^+ \\ 0 & E_{2n} \end{pmatrix} = \begin{pmatrix} A & D^+(x) \\ D(x) & R(x) \end{pmatrix}.$$

Here $D(x) = D + AX$, $R(x) = |x|^2 A + DX^+ + XD^+ + R$.
Since the metric on $I \otimes P_x$ is nondegenerate, $\det R(x) \neq 0$ for all $x \in R^4$.

It follows from the formula

$$(E_{2n}, -D^+(x)R(x)^{-1})\, F' \begin{pmatrix} 0 \\ E_{2n} \end{pmatrix} = 0$$

that the rows of the matrix $(E_{2n}, -D^t(x)\overline{R(x)}^{-1})$ represent the basis of the space $(I \otimes P_x)^\perp$ expressed in the coordinates relative to a basis consisting of the rows of the matrix $\begin{pmatrix} E_{2n} & 0 \\ \overline{X} & E_{2n} \end{pmatrix}$. The corresponding Gram matrix is equal to $A - D^+(x)R(x)^{-1}D(x)$. It is positive semi-definite by the hypothesis.

<u>4.5. Lemma.</u> $\sum_{\mu=1}^{4} \partial_\mu^2 R(x)^{-1} \leq 0$, <u>where</u> $\partial_\mu = \partial / \partial x_\mu$.

Proof. Since $R(x)$ is Hermitian and nondegenerate, it is sufficient to show that $S(x) = -R(x)\left(\sum_{\mu=1}^{4} \partial_\mu^2 R(x)^{-1}\right)R(x) \geq 0$.

We have

$$S(x) = \sum_{\mu=1}^{4} \partial_\mu^2 R(x) - 2\, \partial_\mu R(x) \cdot R(x)^{-1}\, \partial_\mu R(x).$$

We shall show that $S(x) = 4T(A - D^+(x)R(x)^{-1}D(x))$, where the operator T on $(2n \times 2n)$-matrices is defined by the formula

$$T \begin{pmatrix} Z_{11} & Z_{12} \\ Z_{21} & Z_{22} \end{pmatrix} = \begin{pmatrix} Z_{11} + Z_{22} & 0 \\ 0 & Z_{11} + Z_{22} \end{pmatrix}.$$

Clearly T takes semi-definite matrices to semi-definite ones, so the lemma will follow from this property.

Set $\Sigma_\mu = \partial_\mu X$. It follows from the formula (16) and (19) and the property of the Pauli matrices that $\Sigma_\mu \Sigma_\nu^+ + \Sigma_\nu \Sigma_\mu^+ = 2\delta_{\mu\nu} E_{2n}$.
Moreover, $D = \sum_{\mu=1}^{4} D_\mu \Sigma_\mu$, where

$$D_1 = \frac{C - C^+}{2i} \otimes E_2, \quad D_2 = -\frac{C + C^+}{2} \otimes E_2, \quad D_3 = \frac{B - B^+}{2i} \otimes E_2,$$

$$D_4 = \frac{B + B^+}{2} \otimes E_2.$$

It follows from this that

$$R(x) = |x|^2 A + 2 \sum_{\mu=1}^{4} x_\mu D_\mu + R,$$

in particular, $R(x)$ has the form $\begin{pmatrix} Y & 0 \\ 0 & Y \end{pmatrix}$.

Thus we get

$$S(x) = 8A - 4 \sum_{\mu=1}^{4} (x_\mu A + D) R(x)^{-1} (x_\mu A + D_\mu),$$

$$T(A - D^+(x)R(x)^{-1}D(x)) = 2A - \sum_{\mu,\nu=1}^{4} (x_\mu A + D_\mu)R(x)^{-1}(x_\nu A + D_\nu)\, \Sigma_\mu^+ \Sigma_\nu$$

and the required result follows from $T(\Sigma_\mu^+ \Sigma_\nu) = 2\delta_{\mu\nu} E_{2n}$.

4.6. Completion of the proof. On a sphere of large radius M in $\mathbb{R}^4$, we have asymptotically $\partial_\mu R(x)^{-1} = -2 x_\mu M^{-4} A^{-1} + O(M^{-4})$. Computing the integral $\int_{|x| \le M} \sum_{\mu=1}^{4} \partial_\mu R(x)^{-1}$ by Stokes' theorem, we find that $A^{-1} > 0$, from which we get $A > 0$ and we are finished.

4.7. Final remarks. It follows from Theorem 4.1 that we can select a basis $\{e_1, \dots, e_n\}$ in I so that we have $A_n = E_n$, $R_n = (\rho_1, \dots, \rho_n)$ with $0 \le \rho_1 \le \dots \le \rho_n$. This leads to the description of the space of instantons given in the paper [5] for the case of the group $SU(2)$. The orthogonal and symplectic cases can be described analogously.

References.

1. Atiyah M.F., Hitchin N.J., Singer I.M., Self-duality in four-dimensional Riemannian Geometry (Preprint, 1977).
2. Atiyah M.F., Drinfeld V.G., Hitchin N.J., Manin Yu.I., Construction of instantons, Phys. Letters 65A, N3 (1978), 185-187.
3. Drinfeld V.G., Manin Yu.I., Self-dual Yang-Mills fields on the sphere, Funkcional Anal. 12, Vol. 2 (1978), 78-89.
4. Drinfeld V.G, Manin Yu.I. On locally free sheaves on $\mathbb{CP}^3$ associated with the Yang-Mills fields, UMN (Translated as Russian

Math. Surveys), Vol. 2 (1978), 241-242

5. Drinfeld V.G, Manin Yu.I., A description of instantons (Preprint ITEP. N 72, 1978).
6. Beylinson A.A., Coherent sheaves on $\mathbb{P}^N$ and problems of linear algebra, Funkcional Anal. 12, Vol. (1978),
7. Bernstein,I.N., Gelfand,I.M., Gelfand S.I., Algebraic sheaves on $\mathbb{P}^n$ and problems of linear algebra, Funkcional Anal. 12, Vol. (1978),
8. Barth W., Moduli of vector bundles on the projective plane, Inventiones Math., vol. 42 (1977), 63-91.
9. Belavin A.A., Zakharov V.E., Yang-Mills equation as inverse scattering problem (Preprint ITP, Chernogolovka, 1977).
10. Mumford,D., Lectures on Curves on an Algebraic Surface, Annals of Math. Studies 59, Princeton U. Press, Princeton (1966).
11. Hartshorne R., Algebraic Geometry, Springer, Berlin, (1977).
12. Wells, R., Differential Analysis on Complex Manifolds , Prentice-Hall (1973).

Bashkirsky Gosudarstvenny University
V.A. Steklov Mathematical Institute
Academy of Sciences of the USSR
Moscow
USSR

Translated by P.S. Milojević, Noriko Yui and George A. Elliott.

SET THEORETICAL COMPLETE INTERSECTIONS IN CHARACTERISTIC $p > 0$

by Daniel Ferrand.

The main motivation for the remarks presented here is to clarify (for me) some mysterious calculations Hartshorne made in [2] : he showed that the rational curves $C_d \subset \mathbb{P}^3_k$, where $\mathrm{char}(k) = p > 0$, defined parametrically by

$$(u,v) \longrightarrow (u^d, u^{d-1}v, uv^{d-1}, v^d)$$

are the set-theoretical intersection of two surfaces, the equations of which he produced. It seems to me that one way to understand these calculations - i.e., finally, to avoid them - is indicated in the following more general result:

<u>Let</u> C <u>be a smooth curve in</u> $\mathbb{P}^3_k$, <u>where</u> k <u>is algebraically closed and</u> $\mathrm{char}(k) = p > 0$. <u>Assume that</u> C <u>has a linear projection birational to it, and with only cusps as singularities. Then</u> C <u>is a set-theoretical complete intersection</u>.

The proof may be summed up as follows: Let X be the cone in $\mathbb{P}^3_k$ associated to the given projection, and $f:\overline{X} \to X$ its normalization. There exists an effective divisor $\overline{C}$ on $\overline{X}$ mapped isomorphically onto C by f (section 1) . As f is radicial, the Frobenius morphism brings $\overline{C}$ back to an effective divisor on X , whose lifting Y to the ambient space is a surface that $X \cap Y = C$ (as sets).

Unfortunately I do not know how restrictive the hypothesis is; it is at least satisfied for the curves C_d, because C_d is of degree d and has a tangent line with a contact of order $d-1$.

Section 3 contains a criterion for a finite morphism to be the composite of a radicial morphism and an unramified morphism. It is used in Section 4 where we give a straightforward proof of a beautiful result by Cowsik and Nori.

1. The divisor associated with a curve on a cone

The result of this section was the basic tool for early geometers (Cayley, Halphen...); they stated it as: "A curve in space is the partial intersection of a cone and a monoid." Its translation into the language of cycles (theorem of Severi, [3], p. 98) has blunted it a little.

1.1. Fix an algebraically closed field k and a projective space $P = \mathbb{P}_k(V)$ built on a finite dimensional vector space V.

Let $x \in P$ be a closed point defined by a linear form $V \to k$, whose kernel will be denoted by V'. The linear projection with center x is the morphism

$$g: P - \{x\} \to P' = \mathbb{P}_k(V')$$

defined as follows: On $P - \{x\}$, the restriction to V'_P of the canonical quotient $\alpha: V_P \to \underline{O}_P(1)$ is still surjective, and g is the unique morphism such that this quotient is isomorphic to the pull-back $g^*(\alpha'): V'_{P'} \to g^*(\underline{O}_{P'}(1))$ of the canonical quotient on P'. Moreover, every splitting of the exact sequence $0 \to V' \to V \to k \to 0$ gives rise to an isomorphism of P'-schemes

$$\mathbf{V}_{P'}(\underline{O}_{P'}(-1)) \xrightarrow{\sim} P - \{x\}$$

<u>Proposition</u> 1.2 Let C be a smooth curve in P, not containing x, and such that the morphism $C \to g(C)$ induced by g is birational. Let X denote the cone over C with vertex x, and $f: \overline{X} \to X$ its normalization. Then there exists an effective divisor $\overline{C}$ on $\overline{X}$, defined by a section of $f^*(\underline{O}_X(1))$, and such that f induces an isomorphism $\overline{C} \xrightarrow{\sim} C$.

<u>Proof</u>: Let $j: C \to U = X - \{x\}$ denote the immersion of C in the punctured cone, and $C' = g(C)$ the projection of C on P'. Since U is isomorphic to $g^{-1}(C')$, the morphism $U \to C'$ is smooth, and the normalization $\overline{U}$ of U is isomorphic to $U \times_{C'} C$. Consider the following diagram, where $g_C \overline{j} = id$, and $f\overline{j} = j$:

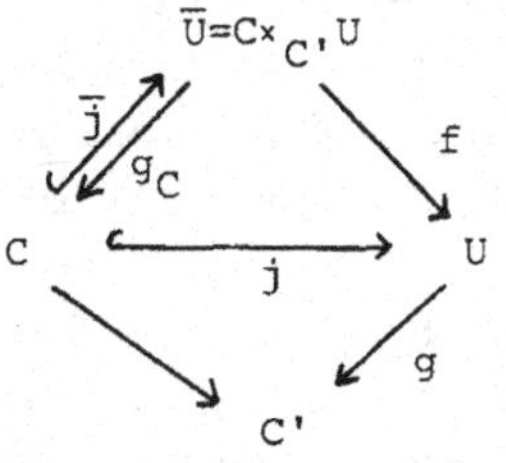

Since $\overline{U}$ is isomorphic to $\mathbf{V}_C(\underline{O}_C(-1))$, $\overline{C} = \overline{j}(C)$ is an effective divisor on $\overline{U}$, defined by a section of $g^*(\underline{O}_C(1)) = f^*(\underline{O}_U(1))$; as $f\overline{j} = j$, f induces an isomorphism from $\overline{C}$ onto C. Now $\overline{U}$ is an open set in the normal scheme $\bar{X}$, and it contains all the points of codimension one in $\bar{X}$, so the section of $\underline{O}_{\bar{U}}(1)$ defining $\bar{C}$ extends to a section of $\underline{O}_{\bar{X}}(1)$.

1.3 The above straightforward construction is usually interpreted - and obscured - in terms of cycles and monoids, as follows:

Since $\bar{X} \to X$ is birational, the section of $\underline{O}_{\bar{X}}(1)$ defining $\bar{C}$ may be considered as a <u>meromorphic</u> section of $\underline{O}_X(1)$, and therefore it defines a (non effective) divisor D on X such that $f^*(D) = \bar{C}$.

Now let $\underline{Z}^1(X)$ denote the group of 1-codimensional cycles, and $cyc: Div(X) \to \underline{Z}^1(X)$ the cycle-map (EGA IV 21.6.7). The theorem of Severi alluded to above is essentially the following equality in $\underline{Z}^1(X)$:

$$cyc(D) = C .$$

Let's make the monoids come into sight: Let N be any effective divisor on X containing the subscheme of X defined by the conductor of $\bar{X} \to X$. Then $M = D+N$ is an effective divisor on X , and this is a monoid. More precisely, N is the divisor associated to an invertible sub-$\underline{O}_X$-module $\underline{O}_X(N)$ of the field of rational functions on X , containing $\underline{O}_{\bar{X}}$. Therefore $\underline{O}_{\bar{X}}(1)$ is contained in $\underline{O}_X(1) \otimes \underline{O}_X(N)$, and the section of $\underline{O}_{\bar{X}}(1)$ associated to $\bar{C}$ furnishes a section of $\underline{O}_X(1) \otimes \underline{O}_X(N)$, which defines M . As a matter of facts, one usually chooses an N such that $\underline{O}_X(N) \cong \underline{O}_X(d)$, in order to be able to lift M to the ambient space (at least when $P = \mathbb{P}^3_k$), and it is the surface thus obtained which is called a monoid.

2. Application to complete intersections

2.1 A morphism of schemes $f:Y \to X$ is said to be <u>radicial</u> if it is injective and if for every $y \in Y$, the residual extension $k(f(y)) \to k(y)$ is radicial (hence trivial in characteristic 0).

The following criterion is useful: $f:Y \to X$ is radicial if and only if the diagonal morphism $\Delta_f: Y \to Y \times_X Y$ is surjective (EGA I 3.7.1).

Let C be an integral curve over an algebraically closed field, and let $f:\bar{C} \to C$ denote its normalization. Then f is radicial if and only if C has only cusps as singularities.

2.2 Let X be a scheme over the finite field $\mathbb{F}_p$. The Frobenius morphism $Fr: X \to X$ is radicial. Conversely, if X is quasi-compact and $f: Y \to X$ is a finite radicial morphism such that $\underline{O}_X \to f^*(\underline{O}_Y)$ is injective, then there exists a power $q = p^n$ of the characteristic such that $f_*(\underline{O}_Y^q) \subset \underline{O}_X$. In other words, if $F = Fr^n$ denotes the n-th iterate of the Frobenius morphism, there exists a morphism $g: X \to Y$ making the following diagram commutative

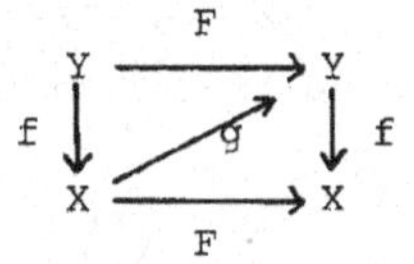

Now, let L be an invertible $\underline{O}_X$-module, and $t: \underline{O}_Y \to f^*(L)$ a section on Y. Then we get a section on X,

$$g^*(t): \underline{O}_X \to g^*f^*(L) = F^*(L) = L^{\otimes q},$$

whose inverse image by f is $F^*(t) = t^q: \underline{O}_Y \to f^*(L^{\otimes q})$.

<u>Proposition</u> 2.3 Let C be a smooth connected curve in $\mathbb{P}^3_k$, where k is algebraically closed, and $\mathrm{char}(k) = p > 0$. Assume the existence of a linear projection with center $x \notin C$, which maps C birationally onto a curve C' with only cusps as singularities. Then C is a set-theoretical complete intersection.

<u>Proof</u>: Keep the notations of 1.2. The morphism $f: \bar{X} \to X$ is radicial because $C \to C'$ is. By 2.2, there exists a power $q = p^n$ of the characteristic, and an effective divisor D on X such that $f^*(D) = q\bar{C}$ and $\underline{O}_X(D) = \underline{O}_X(q)$. The underlying sets of C and D are the same. Moreover, any section of $\underline{O}_X(q)$ comes from a section of $\underline{O}_P(q)$ because X is a divisor in P. Therefore, there exists a surface Y in P such that $X \cap Y = D$.

<u>Remark</u> 2.4 This result is not known if $\mathrm{char}(k) = 0$. The trick above shows at least that $X-C$ is affine: by a theorem of Chevalley (EGA II 6.7.1), this is equivalent to $f^{-1}(X-C)$ being affine; but $f^{-1}(X-C) = \bar{X}-\bar{C}$ because f is radicial, and $\bar{X}-\bar{C}$ is affine because $\bar{C}$ is ample.

3. A factorization for finite morphisms

Proposition 3.1 Let $f:Y \to X$ be a finite morphism between noetherian schemes.

The following conditions are equivalent:

1) There exists a factorization of f as $Y \overset{h}{\to} Z \overset{g}{\to} X$, where g is unramified, h is radicial, and $\underline{O}_Z \to h_*(\underline{O}_Y)$ is injective.

2) The underlying set of the diagonal in $Y \times_X Y$ is open.

Moreover, under these conditions, the factorization is unique.

Remarks 3.2.1 A morphism of finite type $f:Y \to X$ is unramified if and only if its diagonal morphism $\Delta_f : Y \to Y \times_X Y$ is an open immersion; this is, of course, a strictly stronger condition than 2). A finite morphism which is both radicial and unramified is a closed immersion. This implies the unicity in the above factorization.

3.2.2 If, in 3.1, X and Y are integral curves, and if f is birational, then the complement of the diagonal in $Y \times_X Y$ is a finite set of closed points, so the condition 2) is satisfied. If moreover Y is the normalization of X, then Z may be thought of approximately as the curve obtained from X by separating the branches at the multiple points.

Proof of 1 ⇒ 2): Consider the following commutative diagram where $u\Delta_h = \Delta_f$, and where the square is cartesian:

$$\begin{array}{ccccc} Y & \xrightarrow{\Delta_h} & Y \times_Z Y & \xrightarrow{u} & Y \times_X Y \\ & & \downarrow & & \downarrow{\scriptstyle h \times h} \\ & & Z & \xrightarrow[\Delta_g]{} & Z \times_X Z \end{array}$$

Since g is unramified, Δ_g is an open immersion, hence u is an open immersion. Since h is radicial, Δ_h is surjective.

Proof of 2) ⇒ 1): Write $A = \underline{O}_X$, and $B = f_*(\underline{O}_Y)$. We may - and we shall - suppose that $A \to B$ is injective.

To obtain Z, we shall iterate the following construction:

Let U denote the scheme induced by $Y \times_X Y$ on the open set $\Delta_f(Y)$. The direct image R of $\underline{O}_U$ on X is inserted between the following morphisms of A-algebras, where m is surjective and has a nilpotent kernel

$$B \rightrightarrows R \xrightarrow{m} B .$$

Take $C_1 = \mathrm{Ker}(B \rightrightarrows R)$, and $Z_1 = \mathrm{Spec}(C_1)$.

First we show that $h_1 : Y \to Z_1$ is radicial: The construction commutes with flat base change, so we can assume C_1 to be a strictly henselian local ring. Then, it suffices to prove that B is local. For every idempotent $e \in B$, we have, in $B \otimes_A B$,

$$(1-e\otimes 1-1\otimes e)(e\otimes 1-1\otimes e) = 0 .$$

The first factor is invertible in R because its image $1-2e$ is invertible in B, and $\mathrm{Ker}(m)$ is nilpotent. Therefore e is in the local ring C_1; so e is trivial.

This implies that h_1 induces a homeomorphism from Y onto Z_1; hence the condition 2) is satisfied for $g_1 : Z_1 \to X$, and we can iterate the construction.

It remains to prove that $Z_m \to X$ is unramified for m large. Let S_n denote the support of $\Omega^1_{Z_n/X}$ viewed as an A-module. We get a decreasing sequence of closed subsets of the noetherian space X; therefore, we have $S_n = S_{n+1} = \ldots$, for n large enough, and we must show that S_n is empty. Suppose it is not, and let A' denote the strict henselisation of A at a maximal point of S_n. After replacing $B' = A' \otimes_A B$ by $C'_n = A' \otimes_A C_n$, we can assume $A' \to B'$ to be unramified outside the closed points. Let us show that the A'-module B'/A' is then artinian: Since A' is strictly henselian, B' is the direct product of local rings B'_i, and the residual extension of $A' \to B'_i$ is radicial; hence $B'_i \otimes_{A'} B'_i$ is a local ring. The underlying set of the diagonal $\mathrm{Spec}(B'_i) \subset \mathrm{Spec}(B'_i \otimes_{A'} B'_i)$ contains the unique closed point, and is open by condition 2). Therefore it is the whole space, and the morphism $\mathrm{Spec}(B'_i) \to \mathrm{Spec}(A')$ is radicial. But this morphism is unramified outside the closed point, hence it is an isomorphism there, and B'/A' is artinian. This implies that the decreasing sequence C'_m/A' of sub-A'-modules of B'/A' is stationary, hence $C'_m = C'_{m+1} = \ldots$, for m large enough. If $x \in C'_m$, the very construction of C'_{m+1} from C'_m shows that the support of $x\otimes 1-1\otimes x$ in $\mathrm{Spec}(C'_m \otimes_{A'} C'_m)$ is disjoint from the diagonal. Therefore the morphism $\mathrm{Spec}(C'_m) \to \mathrm{Spec}(A')$ is unramified. This contradiction completes the proof.

3.3 This result will be applied, in the next section, to a birational morphism between reduced curves. In this case, the proof is shorter because the A-module B/A is obviously artinian.

4. The theorem of Cowsik-Nori

The following result is the main point in this paper [1] by Cowsik and Nori. Their proof seems a little intricate; here I shall give a proof which is more geometrical and, I hope, easier to follow.

Proposition 4.1 Let k be an algebraically closed field of characteristic $p > 0$, and C a reduced curve in $P = \mathbb{P}^3_k$, with any singularities. Then there exists a curve D in P which is locally a complete intersection, such that $D_{red} = C$.

Proof: Choose a linear projection $g:P - \{x\} \to P'$, with center $x \notin C$, which induces a birational morphism from C onto its image C'. Consider the factorization given by 3.1

$$C \overset{u}{\to} C'' \overset{v}{\to} C' ,$$

where u is radicial and v unramified. The exists a power $q = p^n$ of the characteristic such that, with slight abuses of notation, one has

4.1.1 $$\underline{O}_{C''} = \underline{O}_{C'}[\underline{O}_C^q]$$

Call $U = g^{-1}(C')$ the punctured cone over C, and $F:U \to U^{(q)}$ the n-th power of the relative Frobenius morphism: by definition, one has the following commutative diagram with cartesian square

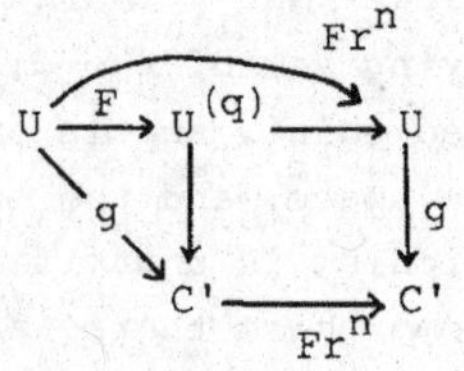

The equality 4.1.1 above means that C'' is the schematic image $F(C)$ of the subscheme C of U. Let $D = F^{-1}F(C) = U \times_{U^{(q)}} C''$. One gets the following factorization of the morphism $C \to C'$:

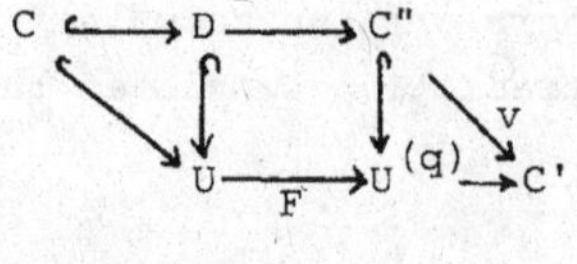

Since C' is a reduced plane curve, $\Omega^1_{C'/k}$ is locally generated by two elements. The exact sequence

$$v^*(\Omega^1_{C'/k}) \to \Omega^1_{C''/k} \to \Omega^1_{C''/C'} \to 0$$

and the fact that v is unramified imply that $\Omega^1_{C''/k}$ is also locally generated by two elements; therefore C'' is locally a complete intersection.

Now $F:U=\mathbb{V}_{C'}(\underline{O}_{C'}(-1)) \to U^{(q)}=\mathbb{V}_{C'}(\underline{O}_{C'}(-q))$ is easily seen to be a complete intersection morphism (EGA IV 19.3); this implies that D is locally a complete intersection. Finally $u:C \to C''$ and $F:D \to C''$ are surjective homeomorphisms, hence $D_{red} = C$.

Bibliography

[1] R.C. Cowsik and M.V. Nori, Curves in characteristic p are Set Theoretic Complete intersections, Inv. Math., (45), 1978, p. 111-114.

EGA A. Grothendieck and J. Dieudonné, Eléments de géométrie algébrique, ch. I: Springer-Verlag, 1971; ch. II to IV: Publ. Math. IHES, Paris.

[2] R. Hartshorne, Complete Intersections in Characteristic $p > 0$, (to appear in Amer. J. Math.).

[3] P. Samuel, Méthodes d'algébre abstraite en géométrie algébrique, Springer-Verlag, 1967.

UER de Mathématiques
Université de Rennes I
Avenue du Général Leclerc
35042 RENNES CEDEX
France

Added in proof. A singular point x of a reduced curve C is called a cusp if a strict henselisation of the local ring $\underline{O}_{C,x}$ is a domain. C has only cusps as singularities if and only if the normalization morphism onto C is radicial (cf. EGA 0_I6.5.10, IV.7.8.3 vii), IV.18.6.12, IV.18.8.15).

INTERSECTION PROPERTIES OF MODULES

by

Hans-Bjørn Foxby [1)]

In this note various intersection properties of a module over a local rings are related to the vanishing of some of the cohomology modules of tensor products of certain complexes of modules.

Throughout M denotes a non-zero f.g. (= finitely generated) module over a local ring (A,m,k) . We introduce the following properties of M .

M has the <u>strong intersection property</u>, if
$\dim N \leqq \operatorname{grade}_A M + \dim(M\otimes N)$ for all f.g. modules N .

M has <u>the intersection property</u>, if
$\dim N \leqq \operatorname{pd}_A M + \dim(M\otimes N)$ for all f.g. modules N .

M <u>has the weak intersection property</u>, if
$\operatorname{depth} M \leqq \operatorname{grade}_A N + \dim(M\otimes N)$ for all f.g. modules N .

Here, dim denotes the usual Krull-dimension, pd_A denotes projective dimension, and for f.g. modules L and K

$$\operatorname{grade}_L K = \text{maximal length of an L-regular sequence in } \operatorname{ann} K$$
$$= \inf\{\ell \mid \operatorname{Ext}^{\ell}(K,L) \neq 0\}, \text{ and}$$

$$\operatorname{depth} L = \operatorname{grade}_L k .$$

If A is regular then M has the strong intersection property, see Serre [10, Chapitre V, Théorème 3]. This is false in general. However one knows of no module of <u>finite</u> projective dimension without the strong intersection property. Indeed M has the strong intersection property if $\operatorname{pd}_A M < \infty$ provided either M is a graded module over a (specially) graded ring (see [9, Théorème 2 (iii)]) or

1) Supported, in part, by SNF (the Danish Natural Science Research Council).

$\text{grade}_A M \leq 1$ (in this case the strong intersection property follows from MacRay [7, Proposition 5.2]).

If A contains a field then M has the intersection property, see Peskine and Szpiro [8, Chapitre II, Théorème (2.1)] and Hochster [3, Corollary 1] or [4]. It is generally believed that M always has the intersection property if $\text{pd}_A M < \infty$, and this is certainly the case if $\text{pd}_A M \leq 2$, see [8, Chapitre II, Théorème (1.3)]. Nevertheless we have introduced also the weak intersection property (which is weaker than the intersection property when $\text{pd}_A M < \infty$) because it is almost the same as the following nice property (even when $\text{pd}_A M = \infty$) (see Theorem 2 below).

M has the <u>regular sequence property</u> if each M-regular sequence is A-regular.

Let $\mathcal{C}$ denote the class of bounded complexes

$$X = 0 \to X^m \to \ldots \to X^n \to 0$$

of f.g. modules, and let $\mathcal{P}$ denote the subclass of $\mathcal{C}$ consiting of complexes X with each X^ℓ f.g. free. For $X \in \mathcal{C}$ we write

$$i(X) = \inf\{\ell \mid H^\ell(X) \neq 0\} \quad \text{and}$$
$$s(X) = \sup\{\ell \mid H^\ell(X) \neq 0\} .$$

The first result is due to Iversen, [6, Theorem 2.4].

<u>Theorem 1. The following are equivalent:</u>

(a) M <u>has the regular sequence property</u>

(b) $\text{grade}_M H \leq \text{grade}_A H$ <u>for all f.g. modules</u> H

(c) $i(M \otimes Y) \leq i(Y)$ <u>for all</u> $Y \in \mathcal{P}$.

The proofs are collected in the last part of this note.

<u>Theorem 2. If M has the regular sequence property then M has the weak intersection property.</u>

<u>If the A_p-module M_p has the weak intersection property for all $p \in \text{Supp } M$ then M has the regular sequence property.</u>

In particular, if A is Cohen-Macaulay and M has the regular sequence property then M has the intersection property.

Now we turn to the intersection property of M, and we assume from now on $pd_A M < \infty$. Furthermore, F will denote a finite resolution of M by f.g. free modules (so $F \cong M$ in the derived category) and F^* denotes the complex $\mathrm{Hom}(F,A)$ $(\in \mathcal{P})$. Finally, we assume that A is a homomorphic image of a Gorenstein local ring (so A admits a dualizing complex), e.g. A is complete or essentially finitely generated over a field.

Theorem 3. The module M has the intersection property if and only if

$$i(F^* \otimes X) \leq s(F^*) + i(X) \quad \text{for all} \quad X \in \mathcal{C}.$$

The next result is an analogue to Theorem 2. The first part is the Result in [1, Section 5], while the second part is closely related to Iversen's result [5, Theorem 3.2].

Theorem 4. The module M has the intersection property if

$$(*) \qquad i(F \otimes X) \leq i(X) \quad \text{for all} \quad X \in \mathcal{C}.$$

If the A_p-module M_p has the intersection property for all $p \in \operatorname{Supp} M$ then $(*)$ holds.

The final result is an analogue to Theorem 3, and its "if"-part has been proved by Iversen [6, Corollary 3.2].

Theorem 5. The module M has the strong intersection property if and only if

$$i(F^* \otimes X) \leq i(F^*) + i(X) \quad \text{for all} \quad X \in \mathcal{C}$$

The proofs.

In the proofs we always work in the derived category of the category of modules over the ring A. The basic notation and some simple facts concerning this derived category are collected in [2, Section 1]. The definitions of depth, dimension, and grade of a complex can be found in Sections 3 and 5 in [2]. We are going to give a lot of references to [2], and a reference to a result in [2] labeled (x,y) will simply be given as [x.y] (instead of [2, (x.y) Result].

For $X \in \mathcal{C}$ we write $X^* = \mathrm{RHom}(X,A)$.

Lemma. For $X, Z \in \mathcal{C}$ there is an equality

$$i(\mathrm{RHom}(X,Z)) = \inf_{\ell}(i(\mathrm{RHom}(H^{\ell}(X),Z))-\ell)$$

Proof. First we prove the inequality $\geqq$. Pick (by [3.4]) $p \in \operatorname{Supp} X \cap \operatorname{Supp} Z$ such that

$$i(\mathrm{RHom}(X,Z)) = \operatorname{depth}_{A_p} Z_p - s(X_p)$$

and write $\ell = s(X_p)$. Then $p \in \operatorname{Supp} H^{\ell}(X)$ and hence

$$\begin{aligned} i(\mathrm{RHom}(H^{\ell}(X),Z)) - \ell &\leqq \operatorname{depth}_{A_p} Z_p - s(X_p) \\ &= i(\mathrm{RHom}(X,Z)) . \end{aligned}$$

Now to prove the other inequality keep ℓ fixed and choose $q \in \operatorname{Supp} H^{\ell}(X) \cap \operatorname{Supp} Z$ such that

$$i(\mathrm{RHom}(H^{\ell}(X),Z)) = \operatorname{depth}_{A_q} Z_q .$$

We have $s(X_q) \geqq \ell$ and hence

$$\begin{aligned} i(\mathrm{RHom}(X,Z)) &\leqq \operatorname{depth}_{A_q} Z_q - s(X_q) \\ &\leqq \operatorname{depth}_{A_q} Z_q - \ell \\ &= i(\mathrm{RHom}(H^{\ell}(X),Z)) - \ell . \quad \square \end{aligned}$$

Proof of Theorem 1. (a) ⇒ (b) and (c) ⇒ (a) are easy as in [6].

To prove (b) ⇒ (c) we proceed as follows.

Write $X = Y^*$, so $Y \cong X^*$ and $M \overset{L}{\otimes} Y \cong \mathrm{RHom}(X,M)$ by [3.15]. From the lemma above and the inequality (b) we get

$$\begin{aligned} i(M \overset{L}{\otimes} Y) &= i(\mathrm{RHom}(X,M)) \\ &= \inf_{\ell}(i(\mathrm{RHom}(H^{\ell}(X),M))-\ell) \\ &\leqq \inf_{\ell}(i(\mathrm{RHom}(H^{\ell}(X),A))-\ell) \\ &= i(\mathrm{RHom}(X,A)) = i(Y) . \quad \square \end{aligned}$$

Proof of Theorem 2. To prove the first part assume that M has the regular sequence property, that N is a f.g. module, $N \neq 0$, and that $g = \mathrm{grade}_A N$ and $\ell = \dim(M \otimes N)$. Then we shall prove depth $M \leqq g + \ell$ be induction on ℓ .

$\underline{\ell = 0}$. $\quad$ depth M = depth RHom(N,M) $\quad$ by [1.7]

$\quad$ = i(RHom(N,M)) $\quad$ by [3.3]

$\quad$ $= \mathrm{grade}_M N \leqq g$ $\quad$ by Theorem 1

$\underline{\ell > 0}$. The case $\ell = 0$ shows in particular depth $M \leqq$ depth A ($= \mathrm{grade}_A k$, while $\dim(M \otimes k) = 0$) , so we will assume $g <$ depth A . Now choose $q \in \mathrm{Supp}\, N$ with $\mathrm{grade}_A A/q = g$. Then $q \neq m$. Now $\dim(M \otimes A/q) \leqq \ell$, so $\dim A/p \leqq \ell$ for all $p \in \mathrm{Supp}(M \otimes A/q)$, and $\dim A/p = \ell$ for only a finite number of these prime ideals p , say for $p_1, \ldots, p_s$ ($s \geqq 0$) . Note that $p_i \neq m$ for all i and choose $a \in m - q \cup p_1 \cup \ldots \cup p_s$. If $N' = A/(q + (a))$ then $\dim(M \otimes N') = \ell - 1$ and $\mathrm{grade}_A N' \leqq g + 1$ (the latter follows from the exact sequence

$$\mathrm{Ext}^g(A/q,A) \xrightarrow{a\cdot} \mathrm{Ext}^g(A/q,A) \to \mathrm{Ext}^{g+1}(N',A) \;) .$$

Therefore the inductive hypothesis gives

$$\text{depth } M \leqq \text{grade } N' + \dim(M \otimes N') \leqq g + 1$$

as required.

To prove the second part of Theorem 2 we are required to show (e.g.) $i(M \overset{L}{\otimes} Y) \leqq i$ when $Y \in P$ and $i = i(Y)$.

If $\dim(M \otimes H^i(Y)) = 0$ then this is easy:

$i(M \overset{L}{\otimes} Y) \leqq \mathrm{depth}(M \overset{L}{\otimes} Y)$ $\quad$ cf. [3.3]

$\quad$ = depth M − pd Y $\quad$ by [3.13.b]

$\quad$ $\leqq \mathrm{grade}_A H^i(Y) - \mathrm{pd}_A Y$

since M is supposed to have the weak intersection property . Now $H^i(Y)$ is a submodule of Coker ($Y^{i-1} \to Y^i$) which is a module of projective dimension $\mathrm{pd}_A Y + i$, and hence we are done in this special case. (Fact: If $H (\neq 0)$ is a submodule of a f.g. module C then $\mathrm{grade}_A H \leqq \mathrm{pd}_A C$ (since if $\mathrm{pd}_A C < \infty$ and $p \in \mathrm{Ass}\, H$ ($\subseteqq$ Ass C) then $\mathrm{grade}_A H \leqq \mathrm{depth}_{A_p} A_p = \mathrm{pd}_{A_p} C_p \leqq \mathrm{pd}_A C$).)

In general pick q minimal in $\mathrm{Supp}(M \otimes H^i(Y))$. Then

$$i(M \overset{L}{\otimes}_A Y) \leq i(M_q \otimes_{A_q} Y_q) \leq i(Y_q) = i \quad . \quad \blacksquare$$

Proof of Theorem 3. Note that $s(F^*) = s(M^*) = \mathrm{pd}\, M$ by [3.16.b].
For $X \in \mathcal{C}$ we have

$$\begin{aligned} i(M^* \overset{L}{\otimes} X) &= \dim A - \dim((M^* \overset{L}{\otimes} X)^\dagger) && \text{by [3.14.d]} \\ &= \dim A - \dim(M \overset{L}{\otimes} X^\dagger) && \text{by [1.2] and [1.4]} \\ &= \dim A - \sup_\ell(\dim(M \otimes H^\ell(X^\dagger)) + \ell) && \text{by [3.12]} \end{aligned}$$

(Fact: If H is a module then $\dim(M \overset{L}{\otimes} H) = \dim(M \otimes H)$. This follows from [3.5]).

Now if M has the intersection property then

$$\begin{aligned} i(M^* \overset{L}{\otimes} X) &\leq \dim A - \sup(\dim H^\ell(X^\dagger) + \ell - \mathrm{pd}\, M) \\ &= \dim A - \dim X^\dagger + \mathrm{pd}\, M && \text{by [3.5]} \\ &= i(X) + s(M^*) \ . \end{aligned}$$

If on the other hand the inequality holds for all $X = N^\dagger$ where N is a f.g. module then

$$\begin{aligned} \mathrm{pd}\, M + \dim(M \otimes N) &= s(M^*) + \dim A - i((M \otimes N)^\dagger) \\ &= s(M^*) + \dim A - i(M^* \overset{L}{\otimes} N^\dagger) \\ &\geq \dim A - i(N^\dagger) = \dim N \quad . \quad \blacksquare \end{aligned}$$

Proof of Theorem 4. To prove the first part assume that (*) holds and let N be a f.g. module, $N \neq 0$, such that $\dim(M \otimes N) = 0$, and we are required to prove $\dim N \leq \mathrm{pd}_A M$. (Then the general case follows easily).

$$\begin{aligned} i(M \overset{L}{\otimes} N^\dagger) &= \mathrm{depth}(M \overset{L}{\otimes} N^\dagger) && \text{by [3.3]} \\ &= \mathrm{depth}\, N^\dagger - \mathrm{pd}\, M && \text{by [3.13.b]} \\ &= \dim A - s(N^{\dagger\dagger}) - \mathrm{pd}\, M && \text{by [3.14.c]} \\ &= \dim A - \mathrm{pd}\, M && \text{since } N^{\dagger\dagger} \cong N \ . \end{aligned}$$

Therefore

$$\mathrm{pd}_A M = \dim A - i(M \overset{L}{\otimes} N^{\dagger}) \geqq \dim A - i(N^{\dagger}) = \dim N \quad .$$

Now we turn to the proof of the second part of Theorem 4 assuming that M_p has the intersection property for all $p \in \mathrm{Supp}\, M$. Let $X \in C$ have $i(X) = i$. If $\dim(M \otimes H^i(X)) = 0$ then

$$\begin{aligned} i(M \overset{L}{\otimes} X) &\leqq \mathrm{depth}(M \overset{L}{\otimes} X) \\ &= \mathrm{depth}\, X - \mathrm{pd}\, M \\ &\leqq \mathrm{depth}\, X - \dim H^i(X) \\ &\leqq i \qquad\qquad \text{by [3.17]} \end{aligned}$$

In general, pick p minimal in $\mathrm{Supp}(M \otimes H^i(X))$. Then

$$i(M \overset{L}{\otimes}_A X) \leqq i(M_p \overset{L}{\otimes}_{A_p} X_p) \leqq i(X_p) = i \ . \qquad \blacksquare$$

<u>Proof of Theorem 5</u>. Replace in proof of Theorem 3 $s(M^*)(= \mathrm{pd}_A M)$ by $i(M^*)(= \mathrm{grade}_A M)$. ∎

<u>Final remarks</u>.

The preceding proofs have been arranged in such a way that they can be used to prove corresponding results for a complex P (in C or in P) in stead of the module M (if such a corresponding result holds, and this is often the case).

For example, the following are equivalent for $P \in P$

(1) $\dim N \leqq \mathrm{pd}_A P + \dim(P \otimes N)$ for all f.g. modules N

(2) $\dim X \leqq \mathrm{pd}_A P + \dim(P \otimes X)$ for all $X \in C$

(3) $i(P^* \otimes X) \leqq s(P^*) + i(X)$ for all $X \in C$.

These inequalities are actually satisfied when A contains a field, see Iversen [5, Theorem 3.2] and the author [2, (4.1) Theorem]. In [2] it is proved that (2) holds even when P is a bounded complex of (not necessarily f.g.) flat modules such that $P \otimes k$ is not exact (provided A contains a field). Note that (1) for $N = A$ is a version of Peskine and Szpiro's New Intersection Theorem, [9, Théorème 1].

If $\mathrm{pd}_A P$ $(= s(P^*))$ in (1), (2), and (3) is replaced by $\mathrm{grade}_A P$ $(= i(P^*))$ then these three conditions are still equivalent.

If A is regular they are satisfied for all $P \in \mathcal{P}$, but in general there is no hope (if $i(P) \neq s(P)$).

References.

1. H.-B. Foxby, Isomorphisms between complexes with applications to the homological theory of modules, Math. Scand. 40(1977), 5-19.

2. H.-B. Foxby, Bounded complexes of flat modules, to appear in J.Pure Appl. Algebra.

3. M. Hochster, The equicharacteristic case of some homological conjectures on local rings, Bull. Amer. Math. Soc. 80 (1974), 683-686.

4. M. Hochster, Topics in the homological theory of modules over commutative rings, (C.B.M.S. Regional Conference Ser. Math. 24) Amer. Math. Soc., Providence, R.I. 1976.

5. B. Iversen, Amplitude inequalities for complexes, Ann. scient. Éc. Norm. Sup. (4 série) 10 (1977), 547-558.

6. B. Iversen, Depth inequalities for complexes, Aarhus Universitet Preprint Series 8 (1977/78).

7. R.E. MacRay, On an application of the Fitting invariants, J. Algebra 2 (1965), 153-169.

8. C. Peskine and L. Szpiro, Dimension projective finie et cohomologie locale, Publ. Math. I.H.E.S. 42 (1973), 49-119.

9. C. Peskine and L. Szpiro, Syzygies et multiplicités, C.R. Acad. Sci. Paris Sér. A 278 (1978), 1421-1424.

10. J.P. Serre, Algrèbre locale. Multiplicités, (Lecture Notes Math. 11) Springer-Verlag, Berlin, Heidelberg, New York, 1965.

Københavns Universitets
Matematiske Institut
Universitetsparken 5
DK-2100 København Ø
Danmark

ON THE THEORY OF ADJOINTS

Silvio Greco

and

Paolo Valabrega

Introduction

There are in the literature several definitions of adjoint divisor to an algebraic curve D contained in a smooth projective surface X.

First of all we find, in order of time, the classical definition due to Brill and Noether (see [B-N], § 7, and [S-R], chap.II, § 2, footnote): they require the passage through any r-fold point P of D, actual or neighbouring, with multiplicity r-1 at least.

Such a definition had some difficulty, because of the following fact (well known to the italian classical geometers): the generic member of the linear system of all adjoints of a given order n passes (if n is large enough) through any r-fold point of D with multiplicity r-1 exactly, so being an adjoint in the sense of Brill and Noether, but, if there are complicated singularities, there are special members of the system with different behaviour (i.e. with multiplicity less than the required one at some neighbouring point of D: see [S], n. 109 and n. 112).

More recently Gorenstein, developping ideas of Zariski, built up a theory of adjoints closely connected to the conductor sheaf of the curve D; precisely in [G] a divisor H is adjoint to D, by definition, if its local equations belong to the stalks of the conductor. It is remarkable that such a definition is the good one to deal properly with the canonical series (and it avoids the problems connected with the bad behaviour at some neighbouring point). Also Zariski in his lectures [Z] considered the above definition relating adjoints and conductor, in his theory of analytic branches over the complex field.

We must also quote Keller's approach to the adjoints (see [K]): he assignes to each place Q centered at a fixed singular point P of D a certain positive integer d_Q, depending on the neighbouring points; then he requires that an adjoint pass, by definition, through the place Q with multiplicity at least d_Q.

Finally we recall that there are other minor variants of the preceding definitions, due for instance to Abhyankar ([A]) or to C.P. Ramamujan (see [R-B]).

The above definitions of adjoint divisor are more or less considered equivalent in the literature, as it is true when D has just ordinary singularities. In the present paper we see that this fact is not true in general and investigate the relations among the three concepts of adjoint above recalled, showing that the classical one is stronger, while the other ones (Gorenstein's and Keller's) are equivalent.

In n.1 we recall a few general facts on blowing up of smooth projective surfaces; in n. 2 we deal with neighbouring points, places and branches of a projective curve on a smooth surface; in n. 3 we discuss some relations between the conductor and the neighbourhoods of a singular point of a curve, stating a characterization of the conductor closely related to adjoints (and based on ideas and results of Northcott: see [M]). In n. 4 we discuss the above quoted definitions of adjoints, proving that the classical one is stronger, while the other ones are equivalent; our main result states that the adjoints in the weak sense are linear combinations, excluded possibly on a hypersurface avoiding all the singularities of D, of adjoints in the strong classical sense. Finally in n. 5 we discuss the behaviour of the generic member of a linear system generated by adjoints in the various senses: it is always a "weak" adjoint and, under some conditions, also a classical adjoint (but we show examples of linear systems of "weak" adjoints containing no "strong" adjoint).

The paper contains also the discussion of several meaningful examples and counterexamples, like the case of ordinary singularities, of the tacnode or of the cusps of higher order.

n. 1 Recalls on blowing up of smooth projective surfaces and divisors.

Once for all we fix an algebraically closed field k of arbitrary characteristic; all the curves and surfaces considered in the present paper will be algebraic schemes over k[(1)].

Let X be a smooth projective surface, i.e. a non singular two-dimensional scheme, together with a closed immersion i: $X \hookrightarrow P^N_k$, fixed once for all. Hence X is isomorphic to a scheme Proj(S), where S is a graded algebra homomorphic image of a ring $k[x_0, x_1, \ldots, x_N]$ with respect to a homogeneous ideal, the x_i's being homogeneous coordinates in P^N_k. Once for all X is supposed irreducible.

Given a closed point P on X, corresponding to the coherent ideal $\underline{M}$, the blowing up of X along $\underline{M}$ (or with center P) is (up to isomorphisms) the scheme $X' = \mathrm{Proj}(\bigoplus_{n=o}^{\infty} \underline{M}^n)$, together with the canonical morphism f: $X' \to X$. The following facts are well known:

(i) $f^{-1}(\underline{M})O_{X'}$ is invertible and f is universal with respect to such a property;

(ii) X-P and $X'-f^{-1}(P)$ are canonically isomorphic;

(iii) X' is a smooth projective surface over k;

(iv) the exceptional divisor E, defined by the sheaf $f^{-1}(\underline{M})O_{X'}$, is a P^1_k;

(v) since $\underline{M}_P$ = maximal ideal of the local ring $O_{X,P}$ is generated by two elements, there is an open affine W = Spec(A) of X, containing P, such that $\underline{M}/_W$ corresponds to a maximal ideal $\underline{m} = (u,v)A$, generated by two elements;

(vi) X' can be described, over W, as the union of two open affines U = Spec(A $[v/u]$) and V = Spec(A $[u/v]$); over any open affine W' of X not containing P, X' is isomorphic to X.

Let now H be an effective Cartier divisor on X; this means that H is a closed subscheme of pure codimension 1 or, equivalently, that there is an affine open covering $(U_i)_{i\in I}$ of X such that on every U_i H has a unique equation $h_i \in A_i = \Gamma(U_i, O_{U_i})$.

---

(1) the majority of results contained in the paper is really valid for any infinite field k, provided that we employ some care in the use of concepts like "point" or "branch" or "multiplicity"; the changes in the proofs being very slight, we deal with the case of an algebraically closed field. On the contrary the situation is quite different when k is finite.

Let P be any point of H; then the blowing up of X with center P induces the blowing up of H with center P, say H', which is, up to isomorphisms, the strict transform of H in X' ([H] ,II,(7.15)). H' is still an effective Cartier divisor on X' ; moreover, if P has multiplicity s for H and if h is a local equation of H in W = Spec(A) (see (v) above), then h/u^s is a local equation for H' in U = Spec(A [v/u]) (for more details see [V] , n.4).

n. 2 Neighbouring points, places and branches of a projective curve.

A curve D on the smooth projective surface X is, once for all, an integral closed subscheme of X, having codimension 1. If $P \in D$ is any r-fold point of D, the blowing up of X with center P induces the blowing up of D with center P, which will be identified with the strict (or proper) transform D' of D in X'. It is easy to see that D' is still a curve on X'. As far as the morphism $g : D' \longrightarrow D$ is concerned, we state the following (well known):

Proposition 2.1 : Let $g : D' \longrightarrow D$ be the blowing up of the curve D with center at the closed point $P \in D$. Then g is a finite morphism.

Proof: Thanks to a well known theorem of Chevalley ([Gr] ,III,(4.4.2)) it is enough to show that g has finite fibers (recall that g is proper).

[Gr] , IV, (19.4.2) says : $g^{-1}(P) = \mathrm{Proj}(G_{O_D}(\underline{I}(P)))$ (equality as schemes), $\underline{I}(P)$ being the Ideal of P on D and $G_{O_D}(\underline{I}(P))$ the associated graded Algebra. Since D has dimension 1, $g^{-1}(P)$ is finite; obviously $g^{-1}(Q)$ is also finite for every $Q \neq P$; hence the claim follows.

The preceding proposition implies obviously that there are only finitely many points of D' over the center P.

Let us now recall the resolution theorem for D on X (see [H] , V, prop. 3.8):

Theorem 2.2 : Let D be a curve on a smooth irreducible projective surface X . Then there exists a finite sequence of blowing up's of X:

$$X = X_0 \overset{f_1}{\longleftarrow} X_1 \longleftarrow \dots \longleftarrow X_{n-1} \overset{f_{n-1}}{\longleftarrow} X_n$$

such that, if

$$D = D_0 \xleftarrow{g_1} D_1 \leftarrow \dots \leftarrow D_{n-1} \xleftarrow{g_{n-1}} D_n = \bar{D}$$

is the corresponding sequence of strict transforms of D, then we have:

1) for every i, f_i has center at a singular point of D_{i-1} ;

2) $\bar{D}$ is smooth (and hence it is the normalization of D).

Remark: the number n of blowing up's considered in the preceding theorem is uniquely determined by the curve D and does not depend on the order one may choose to blow up the points.

Let us now introduce the concept of neighbouring point with the following

Definition 2.3: Let Z be an integral scheme. A neighbouring point Q of Z is a point of a Z-scheme f: Y ⟶ Z obtained from Z by a finite sequence of blowing up's centered at closed points. If f(Q) = P, then we say that Q is a neighbouring point of P.

If f' : Y' ⟶ Z is the composition of another finite sequence of blowing up's and Q' ε Y', we identify Q and Q' if there are open neighbourhoods U,U' of Q,Q' and an isomorphism g : U ⟶ U', compatible with f and f', such that g(Q) = Q'.

The points of Z are special neighbouring points : we call them actual points.

We say that Q belongs to the first neighbourhood of P ε Z if P ≠ Q and if Q lies over P in the blowing up of Z with center at P. We say that Q belongs to the i-th neighbourhood of P if it belongs to the first neighbourhood of some point in the (i-1)-th neighbourhood.

Remark 2.4 : Theorem 2.2 says that every neighbourhood of P ε D contains finitely many points and that there is a neighbourhood containing only simple points. Moreover any neighbouring point of D belongs to some D_i; hence we may consider it also as a point of X_i.

Definition 2.5 : Let H be an effective divisor on X and let Q be a neighbouring point of X. Let Q ε X', where X' is obtained from X by a finite sequence of blowing up's. Let H' be the strict transform of H in X'. We say that Q is a s-fold point for H (or that H passes through Q with multiplicity s) if Q is a s-fold point for H', i.e. if $s = e(O_{H',Q})$ = multiplicity of the ring of H' at Q. If $Q \notin \mathrm{Supp}(H')$ we agree that s = 0.

This applies in particular when H = D (see also [Z], II, § 6).

Example 2.6 : Let (x,y,z) be homogeneous coordinates in the projective plane P^2_k and let D be the projective curve with homogeneous equation $y^2z^3 = x^5$; D has at O = (0,0,1) a double point, with another double point in the first neighbourhood and a simple point in the second neighbourhood; at P_∞ = (0,1,0) D has a triple point, with a double point in the first neighbourhood and a simple point in the second one.

Example 2.7 : The notation being as above, let D be the curve with equation $y^2z^2 = x^4+y^4$; at the origin O = (0,0,1) D has a double point with a double point in the first neighbourhood and then two simple points (we deal with the so called "tac-node").

Example 2.8 : Given the curve D of the example 2.6, the divisor of homogeneous equation yz = 0 passes through O and its neighbouring point in the first order with multiplicity 1; it has the same behaviour in P_∞.

Definition 2.9 : If $P \in D$ and $Q \in \bar{D}$ lies over P, then Q is a place centered at P.

It is easy to see that any r-fold point P is the center of r places at most; if they are r distinct, P is an ordinary singularity.

Definition 2.10 : Let $P = P_0 — P_1 — \dots — P_{m-1} — P_m = Q$ any chain of points such that:

(i) $P \in D$ and P_i belongs to the first neighbourhood of P_{i-1}, for every $i \geq 1$;

(ii) Q is a place with center at P.

Then such a chain is called a "branch" with origin P.

Remark : Let (A,$\underline{m}$) be the local ring of D at P; because of the canonical isomorphism $\hat{\hat{A}} = \hat{\bar{A}}$ there is a one-to-one map between the minimal primes of $\hat{A}$ and the places with center at P (hence with the branches with origin at P); moreover, if $R = O_{X,P}$ and $A = R/(f)$, such minimal primes correspond bijectively to the prime factors of f in $\hat{R}$ (see [Gr], IV_2, (7.8.3), and, for more information on branches, [Gc] and [C]).

Definition 2.11 : Let $P \in D$, let $(A,\underline{m})$ be the local ring of D at P and $Q \in \bar{D}$ a place with center at P having local ring $(B,\underline{n})$. The integer $e(\underline{m}B) = \ell_B(B/\underline{m}B)$ = multiplicity of $\underline{m}B$ in B is called "order" of the branch γ joining P with Q.

It can be shown that, if $\underline{p} \subset \hat{A}$ is the minimal prime corresponding to Q, then the order of γ equals the multiplicity of the local ring $\hat{A}/\underline{p}$ (for more details and generalizations see [C]).

Now we want to introduce (following [K]) a divisor on $\bar{D}$, strictly related with the structure of the singularities of D. Precisely let us consider all the places $Q_1,\ldots,Q_r$ with center at some singular point of D; for every place Q_j there is just one branch γ_j with origin at $P_j \in D$, having the form:

$$P_j = P_{j0} — P_{j1} — \ldots — P_{jn_j} — Q_j,$$

where P_{jh} belongs to D_{jh}, strict transform of $D_{j(h-1)}$ in the blowing up with center $P_{j(h-1)}$. Let us put:

s_{jh} = order of the branch $P_{jh} — \ldots — Q_j$ (with origin on D_{jh})

r_{jh} = multiplicity of P_{jh} as a point of D_{jh}

$d_j = \Sigma_h s_{jh}(r_{jh}-1)$.

Definition 2.12 : With the preceding notations, the divisor $d = \sum_j d_j Q_j$ on $\bar{D}$ is called "the divisor of double points" of D.

Examples : (a) the divisor d for the curve of the example 2.6 consists of the unique place Q_1 lying over O, with coefficient $4 = 2(2-1)+2(2-1)$, and of the unique place Q_2 lying over P_∞, with coefficient $3(3-1)+2(2-1) = 8$.

b) the divisor d for the curve of the example 2.7 (the tacnode) consists of the two places Q_1 and Q_2 lying over O, each with coefficient 2 = 1(2-1)+1(2-1).

Definition 2.13 : Let $H = (U_\alpha, F_\alpha)$ be a Cartier divisor on X and let $f_\alpha = F_\alpha / _{U_\alpha \cap D}$ for every α (so f_α is a rational function on D).

We say that H cuts out on D the divisor d of double points if supp H $\supset$ supp D or

(i) supp H $\not\supset$ supp D (hence $f_\alpha \not\equiv 0$ for every α);

(ii) if $p : \bar{D} \to X$ is the canonical morphism, the divisor $H' = (p^{-1}(U_\alpha), f_\alpha)$ on $\bar{D}$ is greater or equal to d.

Remark : (ii) is equivalent to the following requirement: if $P \in U_\alpha$ is a singular point of D and v is the valuation of $K(D) = K(\bar{D})$ corresponding to a place Q_j with center at P, when $v(f_\alpha) \geq d_j$.

n. 3 The conductor sheaf of a curve and its behaviour in the neighbourhoods.

We recall that the conductor sheaf of the curve D is the sheaf $\gamma_D = \mathrm{Ann}_{O_{\bar{D}}}(O_{\bar{D}}/O_D)$.

Since D is traced on the surface X, we can also consider a conductor on X relative to D (different from the conductor sheaf of X, of course), denoted $\gamma_{X/D}$; it is simply the total inverse image of γ_D on X (for more details see [R-B], n. 5)

Example 3.1 : the conductor γ_D of the curve in the example 2.6 is trivial everywhere, except at O, where the stalk is the ideal generated by $\tilde{x}^2$ and $\tilde{y}$ ($\tilde{x} = x/z$ and $\tilde{y} = y/z$), and P_∞, where the stalk is generated by $\tilde{x}\tilde{z}, \tilde{x}^3, \tilde{z}^2$ ($\tilde{x} = x/y, \tilde{z} = z/y$).

Definition 3.2 : We say that the effective divisor H on X belongs to the conductor of D if $O_X(-H) \subset \gamma_{X/D}$.

The above definition means that, for each $P \in D$, the local equation H_P of H at P belongs to $\gamma_{X/D,P}$.

Now we develop a local teory of conductor for rings A reduced and with finite normalization $\bar{A}$; it will be crucial for the global applications of next sections (for general properties of the conductor of a ring see, for instance, [Z-S], vol. I chap. V, § 5). Once for all we put: γ_A = conductor of A in $\bar{A} = \mathrm{Ann}_{\bar{A}}(\bar{A}/A)$, $\gamma_{B/A}$ = conductor of A in a ring B intermediate between A and $\bar{A} = \mathrm{Ann}_B(B/A)$.

First of all a simple

Lemma 3.3 : *Let A be a domain with finite normalization $\bar{A}$; then either $A = \bar{A}$ or γ_A is not contained in any proper principal ideal of A.*

Proof : Let $a \in A$ be a non invertible element such that $\gamma_A \subset aA$; pick $x \in \gamma_A$, so that $x = ax_1$, with $x_1 \in A$. By definition of conductor we have $(ax_1)\bar{A} \subset A$, hence $ax_1 b \subset \gamma_A \subset aA$, for every $b \in \bar{A}$. Therefore $x_1 b$ belongs to A for each $b \in \bar{A}$, i.e. $x_1 \in \gamma_A \subset aA$; finally $x_1 = ax_2$. We now deduce that $x \in \bigcap_{n\geq 1}^{\infty} a^n A = (0)$, i.e. $\gamma_A = (0)$, which is a contradiction.

We want now to investigate relations between conductor and blowing up, with the aim of stating a result on the structure of the conductor of a ring which descends easily from a result of Northcott on the first neighbourhood and from properties of Gorenstein rings (see [M]).

Lemma 3.4 : *Let A be a one-dimensional ring and Y the blowing up of Spec(A) with center at a closed point. Then we have:*

(i) *the canonical projection $Y \longrightarrow$ Spec(A) is a finite morphism;*

(ii) *Y is affine.*

Proof : (i) follows from prop. 2.1 and (ii) from (i) and the definition of finite morphism (which is affine).

Corollary 3.5 : *Let $(A,\underline{m},K)$ be a reduced local ring having dimension 1 and let B the first neighbourhood of A, in the sense of Northcott-Matlis ([M] page 103). Then Spec(B) coincides, up to isomorphisms, with the blowing up of Spec(A) with center $\underline{m}$.*

Proof : By [M] , 12.2, $\underline{m}B$ is a principal ideal of B generated by a regular element. Hence the following diagram is commutative:

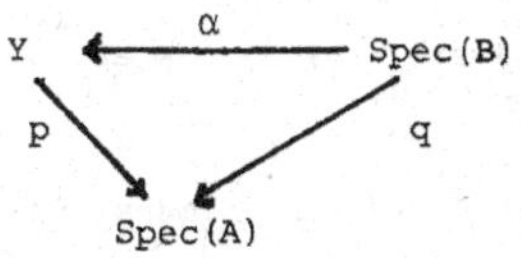

(p and q being the canonical morphisms, Y being the blowing up).

Since p and q are birational, also α is birational, hence dominating. By lemma 3.4, $Y = \mathrm{Spec}(C)$, for some ring C. We have the inclusions: $A \subset B \subset C$; moreover $\underline{m}C = xC$ is principal. BY [M] ,12.1.(2), we have: $\underline{m}^n B = \underline{m}^n$ for some n. Therefore, if $y = x^n$, we get: $yB = \underline{m}^n = \underline{m}^n C = yC$. Since y is regular in B, we obtain that B = C.

Corollary 3.6 : *Let A be a one-dimensional Cohen-Macaulay ring and $\underline{m} \in \mathrm{Spec}(A)$ a closed point, such that $\underline{m}A_{\underline{m}}$ is generated by two elements. Let Spec(B) be the blowing up of Spec(A) with center at $\underline{m}$. If $s = e(A_{\underline{m}})$, we have:*

$$\gamma_{B/A} = \underline{m}^{s-1}.$$

Proof : Since $\underline{m}^r A_{\underline{m}} \cap A = \underline{m}^r$, for each r, we may assume that A is local. By corollary 3.5 B is the first neighourhood of A, hence $\gamma_{B/A} = \underline{m}^{s-1}$ ([M], 13.8).

Given a ring A with total quotient ring K and two sub-A-modules E and F of K , we put: (a) $E:F = (x \in K/xF \quad E)$; (b) $E^{-1} = A:E$. The next two lemmas appear in [D-M] which contains general results on the subject.

Lemma 3.7 : *Let A be a reduced ring and B an overring of A finite over A and such that $A \subset B \subset K$. Then $\gamma_{B/A} = B^{-1}$.*

Proof : Of course $\gamma_{B/A} \subset A:B$. Let now $x \in K$ such that $xB \subset A$. Then $x = x.1 \in A$, hence $x \in A$. It follows that $A:B \subset A$ and finally the claim.

Lemma 3.8 : *Let A be a reduced one-dimensional Gorenstein ring and $A \subset B \subset K$ as in lemma 3.7. Then $(B^{-1})^{-1} = B$.*

Proof : Since B is finite over A, there is a regular element $a \in A$ such that $aB \subset A$. Since $a \in aB$, aB is also a regular ideal of A. Therefore, by [M] , 13.1, pro perties (1) and (7), we have: $((aB)^{-1})^{-1} = aB$, which implies our claim.

Proposition 3.9 : Let A be a one-dimensional Gorenstein ring with total quotient ring K and let $\bar{A}$ be its integral closure in K. If $\bar{A}$ is finite over A and B is any sub-A-algebra of $\bar{A}$ which is Gorenstein, then we have:

$$\gamma_A = \gamma_{B/A}\,\gamma_B.$$

Proof : Since $\gamma_{B/A}\,\gamma_B$ is a regular ideal of A, we have:
$\gamma_{B/A}\,\gamma_B = ((\gamma_{B/A}\,\gamma_B)^{-1})^{-1}$.
Moreover we have:

$$(\gamma_B\,\gamma_{B/A})^{-1} = A:((A:B)(B:\bar{A})) = \qquad \text{by lemma 3.7}$$
$$= (A:(A:B)):(B:\bar{A}) =$$
$$= B:(B:\bar{A}) = \qquad \text{by lemma 3.8}$$
$$= \bar{A} \qquad \text{by lemma 3.8}$$

Finally we get: $\gamma_B\,\gamma_{B/A} = A:\bar{A} = \gamma_A$ (lemma 3.7)

Lemma 3.10 : Let A be a reduced ring having dimension one such that $A = R/\underline{I}$, where R is regular two-dimensional. Let Y = Spec(B) be the blowing up of Spec(A) centered at the closed point $\underline{m}$.
Then we have:

(i) Both A and B are Gorenstein one-dimensional;

(ii) Every localization of A and B at some maximal ideal has embedding dimension ≤ 2.

Proof : Let $\underline{M} \in$ Spec(R) be the closed point of Spec(R) corresponding to $\underline{m}$. Then Spec(B) is a closed reduced subscheme of the blowing up of Spec(R) with center $\underline{M}$. This latter scheme is regular two-dimensional.

Hence (i) follows from [M], 13.2. Moreover (ii) depends on the fact that both Spec(R) and its blowing up have dimension 2.

Proposition 3.11: Let A be a reduced one-dimensional pseudogeometric ring and let us assume that $A = R/\underline{I}$, where R is regular two-dimensional. Let Spec(B) be the blowing up of Spec(A) with center at the closed point $\underline{m}$ and let $s = e(A_{\underline{m}})$. Then we have

$\gamma_A = \underline{m}^{s-1}\gamma_B$ (product in B).

If moreover A is semilocal we have : $\underline{m}B = xB$ and $\gamma_A = x^{s-1}\gamma_B$.

Proof : Since Spec(B) $\longrightarrow$ Spec(A) is finite birational we have: $A \subset B \subset \bar{A}$. Moreover $\bar{A}$ is finite over A because A is pseudogeometric. By lemma 3.10 both A and B are Gorenstein and hence, by proposition 3.9, we have: $\gamma_A = \gamma_{B/A}\gamma_B$. But $\gamma_{B/A} = \underline{m}^{s-1}$ by corollary 3.6.

Corollary 3.12 : Let A be as in 3.11 and let

$A = A_o \to A_1 \to \dots \to A_n = \bar{A}$

be a chain of rings such that A_{i+1} is the ring of the blowing up of Spec(A_i) at the closed point of ideal $\underline{m}_i$. Let then $s_i = e((A_i)_{\underline{m}_i})$.

Then we have:

(i) $\gamma_A = \prod_{i=o}^{n-1} (\underline{m}_i^{s_i-1}\bar{A})$;

(ii) if A is semilocal : $\underline{m}_i A_{i+1} = x_i A_{i+1}$ and $\gamma_A = \prod_{i=o}^{n-1} (x_i^{s_i-1}\bar{A})$.

Remark : If D is a curve on a smooth projective surface X and Spec(A) is an open affine of D, then proposition 3.11 and corollary 3.12 are valid for A. Moreover such results can be employed also for all the curves D_i strict transforms of D_i's in the blowing up's of theorem 2.2, whose sequence desingularizes D.

n. 4 The concept of adjoint divisor to a curve : relations between conductor and effective passage through neighbouring points.

There are in the literature several definitions of adjoint divisor to a projective curve: starting with the classical works of Brill and Noether, we meet the papers of the italian classical geometers and, finally, the more recent results of Gorenstein, Keller and, concerning the analytic branches, of Zariski. When the curve has non ordinary singularities the various definitions raise not so easy problems of equivalence; in the present section we discuss three well known definitions, investigating their relations.

Once for all D is an integral curve on the smooth irreducibile surface $X \subset P_K^N$, while H is an effective (Cartier) divisor on X.

Definition 4.1 (classical : see [B-N] , § 7, or [S-R] , chap, II, § 2, footnote at page 30) : H is adjoint to D (shortly A1) if it passes through every r-fold point P of D, actual or neighbouring, with multiplicity at least r-1 (see def. 2.5).

Definition 4.2 (due to Gorenstein [G]): H is adjoint to D (A2) if it belongs to the conductor sheaf $\gamma_{X/D}$ (see def. 3.2).

Definition 4.3 (due to Keller [K]): H is adjoint to D (A3) if it cuts out the divisor of double points of D (see def. 2.13).

The above definitions occur frequently in the classical and recent literature ; however there are several variants of them, as well as different points of view. Beside Severi's intuitive concept based on virtual passage through the singular points of D ([S] , n. 109 and 112), we have the contributions of Zariski ([Z]), who defines precisely the concept of virtual multiplicity and relates it with the conductor, in the case of analytic branches over the complex field. Moreover Abhyankar in [A] gives other definitions, distinguishing between adjoint and true adjoint (see also [O] , for more details), while some ideas of C.P. Ramanujan in the field have been developped in [R-B] .

In the present section we examine just A1, A2, A3, hoping to dedicate next paper to discuss the point of view of italian geometers and Zariski on the subject.

Before we state our main claim, we want to discuss some examples, useful to enlighten the situation.

Example 4.4 : Let D be a plane projective curve with just ordinary singularities. It is easy to see that a divisor H is A1 if and only if it passes through each r-fold point of D with multiplicity at least r-1 (recall that there is no singular neighbouring point).

It is not easy (but true) that also A2 and A3 are characterized in the same way (see theorems below).

Example 4.5 : Let D be the plane projective curve whose homogeneous equation is $y^2z^3 = x^5$ (same notations of example 2.6). We know that: O is double with a neighbouring double point O_1 (followed by simple points); P_∞ is triple with a neighbouring double point P_1 (followed by simple points).

Moreover the conductor is generated, at O, by $\tilde{x}^2$ and $\tilde{y}$, at P_∞ by $\tilde{x}\tilde{z},\tilde{x}^3,\tilde{z}^2$.

It is now easy to see, with a direct computation, that the divisor whose homogeneus equation is $yz^2 = 0$ is A1, A2, A3 (recall that the coordinate ring of D is, at O, $k[t^2,t^5]$, and $k[t^3,t^5]$ at P_∞).

On the contrary the divisor whose homogeneous equation is $x^3 = 0$ is both A2 and A3 (according with the following theorem 4.6), but it cannot be A1, since it does not pass through the neighbouring point O_1. Exactly the same argument works for all the divisors of equation $x^n = 0$, whenever $n > 3$.

Now we give our first result, relating A2 and A3:

Theorem 4.6 : The effective divisor H is A2 if and only if it is A3.

Proof : Let U = Spec(A) be an open affine of D containing all the singular points of D; then the conductor sheaf γ_D is faithfully described by the ideal γ_A, outside of U being the whole O_D. Let now $P_1,\dots,P_n$ be the singular points of D and $\underline{m}_1$, $\dots,\underline{m}_n$ the corresponding maximal ideals of A. Moreover, for each P_i let $\underline{n}_{i1},\dots,\underline{n}_{is_i}$ be the maximal ideals of $\bar{A}$ lying over $\underline{m}_i$. Our aim is to prove the following equality :

$$\gamma_A = \prod_{i,j} \underline{n}_{ij}^{d_{ij}}$$

were d_{ij} is the coefficient of the place Q_{ij}, corresponding to $\underline{n}_{ij}$, in the divisor of double points of D.

We have:

$$\gamma_A = \prod_{i,j} \underline{n}_{ij}^{a_{ij}},$$

where $\underline{n}_{ij}^{a_{ij}} = (\gamma_A \bar{A}_{\underline{n}_{ij}})^c$ (contraction to $\bar{A}$).

Moreover we have:

$$\gamma_A = \prod_{i,j,h} \underline{m}_{ijh}^{s_{ijh}-1} \bar{A},$$

where $\underline{m}_{ijh}$ is the h-th ideal in the chain of ideals corresponding to the following branch (for h = 0 we agree to have $P_i = P_{ij0}$):

$P_i - \ldots - Q_{ij}$

and s_{ijh} is the multiplicity of the corresponding point.

Hence we get:

$$\underline{n}_{ij}^{a_{ij}} = \prod_h \underline{m}_{ijh}^{s_{ijh}-1} \quad \text{(i and j fixed).}$$

Since $\bar{A}_{\underline{n}_{ij}}$ is a DVR, we have:

$$a_{ij} = \ell(\bar{A}_{\underline{n}_{ij}}/\prod_h \underline{m}_{ijh}^{s_{ijh}-1}) = \sum_h \ell(\bar{A}_{\underline{n}_{ij}}/\underline{m}_{ijh}^{s_{ijh}-1}) =$$

$$= \sum_h (s_{ijh}-1)\ \ell\ (\bar{A}_{\underline{n}_{ij}}/\underline{m}_{ijh}) = \sum_h (s_{ijh}-1) r_{ijh},$$

r_{ijh} being the order of the branch:

$P_{ijh} - \ldots - Q_{ij}$.

So $a_{ij} = d_{ij}$.

Let now H be A2 to D and let h_{P_i} be a local equation of H at $P_i \in D$ $(h_{P_i} \in O_{D,P_i})$

The, by the equality just proved, we have:

$$h_{P_i} \in \prod_j \underline{n}_{ij}^{d_{ij}} \bar{A} \quad \text{(i is fixed)}$$

and therefore H cuts on $\bar{D}$ a divisor not smaller than the divisor of double points.

Conversely, let H be A3; then it follows immediately from the above equality that $h_{P_i} \in \gamma_{P_i}$ for every $P_i \in D$, which proves our claim.

To relate A1 and A2 we need the preliminary.

Definition 4.7: The effective divisor H is special adjoint to D (AS) if it has a point of multiplicity exactly r-1 at every r-fold singular point of D, actual or neighbouring.

Remark: If H is AS, then it is obviously A1.

Lemma 4.8 : Let H be an effective divisor on the smooth projective surface $X \subset P^N_k$ and let $(P_1,\dots,P_n)$ be a finite set of closed points of X. Then there is a hypersurface Z in P^N_k such that:

(i) Z contains no P_i;

(ii) H is principal on the affine open set $X \cap (P^N_k - Z)$.

Proof : Let S be a hyperplane section of X which does not contain the points P_i and let $U = X-S = \mathrm{Spec}(A)$ = open affine of X. Let then $\underline{I} \subset A$ be the invertible ideal of the closed suchscheme H of X restricted to U and let $\underline{m}_i \subset A$ be the maximal ideal corresponding to P_i. If $B = T^{-1}A$ (where $T = A - \bigcup \underline{m}_i$) then $\underline{I}B$ is invertible and hence principal, B being semilocal. Therefore there is an element $g \in T$ suche that $\underline{I}A_g$ is principal. That proves that there exists an open set $V \subset X$ containing $P_1,\dots,P_n$ on which H is principal.

Let us now prove that V can be chosen of the form $X \cap (P^N_k - Z)$, where Z is a suitable hypersurface.

Let $F = X-V$ and let $Z_1,\dots,Z_m$ be forms in the variables $x_0,\dots,x_N$ such that $V = \bigcup (D_+(Z_i) \cap X)$; of course no P_i can be a common zero to $Z_1,\dots,Z_m$. We can assume that the Z_i's have the same degree, raising them, if necessary, to suitable powers. Hence no P_i is a base point of the linear system $\Sigma a_i Z_i = 0$, which means that the generic member of the system does not contain any of the P_i's. Let $Z' = \Sigma a_i Z_i$, with the a_i's generically chosen, i.e. such that Z' contains no P_i. Now Z' vanishes at the whole F and at no P_i. Therefore $X \cap (P^N_k - Z) = U \subset V$ is the required subset, Z being the hypersurface with equation $Z' = 0$.

Lemma 4.9 : Let B be a regular two-dimensional domain, $\underline{M}$ a maximal ideal of B, $\underline{I} \subsetneq \underline{M}$ a prime ideal of height 1, $A = B/\underline{I}$ and $\underline{m} = \underline{M}/\underline{I}$. Let then A' be the blowing up of A with center $\underline{m}$ and assume that $\underline{m}A' = x\underline{A}'$, x being suitable in $\underline{m}$. We have:

(i) $A' = \underline{A}[\underline{m}/x]$ (see [V], n. 2);

(ii) $A' = (B[\underline{M}/y])/\underline{I}'$, y being any inverse image of x in B and $\underline{I}'$ being the strict transform ideal of $\underline{I}$ into $B[\underline{M}/y]$ (see [V-V], n. 1).

Proof : (i) - By the universal property of the blowing up it is easy to see that $A \subset A' \subset A[\underline{m}/x]$. Let now $z \in A[\underline{m}/x]$, so that $zx^n \in \underline{m}^n$ for any large n. Hence $zx^n \in x^n A'$, which means that $zx^n = x^n z'$, for some z' in A'; this implies that

$z = z' \in A'$.

(ii) - Spec($B[\underline{M}/y]$) is a basic open affine of the blowing up of Spec(B) with center at the closed point corresponding to $\underline{M}$; it is easy to check that it induces on the blowing up of Spec(A) the open affine Spec($A[\underline{m}/x]$) (see [V], n. 3), which is the blowing up A' by (i).

Remark : The above lemma is really true under quite general hypotheses (see for more information [V], n. 2 and n. 3 or [V-V], n. 1); but the one-dimensionality of A is essential, to use the fact that the blowing up of A is affine.

Lemma 4.10 : Let A be a domain of dimension 1 and T the multiplicative system complement of the union of finitely many maximal ideals $\underline{m}_o, \ldots, \underline{m}_n$. Than there exists $f \in T$ such that the blowing up of A_f with center at $\underline{m}_o$ is the domain $A_f[\underline{m}_o A_f/x]$, x being suitable in $\underline{m}_o$.

Proof : The blowing up of A_T with center $\underline{m}_o A_T$ is the domain $(A_T)[\underline{m}_o A_T/x]$, where x is suitable in $\underline{m}_o$, since A_T is semilocal; now we have:

$$\underline{m}_o(A_T)[\underline{m}_o A_T/x] = x\,(A_T)[\underline{m}_o A_T/x] = x(A[\underline{m}_o/x])_T.$$

If A' = blowing up of A, we deduce that

$$\underline{m}_o(A')_T = \underline{m}_o(A_T[\underline{m}_o A_T/x]) = x(A_T[\underline{m}_o A_T/x]) = x(A')_T$$

(recall that blowing up and localization commute).

The proceding equalities say that there is $f \in T$ such that:

$$\underline{m}_o(A')_f = x(A')_f = x(A_f[\underline{m}_o A_f/x]),$$

the latter equality depending on the commutativity of blowing up and localization and on lemma 4.9, (i). Hence our claim is proved.

Proposition 4.11 : Let H be an effective divisor on X and U = Spec(B) an open affine of X containing all the singular points of D and such that $H/_U$ has just one equation, say G. Let V = Spec(A) = Spec($B/\underline{I}$) be the induced open affine of D and $A = A_o \to A_1 \to \ldots \to A_{n-1} \to A_n = \bar{A}$ the resolution of A induced by the resolution of D (theorem 2.2), $\underline{m}_o, \ldots, \underline{m}_{n-1}$ being the centers of the blowing up's. If $h \in A$ is a local equation of H on V and we have:

$$h \in \underline{m}_0^{s_0-1} \cdots \underline{m}_i^{s_i-1} A_i - \underline{m}_0^{s_0-1} \cdots \underline{m}_i^{s_i} A_i$$

for all $i = 0,\ldots, n-1$, s_i being the multiplicity of $\underline{m}_i$, then H is AS.

Proof : Let P_i be the point of X_i corresponding to the maximal ideal $\underline{m}_i$, so that s_i = multiplicity of D at P_i. First of all we consider the actual point P_0: our hypothesis for i = 0 says that H passes through P_0 with multiplicity exactly s_0-1. Now we consider the domain B_T obtained semilocalizing B at the (finitely many) singular points of D (P_0 together with some other P_i) and put: $A_{T'} = B_T/\underline{I}B_T$ (T' = image of T). By lemma 4.10 there is a $f \in T'$ such that the blowing up of A_f with center $\underline{m}_0 A_f$ is the domain $(A_f)[\underline{m}_0 A_f/x]$, with x is suitable in $\underline{m}_0$. By lemma 4.9 we have: $(A_f)[\underline{m}_0 A_f/x] = ((B_F)[\underline{M}_0/y])/\underline{I}_1$, where F lifts f and y lifts x, $\underline{I}_1$ being the strict transform of $\underline{I}$, $\underline{M}_0$ = the inverse image of $\underline{m}_0$.

Let now $H^{(1)}$ be the strict transform of H; its local equation on $U_1 = \mathrm{Spec}((B_F)[\underline{M}_0/y])$ is $G_1 = G/y^{s_0-1}$ (see n. 1); hence $h_1 = h/x^{s_0-1}$ is a local equation for $H^{(1)}$ on $\mathrm{Spec}((A_f)[\underline{m}_0/x])$. Our hypothesis for i = 1 says that $H^{(1)}$ has multiplicity s_1-1 exactly at P_1. The same argument being valid for the i-th strict transform, we see that H is AS.

Lemma 4.12 : Let k be an infinite field and let $V_1,\ldots,V_n$ be k-vector spaces. Assume that, for each $i = 1,\ldots,n$, we have v_i and $w_i \in V_i$, both different from 0. Then there exists $a \in k$ such that $av_i \neq w_i$, for every i.

Proof : Assume that v_i and w_i are linearly dependent for $i = 1,\ldots,r$ where r is an integer $\leq n$, linearly independent otherwise. Then we have: $w_i = a_i v_i$ for $i \leq r$. Choose $a \in k^*$ different from all the a_i's; then automatically $av_i \neq w_i$ for all $i = 1,\ldots,n$.

Now we are ready to relate A1 with A2 and A3 by the following

Theorem 4.13 : Let H be an effective divisor on the smooth irreducible projective surface $X \subset P_k^N$ and let D be a curve on X, not component of H. Then we have:

(i) If H is A1 to D, then it is also A2;

(ii) If H is A2 to D, then there exist :

a) a hypersurface $Z \subset P^N_k$ not passing through the singular points of D and the points common to H and D;

b) t hypersurfaces $H_1, \dots, H_t$ of P^N_k, of the same order d, cutting on X divisors AS to D,

such that the divisor cut by H on D coincides, except possibly on Z, with the divisor cut on D by a suitable member of the linear system generated by $H_1, \dots, H_t$.

Proof : Let P be a point of D; once for all we put : H_P = local equation of H at P, Γ_P = stalk at P of the sheaf $\gamma_{X/D}$ (n. 3), $\underline{M}$ = coherent Ideal of P on X, $\underline{M}_P$ = stalk of $\underline{M}$ at P, h_P = image of H_P into $O_{D,P}$, γ_P = stalk at P of γ_D, $\underline{m}$ = coherent Ideal of P on D, $\underline{m}_P$ = stalk of $\underline{m}$ at P.

To prove (i) it is enough to show that $H_P \in \Gamma_P$ for every s-fold singular point P of D. By theorem 2.2 there is a sequence of n blowing up's which gives the desingularization for D; we prove the claim by induction on such a number n (and use from now on the notation of theorem 2.2).

Case n = 1 (here P is the unique singular point of D, center of f_1): by hypothesis H is A1, hence $H_P \in \underline{M}_P^{s-1}$ and $h_P \in \underline{m}_P^{s-1}$; but $\underline{m}_P^{s-1} = \gamma_P$, by proposition 3.11; so $H_P \in \Gamma_P$.

Case n > 1: we assume the claim true for n and show it for n+1. Of course we may assume that P is the center of the first blowing up of the chain of n+1 blowing up's.

Let H' be the strict trasform of H on X_1; so H' is, by the inductive hypothesis, A1 to the curve D_1; this means that, for every $Q \in D_1$, $H'_Q \in \Gamma_Q$ (stalk at Q of γ_{X_1,D_1}). Moreover $H_P \in \underline{M}_P^{s-1}$. Then it is easy to see that, for every Q in the first neighbourhood of P, we have: $h_P \in \underline{m}_P^{s-1} \gamma_Q$; hence by proposition 3.11 $h_P \in \gamma_P$, i.e. $H_P \in \Gamma_P$.

Let us now prove (ii). By lemma 4.8 we may choose an open affine U = Spec(B) on X, such that X-U is the section of X with a hypersurface Z of P^N_k, H is generated on Y by one equation and both the singular points of D and the common points to D and H belong to U.

We put : $D/_U$ = Spec(A), $\bar{H}$ = local equation of H in B, h' = image of $\bar{H}$ in A.

Since the closed immersion $X \subset P^N_k$ is fixed, X can be identified with Proj(S) = = Proj($\oplus S_n$), where $S = \oplus S_n$ is a suitable graded algebra, obtained reducing $k[x_0, \dots x_N]$ modulo a suitable homogeneous ideal. Therefore B is the homogeneous localization of S at some homogeneous element $g \in S_r$ (really g = image of the equation of Z): B =

$= S_{(g)} = (a/g^e \text{ with } a \in S_{re})$.

Let us now consider the desingularization of D given by theorem 2.2; it induces a chain of blowing up's for the ring A:

$$A = A_0 \to A_1 \to \dots \to A_{n-1} \to A_n = \bar{A},$$

where $\bar{A}$ is the normalization of A in its fraction field.

We introduce then $C = A_T$, where T is the multiplicative system complement of the union of those maximal ideals which correspond either to singular points of D or to points common to D and H. Then the preceding chain of blowing up's induces on C the following one:

$$C = C_0 \to C_1 \to \dots \to C_{n-1} \to C_n = \bar{C},$$

where $\bar{C}$ is the normalization of C. Let $h = h'/1$ = image of h' into C.

Let $\underline{m}_i$ be the center of the blowing up $C_i \to C_{i+1}$; then put: $\underline{m}_i C_{i+1} = x_i C_{i+1}$, s_i = multiplicity of $\underline{m}_i$ on C_i. We want to show, by induction on n, that we have:

1) $h = \Sigma h_j$ with

2) $h_j \in \underline{m}_0^{s_0 - 1}$, $h_j \notin \underline{m}_0^{s_0}$, for every j, and

3) $h_j / x_0^{s_0-1} x_1^{s_1-1} \dots x_r^{s_r-1} \in \underline{m}_{r+1}^{s_{r+1}-1}$, $\notin \underline{m}_{r+1}^{s_{r+1}}$, for every j and every $r \leq n-2$.

By hypothesis and by proposition 3.11, we have:

$$h \in \underline{m}_0^{s-1} \gamma_{C_1} = x_0^{s-1} \gamma_{C_1}, \text{ i.e. } h = x_0^{s-1} \bar{h}, \text{ with } \bar{h} \in \gamma_{C_1}.$$

Therefore, by induction, we have: $\bar{h} = \Sigma \bar{h}_j$, where $\bar{h}_j$ fills up conditions 1), 2), 3).

If $\bar{h}_j \notin x_0 C_1$, for every j, then we choose $h_j = x_0^{s-1} \bar{h}_j$, getting the decomposition: $h = \Sigma h_j$, which is the required one.

Let us now assume that some $\bar{h}_j$, for instance $\bar{h}_1$, belongs to $x_0 C_1$. We consider another element $\bar{f} \in \gamma_{C_1}$ which does not belong to $x_0 C_1$ (it exists by lemma 3.3). By our inductive hypothesis we may write that $\bar{f} = \Sigma \bar{f}_j$, where the elements $\bar{f}_j$'s fulfil conditions 2), 3). Moreover at least one among the $\bar{f}_j$'s, say $\bar{f}_1$, does not belong to $x_0 C_1$.

Let us now consider again $\bar{h}_1$ and put, for every i:

$\bar{h}_{1i} = \bar{h}_1/x_1^{s_1-1} \ldots x_i^{s_i-1} \in C_i$ (condition 3)),

z_{1i} = initial form of $\bar{h}_{1i}$ into the associated graded algebra $gr_{\underline{m}_i}(C_i)$ = $= \bigoplus_{0}^{\infty} (m_i^n/\underline{m}_i^{n+1})$.

Similarly we consider the elements $\bar{f}_{1i}$ and their initial forms w_{1i}. Finally we choose, using lemma 4.12, an element $a \in k^*$ such that $aw_{1i} \neq z_{1i}$, for every i. Then we introduce the new element $\bar{g}_1 = a\bar{f}_1$.

By construction the element $\bar{h}_1-\bar{g}_1$ fulfils conditions 2),3) and, moreover, it does not belong to x_0C_1. Therefore the element $h_1 = x_0^{s-1}(\bar{h}_1-\bar{g}_1)$ is the required one. For every $\bar{h}_j \in x_0C_1$ we repeat the argument, obtaining the required decomposition for h.

The step n = 1 can be dealt with using the same arguments.

Therefore we have the equality $h = \Sigma h_j$ in the semilocal ring $C = A_T$. It is now easy to see that the same equality holds in A_f, where f is a suitable element in T; precisely we have: $h'/1 = \Sigma h_j'/1$, where h_j' is an inverse image in A of h_j and the equality concerns elements of A_f. We remark now that A_f is homomorphic image of an affine ring of X, say $B_{f'}$; at the cost of including in the hypersurface Z also a hypersurface cutting f' on Spec(B), we may assume that $h' = \Sigma h_j'$ already in A.

Finally we consider elements $y_j \in B$ such that:

(a) $y_j \rightarrow h_j'$, for every j; (b) $\bar{H} = \Sigma y_j$.

Since $B = S_{(g)}$, with $g \in S_r$, we have: $y_j = Y_j = Y_j/g^d$, where Y_j is an element of $k[x_0,\ldots,x_N]$, homogeneous of degree rd, for every j. Since Spec(B) contains all the singular points of D, Y_j cuts on X a divisor AS by prop. 4.11; since Spec(B) contains all the points common to H and D, H and ΣY_j cut on it the same divisor (they have even the same equation on Spec(B)); hence the theorem is proved.

<u>Example</u> 4.14 : Given the plane projective curve with equation $y^2z^3 = x^5$, we already know that $x^3 = 0$ is A2 but not A1 (see example 4.5). It is not hard to compute that we have the equality: $x^3 = (x^3-(x-z)yz)+(x-z)yz$, which is a sum of divisors AS.

<u>Corollary</u> 4.15 : <u>Let</u> $P \in D$ <u>be a singular point and let</u> $A = O_{D,P}$. <u>Then</u> γ_A <u>is generated by the images of the equations of the divisors</u> A1 <u>to</u> D.

<u>Corollary</u> 4.16 : <u>Assume that all the singular points of</u> D <u>are actual (i.e. or-</u>

dinary singularities). Then the effective divisor H is A1 iff it is A2 iff it is A3.

Proof : It follows from 4.13 (or 3.12) and 4.6.

Remark : Corollary 4.16 now explains rigorously example 4.4.

Corollary 4.17 : There exist divisors H on X which are A1 to D and also AS to D.

Proof : By theorem 4.13 it is enough to show that there is some H which is A2 (and does not contain D).

Let U = Spec(A) be a basic open affine of X, complement of a hyperplane section, containing all the singular points of D. Put : $\underline{I} = \Gamma(U\cap D, \gamma_D)$ = sections on $U\cap D$ of the conductor sheaf. Then $\underline{I}$ is a proper ideal $\neq (0)$ of $\Gamma(U\cap D, O_D)$. If $f \in \underline{I}$, $f \neq 0$ and F lifts it to A, then F = 0 is a curve of U not containing $D\cap U$. Let X = Proj(S) and $A = S_{(z)}$ = homogeneous localization of S at $z \in S_1$; then $F = G/z^m$, where G is a suitable element in S_m. It is now easy to see that the divisor associated to G on X is A2. In fact such a divisor on the basic affine $\mathrm{Spec}(S_{(z)})$, $z \in S_1$, has equation exactly = F.

We proved the preceding theorems for algebraic curves, but the statements are clearly meaningful if we replace everywhere the word "curve" by the word "analytic branch" (in the sense of [Z] , chap. I, II). More or less with the same techniques as above, but with remarkable simplifications, we can prove the following proposition (same notations and terminology of [Z]):

Proposition 4.18 : Let D be an analytic branch and F a positive analytic cycle, defined by an equation f(X, Y) = 0. Then we have:

(i) F is A2 to D iff it is A3;

(ii) If F is A1, then it is A2;

(iii) If F is A2, then there are cycles AS $F_1,\ldots,F_n$ such that F belongs to the linear system generated by the F_i's.

Remarks on the proof : (i) can be shown as in th. 4.6, but we can see it is really already proved in [Z], chap. II, théorème 6.2; (ii) has the same proof of theorem 4.13, (ii); (iii) has the same proof of theorem 4.13 with a simpler machinery, due to the fact that the ring of the branch is already local (so, in particular, we do not need to exclude any hypersurface).

n. 5 Linear systems of adjoint divisors.

In the present section we discuss the behaviour of the generic member of a linear system generated by adjoint divisors in the various meanings of the term.

Proposition 5.1 : *Let* L *be a linear system on* X, D *being not a fixed component of* L. *If the generic element of* L *is* A2 *or* A3, *then all the elements of* L *are so.*

Proof : L has a basis formed with divisors A2; the conductor being an ideal at any point, the claim is now obvious (modulo theorem 4.6).

Lemma 5.2 : *Let* H *be* A2 *to* D *and let* P *be a* s-*fold singular point of* D; *let then* X' *be the blowing up of* X *centered at* P *and* D', H' *the induced strict transforms of* D *and* H. *Then:*

(i) P *is at least* (s-1)-*fold for* H;

(ii) *If* P *is* (s-1)-*fold exactly for* H, *then* H' is A2 to D'.

Proof : (i) follows immediately from theorem 4.13.

To prove (ii), let us put: $R = O_{X,P}$, $\underline{M}$ = maximal ideal of R, $A = O_{D,P} = R/(F)$, $\underline{m} = \underline{M}A$, H_P = equation of H in R. So we have $H_P \in \underline{M}^{s-1}$, $H_P \in \underline{M}^{s}$. Let $P_1,\ldots,P_t$ be the points of D' lying over P and let U = Spec(E) be an open affine of X containing all the P_i's; put then: $\underline{M}S = ZS$, for a suitable $Z \in \underline{M}$. Therefore we can see that H_P/Z^{s-1} is the image into S of an equation of H' in a neighbourhood of the P_i's. Let $B = S/(F')$ be the first neighbourhood of A. Then $\underline{m}B = zB$, where z is the image of Z into B. Moreover, by corollary 3.12, we have: $(h/z^{s-1})z^{s-1} = z^{s-1}h'$, where h = image of H_P into A and $h' \in \gamma_B$; therefore $h/z^{s-1} \in \gamma_B$, which proves the claim.

Lemma 5.3 : *Let* H_1, H_2 *be two linearly equivalent effective divisors on* X *having the same multiplicity* s *at the closed point* $P \in X$.

Let $f: X' \to X$ *be the blowing up with center* P *and let* H'_i *be the strict transform of* H_i. *Then* H'_i *and* H'_2 *are linearly equivalent.*

Proof : Let $(U_1,\dots,U_t)$ be an open covering of X such that H_i is given on U_j by the single equation h_{ij}. Since H_1 and H_2 are linearly equivalent, we have: $H_1 = H_2 + \mathrm{div}(f)$, where f is a rational function, and we may assume that

$h_{1j} = h_{2j}f$, for all $j = 1,\dots,t$.

We may also assume that $P \in U_1$, $P \notin U_j$ for $j \neq 1$ (since P is closed) and that $U_1 = \mathrm{Spec}(B)$ is affine and such that the maximal ideal of B corresponding to P is generated by two elements u,v.

Put: $U'_1 = \mathrm{Spec}(B[u/v])$, $U''_1 = \mathrm{Spec}(B[v/u])$. Then $f^{-1}(U_1) = U'_1 \cup U''_1$ so that $(U'_1, U''_1, U_2,\dots,U_t)$ is an open covering of X'. Then H'_i has the following equations:

h_{i1}/v^s on U'_1, h_{i1}/u^s on U''_1, h_{ij} on U_j if $j \neq 1$.

Now it is immediate to show that $H'_1 = H'_2 + \mathrm{div}(f)$, which is our claim.

Proposition 5.4 : *Let* L *be a linear system on* X, *whose members are all* A2 *to* D. *If* L *contains at least an* AS, *then the generic member of* L *is* AS.

Proof : Let $H_0,\dots,H_t$ be a basis of L, with H_0 AS. If $\underline{a} = (a_0,\dots,a_t) \in k^{t+1}$, we denote $L_{\underline{a}}$ the linear combination $a_0H_0 + \dots + a_tH_t$.

First we consider the behaviour of an actual point $P \in D$, say singular s-fold. Then there is a closed subset $C \subsetneq k^{t+1}$ such that, if $\underline{a} \notin C$, $L_{\underline{a}}$ has P exactly as a (s-1)-fold point. In fact, let $f_0,\dots, f_t$ be the initial forms in $gr(O_{X,P})$ of the equation of $H_0,\dots,H_t$ at P. By lemma 5.2,(i) we can assume that there is an integer $r > 0$ such that:

$\deg f_i = s-1 \qquad$ for $0 \leq i \leq r$

$\deg f_i > s-1 \qquad$ for $r < i \leq t$

Put : $C = ((a_0,\dots,a_t)/ \sum_{i=0} a_if_i = 0)$. Of course C is a closed subset of k^{t+1} with the required property.

Since the generic element of L has a (s-1)-fold point at P, we may assume that H_i has a (s-1)-fold point at P for $i = 0,\dots,t$.

Let $X' \to X$ be the blowing up with center P, and let H'_i be the strict transform of H_i. Then H'_i is A2 to D' (strict transform of D) by lemma 5.2 (ii). Moreover the H_i's are linearly equivalent by lemma 5.3. Let L' be the linear system generated by the H'_i. Since H_O is AS to D, then also H'_O is AS to D'; hence, by induction on the number of blowing up's necessary to resolve D (see theorem 2.2) we may assume that the generic element of L' is AS.

Let now $C' = ((a_O,\ldots,a_t) \varepsilon k^{t+1} / L'_{\underline{a}}$ is AS) and let C be as above. So we have that, if $\underline{a} \notin C \cup C'$, $L_{\underline{a}}$ has multiplicity s-1 at P and that $L'_{\underline{a}}$ is AS to D'. Thus $L_{\underline{a}}$ is AS to D.

Corollary 5.5 (see [S], n. 112) : Let D be a plane projective curve and let L_n be the linear system of all divisors A2 of degree n. If n >> 0, then the generic member of L_n is AS.

Proof : By corollary 4.17 there is a divisor H which is AS. If n_O = deg H and R is a line not passing through any singular point of D, then H+dR is AS with degree n_O+d. Hence for n >> 0 L_n contains an AS; so we are allowed to apply the proposition.

Example 5.6 : proposition 5.3 is false if the linear system does not contain a divisor AS. Consider in fact the curve with equation $y^2z^2 = x^4+y^4$, which has a tacnode at the origin O. The linear system L: $a(x^2z^3-y^5)+b(x^3z^2-y^5) = 0$ is generated by divisors A2 (recall that the conductor at the origin is $\gamma_A = (\tilde{x}^2,\tilde{y}))$; but L contains no A1 (easy direct computation).

Similarly the linear system L': $a(x^2z^3-y^5)+b(y^3z^2-x^5) = 0$ in generated by divisors A2, but the unique A1 of L' is the curve corresponding to a = 0 (and this is not AS).

Example 5.7 : Proposition 5.1 is false for A1. With the same curve of example 5.6, let L be the linear system: $ayz^2+b(yz^2+x^2z-y^3) = 0$.
Then L is generated by divisors AS, but contains $x^2z-y^3 = 0$, which is not A1.

REFERENCES

[A] S. Abhyankar Algebraic Space Curves Univ. de Montréal 1971

[B-N] A. Brill, M. Noether Ueber die algebraischen Functionen und ihre Anwendung in der Geometrie Math. Ann. 7 (1874) 269-310.

[C] C. Cumino Rami analitici e coni tangenti (in preparation)

[D-M] F. De Salvo, M. Manaresi On birational coverings Geometriae Dedicata (to appear)

[G] D. Gorenstein An arithmetic theory of adjoint plane curves Trans.Amer.Math.Soc. 72 (1952) 414-436

[Gc] S. Greco On the theory of branches Proc.Symp.Alg.Geom. Kyoto 1977 311-327.

[Gr] A. Grothendieck E.G.A. Publ. I.H.E.S. n. 4 ... Paris 1960 ...

[H] R. Hartshorne Algebraic Geometry Springer Berlin 1977

[K] O. Keller Vorlesungen über algebraische Geometrie Leipzig 1974

[M] E. Matlis 1-Dimensional Cohen-Macaulay Rings Springer Lecture Notes n. 327 1970

[O] A. Oneto Conduttore e vere aggiunte ad una curva su una superficie Rend. Sem. Mat. Univ. Pol. Torino 1979

[R-B] L. Robbiano, M. Beltrametti Conduttore e curve aggiunte Atti Acc. Ligure 30 (1973) 1-12

[S] F. Severi Trattato di Geometria Algebrica Zanichelli Bologna 1926.

[S-R] J. Semple, L. Roth Introduction to Algebraic Geometry Oxford University Press 1949

[V] P. Valabrega Scoppiamenti, intersezioni complete strette, aggiunte Atti Convegno Geom. Alg. Catania 1978.

[V-V] P. Valabrega, G. Valla Standard Bases and Generators for the strict Transforms (to appear in Boll. U.M.I.)

[Z] O. Zariski Le probléme des modules pour les branches planes Cour de l'Ecole Polytechnique Paris 1973

[Z-S] O. Zariski, P. Samuel Commutative Algebra Van Nostrand New York 1958

Silvio Greco - Paolo Valabrega Istituto Matematico del Politecnico - Torino - Italy

The paper was supported by the C.N.R., while both authors were members of GNSAGA.

INFINITE DIMENSIONAL UNIVERSAL FORMAL GROUP LAWS AND FORMAL A-MODULES.

Michiel Hazewinkel
Dept. Math., Econometric Inst.,
Erasmus Univ. of Rotterdam
50, Burg. Oudlaan,
ROTTERDAM, The Netherlands

1. INTRODUCTION AND MOTIVATION.

Let B be a commutative ring with $1 \in B$. An n-dimensional commutative formal group law over B is an n-tuple of power series $F(X,Y)$ in 2n variables $X_1, \ldots, X_n; Y_1, \ldots, Y_n$ with coefficients in B such that $F(X,0) \equiv X$, $F(0,Y) \equiv Y$ mod degree 2, $F(F(X,Y),Z) = F(X,F(Y,Z))$ (associativity) and $F(X,Y) = F(Y,X)$ (commutativity). From now on all formal group laws will be commutative.

Let A be a discrete valuation ring with finite residue field k. Let $B \in \underline{\underline{Alg}}_A$, the category of commutative A-algebras with 1. A n-dimensional formal A-module over B is a formal group law $F(X,Y)$ over B together with a ring homomorphism $\rho_F: A \to End_B(F(X,Y))$ such that $\rho_F(a) \equiv aX$ mod degree 2 for all $a \in A$. One would like to have a classification theory for formal A-modules which is parallel to the classification theory of formal group laws over $\mathbb{Z}_{(p)}$-algebras. Such a theory is sketched below and details can be found in [2], section 29. As in the case of formal group laws over $\mathbb{Z}_{(p)}$-algebras the theory inevitably involves infinite dimensional objects. Now two important operators for the formal A-module classification theory, viz. ε_q and $\underline{\underline{f}}_\pi$, the analogues of p-typification and Frobenius, are defined by lifting back to the universal case, and, for the moment at least, I know of no other way of defining them, especially if char(A) = p > 0. In case char(A) = 0, cf. also [1]. But by the very nature of the usual definition of infinite dimensional formal group law and formal A-module there cannot exist universal infinite dimensional formal group laws and formal A-modules, so that the definitions of ε_q and $\underline{\underline{f}}_\pi$ break down. In [2], this problem is surmounted by an ad hoc construction which works in the particular case needed (Witt vector like formal A-modules). But this method decidedly lacks elegance. It is the second and main purpose of the present paper to remedy this by showing that after all, in a suitable sense, universal infinite dimensional formal group laws and formal A-modules do exist (and have all the nice properties one could wish for). As a byproduct one obtains then of course such results as liftability to characteristic zero and existence and uniqueness of logarithms also for infinite dimensional formal group laws and formal A-modules.

2. SKETCH OF THE (COVARIANT) CLASSIFICATION THEORY FOR FORMAL GROUP LAWS OVER $\mathbb{Z}_{(p)}$-ALGEBRAS.

Let p be a fixed prime number. Let F(X,Y) be an m-dimensional formal group law over a $\mathbb{Z}_{(p)}$-algebra B. A curve in F over R is simply an m-tuple of power series $\gamma(t)$ in one variable t with coefficients in B such that $\gamma(0) = 0$.

Two curves can be added by means of the formula

$$(2.1)\qquad \gamma(t) +_F \delta(t) = F(\gamma(t),\delta(t))$$

giving us a complete topological abelian group $\mathfrak{C}(F;B)$; the topology is defined by the subgroups of curves $\gamma(t)$ such that $\gamma(t) \equiv 0 \bmod(\text{degree } n)$, $n = 1,2,3, \ldots$ In addition one has operators $\underline{\underline{V}}_n$, $\underline{\underline{f}}_n$, $\langle b\rangle$, for $n \in \mathbb{N}$, $b \in B$.

These are defined as follows

$$(2.2)\qquad \underline{\underline{V}}_n\gamma(t) = \gamma(t^n),\ \langle b\rangle\gamma(t) = \gamma(bt),\ \underline{\underline{f}}_n\gamma(t) = \sum_{i=1}^{n}{}^{F}\, \gamma(\zeta_n^i t^{1/n})$$

where ζ_n is a primitive n-th root of unity. This last definition must be rewritten slightly in case n-th roots of unity make no particular sense over B, cf. [2], section 16 for details.

A curve $\gamma(t)$ is called p-typical if $\underline{\underline{f}}_q\gamma(t) = 0$ for all prime numbers $q \neq p$. The subgroup of p-typical curves, $\mathfrak{C}_p(F;B)$, is complete in the induced topology and stable under $\underline{\underline{f}}_p$ and $\underline{\underline{V}}_p$ and the operators $\langle b\rangle$. Moreover using that B is a $\mathbb{Z}_{(p)}$-algebra there is a projector $\varepsilon_p : \mathfrak{C}(F;B) \to \mathfrak{C}_p(F;B)$ given by the formula

$$(2.3)\qquad \varepsilon_p = \sum_{(n,p)=1} n^{-1}\mu(n)\underline{\underline{V}}_n\underline{\underline{f}}_n$$

where $\mu(n)$ is the Möbius function. We can assemble the operators $\underline{\underline{f}}_p$, $\underline{\underline{V}}_p$, $\langle b\rangle$ into a ring of operators $\mathrm{Cart}_p(B)$ consisting of all sums

$$\sum_{i,j=0}^{\infty} \underline{\underline{V}}_p^i\langle b_{i,j}\rangle\underline{\underline{f}}_p^j$$

with for all i only finitely many $b_{i,j} \neq 0$. For the calculation rules in $\mathrm{Cart}_p(B)$, cf. [2],16.2, 28.1. The subring $\{\sum_{i=0}^{\infty} \underline{\underline{V}}_p^i \langle b_i \rangle \underline{\underline{f}}_p^i\}$ is naturally isomorphic to $W_{p,\infty}(B)$, the ring of Witt vectors over B of infinite length associated to the prime p. Using this we see that $\mathfrak{C}_p(F;B)$ is a module over $W_{p,\infty}(B)[\underline{\underline{f}},\underline{\underline{V}}]$ with calculation rules $\underline{\underline{f}}\underline{\underline{V}} = p$, $\underline{\underline{V}}\underline{\underline{f}} = (0,1,0,\ldots)$, $\underline{\underline{f}}x = x^{\sigma}\underline{\underline{f}}$, $x\underline{\underline{V}} = \underline{\underline{V}}x^{\sigma}$ for all $x \in W_{p,\infty}(R)$, where σ is the Frobenius endomorphism of $W_{p,\infty}(B)$. The functor $F(X,Y) \mapsto \mathfrak{C}_p(F;B)$ turns out to be faithful and its image can be described without much trouble.

3. A CARTIER-DIEUDONNE MODULE CLASSIFICATION THEORY FOR FORMAL A-MODULES (I).

Now let A be a discrete valuation ring with uniformizing element π and finite residue field k of q elements, $q = p^r$. Let K be the quotient field of A. We are going to describe a classification theory for formal A-modules which is completely analogous to the theory sketched in 2 above. In this theory $\underline{\underline{f}}$ gets replaced by $\underline{\underline{f}}_\pi$, $\underline{\underline{V}}$ by $\underline{\underline{V}}_q$, $W_{p,\infty}(B)$ by the appropriate ring of ramified Witt vectors $W^A_{q,\infty}(B)$, $B \in \underline{\underline{\mathrm{Alg}}}_A$, and $\mathfrak{C}_p(F;B)$ by $\mathfrak{C}_q(F;B)$. Of course we should have $\underline{\underline{f}}_\pi\underline{\underline{V}}_q = \pi$, $\underline{\underline{V}}_q\underline{\underline{f}}_\pi = (0,1,0,0,.)$, $x\underline{\underline{V}}_q = \underline{\underline{V}}_q x^{\sigma}$, $\underline{\underline{f}}_\pi x = x^{\sigma}\underline{\underline{f}}_\pi$. In case A is of characteristic zero, $p = u\pi^e$, this shows that $\underline{\underline{f}}_\pi$ and $\underline{\underline{f}}_p$ should be related as

$$[u^{-1}]\underline{\underline{f}}_p = \underline{\underline{f}}_\pi^e \underline{\underline{V}}_p^{er-1} \tag{3.1}$$

Here we shall not discuss the ramified Witt vector functor $W^A_{q,\infty}: \underline{\underline{\mathrm{Alg}}}_A \to \underline{\underline{\mathrm{Alg}}}_A$, cf. [2[, [3], or [4]. It can be most easily obtained by taking q-typical curves in the Lubin-Tate formal group law over A, just as $W_{p,\infty}(-)$ can very nicely be described via the p-typical curves in $\underline{\underline{G}}_m$, the multiplicative formal group law. Alternatively $W^A_{q,\infty}(-)$ can be described via the polynomials

$$X_0^{q^n} + \pi X_1^{q^{n-1}} + \ldots + \pi^n X_n, \quad n = 0,1,2, \ldots \tag{3.2}$$

exactly as $W_{p,\infty}(-)$ is constructed via the Witt polynomials $X_0^{p^n} + pX_1^{p^{n-1}} + \ldots + p^n X_n$.

We shall concentrate on the definition of $\underline{\underline{f}}_\pi$ and the "q-typification" projector $\varepsilon_q: \mathfrak{C}(F;B) \to \mathfrak{C}_q(F;B)$, partly also to illustrate the adagium "do everything first in the universal case", which appears to be particularly effective, in fact even necessary, when dealing with formal A-modules.

Now there seems to be no obvious analogues of the definitions for $\underline{\underline{f}}_p$ and ε_p given in (2.2) and (2.3). Things become better if we restate these definitions in terms of logarithms. Assume therefore that B is torsion free and let $f(X) \in B \otimes \mathbb{Q}[[X]]^m$ be the logarithm of $F(X,Y)$, i.e. $f(X)$ is the unique m-tuple of power series over $B \otimes \mathbb{Q}$ such that $f(X) \equiv X$ mod(degree 2), $F(X,Y) = f^{-1}(f(X) + f(Y))$. Setting

$$(3.3) \qquad f(\gamma(t)) = \sum_{i=1}^{\infty} x_i t^i, \; x_i \in B \otimes \mathbb{Q}^m$$

we then have

$$(3.4) \qquad f(\underline{\underline{f}}_n \gamma(t)) = \sum_{i=1}^{\infty} n x_{ni} t^i$$

$$(3.5) \qquad f(\varepsilon_p \gamma(t)) = \sum_{j=o}^{\infty} x_{p^j} t^{p^j}$$

Now let $(F(X,Y),\rho_F)$ be an m-dimensional formal A-module over $B \in \underline{\underline{\mathrm{Alg}}}_A$. Assume that B is A-torsion free. An A-logarithm for $(F(X,Y), \rho_F)$ is a power series $f(X) \in B \otimes_A K[[X]]^m$ such that $f(X) \equiv X$ mod degree 2 and such that $F(X,Y) = f^{-1}(f(X) + f(Y))$ and $\rho_F(a) = f^{-1}(af(X))$ for all $a \in A$. It is an immediate consequence of the construction of a universal formal A-module below in section 5 that A-logarithms exist. Uniqueness is then easy. Given A-logarithms there are obvious analogues of (3.4) and (3.5) viz.

$$(3.6) \qquad \underline{\underline{f}}_\pi \gamma(t) = f^{-1}\Big(\sum_{i=1}^{\infty} \pi x_{qi} t^i\Big)$$

$$(3.7) \qquad \varepsilon_q \gamma(t) = f^{-1}\Big(\sum_{i=o}^{\infty} x_{q^i} t^{q^i}\Big)$$

It remains of course to prove that the m-tuples of power series thus defined are integral (i.e. that they have their coefficients in B not just in $B \otimes_A K$). This again will be done by proving this to be the case in the universal example, which, fortunately, is defined over the kind of algebra to which the functional equation (integrality) lemma applies. This lemma is our main tool for proving integrality statements. It is remarkably "universally" applicable, cf. also [3] for some other illustrations.

4. THE FUNCTIONAL EQUATION LEMMA.

The ingredients we need are the following

(4.1) $B \subset L$, $\mathfrak{a} \subset B$, $\sigma : L \to L$, p, q, s_1, s_2, ...

Here B is a subring of a ring L, $\mathfrak{a}$ is an ideal in B, σ a ring endomorphism of L, p is a prime number, q is a power of p and the s_i, $i = 1,2,3, \ldots$ are $m \times m$ matrices with coefficients in L. These ingredients are supposed to satisfy the following conditions

(4.2) $p \in \mathfrak{a}$, $\sigma(b) \equiv b^q \bmod \mathfrak{a}$ for all $b \in B$, $\sigma^r(s_i(j,k))\mathfrak{a} \subset B$ for all i,j,k,r

Here $s_i(j,k)$ is the (j,k) entry of the matrix s_i, $j,k \in \{1,\ldots,m\}$.
If $g(X)$ is an m-tuple of power series in $X_1, \ldots, X_n$ with coefficients in L then we denote with $\sigma_* g(X)$ the m-tuple of power series obtained by applying σ to the coefficients of $g(X)$.

4.3. Functional Equation Lemma. Let $f(X) \in L[[X]]^m$ be an m-tuple of power series in m determinates $X_1, \ldots, X_m$ and $\bar{f}(\bar{X}) \in L[[\bar{X}]]^m$ an m-tuple of power series in n indeterminates $\bar{X}_1, \bar{X}_2, \ldots, \bar{X}_n$. Suppose that $f(X) \equiv b_1 X$ mod(degree 2) where b_1 is a matrix with coefficients in B which is invertible (over B). Suppose moreover that

$$f(X) - \sum_{i=1}^{\infty} s_i \sigma_*^i f(X^{q^i}) \in B[[X]]^m, \quad \bar{f}(X) - \sum_{i=1}^{\infty} s_i \sigma_*^i \bar{f}(\bar{X}^{q^i}) \in B[[\bar{X}]]^m \tag{4.4}$$

where X^{q^i} and $\bar{X}^{q^i}$ are short for $(X_1^{q^i}, \ldots, X_m^{q^i})$ and $(\bar{X}_1^{q^i}, \ldots, \bar{X}_n^{q^i})$. Then we have

$$F(X,Y) = f^{-1}(f(X) + f(Y)) \in B[[X;Y]]^m \tag{4.5}$$

$$f^{-1}(\bar{f}(\bar{X})) \in B[[\bar{X}]]^m \tag{4.6}$$

Let $h(\hat{X}) \in B[[\hat{X}]]^m$, $\hat{f}(\hat{X}) = f(h(\hat{X}))$. Then

$$\hat{f}(\hat{X}) - \sum_{i=1}^{\infty} s_i \sigma_*^i \hat{f}(\hat{X}^{q^i}) \in B[[\hat{X}]]^m \tag{4.7}$$

Let $\alpha(\hat{X}) \in B[[\hat{X}]]^m$, $\beta(\hat{X}) \in L[[\hat{X}]]^m$ and $r \in \mathbb{N} = \{1,2,3, \dots\}$. Then

(4.8) $$\alpha(\hat{X}) \equiv \beta(\hat{X}) \bmod \mathfrak{a}^r \iff f(\alpha(\hat{X})) \equiv f(\beta(\hat{X})) \bmod \mathfrak{a}^r$$

For a proof cf. [2], section 10.

5. A UNIVERSAL m-DIMENSIONAL FORMAL A-MODULE.

For each multiindex $\alpha = (n_1,\dots,n_m)$ of length m, $n_i \in \mathbb{N} \cup \{0\}$ let $|\alpha| = n_1 + \dots + n_m$ and $s\alpha = (sn_1,\dots,sn_m)$ for all $s \in \mathbb{N} \cup \{0\}$. For each α such that $|\alpha| \geq 2$ and $i \in \{1,\dots,m\}$ let $U(i,\alpha)$ be an indeterminate. We denote with $\varepsilon(i)$ the multiindex $(0,\dots,0,1,0,\dots,0)$ with 1 in the i-th spot. We set $U(i,\varepsilon(j)) = 0$ if $i \neq j$ and $U(i,\varepsilon(i)) = 1$. For each $\alpha \neq q^r\varepsilon(i)$ for all $r = 1,2, \dots, i \in \{1,\dots,m\}$ we let U_α denote the column vector $U(1,\alpha), \dots, U(m,\alpha)$ and for each $r \in \mathbb{N}$, U_{q^r} denotes the m x m matrix $U_{q^r} = (U(i,q^r\varepsilon(j)))$. Finally let $X^\alpha = X_1^{n_1} \cdot \dots \cdot X_m^{n_m}$. For each multiindex α such that $|\alpha| \geq 1$ we now define the m-vector $a_\alpha(U) \in K[U]^m$ by

(5.1) $$A_\alpha(U) = \sum_{(r_1,\dots,r_t,\beta)} \pi^{-t} U_{q^{r_1}} U_{q^{r_2}}^{(q^{r_1})} \cdots U_{q^{r_t}}^{(q^{r_1+\dots+r_{t-1}})} U_\beta^{(q^{r_1+\dots+r_t})}$$

where the sum is over all $(r_1,\dots,r_t,\underline{\underline{d}})$, $r_i \in \mathbb{N}$, $t \in \mathbb{N} \cup \{0\}$ such that $q^{r_1} q^{r_2} \cdots q^{r_t}\beta = \alpha$ and $\beta \neq q^r\varepsilon(i)$ for all $r \in \mathbb{N}$, $i \in \{1,\dots,m\}$. Here $U_{q^r}^{(q^i)}$ is the matrix obtained from U_{q^r} by raising each of its entries to the q^i-th power. We now define

(5.2) $$f_U^A(X) = \sum_{|\alpha| \geq 1} a_\alpha X^\alpha \in K[U][[X]]^m$$

Now let $L = K[U] \supset A[U] = B$, $\mathfrak{a} = \pi A[U]$, $s_i = \pi^{-1} U_{q^i}$ and $\sigma : L \to L$ the K-algebra endomorphism that sends each $U(i,\alpha)$ into its q-th power. Then the conditions (4.2) hold. Also we have

(5.3) $\quad f_U^A(X) \equiv X \bmod(\text{degree } 2),\ f_U^A(X) - \sum_{i=1}^{\infty} s_i \sigma_*^i f_U^A(X^{q^i}) \in A[U][[X]]^m.$

It follows that if we define

(5.4) $\quad F_U^A(X,Y) = (f_U^A)^{-1}(f_U^A(X)+f_U^A(Y)),\ \rho_U^A(a) = (f_U^A)^{-1}(af_U^A(X))$

then $(F_U^A(X,Y),\rho_U^A)$ is a formal A-module over A[U] (by parts (4.5) and (4.6) of the functional equation lemma 4.3).

5.5. Theorem.

$(F_U^A(X,Y),\rho_U^A)$ is a universal m-dimensional formal A-module.

I.e. if $(G(X,Y),\rho_G)$ is any m-dimensional formal A-module over an A-algebra B then there is a unique A-algebra homomorphism ϕ: A[U] $\to$ B such that $\phi_* F_U^A(X,Y) = G(X,Y)$ and $\phi_* \rho_U^A(a) = \rho_G(a)$ for all $a \in A$.
For a proof cf. [2], section 25.

6. A CARTIER-DIEUDONNE MODULE CLASSIFICATION THEORY FOR FORMAL A-MODULES (2).

For each $n \in \mathbb{N}$, $i \in \{1,\dots,m\}$ let C(n,i) be an indeterminate. Let C_n be the columnvector (C(n,1), ..., C(n,m)). Now consider the curve

(6.1) $$\gamma_C(t) = \sum_{n=1}^{\infty} C_n t^n$$

in the universal formal A-module $(F_U^A(X,Y),\rho_U^A)$ considered as a formal A-module over A[U;C]. This is again the sort of ring to which the functional equation lemma applies. It follows by part (4.7) of lemma 4.3 that the m-tuple of power series in one variable

(6.2) $$f_U^A(\gamma_C(t)) = \sum_{n=1}^{\infty} x_i t^i\ ,\ x_i \in K[U;C]^m$$

satisfies the functional equation condition (4.4). An easy check shows that then the m-tuples of power series

$$\sum_{j=o}^{\infty} x_{q^j} t^{q^j} \qquad \sum_{n=1}^{\infty} \pi x_{qn} t^n$$

also satisfy this condition. It now follows from part (4.6) of the functional equation lemma that

$$\varepsilon_q \gamma_C(t) = (f_U^A)^{-1}\left(\sum_{j=o}^{\infty} x_{q^j} t^{q^j}\right) \tag{6.3}$$

$$\underline{\underline{f}}_\pi \gamma_C(t) = (f_U^A)^{-1}\left(\sum_{n=1}^{\infty} \pi x_{qn} t^n\right) \tag{6.4}$$

have in fact their coefficients in A[U;C].
Now $(F_U^A(X,Y),\rho_U^A,\gamma_C)$ over A[U;C] is (given theorem 5.5) clearly universal for m-dimensional formal A-modules together with a curve.
Let $(F(X,Y),\rho_F)$ be a formal A-module over $B \in \underline{\underline{Alg}}_A$ and let $\gamma(t)$ be a curve in F(X,Y) over B. Let ϕ: A[U;C] $\to$ B be the unique A-algebra homomorphism taking (F_U^A,ρ_U^A) into (F,ρ_F) and $\gamma_C(t)$ into $\gamma(t)$. Then we define

$$\varepsilon_q \gamma(t) = \phi_* \varepsilon_q \gamma_C(t) \tag{6.5}$$

$$\underline{\underline{f}}_\pi \gamma(t) = \phi_* \underline{\underline{f}}_\pi \gamma_C(t) \tag{6.6}$$

It follows immediately that this agrees with the tentative definitions (3.6), (3.7) of section 3 above (if B is A-torsion free so that we have a unique A-logarithm available).
Let $\mathfrak{C}_q(F;B)$ be the image of $\varepsilon_q : \mathfrak{C}(F,B) \to \mathfrak{C}(F;B)$. One now easily proves that ε_q is the identity on $\mathfrak{C}_q(F;B)$ and that $\mathfrak{C}_q(F;B)$ is stable under $\underline{\underline{f}}_\pi$, $\underline{\underline{V}}_q$, $\langle b\rangle$ for all $b \in B$. (Recall that $\underline{\underline{V}}_q\gamma(t) = \gamma(t^q)$, $\langle b\rangle\gamma(t) = \gamma(bt)$). One checks that

$$\underline{\underline{f}}_\pi \underline{\underline{V}}_q = [\pi] \tag{6.7}$$

where $[\pi]$ is the operator induced by the endomorphism $\rho_F(\pi)$ of F(X,Y).
Further

$$\underline{\underline{f}}_\pi \langle b\rangle = \langle b^q\rangle \underline{\underline{f}}_\pi \ , \quad \langle b\rangle \underline{\underline{V}}_q = \underline{\underline{V}}_q \langle b^q\rangle \tag{6.8}$$

We can assemble all these operators into a ring $\mathrm{Cart}_A(B)$

$$\mathrm{Cart}_A(B) = \{ \sum_{i,j=o}^{\infty} \underline{\underline{V}}_q^i \langle b_{i,j}\rangle \underline{\underline{f}}_\pi^j \} \tag{6.9}$$

with for every i only finitely many $b_{i,j} \neq 0$. The subset $\{\sum_{i=0}^{\infty} \underline{\underline{V}}_q^i \langle b_i \rangle \underline{\underline{f}}_\pi^i\}$ turns out to be a subring naturally isomorphic to $W^A_{q,\infty}(B)$, the ring of ramified Witt vectors associated to A with coefficients in B. There results a classification theory of (finite dimensional) formal A-modules in terms of $W^A_{q,\infty}(B)[\underline{\underline{f}}_\pi, \underline{\underline{V}}_q]$ modules which, both in statements and proofs, is completely analogous to the theory for formal group laws over $\mathbb{Z}_{(p)}$-algebras. In particular there is an analogue of Cartier's first theorem. It states that the formal A-module $\hat{W}^A_{q,\infty}(X;Y)$ represents the functor $F \mapsto \mathfrak{C}_q(F;B)$ going from formal A-modules over B to their modules of q-typical curves. Here $\hat{W}^A_{q,\infty}$ is the (infinite dimensional) formal A-module with as A-logarithm the column vector

$$(X_o,\ X_1 + \pi^{-1}X_o^q,\ X_2 + \pi^{-1}X_1^q + \pi^{-2}X_o^{q^2},\ \ldots)$$

As in the case of formal group laws this theorem is important for the proofs of the classification results. This makes it necessary to be able to define ε_q and $\underline{\underline{f}}_\pi$ also for curves in $\hat{W}^A_{q,\infty}$, which can be done by an ad hoc method. It would be nicer to be able to do it also for all other infinite dimensional formal A-modules. It would also be more elegant to be able to extend the classification theory sketched above to all formal A-modules. To do this it is necessary to define ε_q and $\underline{\underline{f}}_\pi$ also in those cases. This, judging from what we did in the finite dimensional case, will involve something like universal infinite dimensional formal A-modules, a gadget which, in terms of the usual definitions, obviously cannot exist. This, the main topic of this paper, is what I take up next.

Before I do so let me remark that the analogy: "formal group laws over $\mathbb{Z}_{(p)}$-algebra" - "formal A-modules" also extends to give a "tapis de Cartier" and related type results for lifting formal A-modules; cf. [2], section 30.

7. "CLASSICAL" INFINITE DIMENSIONAL FORMAL GROUP LAWS AND FORMAL A-MODULES.

Let $(X_i)_{i\in I}$ be a set of indeterminates indexed by an arbitrary index set I. The formal power series ring $B[[X_i;i\in I]]$ is now defined as the ring of all formal (infinite) sums $\sum c_\alpha X^\alpha$ where α runs through all functions

α: $I \to \mathbb{N} \cup \{0\}$ with finite support, i.e. $\text{supp}(\alpha) = \{i \in I | \alpha(i) \neq 0\}$ is finite. We shall call such functions multiindices. Here X^{α} is short for $\prod_{i \in \text{supp}(\alpha)} X_i^{\alpha(i)}$.

One can now consider elements $F(i)(X,Y) \in B[[X_i,Y_i;i \in I]]$ and at first sight one could define an infinite dimensional commutative formal group law as a set of power series $F(i)(X,Y) \in B[[X;Y]]$ indexed by I such that $F(i)(X,Y) \equiv X_i + Y_i \bmod(\text{degree } 2)$, $F(i)(X,Y) = F(i)(Y,X)$ and such that

$$(7.1) \qquad F(i)(X,F(Y,Z)) = F(i)(F(X,Y),Z) \text{ for all } i \in I$$

However, in general this associativity condition (7.1) makes no sense because the calculation of the coefficient of a monomial $X^{\alpha}Y^{\beta}Z^{\gamma}$ in $F(i)(X,F(Y,Z))$ or $F(i)(F(X,Y),Z)$ involves infinite sums of elements of B. The "classical" solution is to require a finite support condition in the following sense.

7.2. Definition.

Let I and J be index sets. Let $f(X)$ be an I-tuple of power series in the indeterminates $X_j, j \in J$. We say that $f(X)$ satisfies the monomials have finite support condition if for all multiindices α: $J \to \mathbb{N} \cup \{0\}$ there are only finity many $i \in I$ such that the coefficient of X^{α} in $f(i)(X)$ is nonzero.

This property is stable under composition and taking inverses in the sense of the following lemma.

7.3. Lemma.

Let I, J, K be index sets. Let $f(X)$ be an I-tuple of power series in the X_j, $j \in J$ and $g(Y)$ a J-tuple of power series in the Y_k, $k \in K$. Suppose that $f(X)$ and $g(Y)$ both satisfy the monomials have finite support condition. Then $f(g(Y))$ is well defined and satisfies the same condition. Further if $f(X) \equiv X \bmod$ degree 2 then $f^{-1}(X)$ is well defined and also satisfies the monomials have finite support condition.

Proof. Write $f(i)(X) = \Sigma\, r_{i,\alpha}X^{\alpha}$, $g(j)(Y) = \Sigma\, s_{j,\beta}Y^{\beta}$. Formally one has

$$(7.4) \qquad f(i)(g(Y)) = \Sigma\, r_{i,\alpha}s_{j_1,\beta_1} \cdots s_{j_t,\beta_t} Y^{\beta_1+\dots+\beta_t}$$

where the sum is over all α and sequences $(j_1,\dots,j_t)$, $(\beta_1,\dots,\beta_t)$ such that

$\beta_i \neq 0$, the zero multiindex, and $j_1 + \ldots + j_t = \alpha$, where $j \in J$ is identified with the multiindex $\varepsilon(j)\colon J \to \mathbb{N} \cup \{0\}$, $j \mapsto 1$, $j' \mapsto 0$ if $j \neq j'$. Given $\gamma\colon K \to \mathbb{N} \cup \{0\}$ there are only finitely many sequences $(\beta_1, \ldots, \beta_t)$ such that $\beta_i \neq 0$ and $\beta_1 + \ldots + \beta_t = \gamma$. For each β_i there are only finitely many j such that $s_{j,\beta_i} \neq 0$; finally $\alpha = j_1 + \ldots + j_t$. It follows that in the sum (7.4) only finitely many coefficients of Y^γ are nonzero (for a given γ). Thus $f(g(Y))$ is welldefined. Also for every γ there are only finitely many α, such that there exist nonzero $s_{j_1,\beta_1}, \ldots, s_{j_t,\beta_t}$ such that $\alpha = j_1 + \ldots + j_t$, $\beta_1 + \ldots + \beta_t = \gamma$. For each α there are only finitely many i such that $r_{i,\alpha} \neq 0$. It follows that the coefficient of Y^γ in $f(i)(g(Y))$ is nonzero for only finitely many i. The second statement of the lemma is proved similarly by comparing coefficients in $f^{-1}(f(X)) = X$. Using these ideas we can now give the "classical" definition of infinite dimensional formal group laws and formal A-modules as follows.

7.5. Definitions.

An (infinite) dimensional formal group law $F(X,Y)$ over B with index set I is an I-tuple of power series $F(X,Y) = (F(i)(X,Y))_{i \in I}$, $F(i)(X,Y) \in B[[X_i, Y_i; i \in I]]$ such that $F(X,Y)$ satisfies the monomials have finite support condition and such that $F(X,0) = X$, $F(0,Y) = Y$, $F(F(X,Y),Z) = F(X,F(Y,Z))$. If moreover $F(X,Y) = F(Y,X)$ the formal group law is said to be commutative. All formal group laws will be commutative from now on. A homomorphism from $F(X,Y)$ with index set I to $G(X,Y)$ with index set J is an J-tuple of power series $\alpha(X)$ in X_i, $i \in I$ with coefficients in B, which satisfies the monomials have finite support condition such that $\alpha(F(X,Y)) = G(\alpha(X), \alpha(Y))$. Finally a formal A-module over $B \in \underline{\underline{\mathrm{Alg}}}_A$ with index set I is a formal group law $F(X,Y)$ over B together with a ring homomorphism $\rho_F\colon A \to \mathrm{End}_B(F(X,Y))$ such that $\rho_F(a) \equiv aX \bmod(\text{degree } 2)$ for all $a \in A$. (This implies of course that all the $\rho_F(a)$ satisfy the monomials have finite support condition). Note that the various formulas above like $F(X,F(Y,Z)) = F(F(X,Y),Z)$ and $\alpha(F(X,Y)) = G(\alpha(X),\alpha(Y)$ make sense because of lemma 7.3.

7.6. It is now immediately obvious that a universal formal group law with infinite index set I cannot exist because there is no predicting for which finitely many $i \in I$ the coefficient of a given monomial $X^{\alpha}Y^{\beta}$ in $F(i)(X,Y)$ will have nonzero coefficient. The way to remedy this is to extend the definition a bit by considering complete topological rings B whose topology is defined by a (filtered) set of ideals $\mathfrak{a}_s$, $s \in S$ such that $\bigcap_s \mathfrak{a}_s = \{0\}$ (so that B is Hausdorff).

7.7. Definition.

Let B be as above in 7.6 and let I and J be index sets. An I-tuple of power series $f(X)$ in X_j, $j \in J$ with coefficients in B is said to be continuous if for all multiindices $\alpha: J \to \mathbb{N} \cup \{0\}$ and all $s \in S$ there are only finitely many $i \in I$ such that the coefficient of X^{α} in $f(i)(X)$ is not in $\mathfrak{a}_s$. It is an immediate consequence of lemma 7.3 that the composite of two continuous sets of power series is welldefined and continuous and that the inverse power series $f^{-1}(X)$ of a continuous power series $f(X)$ such that $f(X) \equiv X$ mod(degree 2) is also welldefined and continuous.

7.8. Definitions.

Let B be as above in 7.6 and let I be an index set. A commutative infinite dimensional formal group law over B is now a continuous I-tuple of power series over B in X_i, Y_i, $i \in I$ such that $F(X,0) = X, F(0,Y) = Y$, $F(F(X,Y),Z) = F(X,F(Y,Z))$, $F(X,Y) = F(Y,X)$. Note that the condition $F(F(X,Y),Z) = F(X,F(Y,Z))$ makes sense again (because it makes sense mod $\mathfrak{a}_s$ for all s and because B is complete). The definitions for homomorphisms and formal A-modules are similarly modified by requiring all I-tuples of power series to be continuous. The definitions of 7.5 correspond to the case of a discretely topologized ring B (defined by the single ideal 0).

8. CONSTRUCTION OF AN INFINITE DIMENSIONAL UNIVERSAL FORMAL GROUP LAW.

8.1. Let R be any ring. Let I be an index set. The first thing to do is to describe the appropriate ring "of polynomials" over which a universal formal group law with index set I will be constructed. For each multiindex $\alpha : I \to \mathbb{N} \cup \{0\}$ (with finite support) such that $|\alpha| \geq 2$ and each $i \in I$ let $U(i,\alpha)$ be an indeterminate. Consider the ring of polynomials $R[U(i,\alpha) \mid i \in I,\ \alpha: I \to \mathbb{N} \cup \{0\},\ |\alpha| \geq 2]$.
Let T be the set of all functions on the set of multiindices on I to the

set of finite subsets of I. For each $\tau \in T$ let $\mathfrak{a}'_\tau \subset R[U]$ be the ideal generated by all the $U(i,\alpha)$ such that $i \notin \tau(\alpha)$. We now denote with $R\langle U;I\rangle$ the completion of $R[U]$ with respect to the topology defined by these ideals, and with $\mathfrak{a}_\tau$ the closure of $\mathfrak{a}'_\tau$ in $R\langle U;I\rangle$ for all $\tau \in T$. If I is a finite set then $R\langle U;I\rangle$ is simply $R[U(i,\alpha)]$ because one of the possible functions τ in this case is $\tau(\alpha) = I$ for all α and then $\mathfrak{a}_\tau = 0$. For each finite subset $\kappa \subset I$ there is a natural surjection $\phi_\kappa : R\langle U;I\rangle \to R\langle U;\kappa\rangle = R[U(i,\alpha) | \operatorname{supp}(\alpha) \cup \{i\} \subset \kappa]$. In fact the kernel is the ideal $\mathfrak{a}_{\tau(\kappa)}$ defined by the function $\tau(\kappa) \in T$, $\tau(\kappa)(\alpha) = \emptyset$ if $\operatorname{supp}(\alpha) \not\subset \kappa$ and $\tau(\kappa)(\alpha) = \kappa$ if $\operatorname{supp}(\alpha) \subset \kappa$. The $\mathfrak{a}_{\tau(\kappa)}$, $\kappa \subset I$, κ finite define another, coarser, topology in $R[U(i,\alpha)]$ which is, however, still Hausdorff. This means that $R\langle U;I\rangle$ is a certain subalgebra of $\varprojlim R\langle U;\kappa\rangle$, which in turn is a proper subalgebra of the projective limit of the polynomial rings in finitely many $U(i,\alpha)$'s over R. For example if $I = \mathbb{N}$ then $\sum_{i=1}^{\infty} U(i,2\varepsilon(i))$ where $\varepsilon(i)$ is the multiindex $\varepsilon(i)(j) = 0$ if $j \neq i$, $\varepsilon(i)(i) = 1$ is an element of $\varprojlim R\langle U;\kappa\rangle$ because for each κ it is a polynomial mod $\mathfrak{a}_{\tau(\kappa)}$. But this element is not an element of $R\langle U;\mathbb{N}\rangle$ because it is not a polynomial modulo $\mathfrak{a}_\tau$ if τ is, e.g., the function $\tau(\alpha) = \operatorname{supp}(\alpha)$.

The R-algebra $R\langle U;I\rangle$ has an obvious freeness property with respect to topological R-algebras B as in 7.6. Let B be such an algebra. And for every $\alpha: I \to \mathbb{N} \cup \{0\}$, $|\alpha| \geq 2$ and $i \in I$ let $b(i,\alpha)$ be an element of B. Suppose that for every α and every $s \in S$ there are only finitely many $b(i,\alpha) \notin \mathfrak{a}_s$. Then there is a unique continuous R-algebra homomorphism $\phi: R\langle U;I\rangle \to B$ such that $\phi(U(i,\alpha)) = b(i,\alpha)$ for all i,α.

8.2. Finite Dimensional Universal Formal Group Laws.

We recall the construction of an m-dimensional universal formal group law in [2], section 11. Let I be a finite set. For each multiindex $\alpha : I \to \mathbb{N} \cup \{0\}$ such that $|\alpha| \geq 2$ and each $i \in I$ let $U(i,\alpha)$ be an indeterminate. Let $\mathbb{Z}[U]$ ($= \mathbb{Z}\langle U;I\rangle$ if I is finite) be the ring of polynomials in these indeterminates.

In addition we define $U(i,\varepsilon(j)) = 1$ if $i = j$ and $= 0$ if $i \neq j$, where $\varepsilon(j)$ is the multiindex $\varepsilon(j)(k) = 0$ if $k \neq j$, $\varepsilon(j)(k) = 1$ if $k = j$. For each multiindex α let U_α be the columnvector $U(i,\alpha)_{i\in I}$. For each prime power $q = p^r$, $r \in \mathbb{N}$, p a prime number we use U_q to denote the matrix

$U_q = (U(i,q\varepsilon(j)))_{i,j\in I}$. Using all this notation we now define for all $\alpha : I \to \mathbb{N} \cup \{0\}$ with $|\alpha| \geq 1$ a column vector a_α with entries in $\mathbb{Q}[U]$ by means of the formula

$$(8.3) \qquad a_\alpha = \sum_{(q_1,\ldots,q_t,\beta)} \frac{n(q_1,\ldots,q_t)}{p_1} \cdots \frac{n(q_{t-1},q_t)}{p_{t-1}} \frac{n(q_t)}{p_t} U_{q_1} U_{q_2}^{(q_1)} \cdots U_{q_t}^{(q_1\ldots q_{t-1})} U_\beta^{(q_1\ldots q_t)}$$

where the sum is over all sequences $(q_1,\ldots,q_t,\beta)$, $t \in \mathbb{N} \cup \{0\}$, $q_i = p_i^{s_i}$, $s_i \in \mathbb{N}$, p_i a prime number, β a multiindex not of the form $\beta = p^r\varepsilon(j)$, $r \in \mathbb{N}$, $j \in I$, p a prime number, such that $q_1\ldots q_t\beta = \alpha$; $U_q^{(r)}$ is the matrix obtained from U_q by raising each of its entries to the power r and $U_\beta^{(s)}$ has the obvious analogous meaning. The numbers $n(q_1,\ldots,q_t)$ are integers which can be chosen arbitrarily subject to the conditions

$$(8.4) \qquad \begin{aligned} n(q_1,\ldots,q_t) &\equiv 1 \bmod p_1^r \text{ if } p_1 = p_2 = \ldots = p_r \neq p_{r+1},\ 1 \leq r \leq t \\ n(q_1,\ldots,q_t) &\equiv 0 \bmod p_2^r \text{ if } p_1 \neq p_2 = \ldots = p_r \neq p_{r+1},\ 2 \leq r \leq t \end{aligned}$$

Sometimes, in order to have reasonable formula for the U's in terms of the a's it is useful to choose the $n(q_1,\ldots,q_t)$ in a very special way, cf. [2] section 5.6 and section 34.4.

We now define

$$(8.5) \qquad f_U(X) = \sum_{|\alpha|\geq 1} a_\alpha X^\alpha, \quad F_U(X,Y) = f_U^{-1}(f_U(X) + f_U(Y))$$

Then, as is proved in section 11 of [2], $F_U(X,Y)$ is a universal formal group law with finite index set I. The integrality of $F_U(X,Y)$ is a consequence of the functional equation lemma 4.3.

8.6. Construction of an Infinite Dimensional Universal Formal Group Law.

Now let I be an infinite index set. Let $\mathbb{Z}\langle U;I\rangle$ be the ring constructed above in 8.1. For each finite subset $\kappa \subset I$ let $\mathbb{Z}\langle U;\kappa\rangle$ be the natural quotient $\mathbb{Z}[U(\alpha,i) \mid \text{supp}(\alpha) \cup \{i\} \subset \kappa]$ of $\mathbb{Z}\langle U;I\rangle$. For each κ let $f_{U,\kappa}(X)$ and $F_{U,\kappa}(X,Y)$ be the power series in X_i, $i \in \kappa$ and X_i, Y_i, $i \in \kappa$ defined by (8.5). Fix a choice of the $n(q_1,\ldots,q_t)$ for all sequences of prime powers

$(q_1,\ldots,q_t)$, $t \in \mathbb{N}$.

For each pair of finite subsets $\kappa, \lambda \subset I$ such that $\kappa \subset \lambda$ we use $\phi_{\lambda,\kappa}$: $\mathbb{Z}\langle U;\lambda\rangle \to \mathbb{Z}\langle U;\kappa\rangle$, $\mathbb{Q}\langle U;\lambda\rangle \to \mathbb{Q}\langle U;\kappa\rangle$, $\mathbb{Q}\langle U;\lambda\rangle[[X_i; i\in\lambda]] \to \mathbb{Q}\langle U;\lambda\rangle[[X_i; i\in\kappa]]$, $\mathbb{Z}\langle U;\lambda\rangle[[X_i,Y_i; i\in\lambda]] \to \mathbb{Z}\langle U;\kappa\rangle[[X_i,Y_i; i\in\kappa]]$ to denote the natural projections ($U(\alpha,i) \mapsto 0$ if $\lambda \supset \mathrm{supp}(\alpha) \cup \{i\} \not\subset \kappa$, $U(\alpha,i) \mapsto U(\alpha,i)$ if $\mathrm{supp}(\alpha) \cup \{i\} \subset \kappa$, $Y_i, X_i \mapsto 0$ if $i \in \lambda \setminus \kappa$, $X_i, Y_i \mapsto X_i, Y_i$ if $i \in \kappa$). Now note that $\phi_{\lambda,\kappa} F_{U,\lambda}(X,Y) = F_{U,\kappa}(X,Y)$ and $\phi_{\lambda,\kappa} f_{U,\lambda}(X) = f_{U,\kappa}(X)$. This means that we can define I-tuples of power series $f_U(X)$ and $F_U(X,Y)$ as follows. For each multiindex $\alpha: I \to \mathbb{N} \cup \{0\}$ and pair of multiindices $\alpha, \beta: I \to \mathbb{N} \cup \{0\}$ and element $i \in I$ consider the finite subsets κ such that $\kappa \supset \mathrm{supp}(\alpha) \cup \mathrm{supp}(\beta) \cup \{i\}$. Now consider the coefficients $e_{\alpha,\kappa}(i)$ and $e_{\alpha,\beta,\kappa}(i)$ of X^α and $X^\alpha Y^\beta$ in $f_{U,\kappa}(i)(X)$ and $F_{U,\kappa}(i)(X,Y)$ respectively. In virtue of the compatibility of the $f_{U,\kappa}(X)$ and $F_{U,\kappa}(X,Y)$ under the $\phi_{\lambda,\kappa}$ the systems of elements $e_{\alpha,\kappa}(i)$ and $e_{\alpha,\beta,\kappa}(i)$ determine welldefined elements $e_\alpha(i)$, $e_{\alpha,\beta}(i)$ in $\varprojlim \mathbb{Q}\langle U;\kappa\rangle$ and $\varprojlim \mathbb{Z}\langle U;\kappa\rangle$ respectively.

We now define $f_U(X)$ and $F_U(X,Y)$ by

$$f_U(i)(X) = \sum_{|\alpha|\geq 1} e_\alpha(i) X^\alpha \qquad F_U(i)(X,Y) = \sum_{\alpha,\beta} e_{\alpha,\beta}(i) X^\alpha Y^\beta$$

I claim that in fact $e_\alpha(i) \in \mathbb{Q}\langle U;I\rangle \subset \varprojlim \mathbb{Q}\langle U;\kappa\rangle$.

Indeed we clearly have

$$(8.7) \quad e_\alpha(i) = \sum U(i,q_1\varepsilon(i_1))U(i_1,q_2\varepsilon(i_2))^{q_1} \cdots U(i_{t-1},q_t\varepsilon(i_t))^{q_1\cdots q_{t-1}} U(i_t,\beta)^{q_1\cdots q_t} d(q_1,\ldots,q_t)$$

where $d(q_1,\ldots,q_t) = p_1^{-1}\cdots p_t^{-1} n(q_1,\ldots,q_t)n(q_2,\ldots,q_t)\ \ldots\ n(q_{t-1},q_t)n(q_t)$ and where the sum is over all sequences $(q_1,\ldots,q_t,\beta)$ as in the sum 8.3 and all $i_1, \ldots, i_t \in I$. Let $\tau \in T$. Because $q_1\cdots q_t\beta = \alpha$ we have that $\mathrm{supp}(\beta) = \mathrm{supp}(\alpha)$. So there are only finitely many i_t such that $U(i_t,\beta) \notin \mathfrak{a}_\tau$, for each of these i_t there are only finitely many i_{t-1} such that $U(i_t,q_t\varepsilon(i_t)) \notin \mathfrak{a}_\tau$, ... , and for each of the i_2 there are only finitely many i_1 such that $U(i_1,q_1\varepsilon(i_2)) \notin \mathfrak{a}_\tau$. Finally there are only finitely many factorizations $q_1 \cdots q_t\beta = \alpha$. It follows that $e_\alpha(i)$ is a polynomial mod $\mathfrak{a}_\tau$ for all τ

proving that $e_\alpha(i) \in \mathbb{Q}\langle U;I\rangle$. Because for every i_1 there are but finitely many i such that $U(i,q_1\varepsilon(i_1)) \notin \mathfrak{a}_T$ it also follows that $e_\alpha(i) \equiv 0 \mod \mathfrak{a}_T$ for all but finitely many i. It follows that $f_U(X)$ is a continuous I-tuple of power series in the sense of definition 7.7. This in turn means that $f_U^{-1}(f_U(X) + f_U(Y))$ makes sense and has its coefficients in $\mathbb{Q}\langle U;I\rangle$. And this finally means that

$$f_U^{-1}(f_U(X) + f_U(Y)) = F_U(X,Y) \tag{8.8}$$

so that $F_U(X,Y)$ has its coefficients in $\mathbb{Z}\langle U;I\rangle \subset \varprojlim \mathbb{Z}\langle U;\kappa\rangle \subset \varprojlim \mathbb{Q}\langle U;\kappa\rangle$.

9. PROOF OF THE UNIVERSALITY OF THE INFINITE DIMENSIONAL UNIVERSAL FORMAL GROUP LAW $F_U(X,Y)$ over $\mathbb{Z}\langle U;I\rangle$.

This proof is in it essentials exactly like the proof in [2], section 11.4 of the universality of the finite dimensional formal group law described in 8.2 above.

If β,α: $I \to \mathbb{N} \cup \{0\}$ are multiindices we write $\alpha > \beta$ if $\alpha(i) \geq \beta(i)$ for all $i \in I$ and $|\alpha| > |\beta|$. We use 0 to denote the multiindex $0(i) = 0$ all $i \in I$. We define $\nu(\alpha) = 1$ unless α is of the form $\alpha = p^r\varepsilon(j)$, $r \in \mathbb{N}$, $j \in I$, p a prime number and $\nu(p^r\varepsilon(j)) = p$. Then $\nu(\alpha)$ is the greatest common divisor of the $\binom{\alpha}{\beta} = \prod_{i\in \mathrm{supp}(\alpha)} \binom{\alpha(i)}{\beta(i)}$ for $0 < \beta < \alpha$.

For each $\alpha > \beta$ choose $\lambda_{\alpha,\beta} \in \mathbb{Z}$ as in [2], 11.3.5 such that

$$\sum_{0<\beta<\alpha} \lambda_{\alpha,\beta}\binom{\alpha}{\beta} = \nu(\alpha) \tag{9.1}$$

Then exactly as in [2], lemma 11.3.7 we have the following lemma

9.2. <u>Lemma</u>. Let α: $I \to \mathbb{N} \cup \{0\}$ be a multiindex, $|\alpha| \geq 2$. For each $0 < \beta < \alpha$ let X_β be an indeterminate and let $X_\beta = X_{\alpha-\beta}$. Then every X_β, $0 < \beta < \alpha$ can be written as a linear expression with coefficients in $\mathbb{Z}$ of the expressions

$$\sum_{0<\beta<\alpha}\lambda_{\alpha,\beta}X_\beta,\quad \binom{\beta+\gamma}{\gamma}X_{\beta+\gamma} - \binom{\gamma+\delta}{\delta}X_{\gamma+\delta},\quad \beta + \gamma + \delta = \alpha,\ \beta,\gamma,\delta > 0$$

9.3. Proof of the Universality of $F_U(X,Y)$.

From formula (8.7) above we see that

$$(9.4) \qquad f_U(X) \equiv X + \sum_{|\alpha|=n} \nu(\alpha)^{-1} U_\alpha X^\alpha \quad \text{mod(degree } n+1,\ U(\beta,j) \text{ with } |\beta| < n)$$

It follows that

$$(9.5) \qquad F_U(X,Y) \equiv X + Y + \sum_{|\alpha|=n} \nu(\alpha)^{-1} U_\alpha X^\alpha + \sum_{|\alpha|=n} \nu(\alpha)^{-1} U_\alpha Y^\alpha - \sum_{|\alpha|=n} \nu(\alpha)^{-1} (X+Y)^\alpha$$

mod(degree n+1, $U(\beta,j)$ with $|\beta| < n$). Now write

$$(9.6) \qquad F_U(i)(X,Y) = X_i + Y_i + \sum_{|\alpha|,|\beta|\geq 1} e_{\alpha,\beta}(i) X^\alpha Y^\beta$$

and define

$$(9.7) \qquad y(i,\alpha) = - \sum_{0<\beta<\alpha} \lambda_{\alpha,\beta} e_{\beta,\alpha-\beta}(i)$$

for all $\alpha: I \to \mathbb{N} \cup \{0\}$, $|\alpha| \geq 2$, $i \in I$. It follows immediately from (9.6) that

$$(9.8) \qquad y(i,\alpha) \equiv U(i,\alpha) \ \text{mod}(U(j,\beta) \text{ with } |\beta| < |\alpha|).$$

Also $y(i,\alpha)$ is a polynomial mod $\mathfrak{a}_\tau$ for all τ, i.e. $y(i,\alpha) \in \mathbb{Z}\langle U;I\rangle$, because (9.7) is a finite sum. From this it follows that we can, so to speak, describe $\mathbb{Z}\langle U;I\rangle$ also as $\mathbb{Z}\langle y;I\rangle$, or, in other words, the $y(i,\alpha)$ are a "free polynomial basis" for $\mathbb{Z}\langle U;I\rangle$ meaning that the images of the $y(i,\alpha)$, $i \in \tau(\alpha)$ are a free polynomial basis for $\mathbb{Z}\langle U;I\rangle/\mathfrak{a}_\tau$ for all τ. Now let $G(X,Y)$ over B, where B is as in 7.6, be any formal group law (in the sense of 7.8) with index set I. We write

$$(9.9) \qquad G(i)(X,Y) = X_i + Y_i + \sum_{|\alpha|,|\beta|\geq 1} b_{\alpha,\beta}(i) X^\alpha Y^\beta$$

We now define a continuous homomorphism $\mathbb{Z}\langle U;I\rangle \to B$ by requiring that

$$(9.10) \qquad \phi(y(i,\alpha)) = - \sum_{0<\beta<\alpha} \lambda_{\alpha,\beta} b_{\beta,\alpha-\beta}(i)$$

for all i,α. This ϕ is welldefined and determined uniquely because of 9.8 and the remarks just below 9.8. The homomorphism is continuous because $G(X,Y)$ is continuous I-tuple of power series in the sense of definition 7.7, and because the sum on the right of (9.10) is finite.
Certainly ϕ is the only possible continuous homomorphism $\mathbb{Z}\langle U;I\rangle \to B$ such that $\phi_* F_U(X,Y) = G(X,Y)$. It remains to show that $\phi(e_{\alpha,\beta}(i)) = b_{\alpha,\beta}(i)$ for all α,β,i. This is obvious if $|\alpha+\beta| = 2$. So by induction let us assume that this has been proved for all α,β with $|\alpha+\beta| < n$. Commutativity and associativity of $F_U(X,Y)$ and $G(X,Y)$ mean that we have relations

$$e_{\alpha,\beta}(i) = e_{\beta,\alpha}(i) \qquad\qquad b_{\alpha,\beta}(i) = b_{\beta,\alpha}(i)$$

$$\binom{\alpha+\beta}{\beta} e_{\alpha+\beta,\gamma}(i) - \binom{\beta+\gamma}{\gamma} e_{\beta+\gamma,\alpha}(i) = Q_{\alpha,\beta,\gamma,i}(e_{\delta,\varepsilon}(j))$$

$$\binom{\alpha+\beta}{\beta} b_{\alpha+\beta,\gamma}(i) - \binom{\beta+\gamma}{\gamma} b_{\beta+\gamma,\alpha}(i) = Q_{\alpha,\beta,\gamma,i}(b_{\delta,\varepsilon}(j))$$

where the $Q_{\alpha,\beta,\gamma,i}$ are certain universal expressions involving only the $e_{\delta,\varepsilon}(j)$, $b_{\delta,\varepsilon}(\gamma)$ with $|\delta+\varepsilon| < |\alpha+\beta+\gamma|$. By induction we therefore know that

$$\phi(\binom{\alpha+\beta}{\beta} e_{\alpha+\beta,\gamma}(i) - \binom{\beta+\gamma}{\gamma} e_{\beta+\gamma,\alpha}(i)) = \binom{\alpha+\beta}{\beta} b_{\alpha+\beta,\gamma}(i) - \binom{\beta+\gamma}{\gamma} b_{\beta+\gamma,\alpha}(i)$$

for all $\alpha,\beta,\gamma > 0$ with $|\alpha+\beta+\gamma| = n$. We also have by the definition of ϕ

$$\phi(\sum_{0<\beta<\alpha} \lambda_{\alpha,\beta} e_{\beta,\alpha-\beta}(i)) = \sum_{0<\beta<\alpha} \lambda_{\alpha,\beta} b_{\beta,\alpha-\beta}(i)$$

for all α, i with $|\alpha| = n$. Using lemma 9.2 it follows that $\phi(e_{\alpha,\beta}(i)) = b_{\alpha,\beta}(i)$ for all α,β,i with $|\alpha+\beta| = n$. With induction this finishes the proof.

9.11. Corollary.

Every infinite dimensional formal group law in the classical sense (cf. definition 7.5) can be lifted to characteristic zero.

Indeed these formal group laws correspond to continuous homomorphisms ϕ: $\mathbb{Z}\langle U;I\rangle \to B$ where B has the discrete topology. This means that $\phi(\mathfrak{a}_\tau) = 0$ for a certain τ and $\mathbb{Z}\langle U;I\rangle/\mathfrak{a}_\tau$ is a ring of polynomials.

9.12. Corollary.

Every infinite dimensional formal group law over a torsion free ring has a unique logarithm.

10. INFINITE DIMENSIONAL UNIVERSAL FORMAL A-MODULES.

Let A be as in section 3 above. Let I be an index set. The construction of an infinite dimensional formal A-module is completely analogous to the constructions of section 8 above. For each finite subset κ let $f_{U,\kappa}(X)$ be the logarithm of the universal formal A-module with index set κ over A $[U(\alpha,i)|\operatorname{supp}(\alpha) \cup \{i\} \subset \kappa,\ |\alpha| \geq 2]$. By taking projective limits of the coefficients we obtain a formal power series $f_U^A(X)$ over $\varprojlim A[U(\alpha,i)|\operatorname{supp}(\alpha) \cup \{i\} \subset \kappa]$ and by making use of the explicit formula (5.1) one shows that in fact the coefficients of $f_U^A(X)$ are in the sub-A-algebra A<U;I> and that $f_U^A(X)$ is a continuous I-tuple of power series. Now let

$$(10.1)\quad F_U^A(X,Y) = (f_U^A)^{-1}(f_U^A(X) + f_U^A(Y)),\ \rho(a) = [a](X) =$$

$$= (f_U^A)^{-1}(af_U^A(X)) \text{ all } a \in A$$

Then $(F_U^A(X,Y),\rho)$ is a formal A-module over A<U;I>. This can be shown either by performing the same projective limit construction with respect to the finite dimensional objects $F_{U,\kappa}^A(X,Y)$, $[a](X)_\kappa$ and observing that the relations (10.1) hold in $\varprojlim A[U(i,\alpha)|\operatorname{supp}(\alpha) \cup \{i\} \subset \kappa]$. This is what we used in section 8 above. Or one can state and prove an appropriate infinite dimensional version of the functional equation lemma. This version is simply obtained by requiring all I-tuples of power series to be continuous. The proof that the formal A-module (10.1) is indeed universal is an entirely straightforward adaptation of the proof in [2], section 25.4 that the finite dimensional formal A-modules described in section 5 above are universal.

10.2. Corollary.

Every infinite dimensional formal A-module in the sense of 7.5 above can be lifted to formal A-module over an A-torsion free A-algebra.

10.3. Corollary.

Every infinite dimensional formal A-module over an A-torsion free algebra B has a unique A-logarithm.

10.4. The A-logarithm $f_U^A(X)$ of the universal formal A-module $F_U^A(X,Y)$ over A<U;I> is of functional equation type, and there does exist a topological analogue of the functional equation lemma 4.3. In the case of A<U;I> and $\mathbb{Z}$ <U;I> this analogue is probably most easily proved by first remarking that the proofs in [2] also work in the infinite dimensional case provided that all the I-tuples of power series involved satisfy the monomials have compact support condition. The topological version alluded to above then results by proving things over A<U;I>/$\mathfrak{a}_\tau$ and $\mathbb{Z}$ <U;I>/$\mathfrak{a}_\tau$ for all τ.

This permits us to define ε_q and $\underline{\underline{f}}_\pi$ for curves in $F_U^A(X;Y)$ and hence by specialization for curves over arbitrary infinite dimensional formal A-modules.

The construction of the infinite dimensional formal group laws $F_U(X,Y)$ over $\mathbb{Z}$ <U;I> and the infinite dimensional universal formal A-modules over A<U;I> also permit us to extend the Cartier-Dieudonné module classification theory of [2], chapter V to cover infinite dimensional case. The proofs are entirely straightforward adaptations of the proofs given in [2].

REFERENCES.

1. V.G. Drinfel'd, Coverings of p-adic Symmetric Domains (Russian), Funk. Analiz i ego Pril. 10(1976), 29-40.
2. M. Hazewinkel, Formal Groups and Applications, Acad. Press (to appear, Sept. 1978).
3. M. Hazewinkel, On Formal Groups: The Functional Equation Lemma and some of its applications. In: Journées dc géométrie algébrique, Rennes, 1978 (to appear, Astérisque).
4. M. Hazewinkel, Twisted Lubin-Tate Formal Group Laws, ramified Witt Vectors and ramified Artin-Hasse Exponentials, Report 7711, Econometric Inst., Erasmus Univ., Rotterdam 1977.

ON THE DUAL OF A SMOOTH VARIETY

by

Audun Holme

§0. Introduction

Recall that if $X \hookrightarrow \mathbb{P}_k^N = \mathbb{P}^N$ is an embedded (reduced and irreducible) projective variety, then the dual (embedded) variety $X^\vee \hookrightarrow \mathbb{P}^{N\vee} = \mathbb{P}^N$ is defined as follows: First, a hyperplane H in $\mathbb{P}^N$ is said to be tangent to X at x if it contains $T_{X,x}$, the embedded Zariski tangent space of X at x. Equivalently, the scheme-theoretic intersection $X_H = X \cap H$ is singular at x. The collection of all tangent hyperplanes at smooth points of X correspond to a subset $U^\vee$ of $\mathbb{P}^{N\vee} = \mathbb{P}^N$; its closure is denoted by $X^\vee$ and referred to as the dual variety of the embedded variety X.

The interplay between the two embedded varieties X and $X^\vee$ constitutes an important part of projective algebraic geometry, classical and modern.

For some of the basic facts one may consult, for instance [Wa 1,2], SGA VII, Exposé XVII, pp. 212-253, [Kl], Chapter IV D. Thus $X^\vee$ is itself a variety; if k is of characteristic zero one has bi-duality, i.e. that $X^{\vee\vee} = X$; and one observes that $X^\vee$ is singular in general even if X is smooth. Moreover, elementary considerations show that

$$N - 1 \geq \dim(X^\vee) \geq N - \dim(X) - 1 .$$

In SGA VII, Exposé XVII §5, N. Katz computes the degree of $X^\vee$ for X smooth under the assumption that $\dim X^\vee = N - 1$. If k is of characteristic zero, this degree is given by a formula in the degrees of the Chern-classes of X, in fact:

$$\deg(X^\vee) = \sum_{i=0}^{n} (i+1)e_{n-i}(X) , \qquad (0.1)$$

where $n = \dim(X)$ and

$$e_j(X) = \deg(c_j(\Omega^1_{X/k})) \quad .$$

Here the degree is taken with respect to the given embedding

$$\phi : X \hookrightarrow \mathbb{P}^N$$

in the usual way: α induces the

$$\phi_* : A(X) \to A(\mathbb{P}^N) = \mathbb{Z}[t]/t^{N+1} \quad ,$$

and for a homogeneous element $c_j \in A^j(X)$ we put

$$\phi_*(c_j) = \deg(c_j)t^{N-n+j} \quad .$$

Two derivations of the formula (0.1) can be found in S.L. Kleimans article [Kl], to which we refere for details, background and notation. Here it is also shown that $\dim(X^\vee) = N - 1$ if and only if the right hand side of (0.1) is non-zero. See also J. Robert's article [Rb].

Formulas for the degree of the dual variety like (0.1) are generalized Plücker-formulas, and there has been a certain ammount of interest in obtaining such formulas. Thus (0.1) holds under the three assumptions that 1. X be non-singular; 2. $\dim(X^\vee) = N - 1$ and 3. k be of characteristic zero.

To generalize (0.1) one may accordingly work in three directions. Thus B. Tessier [Te] found a formula in the case when X is a hypersurface with only isolated singularities, and one of R. Piene's results in [Pi] gives a nice formula for $\deg(X^\vee)$ which holds without the assumption 1, in terms of a certain bundle P on a blow-up of X . Unfortunately one can actually compute only in certain special cases, but even this extends the previously known information considerably. However, the condition 2 is needed.

In the present paper we assume throughout that X is smooth, but not that $\dim(X^{\vee}) = N - 1$. In fact we compute $\dim(X^{\vee})$ by a method similar to the embedding-obstructions in [Ho 1]. Furthermore, we give a formula for $\deg(X^{\vee})$ when k is of characteristic zero, valid for $X^{\vee}$ of arbitrary codimension.

The assumption that X be smooth is not essential, as the reader familiar with [Ho 2], [Jn] or [FM], say, will notice without too much trouble. But for simplicity we consider only the smooth case in this paper.

The present paper is an improved and expanded version of the talk given by the author at the symposium in Kopenhagen 1978. Some of the added material has resulted from several very stimulating and enlightening conversations with A. Landman in Helsinki after the symposium. In particular, the idea behind the proof of Theorem 2.1 is due to him.

§1. Dimension of $X^{\vee}$.

We assume that X is smooth, and approach the dual variety via the linear system of hyperplane sections, following [Kl] pp.360, 361. See also SGA VII and [Rb].

So let $\alpha : D \hookrightarrow X \times \mathbb{P}^N$ be the total space of this linear system with its canonical embedding. Here we have written $\mathbb{P}^{N\vee} = \mathbb{P}^N$. D is a divisor, and we let

$$q : D \to \mathbb{P}^N$$

be the morphism induced by the second projection. Letting R denote its ramified locus, one has

$$X^{\vee} = q(R) .$$

If N denotes the conormal sheaf of the given embedding

$$\phi : X \hookrightarrow \mathbb{P}^N = \mathbb{P}(V) ,$$

we obtain an exact sequence on X ,

$$0 \to N(1) \to V_X \to P^1(O_X(1)) \to 0 .$$

One may show that

$$R \subset X \times \mathbb{P}^N = \mathbb{P}(V_X^\vee)$$

is equal to the closed subscheme

$$\mathbb{P}(N(1)^\vee) \subset \mathbb{P}(V_X^\vee) .$$

Using Porteous' formula and the Adjunction formula, the following expression is derived in [K1]:

$$(1.1) \qquad \alpha_*([R]) = \sum_{i=o}^{n} p^*(c_{n-i}(\Omega^1_{X/k}))[D]^{i+1} ,$$

where $p : X \times \mathbb{P}^N \to X$ denotes the projection. Moreover, since the divisor D is defined by a canonical section of $p^*O_X(1) \otimes q^*O_{\mathbb{P}^N}(1)$, it follows that

$$[D] = p^*c_1(O_X(1)) + q^*c_1(O_{\mathbb{P}^N}(1)).$$

Now put

$$h = p^*(c_1(O_X(1))), \quad t = q^*(c_1(O_{\mathbb{P}^N}(1))), \quad c_i = c_i(\Omega^1_{X/k}) .$$

Then

$$(1.2) \qquad \begin{aligned} \alpha_*([R]) &= \sum_{i=o}^{n} p^*(c_{n-i})(h+t)^{i+1} \\ &= \sum_{i=o}^{n} \sum_{j=o}^{n+1} \binom{i+1}{j} p^*(c_{n-i})h^{i+1-j}t^j \end{aligned}$$

Next, let

$$\psi : X \times \mathbb{P}^N \hookrightarrow \mathbb{P}^N \times \mathbb{P}^N = P$$

be the embedding induced by ϕ. Then the following holds in the Chow-ring $A(P)$:

$$(1.3) \qquad [R] = \psi_*(\alpha_*([R])) = \sum_{i=o}^{n} \sum_{j=o}^{n+1} \binom{i+1}{j} e_{n-i} s^{N+1-j} t^j$$

where

$$e_{n-i} = \deg(c_{n-i}), \qquad s = pr_1^*(c_1(\mathcal{O}_{\mathbb{P}^N}(1)))$$

and where t is identified with $\psi^*(t)$. So in other words, s and t are the pullbacks of the hyperplane classes by pr_1 and pr_2, respectively. Indeed, one observes right away that

$$h = \psi^*(s),$$

and (1.3) follows from (1.2) by the Projection formula.

Reversing the order of summation in (1.3) we get

$$(1.4) \qquad [R] = \sum_{j=o}^{n} \sum_{i=j}^{n} \binom{i+1}{j+1} e_{n-i} s^{N-j} t^{j+1}$$

We now need a lemma which, even though it is quite simple to prove, has been rather usefull in several situations. Let $S \subset P$ be a closed subscheme of pure codimension d, and let s and t be as above. Then

$$(1.5) \qquad [S] = \sum_{i=n_o}^{n_1} a_i s^i t^{d-i}$$

where

$$n_o = \mathrm{Max}\{0, d-N\}$$

$$n_1 = \mathrm{Min}\{N, d\}$$

Lemma 1.6. $\dim(pr_2(S)) \leq r \Leftrightarrow a_i = 0$ *for all* $i \geq \mathrm{Max}\ \{0, d+r+1-N\}$.

Proof. Let P^r denote a r-dimensional linear subspace of $\mathbb{P}^N$ in general position. Then

$$\dim(pr_2(S)) \leq r \Leftrightarrow$$

$$pr_2(S) \cap P^{N-r-1} = \emptyset \Leftrightarrow$$

$$S \cap pr_2^{-1}(P^{N-r-1}) = \emptyset \Leftrightarrow$$

$$[S].[pr_2^{-1}(P^{N-r-1})] = [S]t^{r+1} = 0 \quad .$$

The claim follows from this.

Applying the lemma to the expression (1.3), we get the

Theorem 1.7. *Let* $X \hookrightarrow \mathbb{P}^N_k$ *be an embedded, smooth projective variety. Put*

$$\delta_s = \sum_{i=s}^{n} \binom{i+1}{s+1} e_{n-i} \quad . \tag{1.7.1}$$

Let $N \geq m \geq 0$ *be such that*

$$\delta_m \neq 0 \ , \quad \delta_s = 0 \quad \textit{for all} \quad s < m \quad .$$

Then

$$\dim(X^\vee) = N-1-m \quad .$$

Indeed, as has been pointed out above,

$$X^\vee = pr_2(R) \quad .$$

<u>Remark</u>. Since $\delta_n = e_o = \deg(X)$, we recover the elementary estimate

$$\dim(X^\vee) \geq N-n-1 \ .$$

Clearly this lower limit is attained for X a linear subspace of $\mathbb{P}^N$. A result of <u>A. Landman</u> [Lm 1,2], see [kl] page 363 gives that if X is smooth, non-linear and of dimension ≥ 2, then

$$\dim(X^\vee) \geq N-n-1 \ .$$

Thus the equations

$$(1.8) \qquad \delta_{n-2} = --- = \delta_o = 0$$

imply that $\delta_{n-1} = 0$, and characterize the linear subspaces in $\mathbb{P}^N$ among the smooth projective subvarieties. This is a primitive example of a very interesting problem of classification: To consider certain systems of equations in certain projective invariants, say the degrees of the Chern-classes like (1.8), and try to find <u>canonical forms</u> for the subvarieties of $\mathbb{P}^N$ satisfying the given identities.

We keep the notation from Theorem 1.7, and let H be a hyperplane of $\mathbb{P}^N$ in general position. Put

$$X_H = X \cap H \ .$$

Then we have the

<u>Proposision 1.9</u>. <u>For all</u> $s > 0$,

$$\delta_s(X) = \delta_{s-1}(X_H) \ .$$

<u>Proof</u>. Write $Y = X_H$, and let

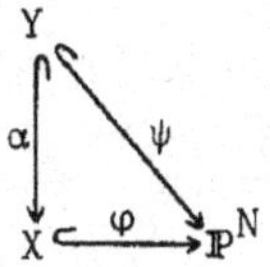

be the diagram of canonial embeddings. The exact sequence

$$0 \to O_Y(-Y) \to \alpha^*\Omega^1_{X/k} \to \Omega^1_{Y/k} \to 0$$

yields

$$\alpha^* c(\Omega^1_{X/k}) = \alpha^*(1-[Y])c(\Omega^1_{Y/k}) \ .$$

Letting $h = [H] \in A(\mathbb{P}^N)$, this gives

$$\psi_*(c(\Omega^1_{Y/k})) = \frac{h}{1-h}\,\varphi_*(c(\Omega^1_{X/k})) \ ,$$

from which one concludes immediately that

$$\begin{aligned} e_o(X) &= e_o(Y) \\ e_1(X) &= e_1(Y) - e_o(X) \\ &\ \vdots \\ e_{n-1}(X) &= e_{n-1}(Y) - e_o(Y) \ . \end{aligned}$$

For $s > 0$ we thus have

$$\begin{aligned} \delta_s(X) &= \sum_{i=s}^{n-1} \binom{i+1}{s+1}(e_{n-i}(Y) - e_{n-i-1}(Y)) + \binom{n+1}{s+1} e_o(Y) \\ &= \sum_{i=s}^{n} \binom{i+1}{s+1} e_{n-i}(Y) - \sum_{i=s}^{n-1} \binom{i+1}{s+1} e_{n-i-1}(Y) \\ &= \sum_{i=s}^{n} \left(\binom{i+1}{s+1} - \binom{i}{s+1}\right) e_{n-i}(Y) \\ &= \delta_{s-1}(Y) \ , \end{aligned}$$

as claimed.

Since

$$[R] = \sum_{j=0}^{n} \delta_j s^{N-j} t^{j+1}$$

is the class of a closed subscheme of $\mathbb{P}^N \times \mathbb{P}^N$, all δ_j's are non negative. But there seems to be no obvious regularity among them.

§2. Degree of $X^\vee$.

Let

$$\gamma : R \longrightarrow X^\vee$$

be the induced morphism. It is well-known that if k is of characteristic zero and if

$$\dim(X^\vee) = N - 1 ,$$

then γ is birational. (See for instance [Wa 2], §3). We assume throughout this section that k is of characteristic zero.

In particular we get that $\deg(X^\vee) = \delta_o$ if $\dim(X^\vee) = N - 1$. But in fact, we have the following more general result:

<u>Theorem 2.1</u>. <u>Let</u> $X \hookrightarrow \mathbb{P}^N$ <u>be an embedded, smooth, projective variety</u>. <u>Assume that</u> $\dim(X^\vee) = N - 1 - m$. <u>Then</u>

$$\deg(X^\vee) = \delta_m .$$

<u>Proof</u>. The claim will follow by repeated application of Proposition 1.9, once we know the following (notation as in §1):

<u>Lemma</u>. <u>If</u> $\delta_o = 0$, <u>then</u>

$$\deg((X_H)^\vee) = \deg(X^\vee) .$$

<u>Proof</u>. Since $\dim(X^\vee) < N - 1$, elementary considerations show that the dual variety of X_H is equal to the cone over $X^\vee$ with vertex in the point which corresponds to H : In the general case $(X_H)^\vee$ is the cone over the subvariety Z in $\mathbb{P}^{N\vee} = \mathbb{P}^N$ which corresponds to those hyperplanes in $\mathbb{P}^N$ which are tangent to X at some point of X_H . But since $\dim(X^\vee) < N - 1$, a hyperplane H' which is tangent to X , must also be tangent to X at some point of X_H , since its locus of contact is at least a curve.

<u>§3. Grassmanians of lines.</u>

Let $X = G(1,N)$, the Grassmanian of lines in $\mathbb{P}^N$, and let $\zeta : X \hookrightarrow \mathbb{P}^{\overline{N}}$ be the Plücker-embedding. A straightforward calculation yields

$$\delta_m = \delta_m(N) =$$

$$\sum_{i=m}^{n} \sum_{j=0}^{\left[\frac{n-i}{2}\right]} \sum_{k=j}^{\left[\frac{n-i}{2}\right]} \sum_{\ell=0}^{j} (-1)^{n-i+j+\ell} 4^{j-\ell} \binom{i+1}{m+1} \binom{i}{j} \binom{N+1}{n-i-j-k} \binom{n-i-j-k}{k-j} \frac{(2(N-1-k+\ell))!}{(N-1-k+\ell)!\,(N-k+\ell)!}$$

As before $n = \dim(X)$, so that here

$$n = 2(N-1) .$$

One obtains the following table of the first few values:

$\delta_m(N)$

m \ N	3	4	5	6	7	8	9	10	11
0	2	0	3	0	4	0	5	0	6
1	2	0	6	0	12	0	20	0	30
2	2	5	12	14	36	30	80	55	150

In fact, using the result of A. Landman [Lm] that $\dim(G(1,2m)^\vee)$ is two less than the hypersurface for $m \geq 2$, we predict the regularly occurring two zeroes in every second column of the table. Moreover, there are many evident regularities, the most evident one being

$$\delta_o(N) = \begin{cases} 0 & \text{for } N \text{ even} \\ \frac{N+1}{2} & \text{for } N \text{ odd} \end{cases} .$$

To prove this from the combinatorical expression for δ_o seems at best quite laborious. Of course the first half follows by geometric means from Landman's result quoted above, and it would be nice to have a similar argument showing the remainder.

References.

[Fu] Fulton, W.: "Rational equivalence on singular varieties." Publications Mathématiques de l'IHES, No. 45, pp 147-167.

[FM] Fulton, W. and MacPherson, R.: "Intersecting cycles on an algebraic variety." Preprint Series, Aarhus University, Denmark. November 1976.

[Ho 1] Holme, A.: "Embedding-obstruction for smooth, projective varieties I." To appear in Advances in Mathematics. Preprint Series, University of Bergen, Norway. June 1974.

[Ho 2] Holme, A.: "Embedding-obstruction for singular algebraic varieties in $\mathbb{P}^N$." Acta Mathematica, 135, pp 155-185 (1975).

[Jn] Johnson, K. W.: "Immersion and embedding of projective varieties." Acta Mathematica, 140, pp 49-74 (1978).

[Kl] Kleiman, S. L.: "The enumerative theory of singularities." In Real and complex singularities, Sijthoff & Noorhoff International Publishers, 1978.

[Lm 1] Landman, A.: "Examples of varieties with 'small' dual varieties." Lecture at Aarhus University, June 24, 1976.

[Lm 2] Landman, A.: "Picard-Lefschetz theory and dual varieties." Lecture at Aarhus University, June 30, 1976.

[Pi] Piene, R.: "Polar classes of singular varieties." Preprint, MIT 1976.

[Rb] Roberts, J.: "A stratification of the dual variety. (Summary of results with indications of proofs)." Preprint (1976).

SGA VII Deligne, P. and Katz, N.: Groupes de monodromie en Géometrie algébrique.
Springer lecture notes in math., 340 (1973).

[Te] Teissier, B.: "Sur divers conditions numériques d'equisingularité des familles de courbes."
Preprint, 1975.

[Wa 1] Wallace, A. H.: Homology theory on algebraic varieties.
International Series of Monographs in Pure and Applied Mathematics. Pergamon Press, 1958.

[Wa 2] Wallace, A. H.: "Tangency and duality over arbitrary fields."
Proc. Lond. Math. Soc. (3) 6 (1956), pp 321-342.

University of Bergen
Allégatan 53-55
N-5014 Bergen
Norway

Symmetric forms and Weierstrass semigroups

Shigeru IITAKA

University of Tokyo

§1. Introduction

First we recall the classical theory of Weierstrass points and semigroups of algebraic curves.

Let C be a complete non-singular algebraic curve defined over $k = \mathbb{C}$, the field of complex numbers. By $\mathrm{Rat}(C)$ we denote the field of rational functions on C. For a point p of C, put

$$\mathbb{L}(*p) = \{\varphi \in \mathrm{Rat}(C) \mid \varphi = 0 \text{ or } \mathrm{div}(\varphi) + mp \geqq 0 \text{ for some } m > 0\}.$$

Here, $\mathrm{div}(\varphi)$ denotes the divisor defined by φ. Letting $\nu_p(\varphi)$ denote the order of φ at p, we have the semigroup

$$SG[p] = \{-\nu_p(\varphi) \mid \varphi \in \mathbb{L}(*p) \setminus \{0\}\} .$$

The following results are well known.

(1) If $g(C) = 0$, i.e., $C = \mathbb{P}^1$, then $SG[p] = \mathbb{N} \sqcup \{0\} = \{0, 1, 2, \ldots\}$.

(2) If $g(C) = 0$, i.e., C is an elliptic curve, then $SG[p] = \{0, 2, 3, \ldots\}$.

(3) If $g(C) \geqq 2$, then $\mathbb{N} \setminus SG[p]$ consists of $i_1, i_2, \ldots, i_g$ where $g = g(C)$ and $i_1 < i_2 < \cdots < i_g$.

(4) If $g(C) \geqq 2$, then letting $w(p) = \sum_{\ell=1}^{g} i_\ell - g(g+1)/2$, one has

$$\sum_{p \in C} w(p) = g(g^2 - 1).$$

Hence the set $\{p \mid w(p) > 0\}$ is a finite set called the set of Weierstrass points of C. The sequence $\{i_1, \ldots, i_g\}$ is said to be the gap sequence at p.

A straightforward generalization of the above $SG[p]$ may be as follows: let V be a complete non-singular algebraic variety and Γ be a prime divisor on V. Define, for any $m \geqq 1$,

$$\mathbb{L}(m\Gamma) = \{\varphi \in \mathrm{Rat}(V), \text{ the field of rational functions of } V \mid \varphi = 0 \text{ or } \mathrm{div}(\varphi) + m\Gamma \geqq 0\}$$

and

$$\mathbb{L}(*\Gamma) = \bigcup_{m=1}^{\infty} \mathbb{L}(m\Gamma).$$

By v_Γ we denote the valuation of $\mathrm{Rat}(V)$ with respect to Γ. Then define

$$SG[\Gamma] = \{-v_\Gamma(\varphi) \mid \varphi \in \mathbb{L}(*\Gamma) \setminus (0)\},$$

which turns out to be a subsemigroup of $\mathbb{N} \cup \{0\}$. However, this semigroup seems not to be well behaved. For instance, $SG[\Gamma]$ is not invariant under birational morphisms.

We come back to the case of curves and generalize $SG[p]$ as follows: let D be an effective divisor on an algebraic curve C and p be a point of $\mathrm{supp}\, D$. Define

$$SG_p^*(D) = \{-\nu_p(\varphi) \mid \varphi \in L(*D) \setminus \{0\}\},$$

which is obviously a semigroup. $SG_p^*(p)$ coincides with $SG[p]$. Moreover, if $\mathrm{supp}\, D \supset \{p\}$, then $SG_p^*(D) = \mathbb{N} \cup \{0\}$. Now, letting $\mathcal{O}(D)$ be the invertible sheaf corresponding to D and ψ be a section of $\mathcal{O}(D)$ satisfying $\mathrm{div}(\psi) = D$, we have an isomorphism:

$$\mathbb{L}(mD) \cong H^0(C, \mathcal{O}(mD))$$

$$\varphi \longmapsto \psi^m \varphi .$$

The gap sequence $\mathrm{seq}(mD)$ associated with mD at p (which is abbreviated as mD-sequence) is $\{a_1, \ldots, a_\ell\}$ which satisfies that $a_1 < \cdots < a_\ell$, $\ell = \dim|mD| + 1$, and

$$a_i = \nu_p(\varphi_i) \text{ for some } \varphi_i \in H^0(C, \mathcal{O}(mD)), \text{ for any } 1 \leqq i \leqq \ell.$$

By definition, we have

$$me - a_1, \ldots, me - a_\ell \in SG_p^*(D),$$

where e is the coefficient at p of D, i.e. $e = \text{mult}_p(D)$.

In [2], the author has defined D-sequence at a point p for any higher dimensional algebraic variety V and an effective divisor D on V by using the notion of symmetric forms and indices.

Here, we shall recall some basic notions concerning symmetric forms and give a fundamental theorem on Wronskian forms (Theorem 1). In the final section, we shall define j-th indices and E-sequences. Especially, we shall prove the additivity of gap-sequences and establish the birational invariance of them. Finally, we shall define Weierstrass semigroups $SG_p(\Lambda)$, $SG_{\Pi,p}(\Lambda)$ and $SG_p^*(D)$ for any linear system Λ and any divisor D. But analogous results for (1), (2), (3), and (4) have not been obtained. One may expect that in the general case, $\mathbb{N} \setminus SG_p^*(D)$ is a finite set and that for a general divisor and a general point p, $SG_p^*(D)$ is of section type, i.e. $SG_p^*(D) = \{r, r+1, r+2, \ldots\}$.

§2. Symmetric forms and symmetric derivation

Let A be a ring and B be a commutative A-algebra. We consider a commutative B-algebra C and an A-linear map δ from C into itself satisfying the following derivation rule:

$$\delta(xy) = \delta(x)y + \delta(y)x.$$

Then for example we have $\delta(x)^a(\delta^2 y)^b(\delta^3 z)^c \in C$ for any $x, y, z \in C$ $a, b, c \in \mathbb{N} \cup \{0\}$; $\delta(1) = 0$ and $\delta|_B : B \longrightarrow \sum_{b \in B} B\delta(b) \subseteq C$ is an A-derivation of B.

It is fairly easy to construct the universal (C, δ) which satisfies the above condition for any given A-algebra B (cf. [2]). The universal couple is denoted $(SF(B/A), d_{B/A})$. And elements of SF(B/A) are called <u>regular</u> (or <u>holomorphic</u>) <u>symmetric</u> <u>form</u> and $d_{B/A}$ is called the <u>symmetric derivation</u> of B over A.

Example 1. Let $A = \mathbb{C}$ and $B = \mathbb{C}[X_1, \ldots, X_n]$. Then

$$SF(B/A) = \mathbb{C}[X_1, \ldots, X_n, Y_{1,1}, \ldots, Y_{1,n}, Y_{2,1}, \ldots, Y_{2,n}, \ldots]$$

is a polynomial in ∞ variables $X_1, \ldots, Y_{m,n}, \ldots,$ and $d = d_{B/A}$ satisfies $dX_j = Y_{1,j}$, $dY_{i,j} = Y_{i+1,j}$ for any $1 \leqq j \leqq n$, $1 \leqq i < \infty$.

Example 2. Let K/k be a field extension and assume $ch(k) = 0$. If K/k has a transcendental base $\{\xi_1, \ldots, \xi_n\}$, then

$$SF(K/k) = K[d\xi_1, \ldots, d\xi_n, d^2\xi_1, \ldots]$$

such that $\{d^i\xi_j\}$ are algebraically independent over K and $d(d^i\xi_j) = d^{i+1}\xi_j$.

For an A-algebra B and for $b_1, \ldots, b_\ell \in B$, define the Wronskian matrix $[d^{i-1}b_j]_{1 \leqq i,j \leqq \ell}$. Its determinant belongs to $SF(B/A)$ and is called the Wronskian form, denoted $W_{B/A}(b_1, \ldots, b_\ell)$.

Theorem 1. If A and B are integral domains such that the fields of fractions have the characteristics zero and A is algebraically closed in B, then

$W_{B/A}(b_1, \ldots, b_\ell) = 0$ if and only if $b_1, \ldots, b_\ell$ are A-linearly independent.

Proof. In the case when $A = \mathbb{C}$ and $B = \mathbb{C}\{z_1, \ldots, z_n\}$, this was proved in [2, Proposition 1.1]. Here, we shall give the proof of Theorem 1, which is quite different from the former one.

Let k be the field of fractions of A. Then

$SF(B/A) \otimes_A k \cong SF(B \otimes_A k/k)$ and

$W_{B/A}(b_1, \ldots, b_\ell) \otimes_A 1 = W_{B\otimes_A k/k}(b_1 \otimes 1, \ldots, b_\ell \otimes 1)$.

Hence we may assume that $A = k$. Further, we can replace B by $k[b_1, \ldots, b_\ell]$. Thus, B is an integral domain finitely generated over

k.

Assuming that $b_1, \ldots, b_\ell$ is k-linearly independent, we derive that $W_{B/k}(b_1, \ldots, b_\ell) \neq 0$. Let $V = \mathrm{Spec}\, B$, which is an affine algebraic variety. If $\dim V = 1$, let p be a general point of V. Then $\mathcal{O}_{V,p}$ is a regular local ring and the completion $\hat{\mathcal{O}}_{V,p} \cong \mathfrak{K}[[t]]$, where $\mathfrak{K} \cong \mathcal{O}_{V,p}/\mathfrak{m}_p$ is the algebraic extension of k. If $W_{\mathfrak{K}[[t]]/k}(b_1, \ldots, b_\ell) = 0$, then $b_1, \ldots, b_\ell$ are $\mathfrak{K}$-linearly dependent. This follows easily (for example, use Proposition 2 in §3). Since $b_1, \ldots, b_\ell \in B$ and since $B_{(0)}$ is the regular extension of k (which follows from the fact that $\mathrm{ch}(k) = 0$ and B/k is the algebraically closed extension), we conclude that $b_1, \ldots, b_\ell$ are k-linearly dependent. Hence $W_{B/k}(b_1, \ldots, b_\ell) \neq 0$ is established.

Now, assuming the case when $\dim V \leq \ell - 1$, we shall prove the case when $\dim V = \ell$. Let $\bar{V}$ be a projective completion of the affine variety V. Then $H = \bar{V} - V$ is the support of the ample divisor. Since $b_1, \ldots, b_\ell \in \Gamma(V, \mathcal{O}_V) \cong \bigcup_{m=1}^{\infty} \Gamma(\bar{V}, \mathcal{O}(mH))$, there exist m, $\beta_0 \in \Gamma(\bar{V}, \mathcal{O}(H))$ such that $\mathrm{div}(\beta_0) = H$, and $\beta_1, \ldots, \beta_\ell \in \Gamma(\bar{V}, \mathcal{O}(mH))$ such that $b_1 = \beta_1/\beta_0^m, \ldots, b_\ell = \beta_\ell/\beta_0^m$. Note that $\beta_1, \ldots, \beta_\ell$ are k-linearly independent. Consider a general hyperplane section W of $\bar{V}$ such that $\mathcal{O}(W) \cong \mathcal{O}(NH)$ with $N > m$. Then we have the exact sequence

$$0 \to \Gamma(\bar{V}, \mathcal{O}(mH) \otimes \mathcal{O}(-W)) \to \Gamma(\bar{V}, \mathcal{O}(mH)) \to \Gamma(W, \mathcal{O}(mH|_W)).$$

Here $\Gamma(\bar{V}, \mathcal{O}(mH) \otimes \mathcal{O}(-W)) = \Gamma(\bar{V}, \mathcal{O}((m-N)H)) = 0$. Hence, letting p be a general point of W, the injection $j : W \to \bar{V}$ induces the ring homomorphism $j^* : \mathcal{O}_{V,p} \to \mathcal{O}_{W,p}$ which satisfies that $j^*(b_1), \ldots, j^*(b_\ell)$ are k-linearly independent. Moreover, letting $C = \mathcal{O}_{W,p}$

$$j^* W_{B/k}(b_1, \ldots, b_\ell) = W_{C/k}(j^*(b_1), \ldots, j^*(b_\ell)).$$

By induction hypothesis, $W_{C/k}(j^*(b_1), \ldots, j^*(b_\ell))$ is not zero and so is $W_{B/k}(b_1, \ldots, b_\ell)$.

If B is a Noetherian local ring and $\mathfrak{m}$ is the maximal ideal of B, we define for $f \in B \setminus (0)$

$$\nu_B(f) = \max\{m \mid f \in \mathfrak{m}^m\}.$$

If $(M, \mathcal{O}_M)$ is a complex manifold and f is a holomorphic function on U which is an open neighborhood of $p \in M$, then $\nu_p(f)$ is defined to be $\nu_{\mathcal{O}_{M,p}}(f)$.

Further, if ω is a holomorphic symmetric form on U, then $\omega = \sum \varphi_I(z)(dz)^I$ where $(z_1, \ldots, z_n)$ is a system of local coordinates around p, $I = \{(m_{1,1}, \ldots), \ldots, (m_{n,1}, \ldots)\}$, and

$$(dz)^I = (dz_1)^{m_{1,1}}(d^2z_1)^{m_{1,2}}\ldots(dz_2)^{m_{2,1}}\ldots(d^iz_j)^{m_{j,i}}\ldots .$$

Define

$$\nu_p(\omega) = \min\{\nu_p(\varphi_I(z)) \mid \omega = \sum \varphi_I(z)(dz)^I\} .$$

The index of ω at p is defined by

$$\rho_p(\omega) = \min\{r \mid \nu_p(d^r\omega) = 0\}.$$

The index is computed by making use of the quadratic transformation at p, i.e. letting $\mu : Q_p(M) \to M$ be a blowing up and q be a general point of $\mu^{-1}(p)$. Then

$$\rho_p(\omega) = \nu_q(\mu^*\omega) \text{ (cf. [2, Proposition 7]).}$$

Moreover, one can prove

$$\rho_p(\omega) = \min\{\nu_p(\omega|_\Gamma) \mid \Gamma \text{ curve on an open neighborhood of } p \text{ on } M \text{ such that } p \in \Gamma\}.$$

Theorem 2 ([2, Theorem 1]). Let $f : M' \to M$ be a birational (i.e. bimeromorphic for general complex manifolds) morphism, ω be a holomorphic symmetric form on M, p be a point of M, and q be a general point of $f^{-1}(p)$. Then

$$\rho_q(f^*\omega) = \rho_p(\omega).$$

Let $\mathcal{L}$ be an invertible sheaf on M and $\mathbb{L}$ be a finite dimensional vector subspace of $H^0(M,\mathcal{L})$. $\mathbb{L}$ defines a linear system $\Lambda = \{\mathrm{div}(\varphi) \mid \varphi \in \mathbb{L}\setminus(0)\}$. Choose a coordinate covering $\{U_\alpha\}$ of M such that $\mathcal{L}|_{U_\alpha}$ is trivial for any α. $\mathcal{L}$ is defined by $\{f_{\alpha\beta}\}$. Let $\{\varphi_\alpha^{(1)}, \ldots, \varphi_\alpha^{(\ell)}\}$ be a base of $\mathbb{L}|_{U_\alpha}$. Thus

$$W(\varphi_\alpha^{(1)}, \ldots, \varphi_\alpha^{(\ell)}) = f_{\alpha\beta}^{\ell} W(\varphi_\beta^{(1)}, \ldots, \varphi_\beta^{(\ell)}) \quad \text{for any } \alpha, \beta.$$

$\{W(\varphi_\alpha^{(1)}, \ldots, \varphi_\alpha^{(\ell)})\}$ is a holomorphic symmetric form with coefficients in $\mathcal{L}^{\otimes\ell}$. This symmetric form is independent of the choice of the base up to constant multiples. We define the Weierstrass form associated with Λ by

$$\omega(M, \Lambda) = \{W(\varphi_\alpha^{(1)}, \ldots, \varphi_\alpha^{(\ell)})\}.$$

The index $\rho_p(\omega(M, \Lambda))$ is written in terms of linear systems as follows: By $\nu_p(\Lambda)$ we denote the multiplicity of a general number of Λ at p.

Case 1. p is not base point of Λ. Let $\Lambda_* = \Lambda(p) = \{D \in \Lambda \mid p \in \Lambda\}$ whose dimenion is $\ell - 2$. Denote ν_1 by $\nu_p(\Lambda_*)$. Consider the blowing up $\mu_1 : M_1 \to M$ at p. Then $\mu_1^*(\Lambda_*)$ has the fixed component $\nu_1 E_1$, where $E_1 = \mu_1^{-1}(p)$. Let p_1 be a general member of $\mu_1^*(\Lambda_*) - \nu_1 E_1$ (which is written as Λ_1). Then p_1 is not a base point of Λ_1. Define $\nu_2 = \nu_{p_1}(\Lambda_1(p_1))$ and repeat the above argument. Thus we obtain $\{\nu_1, \nu_2, \ldots, \nu_{\ell-1}\}$, which is called the Λ-(gap)sequence at p.

Case 2. p is a base point of Λ. Define $\nu_1 = \nu_p(\Lambda)$ and consider the blowing up: $\mu_1 : M_1 \to M$ at p. Then the linear system Λ_1 defined by $\mu_1^*\Lambda = \Lambda_1 + \nu_1 E_1$ does not contain $E_1 = \mu_1^{-1}(p)$. Letting p_1 be a general point of $\mu_1^{-1}(p)$, we arrive at the situation of Case 1. Thus we have $\nu_2, \ldots, \nu_{\ell-1}$. Combining the former ν_1 with these, we obtain the Λ-(gap)sequence $\{\nu_1, \ldots, \nu_\ell\}$ at p.

In the above both cases, we define the Λ-index at p by

$$\rho_p(M, \Lambda) = \sum_{j=1}^{*} \nu_j - \ell(\ell-1)/2,$$

$* = \ell-2$ or $\ell-1$, according to case 1 or case 2, respectively.

From Theorem 2, we deduce the next result (cf. [2, Theorem 2]).

Theorem 3. $\rho_p(M, \Lambda) = \rho_p(\omega(M, \Lambda))$.

§3. Weierstrass semigroups

In this section, we consider finite-dimensional vector subspaces of $B = \mathbb{C}\{z_1, \ldots, z_n\}$ (the ring of convergent power series in n variables) which are abbreviated as vector spaces. $\nu(\varphi)$ denotes $\nu_B(\varphi) = \max\{r \mid \varphi \in \mathfrak{m}^r\}$.

If an ℓ-dimensional vector space E has the base $\{\varphi_1, \ldots, \varphi_\ell\}$ such that $\nu(\varphi_1) < \nu(\varphi_2) < \cdots < \nu(\varphi_\ell)$, then E is said to be the <u>distinguished space</u>. And $\{\varphi_1, \ldots, \varphi_\ell\}$ is said to be the <u>distinguished base</u>.

Proposition 1. If F is a subspace of a distinguished space E, then so is F.

Proof. Let $\{\varphi_1, \ldots, \varphi_\ell\}$ be the distinguished base of E and $\{\psi_1, \ldots, \psi_m\}$ be a base of F. Suppose that $\nu(\psi_1) \leqq \cdots \leqq \nu(\psi_m)$. First we assume that $\nu(\psi_1) = \cdots = \nu(\psi_j) < \nu(\psi_{j+1}) \leqq \cdots$ for some $j \geqq 2$. Then $\psi_i = \sum_{k=r(i)}^{\ell} \lambda_{i,k}\varphi_k$ where $\lambda_{i,r(i)} \neq 0$. Since $\nu(\varphi_1) < \cdots < \nu(\varphi_\ell)$, one has $\nu(\psi_i) = \nu(\varphi_{r(i)})$. Thus if $1 \leqq i \leqq j$, then $r(1) = \cdots = r(j)$ and so one replace ψ_i by $\psi_i/\lambda_{i,r(i)}$. Then $\nu(\psi_1) < \nu(\psi_2 - \psi_1)$ for any $1 \leqq i \leqq j$. Repeating the similar argument, one obtains a distinguished base of F.

The next proposition is obvious.

Proposition 2. If $n = 1$, every vector space is distinguished.

For any vector space $E \subset B$, we shall define the E-index at o denoted $\rho(E)$ by using the induction on ℓ. If $\ell = 1$, $\rho(E)$ is defined to be $\nu(f)$. Let $m = \min\{\nu(f) \mid f \in E \setminus (0)\}$, provided $\ell = \dim E > 0$. If $m = 0$, we have the subspace $F = \{f \mid f = 0$ or $\nu(f) > 0\}$, which has dimension $\ell - 1$. Then, for an f with $\nu(f) = 0$, $E = \mathbb{C}f \oplus F$ and $\rho(E)$ is defined to be $\rho(F)$. If $m > 0$, then letting U be a sufficiently small open neighborhood of o in $\mathbb{C}^n$, we define the blowing up $\mu : V_1 = Q_0(U) \longrightarrow U$ at o. Then letting p be a general point of $\mu^{-1}(o)$ and $(\zeta_1, \zeta_2, \ldots, \zeta_n)$ be a system of local coordinates around p such that $\zeta_1 = 0$ defines $\mu^{-1}(0)$ around p, μ induces a $\mathbb{C}$-algebra homomorphism $\mu^* : \mathcal{O}_{U,o} = B \longrightarrow \mathcal{O}_{V,p} = B'$. If $\{f_1, \ldots, f_\ell\}$ is a base of E, then the $\mu^* f_j$ are divided by ζ_1^m, i.e. $\mu^* f_j = \zeta_1^m g_j$ for certain $g_j \in B'$ and the vector space $E' = \sum \mathbb{C} g_j$ satisfies the former assumption, i.e. $\nu(g_j) = 0$ for a certain g_j. Hence $\rho(E')$ is defined and we define $\rho(E)$ to be $m\ell + \rho(E')$.

By this definition, every E is transformed into a distinguished space by a composition of blowing ups with point center as follows.

Case 1. If there is $f \in E \setminus (0)$ such that $\nu(f) = 0$, then there exist a sequence of complex manifolds V_j and blowing ups μ_j:

$$\begin{array}{ccccccc} V_{\ell-2} & \xrightarrow{\mu_{\ell-2}} & \cdots & \longrightarrow V_1 & \xrightarrow{\mu_1} & V_0 & = U \\ \Downarrow & & & \Downarrow & & \Downarrow & \\ p_{\ell-2} & & & p_1 = p & & p_0 = 0 & \end{array}$$

Case 2. If every $f \in E \setminus (0)$ satisfies that $\nu(f) > 0$, then there exist a sequence of complex manifolds V_j and blowing ups μ_j:

$$\begin{array}{ccccccc} V_{\ell-1} & \longrightarrow & V_{\ell-2} & \longrightarrow \cdots \longrightarrow & V_1 & \longrightarrow & V_0 = U \\ \Downarrow & & \Downarrow & & \Downarrow & & \Downarrow \\ p_{\ell-1} & & p_{\ell-2} & & p_1 = p & & p_0 = 0 \ . \end{array}$$

Letting λ be a composition of those μ_j, $\lambda^* E$ is the distinguished space. In fact, $\lambda^* E$ has a filtration of subspaces:

$$\lambda^*E = F_\ell \supset F_{\ell-1} \supset \cdots \supset F_1 \supset F_0 = \{0\}$$

which satisfies that

1) $\dim F_j = j$,

2) every $f_j \in F_{\ell-j+1} \setminus F_{\ell-j}$ satisfies that $\nu(f_j) < \nu(f_{j+1})$ for all $1 \leq j \leq \ell$.

Thus $\{f_1, \ldots, f_\ell\}$ is the distinguished base of λ^*E.

In this way,

$$\rho(E) \text{ is computed as } \sum \nu(f_j) - \ell(\ell-1)/2.$$

The morphism λ constructed above is said to be <u>distinguished</u>.

Remark. Let $\{\varphi_1, \ldots, \varphi_\ell\}$ be a base of E. Then $\omega(E)$ is a symmetric form defined to be the determinant of the matrix $[d^{i-1}\varphi_j]_{1 \leq i,j \leq \ell}$. And

$$\rho(E) = \rho_o(\omega(E)) \quad \text{by [2, Theorem 2].}$$

Here $\rho_o(\omega)$ is the index of the holomorphic symmetric form at o.

In order to obtain $\{\nu(f_1), \ldots, \nu(f_\ell)\}$ from the E-index, we introduce the j-th E-index $\rho^{(j)}(E)$ for $1 \leq j \leq \ell$. Define $\rho^{(j)}(E)$ to be $\min\{\rho(F) \mid F \subsetneq E \text{ with } \dim F = j\}$. Hence $\rho^{(\ell)}(E) = \rho(E)$. It is evidently true that if F is a more general subspace than a subspace F' with $\dim F = \dim F'$, then $\rho(F) \leq \rho(F')$. Hence, if one chooses a sequence of general subspaces: $E = E_\ell \supset E_{\ell-1} \supset \cdots \supset E_1 \supset (0)$ with $\dim E_j = j$, then $\rho^{(j)}(E) = \rho(E_j)$ for all $1 \leq j \leq \ell$.

We define $\{a_1, \ldots, a_\ell\}$ successively as follows.

$$a_1 = \rho^{(1)}(E), \quad a_1+a_2-1 = \rho^{(2)}(E), \ldots, \quad a_1+a_2+a_3-3 = \rho^{(3)}(E), \ldots$$

$$\sum_{j=1}^{\ell} a_j - \ell(\ell-1)/2 = \rho^{(\ell)}(E) = \rho(E).$$

$\{a_1, \ldots, a_\ell\}$ is called the <u>Weierstrass</u> E-<u>sequence</u> (or gap-sequence, in general) and is denoted by $\mathrm{seq}(E)$. The next result is obvious.

Proposition 3. If $\{f_1, \ldots, f_\ell\}$ is the distinguished base of E, then $\{\nu(f_1), \ldots, \nu(f_\ell)\}$ is the Weierstrass E-sequence.

Let U be an open neighborhood of o in $\mathbb{C}^n$, $f : V \to U$ be a birational morphism, and p be general point of $f^{-1}(o)$. f induces a $\mathbb{C}$-algebra homomorphism $f^* : B = \mathcal{O}_{U,o} \to B' = \mathcal{O}_{V,p}$.

Proposition 4. With the notation above, suppose E is a vector space of B. Then $\rho^{(j)}(E) = \rho^{(j)}(f^*E)$ for all $1 \leq j \leq \ell = \dim E$, hence $\mathrm{seq}(E) = \mathrm{seq}(f^*E)$.

Proof. By definition there exist two sequences $E_1 \subset E_2 \subset \cdots \subset E_\ell \subset E$ and $F_1 \subset F_2 \subset \cdots \subset F_\ell \subset f^*E$ of vector spaces such that $\rho^{(j)}(E) = \rho(E_j)$ and $\rho^{(j)}(f^*E) = \rho(F_j)$ for all $1 \leq j \leq \ell$. Since f^* is injective, there exists G_j ($\subset E$) satisfying $f^*G_j = F_j$ for all $1 \leq j \leq \ell$. By definition of $\{E_j\}$, one has

$$\rho(G_j) \geq \rho(E_j) \quad \text{for all } j,$$

and by the invariance of ρ one has

$$\rho(f^*E_j) = \rho(E_j).$$

On the other hand, $\rho(F_j) = \rho(f^*G_j) \geq \rho(G_j)$, since p may not be general with respect to $\{G_i\}$. Further, by definition of $\{F_j\}$, one has

$$\rho(f^*E_j) \geq \rho(F_j) \quad \text{for all } j.$$

Thus combining these inequalities, one obtains

$$\rho(E_j) = \rho(f^*E_j) \geq \rho(F_j) \geq \rho(G_j) \geq \rho(F_j)$$

and so $\rho(E_j) = \rho(F_j) = \rho(G_j)$ for all j. Hence, we have $\rho^{(j)}(E) = \rho^{(j)}(f^*E)$.

Let $\lambda : V_* \to V$ be a distinguished birational morphism and p_* be $p_{\ell-2}$ (Case 1) or $p_{\ell-1}$ (Case 2). λ induces $\lambda^* : B = \mathcal{O}_{U,o} \to \mathcal{O}_{V_*,p_*}$. Then λ^*E becomes a distinguished space and by Proposition 4, $\rho^{(j)}(\lambda^*E) = \rho^{(j)}(E)$ for all j and so $\mathrm{seq}(\lambda^*E) = \mathrm{seq}(E)$. Since λ^*E is distinguished,

$$\rho^{(j)}(\lambda^*E) = \sum_{i=1}^{j} \nu(f_i) - j(j-1)/2 \quad \text{for all} \quad j,$$

where $\{f_1, \ldots, f_\ell\}$ is the distinguished base of λ^*E. Therefore, $\mathrm{seq}(E) = \{\nu(f_1), \ldots, \nu(f_\ell)\}$.

Proposition 5. If F is a distinguished subspace of E with the distinguished base $\{\psi_1, \ldots, \psi_m\}$, then $\mathrm{seq}(F) = \{\nu(\psi_1), \ldots, \nu(\psi_m)\} \subseteq \mathrm{seq}(E)$.

Proof. Let $\lambda : V_* \to V$ be a distinguished birational morphism and use the previous notation. Then $\mathrm{seq}(F) = \mathrm{seq}(\lambda^*F)$ and $\mathrm{seq}(E) = \mathrm{seq}(\lambda^*E)$. Since λ^*E is distinguished, one has $\mathrm{seq}(\lambda^*F) \subseteq \mathrm{seq}(\lambda^*E)$. Hence $\mathrm{seq}(F) \subseteq \mathrm{seq}(E)$.

If E_1, E_2 are subspaces of E, then $E_1 \cdot E_2$ denotes a subspace spanned by $\{\varphi\psi \mid \varphi \in E_1, \psi \in F_2\}$.

Proposition 6. If $\alpha \in \mathrm{seq}\, E_1$ and $\beta \in \mathrm{seq}\, E_2$, then $\alpha + \beta \in \mathrm{seq}(E_1 \cdot E_2)$

Proof. By a composition of two suitable distinguished birational morphisms, we may assume that E_1 and E_2 are distinguished. Then $\{\alpha\} = \mathrm{seq}\, \mathbb{C}\varphi$ and $\beta = \{\mathrm{seq}\, \mathbb{C}\psi\}$ for some $\varphi \in E_1$ and $\psi \in E_2$. By the last proposition, $\mathbb{C}(\varphi\psi)$ is a distinguished subspace and so $\mathrm{seq}(\mathbb{C}(\varphi\psi)) \subset \mathrm{seq}(E_1 \cdot E_2)$. Since $\alpha + \beta = \nu(\varphi\psi)$, we see that $\alpha + \beta \in \mathrm{seq}(E_1 \cdot E_2)$.

Corollary. Let $\mathcal{E} = \oplus E_j$ be a graded algebra such that the E_j are vector spaces and the product is induced from B. Then both

$$SG(\mathcal{E}) = \bigcup \mathrm{seq}(E_j) \qquad \text{and}$$

$$SG_{II}(\mathcal{E}) = \bigcup \{(je-\alpha)_+ \mid \alpha \in \mathrm{seq}(E_j)\},$$

where $e = \min\{\nu(\varphi) \mid \varphi \in E_1 \setminus (0)\}$, become sub-semigroups of $\mathbb{N} \sqcup \{0\}$. Here x_+ denotes $\max\{x, o\}$.

$SG(\mathcal{E})$ and $SG_{II}(\mathcal{E})$ are called the <u>Weierstrass semigroup</u>

associated $\mathcal{E}$ and Weierstrass semigroup of the second kind associated with $\mathcal{E}$, respectively.

Example 3. Let $E = \mathbb{C}\varphi_1 + \mathbb{C}\varphi_2 \subset \mathbb{C}\{z_1, z_2\}$ where $\varphi_1 = z_1z_2$, $\varphi_2 = z_1^2 + z_2^2$. Then the E-sequence is $\{2, 3\}$. Similarly, the E^m-sequence is $\{2m, 2m+1, \ldots, 3m\}$. Hence $SG(\sum E^m)$ is $\{0, 2, 3, \ldots\}$.

Let Λ be a linear system on a compact complex manifold M corresponding to the ℓ-dimensional vector space $\mathbb{L}$ which is a subspace of $H^0(M, \mathcal{L})$ where $\mathcal{L}$ is an invertible sheaf on M. For any point $p \in M$, define

$$\rho_p^{(j)}(M, \Lambda) = \rho^{(j)}(\mathbb{L}_p) \quad \text{for } 1 \leq j \leq \ell$$

where $\mathbb{L}_p$ is the subspace of $\mathcal{O}_{M,p}$ spanned by germs of elements of $\mathbb{L}$ at p. Further, define

$$\mathrm{seq}_p(M, \Lambda) = \mathrm{seq}(\mathbb{L}_p).$$

Then we have the next result by Theorem 2.

Theorem 4. Let $f : M' \longrightarrow M$ be a birational morphism of compact complex manifolds, Λ be a linear system on M, and o be a point of M. Suppose that p is a general point of $f^{-1}(o)$. Then

$$\rho_p^{(j)}(M', f^*\Lambda) = \rho_o^{(j)}(M, \Lambda) \quad \text{for any } 1 \leq j \leq \ell$$

$= \dim \Lambda + 1$ and

$$\mathrm{seq}_p(M', f^*\Lambda) = \mathrm{seq}_o(M, \Lambda).$$

Finally, letting $(\oplus\mathbb{L}_p^m)$ be the graded algebra defined by $(\oplus\mathbb{L}_p^m)_j = \mathbb{L}_p^j$, we define the Weierstrass semigroups at p associated with Λ

$$SG_p(M, \Lambda) = SG(\oplus\mathbb{L}_p^m), \quad \text{and}$$

$$SG_{II,p}(M, \Lambda) = SG_{II}(\oplus\mathbb{L}_p^m).$$

For convenience, we put $SG(0) = \{0\}$ and consider a divisor D. Define the graded algebra $\mathcal{D}_p$ by $(\mathcal{D}_p)_j = H^0(M, \mathcal{O}(jD))$. Then the

<u>Weierstrass semigroup at</u> p <u>associated with</u> D is the semigroup defined by

$$SG_p^*(D) = SG_{II}(\mathcal{B}_p).$$

If D is a prime divisor Γ and p is a general point of Γ, then $SG_p^*(\Gamma)$ may play some role in the study of M and Γ. One may ask if Γ is a general member of some very ample complete linear system, then is the semigroup $SG_p^*(\Gamma)$ of section type ?

References

[1] S. Iitaka, Symmetric forms and Weierstrass cycles, Proc. Japan Acad., 54 (1978), 101-103.

[2] S. Iitaka, Weierstrass forms associated with linear systems, preprint.

University of Tokyo
Hongo, Tokyo
113 Japan

BIREGULAR THEORY OF FANO 3-FOLDS.

V. A. Iskovskih & V. V. Šokurov

0. Half a century ago, at the International Congress of Mathematicians in Bologna (1928), G. Fano laid down a programme for the study of biregular and birational properties of projective algebraic varities of dimension 3 with canonical curve-sections. Later he developed widely a theory of such varieties in numerous series of articles. His results were summarised by L. Roth [6]. The basic principles of the biregular theory of Fano 3-folds were laid down in [2]. Unfortunately there are many vague points and insufficiently founded statements in Fano's theory.

An attempt was made by the first author to reconstruct Fano's results. He took the ampleness of the anticanonical invertible sheaf as the fundamental property of Fano 3-folds. His results on biregular classification were obtained in [3], [4], [5] under the following two hypotheses:

a. If there exists an invertible sheaf $H \in \mathrm{Pic}\, V$ such that $H^r \simeq K_V^{-1}$ with a natural r, then the linear system $|H|$ contains a smooth surface H.

b. There exists a line on the anticanonical projective model of a Fano 3-fold, excluding Fano 3-holds of index ≥ 2 and $\mathbb{P}^1 \times \mathbb{P}^2$.

Both a) and b) have now been proved by the second author in [8], [9].

So the fundamental problems of the biregular theory of Fano 3-folds, at least in characteristic 0, have proved to be solved and in the main match with classical results.

1. Definition. A smooth complete irreducible algebraic variety V of dimension 3 over a field k will be called a Fano 3-fold if the anticanonical invertible sheaf K_V^{-1} on V is ample.

The field k will be assumed to be algebraically closed with char(k) = 0. The integer $g = g(V) = \frac{-K_V^3}{2} + 1$ will be called the genus of V. By $\Phi_L: V \to P^{\dim|L|}$ we denote the reational map corresponding to an invertible sheaf $L \in \mathrm{Pic}\, V$.

2. Proposition. Let H be an invertible sheaf such that $H^r \simeq K_V^{-1}$ with a natural r. Then the linear system $|H|$ contains

a smooth surface H. For $r = 1$ H is a K3 surface, and for $r = 2$ H is a del Pezzo surface (Th. 1.2 [8]; Cor. 1.5, 1.11 [3]).

3. Corollary. $\mathrm{Pic}\, V \simeq H^2(V,\mathbb{Z})$ is a torsion-free finitely generated Abelian group (Prop. 1.15 [3]).

4. Definition. The maximal integer $r \geq 1$ such that $H^r \approx K_V^{-1}$ for some invertible sheaf $H \in \mathrm{Pic}\, V$ is called the index of the Fano 3-fold V.

5. Proposition. The linear system $|H|$ on a Fano 3-fold V of index r, where $H^r \approx K_V^{-1}$ and $H \in \mathrm{Pic}\, V$, does not have base points, except in the following two cases:

(a) $r = 2$, $H^3 = 1$; $|H|$ has a unique base point;

(b) $r = 1$, H (see §2) contains smooth irreducible curves Z and Y, with Z a curve of genus 0, Y a fibre of an elliptic pencil $|Y|$ on H with $(Z.Y)_H = 1$, and $H_H = \mathcal{O}_H(Z+3Y)$ or $H_H \simeq \mathcal{O}_H(Z+4Y)$; $|H|$ has the unique base curve Z and has no other base points.

Every Fano 3-fold of the type (b) with $H_H \approx \mathcal{O}_H(Z+3Y)$ is the blow up of a Fano 3-fold V of the type (a) with centre a smooth elliptic curve, which is a fibre of the rational map $\Phi_H: V \to \mathbb{P}^2$. (Prop. 3.1, Th. 6.3 [3]; Th. 1.2 [8]). If $H_H \approx \mathcal{O}_H(Z+4Y)$, $V = F \times \mathbb{P}^1$, with F a Del-Pezzo surface of degree 1.

6. Theorem. Let V be a Fano 3-fold of index $r \geq 2$ with $H \in \mathrm{Pic}\, V$ and $H^r \approx K_V^{-1}$. Then:

(i) $r \leq 4$;

(ii) if $r \geq 3$ then we have
for $r = 4$: $\Phi_H: V \tilde{\to} \mathbb{P}^3$ is an isomorphism;
for $r = 3$: $\Phi_H: V \to V_2 \subset \mathbb{P}^4$ is an isomorphism of V with a smooth quadric of $\mathbb{P}^4$;

(iii) if $r = 2$ then a variety V only exists for $1 \leq d \overset{\text{def}}{=} H^3 \leq 7$; for $d \geq 3$, $\Phi_H: V \to V_d \subset \mathbb{P}^{d+1}$ is an embedding of V as a subvariety V_d of degree d in $\mathbb{P}^{d+1}$, with V_d projectively normal, and if $d \geq 4$, V_d is the intersection of the quadrics containing it; conversely, for any $d \geq 3$ every smooth projectively normal 3-fold $V_d \subset \mathbb{P}^{d+1}$, not lying in any hyperplane is a Fano 3-fold, and has index 2, apart from the case $r = 4$, $d = 8$, when V_8 is the image of $\mathbb{P}^3$ in $\mathbb{P}^9$ under the Veronese embedding;

(iv) if $r = 2$ and $3 \leq d \leq 7$ we have
for $d = 7$: V_7 is the projection of the Veronese 3-fold $V_8 \subset \mathbb{P}^9$ from some point of V_8;
for $d = 6$: $V_6 \simeq \mathbb{P}^1 \times \mathbb{P}^1 \times \mathbb{P}^1$ in its Segre embedding;
for $d = 5$: $V_5 \subset \mathbb{P}^6$ is unique up to projective equivalence, and can be realized in either of the two following ways:

(a) as the birational image of a quadric $W \subset \mathbb{P}^4$ under the map defined by the linear system $|\mathcal{O}_W(2) - Y|$ of quadrics passing through a twisted cubic Y;

(b) as the section of the Grassmanian $Gr(2,5)$ of lines in $\mathbb{P}^4$ by 3 hyperplanes in general position;

for $d = 4$: V_4 is any smooth intersection of two quadrics in $\mathbb{P}^5$;
for $d = 3$: V_3 is any smooth cubic hypersurface of $\mathbb{P}^4$;

(v) if $r = 2$ and $d = 1$ or 2 then
for $d = 2$: $\Phi_H: V \to \mathbb{P}^3$ is a double covering with smooth ramification surface $D_4 \subset \mathbb{P}^3$ of degree 4, and any such variety is a Fano 3-fold with $r = 2$ and $d = 2$; every Fano 3-fold with $r = 2$ and $d = 2$ can be realized as a smooth hypersurface of degree 4 in the weighted projective space $\mathbb{P}(x_0,\ldots,x_4)$, where x_4 has degree 2, and the remaining x_i have weight 1;
for $d = 1$: $\Phi_H: V \to \mathbb{P}^2$ is a rational map with a single point of indeterminacy, and with irreducible elliptic fibres, and V can be realized in either of the two following ways:

(a) $\Phi_{K_V}^{-1}: V \to W_4$ is any double cover of the cone W_4 over the Veronese surface $V_4 \subset \mathbb{P}^5$, having smooth ramifacation divisor $D \subset W_4$ cut out on W_4 by a cubic hypersurface not passing through the vertex of the cone;

(b) any smooth hypersurface of degree 6 in the weighted projective space $\mathbb{P}(x_0,\ldots,x_4)$, where x_0, x_1 and x_2 have degree 1, x_3 has degree 2, and x_4 degree 3.

(vi) If V is a Fano 3-fold with $r = 2$ and $d = 1$ or 2 then $\operatorname{Pic} V \simeq \mathbb{Z}$, with H as a generator.

(Th. 1.2 [8]; Prop. 1.12, Th. 4.2, Cor. 6.11 [3]).

7. Proposition. Let V be a Fano 3-fold, let r be the index of V, and let $H = \mathcal{O}_V(H)$ be an invertible sheaf for which $H^r \simeq K_V^{-1}$. Suppose that $|H|$ is without fixed components and base points. Let $\deg \Phi_H$ be the degree of the morphism $\Phi_H: V \to \Phi_H(V)$ ($\deg \Phi_H < \infty$ since H is ample); then $\deg \Phi_H = 1$ or 2 (Cor. 2.2 [3]).

8. Definition. A Fano 3-fold V of index $r = 1$ will be said to be hyperelliptic if its anticanonical map $\Phi_{K_V^{-1}}$ is a morphism and is of degree $\deg \Phi_{K_V^{-1}} = 2$.

9. Theorem. Let V be a hyperelliptic Fano variety, and let $\Phi_{K_V^{-1}}: V \to W \subset \mathbb{P}^{g+1}$ be the corresponding morphism of degree 2. Then W is non-singular and V is uniquely determined by the pair (W,D), where $D \subset W$ is the ramification divisor of $\Phi_{K_V^{-1}}$. The pair (W,D) belongs to one of the following families (and if D is a smooth divisor then for each pair (W,D) there exists a Fano 3-fold V):

(i) $W \simeq \mathbb{P}^3$, and D is smooth hypersurface of degree 6; in this case V can be realized alternatively as a smooth hypersurface of degree 6 in the weighted projective space $\mathbb{P}(x_0,\dots,x_4)$, where $\deg x_i = 1$ for $i = 0,\dots,3$ and $\deg x_4 = 3$.

(ii) $W \simeq V_2$ is a smooth quadric in $\mathbb{P}^4$ and $D \in |\mathcal{O}_{V_2}(4)|$, that is, $D = V_2 \cap V_4$, where V_4 is a quartic of $\mathbb{P}^4$. In this case V can also be realized as a smooth complete intersection in the weighted projective space $\mathbb{P}(x_0,\dots,x_5)$, where $\deg x_i = 1$ for $i = 0,\dots,4$, and $\deg x_5 = 2$: V is the intersection of a quadric cone and a hypersurface of degree 4:

$$F_2(x_0,\dots,x_4) = 0,$$
$$F_4(x_0,\dots,x_5) = 0.$$

(iii) $W \simeq \mathbb{P}(\mathcal{E})$, where $\mathcal{E} = \mathcal{O}_{\mathbb{P}^1}(d_1) \oplus \mathcal{O}_{\mathbb{P}^1}(d_2) \oplus \mathcal{O}_{\mathbb{P}^1}(d_3)$ is a locally free sheaf of rank 3 on $\mathbb{P}^1$, and $d_1 \geq d_2 \geq d_3$ are non-negative intergers; W is a rational scroll in the embedding

$$\Phi_M\colon W \to \mathbb{P}^{d_1+d_2+d_3+2} ,$$

where $M = \mathcal{O}_W(1)$ is the tautological invertible sheaf on W, and the following possibilities occur: $d_1 = d_2 = d_3 = 1$; then $W \simeq \mathbb{P}^2 \times \mathbb{P}^1$ in its Segre embedding, and $D \in |M^4 \otimes L^{-2}|$, where

$$M \simeq p_1^* \mathcal{O}_{\mathbb{P}^2}(1) \otimes p_2^* \mathcal{O}_{\mathbb{P}^1}(1), \quad \text{and} \quad L \simeq p_2^* \mathcal{O}_{\mathbb{P}^1}(1),$$

p_i denoting the projection of $\mathbb{P}^2 \times \mathbb{P}^1$ onto the i'th factor for $i = 1$ and 2;

$d_1 = 2$, $d_2 = d_3 = 1$; $W \subset \mathbb{P}^6$, $\deg W = 4$, and $D \in |\mathcal{O}_W(4)|$;

$d_1 = d_2 = d_3 = 2$; $W \simeq \mathbb{P}^2 \times \mathbb{P}^1$ and the embedding in $\mathbb{P}^8$

is given by the invertible sheaf $M \simeq p_1^* \mathcal{O}_{\mathbb{P}^2}(1) \otimes p_2^* \mathcal{O}_{\mathbb{P}^1}(2)$.

$D \in |p_1^* \mathcal{O}_{\mathbb{P}^2}(4)|$; in this case the 3-fold V is the product $H \times \mathbb{P}^1$, with H a smooth del Pezzo surface with $(K_H \cdot K_H) = 2$, that is a double plane with smooth ramifiacation curve of degree 4.

(iv) If V is a hyperelliptic Fano 3-fold, then $\operatorname{Pic} V \simeq \mathbb{Z}$ if and only if $W \simeq \mathbb{P}^3$ or $W \simeq V_2 \subset \mathbb{P}^4$.

(Th. 7.2, Cor 7.6 [3]).

<u>10. Definition.</u> A Fano 3-fold V will be said to be a variety of <u>the principal series</u> if its anticanonical invertible sheaf K_V^{-1} is very ample. By V_{2g-2} we denote the anticanonical model of such a V defined by the embedding $\Phi_{K_V^{-1}}\colon V \xrightarrow{\sim} V_{2g-2} \subset \mathbb{P}^{g+1}$.

<u>11. Proposition.</u> If a Fano 3-fold V of index 1 (for index > 1 see §6) is not hyperelliptic (for hyperelliptic 3-folds see §9) and the linear system $|K_V^{-1}|$ has no base points (see §5), then

(i) K_V^{-} is very ample; i.e. V is a Fano 3-fold of the principal series;

(ii) $V_{2g-2} \subset \mathbb{P}^{g+1}$ is projectively normal and is, for $g \geq 4$, the intersection of the quadrics and the cubics containing it.

(iii) V_{2g-2} is a smooth variety of degree $-K_V^3 = 2g - 2$ in $\mathbb{P}^{g+1}$, the hyperplane section of which are K3 surfaces, and the curve-sections of which are canonical curves $X_{2g-2} \subset \mathbb{P}^{g-1}$ of genus g.

(iv) Conversely, every smooth algebraic irreducible subvariety V_{2g-2} of dimension 3 and degree $2g - 2$ in $\mathbb{P}^{g+1}$, not lying in any hyperplane, and whose curve-sections $X_{2g-2} = V_{2g-2} \cap \mathbb{P}^{g-1}$ are canonical curves of genus g, will be a Fano 3-fold of the principal series, embedded in $\mathbb{P}^{g+1}$ by its anticanonical map $\Phi_{K^{-1}_{V_{2g-2}}}$

(Prop. 4.4, 1.6 [3]; Prop. 1.7 [4]).

<u>12. Proposition</u>. A Fano 3-fold $V_{2g-2} \subset \mathbb{P}^{g+1}$ of the principal series is a complete intersection only if $g = 3, 4, 5$, where

V_4 is a smooth quartic of $\mathbb{P}^4$,

$V_6 = V_{2.3}$ is a smooth complete intersection of a quadric and a cubic in $\mathbb{P}^5$,

$V_8 = V_{2.2.2}$ is a smooth complete intersection of three quadrics in $\mathbb{P}^6$.

Conversely, every smooth complete intersection indicated above is a Fano 3-fold of the principal series. (Prop 1.3 [4].)

<u>13. Definition</u>. A Fano 3-fold of the principal series $V_{2g-2} \subset \mathbb{P}^{g+1}$ will be called <u>trigonal</u> if its curve-sections $X_{2g-2} = V_{2g-2} \cap \mathbb{P}^{g-1}$ are trigonal canonical curves.

<u>14. Proposition</u>. If $V_{2g-2} \subset \mathbb{P}^{g+1}$ is a non-trigonal Fano 3-fold of the principal series, then V_{2g-2} is the intersection of the quadrics containing it (Prof. 1.7 [4]).

<u>15. Theorem</u>. Let $V_{2g-2} \subset \mathbb{P}^{g+1}$ be a trigonal Fano 3-fold with $g = 5$. Denote by W a closed subscheme of $\mathbb{P}^{g+1}$ which is the intersection of the quadrics containing V_{2g-2}. Then $W \simeq \mathbb{P}(\mathcal{E})$ is a rational scroll of dimension 4, where $\mathcal{E} \simeq \mathcal{O}_{\mathbb{P}^1}(d_1) \oplus \ldots \oplus \mathcal{O}_{\mathbb{P}^1}(d_4)$ is a locally free sheaf of rank 4 on $\mathbb{P}^1$, for non-negative integers $d_1 \geq d_2 \geq d_3 \geq d_4$. Only the following possibilities occur:

n^0	d_1	d_2	d_3	d_4	g	
1	1	1	1	1	6	$V_{10} \simeq V \subset \mathbb{P}^3 \times \mathbb{P}^1$; V: $t_0F_3 + t_1G_3 = 0$, where (t_0, t_1) are homogeneous coordinates of $\mathbb{P}^1$ and F_3, G_4 are cubic forms of $\mathbb{P}^3$.
2	2	1	1	1	7	V_{12} is the blow-up with centre a plane cubic on a smooth cubic hypersurface of $\mathbb{P}^4$.
3	2	2	2	2	10	$V_{18} \simeq S_3 \times \mathbb{P}^1$, where $S_3 \subset \mathbb{P}^3$ is a smooth cubic surface of $\mathbb{P}^3$.

(Th. 2.5, Prop. 2.3 [4]).

<u>16. Theorem</u>. Let $V_{2g-2} \subset \mathbb{P}^{g+1}$ be a Fano 3-fold of the principal series. Then we have one of the following alternatives:

(i) V_{2g-2} contains a line;

(ii) $V_{2g-2} \simeq \mathbb{P}^1 \times \mathbb{P}^2$; or

(iii) V_{2g-2} has index $r \geq 2$.

(Th. 1.2 [9]).

<u>17. Corollary</u>. There exists a line on every Fano 3-fold $V_{2g-2} \subset \mathbb{P}^{g+1}$ of index 1 of the principal series with $\operatorname{Pic} V \simeq \mathbb{Z}$. (Cor. 1.3 [9]).

<u>18. Definition</u>. A Fano 3-fold V will be said to be a variety of the <u>first species</u> if $\operatorname{Pic} V \simeq \mathbb{Z}$. So a Fano 3-fold of the first species is characterized by the conditions: $\operatorname{Pic} V \simeq \mathbb{Z}$ and $K_V^3 < 0$.

<u>19. Theorem</u>. Let $V_{2g-2} \subset \mathbb{P}^{g+1}$ be a Fano 3-fold of the principal series, of the first species and of index 1. There is a line on V_{2g-2} (see §17). Let $\pi_{2Z}: V_{2g-2} \to W \subset \mathbb{P}^{g-6}$ be a double projection (the rational map defined by the linear system $|-K_{V_{2g-2}} - 2Z|$) from a sufficiently general line $Z \subset V_{2g-2}$. Denote by E a hyperplane section of W. Then the following statements are true:

(i) $g \leq 12$;

(ii) if $g = 12$, then $W = W_5 \subset \mathbb{P}^6$ is a Fano 3-fold of the first species of index 2 and degree 5 (possibly with one singular point); the inverse map $\rho_Y: W \to V_{22}$ to π_{2Z} is defined by the linear system $|3E - 2Y|$ where $Y \subset W$ is a normal rational curve of degree 5 in $\mathbb{P}^5$;

(iii) there is no Fano 3-fold of the first species with $g = 11$;

(iv) if $g = 10$, then $W = W_2 \subset \mathbb{P}^4$ is a quadric and $\rho_Y: W \to V_{18}$ is defined by the linear system $|5E - 2Y|$ where Y is a smooth curve of genus 2 and degree 7 in $\mathbb{P}^4$;

(v) if $g = 9$, then $W = \mathbb{P}^3$ and $\rho_Y: \mathbb{P}^3 \to V_{16}$ is defined by the linear system $|7E - 2Y|$ where Y is a smooth curve of genus 3 and degree 7;

(vi) if $g = 8$, then $\pi_{2Z}: V_{14} \to \mathbb{P}^2$ is a rational map whose fibres (on resolving the indeterminacies) are curves of genus 2 and such that the inverse images of lines of $\mathbb{P}^2$ are rational surfaces;

(vii) if $g = 7$, then $\pi_{2Z}: V_{12} \to \mathbb{P}^1$ is a rational map whose general fibre (on resolving the indeterminacy) is a del Pezzo surface of degree 5 with eight points blown up; the variety $\dot{V}_{12}$ is rational and a projection from the line maps it onto a complete intersection of three quadrics containing a smooth rational scroll surface $R_3 \subset \mathbb{P}^4$ in $\mathbb{P}^6$.

(Th. 1.2 [9]; Th. 6.1 [4]).

20. Corollary. Any Fano 3-fold of the first species of index 1 with $g = 7$ or $g \geq 9$ is rational.

21. Table. Fano 3-folds of the first species [4].

n^0	r	H^3	g	p		§	unira-tional	rati-onal
1	4	1	33	0	$\mathbb{P}^3$	6	+	+
2	3	2	28	0	$Q_2 \subset P^4$ is a quadric	6	+	+
3	2	1	5	21	$V_1 \to W_4$ is a double cover of the Veronese cone	6	?	?*
4	2	2	9	10	$V_2 \to \mathbb{P}^3$ is a double cover with ramification in a quartic	6	+	?*
5	2	3	13	5	$V_3 \subset \mathbb{P}^4$ is a cubic	6	+	-
6	2	4	17	2	$V_{2.2} \subset \mathbb{P}^5$ is a complete intersection of two quadrics	6	+	+

n^0	r	H^3	g	p		§	unira-tional	rati-onal
7	2	5	21	0	$V_5 \subset \mathbb{P}^6$ is the section of the Grassmanian $Gr(2,5) \subset \mathbb{P}^9$ by a linear subspace of codimension 3	6	+	+
8	1	2	2	52	$V_2 \to \mathbb{P}^3$ is a double cover with ramification in a sextic	9	?	-
9	1	4	3	30	$V_4 \subset \mathbb{P}^4$ is a quartic	12	?**	-
10	1	4	3	30	$V_4' \to Q_2 \subset \mathbb{P}^4$ is a double cover of a quadric with ramification in a surface of degree 8	9	+	-
11	1	6	4	20	$V_{2.3} \subset \mathbb{P}^5$ is a complete intersection of a quadric and a cubic	12	+	-
12	1	8	5	14	$V_{2.2.2} \subset \mathbb{P}^6$ is a complete intersection of three quadrics	12	+	-
13	1	10	6	10	$V_{10} \subset \mathbb{P}^7$ is the section of the Grassmanian $Gr(2,5) \subset \mathbb{P}^9$ by a linear subspace of codimension 2 and a quadric	4	+	?*
14	1	12	7	5	$V_{12} \subset \mathbb{P}^8$	19	+	+
15	1	14	8	5	$V_{12} \subset \mathbb{P}^9$ is the section of the Grassmanian $Gr(2,6) \subset \mathbb{P}^{14}$ by a linear subspace of codimension 5	19	+	-
16	1	16	9	3	$V_{16} \subset \mathbb{P}^{10}$	19	+	+
17	1	18	10	2	$V_{18} \subset \mathbb{P}^{11}$	19	+	+
18	1	22	12	0	$V_{22} \subset \mathbb{P}^{13}$	19	+	+

r is the index of a Fano 3-fold V (see §4); for $H^3 = d$ see §2, §6; g is the genus of V (see §1); p is the dimension of the intermediate Jacobian of V.

*) Only the non-rationality of a "general" variety of the type indicated has been established [1].

**) The unirationality of some smooth quartics har been proved by B. Segre [7]. The unirationality of a general quartic is unknown.

22. Theorem. For every Fano 3-fold V we have $-K_V^3 \leq 64$.

Sketch proof [5]. The varieties considered in §§ 5,6,9,13 and $\mathbb{P}^1 \times \mathbb{P}^2$ satisfy the above inequality. Therefore one may assume that V is a non-trigonal Fano 3-fold of the principal series of index 1 and not isomorphic to $\mathbb{P}^1 \times \mathbb{P}^2$. In such a case there is a line on the anti-canonical model $V_{2g-2} \simeq V$ (see §16) and hence there is at least a 1-dimensional family of lines on V_{2g-2}. Using Schlessinger's theory of deformations we infer from the above that there is a 2-dimensional family of conics, a 3-dimensional family of curves of degree 3, a 4-dimensional family of curves of degree 4 on V_{2g-2} and so on, a 2-dimensional family of curves of degree 4 passing through a general point of V_{2g-2} and sweeping out the whole of V_{2g-2}.

Let $\sigma': V' \to V \simeq V_{2g-2}$ be the blow-up of V with centre a sufficiently general point $p \in V$, $P = \sigma^{-1}(p)$ and $H^* = \sigma^*(H)$ where $H \in |K_V^{-1}|$. Consider the exact sequence in cohomology

$$0 \to H^0(V', \mathcal{O}_{V'}(H^* - (n+1)P)) \to H^0(V', \mathcal{O}_{V'}(H^* - nP)) \to$$
$$\to H^0(P, \mathcal{O}_P((H^* - nP,P))) \to \dots$$

For $n = 4$ we have $h^0(V', \mathcal{O}_{V'}(H^* - 5P)) = 0$ since $H^* - 5P$ has a negative intersection with a proper inverse image of every curve of degree 4 passing through the point $p \in V$; $h^0(P, \mathcal{O}_P(H^* - 4P,P))) = 15$ since $\mathcal{O}_P((H^* - nP,P)) \simeq \mathcal{O}_P(nL)$ where L is a line on $P \simeq \mathbb{P}^2$. Consequently $h^0(V', \mathcal{O}_V'(H^* - 4P)) \leq 15$. Form this and the exact sequence for $n = 3,2,1,0$ we obtaine the inequality $h^0(V, \mathcal{O}_V(H)) = h^0(V, \mathcal{O}_{V'}(H^*)) \leq \sum_{n=0}^{4} h^0(P, \mathcal{O}_P((H^* - nP,P))) = \sum_{n=0}^{4} h^0(P, \mathcal{O}_P(nL)) = 35$, i.e. $g \leq 33$ or $-K_V^3 \leq 64$.

REFERENCES

1. Beauville A., Variétés de Prym et jacobiennes intermédiaires, Ann. Sci. Ecole norm. super., 1977, 10, No. 3, 309-391.

2. Fano G., Sulle varietà algebriche a tre dimensioni a curve-sezioni canoniche, Comm. Math. Helvetici, 14 (1941-1942), 23-64.

3. Исковских В.А., Трехмерные многообразия Фано I, Известия АН СССР Сер.Матем., Т.42, № 3, 1977, 516-562; English transl. Math. USSR Izvestija (1977).

4. Исковских В.А., Трехмерные многообразия Фано II, Известия АН СССР Сер.Матем., Т.42, № 3, 1978, 506-549.

5. Исковских В.А., Трехмерные многоoьразия Фано III, Известия АН СССР Сер.Матем. (в печати).

6. Roth L., Algebraic threefolds with special regard to problems of rationality, Springer, Berlin-Heidelberg-New York, 1955.

7. Segre B., Variazione continua ad omotopia in geometria algebrica, Ann. mat. pure ed appl., Ser. IV, L (1960), 149-186.

8. Шокуров В.В., Гладкость общего антиканонического дивизора на многообразии Фано, Известия АН СССР, Сер.Матем.(в печати).

9. Шокуров В.В., Существование прямой на многообразии Фано, Известия АН СССР Сер. Матем. (в печати).

Moscow State University
Department of Mathematics
and Mechanics
Moscow
USSR

SINGULARITES RATIONNELLES DU RESULTANT

J.P. JOUANOLOU

Ainsi que le titre l'indique, le propos de cette note, inspirée par une formule "générique" de LASKER ([2], p. 34, l. 24) , qui n'est valable qu'en caractéristique zéro, est d'expliciter aussi complètement que possible la dualité pour le morphisme de désingularisation canonique de la variété résultante, et par suite de décrire l'ensemble des singularités rationnelles de cette dernière.

1. - Rappels et notations.

Soit k un anneau commutatif noethérien, régulier et intègre pour simplifier ; le plus souvent k sera un corps ou l'anneau $\mathbb{Z}$. Etant données n indéterminées $X_1,\ldots,X_n$ et des entiers $d_i \geq 1$ $(1 \leq i \leq n)$, on note

$$f_i = \sum_{|\alpha| = d_i} U_{i,\alpha} X^\alpha$$

le polynôme homogène générique de degré d_i , et on désigne par A l'anneau des coefficients universels

$$A = k[U_{i,\alpha}] \quad (1 \leq i \leq n,\ |\alpha| = d_i) .$$

Par suite, le schéma $S = \operatorname{spec} A$ paramètre les familles $(H_1,\ldots,H_n)$ d'hypersurfaces de P_k^{n-1}, avec $d° H_i = d_i$. On rappelle que l'ensemble de ces familles ayant un point commun dans P_k^{n-1} est une hypersurface intègre, notée T , et définie par l'annulation d'un polynôme géométriquement irréductible à coefficients dans $\mathbb{Z}$, appelé <u>résultant</u> de $f_1, f_2, \ldots, f_n$ et noté R , qui ne dépend pas du choix de l'anneau de base k . En fait, cela ne détermine R qu'au signe près, mais il est possible de le normaliser ([4]) .

Plus précisément, posons

$$C = A[X_1, \ldots, X_n] \quad , \quad B = C / (f_1, \ldots, f_n)$$

et

$$Y = \mathrm{spec}\,(B) \ ,$$

de sorte qu'on a une immersion fermée canonique

$$S = \mathrm{spec}\,(A) \hookrightarrow \mathrm{spec}\,(B) = Y \quad (X_1 = X_2 = \ldots = X_n = 0) \ .$$

L'anneau B étant gradué, on pose $X = \mathrm{proj}\,(B) \subset P_S^{n-1}$, de sorte que

$$X = \{(u_{i,\alpha}) = u \in S \ , \ x \in P_k^{n-1} \mid f_i(u,x) = 0 \quad i = 1,2,\ldots,n\} \ .$$

Le schéma X étant projectif sur S et, comme nous allons le voir, intègre, son image schématique dans S est un sous-schéma fermé intègre de S , et c'est lui que nous avons noté T . On a donc un diagramme commutatif

$$(1.1) \qquad \begin{array}{ccc} X & \xhookrightarrow{\ j\ } & P_S^{n-1} = P \\ g \downarrow & & \downarrow f \\ T & \xhookrightarrow{\ i\ } & S \end{array}$$

dans lequel il est facile de vérifier que g est birationnel.

Précisons pour terminer ces rappels la structure de X . Tout d'abord, posant $V = Y \dot{-} S$, on a une projection canonique

$$V \longrightarrow X$$

qui fait de V le complémentaire de la section nulle du fibré en droites canonique sur X . Par ailleurs, on a une projection canonique

$$S \times (\mathbb{E}^n_k \div 0) \supset X \longrightarrow \mathbb{E}^n_k \div 0 = (\mathbb{E}^n_k)^*$$

$$(u,\xi) \longmapsto \xi$$

qui munit X d'une structure de fibré vectoriel sur $(\mathbb{E}^n_k)^*$. En fait, on a pour tout $i \in [1,n]$, un morphisme surjectif de fibrés vectoriels

$$\theta_i : (\mathbb{E}^n_k)^* \times S^{d_i}(k^n) \longrightarrow (\mathbb{E}^n_k)^* \times k$$

$$(\xi,g) \longmapsto (\xi,g(\xi))$$

au-dessus de $(\mathbb{E}^n_k)^*$ et, posant $F_i = \mathrm{Ker}(\theta_i)$, on voit aussitôt que

$$(1.2) \qquad V = \bigoplus_{i=1}^{n} F_i .$$

A partir de là, des considérations de dimension montrent que T est bien une hypersurface de S . Retenons aussi de ce qui précède que X est lisse sur k , donc constitue une désingularisation naturelle de T , que nous appellerons désingularisation canonique de T .

2. - Désingularisation de T et dualité.

Posons dans la suite

$$\delta = d_1 + d_2 + \ldots + d_n - n .$$

On vérifie tout d'abord l'assertion suivante :

PROPOSITION 2.1. - On a un quasi-isomorphisme (défini sur $\mathbb{Z}$ et pas seulement sur k)

$$g^! \mathcal{O}_T \simeq \mathcal{O}_X(\delta) .$$

Preuve : Comme i est un morphisme d'intersection complète globale défini par l'idéal RA , on a $i^! \mathcal{O}_S \simeq \mathcal{O}_T[1]$. De même, X est une intersection complète globale dans P , de faisceau normal N vérifiant

$$\Lambda^n N \simeq \mathcal{O}_X(-d_1 - d_2 \cdots - d_n) .$$

Pour tout $\mathcal{O}_P$- module localement libre F , on a donc

$$j^! F = j^* F \otimes_{\mathcal{O}_X} (\Lambda^n N)^\vee [n] \simeq j^* F \otimes_{\mathcal{O}_X} \mathcal{O}_X(d_1 + \ldots + d_n)[n]$$

En particulier ,

$$g^! \mathcal{O}_T = g^! i^! \mathcal{O}_S[-1] = j^! f^! \mathcal{O}_S[-1] = j^! \Omega^{n-1}_{P/S}[-n+1-1] ,$$

soit $\quad g^! \mathcal{O}_T \simeq \Omega^{n-1}_{P/S} \otimes_{\mathcal{O}_P} \mathcal{O}_X(d_1 + \ldots + d_n) \simeq \mathcal{O}_X(d_1 + \ldots + d_n - n) = \mathcal{O}_X(\delta)$.

Les isomorphismes ainsi construits lorsque $k = \mathbb{Z}$ fournissent par changement de base le quasi-isomorphisme annoncé.

Explicitons maintenant les foncteurs images directes supérieures $R^i g_*$. Comme X est intersection complète globale des hypersurfaces $f_i = 0$ dans P^{n-1}_S (cela se voit sans peine dans les cartes locales usuelles de P^{n-1}_S) , le complexe de KOSZUL défini par $(f_1, \ldots, f_n)$

$$0 \to \mathcal{O}_P(-d_1 \cdots - d_n) \to \cdots \to \mathcal{O}_P(-d_1) \oplus \cdots \oplus \mathcal{O}_P(-d_n) \xrightarrow[d_o]{(f_1, \ldots, f_n)} \mathcal{O}_P \to 0$$

est une résolution du $\mathcal{O}_P$- module $\mathcal{O}_X$. Posant

$$E = \mathcal{O}_P(-d_1) \oplus \cdots \oplus \mathcal{O}_P(-d_n) ,$$

on en déduit pour tout $a \in \mathbb{Z}$ une suite spectrale

$$(2.2) \quad E_1^{pq} = R^q f_* [(\Lambda^{-p} E)(a)] \Rightarrow E^{p+q} = R^{p+q} f_*(\mathcal{O}_X(a)) = i_* R^{p+q} g_*(\mathcal{O}_X(a)) ,$$

avec la convention $\Lambda^i E = 0$ pour $i < 0$.

Il est clair que $E_1^{pq} = 0$ si $p \notin [-n, 0]$ et la cohomologie des fibrés projectifs montre que $E_1^{pq} = 0$ si $q \neq 0, n-1$.

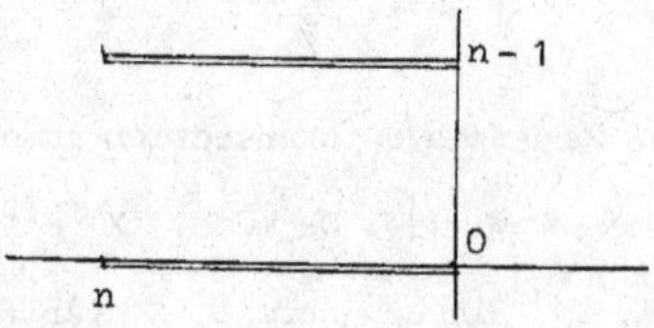

Comme $R^m g_* = 0$ pour $m < 0$, on déduit de (2.2) les conséquences suivantes.

2.3. Pour tout $a \in \mathbb{Z}$, la suite canonique

$$0 \to f_*\mathcal{O}_P(a-d_1 \ldots -d_n) \to \ldots \to f_*\mathcal{O}_P(a-d_1) \oplus \ldots \oplus f_*\mathcal{O}_P(a-d_n) \to f_*\mathcal{O}_P(a) \to \tilde{B}_a \to 0 ,$$

où B_a désigne la composante homogène de degré a du A-module gradué B , est exacte. Autrement dit, on retrouve le fait bien connu que $(f_1,\ldots,f_n)$ est une suite régulière de l'anneau $A[X_1,\ldots,X_n]$.

2.4. Pour $p \geq 2$, la cohomologie en degré p du complexe canonique

$$0 \to R^{n-1}f_*\mathcal{O}_P(a-d_1 \ldots -d_n) \xrightarrow{\delta^\circ} R^{n-1}f_*[(\Lambda^{n-1}E)(a)] \xrightarrow{\delta^1} \ldots \to R^{n-1}f_*E \to R^{n-1}f_*\mathcal{O}_P(a) \to 0$$

↑
degré 0

est isomorphe à $R^{p-1}g_*[\mathcal{O}_X(a))]$. Mais ce complexe est dual, d'après le théorème de dualité pour f , du complexe

$$0 \to f_*\mathcal{O}_P(-a-n) \to f_*\mathcal{O}_P(-a+d_1-n) \oplus \ldots \oplus f_*\mathcal{O}_P(-a+d_n-n) \to \ldots \to f_*\mathcal{O}_P(-a+\delta) \to 0 ,$$

qui n'est autre, par autodualité du complexe de Koszul, que le complexe déduit de (2.3) en remplaçant a par $-a+\delta$, et par suite est une résolution à gauche de $\tilde{B}_{-a+\delta}$. On en déduit :

(2.4.1) Pour $p \geq 1$,

$$R^p g_*\mathcal{O}_X(a) \simeq \operatorname{Ext}_A^{p+1}(B_{\delta-a},A)^\sim$$

(2.4.2) $$\operatorname{Ker} \delta^\circ \simeq (\check{B}_{\delta-a})^\sim .$$

2.5. On a une différentielle injective $s : \operatorname{Ker} \delta^\circ \to \tilde{B}_a$ et une suite exacte

$$0 \to \operatorname{Coker} s \to g_*\mathcal{O}_X(a) \to \operatorname{Ker} \delta^1/\operatorname{Im} \delta^\circ \to 0 ,$$

autrement dit, compte tenu de ce qui précède, on obtient une suite exacte canonique

(2.5.1) $$0 \to \check{B}_{\delta-a} \to B_a \xrightarrow{\text{can}} \Gamma(X,\mathcal{O}_X(a)) \to \mathrm{Ext}^1_A(B_{\delta-a},A) \to 0$$

Exemple 2.6. (a = 0) . Pour $i \geq 1$, on a

$$R^i g_*(\mathcal{O}_X) \simeq \mathrm{Ext}^{i+1}_A(B_\delta,A)^\sim .$$

De plus, on a une suite exacte

$$0 \to (\check{B}_\delta)^\sim \xrightarrow{s} \mathcal{O}_P \xrightarrow{\text{can}} f_*\mathcal{O}_X \to \mathrm{Ext}^1_A(B_\delta,A)^\sim \to 0 ,$$

i.e. $$0 \to \check{B}_\delta \xrightarrow{s} A \to \Gamma(X,\mathcal{O}_X) \to \mathrm{Ext}^1_A(B_\delta,A) \to 0 ,$$

Comme T est irréductible d'équation $R = 0$ dans S et g birationnel, on en conclut

(2.6.1) $$\check{B}_\delta \simeq A \quad \text{et} \quad \mathrm{Im}\, s = RA \quad \text{dans} \quad A .$$

D'autre part, comme X est normal et $f_*\mathcal{O}_X$ fini sur $\mathcal{O}_T$, le support de $\mathrm{Ext}^1_A(B_\delta,A)$ a pour complémentaire dans T l'ensemble des points normaux de T .

Exemple 2.7. $(a = \delta)$ Comme $B_o = A$ on a

$$R^i g_*\mathcal{O}_X(\delta) = 0 \quad (i \geq 1) ,$$

autrement dit, compte tenu de (2.1) , on voit que, dans cette situation particulière, le théorème de GRAUERT - RIEMENSCHNEIDER est valable en toutes caractéristiques. Cela nous permettra dans la suite d'interpréter le morphisme trace $\mathbb{R}g_*(g^!\mathcal{O}_T) \to \mathcal{O}_T$ comme un morphisme de $\mathcal{O}_T$- modules

$$g_*(\mathcal{O}_X(\delta)) \to \mathcal{O}_T ,$$

i.e. comme une application

(2.7.1) $$\Gamma(X,\mathcal{O}_X(\delta)) \to A/RA .$$

Par ailleurs, (2.5.1) fournit une suite exacte

(2.7.2) $$0 \to A \xrightarrow{s} B_\delta \to \Gamma(X,\mathcal{O}_X(\delta)) \to 0$$

que nous expliciterons plus loin.

<u>Remarque</u> 2.8. Il est clair que si $a < \inf(d_1,\ldots,d_n)$, $B_a = A[X_1,\ldots X_n]_a$, donc

$$R^i g_*(\mathcal{O}_X(\delta-a)) = 0 \quad (i \geq 1\ ,\ a < \inf(d_1,\ldots,d_n))\ ,$$

ce qui constitue une généralisation de (2.7).

2.9. Choisissons arbitrairement des polynômes $f_{ij} \in A[X_1,\ldots,X_n]$ tels que

(2.9.1) $$f_i = \sum_{j=1}^{n} X_j f_{ij}\ .$$

Comme $(X_1,\ldots,X_n)$ et $(f_1,\ldots,f_n)$ sont des suites régulières de C, avec $(f_1,\ldots,f_n) \subset (X_1,\ldots,X_n) = \mathfrak{M}$, il résulte d'un lemme de WIEBE [6] (voir aussi [5] (1.2)) que la classe Δ du déterminant

$$\begin{vmatrix} f_{11} & \cdots & f_{1n} \\ & & \\ f_{n1} & \cdots & f_{nn} \end{vmatrix}$$

dans B ne dépend pas du choix des f_{ij} vérifiant (2.9.1) et qu'elle constitue une A-base du sous-A-module de B formé des éléments de B annulés par $\mathfrak{M}$. On notera au passage qu'on peut normaliser les f_{ij} (c'est ce que fait HURWITZ [1]), en imposant que f_{ij} ne dépende que de $X_1,\ldots,X_j$ et alors le choix est unique.

Comme $\Gamma(X,\mathcal{O}_X(a)) = \Gamma(V,\mathcal{O}_V)_a$, il résulte de (2.5.1) que

(2.9.2) $$H^{\circ}_{\mathfrak{M}}(B)_a \simeq \check{B}_{\delta-a} \quad (a \in \mathbb{Z})\ .$$

En particulier,

$$H^{\circ}_{\mathfrak{M}}(B)_a = 0 \quad \text{pour} \quad a > \delta\ ,$$

d'où aussitôt

$$\mathfrak{M}\, H^{\circ}_{\mathfrak{M}}(B)_\delta = 0\ .$$

Comme Δ est homogène de degré δ,

$$H^{\circ}_{\mathfrak{M}}(B)_{\delta} = A\,\Delta .$$

Se ramenant à la situation $k = \mathbb{Z}$, on en déduit que, au signe près, la flèche s de (2.7.2) s'identifie à

(2.9.3)
$$\begin{aligned} \epsilon : A &\longrightarrow B_{\delta} \\ a &\longmapsto a\,\Delta \end{aligned}$$

Enfin, notant J la classe dans B_{δ} du jacobien

$$\det\left(\frac{\partial f_i}{\partial X_j}\right),$$

on vérifie que

(2.9.4)
$$J = d_1 \ldots d_n\,\Delta \text{ dans } B .$$

En effet, il s'agit de le vérifier lorsque $k = \mathbb{Z}$. Alors $J \in A\,\Delta$ et, comme J et Δ sont homogènes de même degré en les X_j et les $U_{i\alpha}$, il existe $m \in \mathbb{Z}$ tel que $J = m\,\Delta$, soit

$$\det\left(\frac{\partial f_i}{\partial X_j}\right) \equiv m \det(f_{ij}) \qquad \mathrm{mod}(X_1,\ldots,X_n) .$$

Il suffit alors de faire la spécialisation $f_i \longmapsto X_i^{d_i}$ pour obtenir (2.9.4).

Par ailleurs, comme X est un T-schéma, $R\Gamma(X,\mathcal{O}_X(a)) = 0$, d'où

(2.9.5)
$$R\,B_a \subset H^{\circ}_{\mathfrak{M}}(B)_a .$$

En particulier, la multiplication par $R : B_{\delta} \to B_{\delta}$ se factorise à travers $H^{\circ}_{\mathfrak{M}}(B)_{\delta} = \mathrm{Im}(\epsilon)$ (2.9.3), donc définit une application A-linéaire

(2.9.6)
$$\omega : B_{\delta} \to A$$

rendant le diagramme

$$\begin{array}{ccc} A & \xrightarrow{\epsilon} & B_\delta \\ & \searrow^{\times R} & \downarrow \omega \\ & & A \end{array}$$

commutatif, i.e. telle que $\omega(\Delta) = R$. En particulier, il résulte de (2.9.4) que

(2.9.8) $$\omega(J) = d_1 \ldots d_n R ,$$

égalité qui caractérise ω d'après (2.6.1) .

PROPOSITION 2.10. (i) <u>L'application transposée</u> $\check{\epsilon} : \check{B}_\delta \to A$ <u>est injective et a pour image</u> RA .

(ii) <u>L'application</u> ω <u>est une A-base de</u> $\check{B}_\delta$.

<u>Preuve</u> . Montrons (i) . La suite exacte (2.7.2) fournit par transposition une suite exacte

$$0 \to \check{B}_\delta \xrightarrow{\check{\epsilon}} A \to \mathrm{Ext}^1_A(\Gamma(X, \mathcal{O}_X(\delta)), A) ,$$

d'où résulte que $A/\mathrm{Im}(\check{\epsilon})$ est annulé par R . D'après (2.6.1) , l'idéal $\mathrm{Im}(\check{\epsilon})$ est principal. Notant a un générateur, on a donc $a|R$. Utilisant l'irréductibilité de R dans A , on voit que $\mathrm{Im}\,\check{\epsilon}$ ne peut être que A ou RA . Si $\check{\epsilon}$ était un isomorphisme, il existerait une application A-linéaire $\lambda : B_\delta \to A$ telle que

$$\lambda(\Delta) = 1 .$$

Or, notant $\mathfrak{a}$ l'idéal de A engendré par les coefficients des f_i , $\Delta \in \mathfrak{a} B_\delta$, d'où $\lambda(\Delta) \in \mathfrak{a}$, ce qui est absurde. On a donc bien $\mathrm{Im}(\check{\epsilon}) = RA$. Comme $\omega(\Delta) = R$, la définition de ϵ montre alors que ω est une base de $\check{B}_\delta$, d'où (ii) .

COROLLAIRE 2.11. <u>On a une suite exacte</u>

$$0 \to A/RA \to \mathrm{Ext}^1_A(\Gamma(X, \mathcal{O}_X(\delta)), A) \to \mathrm{Ext}^1_A(B_\delta, A) \to 0$$

<u>et des isomorphismes</u>

$$\mathrm{Ext}^i_A(\Gamma(X,\mathcal{O}_X(\delta)),A) \xrightarrow{\sim} \mathrm{Ext}^i_A(B_\delta,A) \qquad (i \geq 2) .$$

Pour être complet, signalons l'énoncé suivant et sa démonstration.

PROPOSITION 2.12. (i) <u>Pour</u> $i \geq \delta+1$, B_i <u>est un</u> A/RA - <u>module sans torsion</u>.

(ii) <u>Pour</u> $0 \leq i \leq \delta$, B_i <u>est un</u> A - <u>module sans torsion</u> <u>et l'accouplement</u>

(2.12.1)
$$B_i \underset{A}{\otimes} B_{\delta-i} \to B_\delta$$
$$x \otimes y \longmapsto xy$$

<u>est non dégénéré</u>.

<u>Preuve</u>. Lorsque $i \geq \delta+1$, on a $B_i \simeq \Gamma(X,\mathcal{O}_X(i))$ d'après (2.5.1) , d'où l'assertion (i) car X est intègre et g dominant. Montrons que B_δ est sans A-torsion. La suite exacte (2.7.2)

$$0 \to A \xrightarrow{\varepsilon} B_\delta \xrightarrow{\mathrm{can}} \Gamma(X,\mathcal{O}_X(\delta)) \to 0$$

montre que $\mathrm{Ass}_A(B_\delta) \subset \{(0),(R)\}$. Posant $\mathfrak{P} = (R)$, il s'agit de voir que $\mathfrak{P}$ n'est pas associé à B_δ . Si tel était le cas, il resterait une injection A-linéaire $\alpha : A/\mathfrak{P} \hookrightarrow B_\delta$ de sorte que, comme A est sans torsion, le composé $\gamma = \mathrm{can} \circ \alpha$ serait également injectif. Or, comme g est birationnel, $\Gamma(X,\mathcal{O}_X(\delta))_\mathfrak{P} \simeq A_\mathfrak{P}/\mathfrak{P}A_\mathfrak{P}$, de sorte que $\gamma_\mathfrak{P}$ serait un isomorphisme, et par suite $\varepsilon_\mathfrak{P}$ un monomorphisme direct. De 2.10 (i) résulterait alors que $\check{\varepsilon}_\mathfrak{P}$ serait bijective, donc R inversible dans $A_\mathfrak{P}$, ce qui est absurde. Pour terminer, il suffit de prouver la non-dégénérescence de (2.12.1) , car elle impliquera

$$B_i \subset \mathrm{Hom}_A(B_{\delta-i},B_\delta) \qquad (0 \leq i \leq \delta)$$

d'où l'absence de torsion de B_i , sachant que tel est le cas pour B_δ . Si $i = 0$ ou δ , l'assertion est déjà connue. Si $0 < i < \delta$, soit $x \in B_i$ tel que $x\, B_{\delta-i} = 0$. Il est clair que $\mathfrak{M}^{\delta-i}x = 0$, donc que

$x \in H^{o}_{\mathfrak{M}}(B)_i$. Or nous allons voir qu'on a un isomorphisme de C-modules gradués

(2.12.2) $$H^{o}_{\mathfrak{M}}(B) \simeq \mathrm{Ker}\,\{ H^n_{\mathfrak{M}}(C[-d_1\ldots -d_n]) \xrightarrow{\begin{pmatrix} f_1 \\ \vdots \\ f_n \end{pmatrix}} \bigoplus_i H^n_{\mathfrak{M}}(C[-d_1\ldots \hat{d}_i \ldots -d_n]) \}$$

L'assertion résultera alors de ce que, dans $H^n_{\mathfrak{M}}(C)$, l'accouplement

$$H^n_{\mathfrak{M}}(C)_{i-\delta-n} \otimes (\mathfrak{M}^{\delta-i}/\mathfrak{M}^{\delta-i+1}) \to H^n_{\mathfrak{M}}(C)_{-n}$$

est une dualité parfaite de A-modules (théorème de dualité pour $P^{n-1}_S \to S$) .

Montrons (2.12.1) . Notant $L^{\bullet}$ le complexe de Koszul

$$\begin{array}{ccccc} 0 \to C[-d_1\ldots -d_n] & \xrightarrow{\begin{pmatrix} f_1 \\ \vdots \\ f_n \end{pmatrix}} \ldots \to & C[-d_1] \oplus \ldots \oplus C[-d_n] & \xrightarrow{(f_1,\ldots,f_n)} & C \to 0 \\ \| & & \| & & \| \\ L^{-n} & & L^{-1} & & L^{o} \end{array}$$

qui est une résolution de B , on a deux suites spectrales

(2.12.3) $$'E_1^{p,q} = H^q_{\mathfrak{M}}(L^p) \Rightarrow E^{p+q}$$

(2.12.4) $$''E_2^{p,q} = H^p_{\mathfrak{M}}(H^q L^{\bullet}) \Rightarrow E^{p+q}$$

ayant même aboutissement. L'assertion en résulte aussitôt, compte tenu de ce que $H^j_{\mathfrak{M}}(C) = 0$ pour $j \neq n$.

Enonçons maintenant le résultat principal de ce paragraphe.

PROPOSITION 2.13. (i) <u>L'application</u> $\omega : B_\delta \to A$ <u>est injective</u>.

(ii) <u>Le diagramme</u>

$$\begin{array}{ccccccc} 0 \to A & \xrightarrow{\epsilon} & B_\delta & \xrightarrow{\mathrm{can}} & \Gamma(X,\mathcal{O}_X(\delta)) & \to 0 \\ \mathrm{id}\,\| & & \downarrow \omega & & \downarrow \mathrm{Tr} & \\ 0 \to A & \longrightarrow & A & \xrightarrow{\mathrm{can}} & A/RA & \to 0 \end{array} ,$$

dans lequel Tr est l'application trace de la dualité de Serre-Grothendieck (2.7.1), est commutatif au signe près (i.e. $\mathrm{Tr} \circ \mathrm{can} = \pm \mathrm{can} \circ \omega$) .

Remarque 2.13.1. L'indétermination de signe provient non seulement du choix de R , mais aussi de celui de l'isomorphisme (2.1) .

Preuve de (2.13) . Comme g est un morphisme birationnel entre schémas intègres, le morphisme Tr , qui est un isomorphisme au-dessus d'un ouvert de T , est injectif. L'assertion (i) résulte donc de (ii) . Pour voir (ii) , on peut se limiter au cas $k = \mathbb{Z}$, le cas général s'en déduisant par changement de base. Rappelons tout d'abord que le morphisme trace définit, pour tout $\mathcal{O}_X$- module cohérent F , un isomorphisme de dualité

$$\mathbb{R}g_*\mathbb{R}\,\mathcal{H}om_{\mathcal{O}_X}(F,\mathcal{O}_X(\delta)) \xrightarrow{\sim} \mathbb{R}\,\mathcal{H}om_{\mathcal{O}_T}(\mathbb{R}g_*F,\mathcal{O}_T)$$

composé du morphisme canonique

$$\mathbb{R}g_*\mathbb{R}\,\mathcal{H}om_{\mathcal{O}_X}(F,\mathcal{O}_X(\delta)) \to \mathbb{R}\,\mathcal{H}om_{\mathcal{O}_T}(\mathbb{R}g_*F,\mathbb{R}g_*(\mathcal{O}_X(\delta)))$$

et du morphisme

$$\mathbb{R}\,\mathcal{H}om(\mathrm{id},\mathrm{Tr}) : \mathbb{R}\,\mathcal{H}om_{\mathcal{O}_T}(\mathbb{R}g_*F,\mathbb{R}g_*\mathcal{O}_X(\delta)) \to \mathbb{R}\,\mathcal{H}om_{\mathcal{O}_T}(\mathbb{R}g_*F,\mathcal{O}_T) .$$

Compte tenu du "théorème de Grauert-Riemenschneider" (2.7) , on en déduit, pour $F = \mathcal{O}_X(\delta)$, des isomorphismes

(2.13.2) $$R^i g_*\mathcal{O}_X \xrightarrow{\sim} \mathrm{Ext}^i_{\mathcal{O}_T}(g_*\mathcal{O}_X(\delta),\mathcal{O}_T) ,$$

d'où en particulier, en passant aux sections globales et compte tenu de ce que T est affine, un isomorphisme de $\Gamma(X,\mathcal{O}_X)$ - modules

(2.13.3) $$\Gamma(X,\mathcal{O}_X) \xrightarrow{\sim} \mathrm{Hom}_{A/RA}(\Gamma(X,\mathcal{O}_X(\delta)),A/RA) ,$$

où $\Gamma(X,\mathcal{O}_X)$ opère dans le second membre au moyen de son opération naturelle sur $\Gamma(X,\mathcal{O}_X(\delta))$. Comme (2.13.3) envoie 1 sur Tr , ceci montre en particulier que toute application A-linéaire

$$v : \Gamma(X,\mathcal{O}_X(\delta)) \to A/RA$$

s'écrit de manière unique sous la forme

$$v = \mathrm{Tr} \circ (c_{\Gamma(X,\mathcal{O}_X)}) \ , \text{ avec } \ c \in \Gamma(X,\mathcal{O}_X) \ .$$

Ainsi, notant $\bar{\omega}$ l'unique application A-linéaire rendant le diagramme

$$\begin{array}{ccccccccc} 0 & \to & A & \xrightarrow{\varepsilon} & B_\delta & \xrightarrow{q} & \Gamma(X,\mathcal{O}_X(\delta)) & \to & 0 \\ & & \| & & \downarrow{\scriptstyle \omega} & & \downarrow{\scriptstyle \bar{\omega}} & & \\ 0 & \to & A & \xrightarrow{R} & A & \xrightarrow{p} & A/RA & \to & 0 \end{array}$$

commutatif, il existe $b \in \Gamma(X,\mathcal{O}_X)$ tel que

(2.13.4) $$\bar{\omega} = \mathrm{Tr} \circ (b_{\Gamma(X,\mathcal{O}_X)})$$

et il s'agit de voir que $b = \pm 1$. En tout cas, il résulte de (2.13.4) et de l'injectivité de Tr que $\bar{\omega}$, donc aussi ω , est injective. Ceci montre (i) . Retenons également que, comme $\bar{\omega} \neq 0$ (c'est génériquement un isomorphisme) , $RA \subsetneq \mathrm{Im}\,\omega$. Soit

(2.13.5) $$\begin{cases} B_\delta \xrightarrow[\omega]{\sim} \mathrm{Im}\,\omega \\ RA \subsetneq \mathrm{Im}\,\omega \end{cases}$$

Abordons maintenant la preuve de (ii) proprement dite. On a une suite exacte de A/RA - modules

(2.13.6) $$0 \to A/RA \xrightarrow{\text{can}} \Gamma(X,\mathcal{O}_X) \to \mathrm{Ext}^1_A(B_\delta,A) \to 0$$

déduite de (2.6) . D'autre part, la suite exacte des $\mathrm{Ext}^*_A(B_\delta,.)$ appliquée à $0 \to A \xrightarrow{R} A \to A/RA \to 0$ fournit une autre suite exacte

(2.13.7) $$\begin{array}{ccccccccc} 0 \to \check{B}_\delta & \xrightarrow{R} & \check{B}_\delta & \to & \mathrm{Hom}_A(B_\delta,A/RA) & \to & \mathrm{Ext}^1_A(B_\delta,A) & \to & 0 \\ \uparrow{\scriptstyle \wr\,\omega} & & \uparrow{\scriptstyle \wr\,\omega} & & & & & & \\ A & \xrightarrow{R} & A & & & & & & \end{array}$$

car $\mathrm{Ext}^1_A(B_\delta,A)$ est annulé par R. D'où une nouvelle suite exacte

$$(2.13.8)\qquad 0 \to A/RA \longrightarrow \mathrm{Hom}_A(B_\delta,A/RA) \to \mathrm{Ext}^1_A(B_\delta,A) \to 0\ .$$
$$1 \longmapsto p\circ\omega$$

LEMME 2.13.9. L'application

$$q^* : \mathrm{Hom}_A(\Gamma(X,\mathcal{O}_X(\delta)),A/RA) \to \mathrm{Hom}_A(B_\delta,A/RA)$$
$$u \longmapsto u\circ q$$

est un isomorphisme de A/RA - modules.

On a un diagramme commutatif exact

$$\begin{array}{ccccccc}
 & & & & A/RA & & \\
 & & & & \uparrow \varepsilon^* & & \\
0 \to A/RA & \xrightarrow{1\mapsto p\circ\omega} & \mathrm{Hom}_A(B_\delta,A/RA) & \to & \mathrm{Ext}^1_A(B_\delta,A) & \to & 0 \\
\| & & \uparrow q^* & & & & \\
A/RA & \xrightarrow{1\mapsto\bar\omega} & \mathrm{Hom}_A(\Gamma(X,\mathcal{O}_X(\delta)),A/RA) & & & & \\
 & & \uparrow & & & & \\
 & & 0 & & & &
\end{array}$$

qui montre que $\mathrm{Coker}\, q^*$ est un quotient de $\mathrm{Ext}^1_A(B_\delta,A)$. Or, d'après (2.13.5)

$$\mathrm{Ext}^1_A(B_\delta,A) \xrightarrow{\sim} \mathrm{Ext}^2_A(A/\mathrm{Im}\,\omega,\ A)$$

est annulé par $\mathrm{Im}\,\omega$. Par suite $\mathrm{Coker}\, q^*$, annulé par $\mathrm{Im}\,\omega \supsetneq RA$ (2.13.5) et contenu dans A/RA, est nul, d'où le lemme.

Le lemme (2.13.9) permet d'interpréter (2.13.8) comme une suite exacte

$$\begin{array}{cccccc}
(2.13.10)\quad 0 \to A/RA & \xrightarrow{1\mapsto\bar\omega} & \mathrm{Hom}_A(\Gamma(X,\mathcal{O}_X(\delta)),A/RA) & \to & \mathrm{Ext}^1_A(B_\delta,A) & \to 0 \\
 & & \text{dualité}\ \wr\uparrow & & \uparrow \mathrm{Tr} & \\
 & & & & 1 & \\
 & & \Gamma(X,\mathcal{O}_X) & & &
\end{array}$$

L'élément $b \in \Gamma(X,\mathcal{O}_X)$ défini par (2.13.4) entre donc dans une suite exacte

$$(2.13.11)\qquad 0 \to A/RA \xrightarrow{1\mapsto b} \Gamma(X,\mathcal{O}_X) \to \mathrm{Ext}^1_A(B_\delta,A) \to 0$$

La comparaison de (2.13.11) et (2.13.6) permet de définir une unique application A-linéaire φ rendant le diagramme

$$\begin{array}{ccccccc} 0 \to A/RA & \xrightarrow{1 \mapsto 1} & \Gamma(X,\mathcal{O}_X) & \to & \mathrm{Ext}^1_A(B_\delta,A) & \to 0 & \quad (2.13.6) \\ \| & & \downarrow b & & \downarrow \varphi & & \\ 0 \to A/RA & \xrightarrow{1 \mapsto b} & \Gamma(X,\mathcal{O}_X) & \to & \mathrm{Ext}^1_A(B_\delta,A) & \to 0 & \end{array} \tag{2.13.12}$$

commutatif, d'où une suite exacte

$$(2.13.13)\quad 0 \to \mathrm{Ext}^1_A(B_\delta,A) \xrightarrow{\varphi} \mathrm{Ext}^1_A(B_\delta,A) \to \Gamma(X,\mathcal{O}_X)/b\,\Gamma(X,\mathcal{O}_X) \to 0$$

Pour prouver que $b = \pm 1$, il suffit, comme X est un fibré vectoriel sur $P^{n-1}_{\mathbb{Z}}$, de montrer que b est inversible dans $\Gamma(X,\mathcal{O}_X)$. Supposons par l'absurde que ce ne soit pas vrai, et soit $\mathfrak{P}$ un idéal premier de hauteur 1 de $\Gamma(X,\mathcal{O}_X)$ contenant b . Comme $\Gamma(X,\mathcal{O}_X)$ est fini sur A/RA , $\mathfrak{q} = \mathfrak{P} \cap (A/RA)$ est un idéal premier de hauteur 1 de A/RA . Comme le A/RA - module $\mathrm{Ext}^1_A(B_\delta,A)$ est annulé par $\mathrm{Im}\,\bar{\omega} \neq 0$ (2.13.5) , $\mathrm{Ext}^1_A(B_\delta,A)_{\mathfrak{q}}$ est de longueur finie. Vu l'injectivité de $\varphi_{\mathfrak{q}}$, on en déduit que $\varphi_{\mathfrak{q}}$ est bijective, d'où (2.13.13)

$$\Gamma(X,\mathcal{O}_X)_{\mathfrak{q}} = b\,\Gamma(X,\mathcal{O}_X)_{\mathfrak{q}} ,$$

en contradiction avec le fait que $b \in \mathfrak{P}$.

COROLLAIRE 2.13.14. <u>L'application</u> ω <u>identifie</u> B_δ <u>à un idéal de</u> A <u>conte-nant</u> RA <u>et on a des isomorphismes canoniques</u>

$$\Gamma(X,\mathcal{O}_X(\delta)) \xrightarrow{\sim} \mathrm{Im}\,\omega/RA$$

$$\Gamma(X,\mathcal{O}_X) \xrightarrow{\sim} \mathrm{Hom}_A(\mathrm{Im}\,\omega/RA, A/RA) .$$

Le deuxième isomorphisme provient du théorème de dualité. Si on excepte le cas $d_1 = d_2 = \dots = d_n = 1$, il montre en particulier que $\mathrm{Im}\,\omega \neq A$, car sinon on aurait $A/RA \xrightarrow{\sim} \Gamma(X,\mathcal{O}_X)$, donc T serait normal, en contradiction avec le fait que le morphisme birationnel g a des fibres géométriques finies non réduites à un élément.

Remarque 2.13.15. Ayant normalisé R comme en [4] (nous reviendrons sur ce point au paragraphe suivant), nous choisirons désormais l'isomorphisme (2.1) de manière à rendre le diagramme (2.13) commutatif.

3. - Explicitation du morphisme ω et singularités rationnelles du résultant.

Afin d'expliciter ω, nous allons tout d'abord rappeler, en les généralisant un petit peu, certaines notations et résultats de MACAULAY [4].

3.1. Fixons une fois pour toutes l'ordre des variables $X_1,\dots,X_n$. Etant donnée une partie $K \subset [1,2,\dots,n]$, nous dirons qu'un polynôme $g \in A[X_1,\dots,X_n]$ est réduit par rapport aux X_i $(i \in K)$ si dans sa décomposition canonique

$$g = \sum_\alpha a_\alpha X^\alpha ,$$

on a $a_\alpha = 0$ sauf éventuellement si le multi-indice $\alpha = (\alpha_1,\dots,\alpha_n)$ vérifie

$$\alpha_i \le d_i - 1 \quad (i \in K) .$$

Pour tout $m \in \mathbb{Z}$, nous noterons

(3.1.1) $$R_K C_m$$

le A-module libre formé des $g \in C_m$ qui sont réduits par rapport aux X_i $(i \in K)$. Dans la suite, on prendra le plus souvent K de la forme $[1,p]$ $(1 \le p \le n)$, ou plus généralement de la forme $\sigma([1,p])$, avec $\sigma \in \mathfrak{S}_n$.

Dans l'énoncé suivant, on pose, pour tout $\nu \in [1,n]$,

(3.1.2) $$J^{[\nu]} = (f_1,\dots,f_\nu) \subset C , \quad J^{[\nu]}_m = J^{[\nu]} \cap C_m \quad (m \in \mathbb{Z}) .$$

PROPOSITION 3.2. (i) _Pour tout_ $t \in \mathbb{Z}$, _et tout_ $\nu \in [1,n]$, _l'application_

(3.2.1) $$\varphi_\nu : C_{t-d_1} \oplus R_{[1]} C_{t-d_2} \oplus R_{[1,2]} C_{t-d_3} \oplus \dots \oplus R_{[1,\nu-1]} C_{t-d_\nu} \xrightarrow{(f_1,\dots,f_\nu)} C_t$$

est injective, et $\mathrm{Im}(\varphi_\nu) \subset J^{[\nu]}_t$.

(ii) $J^{[\nu]}_t \cap R_{[1,\nu]} C_t = 0$

(iii) Notant L le corps des fonctions de l'anneau

$$A' = k\,[U_{i,\alpha}]_{1 \leq i \leq \nu}\,, \quad \text{on a}$$

(3.2.2) $$\mathrm{Im}(\varphi_\nu) \underset{A'}{\otimes} L = J_t^{[\nu]} \underset{A'}{\otimes} L$$

(3.2.3) $$(J_t^{[\nu]} \oplus R_{[1,\nu]} C_t) \underset{A'}{\otimes} L = C_t \underset{A'}{\otimes} L\,.$$

Preuve. Pour montrer (i) et (ii) il revient au même de prouver l'injectivité de l'application composée

(3.2.4) $$C_{t-d_1} \oplus R_{[1]} C_{t-d_2} \oplus \ldots \oplus R_{[1,\nu-1]} C_{t-d_\nu} \xrightarrow{\mathrm{pr} \circ \varphi_\nu} C_t / R_{[1,\nu]} C_t\,.$$

Comme les deux membres de (3.2.4) sont des A-modules libres de type fini et A est intègre, il suffit de montrer que l'application qui s'en déduit par la spécialisation

$$f_i \longmapsto X_i^{d_i}$$

est injective. Mais elle est même bijective, car tout monôme non réduit par rapport à l'ensemble des X_i $(i \in [1,\nu])$ s'écrit de manière unique sous la forme

$$X_1^{\alpha_1} \ldots X_\ell^{\alpha_\ell} \ldots X_n^{\alpha_n}$$

avec $\ell \in [1,\nu]$ et $\alpha_i \leq d_i - 1$ $(1 \leq i < \ell)$, $\alpha_\ell \geq d_\ell$; ceci montre d'ailleurs au passage que les deux membres de (3.2.4) ont le même rang. Pour montrer (iii), il suffit de prouver l'assertion analogue en remplaçant A par A' et C par $C' = A'[X_1,\ldots,X_n]$. Comme $f_1,\ldots,f_\nu$ est une suite régulière dans C', la dimension du L - espace vectoriel $(f_1,\ldots,f_\nu)_t \underset{A'}{\otimes} L$ est fournie par le même calcul de séries de Poincaré que celui qui donne la dimension de l'image de l'application déduite de

$$C'_{t-d_1} \oplus \ldots \oplus C'_{t-d_\nu} \xrightarrow{(f_1,\ldots,f_\nu)} C'_t$$

par la spécialisation $f_i \longmapsto X_i^{d_i}$, d'où par ce qui précède

$$\dim_L[(f_1,\ldots,f_\nu)_t \underset{A'}{\otimes} L] = \dim_L \mathrm{Im}(\varphi'_\nu \underset{A'}{\otimes} L) ,$$

où φ'_ν est l'application analogue de φ_ν. Comme l'image de φ'_ν est contenue dans $(f_1,\ldots,f_\nu)_t$, on en déduit (3.2.2). De même, compte tenu de (ii), le fait déjà signalé que $\mathrm{Im}(\varphi'_\nu \underset{A'}{\otimes} L)$ et $(C'_t/R_{[1,\nu]}C'_t) \underset{A'}{\otimes} L$ ont même dimension implique (3.2.3).

3.2.5. Pour toute permutation σ de $[1,n]$, on définit de même des applications A - linéaires

$$\varphi^\sigma_\nu : C_{t-d_{\sigma(1)}} \oplus R_{[\sigma(1)]}C_{t-d_{\sigma(2)}} \oplus\ldots\oplus R_{\sigma([1,\nu-1])}C_{t-d_{\sigma(\nu)}} \xrightarrow{(f_{\sigma(1)},\ldots,f_{\sigma(\nu)})} C_t$$

jouissant de propriétés analogues à celles de φ_ν.

Pour simplifier, lorsque $\nu = n$, nous poserons

$$\varphi^\sigma = \varphi^\sigma_n \quad , \quad J^{[n]} = J = (f_1,\ldots,f_n) .$$

3.3. Notant E le A-module libre de base les monômes en $X_1,\ldots,X_n$ qui ne sont pas réduits par rapport à l'ensemble des variables $X_1,\ldots,X_n$, on définit comme suit des isomorphismes

(3.3.1) $$u^\sigma : E \xrightarrow{\sim} C_{t-d_{\sigma(1)}} \oplus R_{[\sigma(1)]}C_{t-d_{\sigma(2)}} \oplus\ldots\oplus R_{\sigma([1,n-1]}\, C_{t-d_{\sigma(n)}} .$$

Tout monôme de la base canonique de E s'écrit de manière unique sous la forme

$$\mu = X^{\beta_1}_{\sigma(1)} X^{\beta_2}_{\sigma(2)} \ldots X^{\beta_\ell}_{\sigma(\ell)} \ldots X^{\beta_n}_{\sigma(n)} ,$$

avec $\beta_i \leq d_{\sigma(i)} - 1$ $(1 \leq i \leq \ell-1)$, et $\beta_\ell \geq d_{\sigma(\ell)}$. On pose alors

$$u^\sigma(\mu) = \mu / X^{d_{\sigma(\ell)}}_{\sigma(\ell)} \in R_{\sigma([1,\ell-1])}\, C_{t-d_{\sigma(\ell)}} .$$

Nous noterons v^σ le composé

(3.3.2) $$\varphi^\sigma \circ u^\sigma : E \to C_t ,$$

et poserons $\varphi^{id} = \varphi$, $u^{id} = u$, $v^{id} = v$.

Pour tout σ , le morphisme v^σ est injectif et (3.2) a génériquement pour image J_t . On en déduit une factorisation <u>générique</u>

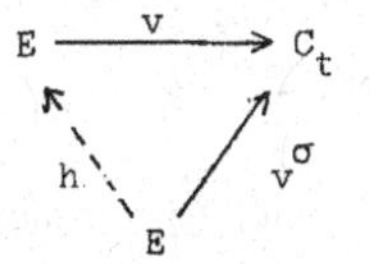

LEMME 3.3.3. <u>La matrice, à coefficients rationnels en les</u> $U_{i,\alpha}$, <u>de</u> h <u>dans la base canonique de</u> E , <u>a des coefficients indépendants de ceux de</u> f_n .

<u>Preuve.</u> Il s'agit de montrer que tout élément de C_t de la forme

(3.3.4) $$y = g_{\sigma(1)}\, f_{\sigma(1)} + \dots + g_{\sigma(n)}\, f_{\sigma(n)} ,$$

où les $g_{\sigma(i)}$ sont réduits par rapport à $X_{\sigma(1)},\dots,X_{\sigma(i-1)}$, peut se mettre sous la forme

(3.3.5) $$\tilde{g}_1\, f_1 + \dots + \tilde{g}_n\, f_n ,$$

avec les $\tilde{g}_i$ réduits par rapport à $X_1,\dots,X_{i-1}$, au moyen de transformations linéaires ne faisant pas intervenir les coefficients de f_n . En fait, nous allons voir que c'est vrai sans tenir compte de l'hypothèse de réduction sur les g_i . Tout d'abord, d'après (3.2.3) , on peut, en inversant seulement les polynômes en les $U_{i,\alpha}$ $(i \neq n)$, écrire de manière unique

$$f_n = k_1 f_1 + \dots + k_{n-1}\, f_{n-1} + \ell ,$$

avec les k_i et ℓ homogènes, k_i réduit en $X_1,\dots,X_{i-1}$ et ℓ réduit en $X_1,\dots,X_{n-1}$. Portant dans (3.3.4) , cela permet d'écrire

(3.3.6) $$y = g'_1\, f_1 + \dots + g'_n\, f_n ,$$

avec g'_n réduit en $X_1,\dots,X_{n-1}$. Utilisant à nouveau (3.2.3) pour $\nu = n-2$, on écrit, en inversant seulement les polynômes en les coefficients de $f_1,\dots,f_{n-2}$,

$$f_{n-1} = k'_1\, f_1 + \dots + k'_{n-2}\, f_{n-2} + \ell' ,$$

où les k_i' et ℓ' sont homogènes et, en particulier, ℓ' réduit en $X_1,..X_{n-2}$. Portant dans (3.3.6), on obtient pour y une expression analogue, avec cette fois g'_{n-1} réduit en $X_1,\dots,X_{n-2}$. Ainsi de suite, on obtient de proche en proche le résultat annoncé.

3.4. Pour les définitions qui suivent, nous supposerons $k = \mathbb{Z}$, le cas général s'en déduisant par le morphisme canonique $\mathbb{Z} \to k$.

Pour tout σ, nous noterons

$$D^{\sigma}(n,t)$$

le déterminant, défini seulement au signe près, faute d'avoir choisi un ordre sur les bases, de l'application

$$\bar{v}^{\sigma} : E \xrightarrow{v_{\sigma}} C_t \to C_t / R_{[1,n]} C_t ,$$

dans les bases canoniques de ces deux A-modules (à vrai dire, comme on travaille sur $\mathbb{Z}$, n'importe quelle base convient tout aussi bien). Avec MACAULAY [3], nous poserons

(3.4.1) $$R(n,t) = \operatorname{pgcd}_{\sigma} D^{\sigma}(n,t) \quad \text{(défini au signe près)}.$$

PROPOSITION 3.4.2. Le polynôme $R(n,t)$ est homogène en les coefficients de chacun des f_i. Son degré d'homogénéité par rapport aux coefficients de f_i est égal au coefficient c_i de x^{t-d_i} dans le développement en série de

$$\frac{(1-x^{d_1}) \dots \widehat{(1-x^{d_i})} \dots (1-x^{d_n})}{(1-x)^n} .$$

Preuve. Les colonnes de la matrice de $\bar{v}^{\sigma}$ sont homogènes (de degré 1) en les coefficients de l'un des f_i et ne dépendent pas des autres. D'où la multi-homogénéité de $D^{\sigma}(n,t)$, et partant celle de $R(n,t)$. Par ailleurs, une généralisation immédiate de (3.3.3) montre que, pour tout couple $\sigma, \sigma' \in \mathfrak{S}_n$, on a

$$v^{\sigma} = v^{\sigma'} \circ w ,$$

où w est une application linéaire à coefficients rationnels indépendants de ceux de $f_{\sigma'(n)}$, donc que

$$D^{\sigma}(n,t) \,/\, D^{\sigma'}(n,t) = \text{dét}(w)$$

est une fraction rationnelle indépendante des coefficients de $f_{\sigma'(n)}$. Choisissant σ' tel que $\sigma'(n) = i$, on en déduit aussitôt que $R(n,t)$ et $D^{\sigma'}(n,t)$ ont même degré par rapport aux coefficients de f_i, à savoir, vu la définition de $\bar{V}^{\sigma'}$,

$$\dim_{\mathbb{Z}} R_{\sigma'[1,n-1]} C_{t-d_i} ,$$

qui est le nombre de monômes homogènes de degré $t-d_i$ réduits par rapport à $X_1,\dots,\hat{X}_i,\dots,X_n$, i.e. le coefficient de x^{t-d_i} dans

$$(1+x+x^2+\dots)(1+x+\dots+x^{d_1-1})\dots\widehat{(1+x+\dots+x^{d_i-1})}\dots(1+x+\dots+x^{d_n-1}) .$$

COROLLAIRE 3.4.3. <u>Choisissant pour tout</u> i <u>une permutation</u> σ_i <u>telle que</u> $\sigma_i(n) = i$, on a

$$R(t,n) = \operatorname*{pgcd}_{i \in [1,n]} D^{\sigma_i}(n,t)$$

En particulier, notant τ la permutation circulaire $(1,2,\dots,n) \longmapsto (2,3,\dots,n,1)$, et posant

$$D^{\tau^p}(n,t) = D^{(p)}(n,t) ,$$

on a

(3.4.4) $$R(n,t) = \operatorname*{pgcd}_{0 \le p \le n-1} D^{(p)}(n,t) ,$$

qui est la définition de [3].

<u>Exemple</u> 3.4.5. MACAULAY prend pour définition du résultant de $f_1,\dots,f_n$

$$R = R(n,\delta+1) .$$

La spécialisation $f_i \longmapsto U_i X_i^{d_i}$ envoie $D(n,t)$ sur

$$\pm\, U_1^{N_1} \dots U_n^{N_n} ,$$

où $N_i = \dim_{\mathbb{Z}} R_{[1,i-1]} \mathbb{C}_{t-d_i}^{d_i}$, d'où résulte, en notant encore pour simplifier U_i le coefficient de X_i dans f_i, que le coefficient de $U_1^{c_1} \ldots U_n^{c_n}$ (3.4.2) dans $R(n,t)$ est ± 1. On normalise habituellement $R = R(n,\delta+1)$ en imposant que ce coefficient soit $+1$. Enfin, il résulte aussitôt de (3.4.2) que R a pour degré

$$d_1 \ldots \hat{d}_i \ldots d_n$$

par rapport aux coefficients de f_i, donc pour degré total

$$d_1 \ldots d_n \left(\sum_{i=1}^{n} \frac{1}{d_i} \right).$$

Exemple 3.4.6. Le polynôme $R(n,\delta)$ a pour degré d'homogénéité

$$d_1 \ldots \hat{d}_i \ldots d_n - 1$$

par rapport aux coefficients de f_i. Il s'agit en effet de déterminer le nombre de suites $(m_1,\ldots,m_n)$ d' entiers vérifiant

$$\begin{cases} 0 \leq m_j \leq d_j - 1 \quad \text{pour} \quad j \neq i & (*) \\ m_1 + \ldots + m_n = d_1 + \ldots + \hat{d}_i + \ldots + d_n - n & (**) . \end{cases}$$

Or chaque choix de m_j vérifiant $(*)$ définit de manière unique un m_n excepté lorsque $m_1 + \ldots + \hat{m}_i + \ldots + m_n \geq d_1 + \ldots + \hat{d}_i + \ldots + d_n - n + 1$, ce qui n'a lieu que lorsque $m_j = d_j - 1$ $(j \neq i)$.

3.4.7. Dans le pas particulier $t = \delta$, $R_{[1,n]} C_\delta$ a pour base $X_1^{d_1-1} \ldots X_n^{d_n-1}$, donc

$$\dim C_\delta = \dim(E) + 1 .$$

Pour tout multi-indice α, avec $\sum_i \alpha_i = \delta$, nous noterons

$$D_\alpha^\sigma (n,\delta)$$

le mineur extrait de la matrice de v^σ en ôtant la ligne correspondant à l'élément X^α de la base canonique de C_δ et nous poserons pour simplifier

$$D_\alpha(n,\delta) = D_\alpha^{id}(n,\delta) .$$

Il est clair que

$$D_\alpha^\sigma(n,\delta) \not\equiv 0 \qquad \text{(spécialiser)} .$$

Pour tout couple $\sigma, \sigma' \in \mathfrak{S}_n$, la factorisation

$$v^\sigma = v^{\sigma'} \circ w$$

utilisée en (3.4.2) montre que, pour tout α ,

$$(3.4.8) \qquad \frac{D_\alpha^\sigma(n,\delta)}{D_\alpha^{\sigma'}(n,\delta)} = \frac{D^\sigma(n,\delta)}{D^{\sigma'}(n,\delta)} = \det(w) .$$

Les égalités déduites de (3.4.8)

$$\frac{D_\alpha^\sigma(n,\delta)}{D^\sigma(n,\delta)} = \frac{D_\alpha(n,\delta)}{D\ (n,\delta)}$$

montrent que, posant

(3.4.9) $\quad R_\alpha(n,\delta) = \operatorname{pgcd}_\sigma D_\alpha^\sigma(n,\delta)$ (défini au signe près) , on a

$$(3.4.10) \qquad \frac{D_\alpha^\sigma(n,\delta)}{D^\sigma(n,\delta)} = \frac{D_\alpha(n,\delta)}{D(n,\delta)} = \pm \frac{R_\alpha(n,\delta)}{R(n,\delta)} \qquad (\sigma \in \mathfrak{S}_n) .$$

On en déduit, en particulier, que $R_\alpha(n,\delta)$ et $R(n,\delta)$ ont même degré d'homogénéité par rapport aux coefficients des divers f_i , et que, choisissant pour tout i une permutation σ_i telle que $\sigma_i(n) = i$, on a

$$(3.4.11) \qquad R_\alpha(n,\delta) = \operatorname{pgcd}_i D_\alpha^{\alpha_i}(n,\delta) .$$

Dans l'énoncé suivant et sa preuve, on pose par abus de notation

$$u(X^\alpha) = u(\overline{X^\alpha}) \quad (u \in \check{B}_\delta) .$$

PROPOSITION 3.5. <u>On peut choisir les déterminations de</u> $R(n,\delta)$ <u>et</u> $R_\alpha(n,\delta)$ <u>de telle sorte que l'application</u> ω (2.9.6) <u>vérifie les égalités</u>

(3.5.1) $$\omega(X^{\alpha}) = R_{\alpha}(n,\delta) \qquad (|\alpha| = \delta) .$$

Remarquons que (3.5.1) détermine entièrement ω . Par ailleurs, nous utiliserons désormais les déterminations indiquées pour $R(n,\delta)$ et les $R_{\alpha}(n,\delta)$.

Preuve de 3.5. On peut supposer $k = \mathbb{Z}$. Par transposition à partir du complexe de KOSZUL associé à $(f_1,\ldots,f_n)$, on obtient une suite exacte

(3.5.2) $$0 \to \check{B}_{\delta} \longrightarrow \check{C}_{\delta} \xrightarrow{\begin{pmatrix} f_1 \\ \vdots \\ f_n \end{pmatrix}} \check{C}_{\delta-d_1} \oplus \ldots \oplus \check{C}_{\delta-d_n} ,$$

$$u \longmapsto (u(X^{\alpha}))_{|\alpha| = \delta}$$

où l'on munit les $\check{C}_j$ de bases duales de celles choisies pour les C_j . Notons M la matrice de l'application

(3.5.3) $$C_{\delta-d_1} \oplus \ldots \oplus C_{\delta-d_n} \xrightarrow{(f_1,\ldots,f_n)} C_{\delta} .$$

Comme $\check{B}_{\delta}$ a pour base ω ,

$$\tilde{\omega} = (\omega(X^{\alpha}))_{|\alpha| = \delta}$$

est caractérisé par les conditions

(3.5.4) $$\begin{cases} {}^{t}M(\tilde{\omega}) = 0 \\ \\ \operatorname{pgcd}_{\alpha} \omega(X^{\alpha}) = 1 . \end{cases}$$

Pour tout $\sigma \in \mathfrak{S}_n$, l'application v^{σ} (3.3.2) se factorise à travers (3.5.3), d'où, notant Q^{σ} la matrice de v^{σ} , l'égalité

(3.5.5) $${}^{t}Q^{\sigma}(\tilde{\omega}) = 0 .$$

On en déduit

$$\frac{\omega(X^{\alpha})}{D^{\sigma}_{\alpha}(n,\delta)} = \pm \frac{\omega(X^{\beta})}{D^{\sigma}_{\beta}(n,\delta)} \qquad (|\alpha| = |\beta| = \delta) ,$$

d'où aussitôt

$$\frac{\omega(X^{\alpha})}{R_{\alpha}(n,\delta)} = \pm \frac{\omega(X_1^{d_1-1} \cdots X_n^{d_n-1})}{R(n,\delta)} \qquad (|\alpha| = \delta) .$$

Une fois choisie une détermination de $R(n,\delta)$, il est donc possible de choisir les $R_{\alpha}(n,\delta)$ de sorte que

(3.5.6)
$$\frac{\omega(X^{\alpha})}{R_{\alpha}(n,\delta)} = \frac{\omega(X_1^{d_1-1} \cdots X_n^{d_n-1})}{R(n,\delta)} ,$$

ce que nous ferons désormais. De plus, soit

$$\Delta = \sum_{|\alpha| = \delta} K_{\alpha} \overline{X}^{\alpha}$$

une décomposition de Δ (2.9) dans B_{δ} . L'application ω est caractérisée par

$$R = \omega(\Delta) = \sum_{|\alpha| = \delta} K_{\alpha}\omega(X^{\alpha}) ,$$

soit, notant W le deuxième membre de (3.5.6) ,

(3.5.7)
$$R = W(\sum_{|\alpha| = \delta} K_{\alpha}R_{\alpha}(n,\delta)) .$$

Par rapport à la totalité des coefficients des f_i , on a les homogénéités :

$$\begin{array}{lll} R & : & \sum_i d_1 \cdots \hat{d}_i \cdots d_n \\ R_{\alpha}(n,\delta) & : & \sum_i d_1 \cdots \hat{d}_i \cdots d_n - n \\ K_{\alpha} & : & n \end{array} .$$

Par suite, écrivant $W = S/T$, avec $\mathrm{pgcd}(S,T) = 1$, les polynômes S et T sont homogènes et

(3.5.8)
$$d^{\circ}(S) = d^{\circ}(T) \leq d^{\circ}R(n,\delta) .$$

Comme R est premier et $d^{\circ}(S) < d^{\circ}(R)$, l'égalité

$$TR = S(\sum_{|\alpha| = \delta} K_{\alpha}R_{\alpha}(n,\delta))$$

montre alors que $W = \pm 1$. Il est clair alors qu'on peut choisir $R(n,\delta)$, et partant les $R_\alpha(n,\delta)$ définis par (3.5.6), de manière à ce que $W = 1$, d'où l'assertion.

COROLLAIRE 3.5.9. (i) On a l'égalité

$$R = \sum_{|\alpha| = \delta} K_\alpha R_\alpha(n,\delta) .$$

(ii) Notant $J(f_1,\ldots,f_n) = \sum_{|\alpha| = \delta} J_\alpha X^\alpha$ la décomposition du jacobien dans C_δ, on a

$$\sum_{|\alpha| = \delta} J_\alpha R_\alpha(n,\delta) = d_1 d_2 \ldots d_n R .$$

Preuve. L'assertion (i) vient d'être démontrée. Quant à (ii), elle en résulte grâce à (2.9.4).

COROLLAIRE 3.5.10. Les polynômes $R_\alpha(n,\delta)$ ($|\alpha| = \delta$) sont premiers dans leur ensemble.

Preuve. N'est autre que la deuxième partie de (3.5.4).

3.6. Faisons le bilan de ce paragraphe. D'après (2.13), le lieu des singularités pires que rationnelles (définies en un sens évident, qui n'est a priori intrinsèque qu'en caractéristique 0) de la variété résultante T est $V(\overline{\omega})$. Nous venons de voir qu'il est défini par les équations

$$R_\alpha(n,\delta) = 0 \quad (|\alpha| = \delta)$$

dans T, mais aussi dans S, car l'équation $R = 0$ résulte des précédentes (3.5.9 (i)). Plus précisément,

$$\text{(3.6.1)} \qquad \operatorname{Im}(\omega) = \sum_{|\alpha| = \delta} A R_\alpha(n,\delta) .$$

4. - Exemples et questions.

Il semble en général difficile de préciser davantage la nature de $V(\omega)$, ou celle de $T^{rat} = T \doteq V(\omega)$, faute de savoir bien expliciter R.

Signalons toutefois quelques cas où cela est possible.

4.1. <u>Cas où</u> $d_1 = d_2 = \dots = d_n = 1$.

Changeant légèrement de notations, posons

$$f_i = u_{i1}X_1+\dots+ u_{in}X_n \ .$$

On sait alors que

$$R = \det(u_{ij}) = J(f_1,\dots,f_n) \ ,$$

de sorte que, vu la caractérisation (2.9.8) , l'application

$$\omega : \ B_o = A \to A$$

est l'identité. On retrouve ainsi le fait bien connu que la variété T d'idéal $\det(u_{ij})$ dans $S = \operatorname{spec} A$ est à singularités rationnelles, donc normale. Rappelons qu'elle n'est pas régulière. Supposant pour simplifier que k est un corps, le lieu singulier T^{sing} , formé des matrices (u_{ij}) de rang $\leq n-2$, est irréductible de dimension

$$(n-2)(2n-(n-2)) = n^2-4 \ ,$$

donc de codimension 3 dans T . (La variété des matrices $p \times q$ de rang $\leq r$ a pour dimension $r(p+q-r)$) .

4.2. <u>Cas</u> $d_1 = d_2 = \dots = d_{n-1} = 1$, $d_n = m \geq 2$.

Posons comme précédemment

$$f_i = u_{i1}X_1+\dots+ u_{in}X_n \quad (1 \leq i \leq n-1)$$

et

$$f_n = f = \sum_{|\alpha| = m} c_\alpha X^\alpha$$

Notant $D_1,\dots,D_n$ les cofacteurs de $v_1,\dots,v_n$ dans la matrice

$$\begin{bmatrix} u_{11} & u_{12} \cdots\cdots & u_{1n} \\ u_{n-1,1} & \cdots\cdots & u_{n-1,n} \\ v_1 & \cdots\cdots & v_n \end{bmatrix},$$

on sait que $R = f(D_1,\ldots,D_n)$. La définition des D_i implique les égalités $f_i(D_1,\ldots,D_n) = 0$ $(1 \leq i \leq n-1)$, et nous allons voir que ω est l'application

(4.2.1)
$$\omega : B_{m-1} \longrightarrow A$$
$$\bar{h} \longmapsto h(D_1,\ldots,D_n)$$

Comme

$$J(f_1,\ldots,f_n) = D_1 \frac{\partial f}{\partial X_1} + \ldots + D_n \frac{\partial f}{\partial X_n},$$ l'assertion résulte aussitôt de (2.9.8) et de la formule d'EULER

$$\sum_i D_i \frac{\partial f}{\partial X_i}(D_1,\ldots,D_n) = m f.$$

Posant $D^\alpha = D_1^{\alpha_1} \ldots D_n^{\alpha_n}$, on a donc $\omega(X^\alpha) = D^\alpha$ $(|\alpha| = m-1)$.

Supposant pour simplifier que k est un corps, la variété $V(\omega)$ est ensemblistement définie par les équations

$$D_1 = \ldots = D_n = 0,$$

donc s'identifie à

$$S^m(k^n) \times \{u : k^n \to k^{n-1} \mid rg(u) \leq n-2\}$$

Elle est donc irréductible de dimension

$$\binom{m+n-1}{m} + (n-2)(n+1),$$

et par suite est une hypersurface de T. Le lieu singulier T^{sing} vérifie les équations

$$D^\alpha = \frac{\partial f}{\partial c_\alpha} = 0,$$

d'où $T^{sing} \subset V(\omega)$. Enfin, comme T est une intersection complète non normale, son lieu normal T^{norm} est tel que $T \div T^{norm}$ soit un fermé purement de codimension un de T, et contenu dans $V(\omega)$.

Avec des notations évidentes, on en déduit les égalités ensemblistes

(4.2.2) $$T^{norm} = T^{rat} = T^{rég} .$$

4.3. <u>Cas</u> $n = 2$.

Notons X et Y les indéterminées et

$$f = u_o X^p + u_1 X^{p-1} Y + \dots + u_p Y^p$$

$$g = v_o X^q + v_1 X^{q-1} Y + \dots + v_q Y^q$$

les polynômes génériques de degré p et q. Identifiant un polynôme homogène de degré $p+q-1$ et la ligne de ses coefficients, rangés suivant les puissances décroissantes de X, on sait que

$$R = \det(X^{q-1}f, X^{q-2}Y f, \dots, Y^{q-1}f, X^{p-1}g, X^{p-2}Y g, \dots, Y^{p-1}g) .$$

Avec des notations analogues, nous allons voir que l'application ω est donnée par

(4.3.1) $$\omega : B_{p+q-2} \to A$$

$$\bar{h} \longmapsto \det(X^{q-2}f, X^{q-3}Yf, \dots, Y^{q-2}f, h, X^{p-2}g, X^{p-3}Yg, \dots, Y^{p-2}g)$$

En effet, il est clair que ce déterminant s'annule lorsque $h \in B_{q-2}f + B_{p-2}g$, donc définit par passage au quotient une application A-linéaire

$$\theta : B_{p+q-2} \to A ,$$

donc, comme ω est une base de $\check{B}_{p+q-2}$, il existe $a \in A$ tel que

$$\theta = a\,\omega .$$

Des considérations d'homogénéité (cf. 3.4.6) impliquent $a \in k$. Faisant la spécialisation $f \longmapsto X^p$, $g \longmapsto Y^q$ dans l'égalité (2.9.7)

$$\theta(\Delta(f,g)) = a\,R(f,g) ,$$

on en déduit que $a = 1$.

Notant T_o le fermé de T défini par

$$\{u_o = u_1 = \dots = u_p = 0\} \cup \{v_o = v_1 = \dots = v_q = 0\} ,$$

la projection $g : X \to T$ est finie au-dessus de $T \dot{-} T_o$, d'où

$$(T \dot{-} T_o) \cap T^{norm} = (T \dot{-} T_o) \cap T^{rat} = (T \dot{-} T_o) \cap T^{rég} .$$

Par ailleurs, il est clair sur les équations que

$$T_o \subset V(\omega) \subset T^{sing} \text{ , d'où}$$

(4.3.2) $$T^{rég} = T^{rat} \subset T \dot{-} T_o .$$

4.4. <u>Questions</u>.

a) Le schéma $V(\omega)$ est-il toujours irréductible ? Cela impliquerait que, lorsque $(p,q) \neq (1,1)$, on a aussi $T^{rég} = T^{norm}$ dans (4.3) , ce qu'on peut vérifier directement lors que $p = q = 2$.

b) De manière générale, il serait intéressant d'étudier les relations entre les schémas T^{rat} , T^{norm} , $T^{rég}$.

REFERENCES

[1] A. HURWITZ : Über die Trägheitsformen eines algebraischen Moduls, Annali di Mathematica pura ed applica (3) 20 (1913) .

[2] E. LASKER : Zur Theorie der Moduln und Ideale, Math. Annalen 60 (1905) .

[3] F.S. MACAULAY : Some formulae on elimination, Proc. Lond. Math. Soc. (1) 35 (1903) .

[4] F.S. MACAULAY : The algebraic theory of modular systems, Cambridge University Press (1916) .

[5] G. SCHEJA , U. STORCH : Über Spurfunktionen bei vollständigen Durschnitten, J. de Crelle 278-279 (1975) .

[6] H. WIEBE : Über homologische Invarianten lokaler Ringe, Math. Annalen 179 (1969) .

Université Louis Pasteur
7, rue René Descartes
F-67084 Strasbourg
France

On the Classification of Non-complete Algebraic Surfaces

Yujiro KAWAMATA

Introduction

Iitaka defined in [4] the logarithmic Kodaira dimension of algebraic varieties which are not necessarily complete and proposed to apply it to the classification theory of algebraic varieties. Following his philosophy we shall study in this paper the structure of non-complete algebraic surfaces of non-negative logarithmic Kodaira dimension defined over the complex number field $\mathbb{C}$.

Let X be a non-singular algebraic surface over $\mathbb{C}$. It is well known that there is a non-singular complete surface $\overline{X}$ which contains X as a Zariski open subset and such that the complement $D = \overline{X} - X$ is a divisor of normal crossing. For a non-negative integer n let $\overline{p}_n(X) = \dim H^0(\overline{X}, n(K+D))$, where K is the canonical sheaf of $\overline{X}$. Then there is an integer (or $-\infty$) $\overline{\kappa}(X)$ such that $an^{\overline{\kappa}(X)} \leq \overline{p}_n(X) \leq bn^{\overline{\kappa}(X)}$ for some $a, b > 0$ and for large n. We know that $\overline{\kappa}(X)$ can take a value in $-\infty$ 0,1 or 2. $\overline{p}_n(X)$ is the logarithmic n-genus and $\overline{\kappa}(X)$ is the logarithmic Kodaira dimension of X. It is easily shown that $\overline{p}_n(X)$ and $\overline{\kappa}(X)$ are proper-birational invariants of X.

In §1 we shall construct a relatively minimal model of X using the theory of an arithmetically effective component of an effective divisor by Zariski [14]. In §2 we shall prove some structure theorems of surfaces with non-negative logarithmic Kodaira dimension following the arguments by Mumford [10]. In §3 we shall show the invariance of the logarithmic Kodaira dimension of surfaces under the deformation, which was conjectured in [6].

The author would like to express his thanks to Prof. Iitaka, the correspondences with whom were encouraging and fruitful, and also to Prof. Lønsted and other members of the Mathematics Institut of the University of Kopenhagen for the hospitality during the meating.

§1. A relatively minimal model

The main result of this section is the following

(1.1) Theorem. Let X, $\bar{X}$, and D as in the introduction. If $\bar{\kappa}(X) \geq 0$, then there exist a non-singular complete surface $\bar{X}_m$, a divisor D_m with coefficients in $\mathbb{Q}$ on $\bar{X}_m$ and a birational morphism $f : \bar{X} \longrightarrow \bar{X}_m$ satisfying the following conditions :

(1) $D_m = \sum_i d_i D_i$, $0 < d_i \leq 1$, where the D_i are prime divisors on $\bar{X}_m$,

(2) $f^*(K_m + D_m)$ is the arithmetically effective component of K+D, where K_m is the canonical sheaf of $\bar{X}_m$.

(1.2) Definition ([14]) Let D be a divisor on a complete surface X with rational coefficients. D is said to be arithmetically effective iff $D \cdot C \geq 0$ for any curve C on X.

(1.3) Theorem (Theorem 7.7 of [14]). Let D be an effective divisor on a surface X (with rational coefficients). Then there is a unique decomposition $D = D^+ + D^-$, where

(1) D^+ and D^- are effective divisors (with rational coefficients),

(2) $D^- = 0$ or the intersection form of the prime components of D^- is negative definit,

(3) D^+ is arithmetically effective,

(4) $D^+ \cdot E = 0$ for every prime component E of D^-.

D^+ (resp. D^-) is said to be the arithmetically effective (resp. negative) component of D.

(1.4) Facts (1) If D is an integral divisor, $H^o(X,D) = H^o(X,[D])$, where [ ] denotes the integral part.

(2) $(nD)^+ = nD^+$ and $(nD)^- = nD^-$ for a positive integer n,

(3) If $D_1 \equiv D_2$ (numerically equivalent), then $D_1^- = D_2^-$. Especially, we can consider the arithmetically negative component of an invertible sheaf whose high multiple has a section.

(4) We can also define D^+ and D^- of D with real coefficients. In this case we can easily prove that they are continuous functions of D, i.e., the coefficients of them are continuous functions of those of D.

(1.5) Proposition If $D_1 > D_2$, then $D_1^+ > D_2^+$.

Proof Put $D_1 = D_2 + D_3$. We shall show $D_1^- < D_2^- + D_3^-$. By Cor. 7.2 of [14] we have only to show that $(D_2^- + D_3^- - D_1^-)\cdot E \leqslant 0$ for every prime component E of D_1^-. The left hand side is equal to $(D_2^- + D_3^- - D_1)\cdot E$, which is not greater than $(D_2 + D_3 - D_1)\cdot E = 0$.

(1.6) We can obtain D^+ from D in the following way. Let $D = \sum_i a_i D_i$, where the D_i are prime components of D. We call it process (i_o) to replace D by a new divisor $D' = \sum_i a'_i D_i$, where $a'_i = a_i$ if $i \neq i_o$ or $D\cdot D_i \geqslant 0$, and $a'_i = a_i - D_i\cdot D / D_i^2$ if $i = i_o$ and $D\cdot D_i < 0$. By (1.5) $D^+ = (D')^+$. We carry out processes (1),$\cdots$, (n), successively, and then after that (1),$\cdots$, (n) again, and so on. D decreases to a limit D^+, which is rational by the theorem of Zariski.

(1.7) Proof of (1.1). We consider all the pairs (Y, C), where Y is a complete surface birational to $\overline{X}$ and C is a divisor on Y with rational coefficients which are positive and not greater than 1. Let E be a curve on Y and assume $(K_Y + C)\cdot E < 0$. We have two cases.

(case 1) E is not a prime component of C. Then, $K\cdot E < 0$ and $E^2 < 0$, hence $K\cdot E = -1$ and $E^2 = -1$ and $E\cdot C \leq 1$. Contract E to a point, and we get a new pair (Y', C'), where C' is the direct image of C : $C' = \mu_* C$, where μ is the contraction. We know $\mu^* C' = C + (C\cdot E)\, E$ and $\mu^* K_{Y'} = K_Y - E$. Hence, $\mu^*(C' + K_{Y'}) = C + K_Y - ((K + C)\cdot E / E^2)\, E$. Thus, this step is exactly the process discussed in (1.6).

(case 2) E is a prime component of C. Then, $C = dE + C_0$, where $0 < d \leq 1$ and E is not a prime component of C_0. We know $E^2 < 0$ and $K\cdot E + E^2 \leq K\cdot E + dE^2 \leq K\cdot E + dE^2 + C_0\cdot E < 0$. Therefore, E is a non-singular rational curve. If $K\cdot E \geq 0$, then d is not smaller than $(K\cdot E + dE^2 + C_0\cdot E)/ E^2$. Thus, the process in (1.6) corresponding to E is carried out by decreasing the d. If $K\cdot E < 0$, then change the d to 0, and we reduce it to (case 1). Remark that (case 1) can occur only a finite number of times. QED.

(1.8) Corollary $\bar{P}_m(X) = H^0(\bar{X}, m(K + D)) = H^0(\bar{X}_m, [m(K_m + D_m)])$.

2. The structure of $(\bar{X}_m, D_m)$

(2.1) Definition A normal singularity of a surface is said to be simple elliptic (resp. simple quasi-elliptic) if the exceptional locus of the minimal resolution consists of a single elliptic curve (resp. a cycle of non-singular rational curves).

A simple elliptic singularity was defined in [11] and proved to have very nice properties. A simple quasi-elliptic singularity was called a 2-dimensional cusp in [5] and has similar properties.

(2.2) Theorem If $\bar{\kappa}(X) = 0$, then $\bar{X}_m$ is a relatively minimal complete surface and there is some integer n such that $[n(K_m + D_m)] = 0$.

(2.3) Theorem. If $\bar{\kappa}(X) = 1$, then some high multiple $|n(K_m + D_m)|$ determines a fiber structure $\pi : \bar{X}_m \longrightarrow C$ over a curve C, which is minimal in the sense of a fiber space. There are two possibilities :

(1) (elliptic case) The general fiber of π is an elliptic curve and $D_m = \sum_i F_i$, where the F_i are distinct fibers. We have

(2.4) $K_m+D_m = \pi^*(K_C+\delta) + \sum_s (m_s-1)E_s + \sum_i F_i$,

where the m_sE_s are the multiple fibers and δ is some divisor on C (see [8]). Moreover, for $n \geq 2$,

(2.5) $\dim H^0([n(K_m+D_m)]) = n(2g-2+t) + \sum_s [n(1-1/m_s)] + \sum_i [n/m_i] + 1-g$,

where g is the genus of C, $t = \deg \delta$ and the m_i are the multiplicities of the fibers F_i.

(2) (quasi-elliptic case) The general fiber of π is a rational curve and $D_m = H + \sum_i d_iF_i$, where H is the horizontal component of D_m and the F_i are fibers. The coefficients in H are 1, the degree of H on C is equal to 2, and H has only normal crossings. For each i, corresponding to whether F_i meats a branch point of H or not, we have

(2.6) $d_i = 1/2\ (1 - 1/m_i)$ or $(1 - 1/m_i)$,

respectively, where m_i is an integer or ∞. We have

(2.7) $K_m+D_m = \pi^*(K_C+\delta) + \sum_i d_iF_i$,

where $t = \deg \delta = 1/2$ (the number of branch points of H). Moreover, for $n \geq 2$,

(2.8) $\dim H^0([n(K_m+D_m)]) = n(2g-2+t) + \sum_i [nd_i] + 1 - g$.

(2.9) Theorem If $\bar{\kappa}(X) = 2$, then the pluricanonical ring R $= \bigoplus_{n\geq 0} H^0(\,n(K+D)\,)$ is finitely generated. We define the canonical model $\overline{X}_c$ of (X, $\overline{X}$, D) to be Proj R. Then, the canonical map $\Phi : \overline{X} \longrightarrow \overline{X}_c$ is a morphism. Denote by D_c the direct image $\Phi_* D$ of D (note that this is only a Weil divisor). Then, a singularity on $\overline{X}_c$ is rational, simple elliptic or simple quasi-elliptic and $\overline{X}_m$ coincides with the minimal resolution of $\overline{X}_c$. Moreover, D_m is determined uniquely by the following conditions :

(1) The direct image of D_m on $\overline{X}_c$ is equal to D_c,

(2) For any curve E_t on $\overline{X}_m$ which contracts to a point on $\overline{X}_c$, $(K_m+D_m)\cdot E_t = 0$.

Let $\mathcal{E}$ be the union of those E_t. Then, $\mathcal{E} \cup \mathrm{Supp}\, D_m$ has at most ordinary double points as singularities. Moreover, we have $(K_m+D_m)^2 > 0$ and $\overline{p}_2 \neq 0$.

The proof of the above theorems is just similar to that by Mumford [10], in which the arithmetical effectivity of the canonical sheaf K was essential. We shall show only the outline of the proof.

(2.10) Proof of (2.2) and (2.3). We assume $(K_m+D_m)^2 = 0$. First, if $[n(K_m+D_m)] = 0$ for some positive integer n, then we get (2.2). In this case, suppose there were an exceptional curve E of first kind on $\overline{X}_m$. Then, we have $D_m \cdot E = 1$. Therfore, E must have already been blown down.

Second, we assume that there is a non-zero member in $|n(K_m+D_m)|$ for a large integer n and we shall prove (2.3). Let $F^o = \sum_i m_i^o E_i$ be a connected component of a member of $|n(K_m+D_m)|$ and put $F = \sum_i m_i E_i$, where $m_i^o = \text{g.c.d.}(m_i^o)\, m_i$. Then we have $E_i \cdot F = E_i \cdot (K_m+D_m) = 0$ for every prime component E_i of F.

(2.11) Lemma Let L be an invertible sheaf on F. We assume that

$\deg L \otimes O_{E_i} = 0$ for every i. Then, $H^0(F, L) \neq 0$ iff $L \cong O_F$. In this case $H^0(O_F) \cong \mathbb{C}$.

Proof is just the same as in Lemma on p.332 of [10].

(2.12) Lemma Put $D_m = D_m' + \sum_i d_i E_i$, where D_m' has no common prime components with F. Then, any prime component E of D_m' which intersects with F has a coefficient 1 in D_m'.

Proof $F \cdot E > 0$ implies $F \cdot (K_m + D_m) > 0$. QED.

(2.13) Corollary The sheaf $\omega = O_{\overline{X}_m}(K_m + D_m' + F) \otimes O_F$ is isomorphic to O_F. We have $D_m \cdot F = 0$ or 2, and accordingly $K \cdot F = 0$ or -2. Hence, $\pi(F) = 1$ or 0.

Proof By (2.12), the formula of ω has a meaning. $\chi(\omega)$ $= \chi(K_m + D_m' + F) - \chi(K_m + D_m') = 1/2(K_m + D_m' + F) \cdot (D_m' + F) - 1/2\, D_m'(K_m + D_m')$ $= 1/2\, D_m \cdot F$. If $D_m \cdot F = 0$, then $\dim H^1(F, \omega) = \dim H^0(F, O_F) \geq 1$. If $D_m \cdot F > 0$, then $\dim H^1(F, \omega) = 0$. Anyway, we have $H^0(F, \omega) \neq 0$. By (2.11), $\omega \cong O_F$. QED.

(2.14) Let F' be an effective divisor with $F' \cdot E_i = 0$ for every i. Then, $F' = nF + F''$, where $n > 0$ and F'' is disjoint to F and effective.

Proof see p.333 of [10].

(2.15) There is a pencil of $\overline{X}_m$ with a fiber F.

Proof If $\pi(F) = 0$, then it is O.K. If $\pi(F) = 1$, we have only to show that $\dim H^0(\overline{X}_m, nF) \geq 2$ for a large integer n, because (2.14). We divide it into two cases :

(case 1) $\overline{p}_g(X) = 0$.

We know that $H^2(\overline{X}_m, n(K_m + D_m') + (n-1)F) = 0$ for large n and $\dim H^1(F, O_F) = 1$. From the exact sequence

$$0 \longrightarrow O_{\overline{X}_m}(n(K_m+D_m') + (n-1)F) \longrightarrow O_{\overline{X}_m}(n(K_m+D_m') + nF) \longrightarrow O_F \longrightarrow 0,$$

we get $H^1(n(K_m+D_m') + nF) \neq 0$. On the other hand,

$\chi(\overline{X}_m, n(K_m+D_m') + nF) = (1/2)nD_m' \cdot (K_m+D_m') + \chi(O_{\overline{X}_m})$, which is non-negative if $\overline{X}_m$ is not a ruled surface of genus greater than one. In this case, the rest of the proof is just as in [10]. If $\overline{X}_m$ is ruled over a curve of genus greater than one, then the general fiber of X is G_m by the addition formula ([7]), hence $\pi(F) = 0$.

(case 2) $\overline{P}_g(X) \neq 0$.

Just the same as in p.334 of [10]. Put $L_n = O_{\overline{X}_m}(-D_m'+nF)/O_{\overline{X}_m}(-D_m')$ $= O_{\overline{X}_m}(nF)/O_{\overline{X}_m}$ and so on. QED.

(2.16) Suppose there were an exceptional curve E of first kind on a fiber. $D_m \cdot F \leqslant 2$ implies that there is some exceptional curve E' such that $D_m E' \leqslant 1$, which is a contradiction.

(2.17) We shall prove (2.4) and (2.7). Only to do is to determine the coefficients in D_m. It was already shown that the coefficients of the horizontal component are 1. Since $f^*(K_m+D_m) \leqslant K + D$, and D is reduced, the horisontal component is normal crossing. The only fractional coefficients occur for rational curves by the construction. Let us determine the coefficient d of a fiber F in the quasi-elliptic case. We only consider the case where D_m intersects with F at distinct two points. The other case can be treated similarly. The graph of curves near F is as follows :

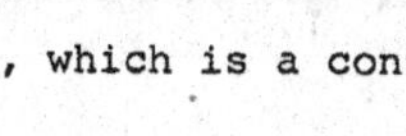

The inverse image of them in $\overline{X}$ is something like

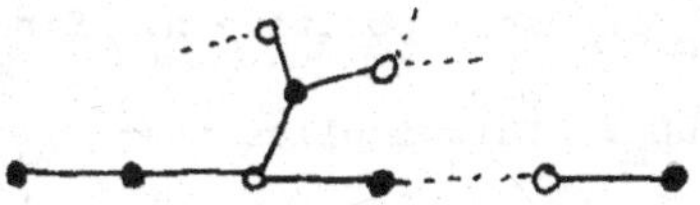

where • denotes a curve in D and o outside of D. First, the branches of the graph are contracted without change of coefficients and we get something like

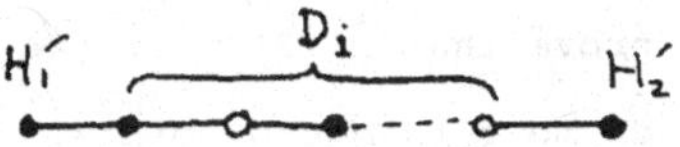

We know $f^*(H_1 + H_2 + F + K_m) = H_1' + H_2' + \sum D_i + K'$. Let $\sum' D_i$ be the sum of those D_i which is in D. Then, d is just the greatest number which satisfies

$$f^*(H_1 + H_2 + dF + K) \leq H_1' + H_2' + \sum' D_i + K'.$$

This is equivalent to $(1 - d)\, F \geq (\sum - \sum')\, D_i$. When we write $F = \sum_i m_i D_i$, then $1 - d = \max_{i \in \Sigma - \Sigma'} (1/m_i, 0)$.

(2.18) The proof of (2.5) and (2.8) is similar to that in [3].

(2.19) <u>Proof</u> of (2.9). By the argument above, $\bar{\kappa}(X) = 2$ iff $(K_m+D_m)^2 > 0$, for from $(K_m+D_m)^2 > 0$ follows $\bar{\kappa}(X) = 2$ by the theorem of Riemann-Roch. We prove first that $\bar{p}_2(X) \neq 0$. Assume the contrary. Put $F = 2D_m - [2D_m]$. First, assume that $\bar{X}_m$ is not birationally equivalent to a ruled surface of genus greater than one. We know that a prime component E of F is a rational curve , and since $E \cdot (K_m+D_m) = 0$, the intersection form of the prime components of F is negative definit, by the Hodge index theorem. If $K_m \cdot E = -1$, then $D_m \cdot E = 1$ and E is already contracted. Thus, $K_m \cdot E \geq 0$. We calculate

$$\chi(\bar{X}_m, 2K_m + [2D_m]) \geq \chi(\mathcal{O}_{\bar{X}_m}) + (K_m+D_m)^2 + 1/2\, K_m \cdot F - 1/2\, F^2 > 0$$

Therefore, $H^2(\bar{X}_m, 2K_m + [2D_m]) \neq 0$. Hence, $H^0(\bar{X}_m, -K_m - 2D_m + F)$

$\neq 0$, i.e., $nF \geqslant n(K_m + 2D_m)$ for some positive n. Since $n(K_m+D_m) \geqslant 0$, for some n, we have $nF \geqslant nD_m$. This implies that $\dim H^0(\overline{X}_m, nD_m) = 1$ for any $n > 0$. On the other hand, since there is a non-zero section s in $H^0(\overline{X}_m, -K_m) \supset H^0(\overline{X}_m, -K_m - [2D_m])$, there is an injection $H^0(\overline{X}_m, n(K_m+D_m)) \longrightarrow H^0(\overline{X}_m, nD_m)$, which is a contradiction.

Next, we assume that $\overline{X}_m$ is birational to a ruled surface of genus greater than one. We shall prove that $\overline{p}_g(X) \neq 0$. We know that the general fiber of X is $\mathbb{P}^1$ minus more than two points. Consider the exact sequence

$$H^0(\overline{X}, K+D) \longrightarrow H^0(D, K_D) \xrightarrow{p} H^1(\overline{X}, K).$$

The map p is the dual of the map $p^* : H^1(\overline{X}, O_{\overline{X}}) \longrightarrow H^1(D, O_D)$. Therfore, we have only to show that p^* is not surjective. Let $D_1, \cdots,$ D_t be the horizontal components of D. If $t \neq 1$, then it is clear. If $t = 1$, then by the Hurwitz formula, $g(D_1) \geqslant tg - t + 1 > g$, where g = genus of X. This shows our assertion.

(2.20) Now let $\mathcal{E}$ be the set of all curves E on $\overline{X}_m$ such that $(K_m+D_m)\cdot E = 0$. By the Hodge index theorem, the intersection form of $\mathcal{E}$ is negative definit. Note that there is no exceptional curve of first kind among $\mathcal{E}$. If $E \in \mathcal{E}$ is not a prime component of D_m, then it is easy to show that $K_m \cdot E = 0$, $D_m \cdot E = 0$ and $E^2 = -2$. These are contracted to rational double points as in [7]. Suppose $E \in \mathcal{E}$ is a prime component of D_m. Put $D_m = D_m' + dE$. Since $E^2 < 0$, $(K_m + E)\cdot E \leqslant (K_m + dE)\cdot E \leqslant (K_m + D_m)\cdot E = 0$. Therefore, $\pi(E) = 0$ or 1, and moreover, $\pi(E) = 1$ iff $d = 1$ and $D_m \cdot E = 0$.

Now we prove that Supp D_m has at most ordinary double points. Note that a prime component of D_m with coefficient less than 1 is a member of $\mathcal{E}$ and hence a non-singular rational curve. Since

$f^{*}(K_m + D_m) \leqslant K + D$, the multiplicity of D_m at any point including an infinitesimally near point cannot be greater than two. Thus, the union of the prime components of D_m with coefficients 1 has at most ordinary double points. We have noly the following four cases left to consider :

(i) Two non-singular rational components of $\mathcal{E}$ meet at a point with multiplicity $n \geqslant 2$,

(ii) Three non-singular rational components of $\mathcal{E}$ meet at a point,

(iii) Two non-singular rational components of $\mathcal{E}$ meet at a point of a component of D_m,

(iv) A non-singular rational component of $\mathcal{E}$ meet a prime component of D_m at a point with multiplicity $n \geqslant 2$.

The case (i) is thrown away as follows. Let E_1 and E_2 be such curves and let d_i be the coefficient of E_i in D_m for $i = 1, 2$. We know that $E_i^2 \leqslant -2$, for $i = 1, 2$. We calculate $0 = (K_m + D_m)\cdot E_1$ $\geqslant K_m \cdot E_1 + d_1 E_1^2 + d_2 n = -2 - (1 - d_1)E_1^2 + d_2 n \geqslant -2 + 2(1-d_1) + d_2 n$ $= -2d_1 + d_2 n$. Thus, $2d_1 \geqslant d_2 n$. Similarly, $2d_2 \geqslant d_1 n$. Therefore, $d_1 = d_2$ and $n = 2$. If $d_1 \neq 1$, then $E_1^2 = E_2^2 = -2$ and $(E_1 + E_2)^2$ $= 0$, which is a contradiction. The other cases are treated in a similar way.

(2.21) Suppose there is a cycle of non-singular rational curves E_1, $\cdots$, E_n in $\mathcal{E}$. The coefficients d_i of E_i in D_m must be 1 forall i.

Proof $0 = (K_m + D_m)\cdot E_i \geqslant 2 - (1 - d_i) E_i^2 + d_{i+1} + d_{i-1}$. Hence, $d_1 = \cdots = d_n$ and, if $d_1 \neq 1$, then $E_i^2 = -2$ for all i. But then, $(\sum_i E_i)^2 = 0$, a contradiction.

It follows also that the cycle has no common points with other components of D_m.

(2.22) We shall contract algebraically $\mathcal{E}$ to points. Let $\mathcal{E}_0$ be a connected component of $\mathcal{E}$ inside of D_m. There are three cases :

(a) $\mathcal{E}_0$ is a tree of non-singular rational curves,

(b) $\mathcal{E}_0$ is a non-singular elliptic curve,

(c) $\mathcal{E}_0$ is a rational curve with an ordinary double point or a cycle of non-singular rational curves.

In case (a) $\mathcal{E}_0$ can be contracted to a rational singular point.

<u>Proof</u> Since the intersection form of $\mathcal{E}_0$ is negative definit, there is a fundamental divisor Z on it. Let $D_m = D_m' + D_m''$, where D_m'' is the sum over the prime components in $\mathcal{E}_0$. Then, $(K_m + Z)\cdot Z$ $= (K_m + D_m'')\cdot Z + (Z - D_m'')\cdot Z \leqslant (K_m + D_m)\cdot Z = 0$. Here the equality holds iff $D_m'\cdot Z = 0$ and $(Z - D_m'')\cdot Z = 0$, i.e., $\mathcal{E}_0$ makes a connected component of D_m and $E_j\cdot Z = 0$ for any prime component E_j of $Z - D_m''$, since $E_j\cdot Z \leqslant 0$ at any rate. Since $Z^2 < 0$, there is at least one prime component E_1 with $d_1 = 1$ in this case. Let d_i be the coefficients of the E_i in D_m as before. We shall classify the graph of $\mathcal{E}_0$ under the assumption that $\mathcal{E}_0$ is a connected component of D_m and there is a E_1 with $d_1 = 1$. Since $(K_m + D_m)\cdot E_i = 0$, the possible graphs are as follows :

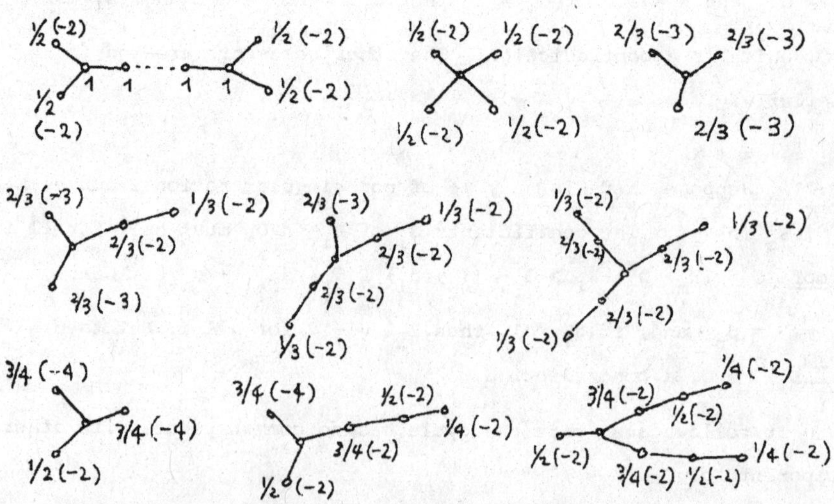

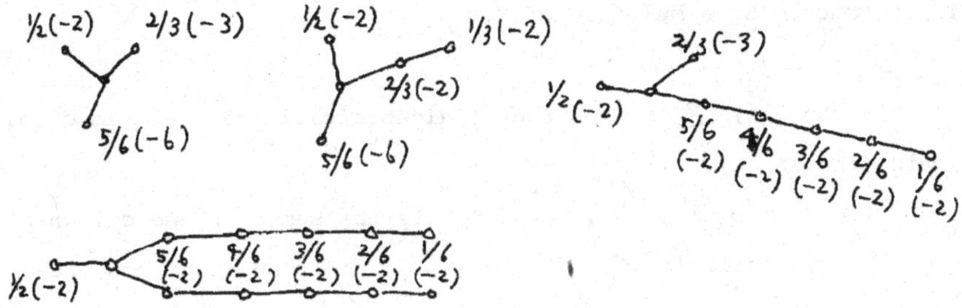

where the integers outside of the brackets are d_i and inside are the self-intersections. We can easily check that each graph determines a rational singularity.

(2.23) In case (b) $\mathcal{E}_0 = D_1$ is contracted algebraically to a simple elliptic singularity.

<u>Proof</u> Consider an exact sequence

$$H^0(\overline{X}_m, K_m + D_1) \xrightarrow{a} H^0(D_1, O_{D_1}) \xrightarrow{b} H^1(\overline{X}_m, K_m).$$

If a is not zero, then D_1 contains no base points of $|K_m + D_1|$, hence of $-|K_m + D_m|$, and can be contracted. If $a = 0$, then the dual $b^*: H^1(\overline{X}_m, O_{\overline{X}_m}) \longrightarrow H^1(D_1, O_{D_1})$ is surjective. Let H be an ample divisor on $\overline{X}_m$. Take an integral multiple, if necessary, and we get

$\deg (H + nD_1)|_{D_1} = 0$ for some positive integer. Since b^* is surjective, there is $L \in \mathrm{Pic}^0(\overline{X}_m)$ such that $(H + nD_1 + L)|_{D_1} = 0$. Put

$M = H + nD_1 + L$. Consider an exact sequence

$$H^0(\overline{X}_m, kM) \longrightarrow H^0(D_1, O_{D_1}) \longrightarrow H^1(\overline{X}_m, kM - D_1),$$

for $k \in \mathbb{N}$. We may assume that $kM - D_1 - K_m$ is arithmetically effective. Since $M^2 = H\cdot(H + nD_1) > 0$, $(kM - D_1 - K_m)^2 > 0$ for a large k. Therfore, by the vanishing theorem of Ramanujan (see [2]), we have $H^1(\overline{X}_m, -kM + D_1 + K_m) = H^1(\overline{X}_m, kM - D_1) = 0$, which shows that

D_1 is contracted by a multiple of M.

(2.24) In case (c) $\mathcal{E}_0$ is contracted algebraically to a simple quasi-elliptic singularity.

Proof Write $D_m = D_m' + D_m''$, where D_m' is the sum of prime components in $\mathcal{E}_0$. Every coefficient in D_m' is 1. Consider an exact sequence

$$H^0(\bar{X}_m, K_m + D_m') \xrightarrow{a} H^0(D_m', O_{D_m'}) \xrightarrow{b} H^1(\bar{X}_m, K_m).$$

If $a \neq 0$, then it is O.K. as in (2.23). If $a = 0$, then the dual $b^* : H^1(\bar{X}_m, O_{\bar{X}_m}) \longrightarrow H^1(D_m', O_{D_m'})$ is surjective. On the other hand, since the restriction of a holomorphic 1-form on $\bar{X}_m$ to D_m' is always zero, the map $H^1(\bar{X}_m, \mathbb{C}) \longrightarrow H^1(D_m', \mathbb{C})$ is zero. These two facts contradict the following commutative diagram

$$\begin{array}{ccc} H^1(\bar{X}_m, \mathbb{Z}) & \longrightarrow & H^1(\bar{X}_m, O_{\bar{X}_m}) \\ \downarrow & & \downarrow \\ H^1(D_m', \mathbb{Z}) & \longrightarrow & H^1(D_m', O_{D_m'}). \end{array}$$

The rest of the proof of (2.9) is just the same as in [9].

§ 3. The deformations

(3.1) Recall briefly the definition of deformations of non-complete algebraic surfaces in [6]. Let $(X_0, \bar{X}_0, D_0)$ be a triple where $\bar{X}_0$ is a non-singular complete algebraic surface over $\mathbb{C}$, D_0 is a reduced divisor on $\bar{X}_0$ of normal crossing and $X_0 = \bar{X}_0 - D_0$. A deformation of it consists of a triple $(X, \bar{X}, D)$ and a base space S with a base point s_0 and a morphism $\pi : \bar{X} \longrightarrow S$, together with an isomorphism

$\psi : (X_o, \bar{X}_o, D_o) \longrightarrow (X, \bar{X}, D) \times_S s_o$, such that π is proper and locally product for $\bar{X}$ and also for D. In particular, π and $\pi|_{D_i}$ are smooth, where D_i is an irreducible component of D. A fiber X_s of $\pi|_X : X \longrightarrow S$ is called a <u>deformation</u> of X_o.

(2.2) <u>Theorem</u>. <u>The logarithmic Kodaira dimension of algebraic surfaces is invariant under deformations</u>.

<u>Proof</u> We may assume that the base space S is a discrete valuation ring having only two points $\{y, s\}$, where y is the generic point and s is the closed point. We shall prove that $\bar{\kappa}(X_y) = d$ iff $\bar{\kappa}(X_s) = d$ for $d = -\infty$, 0 and 2. Since an exceptional curve of first kind on $\bar{X}_s$ can be extended over $\bar{X}$, we can blow down $\bar{X}$ to $\bar{X}_m$, where the special fiber $\bar{X}_{m,s}$ gives a minimal model of $\bar{X}_s$, when $\bar{\kappa}(X_s) \geqslant 0$. In this case, we define D_m as a rational combination of irreducible components of D which gives a minimal model $(\bar{X}_{m,s}, D_{m,s})$ over $\bar{X}_s$. Note that the generic fibers $(\bar{X}_{m,y}, D_{m,y})$ is an intermediate stage from $(\bar{X}_y, D_y)$ to a minimal model of that.

By the upper semi-continuity of cohomology, we get immediately : If $\bar{\kappa}(X_s) = -\infty$, then $\bar{\kappa}(X_y) = -\infty$. and if $\bar{\kappa}(X_y) = 2$, then $\bar{\kappa}(X_s) = 2$.

(2.3) If $\bar{\kappa}(X_y) = -\infty$, then $\bar{\kappa}(X_s) = -\infty$.

<u>Proof</u> By [3], $\bar{X}_s$ is birationally ruled of genus g. First, assume $g = 0$. Suppose $\bar{\kappa}(X_s) \geqslant 0$. Then we have $H^2(\bar{X}_{m,s}, n(K_{m,s}+ D_{m,s})) = 0$, for some $n > 0$, for if $\bar{\kappa}(X_s) > 0$, then it is trivial and if $\bar{\kappa}(X_s) = 0$, then it is because that X_s is rational. By the upper semi-continuity, $H^2(\bar{X}_{m,y}, n(K_{m,y}+ D_{m,y})) = 0$. But on the other hand,

$$\chi(\bar{X}_{m,y}, n(K_{m,y}+D_{m,y})) = 1/2\, n(n-1)(K_{m,y}+ D_{m,y})^2$$

$+ 1/2\ nD_{m,y} \cdot (K_{m,y} + D_{m,y}) + \chi(\mathcal{O}_{X_{m,y}}) > 0$, which is a contradiction.

Second, assume $g \geqslant 1$. Let $a : X_y \longrightarrow A$ be the Albanese map. By the addition formula[7], a general fiber of a is $\mathbb{P}^1$. Therfore, X_s is also a fiber space with $\mathbb{P}^1$ as a fiber. Thus, $\bar{\kappa}(X_s) = -\infty$.

(2.4) If $\bar{\kappa}(X_s) = 0$, then $\bar{\kappa}(X_y) = 0$.

Proof First, assume that the irregularity $q(\bar{X}_s) = 0$. Then, from $[n(K_{m,s} + D_{m,s})] = 0$ follows $[n(K_{m,y} + D_{m,y})] = 0$, hence $\bar{\kappa}(X_y) = 0$. Second, assume $q(\bar{X}_s) \geqslant 1$. Let $a : X_s \longrightarrow A$ be the quasi-Albanese map. By the addition [7], $\bar{q}(A) = q(A) = 1$ and the general fiber of a is an elliptic curve or $\mathbb{G}_m$. In the former case, we have $\kappa(\bar{X}_s) = 1$ and hence $D_{m,s} = 0$. Thus, we reduce the case to [3]. In the latter case, X_y is also an $\mathbb{G}_m$-fiber space over an elliptic curve, and $\bar{\kappa}(X_y) \geqq 0$. By the upper semi-continuity, $\bar{\kappa}(X_y) = 0$.

(2.5) If $\bar{\kappa}(X_s) = 2$, then $\bar{\kappa}(X_y) = 2$.

Proof We know $(K_{m,y} + D_{m,y})^2 = (K_{m,s} + D_{m,s})^2 > 0$. On the other hand, $H^2(\bar{X}_{m,y}, n(K_{m,y} + D_{m,y})) = 0$ by the upper semi-continuity. By the theorem of Riemann-Roch we get the result.

(2.6) If $\bar{\kappa}(X_y) = 0$, then $\bar{\kappa}(X_s) = 0$.

Proof We have only to show that $\bar{\kappa}(X_s)$ cannot be 1. Assume the contrary. Since $(K_{m,y} + D_{m,y})^2 = (K_{m,s} + D_{m,s})^2 = 0$, $K_{m,y} + D_{m,y}$ must be arithmetically effective, i.e., $(\bar{X}_{m,y}, D_{m,y})$ gives a minimal model. Then, $[n(K_{m,s} + D_{m,s})] \equiv [n(K_{m,y} + D_{m,y})] = 0$, a contradiction.

Thus, the proof of (2.2) is complete.

(2.7) Corollary. *If* $\bar{\kappa}(X_o) = 0$ *or* 1, *then* $\bar{P}_n$ *are also deformation invariants*.

<u>Proof</u> Since the set of points of finite order in $Pic^o(\overline{X}_s)$ is discrete, the theorem follows in case $\bar{\kappa}(X_s) = 0$. In case $\bar{\kappa}(X_s) = 1$, this is a consequence of the formulae (2.5) and (2.8).

References

[1] M. Artin : Some numerical criterion for contractibility of curves on algebraic surfaces, Amer. J. Math.,84(1962), 485-496.

[2] E. Bombieri : Canonical models of surfaces of general type, Publ. Math. IHES, 42 (1973), 171-219.

[3] S. Iitaka : Deformations of compact complex surfaces II, Jour. Math. Soc. Jap., 22 (1970), 247-261.

[4] S. Iitaka : On logarithmic Kodaira dimension of algebraic varieties, Complex Analysis and Algebraic Geometry, 1977, Iwanami, Tokyo.

[5] U. Karras : Eigenschaften der lokalen Ringe in zweidimensionalen Spitzen, Math. Ann., 215 (1975), 117-129.

[6] Y. Kawamata : On deformations of compactifiable complex manifolds, Math. ann., 235 (1978), 247-265.

[7] Y. Kawamata : Addition formula of logarithmic Kodaira dimension for morphisms of relative dimension one, Proc. Alg. Geometry in Kyoto, 1977.

[8] K. Kodaira : On compact analytic surfaces II, Ann. of Math., 77 (1963), 563-626.

[9] D. Mumford : The canonical ring of an algebraic surface, appendix to 14 .

[10] D. Mumford : Enriques' classification of surfaces in char p, Global Analysis, 1969, Univ. Tokyo and Univ. Princeton.

[11] K. Saito : Einfach-elliptische Singularitäten, Inv. Math., 23

(1974), 289-325.

[12] F. Sakai : Logarithmic pluricanonical maps of algebraic surfaces, preprint.

[13] P. Wagreich : Elliptic singularities of surfaces, Amer. J. Math., 92 (1970), 419-454.

[14] O. Zariski : The theorem of Riemann-Roch for high multiples of an effective divisor on an algebraic surface, Ann. of Math., 76 (1962), 560-615.

University of Tokyo and Universität Mannheim
Lehrstuhl für Mathematik VI
Universität Mannheim
D-68 Mannheim
Germany

The length of vectors in representation spaces

George Kempf* — The Johns Hopkins University and Princeton University

and

Linda Ness — University of Washington and The Institute for Advanced Study

Let V be a finite dimensional morphic representation of a connected reductive algebriac group G over the complex numbers. Given a maximal compact subgroup K of G, we will fix a Hermitian norm $|| \ ||$ on V so that the action of K on V preserves this norm.

Let v be a vector in V. The purpose of this paper is to study how the length changes as one moves along the orbit $G \cdot v$. Thus we want to examine the function $p_v(g) = ||g \cdot v||^2$ on G. If G_v denotes the stabilizer of v in G, the function p_v on G is invariant on the left by K and on the right by G_v. Hence p_v is constant on double $K - G_v$ cosets of the form $K \cdot g \cdot G_v$.

In this paper we will show that the function p_v has very special properties. We will first prove

Theorem 0.1. a) Any critical point of p_v is a point where p_v obtains its minimum value.

If p_v obtains a minimum value,

b) then the set m where p_v obtains this value consists of a single $K - G_v$ coset and is connected, and

c) the second order variation of p_v at a point of m in any direction not tangent to m is positive.

With this theorem in mind, one may ask, "when does p_v obtain a minimum

---

*Partly supported by NSF contract # MPS75-05578.

value?" If v is a stable vector (i.e., the orbit $G \cdot v$ is closed and $v \neq 0$), then p_v clearly obtains a minimum value. The converse is also true by

Theorem 0.2. The vector v is stable if and only if p_v obtains a minimum value.

The development of the properties of the function p_v is similar to that of Mumford's numerical function (See [8] and [6]). Also the functions p_v for certain unstable vectors v play an interesting role in Borel's treatment of reduction theory [3]. Furthermore, related material may be found in [1,4,7]. We hope that our results may be useful for studying moduli via geometric invariant theory.

In section one, we will deal abstractly with a more general type of function related to the functions like p_v. In the next section, we will apply the previous ideas to prove Theorem 0.1 when G is an algebraic torus. In the third section, we will treat a general reductive group. In the last section, we will prove Theorem 0.2.

§1. Special functions on affine spaces

Let A be a finite dimensional affine space over the real numbers. A special function on A is a finite sum $\sum e^{H_i(a)}$ where the H_i are affine functions on A.

We will begin by studying special functions where A is the real line $\mathbb{R}$. Thus a special function on $\mathbb{R}$ may be written uniquely as $f(x) = \sum a_i e^{\ell_i \cdot x}$, where the a_i's are positive real numbers and the ℓ_i's are distinct real numbers. By calculus, we may deduce

Lemma 1.1. a) The second derivative f'' is never negative.

b) If f'' is zero anywhere, then f is constant.

c) A non-constant special function on $\mathbb{R}$ is a strictly convex Morse function.

Proof. For a), note that $f''(x) = \sum (a_i \cdot \ell_i^2) e^{\ell_i \cdot x}$ is the sum of non-negative terms and, hence, f'' is never negative. For b), if $f''(x) = 0$ for some x, then $a_i \cdot \ell_i^2 = 0$ for each i. This can only happen if $\ell_i = 0$ as $a_i \neq 0$. Therefore, if $f''(x) = 0$ for some x, then f is a constant function. This proves b). Part c) results from a) and b). Q.E.D.

A special function on an affine space is called degenerate if its restriction to any line is constant. We may generalize part a) of the last lemma in

Proposition 1.2. Any non-degenerate special function f on an affine space is a strictly convex Morse function. Hence, if f has a critical point, it must be the unique point where f obtains its minimum value.

Proof. As f is non-degenerate, its restriction to any line is a strictly convex Morse function. Therefore, f is strictly convex and has positive second order variation through its critical points. This proves the first statement. The second statement follows formally from the first. Q.E.D.

To finish our discussion of special functions, we will show how to reduce the general case to the non-degenerate case. The reduction will be achieved by means of

Lemma 1.3. Let f be a special function on an affine space A. There is a unique quotient affine space B of A and a non-degenerate special function on B such that $f = g \circ \pi$ where $\pi : A \to B$ is the quotient mapping.

Proof. If $f = \sum e^{H_i(a)}$ where the H_i's are affine functions on A, there is a maximal affine subspace M_a through any point a of A such that each $H_i|_{M_a}$ is constant. As the M_a's are parallel and have the same dimension, these subspaces $\{M_a\}$ form a quotient affine space, say B. Clearly, f descends to a special function g on B. One may easily check that g is non-degenerate by using the maximality of the M_a. The uniqueness assertion is obvious. Q.E.D.

We may now generalize the last proposition in

Theorem 1.4. Let f be a special function on an affine space A. Then,

a) f is convex,

b) f has no critical points outside of the set m of points where f obtains its minimum value,

c) m is an affine subspace of A,

d) at any point of m, the second order variation of f is positive in any direction not tangent to m.

Proof. By Lemma 1.3, the stated properties may be easily deduced from the Proposition 1.2. Q.E.D.

§2. The toroidal case.

In this section, we will assume that the group G of the introduction is an algebraic torus $T = \{(t_1,\dots,t_n) | t_i \in \mathbb{C} - \{0\}\}$. The maximal compact subgroup K_T consists of the elements of T whose coordinates have absolute value one.

Consider the morphic representation of T on V. Then V may be written uniquely as the direct sum $V = \oplus V_\chi$ where V_χ is the χ-eigenspace of V for the characters χ of T. Recall that a character $\chi: T \to \mathbb{C} - \{0\}$ is a (morphic) homomorphism sending $(t_1,\dots,t_n)$ to $\Pi\, t_i^{m_i}$ where $(m_1,\dots,m_n)$ is a sequence of integers.

Our Hermitian norm $||\ ||$ on V is invariant under K_T if and only if two eigenspaces with distinct characters are perpendicular. With this information, we may easily determine the nature of the functions $p_v(t) = ||t \cdot v||^2$ for any vector v in V.

Lemma 2.1. For any vector v in V, there is a finite set $\Xi(v)$ of elements in Z^n such that

a) $$p_v((t_1,\dots,t_n)) = \sum_{(m_i)\, \varepsilon\, \Xi(v)} (\text{some positive real})\ \Pi |t_i|^{2m_i}$$ and

b) the stabilizer T_v of v in T is given by the system of equations

$$\Pi\, t_i^{m_i} = 1 \quad \text{for} \quad (m_i) \in \Xi(v).$$

Proof. Let $v = \sum v_\chi$ be the eigendecomposition of v where each v_χ is non-zero. Let $\Xi(v)$ be the set of sequences (m_i) corresponding to the characters which occur in this decomposition. As $(t_1,\ldots,t_n)\cdot v = \sum_{(m_i)\in\Xi(v)} \Pi\, t_i^{m_i} \cdot v_{(m_i)}$, the part b) is evident. Using the orthogonality mentioned above, we have

$$p_v\,(t_1,\ldots,t_n) = ||(t_1,\ldots,t_n)\cdot v||^2 = \sum ||\Pi\, t_i^{m_i} \cdot v_{(m_i)}||^2 = \sum ||v_{(m_i)}||^2 \Pi |t_i|^{2m_i}$$

This proves a). Q.E.D.

To prove the desired facts about the function p_v, we may study the induced function p_v' on the quotient Lie group $K_T\backslash T$. Next we will introduce coordinates on the quotient $K_T\backslash T$. Define $x(t_1,\ldots,t_n) = (\log|t_1|,\ldots,\log|t_n|) = (x_1,\ldots,x_n)$ for any $(t_1,\ldots,t_n)$ in T. Thus x defines an isomorphism from $K_T\backslash T$ with the vector space $\mathbb{R}^n$. With these coordinates we may reinterpret the last lemma in

Lemma 2.2. For any vector in V,

a) $p_v'\,(x_1,\ldots,x_n) = \sum_{(m_i)\in\Xi(v)} e^{\text{real} + 2\sum m_i x_i}$ and

b) the locus $\{\sum m_i x_i = 0$ for all (m_i) in $\Xi(v)\}$ is the image of T_v in $K_T\backslash T$

Proof. This is a trivial reformulation of Lemma 2.1. Q.E.D.

We will now apply the theory of section one to the function p_v' on $K_T\backslash T$. By part a) of Lemma 2.2, we know that p_v' is a special function on the vector space $K_T\backslash T = \mathbb{R}^n$. By the proof of Lemma 1.3 and the part b) of Lemma 2.2, we know that the function p'_v comes from a non-degenerate special function on

$\mathbb{R}^n/\{\sum m_i x_i = 0 \mid (m_i) \in \Xi(v)\} = K_T\backslash T/T_v$.

Therefore, by the results of the last section, we have verified the theorem 0.1 when G is an algebraic torus. We will next show how the general case follows from the toroidal case.

§4. The reductive case.

Using E. Cartan's results about the geometry of $K\backslash G$, we will be able to deduce the reductive case from the toroidal case of Theorem 0.1. First we will recall some of Cartan's results.

Let $\mathcal{T}$ be the set of maximal algebraic tori T in the reductive group G such that $K \cap T$ is the maximal compact subgroup K_T of T. Thus $K_T\backslash T$ is a submanifold of $K\backslash G$. We will need to know the Cartan decomposition,

1) $K\backslash G = \bigcup_{T \in \mathcal{T}} K_T\backslash T$

and its infinitesimal form

2) tangent spaces of $K\backslash G$ at $K = \bigcup_{T \in \mathcal{T}}$ tangent space of $K_T\backslash T$ at K_T.

[for instance, 5].

To prove part a) of theorem 0.1, it will suffice to assume that

A) e is the critical point of p_v, as $p_{h\cdot v}(g) = p_v(g\cdot h)$. Let g be any point of G. We want to prove that $p_v(g) \geq p_v(e)$. By 1), $g = k \cdot t$ where $k \in K$ and $t \in T$ for some T in $\mathcal{T}$. By the left K-invariance of p_v, we have $p_v(g) = p_v(t)$. As e must be a critical point of $p_v|_T$, we have the inequality $p_v(t) \geq p_v(e)$ by the toroidal case of the theorem. Hence, $p_v(g) \geq p_v(e)$ which proves part a) of Theorem 0.1.

To prove the last two parts of the theorem, it will suffice to treat the case where

B) p_v obtains its minimum value at e(i.e., $K\cdot G_v \subsetneq m$), because $p_{h\cdot v}(g) = p_v(g\cdot h)$ and $G_{h\cdot v} = hG_vh^{-1}$. Let g be an element of m. If we write $g = k\cdot t$ as before, we find that t is a point of T where $p_v|_T$ obtains its minimum value $p_v(e)$. By the toroidal case, $t \in K_T\cdot T_v$ which is a connected subset containing e. Therefore, $m \subsetneq K \cdot (\bigcup_{T\in\mathcal{T}} K_T\cdot T_v) \subseteq K \cdot G_v$ and $\bigcup_{T\in\mathcal{T}} K_T\cdot T_v$ is connected. Hence, $m = K\cdot G_v (= K \cdot \bigcup_{T\in\mathcal{T}} K_T\cdot T_v)$ is connected because K is also connected as G is. This proves part b) of the theorem 0.1.

To prove part c) of theorem 0.1, it again will suffice to prove the statement about directions through the point e of m. Let p_v' be the function induced by p_v on $K\backslash G$. We want to show that

*) If X is a tangent vector to $K\backslash G$ at K such that that second order variation of p_v' along X is zero, then X is tangent to $K\backslash K \cdot G_v$.

By 2), X is tangent to $K_T\backslash T$ for some T in $\mathcal{T}$. By the toroidal case of the theorem, X must be tangent to $K_T\backslash K_T \cdot T_v$ which is contained in $K\backslash K \cdot G_v$. This completes the proof of theorem 0.1.

§4. Analysis of stability

In this section we will study the relationship between the stability of a vector v in V and the function p_v. We will exclude the trivial case $v = 0$.

Recall the definitions of various notions of stability. The vector is called a) unstable if $\overline{G \cdot v} \ni 0$,

b) not stable if $G \cdot v$ is not closed,

c) stable if $G \cdot v$ is closed and

d) properly stable if v is stable and G_v is connected.

Trivially one has the following criterian:

(†) v is unstable $\Longleftrightarrow$ inf $p_v = 0$.

The Theorem 0.2 will give us a criterian for v being stable in terms of the function p_v. To prove Theorem 0.2, we have to show that,

(*) if v is not stable, then inf p_v is not a value of p_v.

<u>Proof of Theorem 0.2.</u>

Regard G as an algebraic group defined over R such that K is the real locus of G. Then the tori T in $\mathcal{T}$ are exactly the maximal tori of G, which are defined over $\mathbb{R}$.

Assume that the vector v is not stable. Then we may find a parabolic subgroup P of G such that each maximal torus T of P contains a one-parameter algebriac subgroup $\lambda_T: \mathbb{C} - \{0\} \to T$ such that $\lim_{t \to 0} \lambda(t) \cdot v$ exists in V and is a point outside of the orbit $G \cdot v$. See [1] or [6].

Let $\overline{P}$ be the parabolic subgroup of G, which is conjugate to P under the real structure on G. Then $P \cap \overline{P}$ is a subgroup of G which is defined over $\mathbb{R}$ and must contain a maximal torus S defined over $\mathbb{R}$ [2]. As a maximal torus of the intersection of two parabolic subgroups of G is a maximal torus of G [2], S is a maximal torus in the collection $\mathcal{T}$.

By the one-parameter subgroup λ_S of S, we have an action of $\mathbb{C} - \{0\}$ on V such that v is not stable for this action and the maximal compact subgroup of $\mathbb{C} - \{0\}$ preserves the Hermitian norm on V. As the statement (*) for $\mathbb{C} - \{0\}$ implies the statement (*) for G, it will suffice to prove Theorem 0.2 when $G = \mathbb{C} - \{0\}$. Better yet, we may even assume that $\lim_{t \to 0} t * v$ exists in V and does not equal v, where * denotes the action of $\mathbb{C} - \{0\}$ on V.

As in section 2, we may write p'_v on maximal compact $\backslash \mathbb{C} - \{0\} = \mathbb{R}$ uniquely in the form $\sum a_i e^{\ell_i x}$ with positive a_i's and increasing real ℓ_i's. As the above limit exists, the $\lim_{x \to \infty} p_v'(x)$ exists in $\mathbb{R}$ and, hence, the ℓ's are not negative. As the above limit does not equal v, at least one ℓ_i must be positive. Thus, $p_v'(x)$ is a strictly increasing function on $\mathbb{R}$. Hence, p_v' and p_v never obtain a minimum value. This proves the statement (*) and, hence, Theorem 0.2 is true. Q.E.D.

To finish our criteria for stability, it remains to dramatize the meaning of v being properly stable in terms of the function p_v. We will do this in

Theorem 4.1. If v is properly stable, then

a) the induced function p_v' on $K\backslash G$ is a Morse function with one critical point where it obtains its minimum value.

If furthermore p_v obtains its minimum value at the identity e, then

b) the stabilizer G_v is contained in K.

Conversely, if p_v' obtains a minimum at a unique point of $K\backslash G$, then v is properly stable.

Proof. For a), note that Theorems 0.1 and 0.2 say that the only critical points of p_v are situated on a connected $K - G_v$ coset m where p_v obtains its minimum value. As G_v is finite, this double coset m is actually a K-coset. Hence, p_v' has only one critical point on $K\backslash G$ where it obtains its minimum. The last part of Theorem 0.1 means in this case that the second order variation of p_v' in any non-zero direction through its critical point is positive. In other words, p_v' is a Morse function. This proves a).

To prove b), note that a) implies that $K \cdot G_v = K$ as K is the unique critical point of p_v' and p_v' is right invariant under G_v. Therefore, $G_v \subseteq K$. This proves b).

For c), note that Theorem 0.2 shows that v is stable because p_v obtains a minimum value. As we may assume without loss of generality that p_v' obtains its minimum at the point K in $K\backslash G$, we have the inclusion $K \cdot G_v \subseteq K$. Hence, $G_v \leq K$. As G_v is a compact algebraic subgroup of G, it can only be contained in the compact group K if G_v is finite. Therefore, G_v is finite and, hence, v is properly stable. Q.E.D.

Trivially we want to mention a special way to find a vector in the orbit $G \cdot v$ where p_v obtains a minimum value.

Corollary 4.2. Assume that v is properly stable and G_v is contained in K. If $K = \{g \in G \mid gG_vg^{-1} \subseteq K\}$, then p_v obtains its minimum value at e.

Proof. The stablilizer of $g \cdot v$ is gGg^{-1}. By Theorem 4.1, we know that $gG_vg^{-1} \subseteq K$ if p_v has a minimum at g. Therefore the hypothesis means that the coset K is the only possiblity for the set where p_v obtains its minimum. Q.E.D.

An example where this hypothesis is satisfied is when $G = SL(W)$, the induced representation of the finite group G_v on W is irreducible and K consists the elements of G which preserve some Hermitian norm of W. The hypothesis applies because of the uniqueness up to scalar of the Hermitian norm invariant under the action of an irreducible representation.

This kind of example actually occurs in the study of covariants of abelian varieties over C embedded by complete linear systems.

References

[1] D. Birkes, Orbits of linear algebraic groups, Annals of Mathematics 93(1971), 459-475.

[2] A. Borel, Linear Algebraic Groups, Benjamin, New York, 1969.

[3] A. Borel, Introduction aux groupes arithmetiques, Hermann, Paris, 1969.

[4] A. Borel and Harish-Chandra, Arithmetic subgroups of algebraic groups, Annals of Mathematics 75(1962), 485-535..

[5] S. Helgason, Differential Geometry and Symmetric Spaces, Academic Press, New York, 1962.

[6] G. Kempf, Instability invariant theory, Annals of Mathematics, to appear.

[7] G. Mostow and T. Tamagawa, On the compactness of arithmetically defined homogeneous spaces, Annals of Mathematics 76(1962).

[8] D. Mumford, Geometric Invariant Theory, Ergebnisse der Math. (34), Springer-Verlag, Berlin, 1965.

The Johns Hopkins University
Baltimore, Maryland 21218
USA

University of Washington
Seattle, Washington 98105
USA

THE GENERIC PERFECTNESS OF DETERMINANTAL SCHEMES

by H. Kleppe and D. Laksov

We shall in the following article present a proof of the generic perfectness of determinantal schemes. The proof follows closely the one of Eagon and Hochster [1]. In particular we use their idea of introducing a large class of Schubert like schemes and forming what they call a principal radical system of ideals. Moreover, we follow their elegant induction argument showing that the most important schemes in this radical system are perfect. At one single technical point does our proof differ from the one of Eagon and Hochster, namely in the verification of the irreducibility of the basic schemes of the class of ideals considered. This is the crucial property for the class to be radical and they verify it by constructing generic points for the schemes explicitly. We proceed by an inductive method. The justification for having this article printed is that our method can be used with only technical changes to many other interesting classes of ideals like ideals generated by pfaffians of alternating matrices or by determinants of symmetric matrices. In fact, we discovered the method during work on pfaffian ideals described in the article "Deformation and transversality" in this volume. There one can also find a list of other proofs of the perfectness of determinantal schemes.

§ 1. Let R be a noetherian ring (commutative with unit) and let S be an R-algebra. Given an axb-matrix $M = (s(i, j))$ with coordinates in S we denote by R[M] the subalgebra of S generated by the coordinates $s(i, j)$ of M and by $D_p(M)$ the ideal in R[M] generated by all subdeterminants of M of order p. More generally assume that m is an integer such that $m + 1 \leq a$ and $m + 1 \leq b$ and that $0 \leq b_1 < b_2 < \dots < b_m < b$ is a sequence of integers such that $b_i \geq i - 1$ for $i = 1, \dots, m$. Denote by $M(b_i)$ the axb_i-matrix consisting of the b_i first columns of M. Moreover, for each integer $c < b$ denote by $D(M; b_1, \dots b_m; c)$ the ideal in R[M] generated by the elements of $D_{m+1}(M)$, the elements of all the ideals $D_i(M(b_i))$ for $i = 1, \dots, m$ and the elements $\{s(1, 1), \dots, s(1, c)\}$.

Note that all the values $c = 1, \dots, b_1$ give the same ideal.

<u>Example 1</u>. If $b = 2$ then $m = 1$ and we get

$$D(M;\, 1;\, 1) = (s(1,\, 1),\, \ldots,\, s(a,\, 1))\ .$$

If a = 2 then m = 1 and we get

$$D(M;\, b_1;\, b_1) = (s(1,\, 1),\, \ldots,\, s(1,\, b_1),\, s(2,\, 1),\, \ldots\, s(2,\, b_1)\ ,$$
$$s(1,\, j)s(2,\, k) - s(1,\, k)s(2,\, j) \text{ for } j,\, k \geq b_1)\ .$$

<u>Remark 2</u>. (Eagon-Hochster) Assume that $b_n < c < d \leq b_{n+1}$ and put $D = D(M;\, b_1,\, \ldots,\, b_m;\, c)$ and $D_1 = D(M;\, b_1,\, \ldots,\, b_m;\, d)$ and $D_2 = D(M;\, b_1,\, \ldots,\, b_{n-1},\, c,\, b_{n+1},\, \ldots,\, b_m;\, c)$. Denote by Q, Q_1 and Q_2 the radicals of these ideals. Then

(i) $Q = Q_1 \cap Q_2$

(ii) $s(1,\, j)D_2 \subseteq D$ for $j = c + 1,\, \ldots,\, d$.

Indeed the inclusions $D \subseteq D_1$ and $D \subseteq D_2$ imply the inclusion $Q \subseteq Q_1 \cap Q_2$. The opposite inclusion will follow from $D_1 \cap D_2 \subseteq Q$ which again is a consequence of $D_1 \cdot D_2 \subseteq D$. However $D_1 = D + (s(1,\, c+1),\, \ldots,\, s(1,\, d))$ and $D_2 = D + D_n(M')$ where M' is the axc-matrix obtained from the first c columns of M. Hence we easily see that the remaining part of assertion (i) will follow from assertion (ii) and that to prove assertion (ii) it suffices to prove that $s(1,\, j)\det N \in D$ for $j = c + 1,\, \ldots,\, d$ and all nxn-submatrices N of M'.

If the first row of N is taken from the first row of M' there is nothing to prove. If not we expand the (n + 1)x(n + 1)-matrix obtained by adding elements from the first row and j'th column to N along the first row. We obtain $s(1,\, j)\det N + t \in D$ where $t \in (s(1,\, 1),\, \ldots,\, s(1,\, c)) \subseteq D$.

<u>Remark 3</u>. Fix integers $1 \leq p \leq a$ and $1 \leq q \leq b$. Performing elementary row and column operations, we transform the axb-matrix $M = (s(i,\, j))$ into an axb-matrix $M' = (s'(i,\, j))$ where $s'(i,\, j) = s(i,\, j) - s(p,\, j)s(i,\, q)s(p,\, q)^{-1}$ when $p \neq i$ and $q \neq j$ and $s'(p,\, q) = 1$. Denote by $N = (t(i,\, j))$ the (a - 1)x(b - 1)-matrix obtained from M' by deleting row p and column q. If we let $s = s(p,\, q)$ and we assume that s is invertible in S, then

(i) If p = 1 and $b_h < q \leq b_{h+1}$ with $h \geq n$, then

$$D(M;\, b_1,\, \ldots,\, b_m;\, b_n)S + (s(1,\, 1),\, \ldots,\, s(1,\, q-1)) =$$
$$D(M',\, b_1,\, \ldots,\, b_m;\, b_n)S + (s(1,\, 1),\, \ldots,\, s(1,\, q-1)) =$$
$$D(N;\, c_1,\, \ldots,\, c_{m-1};\, c_{n-1})S + ((s(1,\, 1),\, \ldots,\, s(1,\, q-1))\ ,$$

where $c_i = b_i$ for $i = 1,\, \ldots,\, h-1$ and $c_i = b_{i+1} - 1$ for $i = h,\, \ldots,\, m-1$.

(ii) If $p = 2$ and $b_1 < q \leq b_2$ and $I(1) = (s(2, 1), \ldots, s(2, b_1))$ and $I(n) = I(1) + (s(1, q))$ for $n \geq 2$, then

$$D(M; b_1, \ldots, b_m; b_n)S =$$
$$D(M'; b_1, \ldots, b_m; b_n)S + I(n) =$$
$$D(N; c_1, \ldots, c_{m-1}; c_{max(1,n-1)})S + I(n)$$

where $c_i = b_{i+1} - 1$ for $i = 1, \ldots, m - 1$.

Remark 4. In the above remark we see that if all the coordinates $s(i, j)$ of M are algebraically independent then the coordinates $t(i, j)$ of N are algebraically independent and that the coordinates $\{s(p, 1), \ldots, s(p, b), s(1, q), \ldots, s(a, q)\}$ are algebraically independent over $R[N]$ and the polynomial ring in these variables over $R[N]$ is, after localization in s, equal to $R[M]_s$.

In the following assertions we collect some easy and well known results that we shall need in the following.

Lemma 5. Let $x_1, \ldots, x_n = x$ be independent variables over the ring R and let $S = R[x_1, \ldots, x_n]_x$. Moreover, let $m < n$ be an integer and I an ideal of R. Then for every isolated prime ideal Q of $IS + (x_1, \ldots, x_m)$ the following assertions hold,

(i) $Q' = Q \cap R$ is an isolated prime of I in R.

(ii) $Q = Q'S + (t_1, \ldots, t_m)$.

(iii) $htQ = htQ' + m$.

§ 2. Let $X = (x(i, j))$ be an axb-matrix whose coordinates are in S and are algebraically independent over R.

Lemma 6. Let Q be an isolated prime ideal of $D = D(X; b_1, \ldots, b_m; b_n)$. If $a \geq 2$ and $2 \leq p \leq a$ then $s(p, j) \notin Q$ for $b_1 < j \leq b_2$. Moreover if $n = 1$ $s(1, j) \notin Q$ for $b_1 < j \leq b_2$.

Proof. By switching rows we see that it is sufficient to prove the lemma for $p = 2$ and that when $n = 1$ the first and second assertion of the lemma are equivalent.

The lemma holds when $a = 2$ by Example 1. We now proceed by induction on a. First assume that $s(1, j) \in Q$ for $j = 1, \ldots, b$. Then if Z denotes the (a - 1)xb-matrix obtained from X by deleting the first row and if $f : R[X] \to R[Z]$ denotes the resulting natural map, we see that in R[Z] we have

$$f(D) = D(Z; b_1, \ldots, b_m; b_1) .$$

By the induction assumption $z(1, j) \notin f(Q)$ for $b_1 < j \leq b_2$. Since $f(x(2, j)) = z(1, j)$ it follows that $x(2, j) \notin Q$.

Secondly we assume that $x(1, j) \notin Q$ for some j and let q be the smallest such integer. Then $q > b_n$. If $j \leq b_2$ then $n = 1$ and we have finished. On the other hand if $b_h < q \leq b_{h+1}$ and $h \geq n \geq 2$ then by case (i) of Remark 3 with X and Y corresponding to M and N we have

$$D(X; b_1, \ldots, b_m; b_n)_s + (s(1, 1), \ldots, s(1, q-1)) =$$
$$D(Y; c_1, \ldots, c_{m-1}; c_1)R[X]_s + (s(1, 1), \ldots, s(1, q-1)) ,$$

where $s = s(1, q)$ and $c_i = b_i$ for $i = 1, \ldots, h - 1$ and $c_i = b_{i+1} - 1$ for $i = h, \ldots, m - 1$. By Remark 4 and Lemma 5 (i) the ideal $Q' = Q \cap R[Y]$ is an isolated prime of $D(Y; c_1, \ldots, c_{m-1}; c_1)$. Hence by induction assumption $y(1, j) \notin Q'$ for $c_1 < j \leq c_2$. However for $j \leq b_2$ $y(1, j) \equiv x(2, j) \bmod (x(1, 1), \ldots, x(1, q-1))$ and since $(x(1, 1), \ldots, x(1, q-1)) \in Q$ we have that $x(2, j) \notin Q$ for $c_1 < j \leq c_2$ and in particular for $b_1 < j \leq b_2$.

<u>Proposition 7</u>. <u>Let Q be an isolated prime of</u> $D = D(X; b_1, \ldots, b_m; b_n)$. <u>Then the depth of</u> Q <u>is equal to</u>

$$(a - m)(b - m) + \sum_{i=1}^{m} (b_i - i + 1) + n - 1$$

<u>Proof</u>. We proceed by induction on b starting from the case b = 2 of Example 1.

By Lemma 6 we have that $x = x(2, b_2) \notin Q$ and by the case (ii) of Remark 3 with X and Y corresponding to M and N and $q = b_2$

$$D(X; b_1, \ldots, b_m; b_1)R[X]_x =$$
$$D(Y; c_1, \ldots, c_{m-1}; c_1)R[X]_x + (x(2, 1), \ldots, x(2, b_1))$$

and if $n \geq 2$

$$D(X; b_1, \ldots, b_m; b_n)R[X]_x =$$
$$D(Y; c_1, \ldots, c_{m-1}; c_{n-1})R[X]_x + (x(2, 1), \ldots, x(2, b_1), x(1, b_2))$$

where $c_i = b_{i+1} - 1$ for $i = 1, \ldots, m - 1$.

From Remark 4 and Lemma 5 (i) it follows that $Q' = Q \cap R[X]$ is an isolated prime ideal of $D(Y; c_1, \ldots, c_{m-1}; c_{n-1})$ if $n \geq 2$ and of $D(Y; c_1, \ldots, c_{m-1}; c_1)$ otherwise. Hence by the induction assumptation the height of Q' is

$$(a - m)(b - m) + \sum_{i=1}^{m-1} (c_i - i + 1) + \max (0, n - 2) .$$

This number differs from the asserted height of Q by b_1 if $n = 1$ and by $b_1 + 1$ if $n > 1$. The proposition now follows from Lemma 5 (iii).

Lemma 8. Assume that $b_{r+1} > r$ and let $2 \leq i_1 < \ldots < i_r \leq a$ and $j_1 < \ldots < j_r \leq b$ be two sequences of integers where $b_k < j_k \leq b_{k+1}$ for $k = 1, \ldots, r$. Denote by $X(i_1, \ldots, i_r, j_1, \ldots, j_r)$ the rxr-matrix with coordinates $x(i_p, j_q)$. Let Q be an isolated prime ideal of $D = D(X; b_1, \ldots, b_m; b_n)$. Then $\det (X(i_1, \ldots, i_r, j_1, \ldots j_r)) \notin Q$.

Proof. By switching rows we see that we may assume that $i_1 = 2$. We proceed by induction on r. By lemma 6 the assertion holds when $r = 1$. Let $x = x(2, j_1)$. From Remark 3 part (ii) with $q = i_1$ and with X and Y corresponding to M and N we have

$$D(X; b_1, \ldots, b_m; b_n)R[X]_x =$$
$$D(Y; c_1, \ldots, c_m; c_{\max(1, n-1)})R[X]_x + I(n)$$

where $c_i = b_{i+1} - 1$ for $i = 1, \ldots, m - 1$. Moreover $\det X(i_1, \ldots, i_r, j_1, \ldots, j_r) = \det Y(i_2 - 1, \ldots, i_r - 1, j_2 - 1, \ldots, j_r - 1) = y$. By Remark 4 and Lemma 5 (i) the ideal $Q' = Q \cap R[Y]$ is an isolated prime ideal of $D' = D(Y; c_1, \ldots, c_m; c_{\max(1,n-1)})$. Hence $y \notin D'$ and the proposition is proved.

Proposition 9. Let $2 \leq i_1 < \ldots < i_m \leq a$ and $j_1 < \ldots < j_m \leq b$ be two sequences of different integers such that $b_k < j_k \leq b_{k+1}$ for $k = 1, \ldots, m$. Denote by x_r the determinant of the rxr-matrix with coordinate $x(i_k, j_l)$ for $l, k = 1, \ldots, r$ and by x the product of those x_r where $b_r > r - 1$. We define $m + 2$ sets as follows

$$B_{m+1} = \{x(i, j) | i \notin \{i_1, \ldots, i_m\}, j \notin \{j_1, \ldots, j_m\}\}$$
$$B_k = \{x(i_k, 1), \ldots, x(i_k, b_k)\} \setminus \{x(i_k, j_1), \ldots, x(i_k, j_{k-1})\}$$
$$B_o = \{x(1, j_k) | j_k \leq b_n\} .$$

Finally let S be the polynomial ring over R in the

$$ab - [(a - m)(b - m) + \sum_{i=1}^{m} (b_i - i + 1) + n - 1]$$

variables of X not in the set $\bigcup_{i=o}^{m+1} B_i$ and let $D = D(X; b_1, \ldots, b_m; b_n)$.

Then the principal open subset V_x of $V = \text{Spec}\,(R[X]/D)$ is dense and if R is an integral domain its coordinate ring is S_x.

Note that if $b_k = k - 1$ then B_k is empty.

Proof. We conclude immediately from Lemma 8 that x is not in any isolated prime ideal of D and consequently that V_x is dense in V. If $x(p, q) \in B_{m+1}$ then we add row p and column q to x_m and expand the resulting $(m + 1)x(m + 1)$-matrix along column q. We obtain a relation

$$x(p, q)x_m + f_{m,p,q} \in D$$

where $f_{m,p,q}$ is a polynomial in R[X] not containing x(p, q).
If $x(p, q) \in B_k$, we add row p and column q to the matrix x_{k-1}. The resulting kxk-matrix we expand along the q'th column. We then obtain a relation

$$x(p, q)x_{k-1} + f_{k,p,q} \in D$$

where $f_{k,p,q}$ is a polynomial not containing x(p, q).
The set $\bigcup_{i=1}^{m+1} B_i$ contains exactly the coordinates $\{x(1, j) | j \notin \{j_1, \ldots, j_m\}\}$ from the first row. Consequently the elements that are not in this set but in the set $\{x(1, 1), \ldots, x(1, b_n)\}$ are B_o.
Let I be the ideal in $R[X]_x$ generated by the elements

$$\{x(i, j) + f_{k,i,j}x_k^{-1} | x(i, j) \in \bigcup_{k=1}^{m+1} B_k\} \cup \{x(i, j) | x(i, j) \in B_o\} .$$

Then $R[X]_x/I \cong S_x$. Consider the natural surjective map

$$R[X]_x/I \to R[X]_x/D .$$

By Proposition 7 we have that $\text{ht}Q = \dim R[X] - \dim S_x = \text{ht}I$ for any isolated prime ideal Q containing D, and if S_x is integral this surjection is also injective.

Corollary 10. If R is integral, then the radical of the ideal $D(X; b_1, \ldots, b_m; b_n)$ is a prime ideal in R[X].

§ 3. Recall that an ideal I in a ring S is called perfect if the homological dimension dhS/I of S/I is equal to the length of the longest S-sequence with elements from I (the grade of I). The aim of this section is to prove, following the method of Eagon and Hochster, the following result.

Theorem 11. If R is an integral domain and $c = b_n$ or $c = b_n + 1$ for some n, then $D = D(X; b_1, \ldots, b_m; c)$ is a perfect ideal in R[X] of height

$$(a - m)(b - m) + \sum_{i=1}^{m} (b_i - i + 1) + n - 1 + c - b_n .$$

When $c = b_n$ then D is a prime ideal.

One of the main features of Eagon and Hochsters proof is that it demonstrates the importance of Corollary 10. They show that, on the one hand, Corollary 10 together with Remark 2 are the properties of the system of ideals $D(X; b_1, \ldots, b_m; c)$ that implies that they, in Eagon and Hochsters terminology, form a principal radical system and consequently that they are radical. On the other hand they show that Theorem 11 is an immediate consequence of Corollary 10, the height formula of Proposition 7 and the property that the ideals are radical.

The arguments given by Eagon and Hochster are so short and elegant that it is both more instructive and more economical to repeat them than to refer to their article.

Proposition 12. If R is an integral domain, then the ideals $D(X; b_1, \ldots, b_m; c)$ are radical.

Proof. We shall perform descending induction on the lattice of ideals of the form $D(X; b_1, \ldots, b_m; c)$. The maximal such ideal $D(X; b - m, \ldots, b - 1; b - 1)$ is obviously radical.

Let Q be the radical of $D(X; b_1, \ldots, b_m; c)$, and suppose all the ideals $D(X, b_1', \ldots, b_h', c')$ that contains $D(X; b_1, \ldots, b_m; c)$ are radical. Then $Q \subset Q_1 = D_1 = D + (x)$ and consequently $Q = D + x(Q : x)$. If $D \neq Q$ then, since D is a homogenous ideal in R[X] and consequently $(D : Q)$ is homogenous, we have that $(D : Q) + (x)$ is a proper ideal. Hence localizing in a prime ideal containing $(D : Q) + (x)$ we may assume that $P \neq Q$ in a local ring. If $c = b_n$ for some n then Q is prime by Corollary 10 so that $(Q : x) = Q$. Hence we obtain an equality $Q = D + xQ$ which, by Nakayamas lemma, contradicts the assumption $D \neq Q$. If $b_n < c < b_{n+1}$ then by Remark 2 we have that $xD_2 \subseteq D$ and consequently that

$D_2 = (D : x) = (Q : x) = Q_2$. From the above we then obtain $Q = D + x(Q:x) = D + xD_2 \subseteq D$ which again contradicts the assumption $D \neq Q$.

Lemma 13. Let $D = D(X; b_1, \ldots, b_m; c)$. Then the $\mathbb{Z}$-module $\mathbb{Z}[X]/D$ is faithfully flat.

Proof. As a consequence of Proposition 13 and Corollary 10 we have that D is prime when $c = b_n$ for some n and by Remark 2 (i) with $d = b_{n+1}$ we have that $D = D_1 \cap D_2$ is the intersection of two primes if $b_n < c < b_{n+1}$. Hence for any c the $\mathbb{Z}$-module $\mathbb{Z}[X]/D$ has no torsion.

Proof of Theorem 11. As a consequence of Lemma 13 we may assume that R is a field (see e.g. [1], Prop. 20, p. 1038). We proceed by descending induction on the lattice of ideals of the form $D(X; b_1, \ldots, b_m; c)$ where $c = b_n$ or $c = b_n + 1$ for some n. Again the theorem holds for the maximal ideal $D(X; b - m, \ldots, b - 1; b - 1)$ of this form.

If $c = b_n$ then D is prime by Corollary 10 and Proposition 12. Hence $x = x(1, b_n + 1)$ is not a zero divisor in $R[X]/D$ and since $D_1 = D + (x)$ is perfect of grade

$$h = (a - m)(b - m) + \sum_{i=1}^{m} (b_i - i + 1) + n$$

by induction, we have that D is perfect of grade $h - 1$.

If $c = b_n + 1$, then by Remark 2 we have that $D = D_2 \cap D_3$. By induction assumption D_2 and D_3 are perfect of grade h and $D_2 + D_3 = D(X; b_1, \ldots, b_{n-1}, c, b_{n+1}, \ldots, b_m; b_{n+1})$ is perfect of grade $h + 1$. An easy lemma (see e.g. [1], Prop. 18, p. 1037) shows that $D = D_2 \cap D_3$ is perfect of grade h.

Together with standard results on free resolutions of modules (see e.g. [2], Prop. 4, p. 161, or [3], Corollary 8, p. 160) we obtain from Theorem 11

Theorem 14. Let R be an integral domain and $M = (s(i, j))$ an $a{\times}b$-matrix with entries in an R-algebra S. Let $c = b_n$ or $c = b_n + 1$ for some integer n. Assume that the ideal $D = D(M; b_1; \ldots, b_m; c)$ in $R[M]$ is of grade

$$(a - m)(b - m) + \sum_{i=1}^{m} (b_i - i + 1) + n - 1 + c - b_n .$$

Then D is perfect.

§ 4. Denote by $D_c(a, b)$ the generic determinantal scheme of all axb-matrices whose cxc-minors vanish. Then $D_c(a,b) = \mathrm{Spec}(R[X]/D(X;0,\ldots, c-2;0))$. The most important geometric properties of the schemes $D_c(a, b)$ we collect in the following result.

Theorem 15. Assume R is a field. With the above notation the following assertions hold:

(i) The scheme $D_c(a, b)$ is integral, Cohen-Macauly and of codimension $(a - c + 1)(b - c + 1)$ in $\mathbb{A}_k^{a\cdot b}$.

(ii) The scheme $D_{c-1}(a, b)$ is the singular locus of $D_c(a, b)$.

Proof. Property (i) follows immediately from Theorem 11. To prove property (ii) we denote by $f_1, \ldots, f_N$ the minors of X of order c. The entries $\partial f_k/\partial x(i, j)$ of the Jacobian matrix are either minors of X of order $c - 1$ or zero. Consequently the scheme $D_{c-1}(a, b)$ is contained in the singular locus. Conversely let P be a point of $D_c(a, b) \setminus D_{c-1}(a, b)$ and assume that the determinant x of the $(c - 1)x(c - 1)$-submatrix of X taken from rows $i_1 < \ldots < i_{c-1}$ and columns $j_1 < \ldots < j_{c-1}$ does not vanish on P. We may reorder rows and columns of X arbitrarily without changing $D_c(a, b)$ and $D_{c-1}(a, b)$. Hence we may assume that $j_k = k$ for $k = 1, \ldots, c - 2$. Then by Proposition 9 for the case $b_i = i - 1$ for $i = 1, \ldots, c - 1$ and $b_i = b$ we have that the principal open subset $D_c(a, b)_x$ is regular. Hence every point of $D_c(a, b) \setminus D_{c-1}(a, b)$ is regular.

References

[1] Eagon, J.A. & Hochster, M., "Cohen-Macauley rings, invariant theory and the generic perfection of determinantal loci". Amer. J. Math. 93 (1971), 1020-1058.

[2] Eagon, J.A. & Northcott, D.G., "Generically acyclic complexes and generically perfect ideals". Proc. Roy Soc., A, 299 (1967), 147-172.

[3] Kempf, G. & Laksov, D., "The determinantal formula of Schubert calculus". Acta Math. 132 (1974), 153-162.

University of Oslo
Blindern, Oslo 3
Norway

Institut Mittag-Leffler
Auravägen 17
S-18262 Djursholm
Sweden

On Weierstrass points and automorphisms of curves of genus three

by

Akikazu Kuribayashi and Kaname Komiya

The aim of this paper is to give the group G of automorphisms of a curve of genus three. Let H be the cyclic subgroup of maximum order of G. Let $n = \#G$ be the order of G and let $m = \#H$ be the order of H. In §1 we study Riemann surfaces which are cyclic coverings of the Riemann sphere. If the order m of the cyclic group is greater than 4, we obtain Theorem 1 using mainly the Riemann-Hurwitz relation which is classical but powerful. In §2 and §3 we study Riemann surfaces which are defined by

$$y^3 = x(x-1)(x-t_1)(x-t_2) \quad \text{and} \quad y^4 = x(x-1)(x-t).$$

We obtain Theorem 2 and Theorem 3 by considering Weierstrass points of these Riemann surfaces and canonical models of them in $\mathbb{P}^2$. In §4 we consider special Riemann surfaces which are characterized by Weierstrass points. We obtain concretely the groups of automorphisms of these Riemann surfaces and the equations by which these surfaces are defined. In §5 we study Riemann surfaces which are cyclic coverings of complex tori. We obtain Theorem 4 where some new curves come forward.

The main result is as follows:

The order of group		The equation of Riemann surface
(1) 168		$y^3 + x + x^3y = 0$
(2) 96		$y^4 = x(x^2-1)$
(3) 48	(a)	$y^2 = x(x^6-1)$
	(b)	$y^3 = x^4 - 1$
(4) 32		$y^2 = x^8 - 1$
(5) 24		$x^4 + y^4 + 2a(x^2y^2 + x^2 + y^2) + 1 = 0$

(6) 16 (a) $y^2 = x^8 + 2ax^4 + 1$

(b) $y^4 = x(x-1)(x-t)$

(7) 14 $y^2 = x^7 - 1$

(8) 12 $y^2 = x(x^3-1)(x^3-t)$

(9) 9 $y^3 = x(x^3-1)$

(10) 8 (a_1) $y^2 = x(x^2-1)(x^2-t)(x^2-1/t)$

(a_2) $y^2 = (x^2-1)(x^2-t_1)(x^2-t_2)(x^2-t_1/t_2)$

(b) $x^4+ y^4+ 2ax^2y^2+ 2b(x^2+ y^2) + 1 = 0$

(11) 6 (b_1) $y^3 = x(x-1)(x-t)(x-(1-t))$

(b_2) $a(x^4+ y^4+ 1) + b(x^3y + y^3x + x^3+ y^3+ x + y) + c(x^2y^2+ x^2+ y^2) = 0.$

We have listed above Riemann surfaces whose groups of automorphisms are not less than 6 in the order. However, in the case of the order of the group is 4, we have to consider Riemann surfaces which are normal coverings of order 4 over the sphere and tori:

(12) 4 (a) $y^2 = x(x^2-1)(x^2-t_1)(x^2-t_2)$

(b_1) $x^4+ y^4+ 2ax^2y^2+ 2bx^2+ 2cy^2+ 1 = 0$

(b_2) $a(x^4+y^4+1)+ b(x^3y-y^3x)+ cx^2y^2+ d(x^2+y^2) = 0.$

In the case of the order of the group is 3, we have

(13) 3 (b_1) $y^3 = x(x-1)(x-t_1)(x-t_2)$

and moreover we have to take Riemann surfaces which are cyclic coverings of order 3 over tori:

(b_2) $a(x^4+y^4+1) + b(x^3y+y^3+x) + c(y^3x+x^3+y) + d(x^2y^2+x^2+y^2) = 0.$

In the case of the order of the group is 2, we have

(14) 2 (a) $y^2 = x(x-1)(x-t_1)\cdots(x-t_5)$

and moreover we have to take Riemann surfaces which are double coverings of tori: (b) $y^4+(a_0x^2+a_1x+b)y^2+ x^4+a_2x^3+a_3x^2+a_4x+1 = 0.$

In each case, we should take parameters in general.

R. Tsuji studied also on the same subject. However, it seems to us that his paper [9] leaves something to be desired. Because he did not consider the Riemann surfaces which are coverings of tori.

§0 Preliminaries

We summarize here some fundamental lemmas which will be needed later.

Lemma 1 [Riemann-Hurwitz; 3]. Let R be an n-sheeted normal covering of R_0 ($R_0 = R/G$), where the group of automorphisms of R is denoted by G. Let $p_1, \dots, p_r$ be the points of R_0 over which the ramification occurs. For each p_j $(1 \leq j \leq r)$ there is an integer ν_j so that above p_j there are n/ν_j branch points each of multiplicity ν_j. Then we have

$$2g - 2 = n(2g_0 - 2) + V$$

where the total ramification V is given by

$$V = n\ \Sigma_{j=1}^{r}(1 - \nu_j^{-1})$$

and g, g_0 are the genuses of R, R_0 respectively.

Lemma 2. Let H be a cyclic subgroup of the group of automorphisms of R. Let g' be the genus of R/H and let m be the order of H, then we have

(i) if $g' \geq 2$, then $m \leq g - 1$

(ii) if $g' = 1$, then $m \leq 2(g - 1)$

(iii) if $g' = 0$, and moreover

(a) if $r \geq 5$, then $m \leq 2(g - 1)$

(b) if $r = 4$, then $m \leq 6(g - 1)$

(c) if $r = 3$, then $m \leq 10(g - 1)$.

Lemma 3. Let K be a Galois extention of a field k. Suppose that the Galois group is cyclic and of order n, and k contains a primitive n-th root of unity ζ. Then for every generator σ of the Galois group, there exists an element y of K such that

$$\sigma(y) = \zeta y,\ K = k(y) \quad \text{and} \quad y^n \in k.$$

§1 Riemann surfaces which are cyclic coverings of order m > 4 over the Riemann sphere

Let K, K' be the meromorphic function fields of R, R' = R/H respectively. By Lemma 3 K' is the rational function field $\mathbf{C}(x)$ and the equation of R is given by

$$y^m = (x - a_1)^{n_1}(x - a_2)^{n_2}\cdots(x - a_r)^{n_r}$$

where $\Sigma_{i=1}^r n_i$ is divided by m and $1 \leq n_i < m$.

Hereafter we assume that R is of genus 3, and so by Lemma 2, $m \leq 20$.

If m = 2, then we have

$$y^2 = (x - a_1)(x - a_2)\cdots(x - a_8) .$$

If m = 3, then we have two types

$$y^3 = (x - a_1)(x - a_2)(x - a_3)(x - a_4)(x - a_5)^2 \text{ and}$$

$$y^3 = (x - a_1)^2(x - a_2)^2(x - a_3)^2(x - a_4)^2(x - a_5).$$

However by a birational transformation $X = x,\ Y = y^2(x - a_5)^{-1}$ we see that the former is conformally equivalent to the latter. Therefore, after this we always except the latter type from our list.

If m = 4, then by Lemma 1 the values which ν_i may be able to take are 2 and 4. Therefore we have following:

	ν_1	ν_2	ν_3	ν_4	ν_5	ν_6
(i)	2	2	2	2	2	2
(ii)	2	2	2	4	4	
(iii)	4	4	4	4		

In case (i), if there were such a Riemann surface, the equation becomes to be reducible. Therefore there does not exist such a Riemann surface.

In case (ii), we have a single type of Riemann surfaces :

$$y^4 = (x - a_1)^2(x - a_2)^2(x - a_3)^2(x - a_4)(x - a_5)$$

It is easy to see that this is hyperelliptic. In fact, we normalize the equation as follows :

$$y^4 = x^2(x - 1)^2(x - t_1)^2(x - t_2)$$

By a birational transformation $x = Y^2 + t_2,\ y = YX^{-1}$

it becomes to

$$Y^2X^{-2} = Y(Y^2 + t_2)(Y^2 + t_2 - 1)(Y^2 + t_2 - t_1).$$

Thus we obtain

$$\eta^2 = \xi(\xi^2 - 1)(\xi^2 - \alpha_1)(\xi^2 - \alpha_2).$$

In case (iii), we have two types

(1) $y^4 = (x - a_1)(x - a_2)(x - a_3)(x - a_4)$,

(2) $y^4 = (x - a_1)(x - a_2)(x - a_3)^3(x - a_4)^3$.

It is easy to see that (1) is elliptic-hyperelliptic and (2) is hyperelliptic. In fact, (2) will be normalized to

$$y^4 = x^3(x - 1)(x - t).$$

By a birational transformation $x = Y,\ y = XY$

it becomes to

$$X^4 Y = (Y - 1)(Y - t).$$

Thus we obtain

$$\eta^2 = \xi^8 + \alpha\xi^4 + 1.$$

If m = 6, then by Lemma 1 the values which ν_i may be able to take are 2, 3 and 6. Therefore we have following :

	ν_1	ν_2	ν_3	ν_4	ν_5
(i)	2	2	2	2	3
(ii)	2	2	6	6	
(iii)	2	3	3	6	
(iv)	3	3	3	3	

In case (i), by Lemma 2 (iii) (a) we see that there does not exist such a Riemann surface.

In case (ii), we have a single type of Riemann surfaces :

$$y^6 = (x - a_1)^3(x - a_2)^3(x - a_3)(x - a_4)^5.$$

It is easy to see that this is hyperelliptic. In fact, we normalize the equation as follows :

$$y^6 = x^3(x - 1)^3(x - t).$$

By a birational transformation $x = Y^3 + t,\ y = YX^{-1}$

it comes to

$$Y^2X^{-2} = Y(Y^3 + t)(Y^3 + t - 1).$$

Thus we obtain

$$\eta^2 = \xi(\xi^3 - 1)(\xi^3 - \alpha).$$

In case (iii), we have a single type of Riemann surfaces :

$$y^6 = (x - a_1)^3(x - a_2)^2(x - a_3)^2(x - a_4)^5.$$

We normalize the equation as follow :

$$y^6 = x^3(x - 1)^2(x - t)^2.$$

We have a basis of differentials of the first kind

$$\omega_1 = y^{-1}dx, \quad \omega_2 = y^{-2}xdx, \quad \omega_3 = y^{-5}x^2(x-1)(x-t)dx.$$

Put

$$X = xy^{-1}, \; Y = y^{-4}x^2(x-1)(x-t) \quad ; \quad x = X^2Y^{-2}, \; y = XY^{-2}.$$

Then we have as a canonical model

$$(X^2 - Y^2)(X^2 - tY^2) + YZ^3 = 0.$$

Thus we have an equation for the Riemann surface

$$y^3 = (x^2 - 1)(x^2 - t).$$

We can rewrite this equation in

$$\eta^3 = \xi(\xi - 1)(\xi - \tau)(\xi - (1 - \tau)).$$

In case (iv), if there were such a Riemann surface, the equation becomes to be reducible. Therefore there does not exist such a Riemann surface.

If $m = 7$, we have four types

(1) $y^7 = (x - a_1)(x - a_2)(x - a_3)^5$

(2) $y^7 = (x - a_1)(x - a_2)^2(x - a_3)^4$

(3) $y^7 = (x - a_1)(x - a_2)^3(x - a_3)^3$

(4) $y^7 = (x - a_1)^2(x - a_2)^2(x - a_3)^3.$

We see that (1) is birationally equivalent to (4). In fact, we have a birational transformation

$$x = X, \quad y = Y^{-3}(X - a_1)(X - a_2)(X - a_3)^2.$$

We see also that (3) is birationally equivalent to (4). In fact, we have a birational transformation

$$x = X, \quad y = Y^{-2}(x - a_1)(x - a_2)(x - a_3).$$

Now then, (1) is hyperelliptic. In fact, we normalize the equation as follows :

$$y^7 = x(x - 1).$$

It is easy to see that this equation becomes to

$$\eta^2 = \xi^7 - 1.$$

As for (2), we normalize the equation as follows :

$$y^7 = x^2(x-1).$$

We have a basis of differentials of the first kind

$$\omega_1 = y^{-3}dx, \quad \omega_2 = y^{-5}xdx, \quad \omega_3 = y^{-6}xdx.$$

Put

$$x = -X^3Y^{-2}, \quad y = XY^{-1}.$$

Then we have as a canonical model

$$X^3Y + Y^3Z + Z^3X = 0.$$

If $m = 8$, then the values which ν_i may be able to take are 2,4 and 8. Therefore we have following :

	ν_1	ν_2	ν_3	ν_4	ν_5
(i)	2	2	2	2	2
(ii)	2	2	4	4	
(iii)	4	8	8		

In case (i) and (ii), if there were such a Riemann surface, the equation becomes to be reducible. Therefore there does not exist such a Riemann surface.

In case (iii), we have three types

(1) $y^8 = (x-a_1)^2(x-a_2)^3(x-a_3)^3$

(2) $y^8 = (x-a_1)(x-a_2)(x-a_3)^6$

(3) $y^8 = (x-a_1)(x-a_2)\ (x-a_3)^5.$

By a birational transformation

$$x = X, \quad y = (X-a\ )^{-2}(X-a\ )^{-1}(X-a\ )^{-1}Y,$$

we see that (2) is conformally equivalent to (1).

It is easy to see that (2) is hyperelliptic. In fact, we normalize the equation as follows :

$$y^8 = x(x-1)$$

and this becomes to

$$\eta^2 = \xi^8 - 1.$$

As for (3), we normalize the equation as follows :

$$y^8 = x^2(x-1).$$

We have a basis of differentials of the first kind

$$\omega_1 = y^{-3}dx, \quad \omega_2 = y^{-6}xdx, \quad \omega_3 = y^{-7}xdx.$$

Put

$$x = -X^{-1}Y^4, \quad y = Y.$$

Then we have as a canonical model

$$X^3Z + XZ^3 + Y^4 = 0.$$

Remark : This is equivalent to

$$X^4 + Y^4 + Z^4 = 0.$$

If m = 9, then the values which ν_i may be able to take place are 3 and 9. Therefore we have following :

	ν_1	ν_2	ν_3
(i)	3	9	9

Then we have three types

(1) $y^9 = (x - a_1)^3(x - a_2)(x - a_3)^5$

(2) $y^9 = (x - a_1)^3(x - a_2)^2(x - a_3)^4$

(3) $y^9 = (x - a_1)^6(x - a_2)(x - a_3)^2.$

We normalize these equations as follows :

(1)' $y^9 = x^3(x - 1)$

(2)' $y^9 = x^3(x - 1)^4$

(3)' $y^9 = x^6(x - 1).$

By a birational transformation $x = X, \quad y = X^3(X - 1)Y^2$ (1)' is conformally equivalent to (2)' and (3)' is conformally equivalent to $y^9 = x^3(x - 1)^5$ which is equivalent to (1)'.

Now then, as for (1), we have a basis of differentials of the first kind

$$\omega_1 = y^{-5}xdx, \quad \omega_2 = y^{-7}x^2dx, \quad \omega_3 = y^{-8}x^2dx.$$

Put

$$x = -X^3Y^{-2}, \quad y = -XY^{-1}.$$

Then we have as a canonical model

$$X^3Y + Y^3Z + Z^4 = 0.$$

In affine coordinates we have easily

$$y^3 = x(x^3 - 1).$$

If $m = 10$, then the values which ν_i may be able to take are 2, 5 and 10. Therefore we have following :

	ν_1	ν_2	ν_3	ν_4
(i)	2	2	2	10
(ii)	5	5	5	

Obviously there does not exist such a Riemann surface.

If $m = 12$, then the values which ν_i may be able to take are 2, 3, 4, 6 and 12. Therefore we have following :

	ν_1	ν_2	ν_3	ν_4
(i)	2	2	2	6
(ii)	2	2	3	3
(iii)	2	12	12	
(iv)	3	3	6	
(v)	3	4	12	
(vi)	4	4	6	

It is easy to see that in each case of (i),(ii),(iv) and (vi), there does not exist such a Riemann surface.

In case (iii), we have two types

(1) $\qquad y^{12} = (x - a_1)^6(x - a_2)(x - a_3)^5$

(2) $\qquad y^{12} = (x - a_1)^6(x - a_2)^7(x - a_3)^{11}$.

However obviously (1) is conformally equivalent to (2). Furthermore, (1) is hyperelliptic. In fact, we normalize the equation as follows :

$$y^{12} = x^6(x - 1).$$

By a birational transformation $x = X^{-2}Y$, $y = X^{-1}Y$; it becomes to

$$Y^6 = X^{-2}Y - 1.$$

Thus we obtain

$$\eta^2 = \xi(\xi^6 - 1).$$

In case (v), we have two types

(1) $\qquad y^{12} = (x - a_1)^4(x - a_2)^3(x - a_3)^5$

(2) $\qquad y^{12} = (x - a_1)^4(x - a_2)^9(x - a_3)^{11}$.

We normalize (1) and (2) as follows :

(1)' $y^{12} = x^4(x - 1)^3$

(2)' $y^{12} = x^4(x - 1)^9$.

By a birational transformation $x = X,\ \ y = X^2(X - 1)^2 Y^5$ (2)' is equivalent to (1)'. Furthermore, by a birational transformation $x = (1 - X)^{-1},\ \ y = \zeta X^{-1}(1 - X)^{-1}Y^5,\ \zeta^{12} = -1$, (1)' is equivalent to

$$y^{12} = x^3(x - 1).$$

For this surface, we have a basis of differentials of the first kind

$$\omega_1 = y^{-7}x\,dx,\quad \omega_2 = y^{-10}x^2dx,\quad \omega_3 = y^{-11}x^2dx.$$

Put

$$x = -X^4Y^{-3},\quad y = \eta XY^{-1},\ \eta^4 = -1.$$

Then we have as a canonical model

$$X^4 + Y^3Z + Z^4 = 0.$$

In affine coordinates we have

$$y^4 = x^3 - 1.$$

If m = 14, then the values which ν_i may be able to take are 2, 7 and 14. Therefore we have following :

	ν_1	ν_2	ν_3
(i)	2	7	14

We have a single type of Riemann surfaces :

$$y^{14} = (x - a\ )^7(x - a\ )^2(x - a\ )^5.$$

It is easy to see that this is hyperelliptic. In fact, we normalize the equation as follows

$$y^{14} = x^7(x - 1)^2.$$

By a birational transformation

$$X = x^{-3}(x-1)^{-1}y^6,\ Y = x^{-3}(x-1)^{-2}y^7\ ;\quad x = Y^2,\ y = X^{-1}Y$$

it becomes to

$$Y^2 = X^{-7} + 1\quad,\ \text{i.e.,}\quad \eta^2 = \xi^7 - 1.$$

If m = 15,16,18 and 20, there are no Riemann surfaces which satisfy the conditions.

Remark. By Lemma 1, we need not consider the case of m = 5,11,13,17 and 19.

Summarizing above we obtain following theorem :

Theorem 1. We can classify Piemann surfaces which are cyclic coverings of the Riemann sphere as follows :

(i) Hyperelliptic Case.

(1) $m = 2$ $\quad y^2 = x(x - 1)(x - \alpha_1)\cdots(x - \alpha_5)$

(2) $m = 4$ (a) $y^2 = x^8 + \alpha x^4 + 1$

(b) $y^2 = x(x^2 - 1)(x^2 - \alpha_1)(x^2 - \alpha_2)$

(3) $m = 6$ $\quad y^2 = x(x^3 - 1)(x^3 - \alpha)$

(4) $m = 8$ $\quad y^2 = x^8 - 1$

(5) $m = 12$ $\quad y^2 = x(x^6 - 1)$

(6) $m = 14$ $\quad y^2 = x^7 - 1$

(ii) Non-hyperelliptic Case.

(1) $m = 3$ $\quad y^3 = x(x - 1)(x - \alpha_1)(x - \alpha_2)$

(2) $m = 4$ $\quad y^4 = x(x - 1)(x - \alpha)$

(3) $m = 6$ $\quad y^3 = x(x - 1)(x - \alpha)(x - (1-\alpha))$

(4) $m = 7$ $\quad y^3 + yx^3 + x = 0$

(5) $m = 8$ $\quad y^4 = x(x^2 - 1)$

(6) $m = 9$ $\quad y^3 = x(x^3 - 1)$

(7) $m = 12$ $\quad y^4 = x^3 - 1$

Remark. In non-hyperelliptic case, each equation becomes to a projective canonical model of the Riemann surface. For example, in (1)

$$Y^3Z = X(X - Z)(X - \alpha_1 Z)(X - \alpha_2 Z)$$

is the projective canonical model.

Remark. It is well-known that the groups of automorphisms of hyperelliptic Riemann surfaces are given by the hyperelliptic involution which can commute to any other automorphisms and by the groups of transformations of the Riemann sphere which has some distinquished points. Hence we obtain

Corollary. In hyperelliptic case, we have

(1) $m = 2$ $\quad$ the order of the group = 2

(2) $m = 4$ (a) the order of the group = 16

(b) the order of the group = 4

(3) $m = 6$ the order of the group = 12

(4) $m = 8$ the order of the group = 32

(5) $m = 12$ the order of the group = 48

(6) $m = 14$ the order of the group = 14.

In hyperelliptic case, branch points play an important role in an investigation of the group of automorphisms. However we notice, now, that it is not branch points but Weierstrass points that play the important part in automorphisms of Riemann surfaces.

§2 Automorphisms of Riemann surfaces defined by

$$y^3 = x(x-1)(x-t_1)(x-t_2)$$

On our study the following lemma is fundamental :

Lemma 4. Let R be a non-hyperelliptic Riemann surface of genus g ($g \geq 3$) and let σ be an automorphism of R. Let R_0 be a canonical model of R in the projective space $\mathbb{P}^{g-1}$. Then σ is obtained as a projective transformation of $\mathbb{P}^{g-1}$ restricted to R_0.

As we have see in §1, a canonical model of our Riemann surface is

$$Y^3 Z = X(X-Z)(X-t_1 Z)(X-t_2 Z).$$

Put

$$X' = a_{11} X + a_{12} Y + a_{13} Z$$

$$Y' = a_{21} X + a_{22} Y + a_{23} Z$$

$$Z' = a_{31} X + a_{32} Y + a_{33} Z.$$

Then we must have

$$Y'^3 Z' = X'(X'-Z')(X'-t_1 Z')(X'-t_2 Z').$$

We must decide 3x3 matrix

$$\begin{pmatrix} a_{11} & a_{12} & a_{13} \\ a_{21} & a_{22} & a_{23} \\ a_{31} & a_{32} & a_{33} \end{pmatrix}$$

such that this condition is satisfied, i.e., we have to solve the following simultaneous algebraic equations with 9 unknowns :

(1) $a_{21}^3 a_{31} - a_{11}^4 + (t_1+t_2+1)a_{11}^3 a_{31} - (t_1+t_2+t_1t_2)a_{11}^2 a_{31}^2 +$
$t_1t_2 a_{11} a_{31}^3 = -k$

(2) $a_{22}^3 a_{32} - a_{12}^4 + (t_1+t_2+1)a_{12}^3 a_{32} - (t_1+t_2+t_1t_2)a_{12}^2 a_{32}^2 +$
$t_1t_2 a_{12} a_{32}^3 = 0$

(3) $a_{23}^3 a_{33} - a_{13}^4 + (t_1+t_2+1)a_{13}^3 a_{33} - (t_1+t_2+t_1t_2)a_{13}^2 a_{33}^2 +$
$t_1t_2 a_{13} a_{33}^3 = 0$

(4) $a_{21}^3 a_{32} + 3a_{21}^2 a_{22} a_{31} - 4a_{11}^3 a_{12} + (t_1+t_2+1)(a_{11}^3 a_{32} + 3a_{11}^2 a_{12} a_{31})$
$- 2(t_1+t_2+t_1t_2)(a_{11}^2 a_{31} a_{32} + a_{11} a_{12} a_{31}^2) + t_1t_2(a_{12} a_{31}^3 +$
$3a_{11} a_{31}^2 a_{32}) = 0$

(5) $a_{22}^3 a_{33} + 3a_{22}^2 a_{23} a_{32} - 4a_{12}^3 a_{13} + (t_1+t_2+1)(a_{12}^3 a_{33} + 3a_{12}^2 a_{13} a_{32})$
$- 2(t_1+t_2+t_1t_2)(a_{11}^2 a_{31} a_{32} + a_{11} a_{12} a_{31}^2) + t_1t_2(a_{13} a_{32}^3 +$
$3a_{12} a_{32}^2 a_{33}) = k$

(6) $a_{21}^3 a_{33} + 3a_{21}^2 a_{23} a_{31} - 4a_{11}^3 a_{13} + (t_1+t_2+1)(a_{11}^3 a_{33} + 3a_{11}^2 a_{13} a_{31})$
$- 2(t_1+t_2+t_1t_2)(a_{11}^2 a_{31} a_{33} + a_{11} a_{13} a_{31}^2) + t_1t_2(a_{31} a_{13}^3 +$
$3a_{11} a_{31}^2 a_{33}) = k(t_1+t_2+1)$

(7) $a_{22}^3 a_{31} + 3a_{21} a_{22}^2 a_{32} - 4a_{11} a_{12}^3 + (t_1+t_2+1)(a_{12}^3 a_{31} + 3a_{11} a_{12}^2 a_{31})$
$- 2(t_1+t_2+t_1t_2)(a_{11} a_{12} a_{32}^2 + a_{12}^2 a_{31} a_{32}) + t_1t_2(a_{11} a_{32}^3 +$
$3a_{12} a_{31} a_{32}^2) = 0$

(8) $a_{23}^3 a_{32} + 3a_{22} a_{23}^2 a_{33} - 4a_{12} a_{13}^3 + (t_1+t_2+1)(a_{13}^3 a_{32} + 3a_{12} a_{13}^2 a_{33})$
$- 2(t_1+t_2+t_1t_2)(a_{12} a_{13} a_{33}^2 + a_{13}^2 a_{32} a_{33}) + t_1t_2(a_{12} a_{33}^3 +$
$3a_{13} a_{32} a_{33}^2) = 0$

(9) $a_{23}^3 a_{31} + 3a_{21} a_{23}^2 a_{33} - 4a_{11} a_{13}^3 + (t_1+t_2+1)(a_{13}^3 a_{31} + 3a_{11} a_{13}^2 a_{33})$
$- 2(t_1+t_2+t_1t_2)(a_{13}^2 a_{31} a_{33} + a_{11} a_{13} a_{33}^2) + t_1t_2(a_{11} a_{33}^3 +$
$3a_{13} a_{31} a_{33}^2) = kt_1t_2$

(10) $3(a_{21} a_{22}^2 a_{31} + a_{21}^2 a_{22} a_{32}) - 6a_{11}^2 a_{12}^2 + 3(t_1+t_2+1)(a_{11} a_{12}^2 a_{31} +$
$a_{11}^2 a_{12} a_{32}) - (t_1+t_2+t_1t_2)(a_{11}^2 a_{32}^2 + a_{12}^2 a_{31}^2 + 4a_{11} a_{12} a_{31} a_{32})$
$+ 3t_1t_2(a_{11} a_{31} a_{32}^2 + a_{12} a_{31}^2 a_{32}) = 0$

(11) $3(a_{22} a_{23}^2 a_{32} + a_{22}^2 a_{23} a_{33}) - 6a_{12}^2 a_{13}^2 + 3(t_1+t_2+1)(a_{12} a_{13}^2 a_{32} +$
$a_{12}^2 a_{13} a_{33}) - (t_1+t_2+t_1t_2)(a_{12}^2 a_{33}^2 + a_{13}^2 a_{32}^2 + 4a_{12} a_{13} a_{32} a_{33}) +$

$+ 3t_1t_2(a_{12}a_{32}a_{33}^2+a_{13}a_{32}^2a_{33}) = 0$

(12) $3(a_{21}a_{23}^2a_{31}+a_{21}^2a_{23}a_{33}) - 6a_{11}^2a_{13}^2 + 3(t_1+t_2+1)(a_{11}a_{13}^2a_{31}+$
$a_{11}^2a_{13}a_{33}) - (t_1+t_2+t_1t_2)(a_{11}^2a_{33}^2+a_{13}^2a_{31}^2+4a_{11}a_{13}a_{31}a_{33})$
$+ 3t_1t_2(a_{11}a_{31}a_{33}^2+a_{13}a_{31}^2a_{33}) = -k(t_1+t_2+t_1t_2)$

(13) $3(2a_{21}a_{22}a_{23}a_{31}+a_{21}^2a_{23}a_{32}+a_{21}^2a_{22}a_{33}) - 12a_{11}^2a_{12}a_{13} +$
$3(t_1+t_2+1)(2a_{11}a_{12}a_{13}a_{31}+a_{11}^2a_{13}a_{32}+a_{11}^2a_{12}a_{33}) -$
$2(t_1+t_2+t_1t_2)(a_{11}^2a_{32}a_{33}+a_{12}a_{13}^2a_{31}+2a_{11}a_{12}a_{31}a_{33}+2a_{11}a_{13}a_{31}a_{32})$
$+3t_1t_2(2a_{11}a_{31}a_{32}a_{33}+a_{12}a_{31}^2a_{33}+a_{13}a_{31}^2a_{32}) = 0$

(14) $3(2a_{21}a_{22}a_{23}a_{32}+a_{22}^2a_{23}a_{31}+a_{21}a_{22}^2a_{33}) - 12a_{11}a_{12}^2a_{13} +$
$3(t_1+t_2+1)(2a_{11}a_{12}a_{13}a_{32}+a_{12}^2a_{13}a_{31}+a_{11}a_{12}^2a_{33}) -$
$2(t_1+t_2+t_1t_2)(a_{12}^2a_{31}a_{33}+a_{11}a_{13}a_{32}^2+2a_{11}a_{12}a_{32}a_{33}+2a_{12}a_{13}a_{31}a_{32})$
$+3t_1t_2(2a_{31}a_{32}a_{33}a_{12}+a_{13}a_{31}a_{32}^2+a_{11}a_{32}^2a_{33}) = 0$

(15) $3(2a_{21}a_{22}a_{23}a_{33}+a_{22}a_{23}^2a_{31}+a_{21}a_{23}^2a_{32}) - 12a_{21}a_{22}a_{23}^2 +$
$3(t_1+t_2+1)(2a_{11}a_{12}a_{13}a_{33}+a_{12}a_{13}a_{31}^2+a_{11}a_{13}^2a_{32}) -$
$2(t_1+t_2+t_1t_2)(a_{13}^2a_{31}a_{32}+a_{11}a_{12}a_{33}^2+2a_{12}a_{13}a_{31}a_{33}+2a_{11}a_{13}a_{32}a_{33})$
$+3t_1t_2(2a_{13}a_{31}a_{32}a_{33}+a_{11}a_{32}a_{33}^2+a_{12}a_{31}a_{33}^2) = 0.$

Here, of course, k is a non-zero constant.

The Wronskian W of our Riemann surface is

$$W = y''(dx/y)^6.$$

Put

$$\gamma(x) = x(x-1)(x-t_1)(x-t_2).$$

Then we obtain

$$\begin{aligned}\operatorname{div} W &= \operatorname{div}\,(3\gamma\gamma'' - 2\gamma'^2) + P_0 + P_1 + P_{t_1} + P_{t_2} + 20P_\infty \\ &= P_{s_1} + P'_{s_1} + P''_{s_1} + \cdots + P_{s_6} + P'_{s_6} + P''_{s_6} \\ &\quad + P_0 + P_1 + P_{t_1} + P_{t_2} + 2P_\infty.\end{aligned}$$

It is easy to see that s_i $(1\leq i\leq 6)$ are different from $0, 1, t_1$ and t_2 and the following three cases take place in the equation

$$F(x) = 3\gamma(x)\gamma''(x) - 2\gamma'^2(x) = 0.$$

(i) There are no double roots.

(ii) There are only one double roots.

(iii) There are two distinct double roots.

For cases (i) and (iii), however, it is easy to see that every automorphism fixes the point P_∞, i.e., we must have

$$\begin{pmatrix}0\\ \lambda\\ 0\end{pmatrix} = \begin{pmatrix}a_{11} & a_{12} & a_{13}\\ a_{21} & a_{22} & a_{23}\\ a_{31} & a_{32} & a_{33}\end{pmatrix}\begin{pmatrix}0\\ 1\\ 0\end{pmatrix}.$$

Here λ is an arbitrary non-zero constant. Hence we have

$$a_{12} = 0, \quad a_{22} = \lambda, \quad a_{32} = 0.$$

From (7) in the equations we obtain

$$a_{31} = 0$$

and from (8), $a_{23} = 0$ or $a_{33} = 0$. However from (5) we must have $a_{33} \neq 0$. From (13) we have $a_{21} = 0$. Further, we may assume that $a_{33} = 1$.

Then the fifteen equations become to

(1) $a_{11}^4 = k$

(3) $-a_{13}^4 + (t_1+t_2+1)a_{13}^3 - (t_1+t_2+t_1t_2)a_{13}^2 + t_1t_2a_{13} = 0$

(5) $a_{22}^3 = k$

(6) $-4a_{11}^3a_{13} + (t_1+t_2+1)a_{11}^3 = k(t_1+t_2+1)$

(9) $-4a_{11}a_{13}^3 + 3(t_1+t_2+1)a_{11}a_{13}^2 - 2(t_1+t_2+t_1t_2)a_{11}a_{13} + t_1t_2a_{11} = kt_1t_2.$

(12) $-6a_{11}^2a_{13}^2 + 3(t_1+t_2+1)a_{11}^2a_{13} - (t_1+t_2+t_1t_2)a_{11}^2 = -k(t_1+t_2+t_1t_2).$

Example 1. $y^3 = x(x^3 - 1)$

We have $t_1+t_2+t_1t_2 = 0$, $t_1+t_2+1 = 0$ and $t_1t_2 = 1$. Hence, we have

(1) $a_{11}^4 = k$ (5) $a_{22}^3 = k$

(6) $a_{13} = 0$ (9) $a_{11} = k.$

Thus we obtain

$$\sigma = \begin{pmatrix}\zeta_1 & 0 & 0\\ 0 & \zeta_2 & 0\\ 0 & 0 & 1\end{pmatrix}.$$

Here $\zeta_1^3 = 1$ and $\zeta_1 = \zeta_2^3$. We see that the order of the group of automorphisms of this Riemann surface is exactly 9.

Example 2. $y^3 = x(x-1)(x-t)(x-(1-t))$

We have $t_1+t_2+1 = 2$, $t_1+t_2+t_1t_2 = 1+t(1-t)$. From (6) and (1)

$$a_{11} = 1 - 2a_{13} .$$

Hence, if $a_{13} = 0$, then $a_{11} = 1$. We assume that $a_{13} \neq 0$. From (9) and (1) we have

$$a_{13} = 1$$

if t is not equal to $(1\pm i)/2$. If $t = (1\pm i)/2$, then the equation of the Riemann surface becomes to

$$y^3 = x^4 - 1$$

which will be studied in §3 and it is known that the surface has the group of automorphisms of order 48. However in case $t \neq (1\pm i)/2$ our Riemann surface has for automorphisms which fix P_∞

$$\begin{pmatrix} 1 & 0 & 0 \\ 0 & \zeta_1 & 0 \\ 0 & 0 & 1 \end{pmatrix}, \qquad \begin{pmatrix} -1 & 0 & 1 \\ 0 & \zeta_2 & 0 \\ 0 & 0 & 1 \end{pmatrix} \quad (\zeta_1^3 = \zeta_2^3 = 1).$$

Furthermor, we know the equation $F(x) = 0$ for this surface has no double root. In fact, if there is a double root s, then we have $s = 1/2$ from the existence of the latter automorphisms and hence $t = (1\pm i)/2$. Therefore the order of the group is 6.

Now, in general, as for the group G' of automorphisms which fix P_∞, if $t_1+t_2+1 = 0$, then $a_{13} = 0$ and except for the case of Example 1, we have $a_{11} = 1$. Hence the order of the group G' is 3. If $t_1+t_2+1 \neq 0$ and $t_1+t_2+t_1t_2 = 0$, then $a_{13} = 0$ or $(t_1+t_2+1)/2$ from (12), and we see that the order of G' is 3. Assume that $t_1+t_2+1 \neq 0$ and $t_1+t_2+t_1t_2 \neq 0$. Then, if $a_{13} = 0$, we have

$$\sigma = \begin{pmatrix} 1 & 0 & 0 \\ 0 & \zeta & 0 \\ 0 & 0 & 1 \end{pmatrix} \quad (\zeta^3 = 1).$$

If $a_{13} \neq 0$, then from (3) we have $a_{13} = 1$ or t_1 or t_2 and then we know $t_1 + t_2 = 1$ or $t_1 - t_2 = 1$ from $\{\sigma P_0, \sigma P_1, \sigma P_{t_1}, \sigma P_{t_2}\} = \{P_0, P_1, P_{t_1}, P_{t_2}\}$. Riemann surfaces for these t_i are birationally equivalent to those of

Example 2.

Summarizing above, we obtain following theorem:

Theorem 2. We can classify Riemann surfaces defined by

$$y^3 = x(x-1)(x-t_1)(x-t_2)$$

into four classes, i.e.,

(1) $y^3 = x^4 - 1$, $\#(G) = 48$

(2) $y^3 = x(x^3-1)$, $\#(G) = 9$

(3) $y^3 = x(x-1)(x-t)(x-(1-t))$, $t \neq (1\pm i)/2$, $\#(G) = 6$

(4) the remainder, $\#(G') = 3$ for general surfaces.

Here $\#(G)$ and $\#(G')$ mean orders of the group G of all automorphisms and of the group G' of automorphisms which fix P_∞ respectively.

Remark. From results of §5, we know a surface which has an automorphism σ such that $\sigma P_\infty \neq P_\infty$ is (1). Hence, in (4), we have $\#(G) = 3$ for every surface which is not birationally equivalent to one of (1),(2) and (3).

Remark. A surface $y^3 = x(x-1)(x-t_1)(x-t_2)$ is equivalent to (1) if and only if $\{t_1, t_2\} = \{(1\pm i)/2, (1\mp i)/2\}$, $\{\pm i, 1\pm i\}$.

§3 Automorphisms of Riemann surfaces defined by

$$y^4 = x(x-1)(x-t)$$

As we have seen in §2, a canonical model of our Riemann surface is

$$Y^4 + XZ(X-Z)(X-tZ) = 0.$$

We have to solve the following simultaneous algebraic equations with 9 unknowns :

(1) $a_{21}^4 + a_{11}^3 a_{31} - (t+1)a_{11}^2 a_{31}^2 + ta_{31}^3 a_{11} = 0$

(2) $a_{12}^3 a_{32} + a_{22}^4 - (t+1)a_{12}^2 a_{32}^2 + ta_{32}^3 a_{12} = k$

(3) $a_{23}^4 + a_{13}^3 a_{33} - (t+1)a_{13}^2 a_{33}^2 + ta_{33}^3 a_{13} = 0$

(4) $4a_{21}^3 a_{22} + a_{11}^3 a_{32} + 3a_{11}^2 a_{12} a_{31} - 2(t+1)(a_{11}^2 a_{31} a_{32} + a_{11} a_{12} a_{31}^2) +$
$t(a_{31}^3 a_{12} + 3a_{31}^2 a_{32} a_{11}) = 0$

(5) $4a_{22}^3 a_{23} + a_{12}^3 a_{33} + 3a_{12}^2 a_{13} a_{32} - 2(t+1)(a_{12}^2 a_{32} a_{33} + a_{12} a_{13} a_{32}^2) +$

$t(a_{32}^3 a_{13} + 3a_{32}^2 a_{33} a_{12}) = 0$

(6) $4a_{21}^3 a_{23} + a_{11}^3 a_{33} + 3a_{11}^2 a_{13} a_{31} - 2(t+1)(a_{11}^2 a_{31} a_{33} + a_{11} a_{13} a_{31}^2) +$
$t(a_{31}^3 a_{13} + 3a_{31}^2 a_{33} a_{11}) = k$

(7) $4a_{21} a_{22}^3 + a_{12}^3 a_{31} + 3a_{11} a_{12}^2 a_{32} - 2(t+1)(a_{11} a_{12} a_{32}^2 + a_{12}^2 a_{31} a_{32}) +$
$t(a_{32}^3 a_{11} + 3a_{31} a_{32}^2 a_{12}) = 0$

(8) $4a_{22} a_{23}^3 + a_{13}^3 a_{32} + 3a_{12} a_{13}^2 a_{33} - 2(t+1)(a_{12} a_{13} a_{33}^2 + a_{13}^2 a_{32} a_{33}) +$
$t(a_{33}^3 a_{12} + 3a_{32} a_{33}^2 a_{13}) = 0$

(9) $4a_{21} a_{23}^3 + a_{13}^3 a_{31} + 3a_{11} a_{13}^2 a_{33} - 2(t+1)(a_{13}^2 a_{31} a_{33} + a_{11} a_{13} a_{33}^2) +$
$t(a_{33}^3 a_{11} + 3a_{31} a_{33}^2 a_{13}) = kt$

(10) $6a_{21}^2 a_{22}^2 + 3(a_{11} a_{12}^2 a_{31} + a_{11}^2 a_{12} a_{32}) - (t+1)(a_{11}^2 a_{32}^2 + a_{12}^2 a_{31}^2 +$
$4a_{11} a_{12} a_{31} a_{32}) + 3t(a_{31} a_{32}^2 a_{11} + a_{31}^2 a_{32} a_{12}) = 0$

(11) $6a_{22}^2 a_{23}^2 + 3(a_{12} a_{13}^2 a_{32} + a_{12}^2 a_{13} a_{33}) - (t+1)(a_{12}^2 a_{33}^2 + a_{13}^2 a_{32}^2 +$
$4a_{12} a_{13} a_{32} a_{33}) + 3t(a_{32} a_{33}^2 a_{12} + a_{32}^2 a_{33} a_{13}) = 0$

(12) $6a_{21}^2 a_{23}^2 + 3(a_{11} a_{13}^2 a_{31} + a_{11}^2 a_{13} a_{33}) - (t+1)(a_{11}^2 a_{33}^2 + a_{13}^2 a_{31}^2 +$
$4a_{11} a_{13} a_{31} a_{33}) + 3t(a_{31}^2 a_{33} a_{13} + a_{31} a_{33}^2 a_{11}) = -k(t+1)$

(13) $12a_{21}^2 a_{22} a_{23} + 3(2a_{11} a_{12} a_{13} a_{31} + a_{11}^2 a_{13} a_{32} + a_{11}^2 a_{12} a_{33})$
$- 2(t+1)(a_{11}^2 a_{32} a_{33} + a_{12} a_{13} a_{31}^2 + 2a_{11} a_{12} a_{31} a_{33} + 2a_{11} a_{13} a_{31} a_{32})$
$+ 3t(2a_{31} a_{32} a_{33} a_{11} + a_{31}^2 a_{33} a_{12} + a_{31}^2 a_{32} a_{13}) = 0$

(14) $12a_{21} a_{22}^2 a_{23} + 3(2a_{11} a_{12} a_{13} a_{32} + a_{12}^2 a_{13} a_{31} + a_{11} a_{12}^2 a_{33})$
$- 2(t+1)(a_{12}^2 a_{31} a_{33} + a_{11} a_{13} a_{32}^2 + 2a_{11} a_{12} a_{32} a_{33} + 2a_{12} a_{13} a_{31} a_{32})$
$+ 3t(2a_{31} a_{32} a_{33} a_{12} + a_{31} a_{32}^2 a_{13} + a_{32}^2 a_{33} a_{11}) = 0$

(15) $12a_{21} a_{22} a_{23}^2 + 3(2a_{11} a_{12} a_{13} a_{33} + a_{12} a_{13}^2 a_{31} + a_{11} a_{13}^2 a_{32})$
$- 2(t+1)(a_{13}^2 a_{31} a_{32} + a_{11} a_{12} a_{33}^2 + 2a_{12} a_{13} a_{31} a_{33} + 2a_{11} a_{13} a_{32} a_{33})$
$+ 3t(2a_{31} a_{32} a_{33} a_{13} + a_{32} a_{33}^2 a_{11} + a_{31} a_{33}^2 a_{12}) = 0.$

Now, we consider the Hessian $H(X,Y,Z)$ of the curve

$$F(X,Y,Z) = Y^4 + X^3Z - (t+1)X^2Z^2 + tXZ^3 = 0 :$$

$$H(X,Y,Z) = \begin{vmatrix} F_{XX} & F_{XY} & F_{XZ} \\ F_{YX} & F_{YY} & F_{YZ} \\ F_{ZX} & F_{ZY} & F_{ZZ} \end{vmatrix}$$

$$= AY^2(X-k_1Z)(X-k_2Z)(X-k_3Z)(X-k_4Z) = 0.$$

Here, A and $k_1, \cdots, k_4$ are suitable constants. Hence, we see that P_0,

P_1, P_t and P_∞ over the points 0, 1, t and ∞ on the Riemann sphere are Weierstrass points of degree 2 and lie on the straight line $Y = 0$.

First, consider the case where $k_1 = k_2$ and $k_3 = k_4$, i.e., the number of Weierstrass points is twelve. We can represent the case by the permutations of P_0, P_1, P_t and P_∞. They are

$$\text{(Ia)} \quad \begin{pmatrix} P_0 & P_1 & P_\infty & P_t \\ P_0 & P_1 & P_t & P_\infty \end{pmatrix} \quad \begin{pmatrix} P_0 & P_1 & P_\infty & P_t \\ P_1 & P_0 & P_\infty & P_t \end{pmatrix} \quad \begin{pmatrix} P_0 & P_1 & P_\infty & P_t \\ P_t & P_\infty & P_0 & P_1 \end{pmatrix} \quad \begin{pmatrix} P_0 & P_1 & P_\infty & P_t \\ P_\infty & P_t & P_1 & P_0 \end{pmatrix}.$$

$$\text{(Ib)} \quad \begin{pmatrix} P_0 & P_1 & P_\infty & P_t \\ P_0 & P_\infty & P_1 & P_t \end{pmatrix} \quad \begin{pmatrix} P_0 & P_1 & P_\infty & P_t \\ P_\infty & P_0 & P_t & P_1 \end{pmatrix} \quad \begin{pmatrix} P_0 & P_1 & P_\infty & P_t \\ P_1 & P_t & P_0 & P_\infty \end{pmatrix} \quad \begin{pmatrix} P_0 & P_1 & P_\infty & P_t \\ P_t & P_1 & P_\infty & P_0 \end{pmatrix}.$$

$$\text{(Ic)} \quad \begin{pmatrix} P_0 & P_1 & P_\infty & P_t \\ P_0 & P_t & P_\infty & P_1 \end{pmatrix} \quad \begin{pmatrix} P_0 & P_1 & P_\infty & P_t \\ P_t & P_0 & P_1 & P_\infty \end{pmatrix} \quad \begin{pmatrix} P_0 & P_1 & P_\infty & P_t \\ P_\infty & P_1 & P_0 & P_t \end{pmatrix} \quad \begin{pmatrix} P_0 & P_1 & P_\infty & P_t \\ P_1 & P_\infty & P_t & P_0 \end{pmatrix}.$$

Next, consider following permutations :

$$\text{(II)} \quad \begin{matrix} \begin{pmatrix} P_0 & P_1 & P_\infty & P_t \\ P_0 & P_\infty & P_t & P_1 \end{pmatrix} & \begin{pmatrix} P_0 & P_1 & P_\infty & P_t \\ P_0 & P_t & P_1 & P_\infty \end{pmatrix} & \begin{pmatrix} P_0 & P_1 & P_\infty & P_t \\ P_\infty & P_0 & P_1 & P_t \end{pmatrix} & \begin{pmatrix} P_0 & P_1 & P_\infty & P_t \\ P_1 & P_\infty & P_0 & P_t \end{pmatrix} \\ \begin{pmatrix} P_0 & P_1 & P_\infty & P_t \\ P_t & P_0 & P_\infty & P_1 \end{pmatrix} & \begin{pmatrix} P_0 & P_1 & P_\infty & P_t \\ P_1 & P_t & P_\infty & P_0 \end{pmatrix} & \begin{pmatrix} P_0 & P_1 & P_\infty & P_t \\ P_\infty & P_1 & P_t & P_0 \end{pmatrix} & \begin{pmatrix} P_0 & P_1 & P_\infty & P_t \\ P_t & P_1 & P_0 & P_\infty \end{pmatrix} \end{matrix}.$$

Last, consider remaining permutations :

$$\text{(III)} \quad \begin{pmatrix} P_0 & P_1 & P_\infty & P_t \\ P_0 & P_1 & P_\infty & P_t \end{pmatrix} \quad \begin{pmatrix} P_0 & P_1 & P_\infty & P_t \\ P_1 & P_0 & P_t & P_\infty \end{pmatrix} \quad \begin{pmatrix} P_0 & P_1 & P_\infty & P_t \\ P_\infty & P_1 & P_t & P_0 \end{pmatrix} \quad \begin{pmatrix} P_0 & P_1 & P_\infty & P_t \\ P_t & P_\infty & P_1 & P_0 \end{pmatrix}.$$

From these permutations we obtain automorphisms of curves as follows ; Put

$$P_\infty = \begin{pmatrix} 1 \\ 0 \\ 0 \end{pmatrix}, \quad P_0 = \begin{pmatrix} 0 \\ 0 \\ 1 \end{pmatrix}, \quad P_1 = \begin{pmatrix} 1 \\ 0 \\ 1 \end{pmatrix}, \quad P_t = \begin{pmatrix} t \\ 0 \\ 1 \end{pmatrix}.$$

By $\sigma(P_0) = P_0$, we have

$$a_{13} = 0, \quad a_{23} = 0, \quad a_{33} = 1.$$

By $\sigma(P_1) = P_1$, we have

$$a_{11} = \lambda, \quad a_{21} = 0, \quad a_{31} = \lambda - 1.$$

By $(P_\infty) = P_t$, we have

$$\begin{pmatrix}\mu t\\0\\\mu\end{pmatrix} = \begin{pmatrix}\lambda & a_{12} & 0\\0 & a_{22} & 0\\\lambda-1 & a_{32} & 1\end{pmatrix}\begin{pmatrix}1\\0\\0\end{pmatrix}.$$

Hence, we have $\lambda = t(t-1)^{-1}$. By P_t being transformed to P_∞, we have

$$\begin{pmatrix}\mu\\0\\0\end{pmatrix} = \begin{pmatrix}t(t-1)^{-1} & a_{12} & 0\\0 & a_{22} & 0\\(t-1)^{-1} & a_{32} & 1\end{pmatrix}\begin{pmatrix}t\\0\\1\end{pmatrix}.$$

Hence, we have $t = 1/2$ and so

$$\sigma = \begin{pmatrix}-1 & a_{12} & 0\\0 & a_{22} & 0\\-2 & a_{32} & 1\end{pmatrix}.$$

Substituting these values into the fifteen equations, we have from (4)

$$a_{32} = 2a_{12}$$

and from (8) $a_{12} = 0$ and so $a_{32} = 0$. Thus from (2), $a_{22}^4 = 1$ and we have

$$\sigma = \begin{pmatrix}-1 & 0 & 0\\0 & \zeta & 0\\-2 & 0 & 1\end{pmatrix} \quad (\zeta^4 = 1).$$

Moreover, $t = 1/2$ informs us that the Hessian of the curve is

$$cY^2(2X^2 - 2XZ + Z^2)^2 = 0.$$

Here c is a suitable constant. This shows us that the number of Weierstrass points is exactly twelve.

By the same way it is seen that $t = 1/2$ for (Ia), $t = 2$ for (Ib) and $t = -1$ for (Ic). All (I) around, the curves are all conformally equivalent. In each case, we have 4 automorphisms.

Next, consider permutations (II).

By $\sigma(P_0) = P_0$, we have

$$a_{13} = 0, \quad a_{23} = 0, \quad a_{33} = 1.$$

By $\sigma(P_1) = P_\infty$, we have

$$a_{11} = \lambda, \quad a_{21} = 0, \quad a_{31} = -1.$$

By $\sigma(P_\infty) = P_t$, we have

$$\mu = -1, \quad \lambda = -t.$$

By $\sigma(P_t) = P_1$, we have

$$t = (1+\sqrt{3}i)/2 \text{ or } (1-\sqrt{3}i)/2.$$

Substituting these values into the fifteen equations, we have from (4) $a_{32} = ta_{12}$. From (2) we have $a_{22}^4 = k$. From (5) we have $a_{12} = a_{23} = 0$. From (9) we have $k = -t$. Thus we have finally,

$$\sigma = \begin{pmatrix} -t & 0 & 0 \\ 0 & \zeta & 0 \\ -1 & 0 & 1 \end{pmatrix} \qquad (\zeta^4 = -t).$$

Moreover, in the Hessian of the curve

$$Y^2[-3X^4 + 4(t+1)X^3Z - 2(2t^2+t+2)X^2Z^2 + 4t(t+1)XZ^3 - 3t^2Z^4] = 0,$$

if we put $t = (1\pm\sqrt{3}i)/2$, then we see that the number of Weierstrass points is exactly 20.

By the same way, it is seen that all (II) around t is $(1\pm\sqrt{3}i)/2$, and the curves are all coincident with each other. In each case, we have 4 automorphisms.

Last, consider permutations (III). By $\sigma(P_0) = P_t$, we have

$$a_{13} = t, \quad a_{23} = 0, \quad a_{33} = 1.$$

By $(P_1) = P_\infty$, we have

$$\begin{pmatrix} \lambda \\ 0 \\ 0 \end{pmatrix} = \begin{pmatrix} a_{11} & a_{12} & t \\ a_{21} & a_{22} & 0 \\ a_{31} & a_{32} & 1 \end{pmatrix} \begin{pmatrix} 1 \\ 0 \\ 1 \end{pmatrix}.$$

Here λ is a non-zero value. From this we have

$$a_{11} = -t + \lambda, \quad a_{21} = 0, \quad a_{31} = -1$$

and by $\sigma(P_\infty) = P_1$, we have

$$\begin{pmatrix} \mu \\ 0 \\ \mu \end{pmatrix} = \begin{pmatrix} \lambda-t & a_{12} & t \\ 0 & a_{22} & 0 \\ -1 & a_{32} & 0 \end{pmatrix} \begin{pmatrix} 1 \\ 0 \\ 0 \end{pmatrix}.$$

In this case, P_∞ goes automatically to P_1 and we see that we can take t arbitrarily.

Substituting these values into the fifteen equations, we have from

(4) and (5)

$$a_{12} = a_{32} = 0$$

and from (2) we have $a_{22}^4 = k$. Further, from (6) we have $k = (t-1)^2$. Thus we have finally,

$$\sigma = \begin{pmatrix} -1 & 0 & t \\ 0 & \zeta & 0 \\ -1 & 0 & 1 \end{pmatrix} \quad (\zeta^4 = (t-1)^2).$$

By the same way as above we see that all (III) around, t is arbitrary and in each case, we have 4 automorphisms.

Remark. In the Hessian of our curve, if we have $k_1 = k_2$ and $k_3 \neq k_4$, then the number of Weierstrass points is exactly 16 and there exist 4 Weierstrass points of degree 2 besides P_0, P_1, P_t and P_∞. They are on a straight line. However, it is easy to see that we cannot transform the line on which P_0, P_1, P_t, P_∞ are to the line on which these 4 Weierstrass points are.

Now, we have to reckon up the order of the group of automorphisms of the Riemann surface defined by $y^4 = x(x-1)(x+1)$. This is birationally equivalent to $x^4 + y^4 + 1 = 0$, the canonical model of which is

$$F(X,Y,Z) = X^4 + Y^4 + Z^4 = 0$$

and the Hessian of which is

$$H(X,Y,Z) = cX^2Y^2Z^2 = 0.$$

Here c is a suitable constant. By the same idea used in (I), and by the fact that a projective transformation transforms a straight line into a straight line, we get the order of the group is 96. Indeed, put

$$L_1 : X = 0, \quad L_2 : Y = 0, \quad L_3 : Z = 0.$$

If $L_0 \to L_0$, $L_1 \to L_1$ and $L_2 \to L_2$, then we have

$$\sigma = \begin{pmatrix} \zeta & 0 & 0 \\ 0 & \zeta' & 0 \\ 0 & 0 & 1 \end{pmatrix}.$$

If $L_0 \to L_0$, $L_1 \to L_2$ and $L_2 \to L_1$, then we have

$$\sigma = \begin{pmatrix} \zeta & 0 & 0 \\ 0 & 0 & 1 \\ 0 & \zeta' & 0 \end{pmatrix}.$$

If $L_0 \to L_2$, $L_1 \to L_1$ and $L_2 \to L_0$, then we have

$$\sigma = \begin{pmatrix} 0 & 0 & 1 \\ 0 & \zeta & 0 \\ \zeta' & 0 & 0 \end{pmatrix}.$$

If $L_0 \to L_1$, $L_1 \to L_0$ and $L_2 \to L_2$, then we have

$$\sigma = \begin{pmatrix} 0 & \zeta & 0 \\ \zeta' & 0 & 0 \\ 0 & 0 & 1 \end{pmatrix}.$$

If $L_0 \to L_1$, $L_1 \to L_2$ and $L_2 \to L_0$, then we have

$$\sigma = \begin{pmatrix} 0 & 0 & 1 \\ \zeta & 0 & 0 \\ 0 & \zeta' & 0 \end{pmatrix}.$$

If $L_0 \to L_2$, $L_1 \to L_0$ and $L_2 \to L_1$, then we have

$$\sigma = \begin{pmatrix} 0 & \zeta & 0 \\ 0 & 0 & 1 \\ \zeta' & 0 & 0 \end{pmatrix}.$$

Here in each case, ζ and ζ' are any fourth roots of unity. Thus, finally we see that the order of the group is 96.

Next, we have to reckon up the order of the group of automorphisms of the Riemann surface defined by

$$y^4 = x(x-1)(x-\omega), \quad \omega = (1 \pm \sqrt{3}i)/2.$$

This is birationally equivalent to $y^4 = x^3 - 1$, the canonical model of which is

$$F(X,Y,Z) = X^3Z - Y^4 - Z^4 = 0$$

and the Hessian of which is

$$H(X,Y,Z) = cY^2X(X+2Z)(X+2\zeta Z)(X+2\zeta^2 Z) = 0.$$

Here c is a suitable constant and $\zeta^3 = 1$, $\zeta \neq 1$. Therefore there are exactly 4 Weierstrass points. From this we can count up the order of the group by the same method as above. However, in this case, we have already got the order of the group. Indeed, permutations (II) and (III) shows us that the order is exactly 48.

Summarizing above, we obtain following theorem :

Theorem 3. We can classify Riemann surfaces defined by

$$y^4 = x(x-1)(x-t)$$

into three classes, i.e.,

(1) $y^4 = x(x-1)(x-\alpha)$, $\alpha = -1,2,1/2$, $\#(G) = 96$

(2) $y^4 = x(x-1)(x-\omega)$, $\omega = (1\pm\sqrt{3}i)/2$, $\#(G) = 48$

(3) the remainder, $\#(G) = 16$.

Here $\#(G)$ means the order of the group G of automorphisms of the Riemann surface.

Remark. It is easy to see that a curve $y^4 = x(x-1)(x-t)$ is birationally equivalent to $x^4 + y^4 + 2ax^2y^2 + 1 = 0$.

§4 Automorphisms of Riemann surfaces which are characterized by Weierstrass points

It is well known that if we put N the number of Weierstrass points of Riemann surfaces of genus g, then we have $2g+2 \leq N \leq (g-1)g(g+1)$. $N = 2g+2$ occurs if and only if R is hyperelliptic. If R is non-hyperelliptic of genus 3, $N = 12$ is the minimum and $N = 24$ is the maximum and there exist two Riemann surfaces which attain $N = 12$ [8];

(1) $x^4 + y^4 = 1$,

(2) $x^4 + y^4 + 3(x^2y^2 + x^2 + y^2) + 1 = 0$.

Almost all Riemann surfaces attain $N = 24$ and many of them have no non-trivial automorphisms.However, a Riemann surface defined by

(3) $y^3 + x + x^3y = 0$

which has the group of automorphisms of maximum order 168, attain $N = 24$ [4,6].

We must investigate the group of these Riemann surfaces. As for (1) we already see the structure of the group. Therefore, we start with (2). A canonical model of R is

$$F(X,Y,Z) = X^4 + Y^4 + Z^4 + 3(X^2Y^2 + Y^2Z^2 + Z^2X^2) = 0$$

and the Hessian of the curve is

$$H(X,Y,Z) = (X^2 + Y^2)(Y^2 + Z^2)(Z^2 + X^2).$$

Let L_j $(1\leq j\leq 6)$ be straight lines as follows

$$L_1: X = iY, \quad L_3: Y = iZ, \quad L_5: Z = iY,$$

$$L_2: X = -iY, \quad L_4: Y = -iZ, \quad L_6: Z = -iY.$$

The Weierstrass points are given as follows :

$$P_{11} = \begin{pmatrix} i \\ 1 \\ 1 \end{pmatrix} \quad P_{12} = \begin{pmatrix} -i \\ -1 \\ 1 \end{pmatrix} \quad P_{13} = \begin{pmatrix} -1 \\ i \\ 1 \end{pmatrix} \quad P_{14} = \begin{pmatrix} 1 \\ -i \\ 1 \end{pmatrix} \quad \text{on } L_1.$$

$$P_{21} = \begin{pmatrix} -i \\ 1 \\ 1 \end{pmatrix} \quad P_{22} = \begin{pmatrix} i \\ -1 \\ 1 \end{pmatrix} \quad P_{23} = \begin{pmatrix} 1 \\ i \\ 1 \end{pmatrix} \quad P_{24} = \begin{pmatrix} -1 \\ -i \\ 1 \end{pmatrix} \quad \text{on } L_2.$$

$$P_{31} = \begin{pmatrix} 1 \\ i \\ 1 \end{pmatrix} \quad P_{32} = \begin{pmatrix} -1 \\ i \\ 1 \end{pmatrix} \quad P_{33} = \begin{pmatrix} -i \\ i \\ 1 \end{pmatrix} \quad P_{34} = \begin{pmatrix} i \\ i \\ 1 \end{pmatrix} \quad \text{on } L_3.$$

$$P_{41} = \begin{pmatrix} 1 \\ -i \\ 1 \end{pmatrix} \quad P_{42} = \begin{pmatrix} -1 \\ -i \\ 1 \end{pmatrix} \quad P_{43} = \begin{pmatrix} -i \\ -i \\ 1 \end{pmatrix} \quad P_{44} = \begin{pmatrix} i \\ -i \\ 1 \end{pmatrix} \quad \text{on } L_4.$$

$$P_{51} = \begin{pmatrix} -i \\ -i \\ 1 \end{pmatrix} \quad P_{52} = \begin{pmatrix} -i \\ i \\ 1 \end{pmatrix} \quad P_{53} = \begin{pmatrix} -i \\ -1 \\ 1 \end{pmatrix} \quad P_{54} = \begin{pmatrix} -i \\ 1 \\ 1 \end{pmatrix} \quad \text{on } L_5.$$

$$P_{61} = \begin{pmatrix} i \\ i \\ 1 \end{pmatrix} \quad P_{62} = \begin{pmatrix} i \\ -i \\ 1 \end{pmatrix} \quad P_{63} = \begin{pmatrix} i \\ 1 \\ 1 \end{pmatrix} \quad P_{64} = \begin{pmatrix} i \\ -1 \\ 1 \end{pmatrix} \quad \text{on } L_6.$$

Thus distinct Weierstrass points are 12 in all.

Case I: $\sigma(L_1) = L_1$ (naturally it follows $\sigma(L_2) = L_2$).

Obviously, we have from $\sigma(L_1) = L_1$

$$\sigma = \begin{pmatrix} a_{11} & a_{12} & 0 \\ a_{21} & a_{22} & 0 \\ a_{31} & a_{32} & 1 \end{pmatrix}.$$

If $P_{11} \to P_{11}$ and $P_{21} \to P_{21}$, then we have

$$\sigma = \begin{pmatrix} 1 & 0 & 0 \\ 0 & 1 & 0 \\ 0 & 0 & 1 \end{pmatrix}.$$

If $P_{11} \to P_{12}$ and $P_{21} \to P_{22}$, then we have

$$\sigma = \begin{pmatrix} -1 & 0 & 0 \\ 0 & -1 & 0 \\ 0 & 0 & 1 \end{pmatrix}.$$

If $P_{11} \to P_{13}$ and $P_{21} \to P_{24}$, then we have

$$\sigma = \begin{pmatrix} 0 & -1 & 0 \\ 1 & 0 & 0 \\ 0 & 0 & 1 \end{pmatrix}.$$

If $P_{11} \to P_{14}$ and $P_{21} \to P_{23}$, then we have

$$\sigma = \begin{pmatrix} 0 & 1 & 0 \\ -1 & 0 & 0 \\ 0 & 0 & 1 \end{pmatrix}.$$

It is easy to see that the other permutations of P's cannot occur and we have 4 automorphisms.

Case II: $\sigma(L_1) = L_2$ (naturally it follows $\sigma(L_2) = L_1$).

We have by the same way as above 4 automorphisms:

$$\begin{pmatrix} -1 & 0 & 0 \\ 0 & 1 & 0 \\ 0 & 0 & 1 \end{pmatrix}, \begin{pmatrix} 1 & 0 & 0 \\ 0 & -1 & 0 \\ 0 & 0 & 1 \end{pmatrix}, \begin{pmatrix} 0 & 1 & 0 \\ 1 & 0 & 0 \\ 0 & 0 & 1 \end{pmatrix}, \begin{pmatrix} 0 & -1 & 0 \\ -1 & 0 & 0 \\ 0 & 0 & 1 \end{pmatrix}.$$

Case III: $\sigma(L_1) = L_3$ (naturally it follows $\sigma(L_2) = L_4$).

We have by the same way as above 4 automorphisms:

$$\begin{pmatrix} 0 & 0 & 1 \\ 1 & 0 & 0 \\ 0 & 1 & 0 \end{pmatrix}, \begin{pmatrix} 0 & 0 & 1 \\ -1 & 0 & 0 \\ 0 & -1 & 0 \end{pmatrix}, \begin{pmatrix} 0 & 0 & 1 \\ 0 & -1 & 0 \\ 1 & 0 & 0 \end{pmatrix}, \begin{pmatrix} 0 & 0 & 1 \\ 0 & 1 & 0 \\ -1 & 0 & 0 \end{pmatrix}.$$

Case IV: $\sigma(L_1) = L_4$ (naturally it follows $\sigma(L_2) = L_5$).

We have by the same way as above 4 automorphisms:

$$\begin{pmatrix} 0 & 0 & 1 \\ -1 & 0 & 0 \\ 0 & 1 & 0 \end{pmatrix}, \begin{pmatrix} 0 & 0 & 1 \\ 1 & 0 & 0 \\ 0 & -1 & 0 \end{pmatrix}, \begin{pmatrix} 0 & 0 & 1 \\ 0 & 1 & 0 \\ 1 & 0 & 0 \end{pmatrix}, \begin{pmatrix} 0 & 0 & 1 \\ 0 & -1 & 0 \\ -1 & 0 & 0 \end{pmatrix}.$$

Case V: $\sigma(L_1) = L_5$ (naturally it follows $\sigma(L_2) = L_6$).

We have by the same way as above 4 automorphisms:

$$\begin{pmatrix} 0 & 1 & 0 \\ 0 & 0 & 1 \\ 1 & 0 & 0 \end{pmatrix}, \begin{pmatrix} 0 & -1 & 0 \\ 0 & 0 & 1 \\ -1 & 0 & 0 \end{pmatrix}, \begin{pmatrix} 1 & 0 & 0 \\ 0 & 0 & 1 \\ 0 & -1 & 0 \end{pmatrix}, \begin{pmatrix} -1 & 0 & 0 \\ 0 & 0 & 1 \\ 0 & 1 & 0 \end{pmatrix}.$$

Case VI: $\sigma(L_1) = L_6$ (naturally it follows $\sigma(L_2) = L_5$).

We have by the same way as above 4 automorphisms:

$$\begin{pmatrix} 0 & 1 & 0 \\ 0 & 0 & 1 \\ -1 & 0 & 0 \end{pmatrix}, \begin{pmatrix} 0 & -1 & 0 \\ 0 & 0 & 1 \\ 1 & 0 & 0 \end{pmatrix}, \begin{pmatrix} 1 & 0 & 0 \\ 0 & 0 & 1 \\ 0 & 1 & 0 \end{pmatrix}, \begin{pmatrix} -1 & 0 & 0 \\ 0 & 0 & 1 \\ 0 & -1 & 0 \end{pmatrix}.$$

Summarizing above we obtain 24 automorphisms in all.

Now we treat of (3). A canonical model of R is given by

$$F(X,Y,Z) = XY^3 + YZ^3 + ZX^3 = 0$$

and the Hessian of the curve is given by

$$H(X,Y,Z) = 5X^2Y^2Z^2 - X^5Y - Y^5Z - Z^5X = 0.$$

Hence we see that the Weierstrass points are given as follows:

$$A_0 = \begin{pmatrix} 0 \\ 0 \\ 1 \end{pmatrix} \quad A_1 = \begin{pmatrix} 0 \\ 1 \\ 0 \end{pmatrix} \quad A_\infty = \begin{pmatrix} 1 \\ 0 \\ 0 \end{pmatrix} \quad P_{ij} = \begin{pmatrix} x_{ij} \\ y_{ij} \\ 1 \end{pmatrix} \quad (1 \leq i \leq 3,\ 1 \leq j \leq 7).$$

Here x_{ij} are the roots of $(x^7)^3 + 289(x^7) - 57x^7 - 1 = 0$. There are exactly 3 real roots which we denote by $\omega_1, \omega_2, \omega_3$. Then x_{ij} are written by $\zeta^{j-1}\omega_i$, where $\zeta = \exp(2\pi i/7)$. Furthermore, we know that x-coordinates of Weierstrass points $P(x,y,1)$ which are satisfied by the equation $xy + x + y = 0$ are the roots of

$$x^3 + 2x^2 - x - 1 = 0.$$

These are real and hence ω_1, ω_2, ω_3. Therefore we see that if we put $y_i = -\omega_i/(\omega_i+1)$, then y_{ij} is written by $\zeta^{3j-3}y_i$.

We can construct the group of automorphisms ;

Case I: $\sigma(A_k) = A_k$ $(k=0,1,\infty)$. We have

$$\sigma = \begin{pmatrix} \zeta & 0 & 0 \\ 0 & \zeta^3 & 0 \\ 0 & 0 & 1 \end{pmatrix} \quad (\zeta = \exp(2\pi i/7)).$$

and for each i (i=1,2,3), P_{i1} goes to P_{i2} and P_{i2} goes to P_{i3},..., and P_{i7} goes to P_{i1}. Thus σ generates a cyclic subgroup H of order 7.

Case II: $\tau(A_0) = A_1$, $\tau(A_1) = A_\infty$, $\tau(A_\infty) = A_0$. We have

$$\tau = \begin{pmatrix} 0 & 1 & 0 \\ 0 & 0 & 1 \\ 1 & 0 & 0 \end{pmatrix}.$$

Then, by τ, $\{P_{ij}\}$ goes to $\{P_{ij}\}$ and τ generates a cyclic subgroup of order 3.

Case III: $\lambda(A_\infty) = P_{11}$. We have

$$\lambda = \begin{pmatrix} \omega_1 & y_1 & 1 \\ y_1 & 1 & \omega_1 \\ 1 & \omega_1 & y_1 \end{pmatrix}.$$

λ is an involution. If we put $\lambda_{ij}(A_\infty) = P_{ij}$, then λ_{ij} are written in a form $\sigma^k\tau^{k'}\lambda$ with suitable integers k and k'. It is easy to see that the group of automorphisms, G, is resolved into

$$G = H + H\tau + H\tau^2 + \Sigma H\lambda_{ij}$$

which shows that the order of G is 168.

Remark. For the proof of λ being automorphism, it is sufficient to show that the coefficients of X^4, X^3Y, X^2Y^2, X^2YZ are equal to 0.

Remark. The roots of $x^3 + 2x^2 - x - 1 = 0$ are $(\zeta+\zeta^6)^{-1}$, $(\zeta^2+\zeta^5)^{-1}$ and $(\zeta^3+\zeta^5)^{-1}$. If we put $\omega_1 = (\zeta+\zeta^6)^{-1} = \zeta+\zeta^2+\zeta^5+\zeta^6$, then $y_1 = \zeta^2+\zeta^5$.

Riemann surfaces defined by (1) and (2) are special cases of the following:

$$(\#) \qquad x^4 + y^4 + 2ax^2y^2 + 2bx^2 + 2cy^2 + 1 = 0$$

where $a^2, b^2, c^2 \neq 1$ and $1 + 2abc - a^2 - b^2 - c^2 \neq 0$.

For this surface we have following lemma:

Lemma 5 [8]. A non-hyperelliptic Riemann surface of genus 3, which has two elliptic-hyperelliptic involutions σ_1, σ_2 such as $\sigma_1\sigma_2 = \sigma_2\sigma_1$ is given by (#).

Now, we investigate the group of automorphisms of Riemann surfaces defined by (#).

First, if $a = b = c = 0$, then as we have seen already the order of the group is 96.

Second, if any two of a, b and c are zero, then it is reduced to the case treated in Theorem 3. Thereforc the order of the group is determined, i.e., 96 or 48 or 16.

Third, if $a = b = c \neq 0$, then the order of the group is exactly 24. Indeed, the automorphisms obtained for $a = b = c = 3/2$ are precisely automorphisms of our Riemann surface. Therefore the order of our group is equal to or greater than 24. However, if it is greater than 24 it must be 96 or 48. Obviously it is not 96 and §5 (iii) (1) shows that 48 must also be rejected.

Fourth, if $a \neq 0$, and $b = c \neq 0$, then the order of the group is 8 in general. Indeed, we see easily that the surface has at least 8 automorphisms, i.e., following three generate a subgroup of order 8:

$$\sigma_1(x,y) = (x,-y), \quad \sigma_2(x,y) = (-x,y), \quad \sigma_3(x,y) = (y,x).$$

Therefore the order of our group is equal to or greater than 8. However if it is greater than 8 it must be 16 or 24 Or 32 or 48 or 96. Obviously it is not 96, 48, 32 and 24. Further, §5 (ii)(2) shows that 16 must also be rejected.

If $a = 0$, and $b = c \neq 0$, then the situation is the same as above.

Fifth, if $abc \neq 0$ and a, b, c are different from each other, then the order of the group is 4. By the same reasoning as above we see easily this fact.

If $a = 0$, and $bc \neq 0$ and b, c are different from each other, the situation is the same as above.

In each case a, b and c are taken in general.

§5 Riemann surfaces having no automorphisms of order > 4 and cyclic coverings of tori

Let G be the group of automorphisms of a Riemann surface R of genus 3 and let n be the order of G. Let H be a cyclic subgroup of G which has maximal order and let m be the order. It is easy to see that if $m > 4$, then R/H is the Riemann sphere and we have studied these Riemann surfaces hitherto. Therefore it remains to study for m = 2, 3 and 4. However even for m = 2, 3 and 4, if R/H is the Riemann sphere, we have studied already to a certain extent.

First, assume that n is greater than 4 and $m \leq 4$. Then by Lemma 1, §0 we see that R/G is the Riemann sphere since the number r of ramification points is greater than one. Further we see that $r \geq 3$ by Lemma 1. If such a Riemann surface exists it must satisfy following:

(i) if $r \geq 5$, then $n \leq 8$ and

(1) $n = 8$ $\quad \nu_1 = 2,\ \nu_2 = 2,\ \nu_3 = 2,\ \nu_4 = 2,\ \nu_5 = 2$

(2) $n = 6$ $\quad \nu_1 = 2,\ \nu_2 = 2,\ \nu_3 = 2,\ \nu_4 = 3,\ \nu_5 = 3$

(ii) if $r = 4$, then $n \leq 24$ and

(1) $n = 24$ $\quad \nu_1 = 2,\ \nu_2 = 2,\ \nu_3 = 2,\ \nu_4 = 3$

(2) $n = 16$ $\quad \nu_1 = 2,\ \nu_2 = 2,\ \nu_3 = 2,\ \nu_4 = 4$

(3) $n = 12$ $\quad \nu_1 = 2,\ \nu_2 = 2,\ \nu_3 = 3,\ \nu_4 = 3$

(4) $n = 8$ $\quad \nu_1 = 2,\ \nu_2 = 2,\ \nu_3 = 4,\ \nu_4 = 4$

(5) $n = 6$ $\quad \nu_1 = 3,\ \nu_2 = 3,\ \nu_3 = 3,\ \nu_4 = 3$

(iii) if $r = 3$, then $n \leq 48$ and

(1) $n = 48$ $\quad \nu_1 = 3,\ \nu_2 = 3,\ \nu_3 = 4$

(2) $n = 24$ $\quad \nu_1 = 3,\ \nu_2 = 4,\ \nu_3 = 4$

(3) $n = 16$ $\quad \nu_1 = 4,\ \nu_2 = 4,\ \nu_3 = 4.$

We shall solve the problem on the existence or non-existence of these Riemann surfaces which satisfy above conditions in order.

It may be convenient to give following lemmas:

Lemma 6. Let R be a Remann surface of genus g ($\geqq 3$) and let σ be an automorphism of R such that $R/\langle\sigma\rangle$ be a torus. We denote by ν_P the ramification multiplicity of a branch point P of the covering π: R $\to$ $R/\langle\sigma\rangle$. Then the divisor $\Sigma(\nu_P-1)P$ is canonical.

Proof. Let ω be a differential of the first kind on the torus $R/\langle\sigma\rangle$. We may regard ω as a differential on R and we have

$$\mathrm{div}_R(\omega) = \pi^{-1}\mathrm{div}_{R/\langle\sigma\rangle}(\omega) + \Sigma(\nu_P-1)P = \Sigma(\nu_P-1)P.$$

Lemma 7. Let R be a non-hyperelliptic Riemann surface of genus 3. Assume that R has an automorphism σ of order 4 and σ has fixed points on R. Then we see that fixed points by σ are four in all. If we denote them by P_1, P_2, P_3 and P_4, then $\Sigma_{i=1}^{4}P_i$ and $4P_i\,(1\leqq i\leqq 4)$ are canonical.

Proof. It is easy by Lemma 6 and the results in §3.

Lemma 8 [1,2,5]. Let R be of genus 3 and be 2-hyperelliptic. Then R is hyperelliptic and elliptic-hyperelliptic.

Corollary. Every involutions on non-hyperelliptic Riemann surfaces of genus 3 are elliptic-hyperelliptic.

(i) (1) $n = 8,\quad \nu_1 = 2,\quad \nu_2 = 2,\quad \nu_3 = 2,\quad \nu_4 = 2,\quad \nu_5 = 2$

As we see in §2, we have no automorphism of order 8 in this case. Further we see that every automorphism of order 4 has no fixed points since $\nu_i = 2$ $(1\leqq i\leqq 5)$. Hence, if there were an automorphism σ of order 4, then the genus of $R/\langle\sigma\rangle$ is one (cf. Case m = 4 in §1).

First, consider hyperelliptic case:

Let σ_0 be the hyperelliptic involution. The number of points fixed by σ_0 is 8 since they are Weierstrass points of R. Now, branch points over R/G are 20 in all. Therefore there exist exactly three elliptic-hyperelliptic involutions σ_1, σ_2 and σ_3. We see that each of $\sigma_0\sigma_1$, $\sigma_0\sigma_2$ and $\sigma_0\sigma_3$ is a 2-hyperelliptic involutions [1]. Hence the group of automorphisms, $G = \{1,\sigma_0,\sigma_1,\sigma_2,\sigma_3,\sigma_0\sigma_1,\sigma_0\sigma_2,\sigma_0\sigma_3\}$ is commutative. Let $P_1,\ldots,P_8$ be Weierstrass points and let $Q_1,\ldots,Q_4$ be points fixed by σ_1. Since $\sigma_0 Q_i = Q_j (i \neq j)$, we may put $\sigma_0 Q_1 = Q_2$ and $\sigma_0 Q_3 = Q_4$. Then we obtain

$$y^2 = (x-1)(x+1)(x-t_1)\cdots(x-t_6)$$

with $\mathrm{div}(x) = Q_3 + Q_4 - Q_1 - Q_2$ and $\mathrm{div}(y) = \Sigma_{i=1}^{8} P_i - 4(Q_1 + Q_2)$. Here $x(P_1) = 1$, $x(P_2) = -1$ and $x(P_{i+2}) = t_i$ $(1 \leq i \leq 6)$. We see that $\sigma_2 Q_1 = Q_3$, $\sigma_2 Q_2 = Q_4$ and $\sigma_1(P_i) = P_j$ $(i \neq j)$. Thus we have

$$\sigma_0 x = x,\ \sigma_0 y = -y;\quad \sigma_1 x = -x,\ \sigma_1 y = y;\quad \sigma_2 x = \alpha/x,\ \sigma_2 y = \beta y/x^4.$$

Since $\sigma_2^2 = 1$, we have $\beta^2 = \alpha^4$. Operating σ_2 to the equation, we have $\beta^2 = -t_1 t_2\cdots t_6$. σ_1 and σ_2 give permutations of x-coordinates of P_i $(1 \leq i \leq 8)$. Therefore $\alpha = t_1$ and we obtain for the equation of R

$$y^2 = (x^2-1)(x^2-t_1^2)(x^2-t_2^2)(x^2-t_1^2/t_2^2).$$

This Riemann surfaces has the group of automorphisms of order 8 in general.

Second, consider non-hyperelliptic case:

Every involution is elliptic-hyperelliptic by Lemma 8 and so each of them has 4 fixed points. Since the order of G is 8, there exists one automorphism σ of order 4. The index of the subgroup H generated by σ is 2. Hence H is normal. Put $\sigma_1 = \sigma^2$. Then we have $\tau\sigma^2\tau^{-1} = \sigma^2$ for every τ of G. The number of branch points is 20 in all, Therefore there must be 4 involutions except for σ_1. Let σ_2 be one of them. Then we have $\sigma_1\sigma_2 = \sigma_2\sigma_1$ $(= \sigma_3)$. We get $x^4 + y^4 + 2ax^2y^2 + 2bx^2 + 2cy^2 + 1 = 0$, by Lemma 5, for the equation of R. To be more precise, let $Q_1^{(i)},\ldots,Q_4^{(i)}$ $(1 \leq i \leq 3)$ be points fixed by σ_i, and put

$$\mathrm{div}(x) = \Sigma_{j=1}^{4} Q_j^{(2)} - \Sigma_{j=1}^{4} Q_j^{(1)},\quad \mathrm{div}(y) = \Sigma_{j=1}^{4} Q_j^{(3)} - \Sigma_{j=1}^{4} Q_j^{(1)}$$

and so

$$\sigma_1: (x,y) \to (-x,-y),\quad \sigma_2: (x,y) \to (-x,y)$$

$$\sigma_3: (x,y) \to (x,-y).$$

Now, we have $G = H + \sigma_2 H$ and every element of $\sigma_2 H$ is an involution. We see that $\sigma_2\sigma \neq \sigma\sigma_2$. Hence we get $\sigma_2\sigma\sigma_2 = \sigma^3$ and so $\sigma\sigma_1\sigma_2\sigma^{-1} = \sigma_2$. Therefore, we have

$$\sigma(\Sigma_{j=1}^4 Q_j^{(1)}) = \Sigma_{j=1}^4 Q_j^{(1)}, \quad \sigma(\Sigma_{j=1}^4 Q_j^{(3)}) = \Sigma_{j=1}^4 Q_j^{(2)},$$
$$\sigma(\Sigma_{j=1}^4 Q_j^{(2)}) = \Sigma_{j=1}^4 Q_j^{(3)}.$$

Hence we can put $\sigma(x,y) = (\alpha y, \beta X)$. We see then that $\alpha\beta = -1$ since $\sigma^2 = \sigma_1$. By operating σ to our equation, we have

$$\alpha^4 = \beta^4 = 1 \quad \text{and} \quad \alpha^2 b = c.$$

Thus after some change of notation if necessary, our equation becomes to

$$x^4 + y^4 + 2ax^2y^2 + 2bx^2 + 2by^2 + 1 = 0.$$

The order of this Riemann surface is exactly 8 in general.

(i) (2) $n = 6$, $\nu_1 = 2$, $\nu_2 = 2$, $\nu_3 = 2$, $\nu_4 = 2$, $\nu_5 = 3$.

We see that if there is such a Riemann surface, it must be non-hyperelliptic. In fact, if not, there is the hyperelliptic involution σ. By assumption we have an element τ of G whose order is 3. Then we have $\sigma\tau = \tau\sigma$ because the hyperelliptic involution commute with every automorphism. Hence $\sigma\tau$ is an element of G whose order is 6 and so we see in §1 that there does not exist such a Riemann surface.

Now assume that R is non-hyperelliptic. Every involution must be elliptic-hyperelliptic and so each of them has 4 fixed points. Hence there are 3 involutions σ_1, σ_2 and σ_3.

Let P_1 and P_2 be branch points whose ramification multiplicity is three. Let τ be the automorphism of order 3, by which P_1 and P_2 are fixed. We see that $\sigma_1\tau\sigma_1 = \tau^2$ and $R/\langle\tau\rangle$ is a torus. By Lemma 6 we see that the divisor $2(P_1 + P_2)$ is canonical. We have

$$G = \{1, \tau, \tau^2, \sigma_1, \tau\sigma_1, \tau^2\sigma_1\}.$$

We may put $\sigma_2 = \tau\sigma_1$ and $\sigma_3 = \tau^2\sigma_1$. Let $\{Q_i^{(1)}\}, \{Q_i^{(2)}\}, \{Q_i^{(3)}\}$ be fixed points of $\sigma_1, \sigma_2, \sigma_3$ respectively. Then by Lemma 6, $\Sigma_{i=1}^4 Q^{(1)}, \Sigma_{i=1}^4 Q^{(2)}$, $\Sigma_{i=1}^4 Q_i^{(3)}$ are canonical. By $\sigma_1\sigma_2\sigma_1 = \sigma_3$, $\sigma_1 Q_i^{(2)} = Q_j^{(3)}$ and so

$\sigma_1(\Sigma_{i=1}^4 Q_i^{(2)}) = \Sigma_{i=1}^4 Q_i^{(3)}$. By $\tau\sigma_1\tau^{-1} = \sigma_3$, $\tau Q_i^{(1)} = Q_k^{(3)}$ and so

$$\tau^{-1}(\Sigma_i Q_i^{(1)}) = \Sigma_i Q_i^{(2)}, \quad \tau^{-1}(\Sigma_i Q_i^{(2)}) = \Sigma_i Q_i^{(3)},$$
$$\tau^{-1}(\Sigma_i Q_i^{(3)}) = \Sigma_i Q_i^{(1)}.$$

Futher we have $\sigma_1 P_1 = P_2$. Put

$$\operatorname{div}(x) = \Sigma_{i=1}^4 Q_i^{(2)} - 2(P_1 + P_2),$$
$$\operatorname{div}(y) = \Sigma_{i=1}^4 Q_i^{(3)} - 2(P_1 + P_2).$$

We have $\sigma_1 x = \alpha y$ since $\operatorname{div}(\sigma_1 x) = \sigma_1^{-1}\operatorname{div}(x) = \operatorname{div}(y)$. By the same way we have $\sigma_1 y = \beta x$. Further we see $\alpha\beta = 1$ since $\sigma_1^2 = 1$. Hence we have $\sigma_1(\alpha y) = \alpha\beta x = x$. Therefore rewrite y instead of αy. We have

$$\sigma_1 x = y \quad \text{and} \quad \sigma_1 y = x.$$

Put

$$\operatorname{div}(z) = \Sigma_{i=1}^4 Q_i^{(1)} - 2(P_1 + P_2).$$

By the same way as above, we see that $\sigma_1 z = -z$.

We may put

$$\tau x = \alpha y, \quad \tau y = \beta z, \quad \tau z = \gamma x.$$

with suitable constants α, β and γ. We have $\alpha\beta\gamma = 1$ since $\tau^3 = 1$, and we have $\alpha = -1$ since $\sigma_1\tau\sigma_1 = \tau^2$. Therefore we have $\tau x = -y$.

The vector space $L(2P_1+2P_2)$ is three dimensional, and 1, x and y are a basis of the space. z is expressed by $ax + by + c$ with a suitable constants a, b and c. It is easy to see that $a = -b$ and $c = 0$. Therefore rewrite z instead of z/a. Then we have $z = x - y$. Hence we have $\tau y = x - y$.

Now, put

$$x' = \frac{x - 2y + 1}{x + y + 1}, \qquad y' = \frac{-2x + y + 1}{x + y + 1}.$$

Then we have

$$\sigma_1 : (x',y') \to (y',x'), \qquad \tau : (x',y') \to (y'/x', 1/x').$$

Rewrite x,y instead of x',y'. We have for the equation of R

$$f(x,y) = \Sigma\, a_{ij}x^i y^j.$$

Here i and j run from 0 to 4 and $i + j \leq 4$. Indeed, 1, x, y are a basis of L(K) where K is a canonical divisor, and so the function field of R is given by $\mathbb{C}(x,y)$. Moreover the dimension of L(4K) is equal to 14.

In the homogeneous coordinates, σ_1 and τ are represented by

$$\sigma_1 \begin{pmatrix} X \\ Y \\ Z \end{pmatrix} = \begin{pmatrix} 0 & 1 & 0 \\ 1 & 0 & 0 \\ 0 & 0 & 1 \end{pmatrix} \begin{pmatrix} X \\ Y \\ Z \end{pmatrix}, \quad \tau \begin{pmatrix} X \\ Y \\ Z \end{pmatrix} = \begin{pmatrix} 0 & 1 & 0 \\ 0 & 0 & 1 \\ 1 & 0 & 0 \end{pmatrix} \begin{pmatrix} X \\ Y \\ Z \end{pmatrix}.$$

The equation of R in the homogeneous coordinates must be invariant by the symmetric group G of degree 3. Therefore we obtain

$$(*) \qquad A(X^4+ Y^4+ Z^4) + B(X^3Y + Y^3Z + Z^3X + X^3Z + Z^3Y + Y^3X) + C(X^2Y^2+ Y^2Z^2+ Z^2X^2) = 0.$$

Here A, B and C are suitable constants. If $B = C = 0$ and $A \neq 0$, then the order of the group is 96. If $B = 0$ and $AC \neq 0$, then the order of the group is 24 in general. Since our Riemann surface is not hyperelliptic, the order of the group cannot take the value 12. Thus for general A, B, C, (*) is the desired equation of our Riemann surface.

(ii) (1) $\quad n = 24, \quad \nu_1 = 2, \quad \nu_2 = 2, \quad \nu_3 = 2, \quad \nu_4 = 3$

We see that R must be non-hyperelliptic by the same reason as in (2) of (i). Let H be a 2-Sylow subgoup of order 8. The covering π: $R \to R/H$ is of type $\nu'_1 = \nu'_2 = \ldots = \nu'_5 = 2$ since each ramification multiplicity ν'_i $(1 \leqq i \leqq 5)$ of π is smaller or equall to 2. Therefore by the same reason as in (1) of (i), H has an automorphism σ of order 4. An automorphism τ of order 3 has five fixed points if $R/\langle\tau\rangle$ is of genus 0 and has two fixed points if $R/\langle\tau\rangle$ is of genus 1. The branch points such as $\nu = 3$ are eight in all. Hence there are four groups of order 3 each of which has two fixed points. Let $\langle\tau\rangle$ be one of them. If we put $\sigma_1 = \sigma^2$, then we have $\tau\sigma_1\tau^{-1} \neq \sigma_1$. Indeed, let $Q_i^{(1)}$ $(1 \leqq i \leqq 4)$ be fixed points by σ_1. If $\tau\sigma_1\tau^{-1} = \sigma_1$, then we have $\tau Q_i^{(1)} = Q_j^{(1)}$ $(i \neq j)$. This is a contradiction. Put $\tau\sigma_1\tau^{-1} = \sigma_2$ and take $H = \langle\sigma\rangle + \sigma_2\langle\sigma\rangle$ as a 2-Sylow subgroup. Then we have $\sigma_1\sigma_2 = \sigma_2\sigma_1$ and we obtain by Lemma 5 as (1) of (i)

$$x^4+ y^4+ 2ax^2y^2+ 2bx^2+ 2by^2+ 1 = 0.$$

Let $Q_i^{(2)}$, $Q_i^{(3)}$ $(1 \leq i \leq 4)$ be fixed points by σ_2, $\sigma_3 (= \sigma_1\sigma_2)$ respectively. Since $\tau Q_i^{(2)}$ is a branch point of $\tau\sigma_2\tau^{-1} = \sigma_3$ we have

$$\tau(\Sigma_{i=1}^4 Q_i^{(1)}) = \Sigma_{i=1}^4 Q_i^{(2)} \text{ and } \tau(\Sigma_{i=1}^4 Q_i^{(2)}) = \Sigma_{i=1}^4 Q^{(3)}.$$

Thus we have

$$\mathrm{div}(\tau x) = \tau^{-1}(\Sigma_i Q_i^{(2)} - \Sigma_i Q_i^{(1)}) = \Sigma_i Q_i^{(1)} - \Sigma_i Q_i^{(3)},$$
$$\mathrm{div}(\tau y) = \tau^{-1}(\Sigma_i Q_i^{(3)} - \Sigma_i Q_i^{(1)}) = \Sigma_i Q_i^{(2)} - \Sigma_i Q_i^{(3)}.$$

Hence we see that $\tau(x) = \alpha/y$, and $\tau(y) = \beta x/y$. Operating to our equation we have

$$\alpha^4 = \beta^4 = 1, \quad b\beta^2 = a, \quad a\alpha^2\beta^2 = b \quad \text{and} \quad b\alpha^2 = b.$$

If $b = 0$, then $a = 0$ and as we see in §3, in this case R has the group of automorphisms whose order is 96. Therefore $b \neq 0$ and the equation becomes to

$$x^4 + y^4 + 2a(x^2y^2 + x^2 + y^2) + 1 = 0$$

and the order of the group of this Riemann surface is exactly 24 in general as we see in §4.

(ii) (2) $n = 16$, $\nu_1 = 2$, $\nu_2 = 2$, $\nu_3 = 2$, $\nu_4 = 4$

Let P_1, P_2, P_3 and P_4 be branch points of multiplicity 4 and let $H = \langle\sigma\rangle$ be a cyclic group of order 4 and P_1 be fixed by σ. We see that σ^2 is an elliptic-hyperelliptic involution. In fact, if not, σ^2 would be the hyperelliptic involution and so P_1, P_2, P_3 and P_4 would be Weierstrass points and the other 4 Weierstrass points were among branch points of multiplicity 2. This is a contradiction. Since σ fixes P_1, P_2 P_3 and P_4, H must be a normal subgroup and we have $\sigma^2\rho = \rho\sigma^2$ for every ρ of G since σ^2 is the only one element which is of order 2 in H.

First, consider the hyperelliptic case:

Let σ_0 be the hyperelliptic involution and let $Q_i (1 \leq i \leq 8)$ be Weierstrass points. Further we may put $\sigma_0 P_1 = P_2$ and $\sigma_0 P_3 = P_4$. The equation of R is given by

$$y^2 = (x-a_1)(x-a_2)\cdots(x-a_8)$$

with $\mathrm{div}(x) = P_3 + P_4 - (P_1 + P_2)$, $\mathrm{div}(x-a_i) = 2Q_i - (P_1 + P_2)$ and $\mathrm{div}(y) = \Sigma_{i=1}^8 Q_i - 4(P_1 + P_2)$. Then we have

$$\operatorname{div}(\sigma x) = \sigma^{-1}\operatorname{div}(x) = P_3 + P_4 - (P_1 + P_2).$$

Hence we have $\sigma x = \alpha x$, $\alpha^4 = 1$, say, $\alpha = i$. further we have $\sigma y = y$ since we have $\Sigma_{i=1}^{8} Q_i = \Sigma_{\nu=0}^{3} \sigma^{\nu} Q_1 + \Sigma_{\nu=0}^{3} \sigma^{\nu} Q_5$. Put $\sigma^{\nu} Q_1 = Q_{1+\nu}$ and $\sigma^{\nu} Q_5 = Q_{5+\nu}$ $(0 \leqq \nu \leqq 3)$. Then we have

$$ia_1 = a_2, \quad -a_1 = a_3, \quad -ia_1 = a_4,$$

$$ia_5 = a_6, \quad -a_5 = a_7, \quad -ia_5 = a_8.$$

Then we obtain

$$y^2 = (x^4 - a_1^4)(x^4 - a_5^4)$$

and finally we obtain

$$y^2 = x^8 + 2ax^4 + 1.$$

Further R has one more involution σ': $(x,y) \to (\gamma/x, \gamma^2 y/x^4)$, $\gamma^4 = 1$ and we have

$$G = H + \sigma_0 H + \sigma' H + \sigma_0 \sigma' H.$$

Second, consider non-hyperelliptic case.

There exist 6 elliptic-hyperelliptic involutions besides $\sigma_1 = \sigma^2$. We denote one of them by σ_2. Let $Q_1^{(2)}, Q_2^{(2)}, Q_3^{(2)}$ and $Q_4^{(2)}$ be fixed points by σ_2 and $Q_1^{(3)}, Q_2^{(3)}, Q_3^{(3)}$ and $Q_4^{(3)}$ be fixed points by $\sigma_1\sigma_2$. Put as in (2) of (i)

$$\operatorname{div}(x) = \Sigma_i Q_i^{(2)} - \Sigma_i P_i, \quad \operatorname{div}(y) = \Sigma_i Q_i^{(3)} - \Sigma_i P_i.$$

We obtain as an equation of R

$$x^4 + y^4 + 2ax^2y^2 + 2bx^2 + 2cy^2 + 1 = 0.$$

Now, we have $\sigma_2\sigma\sigma_2 = \sigma$ or σ^3 by $\sigma_2 H \sigma_2 = H$. Assume that $\sigma_2\sigma\sigma_2 = \sigma$. Then by $\sigma(\Sigma_i Q_i^{(2)}) = \Sigma_i Q_i^{(2)}$ and $\sigma(\Sigma_i Q_i^{(3)}) = \Sigma_i Q_i^{(3)}$, we have

$$\sigma : (x,y) \to (\alpha x, \beta Y).$$

However $\sigma^2(x,y) = (-x,-y)$. Hence $\alpha^2 = \beta^2 = -1$. Operating σ to the equation we have $b = c = 0$ and we obtain finally

$$x^4 + y^4 + 2ax^2y^2 + 1 = 0.$$

However the Riemann surface defined by this equation has the group of automorphisms whose order is given in (3) of Thorem 3 in §3. We have

$$G = H + \sigma_2 H + \sigma' H + \sigma'\sigma_2 H.$$

Here σ_2: $(x,y) \to (-x,y)$ and σ': $(x,y) \to (y,x)$.

Assume that $\sigma_2\sigma\sigma_2 = \sigma^3$. Then by the same way as in (2) of (i), we have

$$\sigma : (x,y) \to (\alpha y, \beta x), \quad \alpha^2 = 1, \ \alpha\beta = -1$$

and $b = c$. σ can be represented in homogeneous coordinates by

$$\sigma : (X,Y,Z) \to (\alpha Y, -\alpha^{-1}X, Z)$$

and P_i is fixed by σ. Hence we see that by σ

$$(k,1,0) \to (\alpha, -\alpha^{-1}k, 0).$$

Here $k^4 + 2ak^2 + 1 = 0$. Thus we have $k^2 = -\alpha^2 = -1$ and so $a = 1$ contrary to our assumption. Therefore this case does not occur.

(ii) (3) $n = 12$, $\nu_1 = 2$, $\nu_2 = 2$, $\nu_3 = 3$, $\nu_4 = 3$

By the same reason as in (2) of (i), we see that R must be non-hyperelliptic. By the way there are 12 branch points of multiplicity 2. Hence there are 3 elliptic-hyperelliptic involutions. There are 8 branch points of multiplicity 3. Hence there are 4 groups of automorphisms of order 3, each of which has 2 fixed points. Let $H = \{1, \sigma_1, \sigma_2, \sigma_3\}$ be 2-Sylow subgroup of G. Then H is normal and σ_1, σ_2 and σ_3 are involutions since there are exactly 3 involutions in G. Therefore by Lemma 5, we have as an equation of R

$$x^4 + y^4 + 2ax^2y^2 + 2bx^2 + 2cy^2 + 1 = 0.$$

Further we see that

$$\tau(\Sigma_i Q_i^{(1)}) = \Sigma_i Q_i^{(2)}, \quad \tau(\Sigma_i Q_i^{(2)}) = \Sigma_i Q_i^{(3)}, \quad \tau(\Sigma_i Q_i^{(3)}) = \Sigma_i Q_i^{(1)}$$

or

$$\tau(\Sigma_i Q_i^{(1)}) = \Sigma_i Q_i^{(3)}, \quad \tau(\Sigma_i Q_i^{(3)}) = \Sigma_i Q_i^{(2)}, \quad \tau(\Sigma_i Q_i^{(2)}) = \Sigma_i Q_i^{(1)}$$

since H is normal and we cannot have $\tau(\Sigma_i Q_i^{(j)}) = \Sigma_i Q_i^{(j)}$. Here $Q_1^{(j)}, \ldots, Q_4^{(j)}$ are fixed points by σ_j $(1 \leq j \leq 3)$.

Thus by the same way as in (1) of (ii), we have in the former case

$$\tau : (x,y) \to (\alpha/y, \beta x/y)$$

and we see that

$$\alpha^4 = \beta^4 = 1, \quad \beta^2 c = a, \quad \alpha^2\beta^2 a = 1$$

and finally we obtain as an equation of R

$$x^4 + y^4 + 2a(x^2y^2 + x^2+y^2) + 1 = 0.$$

We have also the same equation in the latter case for τ. However the order of the group of this Riemann surface is equal to or greater than 24 as we see in §4.

Thus we can conclude that there exists no such a Riemann surface. Riemann surfaces which have the group of order 12 are only hyperelliptic.

(ii) (4) $n = 8,\ \nu_1 = 2,\ \nu_2 = 2,\ \nu_3 = 4,\ \nu_4 = 4$

There are 4 branch points of degree 4. Let P_1, P_2, P_3 and P_4 be these points. Let $H = \langle\sigma\rangle$ be a cyclic group of order 4 and P_1 be fixed by σ. Obviously H is normal. Therefore we have $\sigma_1\rho = \rho\sigma_1$ for every ρ of G. Here $\sigma_1 = \sigma^2$.

First, consider the hyperelliptic case :

Let σ_0 be the hyperelliptic involution. We divide into tow cases :

(a) $\sigma_1 \neq \sigma_0$. Therefore $R/\langle\sigma_1\rangle$ is elliptic and by the same way as in (2) of (ii) we have as an equation of R

$$y^2 = x^8 + 2ax^4 + 1.$$

This Riemann surface has one more involution which does not belong to $H + \sigma_0 H$ (see (2) of (ii)). We must shut out this case.

(b) $\sigma_1 = \sigma_0$. σ has two fixed points. We denote them by P_1 and P_2. Let σ' be an automorphism of order 4 which fixes the point P_3. Then σ' fixes also P_4 and we see that $\sigma'^2 = \sigma_0$. Therefore, we see that $P_1,\dots,P_4$ are Weierstrass points. Let $P_5,\dots,P_8$ be the remaining Weierstrass points. The equation of R is given by

$$y^2 = x(x-1)(x-a_1)(x-a_2)\cdots(x-a_5)$$

with

$$\operatorname{div}(x) = 2P_2 - 2P_1,\quad \operatorname{div}(x-1) = 2P_3 - 2P_1,$$

$$\operatorname{div}(y) = \Sigma_{i=2}^{8} P_i - 7P_1.$$

Then we have

$$\sigma : (x,y) \to (\alpha x, \beta y),\quad \alpha^2 = 1,\quad \beta^2 = -1.$$

Operating σ to our equation we have

$$y^2 = -\alpha x(x-\alpha)(x-\alpha a_1)\cdots(x-\alpha a_5).$$

Therefore we have

$$y^2 = x(x^2-1)(x^2-t_1^2)(x^2-t_2^2).$$

We may consider that

$$\sigma_0 : (x,y) \to (x,-y), \quad \sigma : (x,y) \to (-x,iy).$$

As for σ' we have

$$\sigma'^{-1}P_1 = P_2, \quad \sigma'^{-1}P_2 = P_1 \text{ and } \sigma'P_3 = P_3$$

since $\sigma'^{-1}H\sigma' = H$. Therefore we have

$$\sigma' : (x,y) \to (\alpha/x, \beta y/x\ ), \quad \beta^2/\alpha^4 = -1.$$

Hence we have

$$y^2 = \alpha^{-3}t_1^2t_2^2(x^2-\alpha^2)(x^2-\alpha^2/t_1^2)(x^2-\alpha^2/t_2^2).$$

We see that $\alpha = 1$ by $\sigma'P_3 = P_3$. Hence $t_1^2t_2^2 = 1$. Finally we obtain

$$y^2 = x(x^2-1)(x^2-t^2)(x^2-1/t^2).$$

This Riemann surface has the group of automorphisms whose order is exactly 8 in general.

Second, consider non-hyperelliptic case :

$R/\langle\sigma_1\rangle$ is elliptic and P_1,P_2,P_3,P_4 are fixed by σ_1. There are two involutions besides σ_1 since there are 8 branch points besides $P_1,\ldots,P_4$. Let σ_2 be one of them. We have $\sigma_1\sigma_2 = \sigma_2\sigma_1$. By the same way as in (2) of (ii), we obtain as an equation of R

$$x^4+ y^4+ 2ax^2y^2+ 1 = 0.$$

However this surface does not have the group of order 8 but have one of order 96 or 48 or 16 depending on a.

(ii) (5) $n = 6, \quad \nu_1 = 3, \quad \nu_2 = 3, \quad \nu_3 = 3, \quad \nu_4 = 3$

Since $n = 6$ there must be at least one involution. It must be 2-hyperelliptic. By Lemma 8 R must be hyperelliptic. Hence we see that there does not exist such a Riemann surface.

(iii) (1) $n = 48, \quad \nu_1 = 3, \quad \nu_2 = 3, \quad \nu_3 = 4$

By the same reason as (2) of (i), we see that R must be non-hyperelliptic. Obviously, there does not exist Weierstrass points on non-branch points. Moreover, Weierstrass points are not on the branch points of multiplicity 3. Therefore they must be on the branch points of multiplicity 4. Thus the number of Weierstrass points are 12, say, $P_1, P_2, \ldots, P_{12}$. By Lemma 7 we see that $P_1 + \cdots + P_4$ and $P_5 + \cdots + P_8$ and $P_9 + \cdots + P_{12}$ are canonical divisors. We see that such a Riemann surface is given by

$$y^4 = x(x^2-1)$$

([8] p.761,Th.2) and has the group of automorphisms of order 96 as we see in §3.

(iii) (2) $n = 24, \quad \nu_1 = 3, \quad \nu_2 = 4, \quad \nu_3 = 4$

By the same reason as above, we see that R must be non-hyperelliptic. Let P_1 be a branch point of multiplicity 4 and let $H = \langle\sigma\rangle$ be a cyclic subgroup of order 4 which fixes the point P_1. Let P_1, P_2, P_3 and P_4 be fixed points of σ^2. By Lemma 7, $4P_i$ and $\Sigma_{i=1}^{4} P_i$ are canonical divisors. We have as an equation of R

$$y^4 = x(x-1)(x-t)$$

with

$$\mathrm{div}(x) = 4P_2 - 4P_1, \quad \mathrm{div}(y) = P_2 + P_3 + P_4 - 3P_1.$$

However this surface has the group of automorphisms of order 96 or 48 or 16 as we see in Theorem 3.

(iii) (3) $n = 16, \quad \nu_1 = 4, \quad \nu_2 = 4, \quad \nu_3 = 4$

We see that R has an automorphism σ of order 4 which has a fixed point. If $R/\langle\sigma\rangle$ is the Riemann sphere, we have already studied and we see that there is no such a Riemann surface by Th.1 and (2) of (ii). Therefore it is sufficient only to consider the case where $R/\langle\sigma\rangle$ is a torus. However this case does not occur by Lemma 1.

Summarizing above, we obtain following theorem:

Theorem 4.

(i) (1) $n = 8$, $(2,2,2,2,2)$.

hyperelliptic : $y^2 = (x^2-1)(x^2-t_1)(x^2-t_2)(x^2-t_1/t_2)$

non-hyperelliptic : $x^4 + y^4 + 2ax^2y^2 + 2b(x^2+y^2) + 1 = 0$

(2) $n = 6$, $(2,2,2,2,3)$.

hyperelliptic : non-existent

non-hyperelliptic : $A(X^4+Y^4+Z^4) + B(X^3Y+Y^3Z+Z^3X+X^3Z+Z^3Y+Y^3X) + C(X^2Y^2+Y^2Z^2+Z^2X^2) = 0$

where $ABC \neq 0$.

(ii) (1) $n = 24$, $(2,2,2,3)$.

hyperelliptic : non-existent

non-hyperelliptic : $x^4 + y^4 + 2a(x^2y^2 + x^2 + y^2) + 1 = 0$

(2) $n = 16$, $(2,2,2,4)$.

hyperelliptic : $y^2 = x^8 + 2ax^4 + 1$ $(a \neq 0,1)$

non-hyperelliptic : $x^4 + y^4 + 2ax^2y^2 + 1 = 0$ $(a \neq 0)$

(3) $n = 12$, $(2,2,3,3)$.

hyperelliptic : non-existent

non-hyperelliptic : non-existent

(4) $n = 8$, $(2,2,4,4)$.

hyperelliptic : $y^2 = x(x^2-1)(x^2-t)(x^2-1/t)$

non-hyperelliptic : non-existent

(5) $n = 6$, $(3,3,3,3)$.

hyperelliptic : non-existent

non-hyperelliptic : non-existent

(iii) (1) $n = 48$, $(3,3,4)$

hyperelliptic : non-existent

non-hyperelliptic : non-existent

(2) $n = 24$, $(3,4,4)$

hyperelliptic : non-existent

non-hyperelliptic : non-existent

(3) $n = 16$, $(4,4,4)$

hyperelliptic : non-existent

non-hyperelliptic : non-existent.

Now, assume that n is equal to 4. Then there exists a Riemann surfaces which is a normal covering of order 4 of the sphere. The group has no fixed points. The equation is given by

$$x^4 + y^4 + 2ax^2y^2 + 2bx^2 + 2cy^2 + 1 = 0$$

as we see in §4. We have one more type of Reimann surfaces which are cyclic coverings of order 4 of tori. The equation is given by

$$a(x^4 + y^4 + 1) + b(x^3y - y^3x) + cx^2y^2 + d(x^2 + y^2) = 0.$$

In fact, let σ be the automorphism. Then $R_0 = R/\langle\sigma^2\rangle$ is a torus. Let p_1, p_2, p_3 and p_4 are points of R_0 over which the ramification occurs. There is a function on R_0 and a divisor 2D such that $\mathrm{div}(f) = p_1 + \cdots + p_4 - 2D$ and if we adjoint $f^{1/2}$ to the elliptic function field we obtain the function field of R ([5,7]). There exists a point q such that $2q \sim D$. Let Q_1 and Q_2 be the two points over q and let $P_1, \cdots, P_4$ be the branch points over $p_1, \ldots, p_4$. Put

$$\mathrm{div}(x) = 2Q_1 + 2Q_2 - (P_1 + \cdots + P_4), \quad \mathrm{div}(y) = 2\sigma(Q_1 + Q_2) - (P_1 + \cdots + P_4).$$

Then we have as before $\sigma(x) = y$ and $\sigma(y) = -x$. By the same method as before we obtain the equation.

Assume that n is equal to 3. Then we have in §2

$$y^3 = x(x-1)(x-t_1)(x-t_2).$$

We have one more type of Riemann surfaces which are cyclic coverings of order 3 of a torus. We shall constract the equation. Let $\pi: R \to R/\langle\tau\rangle$ be the covering. Let p_1 and p_2 be points of $R/\langle\tau\rangle$ over which the ramification occurs. There is a function f on $R/\langle\tau\rangle$ and a point q such that $\mathrm{div}(f) = p_1 + 2p_2 - 3q$ and if we adjoint to the elliptic function field, $f^{1/3}$ then we obtain the function field of R. Let Q_1, Q_2 and Q_3 be points of R over the point q. We select Q_1', Q_1'' and Q_1''' such that $Q_1 + Q_1' + Q_1'' + Q_1'''$ are canonical. Then $\ell(Q_1' + Q_1'' + Q_1''') = 2$ and so we select Q's as

Q's are different from P_1, P_2, Q_1, Q_2, Q_3. Here P_1, P_2 are over p_1, p_2. Put

$$\operatorname{div}(x) = Q_1 + Q_2 + Q_3 - (P_1 + 2P_2),$$

$$\operatorname{div}(y) = Q_1 + Q_1' + Q_1'' + Q_1''' - (2P_1 + 2P_2).$$

Then we see that 1, x, and y are linearly independent. From this we see that y, τy and $\tau^2 y$ are linearly independent. Thus, if we put

$$x' = \tau y/y \quad \text{and} \quad y' = \tau^2 y/y,$$

then τ is represented by

$$\begin{pmatrix} 0 & 1 & 0 \\ 0 & 0 & 1 \\ 1 & 0 & 0 \end{pmatrix}$$

in the projective space $\mathbb{P}^2$. By the same way as in (2) of (i) we obtain the equation

$$A(X^4+Y^4+Z^4) + B(X^3Y+Y^3Z+Z^3X) + C(X^3Z+Z^3Y+Y^3X) + D(X^2Y^2+Y^2Z^2+Z^2X^2) = 0.$$

Finally assume that n is equal to 2. Then we have in §1

$$y^2 = x(x-1)(x-t_1)\cdots(x-t_5).$$

We have one more type of Riemann surfaces which is a cyclic covering of order 2 of a torus. We can construct the equation by the same way as n = 3. We have this time

$$\sigma(x) = x, \quad \sigma(y) = -y.$$

and the equation is given by

$$y^4 + y^2(a_0x^2+a_1x+b) + x^4+a_2x^3+a_3x^2+a_4x+a_5 = 0.$$

References

[1] R.D.M.Accola, Riemann surfaces with automorphism groups admitting partitions, Proc. Amer. Math. Soc. 21(1969) 477-482.

[2] H.M.Farkas, Automorphisms of compact Riemann surfaces and vanishing theta constants, Bull. Amer. Math. Soc. 73(1967) 231-232.

[3] A.Hurwitz, Über algebraischen Gebilde mit eindeutige Transformationen in sich, Math. Ann. 41(1893), 403-442.

[4] F.Klein, Mathematische Abhandlungen, III Springer, Berlin (1923) 106-108.

[5] K.Komiya, On families of curves of genus 3 with involutions. Mem. of The Faculty of Lib. Arts & Education, Yamanashi Univ. 26(1975) 4-9.

[6] A.Kuribayashi, On analytic families of Riemann surfaces with non-trivial automorphisms, Nagoya Math. J. 28(1966) 119-165.

[7] ―――――――― , On the equations of compact Riemann surfaces of genus 3 and the generalized Teichmuller spaces. J. Math. Soc. Japan 28(1976) 712-736.

[8] A.Kuribayashi-K.Komiya, On Weierstrass points of non-hyperelliptic compact Riemann surfaces of genus three, Hiroshima Math. J. 7(1977) 743-768.

[9] R.Tsuji, Conformal automorphisms of a compact bordered Riemann surfaces of genus 3, Kodai Math. Sem. Rep. 27(1976) 271-290.

[10] A.Weil, Foundations of algebraic geometry, A.M.S. Coll. 29, New York, 1946.

[11] ――――,Sur les courbes algébriques et les variétés qui s'en déduisent, Hermann, Paris, 1948.

Department of Mathematics
Chuo University, Japan
and
Department of Mathematics
Yamanashi University, Japan

Errata to the preceeding paper.

(I) p.2. (11) 6 (b_2) must be

$$a(x^4+y^4+1)+b(x^3y+y^3x+x^3+y^3+x+y)+c(x^2y^2+x^2+y^2)$$
$$+d(x^2y+xy^2+xy)=0$$

(II) p.2. (12) 4 (b_2) is included in (10) 8 (b).

In fact, (12) 4 (b_2) is transformed into

$$a'x_1^4+b'y_1^4+c'+d'x_1^2y_1^2+e'x_1y_1=0$$

by the transformation

$$x=x_1+y_1$$
$$y=-i(x_1-y_1)$$

Here we remark that $a'b'c'$ is not equal to zero. Furthermore, we see that this is transforemed into

$$x^4+y^4+1+ax^2y^2+bxy=0.$$

Therefore this has an involution

$$\begin{pmatrix} 0 & 1 & 0 \\ 1 & 0 & 0 \\ 0 & 0 & 1 \end{pmatrix}$$

in addition to the automorphism

$$\begin{pmatrix} i & 0 & 0 \\ 0 & -i & 0 \\ 0 & 0 & 1 \end{pmatrix}$$

of order 4.

It is easy to see that this is equivalent to (10) 8 (b).

(III) p.2. (13) 3 (b_2) is included in (11) 6 (b_2).

In fact, we denote our equation by

$$a(X^4+Y^4+Z^4)+b(X^3Y+Y^3Z+Z^3X)+c(X^3Z+Z^3Y+Y^3X)$$
$$d(X^2Y^2+Y^2Z^2+Z^2X^2)+e(X^2YZ+XY^2Z+XYZ^2)=0.$$

By the transformation

$$X = X_1 + Y_1 + Z_1$$

$$Y = \omega X_1 + \omega^2 Y_1 + Z_1$$

$$Z = \omega^2 X_1 + \omega Y_1 + Z_1$$

the equation is transformed into

$$a'X_1^3 Z_1 + b'Y_1^3 Z_1 + c'Z_1^4 + d'X_1^2 Y_1^2 + e'X_1 Y_1 Z_1^2 = 0.$$

Here we must remark that a'b'c'd' is not equal to zero. Further, we see that this is transformed into

$$X^3 Z + Y^3 Z + Z^4 + bX^2 Y^2 + cXYZ^2 = 0 \quad (b \neq 0).$$

This has an involution

$$\begin{pmatrix} 0 & 1 & 0 \\ 1 & 0 & 0 \\ 0 & 0 & 1 \end{pmatrix}$$

in addition to the automorphism

$$\begin{pmatrix} \omega & 0 & 0 \\ 0 & \omega^2 & 0 \\ 0 & 0 & 1 \end{pmatrix}$$

of order 3.

Thus we have a following proposition:

Riemann surfaces of genus three which is non-hyperelliptic and three sheeted covering of tori have S_3 as a subgroup of automorphism group.

(I) and (III) are informed by L.Vermeulen to us. We thank to him very much.

A. Kuribayashi
K. Komiya

Deformation and transversality

by Dan Laksov.

Institut Mittag-Leffler
Auravägen 17
S-18262 Djursholm
Sweden

§1. Deformations and transversality.

The motivation behind the investigations that we are going to describe in this lecture, comes from the idea to use transversality results to deform algebraic varieties. To be more precise we introduce the following notation:

Let

$$F: G\times X \longrightarrow M$$

be a morphism of smooth connected algebraic varieties G, X and M. Moreover, let D be a closed subscheme of M. Given a rational point g of G we shall denote by F_g the morphism

$$F|(g\times X): \quad (g\times X) \cong X \longrightarrow M.$$

The goal of most transversality results is to give conditions on F such that the morphisms F_g are transversal to D for "nearly all" rational points g of G. Consider instead of F the projection

$$q: F^{-1}(D) \longrightarrow G.$$

Then clearly $q^{-1}(g) \cong F_g^{-1}(D)$ and we see that the transversality results will relate a stratification of D into singular loci $\Phi = D_0 \subset D_1 \subset \cdots \subset D_c = D$ where D_{i-1} is the singular locus of D_i for $i=1,\cdots,c$, with a similar stratification into singular loci of the general member of the family q.

Look at the situation in the opposite way. We see that if we are given the varieties X and M and a morphism

$$f: X \longrightarrow M,$$

then to find a family of varieties containing the member $f^{-1}(D)$ and whose general member has a stratification like that of D, it is sufficient to find a variety G and a morphism F satisfying the conditions of an appropriate transversality theorem and such that $F_e = f$ for some distinguished point e of G.

In a previous article [22] we proved a rather strong transversality result which generalizes Bertinis' theorem in the form given by S.L. Kleiman [17] and which, in the case X and M are both affine spaces, makes the construction of a corresponding variety G and morphism F easy. In this way every subscheme Y of an affine space $\mathbb{A}^p$, which is the inverse of some subvariety D of another affine space $\mathbb{A}^q$ by a morphism f of affine spaces, will be a member of a family whose general member has the same stratification into singular loci as D.

However, our aim being to construct deformations of the variety $Y = f^{-1}(D)$ we want the family q to be flat in a neighbourhood of $Y = q^{-1}(e)$. One of the main observations of the mentioned article [22] was that, when D is Cohen-Macaulay, a property that we shall later see is satisfied in several interesting cases, then q is flat in a neighbourhood of Y when the natural condition

$$\mathrm{codim}(\, f^{-1}(D),\, X) = \mathrm{codim}(D,G)$$

is satisfied.

The statements of the results of the article [22] are rather lenghty so we shall here content ourselves by stating a particular case, whose applications will concern us in the following.

Theorem 1. Let

$$\Phi = D_0 \subseteq D_1 \subseteq \cdots \subseteq D_c = D$$

be a sequence of irreducible subspaces of the q-dimensional affine space $\mathbb{A}^q$ and assume that D is a Cohen-Macaulay scheme. Moreover let

$$f\colon \mathbb{A}^p \longrightarrow \mathbb{A}^q$$

be a morphism of affine spaces such that

$$\mathrm{codim}(f^{-1}(D),\, \mathbb{A}^p) = \mathrm{codim}(D,\, \mathbb{A}^q).$$

Then there exists a morphism

$$F\colon \mathbb{A}^{(p+1)q} \times \mathbb{A}^p \longrightarrow \mathbb{A}^q$$

and an open subset G of $\mathbb{A}^{(p+1)q}$, containing the origin e, such that for each rational point $g \in G$ the following four assertions hold:

(i) $$F_e = f.$$

(ii) Each scheme in the stratification

$$\Phi = F_g^{-1}(D_0) \subsetneqq F_g^{-1}(D_1) \subsetneqq \cdots \subsetneqq F_g^{-1}(D_c) = F_g^{-1}(D),$$

satisfies

$$\operatorname{codim}(F_g^{-1}(D_i), \mathbb{A}^p) = \operatorname{codim}(D_i, \mathbb{A}^q)$$

(and is empty if $\operatorname{codim}(D_i, \mathbb{A}^q) > p$).

(iii) The scheme $F_g^{-1}(D_{i-1})$ is the singular locus of $F_g^{-1}(D)$ for $i=1,\cdots,c$.

(iv) The projection

$$q: F^{-1}(D) \longrightarrow \mathbb{A}^{(p+1)q},$$

is flat over G.

§2. Determinantal varieties.

To illustrate the usefulness of Theorem 1 we shall give an application to determinantal varieties, which was the starting point of the investigations described above.

We first introduce some notation and recall the main properties that the determinantal varieties satisfy.

Denote by $M(a,b) = \mathbb{A}^{ab}$ the affine space of all $a \times b$-matrices and by $D_c(a,b)$ the generic determinantal scheme of $a \times b$-matrices whose minors of order c all vanish. Let $R= k[x_{1,1},\cdots,x_{a,b}]$ be the polynomial ring in ab variables over the field k. Then $M(a,b)=$ Spec R and $D_c(a,b)$ is defined by the ideal in R generated by all subdeterminants of order c of the $a \times b$-matrix $(x_{i,j})$.

The following properties hold:

(i) The scheme $D_c(a,b)$ is intergral and

$$\operatorname{codim}(D_c(a,b), M(a,b)) = (a-c+1)(b-c+1).$$

(ii) The scheme $D_{c-1}(a,b)$ is the singular locus of $D_c(a,b)$.

(iii) The scheme $D_c(a,b)$ is Cohen-Macaulay.

Let A be a k-algebra and $M=(a_{i,j})$ an $a\times b$-matrix with coordinates in A. Denote by $D_c(M)$ the closed subscheme of X= Spec A defined by the ideal generated by all subdeterminants of order c of M. Then there is a natural morphism

$$f\colon X\longrightarrow M(a,b)$$

such that $f^{-1}D_c(a,b)=D_c(M)$.

We say that $D_c(M)$ is a determinantal subscheme of X if it is of pure codimension $(a-c+1)(b-c+1)$ in X.

When $D_c(M)$ is determinantal and $X=\mathbb{A}^p$, we see that we are in a situation where Theorem 1 applies. Appropriately reformulated that theorem asserts the following:

Theorem 2. Assume that $D_c(M)$ is a determinantal subvariety of $\mathbb{A}^p=\operatorname{Spec} k[x_1,\cdots,x_p]$. Then there exists a faithfully flat morphism

$$q\colon V\longrightarrow W$$

from an algebraic variety V to an open subset W of an affine space, such that $q^{-1}(e)=D_c(M)$ for some rational point e of W and such that for every rational point g of W the following assertions hold:

(i) There exists an $a\times b$-matrix M(g) with coordinates in the polynomial ring $k[x_1,\cdots,x_p]$ such that $q^{-1}(g)$ is isomorphic to $D_c(M(g))$.

(ii) Each scheme in the sequence

$$\phi=D_0(M(g))\subseteq D_1(M(g))\subseteq\cdots\subseteq D_c(M(g))$$

is a determinantal subscheme of the affine space $\mathbb{A}^p$, that is $\operatorname{codim}(D_i(M(g)),\mathbb{A}^p)=(a-i+1)(b-i+1)$ for $i=1,\cdots,c$ (and empty if $(a-i+1)(b-i+1)>p$).

(iii) The scheme $D_{i-1}(M(g))$ is the singular locus of the scheme $D_i(M(g))$ for $i=1,\cdots,c$.

This result was found independently by M. Schaps [26] and the author [22] It is in a very precise way the best possible result. Indeed, the stratification into singular loci of a general member of the families described above are similar to the stratifications of the generic determinantal varieties and the latter varieties were proved by T. Svanes to be rigid [27] (except those in codimension one), that is, they can not be deformed non-trivially at all.

§3. Scemes of codimension two.

Considering the variety of situations, particularly in enumerative geometry, where determinantal varieties appear, the above results about their deformation are of significance in themselves. The interest in these results in this connection, came however from an attempt to deform curves in $\mathbb{A}^3$, or more generally to deform schemes of codimension two in affine spaces. The reason why determinantal schemes are involved here is suggested by the following result, which was first proved by L. Burch [4]:

Let R be a regular local ring and I an ideal of pure codimension (or equivalently grade) two in R. Then R/I is Cohen-Macaulay if and only if I is of the form $D_{n-1}(M)$ for some $(n-1)\times n$-matrix M with entries in R.

This result was extended to the case when R is a polynomial ring by M. Schaps [26], but was already known, in particular cases, by D. Hilbert. As a consequence of this extension one obtains from Theorem 2 deformations of codimension two Cohen-Macaulay subschemes of affine spaces, that are, as was mentioned in the previous section, the best possible. The most striking consequence of the results obtained is the following assertion:

Every Cohen-Macaulay scheme of codimension two in an affine space $\mathbb{A}^p$ can be deformed into a scheme whose singular locus has codimension at least 6. In particular, if $p < 6$, then the scheme can be smoothed.

§4. Schemes of codimension three.

Turning our attention to subschemes of codimension three in affine spaces, it is clear from the outset that the situation is much more complicated than in the codimension two case described above. Whereas we have seen that points in the plane can be smoothed and it is well known (see e.g. [9] or [10]) that the Hilbert scheme of such points is irreducible, it was shown by T. Iarrobino [13] that most points in $\mathbb{A}^3$ can not be smoothed and that the Hilbert scheme for such points has lots of components.

However, if we restrict our attention to those subschemes that are Gorenstein, the situation is more similar to the codimension two case. Then, corresponding to Burch s theorem, one has the following result by D. A. Buchsbaum and D. Eisenbud [2]:

Let R be a regular local ring and I an ideal of codimension three. Then R/I is Gorenstein if and only if I is generated by the subpfaffians of order 2c of an alternating $(2c+1)\times(2c+1)$-matrix with entries in R.

Recall that an $a\times a$-matrix $A=(a_{i,j})$ with entries in R is alternating if $a_{i,j}=-a_{i,j}$ and $a_{i,i}=0$ for all i and j and that the determinant of such a matrix is the square of a uniquely determined element in R called the pfaffian of A (see e.g. [1]).

One half of the above result holds when R is not necessarily local. Let $A=(a_{i,j})$ be an alternating $(2c+1)\times(2c+1)$-matrix and denote by p_i the pfaffian obtained by deleting row and column number i in A. Assume that $I=(p_1,\cdots,p_{2c+1})$. Let $B\colon R^{2c+1}\longrightarrow R$ be the map defined by $B(r_1,\cdots,r_{2c+1})=\sum_{i=1}^{2c+1}(-1)^{i+1}r_ip_i$. Then $B\,A=0$ because $B(a_{i,1},\ \ ,a_{i,2c+1})=\sum_{j=1}^{2c+1}(-1)^{j+1}\,a_{i,j}p_j$ is the expansion along the first row of the alternating $(2c+1)\times(2c+1)$-matrix having the vector $(0,a_{i,1},\cdots,a_{i,2c+1})$ as first row and having A in the bottom right corner. Moreover $A\,B^t=-A^t\,B^t=0$. Correspondingly we have a complex,

$$0 \longrightarrow R \xrightarrow{B^t} R^{2c+1} \xrightarrow{A} R^{2c+1} \xrightarrow{B} R \longrightarrow R/I \longrightarrow 0$$

The easy part of Buchsbaum and Eisenbuds proof shows that this sequence is exact and consequently, not only proves that R/I is Gorenstein, but also that $Ext_R^3(R/I,R)$ is a cyclic R-module.

Unfortunately the converse of the Buchsbaum-Eisenbud result does not hold when R is not local. We shall next give an example, due to H. Kleppe and the author, where R is a polynomial ring in seven variables and I is an ideal of codimension three such that R/I is regular, but $Ext_R^3(R/I,R)$ is not cyclic. Then, by the above, I can not be the ideal of 2c-pfaffians of any alternating $(2c+1)\times(2c+1)$-matrix.

<u>Example.</u> Let

$$M=\begin{bmatrix} x_{1,1}, & x_{1,2}, & x_{1,3}, & x_{1,4} \\ x_{2,1}, & x_{2,2}, & x_{2,3}, & x_{1,1}+1 \end{bmatrix} = (r_{i,j}),$$

be a matrix whose coordinates, except the (2,4)-coordinate, are independent variables and let R be the polynomial ring in these variables. If P is a prime ideal in R containing I and not $r_{1,1}$, then

$I_P = (r_{2,2}-r_{2,1}r_{1,2}r_{1,1}^{-1},\ r_{2,3}-r_{2,1}r_{1,3}r_{1,1}^{-1},\ r_{2,4}-r_{2,1}r_{1,4}r_{1,1}^{-1})$ and if P contains $r_{1,1}$ then

$I_P = (r_{1,1}-r_{2,1}r_{1,4}r_{2,4}^{-1},\ r_{1,2}-r_{2,2}r_{1,4}r_{2,4}^{-1},\ r_{1,3}-r_{2,3}r_{1,4}r_{2,4}^{-1})$.

Hence in both cases, $(R/I)_P$ is regular of dimension five. Since I is also generated by the maximal minors of a 2×4-matrix and hence is determinental, we have explicit free resolutions of I bearing the names of Eagon-Northcott [7] and Buchsbaum-Rim [3]. The Eagon-Northcott resolution is of the form

$$0 \longrightarrow R^3 \xrightarrow{N} R^8 \longrightarrow R^6 \longrightarrow R \longrightarrow R/I \longrightarrow 0$$

where N is the matrix

$$\begin{bmatrix} r_{1,1}, & -r_{1,2}, & r_{1,3}, & -r_{1,4}, & 0, & 0, & 0, & 0 \\ r_{2,1}, & -r_{2,2}, & r_{2,3}, & -r_{2,4}, & r_{1,1}, & -r_{1,2}, & r_{1,3}, & -r_{1,4} \\ 0, & 0, & 0, & 0, & r_{2,1}, & -r_{2,2}, & r_{2,3}, & -r_{2,4} \end{bmatrix}$$

Hence $E = \mathrm{Ext}_R^3(R/I,R)$ is the quotient of R^3 by the module generated by the elements $r_{1,i}e_1+r_{2,i}e_2$ and $r_{1,i}e_2+r_{2,i}e_3$ for $i=1,\cdots,4$. Let $f_1=e_1$, $f_2=e_3$ and $f=r_{1,4}f_1-r_{2,1}f_2$. An easy computation shows that $e_2=f$ and that E is the quotient of R^2 by the module generated by the elements $r_{1,i}f_1-r_{2,i}f$ and $-r_{1,i}f+r_{2,i}f_2$ for $i=1,\cdots,4$.

Let $J=(r_{1,2}, r_{1,3}, r_{2,2}, r_{2,3}, r_{1,1}r_{2,4}-r_{2,1}r_{1,4})$ and let $x=r_{1,1}$, $y=r_{2,1}$ and $z=r_{1,4}$. Then $S=R/J=k[x,y,z]/(x(x+1)-yz)$ and E/JE is the quotient of S^2 by the module F generated by $\{xf_1-yf,\ zf_1-(x+1)f,\ -xf+yf_2,\ -zf+(x+1)f_2\}$, where $f=zf_1-yf_2$. The following four relations

$$xf_1-yf=zf_1-yzf_1+y^2f_2=-x^2f_1+y^2f_2$$
$$zf_1-(x+1)f=-xf+yf_2$$
$$-xf+yf_2=-xzf_1+yxf_2+yf_2=y(-zf+(x+1)f_2)$$
$$-zf+(x+1)f_2=-z^2f_1+zyf_2+(x+1)f_2=-z^2f_1+(x+1)^2f_2,$$

shows that F is generated by $x^2f_1-y^2f_2$ and $z^2f_1-(x+1)^2f_2$. To prove that **E** is not cyclic it is clearly sufficient to show that the quotient S^2/F is not cyclic. However S^2/F is a locally free S module of rank one. This follows from R being Gorenstein, or is seen by an easy computation. Hence to prove that F is not cyclic, it is sufficient to prove that it is not free. This we achive by an elegant argument pointed out to us by G. Ellingsrud and S. A. Strömme. Let Z be the subscheme of $\mathbb{P}^3$ defined by the ideal $x(x+w)-zy$ and let $U\cong\mathbb{A}^3$ be the principal open set $\mathbb{P}^3\setminus V(w)$. Clearly Z is isomorphic to $\mathbb{P}^1\times\mathbb{P}^1$ imbedded in $\mathbb{P}^3$ via the Segre map, and under this isomorphism Spec $S=Z\cap U$ is isomorphic to $\mathbb{P}^1\times\mathbb{P}^1\setminus\Delta$. Moreover the line $L_1=V(x,y)$ is mapped onto $(0:1)\times\mathbb{P}^1$ and $L_2=V(x+w,y)$ onto $\mathbb{P}^1\times(1:0)$.

Denote by $\mathcal{F}$ the cokernel of the map

$$\begin{bmatrix} x^2, & -y^2 \\ z^2, & -(x+1)^2 \end{bmatrix}:\mathcal{O}_Z(-2)^2\longrightarrow\mathcal{O}_Z^{\ 2}.$$

Then $\mathcal{F}|U = S^2/F$ and putting

$$\alpha = (z^2, -(x+1)^2) \quad \text{and} \quad \beta = \begin{bmatrix} x^2, & 0 \\ z^2, & 0 \end{bmatrix}$$

we have exact sequences

$$0 \longrightarrow \mathcal{O}(-2)|L_1 \xrightarrow{\alpha} \mathcal{O}_{L_1}^2 \longrightarrow \mathcal{O}|\mathcal{F} \longrightarrow 0 \quad \text{and}$$

$$0 \longrightarrow \mathcal{O}(-4)|L_2 \xrightarrow{(z^2, -x^2)} \mathcal{O}(-2)^2|L_2 \longrightarrow \mathcal{O}_{L_2}^2 \longrightarrow \mathcal{F}|L_2 \longrightarrow 0.$$

We conclude that $\deg \mathcal{F}|L_1 = 2$ and that $\deg \mathcal{F}|L_2 = 0$ and consequently that $\mathcal{F}$ is not in the kernel $\mathbb{Z}(1,1)$ of the map

$$\mathbb{Z} \times \mathbb{Z} = \text{Pic}(Z) \longrightarrow \mathbb{Z} = \text{Pic } U.$$

Hence S^2/F is not zero in Pic U, that is, it is not free.

From the above example we see that to construct deformations of codimension three Gorenstin schemes it does not quite suffice to construct deformations of pfaffian schemes. On the other hand we do not know of any example of a codimension three Gorenstein scheme which does not have deformations of the same type as those constructed for pfaffian schemes is the next section. For example we prove in the following section that every pfaffian scheme of codimension three in an affine space of dimension strictly less than ten can be smoothed. We ignore if the same is true for codimension three Gorenstein schemes.

Another similar question is wether the part of the Hilbert scheme coming from codimension three Gorenstein schemes is irreducible. We believe that the part coming from codimension three pfaffians can be determined in the same way that G. Ellingsrud [9] determined those for codimension two determinental schemes.

§5. Deformation of pfaffian schemes.

The previous section suggests the importance of finding deformations of pfaffian schemes. We shall now define such schemes in general and give their main properties.

Denote by $M(a)$ the $a(a-1)/2$-dimensional affine space of all alternating $a \times a$-matrices and by $P_{2c}(a)$ the generic pfaffian scheme of $a \times a$-matrices whose pfaffians of order 2c vanish. Let R be the polynomial ring in the a(a-1)/2 independent variables $x_{i,j}$ for $1 \leq i < j \leq a$ over a field and let M be the alternating matrix having the variables $x_{i,j}$ as coordinates above the diagonal. Then M(a)= Spec R and $P_{2c}(a)$ is defined by the ideal in R generated by the pfaffians of all alternating $2c \times 2c$-submatrices of M.

The following three properties hold:

(i) The scheme $P_{2c}(a)$ is integral and

$$\operatorname{codim}(P_{2c}(a), M(a)) = (a-2c+2)(a-2c+1)/2.$$

(ii) The scheme $P_{2(c-1)}(a)$ is the singular locus of $P_{2c}(a)$.

(iii) The scheme $P_{2c}(a)$ is Gorenstein.

Except for the assertion that $P_{2c}(a)$ is reduced, the properties (i) and (ii) are classical (see e.g. T.G. Room [25] or more recently H. Kleppe [18]). As for the remaining properties see section 6 below.

Let A be a k-algebra and N= $(a_{i,j})$ an alternating $a \times a$-matrix with coordinates in A. Denote by $P_{2c}(N)$ the closed subscheme of X= Spec A defined by the ideal generated by the pfaffians of all $2c \times 2c$-submatrices of N formed from the same 2c rows and columns. Then there is a natural morphism

$$f: X \longrightarrow M(a)$$

such that $f^{-1}P_{2c}(a) = P_{2c}(N)$.

We say that $P_{2c}(N)$ is a pfaffian subscheme of X if it is of pure codimension (a-2c+2)(a-2c+1)/2 in X.

When $P_{2c}(N)$ is pfaffian and X= $\mathbb{A}^p$ we see that we are exactly in a situation where Theorem 1 applies. Appropriately reformulated that theorem asserts the following:

Theorem 3. Assume that $P_{2c}(N)$ is a pfaffian subvariety of $\mathbb{A}^p = \operatorname{Spec} k[x_1,\cdots,x_p]$. Then there exists a faithfully flat morphism

$$q: V \longrightarrow W$$

from an algebraic variety V to an open subset W of an affine space, such that $q^{-1}(e) = P_{2c}(N)$ for some rational point e of W and such that for every rational point g of W the following assertions hold:

(i) There exists an alternating a a-matrix N(g) with coordinates in the polynomial ring $k[x_1,\cdots,x_p]$ such that $q^{-1}(g)$ is isomorphic to $P_{2c}(N(g))$.

(ii) Each sheme in the sequence

$$\emptyset = P_0(N(g)) \subsetneqq P_2(N(g)) \subsetneqq \cdots \subsetneqq P_{2c}(N(g))$$

is a pfaffian subscheme of the affine space $\mathbb{A}^p$, that is

$$\operatorname{codim}(P_{2i}(N(g)), \mathbb{A}^p) = (a-2i+2)(a-2i+1)/2 \text{ for } i=1,\cdots,c.$$

(iii) The scheme $P_{2(i-1)}(N(g))$ is the singular locus of the scheme $P_{2i}(N(g))$ for $i = 1,\cdots,c$.

It would be interesting to know if the generic pfaffian schemes in codimension at least two have rigid singularities. If so, the stratification of the deformed pfaffians into singular loci described in Theorem 3 would be the best possible.

§6. The Cohen-Macaulay property.

From what we have seen the flatness of the families we construct by the transversality results follow from the Cohen-Macaulay property of the schemes involved. This property is thus crucial, but is difficult to verify in the most interesting applications.

For determinantal schemes there has during the last six years appeared several different proofs that they are Cohen-Macaulay. We shall here mention the main contributions.

(i) The first proof was given by J.A. Eagon and M. Hochster [6]. They used induction on a large class of Schubert type determinantal schemes and needed to construct generic points for several such schemes. A simplification of their proof obtained by H. Kleppe and the author will be presented in this volume. Our proof avoids the explicit construction of generic points and has the advantage that it can be used, almost without any modifications, to pfaffians and to determinants of symmetric matrices.

(ii) A global algebraic proof was found simultaneously by M. Hochster [11], D. Laksov [21] and C. Musili [24].

(iii) A beautiful geometric approach was found by G.R. Kempf [15]. He related the Cohen-Macaulay property to vanishing theorems for certain line bundles on homogenous spaces and even proved that the determinantal varieties have rational singularities. Later Kempf [16] generalized the methods vastely to quotients of reductive groups by parabolic subgroups.

(iv) In the characteristic zero case A. Lascoux [23] refined Kempfs geometric approach by introducing "Schur modules" and succeded in finding all the syzygies of the determinantal ideals.

(v) The determinants have long been known to be invariants under the general linear group acting on a regular ring (see H. Weyl [28] in characteristic zero and C. De Cocini and C. Procesi [5] in general). Hence it follows from M. Hochster and J.L. Roberts result [12] that the invariants of a reductive group acting on a regular local ring are Cohen-Macaulay that the determinantal ideals have this property in characteristic zero.

(vi) It was noted by G. Eisenreich [8] that if a certain natural assertion about the first syzygies of determinantal ideals hold, then they are Cohen-Macaulay. The argument given by Eisenreich to support this assertion is incomplete. However, T. Jozefiak and H.A. Nielsen have pointed out that, at least in characteristic zero, the assertion is correct. Their argument is based upon certain vanishing results that are stronger than the Cohen-Macaulay property. As Eisenreichs idea also could be used to prove that pfaffians are Cohen-Macaulay it would be of interest to have an elementary proof of his assertion valid in all characteristics.

Corresponding to the proofs mentioned above of the determinantal varieties being Cohen-Macaulay, there has during the last year been announced similar proofs for the pfaffians being Gorenstein.

(i) A proof has been given by H. Kleppe and the author based on our simplified version of Eagon and Hochsters proof presented in this volume. Another proof following more closely the ideas of Eagon and Hochster was independently given by V. Marinov (Thesis, Bulgarian academy of sciences).
(ii) A proof similar to the one mentioned in (ii) above has been given by the author.
(iii) A proof similar to the one mentioned in (iii) above can probably be worked out using ideas of V. Lakshimilbai, C. Musili and C.S. Seshadri [19] and [20].
(iv) The method of Lascoux has been applied to pfaffians by T. Jozefiak and T. Pragasz [14]. In characteristic zero they find all the syzygies and construct rational resolutions of the pfaffian ideals.
(v) The pfaffians are invariants under the symplectic group acting on a regular ring. This is classical in characteristic zero (see H. Weyl [28]) and follows in all characteristics form the work [5] of De Concini and C. Procesi. Hence it follow from the result of Hochster and Roberts mentioned in (v) above, that in characteristic zero the pfaffians are Cohen-Macaulay.

§7. Applications to other schemes.

We have seen that the most important feature of eur construction of deformations of a subscheme of an affine space $\mathbb{A}^p$, is to have a morphism

$$f: \mathbb{A}^p \longrightarrow \mathbb{A}^q$$

and a Cohen-Macaulay subscheme D of $\mathbb{A}^q$ having a nice stratification into singular loci such that $f^{-1}(D) = X$ and such that $\operatorname{codim}(X, \mathbb{A}^p) = \operatorname{codim}(D, \mathbb{A}^q)$. Examples of such schemes, like the two we have already encountered, often appear in connection with rank conditions on certain matrices and more generally as invariants under the classical groups ar even of general reductive groups. In general it is not clear however, if the deformations constructed have any interesting applications. A more interesting line of investigation is to decide how many deformations that can be obtained from transversality results. For examples of such investigations the reader can consult M. Schaps article in this volume. Another question that may be worhtwhile considering is wether the deformations obtained by our methods " in generic situations " are the best possible.

References.

[1] Bourbaki, N.. Elements de mathematiques, Algèbre chap. 9. Hermann 1959.

[2] Buchsbaum, D.A. & Eisenbud, D., "Algebra structures for finite free resolutions and some structure theorems for ideals of codimension 3". Amer. J. Math. 99 (1977), 447-485.

[3] Buchsbaum, D.A. & Rim, D.S., "A generalized Koszul complex I". Trans. Amer. Math. Soc. 111 (1964), 183-196.

[4] Burch, L., "On ideals of finite homological dimension in local rings". Proc. Cambridge Phil. Soc. 64 (1968), 941-952.

[5] De Concini, C. & Proceci, C., "A characteristic free approach to invariant theory". Advances in Math. 21 (1976), 330-354.

[6] Eagon, J.A. & Hochster, M., "Cohen-Macaulay rings, invariant theory and the generic perfection of determinantal loci". Amer. J. Math. 93 (1971), 1020-1058.

[7] Eagon, J.A. & Nothcott, D.G. "Ideals defined by matrices and a certain complex associated with them". Proc. Roy. Soc. Ser. A. 269 (1962), 188-204.

[8] Eisenreich, G., "Zur perfectheit von Determinantideale". Beiträge zur Algebra und Geometrie 3 (1974), 49-54.

[9] Ellingsrud, G., "Sur le schema de Hilbert des variétés de codimension 2 dans $\mathbb{P}^e$ a cone de Cohen-Macaulay". Annales Sci. de l'Ecole Normale Sup. 4e ser. 8 (1975), 423-432.

[10] Fogarty, J., "Algebraic families on an algebraic surface". Amer. J. Math. 90 (1968), 511-521.

[11] Hochster, M., "Grassmannians and their Schubert varieties are arithmetically Cohen-Macaulay". J. Algebra 25 (1973), 40-57.

[12] Hochster, M. & Roberts, J.L., "Rings of invariants of reductive groups acting on regular rings are Cohen-Macaulay". Advances in Math. 13 (1974), 115-175.

[13] Iarrobino, A., "Reducibility of the family of 0-dimensional schemes on a variety". Invent. Math. 15 (1972), 72-77.

[14] Jozefiak, T. & Pragacz, P., "Syzygies de pfaffiens". Comptes Rendus 287 (1978), 89-91.

[15] Kempf, G.R., "Vanishing theorems for flag manifolds". Amer. J. Math. 98 (1976), 325-331.

[16] Kempf, G.R., "Linear systems on homogenous spaces". Ann. of Math. 103 (1976), 557-591.

[17] Kleiman, S.L., "The transversality of a general translate". Compositio Math. 28 (1974), 287-297.

[18] Kleppe, H., "Deformation of schemes defined by vanishing of pfaffians". J. Algebra 53 (1978), 84-92.

[19] Lakshmibai, V. & Seshadri, C.S., "Geometry of G/P-II". Proc. Indian Acad. Sci. 87A (1978), 1-54.

[20] Lakshimibai, V., Musili, C. & Seshadri, C.S., "Geometry of G/P". Preprint.

[21] Laksov, D., "The arithmetic Cohen-Macaulay character of Schubert schemes". Acta Math. 129 (1972), 1-9.

[22] Laksov, D., "Deformation of determinantal schemes". Compositio Math. 30 (1975), 273-292.

[23] Lascoux, A.. Thesis, Paris 1977.

[24] Musili, C., "Postulation formula for Schubert varieties". Journ. Indian Math. Soc. 36 (1972), 143-171.

[25] Room, T.G.. The geometry of determinantal loci. Cambr. Univ. Press 1938.

[26] Schaps, M., "Deformation of Cohen-Macaulay schemes of codimension 2 and non-singular deformation of space curves". Amer. J. Math. 99 (1977), 669-685.

[27] Svanes, T., "Coherent cohomology on flag manifolds and rigidity". Advances in Math. 14 (1974), 369-453.

[28] Weyl, H.. The classical groups. Princeton Univ. Press 1946.

FINITE GENERATIONS OF LIFTED P-ADIC HOMOLOGY WITH COMPACT SUPPORTS. GENERALIZATION OF THE WEIL CONJECTURES TO SINGULAR, NON-COMPLETE ALGEBRAIC VARIETIES.

by Saul Lubkin

Department of Mathematics
University of Rochester
Rochester, NY 14627

CHAPTER 1.

Lifted p-Adic Homology With Compact Supports.

Let $\mathcal{O}$ be a complete discrete valuation ring having a quotient field K of characteristic zero, and with residue class field k. Let $\underline{A}$ be an $\mathcal{O}$-algebra and let $A = \underline{A} \otimes_{\mathcal{O}} k$. Let C be a scheme of finite presentation over A_{red} such that C is polynomially properly embeddable over A (see below - e.g., it suffices that C be quasiprojective over A). Then in this chapter we define the lifted p-adic homology groups with compact supports

(1) $H_h^c(C,(\underline{A}\dagger) \otimes_{\mathbb{Z}} \mathbb{Q})$, all integers h. In the special case that C is simple over A_{red} with fibers of constant dimension N and liftable over $\underline{A}$, these are canonically isomorphic to $H^{2N-h}(\underline{C}, \Gamma^*_{\underline{A}}(\underline{C})\dagger \otimes_{\mathbb{Z}} \mathbb{Q})$, all integers h, where $\underline{C}$ is any simple and separated lifting of finite presentation over $\underline{A}$. (In general the groups (1) are defined to be $H^{2N-h}(X,X-C,(\Gamma^*_{\underline{A}}(\underline{X})\dagger) \otimes_{\mathbb{Z}} \mathbb{Q})$, where $\underline{X}$ is simple and separated of finite presentation over $\underline{A}$, with fibers of constant dimension N over $\underline{A}$, $X=\underline{X} \times_{\mathcal{O}} k$ and C is closed in X.) These groups are shown to be a functor with respect to maps of reduced schemes over A_{red}.

The research for this chapter was partly done at Harvard University in Spring, 1970, which portion was supported by NSF grants and a Sloan Foundation grant, and partly done in Spring, 1978, at the University of Rochester.

We now begin.

Let $\mathcal{O}$ be a complete discrete valuation ring having quotient field K and residue class field k. Let $\underline{A}$ be an $\mathcal{O}$-algebra and let $A = \underline{A} \otimes_{\mathcal{O}} k$.

We first consider the following question. Let C be a reduced prescheme over Spec(A), and let $\underline{X}$, resp: $\underline{D}$, be preschemes simple of finite presentation over Spec($\underline{A}$), such that C is A-isomorphic to a closed subprescheme of $X = \underline{X} \times_{\underline{A}} A$, resp.: of $D = \underline{D} \times_{\underline{A}} A$, and such that $X - C$, resp: $D - C$, are quasicompact. Suppose that the dimensions of all the fibers of $\underline{X}$, resp: $\underline{D}$, over $\underline{A}$ are all equal to the same integer N, respectively M. Then we find conditions under which we construct canonical isomorphisms:

$$(1) \quad H^{2N-h}(X, X-C, (\Gamma^*_{\underline{A}}(\underline{X})\dagger) \otimes_{\mathbb{Z}} \mathbb{Q}) \approx$$

$$H^{2M-h}(D, D-C, (\Gamma^*_{\underline{A}}(\underline{D}))\dagger \otimes_{\mathbb{Z}} \mathbb{Q}), \quad \text{all integers } h.$$

Lemma 1. C, $\underline{X}$ and $\underline{D}$ as above, suppose that there exists an A-map $f: X_{red} \to D_{red}$ such that the restriction of f induces the identity isomorphism from $C \subset X_{red}$ onto $C \subset D_{red}$. Let ∇ denote the image of C under the mapping: $C \to X \times_A D$ such that the composites with the first and second projections are the inclusions: $C \to X$ and $C \to D$ respectively. Then we have isomorphisms:

$$(2) \quad H^{2M+2N-h}(X\times_A D, X\times_A D-\nabla, (\Gamma^*_{\underline{A}}(\underline{X}\times_{\underline{A}}\underline{D})\dagger) \otimes_{\mathbb{Z}} \mathbb{Q}) \approx$$

$$H^{2N-h}(X, X-C, \Gamma^*_{\underline{A}}(\underline{X})\dagger \otimes_{\mathbb{Z}} \mathbb{Q}).$$

Sketch of Proof: Consider the composite mapping:

$$(3) \quad H^{2N-h}(X, X-C) \xrightarrow[\text{by projection)}]{\text{(map induced}} H^{2N-h}(X\times_A D, X\times_A D - C\times_A D)$$

$$\xrightarrow[u_{X\times_A D,\, \Gamma_f}]{\text{cupping with}} H^{2N+2M-h}(X\times_A D, X\times_A D-\nabla),$$

where $u_{X\times_A D, \Gamma_f} \in H^{2M}(X\times_A D, X\times_A D - \Gamma_f)$ is the canonical class of Γ_f on $X \times_A D$. ([2]). This gives an $(\underline{A}\dagger \otimes_{\mathbb{Z}} \mathbb{Q})$-homomorphism between the groups that we wish to prove isomorphic.

By the second Leray spectral sequence of relative hypercohomology ([1]) we have the first quadrant spectral sequence

$$(4) \quad E_2^{p,q} = H^p(X\, H_X^q(X, X-C, (\Gamma^*(\underline{X})\dagger) \underset{\underline{A}}{} \underset{\mathbb{Z}}{\otimes} \mathbb{Q}))$$

$$\Longrightarrow H^n(X, X-C, (\Gamma^*(\underline{X})\dagger) \underset{\underline{A}}{} \underset{\mathbb{Z}}{\otimes} \mathbb{Q}),$$

and a similar second Leray spectral sequence of relative hypercohomology call it (5), abutting at the first groups in equation (2). The construction of the composite mapping (3) defines a mapping from the spectral sequence (4) into the spectral sequence (5). Therefore the proof that the composite mapping (3) is an isomorphism becomes a local problem--i.e., it sufficies to prove the analogous assertion for any collection of open subsets of C that are an open base for the topology of C. But an open base for the topology of C consists of those open subsets C' of C, such that there exists an open neighborhood X' of C' in X, such that, if $f' = f|X'$, then $\Gamma_{f'}$ is globally regularly embedded ([2]) as $\mathcal{O}$-space in $X \underset{A}{\times} D$, and equation (1) has been proved in that case in [2].

Remark: The proof of Lemma 1 shows more generally, "Let $C, \underline{X}$ and $\underline{D}$ be as above, and suppose that we have an A-map $\pi: D_{red} \to X_{red}$ that induces the identity from C onto C, and also that there exists a reduced closed subscheme of finite presentation E of D such that, all points of E that are generic in their fiber over Spec(A) are simple points of E over Spec(A_{red}), and of codimension $M-N$ on D_{red}, and such that $\pi^{-1}(C) \cap E$ is the closed subset C of D_{red}, the intersection being in general position and transverse regular (as defined in [3], Proposition II.5.2, bottom of page 231). Then equation (1) holds."

Lemma 2. If there exist A-maps: $X_{red} \to D_{red}$ and $D_{red} \to X_{red}$ that both induce the identity mapping of C, then one can establish isomorphisms as in equation (1).

Proof: By Lemma 1 we have the isomorphisms (2). Lemma 1 with "$\underline{D}$" and "$\underline{X}$" interchanged completes the proof. Q.E.D.

For the moment, Lemma 2 above will suffice for the next set of applications.

Next we define a category which we denote $C'_{0,\underline{A}}$. The objects of $C'_{0,\underline{A}}$ are the pairs $(C,\overline{\underline{X}})$ where C is a reduced scheme over A and $\overline{\underline{X}}$ is proper and of finite presentation over Spec($\underline{A}$), and such that C is a locally closed sub A-scheme of $\overline{X} = \overline{\underline{X}} \times_0 k$, and such that C is contained in some dense open subscheme $\underline{X}$ of $\overline{\underline{X}}$ such that $\underline{X}$ is simple of finite presentation over $\underline{A}$, such that $\underline{X} - C$ is quasicompact and such that all the connected components of all the fibers of $\underline{X}$ over points of Spec($\underline{A}$) are of the same constant dimension N.

Example. If C is simple over Spec(A_{red}), such that there exists $\underline{C}$ simple of finite presentation over Spec($\underline{A}$) such that $C = (\underline{C} \times_0 k)_{red}$, and such that there exists $\overline{\underline{X}}$ proper of finite presentation, all the connected components of the fibres of which over Spec($\underline{A}$) have the same dimension, such that $\underline{C}$ is an open subscheme of $\overline{\underline{X}}$, then $(C,\overline{\underline{X}}) \in C'_{0,\underline{A}}$. (Therefore the reader will verify that lifted p-adic homology with compact supports, as defined below, generalizes the hypercohomology of a flat lifting with coefficients in the † of the sheaves of differential forms, when such exists, of a simple scheme).

Given two such objects $(C,\overline{\underline{X}})$ and $(D,\overline{\underline{Y}})$ in $C'_{0,\underline{A}}$, a map from $(C,\overline{\underline{X}})$ into $(D,\overline{\underline{Y}})$ is a pair (ι,f) Where

(5) ι: $C \to D$ is a proper map over A_{red}, and where

(6) f: $X_{red} \to \overline{Y}_{red}$ is an A-map extending ι,

where X is a "sufficiently small" open neighborhood of C in $\overline{X} = \overline{\underline{X}} \times_0 k$ and where $\overline{Y} = \overline{\underline{Y}} \times_0 k$ (two such "f"'s, defined on different open neighborhoods "X" of C in $\overline{X}$, are considered to be the same if they agree on some smaller neighborhood of C in $\overline{X}$).

Proposition 3. For each object $(C,\underline{\overline{X}}) \in C'_{\mathcal{O},\underline{A}}$, define

$$H^c_h(C,\underline{\overline{X}},\underline{A}^{\dagger} \underset{\mathbb{Z}}{\otimes} \mathbb{Q}) = H^{2N-h}(X,X-C,(\Gamma^*_{\underline{A}}(\underline{X})^{\dagger}) \underset{\mathbb{Z}}{\otimes} \mathbb{Q}),$$

where $\underline{X}$ is some open neighborhood of C in $\underline{\overline{X}}$ such that $\underline{X}$ is simple over Spec($\underline{A}$) and such that the dimensions of all the connected components of all fibers of $\underline{X}$ over Spec($\underline{A}$) are equal to some constant integer N. Assume for simplicity that the ring $\underline{A}$ is normal (i.e., is isomorphic to the direct product of finitely many integral domains, each of which is integrally closed in its field of quotients). (We'll show later how to remove this hypothesis). Then the assignment:

$$(C,\underline{\overline{X}}) \rightsquigarrow H^c_h(C,\underline{\overline{X}}, (\underline{A}^{\dagger}) \underset{\mathbb{Z}}{\otimes} \mathbb{Q})$$

is in a natural way a covariant functor from the category $C'_{\mathcal{O},\underline{A}}$ into the category of $(\underline{A}^{\dagger}) \underset{\mathbb{Z}}{\otimes} \mathbb{Q}$ -modules, all integers h.

Sketch of Proof: The proof in many ways resembles that of III. 1.7, pg 249, of [3]. (That theorem is basically the special case of this Proposition in which $\underline{A} = \mathcal{O}$, and we restrict attention to those $(C,\underline{\overline{X}}) \in C'_{\mathcal{O},\underline{A}}$ such that $C = \underline{X} \underset{\mathcal{O}}{\times} k$).

The key step is, if (ι,f): $(C,\underline{\overline{X}}) \to (D,\underline{\overline{Y}})$ is a map in $C'_{\mathcal{O},\underline{A}}$, then we must construct

$$H^c_h(\iota,f,\underline{A}^{\dagger} \underset{\mathbb{Z}}{\otimes} \mathbb{Q}): H^c_h(C,\underline{\overline{X}},(\underline{A}^{\dagger}) \underset{\mathbb{Z}}{\otimes} \mathbb{Q}) \to H^c_h(D,\underline{\overline{Y}}, (\underline{A}^{\dagger}) \underset{\mathbb{Z}}{\otimes} \mathbb{Q}),$$

a homomorphism of $\underline{A}^{\dagger} \underset{\mathbb{Z}}{\otimes} \mathbb{Q}$ -modules, all integers h.

First, we can easily reduce to the case $\underline{A}$ an integral domain, which we assume. Also, it is easy to reduce to the case in which $\underline{\overline{X}}$ and $\underline{\overline{Y}}$ are connected, which we assume. Let $\underline{X}$, respectively: $\underline{Y}$, be an open neighborhood of C in $\underline{\overline{X}}$, respectively D in $\underline{\overline{Y}}$, such that $\underline{X}$ and $\underline{Y}$ are simple over Spec($\underline{A}$), such that C, respectively D, is closed in $\underline{X}$, respectively $\underline{Y}$, such that there exists an integer N, respectively: M, such that all of the connected components of fibers of $\underline{X}$, respectively: $\underline{Y}$, over Spec($\underline{A}$) are of dimension N, respectively: M. Let $X = \underline{X} \times_{\mathcal{O}} k$, $Y = \underline{Y} \times_{\mathcal{O}} k$, $\bar{X} = \bar{\underline{X}} \times_{\mathcal{O}} k$, $\bar{Y} = \bar{\underline{Y}} \times_{\mathcal{O}} k$. Replacing $\underline{X}$ by an open subset if necessary, we can assume that the map f in (6) is defined on X. Then by Lemma 2 with $\underline{D} = \underline{X} \times_{\underline{A}} \underline{Y}$, we have isomorphisms

$$(7) \qquad H^{2N-h}(X, X-C, (\Gamma^*_{\underline{A}}(\underline{X})\dagger) \otimes_{\mathbb{Z}} \mathbb{Q} \approx H^{2N+2M-h}(X \times_A Y,$$

$$X \times_A Y - \Gamma_\iota, (\Gamma^*_{\underline{A}}(\underline{X})\dagger) \otimes_{\mathbb{Z}} \mathbb{Q}) \quad , \quad \text{all integers } h,$$

where Γ_ι is the graph of the map (5). (The maps, the projection: $X_{red} \times_A Y_{red} \to X_{red}$ and $x \to (x,f(x)): X_{red} \to X_{red} \times_A Y_{red}$ fulfill the requirements of Lemma 2).

Since the map $\iota: C \to D$ is a <u>proper</u> map over A (see (5)), since Γ_ι is D-isomorphic to C and since D is closed in Y it follows that both Γ_ι and $\bar{X} \times_A Y$ are proper over Y. Therefore Γ_ι is <u>closed</u> in $\bar{X} \times_A Y$. Therefore we have the excision isomorphism ([3], I. 6.4, pps. 146-147),

$$(8) \qquad H^{2N+2M-h}(X \times_A Y, \bar{X}\times_A Y-\Gamma_\iota, (\Gamma^*_{\underline{A}}(\underline{X}\times_{\underline{A}}\underline{Y})\dagger) \otimes_{\mathbb{Z}} \mathbb{Q})$$

$$\overset{\approx}{\leftarrow} H^{2N+2M-h}(\bar{X}\times_A Y, \bar{X}\times_A Y-\Gamma_\iota, ('\Gamma^*_{\underline{A}}(\bar{\underline{X}}\times_{\underline{A}}\underline{Y})\dagger) \otimes_{\mathbb{Z}} \mathbb{Q})$$

all integers h, where

$$'\Gamma^j_{\underline{A}}(\bar{\underline{X}}\times_{\underline{A}}\underline{Y}) = \begin{cases} \Gamma^j_{\underline{A}}(\bar{\underline{X}}\times_{\underline{A}}\underline{Y}), & j \leq N+M, \\ 0 \quad , & j \geq N+M+1, \end{cases}$$

all integers $j \geq 0$. The restriction maps the right side of equation (8) into $H^{2N+2M-h}(\bar{X}\times_A Y, \bar{X}\times_A Y-\bar{X}\times_A D, ('\Gamma^*_{\underline{A}}(\bar{\underline{X}}\times_{\underline{A}}Y)\dagger) \otimes_{\mathbb{Z}} \mathbb{Q})$, which we prefer writing as

(9) $$H^{2N+2M-h}(\overline{X}\underset{A}{\times}(Y,Y-D),\ ('\Gamma^*(\overline{X}\underset{\underline{A}}{\times}Y))^{\dagger}\underset{\mathbb{Z}}{\otimes}\mathbb{Q}).$$

Since $\overline{X}$ is proper of finite presentation over $\underline{A}$, there exists a blowing up $\overline{\underline{X}}'$ of $\overline{\underline{X}}$ that is projective over $\mathrm{Spec}(\underline{A})$. Then we have the natural mapping, induced by the morphism: $\overline{\underline{X}}'\to\overline{\underline{X}}$,

(10) $$H^{2N+2M-h}(\overline{X}\underset{A}{\times}(Y,Y-D))\to H^{2N+2M-h}(\overline{X}'\underset{A}{\times}(Y,Y-D)),$$

where $\overline{X}'=\overline{\underline{X}}'\underset{\mathcal{O}}{\times}k$ (and where the coefficients are in $('\Gamma^*_{\underline{A}}(\overline{\underline{X}}\underset{\underline{A}}{\times}\underline{Y})^{\dagger})\underset{\mathbb{Z}}{\otimes}\mathbb{Q}$ and in $(('\Gamma^*_{\underline{A}}(\overline{\underline{X}}'\underset{\underline{A}}{\times}\underline{Y})^{\dagger})\underset{\mathbb{Z}}{\otimes}\mathbb{Q})$ respectively). Since $\overline{\underline{X}}'$ is projective over $\underline{A}$, there exists a finite map (meaning a map of module finite presentation)

$$\overline{\underline{X}}'\to\mathbb{P}^N(\underline{A}).$$

Since $\overline{\underline{X}}'$ is connected and finite over the normal scheme $\mathbb{P}^N(\underline{A})$, there exists $\overline{\underline{X}}''$ finite over $\overline{\underline{X}}'$ such that $\overline{\underline{X}}''$ is normal, and such that $\overline{\underline{X}}''$ is Galois over $\mathbb{P}^N(\underline{A})$. Let d be the degree of the finite covering $\overline{\underline{X}}''$ of $\overline{\underline{X}}'$ (thus $d=[K(\overline{\underline{X}}''):K(\overline{\underline{X}}')]$), and let G be the Galois group of $\overline{\underline{X}}''$ over $\mathbb{P}^N(\underline{A})$. Then we have the natural mapping from the group on the right of equation (10) into $H^{2N+2M-h}(\overline{X}''\times(Y,Y-D))$ (where $\overline{X}''=\overline{\underline{X}}''\underset{\mathcal{O}}{\times}k$, and the coefficients are in $'\Gamma^*_{\underline{A}}(\overline{\underline{X}}''\underset{\underline{A}}{\times}\underline{Y})^{\dagger}\underset{\mathbb{Z}}{\otimes}\mathbb{Q}$).

Let

(11) $$H^{2N+2M-h}(\overline{X}'\underset{A}{\times}(Y,Y-D))\xrightarrow[\text{natural mapping}]{(1/d)\ \text{times}}H^{2N+2M-h}(\overline{X}''\underset{A}{\times}(Y,Y-D)),$$

denote $1/d$ times the natural mapping, all integers h. Since $\overline{X}''$ is finite, and therefore affine, over $\mathbb{P}^N(A)$, by Corollary II.3.1.2, pg. 191 of [3] we have the isomorphisms:

(12) $$H^{2N+2M-h}(\overline{X}''\underset{A}{\times}(Y-D),('\Gamma^*(\overline{\underline{X}}''\underset{\underline{A}}{\times}\underline{Y})^{\dagger})\underset{\mathbb{Z}}{\otimes}\mathbb{Q})$$
$$\approx H^{2N+2M-h}(\mathbb{P}^N(A)\underset{A}{\times}(Y-D),$$
$$(\pi\underset{A}{\times}\mathrm{id}_{\underline{Y}})_*(('\Gamma^*_{\underline{A}}(\overline{\underline{X}}''\underset{\underline{A}}{\times}\underline{Y})^{\dagger})\underset{\mathbb{Z}}{\otimes}\mathbb{Q}),$$

where $\pi\colon \underline{\overline{X}}" \to \mathbb{P}^N(\underline{A})$ is the morphism. Since $\mathbb{P}^N(\underline{A})$ is normal and since $\underline{\overline{X}}"$ is a Galois finite covering of $\mathbb{P}^N(\underline{A})$ with Galois group G, it is easy to see, by arguments similar to those in the proof of Theorem II.4.5, pgs. 209-210 of [3], that

$$\pi_*(\mathcal{O}_{\underline{\overline{X}}"})^G = \mathcal{O}_{\mathbb{P}^N(\underline{A})}.$$

It follows (by methods not unlike those in the proof of II.4.5 of [3]) that

$$(13)\quad (\pi \underset{\underline{A}}{\times} \mathrm{id}_{\underline{Y}})_*(('\Gamma^*_{\underline{A}}(\underline{\overline{X}}"\underset{\underline{A}}{\times}\underline{Y})\dagger) \underset{\mathbb{Z}}{\otimes} \mathbb{Q})^G$$
$$\approx (\Gamma^*_{\underline{A}}(\mathbb{P}^N(\underline{A}))\dagger) \underset{\mathbb{Z}}{\otimes} \mathbb{Q}.$$

The assignment: $\alpha \to \sum_{g\in G} \alpha\cdot g$ in each stalk defines a mapping of cochain complexes of sheaves of $(\underline{A}\dagger) \underset{\mathbb{Z}}{\otimes} \mathbb{Q}$ modules over $\mathbb{P}^N(A)$ from

$$(\pi \underset{\underline{A}}{\times} \mathrm{id}_{\underline{Y}})_*('\Gamma^*_{\underline{A}}(\underline{\overline{X}}"\underset{\underline{A}}{\times}\underline{Y})\dagger) \underset{\mathbb{Z}}{\otimes} \mathbb{Q}) \qquad \text{into}$$

$$((\pi \underset{\underline{A}}{\times} \mathrm{id}_{\underline{Y}})_*(\Gamma^*_{\underline{A}}(\underline{\overline{X}}"\underset{\underline{A}}{\times}\underline{Y})\dagger) \underset{\mathbb{Z}}{\otimes} \mathbb{Q}))^G .$$

Combining this observation with equation (13) we obtain the transfer mapping from the right side of equation (12) into

$$(14)\quad H^{2N+2M-h}(\mathbb{P}^N(A)\underset{A}{\times}(Y,Y-D),\ \Gamma^*_{\underline{A}}(\mathbb{P}^N(\underline{A})\underset{\underline{A}}{\times}\underline{Y})\dagger \underset{\mathbb{Z}}{\otimes} \mathbb{Q}).$$

Before completing the indication of proof of Proposition 3, we digress to prove a Lemma and Corollary.

One can deduce from Corollary II.4. 4.1, pg. 206 of [3], (or rather, from its generalization with "$\underline{A}$" replacing "$\mathcal{O}$", given in [2]), the following Lemma.

Lemma 4. If Y is an $\mathcal{O}$-space of finite presentation over A, if D is a closed subset of Y and if N is any non-negative integer then there are induced canonical isomorphisms,

(15) $$H^h(\mathbb{P}^N(A)\underset{A}{\times}(Y,Y-D),\Gamma^*_{\underline{A}}(\mathbb{P}^N(A)\underset{A}{\times}Y)\dagger \underset{\mathbb{Z}}{\otimes} \mathbb{Q}) \approx \bigoplus_{i=0}^{N} H^{h-2i}(Y,Y-D,(\Gamma^*_{\underline{A}}(Y)\dagger) \underset{\mathbb{Z}}{\otimes} \mathbb{Q}).$$

Proof: We first define a natural mapping from the right side of equation (15) into the left side. We must build mappings: $H^{h-2i}(Y,Y-D) \to H^h(\mathbb{P}^N(A)\underset{A}{\times}(Y,Y-D))$, $0 \le i \le N$, $h \ge 0$. In fact, take the composite:

$$H^{h-2i}(Y,Y-D) \xrightarrow{p_1^*} H^{h-2i}(\mathbb{P}^N(A)\underset{A}{\times}(Y,Y-D)) \xrightarrow[p_2^*(u_i)]{\text{cupping with}} H^h(\mathbb{P}^N(A) \underset{A}{\times}(Y,Y-D)),$$

where p^*_1 and p_2^* are induced by projections and where

$u_i \in H^{2i}(\mathbb{P}^N(A),\Gamma^*_{\underline{A}}(\mathbb{P}^N(A))\dagger \underset{\mathbb{Z}}{\otimes} \mathbb{Q})$ is the cohomology class of any linear subspace of $\mathbb{P}^N(A)$ of dimension $N-i$ (i.e., is the image of $u_{\mathbb{P}^N(A),\mathbb{P}^{N-i}(A)} \in H^{2i}(\mathbb{P}^N(A),\mathbb{P}^N(A)-\mathbb{P}^{N-i}(A)))$. This defines a canonical mapping as indicated. Considering the cohomology sequences of the pairs $(Y,Y-D)$ and $(\mathbb{P}^N(A)\underset{A}{\times}(Y,Y-D))$ ([3], Theorem I.6.3, pgs. 140-141), and the Five Lemma, the proof of Lemma 4 reduces to the case $D=\emptyset$ We have

(16) $\mathbb{A}^N(A) = \mathbb{P}^N(A)-\mathbb{P}^{N-1}(A) \subset \mathbb{P}^N(A)-\mathbb{P}^{N-2}(A) \subset \mathbb{P}^N(A)-\mathbb{P}^{N-3}(A) \subset \ldots,$

where "$\mathbb{P}^i(A)$" denotes the closed subset, $(T_0 = T_1 = \ldots = T_{N-i-1}=0)$ of $\mathbb{P}^N(A)$ and $\mathbb{A}^N(A)$ is affine N-space over A. Let $\mathbb{P}^i$ stand for $\mathbb{P}^i(A)$. Then exicising the closed subset "$T_i=0$" we see that the restriction is an isomorphism

(17) $$H^h(\mathbb{P}^N-\mathbb{P}^{N-(i+1)},\ \mathbb{P}^N-\mathbb{P}^{N-i})\underset{A}{\times}Y) \approx H^h((\mathbb{A}^N_i,\mathbb{A}^N_i-(T_0=\ldots=T_{i=1}=0))\underset{A}{\times}Y),$$

where $\mathbb{A}^N_i$ denotes the affine open subset, "$T_i \neq 0$" of $\mathbb{P}^N(A)$, so that $A^N_i \approx \mathrm{Spec}(A[T_0,\ldots,T_{i-1},T_{i+1},\ldots,T_N])$, $0 \le i \le N$. The cohomology sequence of a triple ([3], pg. 141, Note following

Theorem I.6.3, equation (*)), and iteration of equation (16), together with equation (17) and Corollary II.4.4.1, pg. 206, of [3] (as generalized in [2] with "$\underline{A}$" replacing "$\mathcal{O}$") completes the proof of Lemma 4.

Corollary 4.1. The hypotheses being as in Lemma 4, there is induced a natural mapping

$$(18) \qquad H^h(\mathbb{P}^N(A) \underset{A}{\times} (Y,Y-D)) \to H^{h-2N}(Y,Y-D).$$

Completion of the sketch of proof of Proposition 3: Define $H_h^c(\iota,f,(\underline{A}\dagger)\underset{\mathbb{Z}}{\otimes}\mathbb{Q})$ to be the composite of: the isomorphism (7), followed by the isomorphism (8), the restriction mapping into (9), the natural mapping (10), followed by the mapping (11), the isomorphism (12), the transfer mapping (14) and the mapping (18). It remains to show that, (1) If $(\iota,f)\colon (C,\underline{\overline{X}}) \to (D,\underline{\overline{Y}})$ and $(j,g)\colon (D,\underline{\overline{Y}}) \to (E,\underline{\overline{Z}})$ are maps in $C'_{\mathcal{O},\underline{A}}$, then $H_h^c(j,g) \circ H_h^c(\iota,f) = H_h^c(j\circ\iota, g\circ f)$, all integers h, and (2) If $(id_C, id_{\underline{\overline{X}}})$ is the identity of $(C,\underline{\overline{X}})$ in $C'_{\mathcal{O},\underline{A}}$ then $H_h^c(id_C,id_{\underline{\overline{X}}})$ is the identity of $H_h^c(C,\underline{\overline{X}},(\underline{A}\dagger)\underset{\mathbb{Z}}{\otimes}\mathbb{Q})$, all integers h. (To prove equation (1), one first considers the case in which $\underline{\overline{X}}$ and $\underline{\overline{Y}}$ are projective spaces over $\underline{A}$, and argues similarly to the proof of Theorem III.1.5 of [3] , pgs. 244-246. (The proof is in some ways easier, since one doesn't have to argue with Poincaré duality.)) (To prove equation (2), one notes that it suffices to show that $H_h^c(id_C,id_{\underline{\overline{X}}})$ is an isomorphism for all integers h. And, by using the second Leray spectral sequence of relative hypercohomology [1] this problem becomes local in C. By such techniques, one reduces to the case: C an affine scheme, $\underline{\overline{X}}$ = a projective space over $\underline{A}$, and then an explicit computation is made).

Corollary 3.1. The same hypotheses as Proposition 3, but delete the hypothesis that "$\underline{A}$ is normal." Let $C''_{\mathcal{O},\underline{A}}$ be the full subcategory of $C'_{\mathcal{O},\underline{A}}$ generated by those pairs $(C,\underline{\overline{X}}) \in C'_{\mathcal{O},\underline{A}}$

such that $\underline{\overline{X}} \approx \mathbb{P}^N(\underline{A})$, there exists an integer $N \geq 0$. Then the assignment, $(C,\underline{\overline{X}}) \rightsquigarrow H_h^c(C,\underline{\overline{X}},(\underline{A}\dagger)\otimes_{\mathbb{Z}}\mathbb{Q})$, $h \geq 0$, is a functor on the category $\mathcal{C}''_{\mathcal{O},\underline{A}}$.

Proof: Similar to (but easier than) Proposition 3; simply delete equations (10),(11),(12),(13) and (14) from the proof of Proposition 3.

Remark: For a better generalization of Proposition 3 beyond the normal case, see Remark 2 following Proposition 7 below.

Proposition 5. The hypotheses being in Proposition 3, let $(C,\underline{\overline{X}})$ and $(C,\underline{\overline{X}}')$ be two objects in the category $\mathcal{C}'_{\mathcal{O},\underline{A}}$ having the same C. Then there are induced canonical isomorphisms of $(\underline{A}\dagger)\otimes_{\mathbb{Z}}\mathbb{Q}$ -modules

$$H_h^c(C,\underline{\overline{X}},(A\dagger)\otimes_{\mathbb{Z}}\mathbb{Q}) \approx H_h^c(C,\underline{\overline{X}}',(A\dagger)\otimes_{\mathbb{Z}}\mathbb{Q}),$$

all integers h.

Sketch of Proof: We have maps in the category $\mathcal{C}'_{\mathcal{O},\underline{A}}$,

$$(id_C,\pi_1): (C,\underline{\overline{X}} \times_{\underline{A}} \underline{\overline{X}}') \to (C,\underline{\overline{X}}) \quad \text{and}$$

$$(id_C,\pi_2): (C,\underline{\overline{X}} \times_{\underline{A}} \underline{\overline{X}}') \to (C,\underline{\overline{X}}'),$$

where π_1 and π_2 are the canonical projections. Therefore we need only show that both of these maps induce isomorphisms on lifted p-adic homology with compact supports, e.g., the first of these. Let $\underline{X}$ and $\underline{X}'$ be as in the proof of Proposition 3, and $X = \underline{X} \otimes_{\mathcal{O}} k$, $X' = \underline{X}' \otimes_{\mathcal{O}} k$. Using second Leray spectral sequences of relative hypercohomology ([1]), we have the first quadrant cohomological spectral sequence

$$E_2^{p,q} = H^p(X,X-C,\mathbb{H}^q((\Gamma^*_{\underline{A}}(\underline{X})\dagger)\otimes_{\mathbb{Z}}\mathbb{Q})) \Rightarrow H^n(X,X-C,(\Gamma^*_{\underline{A}}(\underline{X})\dagger)\otimes_{\mathbb{Z}}\mathbb{Q})$$

and a similar spectral sequence abutting at $H^n(X \times_A X', X \times_A X'-C, (\Gamma^*_{\underline{A}}(\underline{X} \times_{\underline{A}} \underline{X}')\dagger) \otimes_{\mathbb{Z}} \mathbb{Q})$. Therefore, tracing the proof of Proposition 3 and of Lemma 1, (notice that all the maps constructed generalize to the cochain level (as defined in [3], Chapter I), and therefore induce maps of spectral sequences), this problem is local in C. Let N, respectively: N', be the dimension of the fibers of $\underline{X}$, respectively: $\underline{X}'$, over $\underline{A}$.

Case I. $N = N'$. Then if we replace C by a small enough open subset, then there exists $\underline{E}$ over $\underline{A}$ such that C is A-isomorphic to a closed subset of $E = \underline{E} \times_{\underline{0}} k$ and such that there exist étale $\underline{A}$-mappings of finite presentation: $\underline{E} \to \underline{X}$ and $\underline{E} \to \underline{X}'$ both of which induce the identity on C. But then, by the first cohomology theorem in [2], we have that these mappings induce isomorphisms,

$$H^{2N-h}(X, X-C, (\Gamma^*_{\underline{A}}(\underline{X})\dagger) \otimes_{\mathbb{Z}} \mathbb{Q}) \overset{\approx}{\to} H^{2N-h}(E, E-C, (\Gamma^*_{\underline{A}}(\underline{E})\dagger) \otimes_{\mathbb{Z}} \mathbb{Q}),$$

$$H^{2N-h}(X', X'-C, (\Gamma^*_{\underline{A}}(\underline{X}')\dagger) \otimes_{\mathbb{Z}} \mathbb{Q}) \overset{\approx}{\to} H^{2N-h}(E, E-C, (\Gamma^*_{\underline{A}}(\underline{E})\dagger) \otimes_{\mathbb{Z}} \mathbb{Q}),$$

all integers h. (And it is easy to see that the isomorphisms thus obtained between $H^{2N-h}(X, X-C)$ and $H^{2N}(X', X'-C)$, $h \in \mathbb{Z}$, coincide with the ones indicated above.)

Case II. $N \neq N'$, say $N' > N$. Let $\underline{D} = \underline{X} \times_{\underline{A}} \mathbb{P}^{N'-N}(\underline{A})$ and let $C \to X \times_A \mathbb{P}^{N'-N}(A)$ be the map whose first coordinate is the inclusion: $C \hookrightarrow X$ and whose second coordinate is the composite of the structure map: $C \to \mathrm{Spec}(A)$ with the map $\mathrm{Spec}(A) \approx$ (the section $(T_1 = \ldots = T_{N'-N} = 0)$ of $\mathbb{P}^{N'-N}(A)) \subset \mathbb{P}^{N'-N}(A)$. Then Lemma 2 applies (since we have the projection mapping: $\underline{X} \times_{\underline{A}} \mathbb{P}^{N'-N}(\underline{A}) \to \underline{X}$ and the mapping: $\underline{X} \to \underline{X} \times_{\underline{A}} \mathbb{P}^{N'-N}(\underline{A})$ whose coordinates are the identity of $\underline{X}$ and the composite $\underline{X} \to \mathrm{Spec}(\underline{A}) \approx$ (the section $(T_1 = \ldots = T_{N'-N} = 0)$ of $\mathbb{P}^{N'-N}(\underline{A})) \hookrightarrow \mathbb{P}^{N'-N}(\underline{A})$ both of which induce the identity on C), whence by Lemma 2

$$H^{2N-h}(X,X-C) \approx H^{2N-h}(X \times_{\underline{A}} \mathbb{P}^{N'-N}(A), X \times_{\underline{A}} \mathbb{P}^{N'-N}(A) - C),$$

all integers h. But by Case I, with $\overline{X} \times_{\underline{A}} \mathbb{P}^{N'-N}(\underline{A})$ replacing $\underline{\overline{X}}$, we have that

$$H^{2N-h}(X \times_{\underline{A}} \mathbb{P}^{N'-N}(A), X \times_{\underline{A}} \mathbb{P}^{N'-N}(A) - C) \approx H^{2N-h}(X,X-C),$$

all integers h.

Definition 1. Let $\underline{A}$ be a ring and let A be a quotient ring of $\underline{A}$. A scheme C over A is properly embeddable over A if there exists $\underline{\overline{X}}$ proper of finite presentation over $\underline{A}$ such that C is A-isomorphic to a locally closed subscheme of $\overline{X} = \underline{\overline{X}} \times_{\underline{A}} A$, such that $\overline{X}-C$ is quasicompact and such that C is contained in the set of simple points of $\underline{\overline{X}}$ over $\underline{A}$.

Remark 1. (When this is the case, it is easy to see that $\underline{\overline{X}}$ can be taken so that there exists $\underline{X}$ open in $\underline{\overline{X}}$ and simple over $\underline{A}$ such that C is closed in $\underline{X}$ and such that the connected components of fibers of $\underline{X}$ over $\mathrm{Spec}(\underline{A})$ all have the same dimension).

Remark 2. By a well-known theorem of Nagata [4], it is easy to show that, C is properly embeddable over $\underline{A}$ iff C is A-isomorphic to a closed subscheme of a scheme $\underline{X}$ that is separated and simple of finite presentation over $\mathrm{Spec}(\underline{A})$, and such that $X-C$ is quasicompact where $X = \underline{X} \times_{\mathcal{O}} k$.

Definition 1 is relative to $\underline{A}$ as well as A. Definition 2 below is independent of $\underline{A}$, depends on the $\mathcal{O}$-algebra structure of A, and appears to be more restrictive.

Definition 2. Let $\mathcal{O}$ be a ring, let A be an $\mathcal{O}$-algebra and let C be a scheme over Spec(A). Fix a polynomial algebra $\underline{P}$ over $\mathcal{O}$ (in possibly infinitely many variables. I.e., $\underline{P} \approx \mathcal{O}[(T_i)_{i \in I}]$ as $\mathcal{O}$ algebras, there exists some set I) and an epimorphism of $\mathcal{O}$-algebras: $\underline{P} \to A$. Then we say that C is polynomially properly embeddable over A iff C is properly embeddable over $\underline{A}$ in the sense of Definition 1. (It is easy to see that this definition is independent of the choice of such a $\underline{P}$ and of an epimorphism: $\underline{P} \to \underline{A}$, since polynomial algebras are projective objects in the category of $\mathcal{O}$-algebras).

Examples 1. If C is quasiprojective over A, then C is properly embeddable over $\underline{A}$ for $\underline{A}$ as in Definition 1 (take $\overline{\underline{X}} =$ a suitable $\mathbb{P}^N(\underline{A})$) (and therefore C is also polynomially properly embeddable over A).

2. If C is simple over A_{red} and liftable over $\underline{A}$, in the sense that there exists $\underline{X}$ separated and simple of finite presentation over $\underline{A}$ such that $X \approx \underline{X} \times_{\underline{A}} A$, then C is properly embeddable over A. The main theorem is Theorem 6.

We return to the notation $\mathcal{O}, K, k, A, \underline{A}$ as in the beginning of the chapter.

Theorem 6. Suppose that the ring $\underline{A}$ is normal (respectively: Make no additional hypothesis on $\underline{A}$). Then let $C_{\mathcal{O},\underline{A}}$ be the category having for objects all schemes C over Spec(A) that are properly embeddable over $\underline{A}$ (respectively: that are polynomially properly embeddable over A), and for maps all proper A-maps. Then we have functors, "lifted p-adic homology with compact supports",

$$C \rightsquigarrow H_h^c(C, (\underline{A}^\dagger) \otimes_{\mathbb{Z}} \mathbb{Q}), \quad \text{all integers } h,$$

from the category $C_{\mathcal{O},\underline{A}}$ into the category of $(\underline{A}^\dagger) \otimes_{\mathbb{Z}} \mathbb{Q}$ -modules.

Sketch of Proof: (We sketch the case in which $\underline{A}$ is normal. The other case is covered in Remark 2 following Proposition 7 below.) For every object $C \in \mathcal{C}_{0,\underline{A}}$ by definition there exists $\underline{\overline{X}}$ such that $(C,\underline{\overline{X}}) \in \mathcal{C}'_{0,\underline{A}}$. Then by Proposition 3 we have

$$H^h_c(C,\underline{\overline{X}},(A\dagger) \otimes_{\mathbb{Z}} \mathbb{Q}), \quad \text{all integers } h.$$

But by Proposition 5 these groups are independent, up to canonical isomorphisms, of $\underline{\overline{X}}$. So define

$$(19) \quad H^h_c(C,(\underline{A}\dagger) \otimes_{\mathbb{Z}} \mathbb{Q}) = H^h_c(C,\underline{\overline{X}},(A\dagger) \otimes_{\mathbb{Z}} \mathbb{Q}),$$

for any such $\underline{\overline{X}}$, all integers h. The Propositions 3 and 5 prove that $H^h(C,(A\dagger) \otimes_{\mathbb{Z}} \mathbb{Q})$ is a functor on the category $\mathcal{C}_{0,\underline{A}}$, all integers h.

Proposition 7. Let $C \in \mathcal{C}_{0,\underline{A}}$ and let D be a reduced closed A-subscheme of C such that $D \in \mathcal{C}_{0,\underline{A}}$ (i.e., such that $C-D$ is quasicompact). Let U be the open subset $C-D$ of C. Then there is induced a homomorphism of $(\underline{A}\dagger) \otimes_{\mathbb{Z}} \mathbb{Q}$ -modules, which we call the restriction,

$$H^c_h(C,(\underline{A}\dagger) \otimes_{\mathbb{Z}} \mathbb{Q}) \to H^c_h(U,(\underline{A}\dagger) \otimes_{\mathbb{Z}} \mathbb{Q}),$$

all integers h. Moreover, we have a long exact sequence:

$$(20) \cdots \xrightarrow{\partial_{h+1}} H^c_h(D,(\underline{A}\dagger) \otimes_{\mathbb{Z}} \mathbb{Q}) \xrightarrow{H^c_h(\iota)} H^c_h(C,(\underline{A}\dagger) \otimes_{\mathbb{Z}} \mathbb{Q})$$

$$\xrightarrow{\text{restriction}} H^c_h(U,(A\dagger) \otimes_{\mathbb{Z}} \mathbb{Q}) \xrightarrow{\partial_h} H^c_{h-1}(D,(\underline{A}\dagger) \otimes_{\mathbb{Z}} \mathbb{Q}) \to \cdots ,$$

where ι is the inclusion: $D \to C$.

Proof: Choose $\underline{\overline{X}}$ such that $(C,\underline{\overline{X}}) \in \mathcal{C}'_{0,\underline{A}}$, and let $\underline{X}$ be as in the proof of Proposition 3. Then

$$H^c_h(D,(\underline{A}\dagger) \otimes_{\mathbb{Z}} \mathbb{Q}) = H^{2N-h}(X,X-D), \quad H^c_h(C,\underline{A}\dagger \otimes_{\mathbb{Z}} \mathbb{Q}) = H^{2N-h}(X,X-C)$$

and $H_h^c(U,(\underline{A}\dagger)\otimes_{\mathbb{Z}}\mathbb{Q}) = H^{2N-h}(X-D,X-C)$, all integers h. Therefore the indicated long exact homology sequence is the cohomology sequence of the triple ([3],I.6, Note, pgs. 141-142) (X,X-D,X-C) with coefficients in $(\Gamma^*_{\underline{A}}(\overline{\underline{X}})\dagger)\otimes_{\mathbb{Z}}\mathbb{Q}$.

Remarks 1. The hypotheses being as in Theorem 6, let M be any module over the ring $(\underline{A}\dagger)\otimes_{\mathbb{Z}}\mathbb{Q}$. Then for every $C \in C_{0,\underline{A}}$ we define the lifted p-adic homology with compact supports of C with coeffients in M as follows. Fix any $\overline{X}$ such that $(C,\overline{X}) \in C'_{0,\underline{A}}$, and let $\underline{X}$ be as in the proof of Proposition 3. Let $\mathfrak{U}$ be a finite set of affine open subsets of $\underline{X}$ that is a covering (in the sense of [3], I.5, pg. 127), and define

$$(21)\quad H_h^c(C,M) = H^{2N-h}(C^*(\mathfrak{U},(X,X-C),(\Gamma^*_{\underline{A}}(\underline{X})\dagger)\otimes_{\mathbb{Z}}\mathbb{Q})\otimes_{((\underline{A}\dagger)\otimes_{\mathbb{Z}}\mathbb{Q})} M),$$

all integers h (where $C^*(\mathfrak{U}, X,X-C,\ )$ is as defined in [3], I. 6, pg. 144). Then since $C^* = C^*(\mathfrak{U},(X,X-C),(\Gamma^*_{\underline{A}}(\underline{X})\dagger)\otimes_{\mathbb{Z}}\mathbb{Q})$ is flat over $\underline{A}$, and since $H^{2N-h}(C^*) \approx H^{2N-h}(X,X-C,(\Gamma^*_{\underline{A}}(\underline{X})\dagger)\otimes_{\mathbb{Z}}\mathbb{Q})$, all integers h (by [3], Theorem I. 6.7, pg. 152 (as generalized in [2])), we have the universal coefficients spectral sequence (see [5], Chapter V, shortly after the definition of "percohomology"), a homological spectral sequence confined to the region: $p \geq 0$, $2N-M \leq q \leq 2N$ where M is an integer such that $H^i(X,X-C,(\Gamma^*_{\underline{A}}(\underline{X})\dagger)\otimes_{\mathbb{Z}}\mathbb{Q}) = 0$, all integers $i \geq M+1$,

$$(22)\quad \mathrm{Tor}_p^{(\underline{A}\dagger)\otimes_{\mathbb{Z}}\mathbb{Q}}(H_q^c(C,(\underline{A}\dagger)\otimes_{\mathbb{Z}}\mathbb{Q}),M) \Longrightarrow H_n^c(C,M).$$

Using these "universal coefficients spectral sequences" the proofs of Lemma 1 and of Propositions 3 and 5 show that the definition (21) is independent of all choices (i.e. of $\overline{X}$ and $\mathfrak{U}$), and is a functor on the category $C_{0,\underline{A}}$, all integers h, all $(\underline{A}\dagger) \otimes_{\mathbb{Z}} \mathbb{Q}$ -modules M. And of course also the long exact sequence (20) of Proposition 7 goes through with "M" replacing "$(A\dagger) \otimes_{\mathbb{Z}} \mathbb{Q}$."

Remark 2. We now prove the parenthetical case of Theorem 6 (and also Proposition 7). Let $\underline{P}$ be any polynomial algebra over $\mathcal{O}$ such that A is isomorphic to a quotient $\mathcal{O}$-algebra of $\underline{P}$, and fix any $\mathcal{O}$-homomorphism: $\underline{P} \to \underline{A}$ lifting the epimorphism: $\underline{P} \to A$. Let $P = \underline{P} \otimes_{\mathcal{O}} k$. Then $\underline{P}$ being a polynomial algebra is normal, and therefore $\underline{P}$ and P obey the hypotheses of the non-parenthetical case of Theorem 6, so we have the functors on the category $C_{\mathcal{O},\underline{P}}$: $C \rightsquigarrow H^c_h(C, \underline{P}\dagger \otimes_{\mathbb{Z}} \mathbb{Q})$, and by Remark 1 above even the functors: $C \rightsquigarrow H^c_h(C,M)$, all integers h, all $(P\dagger) \otimes_{\mathbb{Z}} \mathbb{Q}$ -modules M, on the category $C_{\mathcal{O},\underline{P}}$. But the category $C_{\mathcal{O},\underline{A}}$ (with polynomially properly embeddable objects) is a full subcategory of $C_{\mathcal{O},\underline{P}}$ (with properly embeddable objects over $\underline{P}$). If now M is any $\underline{A}\dagger \otimes_{\mathbb{Z}} \mathbb{Q}$ -module, then regarding M as a $(\underline{P}\dagger) \otimes_{\mathbb{Z}} \mathbb{Q}$ -module, the restriction of the functor $C \rightsquigarrow H^c_h(C,M)$ to the full subcategory $C_{\mathcal{O},\underline{A}}$ of $C_{\mathcal{O},\underline{P}}$ proves Remark 1 (for the parenthetical hypotheses of Theorem 6). And the special case $M = (\underline{A}\dagger) \otimes_{\mathbb{Z}} \mathbb{Q}$ proves the parenthetical part of Theorem 6. Similarly for Proposition 7. Q.E.D.

Remark 3. Let $D \in C_{\mathcal{O},\underline{A}}$ and let d be the largest dimension of fibers of D over A. Then by theorems in [2], $H^c_h(D,(\underline{A}\dagger) \otimes_{\mathbb{Z}} \mathbb{Q}) = 0$ for $h \geq 2d + 1$. By the universal coefficients spectral sequence (Remark 1 above, equation (22)), it follows that $H^c_h(D,M) = 0$ for all $(\underline{A}\dagger) \otimes_{\mathbb{Z}} \mathbb{Q}$ -modules M, all integers $h \geq 2d + 1$.

4. If we take $\underline{A} = \mathcal{O}$, and if $D \in C_{\mathcal{O},\mathcal{O}}$, then it is easy to see that $H^c_h(D,K) = 0$, all integers h such that either $h < 0$ or $h > 2d$, $d = \dim D$. However,

Example. Let $\underline{A}$ be an $\mathcal{O}$-algebra that is simple over $\mathcal{O}$ and that is a local ring of dimension $n + 1 \geq 2$ and let D be the closed point of Spec(A). Then $D \in C_{\mathcal{O},\underline{A}}$ and $H^{c}_{-n}(D,(\underline{A}\dagger) \otimes_{\mathbb{Z}} \mathbb{Q}) \approx H^{n}(\mathrm{Spec}(A),\mathrm{Spec}(A)-D,\mathcal{O}_{\underline{\mathrm{Spec}(A)}}\dagger) \otimes_{\mathbb{Z}} \mathbb{Q} \neq 0$. Therefore negative homology groups with compact supports need not always vanish if $\underline{A} \neq \mathcal{O}$.

CHAPTER 2

Finite Generation of Lifted p-Adic Homology With Compact Supports.

Let $\mathcal{O}$ be a complete discrete valuation ring having a quotient field of characteristic zero, and with residue class field k. Let C be an algebraic variety over k that is properly embeddable (see Chapter 1) over $\mathcal{O}$ (e.g., it suffices that C be quasi-projective), and let K be the quotient field of $\mathcal{O}\dagger$. Then in this chapter we prove that the lifted p-adic homology with compact supports, $H^{c}_{h}(C,K)$, as defined in Chapter 1, is finite dimensional over K, all integers h. In consequence if C is simple over k and embeddable ([6]) over $\mathcal{O}$, then the lifted p-adic cohomology of C $H^{h}(C,K)$, as defined in [6], is finite dimensional over K, all integers h. (Therefore if C is simple of finite type over k, and if C should admit the simple lifting $\underline{C}$ over $\mathcal{O}$, then the groups $H^{h}(C,(\Gamma^{*}_{\mathcal{O}}(\underline{C})\dagger) \otimes_{\mathcal{O}} K)$, $h \geq 0$, as defined in [3], are finite dimensional over K, all integers h). (The analogous theorems for q-adic homology with compact supports, and for q-adic cohomology, about finite generation, can also be proved by the same method).

The research for this chapter was begun at Harvard University in Spring, 1970. That portion was partially supported by an

NSF Postdoctoral Fellowship and a Sloane Foundation grant. This research was completed at the University of Rochester in Spring, 1978.

Let $\mathcal{O}$, K, k, $\underline{A}$, A, and $C_{\mathcal{O},\underline{A}}$ be as in Chapter 1.

Lemma 1. Suppose that $\underline{A}$ is normal and let C, D $\in C_{\mathcal{O},\underline{A}}$ be such that C is proper over Spec(A), and let f: C $\to$ D be an A-map. (Therefore f is a map in $C_{\mathcal{O},\underline{A}}$). Suppose that there exists $\underline{\bar{X}}$ and $\underline{\bar{Y}}$ proper of finite presentation over $\underline{A}$, such that $\bar{X}_K = \underline{\bar{X}} \underset{\mathcal{O}}{\times} K$ is simple over $\underline{A} \underset{\mathcal{O}}{\otimes} K$, such that C(respectively: D) is a closed A-subscheme of $\bar{X} = \underline{\bar{X}} \underset{\mathcal{O}}{\times} k$ (respectively: of $\bar{Y} = \underline{\bar{Y}} \underset{\mathcal{O}}{\times} k$), such that D is contained in the set $\bar{Y}$ of $\underline{\bar{Y}}$ of simple points of $\underline{\bar{Y}}$ over Spec($\underline{A}$) and such that the dimension of all the connected components of $\underline{\bar{X}}$ (respectively: $\underline{\bar{Y}}$) in their fibers over Spec($\underline{A}$) is equal to a constant integer N (respectively M). Let $Y = \underline{Y} \underset{\mathcal{O}}{\times} k$. Suppose also that there exists $\underline{\bar{D}}$ a closed $\underline{A}$-subscheme of $\underline{\bar{Y}}$ such that $D = (\underline{\bar{D}} \underset{\mathcal{O}}{\times} k)_{red}$, and that there exists $\underline{f}: \underline{\bar{X}} \to \underline{\bar{D}}$ a mapping of $\underline{A}$-schemes such that the restriction of $\underline{f}$ to C is f. Then there are induced a homomorphism of $\underline{A}\dagger \underset{\mathcal{O}}{\otimes} K$ -modules

$$\text{(1)}\quad H^{2N-h}(\bar{X},\bar{X}-C,('\Gamma^*_{\underline{A}}(\underline{\bar{X}})\dagger) \underset{\mathcal{O}}{\otimes} K)$$

$$\longrightarrow H^{2M-h}(Y,Y-D,(\Gamma^*_{\underline{A}}(\underline{Y})\dagger) \underset{\mathcal{O}}{\otimes} K) = H^c_h(D,(\underline{A}\dagger) \underset{\mathcal{O}}{\otimes} K),$$

all integers h. Suppose, in addition, that <u>any one</u> of the following three technical conditions holds: <u>Either</u> (a) $\underline{A} = \mathcal{O}$ or (b) the image of: $u_{\underline{\bar{X}} \underset{A}{\times} \underline{Y},\, \Gamma_{\underline{f}}}$ (as defined below)) in (b1) $H^{2M}(\underline{\bar{X}} \underset{\underline{A}}{\times} (\underline{Y},\underline{Y}-\underline{\bar{D}}),\ '\Gamma^*_{A}(\underline{\bar{X}} \underset{\underline{A}}{\times} \underline{Y}) \underset{\mathcal{O}}{\otimes} K)$ comes from an element of

$$\text{(b2)}\quad \bigoplus_{i=0}^{2M} [H^i(\underline{\bar{X}},\ '\Gamma^*_{\underline{A}}(\underline{\bar{X}}))] \underset{\underline{A}}{\otimes} [H^{2M-i}(\underline{Y},\underline{Y}-\underline{\bar{D}},\Gamma^*_{\underline{A}}(\underline{Y}))] \underset{\mathcal{O}}{\otimes} K;$$

or that

(c) $H^h(\underline{X}, '\Gamma^*_{\underline{A}}(\underline{X})\dagger \underset{0}{\otimes} K)$ is finitely generated as $(\underline{A}\dagger) \underset{0}{\otimes} K$ -module, all integers h, $0 \le h \le 2M$. Moreover, if the ring $\underline{A}\dagger$ is Noetherian, then the image of the homomorphism (1) is finitely generated as $((\underline{A}\dagger) \underset{0}{\otimes} K)$ -module.

Note. The proof of the Lemma shows that, if the hypothesis "$\underline{f}: \underline{X} \to \underline{D}$" is weakened to: "$\underline{f} : \underline{X} \to \underline{Y}$", then one can still construct the mapping (1).

Proof. The construction of the map (1) is similar to the proof of Proposition 3 of Chapter 1. First, notice that since $\overline{X}_K$ and $Y_K (= \underline{Y} \underset{0}{\times} K)$ are simple of finite presentation over $A_K = \underline{A} \underset{0}{\otimes} K$, by [2] (or, in the case $\underline{A} = 0$, by [3], II.5, pg. 231, just before Prop. 2), we have the canonical class

$$(2) \qquad u_{\overline{X}_K \underset{A_K}{\times} Y_K, \Gamma_{f_K}} \in H^{2M}(\overline{X}_K \underset{A_K}{\times} Y_K, \ \overline{X}_K \underset{A_K}{\times} Y_K - \Gamma_{f_K}, \Gamma^*_{A_K}(X_K \underset{A_K}{\times} Y_K)),$$

where $f_K = \underline{f} \underset{0}{\times} K$ and $\Gamma_{f_K} \subset \overline{X}_K \underset{A_K}{\times} Y_K$ is the graph of f_K.

But

$$(3) \qquad H^{2M}(\overline{\underline{X}} \underset{\underline{A}}{\times} \underline{Y}, \overline{\underline{X}} \underset{\underline{A}}{\times} \underline{Y} - \Gamma_{\underline{f}}, '\Gamma^*_{\underline{A}}(\overline{\underline{X}} \underset{\underline{A}}{\times} \underline{Y}) \underset{0}{\otimes} K)$$

$$\approx H^{2M}(\overline{X}_K \underset{A_K}{\times} Y_K, \overline{X}_K \underset{A_K}{\times} Y_K - \Gamma_{f_K}, \Gamma^*_{A_K}(\overline{X}_K \underset{A_K}{\times} Y_K))$$

(since $X_K \underset{A_K}{\times} Y_K = (\overline{\underline{X}} \underset{\underline{A}}{\times} \underline{Y}) \underset{0}{\times} K$ and the objects are of finite presentation over $\underline{A}$ see, e.g., [2]). Therefore the element (2) defines an element, call it "$u_{\overline{\underline{X}} \underset{\underline{A}}{\times} \underline{Y}, \Gamma_{\underline{f}}}$", in the left side of equation (3). We also use the same notation for the image of this element in $H^{2M}(\overline{X} \underset{A}{\times} Y, \overline{X} \underset{A}{\times} Y - \Gamma_{\underline{f}}, \Gamma^*_{\underline{A}}(\overline{\underline{X}} \underset{\underline{A}}{\times} \underline{Y})\dagger \underset{0}{\otimes} K)$.

We define the $(\underline{A}\dagger \underset{0}{\otimes} K)$ -homomorphism (1), each integer h, to be the composite of the sequence:

$$(4) \qquad H^{2N-h}(\overline{X}, \overline{X}-C, '\Gamma^*_{\underline{A}}(\underline{\overline{X}})\dagger \underset{0}{\otimes} K) \xrightarrow{\pi_1^*}$$

$$H^{2N-h}((\overline{X},\overline{X}-C) \underset{A}{\times} Y, '\Gamma^*_{\underline{A}}(\underline{\overline{X}\times Y})^\dagger \underset{\mathcal{O}}{\otimes} K) \xrightarrow[u_{\underline{\overline{X}\times Y}},\Gamma_{\underline{f}}]{\text{cupping with}}$$

$$H^{2N+2M-h}(\overline{X}\underset{A}{\times}Y, \overline{X}\underset{A}{\times}Y - \Gamma_f, '\Gamma^*_{\underline{A}}(\underline{\overline{X}\times Y})^\dagger \underset{\mathcal{O}}{\otimes} K) \xrightarrow{\text{restriction}}$$

$$H^{2N+2M-h}(\overline{X}\underset{A}{\times}Y, \overline{X}\underset{A}{\times}Y - \overline{X}\underset{A}{\times}D, '\Gamma^*_{\underline{A}}(\underline{\overline{X}\times Y})^\dagger \underset{\mathcal{O}}{\otimes} K) =$$

$$H^{2N+2M-h}(\overline{X}\underset{A}{\times}(Y,Y-D)) \xrightarrow[\text{natural map}]{}$$

$$H^{2N+2M-h}(\overline{X}'\underset{A}{\times}(Y,Y-D)) \xrightarrow[\text{map}]{(1/d)\cdot\text{natural}}$$

$$H^{2N+2M-h}(\overline{X}''\underset{A}{\times}(Y,Y-D)) \xrightarrow{\alpha \to \sum_{g\in G} \alpha g}$$

$$H^{2N+2M-h}(\mathbb{P}^N(A)\underset{A}{\times}(Y,Y-D)) \xrightarrow[\text{of Chapter 1}]{\text{map of Cor. 4.1}}$$

$$H^{2M-h}(Y,Y-D,\Gamma^*_{\underline{A}}(\underline{Y})^\dagger \underset{\mathcal{O}}{\otimes} K),$$

where $\underline{\overline{X}}'$, $\underline{\overline{X}}''$, $\overline{X}'$, $\overline{X}''$, d and G are constructed as in the proof of Proposition 3 of Chapter 1.

Considering the second mapping and the fifth group in the sequence (4), we see that the image of any element $x \in H^{2N-h}(\overline{X},\overline{X}-C, '\Gamma^*_{\underline{A}}(\underline{\overline{X}})^\dagger \underset{\mathcal{O}}{\otimes} K)$ under the mapping (1) depends only on the value of the image of $\pi_1^*(x)$ in $H^{2N-h}(\overline{X}\underset{A}{\times}Y, '\Gamma^*_{\underline{A}}(\underline{\overline{X}})^\dagger \underset{\mathcal{O}}{\otimes} K)$ after cupping with the image of $u_{\underline{\overline{X}\times Y}}, \Gamma_{\underline{f}}$ in

$$(5)\quad H^{2M}(\overline{X}\underset{A}{\times}(Y,Y-D), ('\Gamma^*_{\underline{A}}(\underline{\overline{X}\times Y})^\dagger) \underset{\mathcal{O}}{\otimes} K).$$

Condition (a) implies condition (b). Therefore it suffices to prove the Lemma if either (b) or (c) holds.

Case 1. Condition (b) holds. Then by condition (b) the image $u_{\underline{\overline{X}\times Y}},\Gamma_{\underline{f}}$ in the group (5) can be written in the form

$$(6)\quad \sum_{i=0}^{2M} \sum_{j=1}^{B} \pi_1^*(e_{ij}) \cup \pi_2^*(f_{ij}),$$

where B is an integer ≥ 1, and where $e_{ij} \in H^i(\underline{X}, '\Gamma^*_{\underline{A}}(\underline{X})\dagger \underset{0}{\otimes} K)$ and $f_{ij} \in H^{2M-i}(Y, Y-D, (\Gamma^*_{\underline{A}}(\underline{Y})\dagger) \underset{0}{\otimes} K)$, $1 \leq j \leq B$, $0 \leq i \leq 2M$. Therefore in this case the image of the element x in the fifth group of the sequence (4) can be written as

$$(7) \quad \sum_{i=0}^{2M} \sum_{j=1}^{B} [\pi_1^*(x \cup e_{ij})] \cup \pi_2^*(f_{ij}).$$

Considering the maps leaving the fifth, sixth and seventh groups in equation (4), it follows that image of x under the composite mapping (4) depends only on the images, for all integers i,j such that $0 \leq i \leq 2M$, $1 \leq j \leq B$,

$$(8) \quad \alpha_{ij}(x) \in H^{2N-h+i}(\mathbb{P}^N(A), \Gamma^*_{\underline{A}}(\mathbb{P}^N(\underline{A}))\dagger \underset{0}{\otimes} K)$$

of the elements: $\pi_1^*(x \cup e_{ij})$ under the composite mappings:

$$H^{2N-h+i}(\overline{X}, ('\Gamma^*_{\underline{A}}(\overline{X})\dagger) \underset{0}{\otimes} K) \to H^{2N-h+i}(\overline{X}', ('\Gamma^*_{\underline{A}}(\overline{X}')\dagger) \underset{0}{\otimes} K)$$

$$\to H^{2N-h+i}(\overline{X}'', ('\Gamma^*_{\underline{A}}(\overline{X}'')\dagger) \underset{0}{\otimes} K) \to H^{2N-h+i}(\mathbb{P}^N(A),$$

$(\Gamma^*_{\underline{A}}(\mathbb{P}^N(\underline{A}))\dagger) \underset{0}{\otimes} K)$, for $0 \leq i \leq 2M$. In fact, considering the last mapping in the sequence (4), the image of $x \in H^{2N-h}(\overline{X}, \overline{X}-C, ('\Gamma^*_{\underline{A}}(\overline{X})\dagger) \underset{0}{\otimes} K)$ under the mapping (1) depends actually only on thôse $\alpha_{ij}(x)$ in equation (8) such that $i = h$ - i.e., only on the value of the elements $\alpha_{h,j}(x) \in H^{2N}(\mathbb{P}^N(A), \Gamma^*_{\underline{A}}(\mathbb{P}^N(\underline{A}))\dagger \underset{0}{\otimes} K)$, $1 \leq h \leq B$. Since (by Lemma 4 of Chapter 1 with $\underline{Y} = \mathrm{Spec}(\underline{A})$) this latter group is isomorphic to $(\underline{A}\dagger) \underset{0}{\otimes} K$, it follows that for each integer h, $1 \leq h \leq B$, the assignment: $x \to \alpha_{h,j}(x)$ is a homomorphism of $(\underline{A}\dagger) \underset{0}{\otimes} K$ -modules from the $(\underline{A}\dagger) \underset{0}{\otimes} K$ -module

(9) $H^{2N-h}(\overline{X}, \overline{X}-C, ('\Gamma^*_{\underline{A}}(\overline{X})\dagger) \underset{0}{\otimes} K)$, into $(\underline{A}\dagger) \underset{0}{\otimes} K$, and that if $\alpha_h(x) = (\alpha_{hj}(x))_{1 \leq j \leq B}$ is the mapping from the group (9) into $((\underline{A}\dagger) \underset{0}{\otimes} K)^B$, then the image of x under (1) is completely determined by the image of x under α_h, all elements x of the group (9).

Therefore the image of the homomorphism (1) is naturally $(\underline{A}\dagger) \underset{0}{\otimes} K$ -isomorphic to a quotient module of the image of α_h, and therefore is finitely generated as $(\underline{A}\dagger) \underset{0}{\otimes} K$ - module, Q.E.D.

Case 2. Condition (c) holds. We have seen that the image of an element $\in H^{2N-h}(\overline{X},\overline{X}-C,('\Gamma^*_{\underline{A}}(\underline{\overline{X}})\dagger) \underset{0}{\otimes} K)$ under the mapping (1) depends only on the image of $\pi_1^*(x)$ in $H^{2N-h}(\overline{X}\underset{\underline{A}}{\times}Y, '\Gamma^*_{\underline{A}}(\underline{\overline{X}})\dagger \underset{0}{\otimes} K)$. Therefore the image of x under (1) depends only on the image of x in $H^{2N-h}(\overline{X},('\Gamma^*_{\underline{A}}(\underline{\overline{X}})\dagger) \underset{0}{\otimes} K$. Therefore the image of (1) is naturally $(\underline{A}\dagger) \underset{0}{\otimes} K$ -isomorphic to a quotient module of the image of the restriction map:

$$H^{2N-h}(\overline{X},\overline{X}-C,('\Gamma^*_{\underline{A}}(\underline{\overline{X}})\dagger) \underset{0}{\otimes} K) \longrightarrow$$

$$H^{2N-h}(\overline{X},('\Gamma^*_{\underline{A}}(\underline{\overline{X}}))\dagger \underset{0}{\otimes} K).$$

But by hypothesis (c) the range of this latter mapping is finitely generated. Q.E.D.

Remark: If one drops the hypothesis that "the ring $\underline{A}\dagger$ is Noetherian", then the proof of Lemma 1 shows that the image of the mapping (1) is a submodule of a finitely generated $(\underline{A}\dagger) \underset{0}{\otimes} K$ -module. Also, if M is any finitely generated $(\underline{A}\dagger) \underset{0}{\otimes} K$ -module, then the image of the analogous mapping

$$(1_M)\quad H^{2N-h}(\overline{X},\overline{X}-C,M) \longrightarrow H_h^c(D,M),$$

with coefficients in M, is a submodule of a finitely generated $(\underline{A}\dagger) \underset{0}{\otimes} K$ -module, all integers h.

Corollary 1.1. The hypotheses being as in the Note to Lemma 1, if U is any open subset of C such that $U \in C_{0,\underline{A}}$ (i.e., such that U is quasicompact) and such that U is contained in the simple points of $\underline{\overline{X}}$, and if V is an open subset of D such that $V \in C_{0,\underline{A}}$ (i.e., such that V is quasicompact), such that f maps U into V, and such that the restriction $f_U : U \to V$ is proper over A, then the following diagram is commutative

$$\begin{array}{ccccc}
H^{2N-h}(\bar{X},\bar{X}-C,\Gamma^{*}_{\underline{A}}(\underline{\bar{X}})\dagger\otimes_{\mathcal{O}} K) & \xrightarrow[\text{of Lemma 1}]{\text{map (1)}} & H^{2M-h}(Y,Y-D,\Gamma^{*}_{\underline{A}}(\underline{Y})\dagger\otimes_{\mathcal{O}} K) & = & H^{c}_{h}(D,\underline{A}\dagger\otimes_{\mathcal{O}} K) \\
\downarrow \text{restriction} & & \downarrow \text{restriction} & & \downarrow \text{restriction} \\
H^{2N-h}(X_{o},X_{o}-U,\Gamma^{*}_{\underline{A}}(\underline{X})\dagger\otimes_{\mathcal{O}} K) & \rightarrow & H^{2M-h}(Y_{o},Y_{o}-V,\Gamma^{*}_{\underline{A}}(\underline{Y})\dagger\otimes_{\mathcal{O}} K) & = & H^{c}_{h}(V,\underline{A}\dagger\otimes_{\mathcal{O}} K) \\
\| & & & \nearrow & \\
H^{c}_{h}(U,\underline{A}\dagger\otimes_{\mathcal{O}} K) & & \xrightarrow{\quad H^{c}_{h}(f_{U}) \quad} & &
\end{array}$$

all integers h, where $\underline{X}_{o}$ (resp: $\underline{Y}_{o}$) is any open neighborhood of U (resp: V) in $\bar{X}$ (resp: $\bar{Y}$) that is simple of finite presentation over $\underline{A}$ such that if $X_{o} = \underline{X}_{o} \times_{\mathcal{O}} K$ (resp: $Y_{o} = \underline{Y}_{o} \times K$) then U (resp: V) is closed in X_{o} (resp: Y_{o}).

Proof: The three "equalities" in the diagram are the definition of H^{c}_{h}; commutativity of the upper right square is the definition of the "restriction" mapping in H^{c}_{h}; and commutativity of the bottom square is by definition of $H^{c}_{h}(f_{U})$. Commutativity of the upper left square follows from the definition of the map (1) of Lemma 1 and the definition (see the Proof of Proposition 3 of Chapter 1) of $H^{c}_{h}(f_{U})$.

Lemma 2. Let $\bar{X}$ and C be as in the hypotheses of Lemma 1. Let $(C',\underline{\bar{X}}') \in C_{\mathcal{O},\underline{A}}$ be such that C' is closed in $\bar{X}' = \underline{\bar{X}}' \times_{\mathcal{O}} k$, and such that we have a mapping $\rho: \underline{\bar{X}} \to \underline{\bar{X}}'$ of $\underline{A}$-schemes that maps C into C'. Suppose that we have $\underline{X}$ an open subset of $\underline{\bar{X}}$ that is simple of finite presentation over $\underline{A}$, and $\underline{\bar{E}}$ a closed subscheme of $\underline{\bar{X}}$ of finite presentation over $\underline{A}$, of Krull codimension $N - N'$, such that $\underline{\bar{E}} \cap \rho^{-1}(C') \subset C$ as sets, and such that if $\bar{E} = \underline{\bar{E}} \times_{\mathcal{O}} k$, $X = \underline{X} \times_{\mathcal{O}} k$, $E = X \cap \bar{E}$ and $U = X \cap C$, then E_{red}, U, and the closed subscheme $\rho^{-1}(C') \cap X$ of X_{red} are simple over $\mathrm{Spec}(A_{red})$, and E_{red} and $\rho^{-1}(C') \cap X$ are in general position and intersect transverse regularly ([3], Proposition II.5.2, bottom of pg. 231) over $\mathrm{Spec}(A_{red})$. Suppose also that E is dense in $\bar{E}$, and that the generic points of fibers of

$E_K = \underline{\overline{E}} \times_{0} K$ over Spec$(\underline{A} \otimes_{0} K)$ are simple over Spec$(\underline{A} \otimes_{0} K)$. Suppose that we have $\underline{X}'$ open in $\underline{\overline{X}}'$ of finite presentation over Spec$(\underline{A})$ such that ρ maps $\underline{X}$ into $\underline{X}'$, and such that $\underline{X}$ is simple over $\underline{X}'$, and such that if $U' = C' \cap \underline{X}'$, then $\rho^{-1}(U') \cap \underline{X} = U$, and the restriction τ of ρ to U is a proper mapping from U into U'. Suppose, for simplicity, that the A-mapping $\tau: U \to U'$ is an <u>isomorphism</u> from U onto U'. Then the following diagram is commutative:

$$\begin{array}{ccccc}
H^{2N-h}(\overline{X},\overline{X}-C,('\Gamma^*_{\underline{A}}(\underline{\overline{X}})\dagger)\otimes_{0} K) & \xrightarrow{\text{restriction}} & H^{2N-h}(X,X-U,(\Gamma^*_{\underline{A}}(\underline{X})\dagger)\otimes_{0} K) & = & H^c_h(U,(\underline{A}\dagger)\otimes_{0} K) \\
\uparrow \beta & & \uparrow \beta_0 & & \uparrow H^c_h(\tau,(A\dagger)\otimes_{0} K) \\
H^{2N'-h}(\overline{X}',\overline{X}'-C',(\Gamma^*_{\underline{A}}(\underline{\overline{X}}')\dagger)\otimes_{0} K) & \xrightarrow{\text{restriction}} & H^{2N'-h}(X',X'-U',(\Gamma^*_{\underline{A}}(\underline{X}')\dagger)\otimes_{0} K) & = & H^c_h(U',(\underline{A}\dagger)\otimes_{0} K) \\
\| & & \| & \diagup\!\!\diagup & \\
H^c_h(C',(\underline{A}\dagger)\otimes_{0} K) & \xrightarrow{\text{restriction}} & H^c_h(U',(\underline{A}\dagger)\otimes_{0} K) & &
\end{array}$$

where $X' = \underline{X}' \times_{0} k$.

<u>Remark</u>: The hypothesis that "the A-mapping $\tau: U \to U'$ is an isomorphism from U onto U'" can be <u>eliminated</u>, leaving a true statement (which we shall not prove). (The proof is similar).

<u>Proof</u>: We first must define the mapping β. Since the generic points of fibers of E_K over Spec(A_K) are simple over Spec(A_K) (where $A_K = \underline{A} \otimes_{0} K$), by [2] we have the canonical class $u_{X_K,E_K} \in H^{2(N-N')}(X_K, X_K-E_K, \Gamma^*_{A_K}(X_K))$, where $X_K = \underline{\overline{X}} \times_{0} K$. This latter cohomology group is isomorphic to $H^{2(N-N')}(\underline{\overline{X}},\underline{\overline{X}}-\underline{\overline{E}},'\Gamma^*_{\underline{A}}(\underline{X}) \otimes_{0} K)$. Let u_{X_K,E_K} denote the image of that element in this latter group, and let $u_{\underline{\overline{X}},\underline{\overline{E}}}$ denote the image in $H^{2(N-N')}(\overline{X},\overline{X}-\overline{E},(\Gamma^*(\underline{\overline{X}})\dagger)\otimes_{0} K)$.

Then let β be the composite:

$$H^{2N'-h}(\overline{X}',\overline{X}'-C',(\Gamma^*_{\underline{A}}(\underline{\overline{X}}')\dagger)\otimes_{0} K) \xrightarrow{\rho^*} H^{2N'-h'}(\overline{X},\overline{X}-\rho^{-1}(C'),('\Gamma^*_{\underline{A}}(\underline{\overline{X}})\dagger)\otimes_{0} K)$$

$$\xrightarrow[u_{\underline{X},\underline{E}}]{\text{cupping with}} H^{2N-h}(\overline{X},\overline{X}-C,({}'\Gamma^*_{\underline{A}}(\underline{X})\dagger)\underset{\mathcal{O}}{\otimes}K).$$

Then if we define β_o similarly, then the upper left square clearly commutes. And, by the Remark following Lemma 1 of Chapter 1, it is easy to see that the upper right square commutes. Commutativity of the bottom square and triangle are by definition. Q.E.D.

Theorem 3. Let D be an algebraic variety (= scheme of finite type) over k that is properly embeddable over $\mathcal{O}$. Then the p-adic homology with compact supports $H^c_h(D,K)$ of D with coefficients in K is finitely generated as K vector space, all integers h.

Proof: The proof is by induction on the dimension of D. If

If U' is a dense open subset of D, then by Proposition 7 of Chapter 1 we have the exact sequence:

$$(10)\ldots\xrightarrow{\partial_{h+1}} H^c_h(D-U',K)\xrightarrow{H^c_h(\iota)} H^c_h(D,K)\xrightarrow{\text{restriction}} H^c_h(U',K)\xrightarrow{\partial_h} H^c_{h-1}(D-U',K)\to\cdot$$

Since dim(D-U') < dim(D), by the inductive assumption to prove the theorem for D it suffices to prove it for any variety birationally equivalent to D. Therefore the theorem easily reduces to the case in which D is irreducible, projective and a hypersurface. Then choose $\underline{\overline{D}}$ proper and flat over Spec($\mathcal{O}$), and an irreducible, projective hypersurface over Spec($\mathcal{O}$), such that D is k isomorphic to $\underline{\overline{D}}\underset{\mathcal{O}}{\times}k$, and such that the general fiber, $D_K=\underline{\overline{D}}\underset{\mathcal{O}}{\times}K$, is simple over K (this is easily done by the Jacobian criterion). Since $\underline{\overline{D}}$ is projective over $\mathcal{O}$, $\underline{\overline{D}}$ is $\mathcal{O}$-isomorphic to a closed subscheme of $\mathbb{P}^{N'}(\mathcal{O})$ for some integer N'. Let x be a point of $\underline{\overline{D}}$ that is in D and that is a simple point of $\underline{\overline{D}}$ over $\mathcal{O}$. Then there exists an open neighborhood $\underline{W}$ of x in $\mathbb{P}^{N'}(\mathcal{O})$ and d functions $t_1,\ldots,t_d\in\Gamma(\underline{W},\mathcal{O}_{\underline{W}})$ where d = dim D, such that if $\underline{U} = \underline{W} \cap \underline{\bar{D}}$ and if $t'_1,\ldots,t'_d$ are the images of $t_1,\ldots,t_d$ in $\Gamma(\underline{U},\mathcal{O}_{\underline{U}})$, and if we let $s_i = t_i \underset{\mathcal{O}}{\otimes} 1 - 1 \underset{\mathcal{O}}{\otimes} t'_i \in \Gamma(\underline{W} \underset{\mathcal{O}}{\times} \underline{U},\mathcal{O}_{\underline{\bar{X}}})$, $1 \leq i \leq d$, where $\underline{\bar{X}} = \mathbb{P}^{N'}(\mathcal{O}) \underset{\mathcal{O}}{\times} \underline{\bar{D}}$, then the closed subset: $(s_1 = \cdots = s_d = 0)$ of $\underline{W} \underset{\mathcal{O}}{\times} \underline{U}$ intersected with $\underline{U} \underset{\mathcal{O}}{\times} \underline{U}$ is the diagonal of $\underline{U} \underset{\mathcal{O}}{\times} \underline{U}$. Let $\underline{\bar{E}}$ be the closure in $\underline{\bar{X}}$ of the closed subset $(s_1 = \cdots = s_d = 0)$ of $\underline{W} \underset{\mathcal{O}}{\times} \underline{U}$,

let $\underline{\bar{C}}$ be the intersection of $\underline{\bar{E}}$ and $\underline{\bar{D}} \underset{\mathcal{O}}{\times} \underline{\bar{D}}$ in $\underline{\bar{X}}(=\mathbb{P}^{N'}(\mathcal{O}) \underset{\mathcal{O}}{\times} \underline{\bar{D}})$, let $C = (\underline{\bar{C}} \underset{\mathcal{O}}{\times} k)_{red}$, let $\underline{\bar{X}}' = \underline{\bar{Y}} = \mathbb{P}^{N'}(\mathcal{O})$, let $\underline{f}:\underline{\bar{X}} \to \underline{\bar{D}}$ be the second projection and let $C' = D$. Let U' be the open subset $\underline{U} \underset{\mathcal{O}}{\times} k$ of D, let $V = U'$ and let U be the diagonal of $U' \underset{k}{\times} U'$. Then U is an open subset of C and the restriction τ of the first projection is an isomorphism from U onto U'. Let $\underline{X}_o = \underline{W} \underset{\mathcal{O}}{\times} \underline{U}$, $\underline{Y}_o = \underline{W}$, $\underline{X}' = \underline{W}$, $X_o = \underline{X}_o \underset{\mathcal{O}}{\times} k$, $Y_o = \underline{Y} \underset{\mathcal{O}}{\times} k$, $X' = \underline{X}' \underset{\mathcal{O}}{\times} k$. Then all the hypotheses of Lemmae 1 and 2, and of Corollary 1.1 hold, so that we have the mapping γ, β, β_0 and f_U of Lemma 1, 2,2, and of Corollary 1.1 respectively, (notice that f_U is an isomorphism) and we have the commutative diagram (the notation "Γ^{*+}_{K}" is as in [3]):

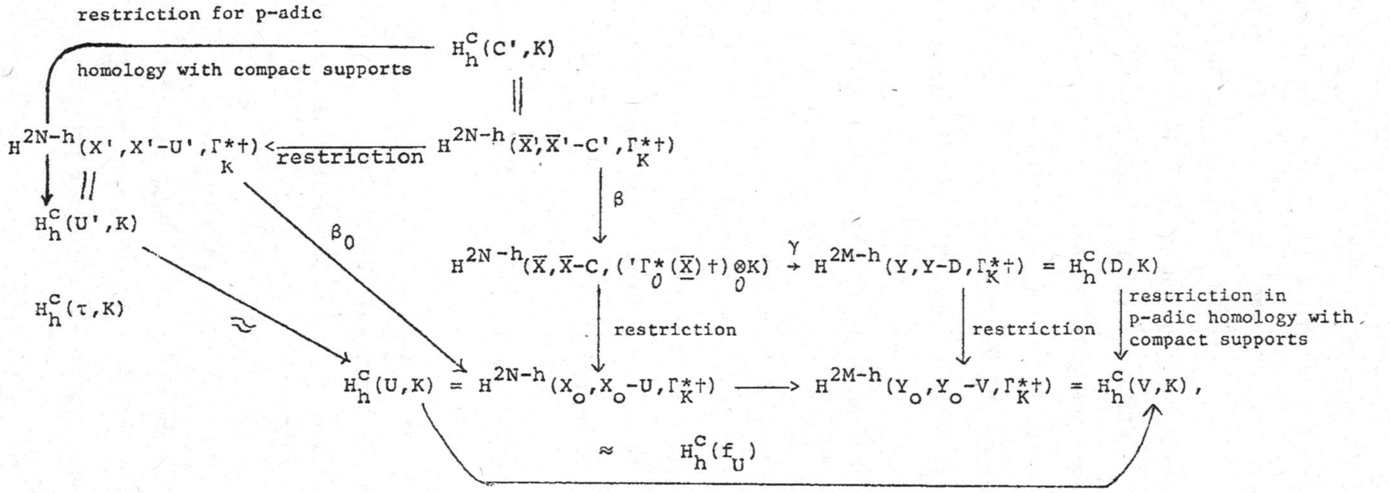
restriction for p-adic
homology with compact supports
$H^c_h(C',K)$
$H^{2N-h}(X',X'-U',\Gamma^*_K\dagger)$
restriction
$H^{2N-h}(\overline{X}',\overline{X}'-C',\Gamma^*_K\dagger)$
β
$H^c_h(U',K)$
β_0
$H^{2N-h}(\overline{X},\overline{X}-C,('\Gamma^*_0(\underline{\overline{X}})\dagger)\otimes_0 K)$
γ
$H^{2M-h}(Y,Y-D,\Gamma^*_K\dagger) = H^c_h(D,K)$
$H^c_h(\tau,K)$
$\approx$
restriction
restriction
restriction in
p-adic homology with
compact supports
$H^c_h(U,K) = H^{2N-h}(X_o,X_o-U,\Gamma^*_K\dagger)$
$H^{2M-h}(Y_o,Y_o-V,\Gamma^*_K\dagger) = H^c_h(V,K)$,
$\approx$
$H^c_h(f_U)$

all integers h. (Commutativity of the three squares to the bottom right is Corollary 1.1; commutativity of the other squares is Lemma 2). By Lemma 1, the image of γ is a finite dimensional K-vector space. Diagram chasing in the above diagram, it follows that the restriction mapping: $H_h^c(C',K) \to H_h^c(U',K)$ has a finite dimensional image, all integers h. But $C' = D$. Then considering the long exact sequence (10), we have that the mappings "restriction" in equation (10) have finite dimensional images. Since by the inductive assumption $H_h^c(D-U',K)$ is finite dimensional over K, from the exact sequence (10) we deduce that $H_h^c(D,K)$ is finite dimensional, all integers h. Q.E.D.

The proof of Theorem 3 shows equally well

Corollary 3.1. Suppose that the $\mathcal{O}$-algebra $\underline{A}$ is a discrete valuation ring $\mathcal{O}'$ such that the pre-image of the maximal ideal of $\mathcal{O}'$ in $\mathcal{O}$ is the maximal ideal of $\mathcal{O}$. Let k' be the residue class field of $\mathcal{O}'$ and let K' be the quotient field of $\mathcal{O}'\dagger^{\mathcal{O}}$. Then the p-adic homology with compact supports $H_h^c(D,K')$ is finite dimensional as K'-vector space, all integers h.

Remark 1: One might hope to prove the analogue of Theorem 3 with "$\underline{A}$", an arbitrary $\mathcal{O}$-algebra, replacing "$\mathcal{O}$". This is in general false, even if D is proper over A_{red} and $\underline{A}$ is a regular local ring, even a simple $\mathcal{O}$-algebra.

Counterexample: Take $\underline{A}$ a simple local $\mathcal{O}$-algebra of Krull dimension $n + 1 \geq 2$ and let D = (the closed point of Spec($\underline{A}$)). Then $H_{-n}^c(D,(\underline{A}\dagger) \otimes_{\mathcal{O}} K)$ is isomorphic to $H^n(\mathrm{Spec}(A),\mathrm{Spec}(A)-\{\text{closed point}\}, \mathcal{O}_{\mathrm{Spec}(\underline{A})}\dagger) \otimes_{\mathcal{O}} K$, where $n = \dim A$, and is not finitely generated over $\underline{A}\dagger \otimes_{\mathcal{O}} K$, see the Example following the four Remarks after Proposition 7 of Chapter 1.

Remark 2. Under the hypotheses of Lemma 1, the proof of Lemma 1 can be refined slightly to prove a bit more. Namely, that the composite of the natural mapping:

$$H^{2N-h}(\overline{X},\overline{X}-C,{}'\Gamma^{\star}_{\underline{A}}(\overline{X})\dagger) \to$$
$$H^{2N-h}(\overline{X},\overline{X}-C,({}'\Gamma^{\star}_{\underline{A}}(\overline{X})\dagger) \underset{\mathcal{O}}{\otimes} K)$$

with the mapping (1) of Lemma 1 is such that, the image of that composite mapping is finitely generated as $(\underline{A}\dagger)$-module.

CHAPTER 3

Generalization of the Weil Conjectures to Singular, Non-Complete Varieties.

Let $\mathcal{O}$ be a c.d.v.r. of mixed characteristic with quotient field K and residue class field k. Let C be an algebraic variety over k that is properly embeddable over $\mathcal{O}$. (See chapter 1. It suffices that C be quasi-projective). If $F:\mathcal{O}\to\mathcal{O}$ induces the p'th power endomorphism of k, $p=\text{char}(k)$, then we define the zeta matrices $W^h(C)$, $0\le h\le 2\dim C$. These generalize the zeta function of varieties over finite fields. This uses the author's lifted p-adic homology with compact supports $H^c_h(C,K)$ (See Chapters 1 and 2). (More generally, if $\underline{A}$ is an $\mathcal{O}$-algebra and $F:\underline{A}\to\underline{A}$ induces the p'th power endomorphism of A_{red}, where $A=\underline{A}\underset{\mathcal{O}}{\otimes}k$, and if C is a reduced scheme over A_{red} that is polynomially properly embeddable (see chapter 1) over A_{red}, then we define the zeta endomorphism ζ_h of the lifted p-adic homology with compact supports $H^c_h(C,(\underline{A}\dagger)\underset{\mathcal{O}}{\otimes}K)$, (chapter 1) all integers h. These essentially determine the zeta matrices of all the algebraic varieties (= fibers over $\mathrm{Spec}(A_{red})$) in the algebraic family C).

If the field k is finite then a generalization of the First Weil Conjecture, "Lefschetz Theorem", is stated and proved for C, p-adically in chapter 3, and q-adically in chapter 4. Also, a generalization of the Third Weil Conjecture, "Riemann Hypothesis", is stated for C (conjectured p-adically in chapter 3, and proved q-adically in chapter 4, $q \neq \mathrm{char}\ k$).

The research for this chapter was done at Berkeley in 1968-9, parially supported by an NSF grant, and at Harvard in 1969-70, supported by an NSF Postdoctoral Fellowship and a Sloane Foundation grant.

Let $\mathcal{O}$ be a complete discrete valuation ring with residue class field k and quotient field K, such that K is of characteristic zero. Then we define categories $C_{\mathcal{O}}$ and $C_{\mathcal{O}}^{\mathrm{normal}}$.

The <u>objects</u> in $C_{\mathcal{O}}^{\mathrm{normal}}$ (resp: $C_{\mathcal{O}}$) are the pairs $(\underline{A}, C)$ where $\underline{A}$ is a normal (resp: an) $\mathcal{O}$-algebra and where C is a reduced scheme over $\mathrm{Spec}(A_{red})$ where $A = \underline{A} \otimes_{\mathcal{O}} k$, such that C is properly embeddable (respectively: properly polynomially embeddable) over $\underline{A}$ (respectively: over A_{red}). (See chapter 1.) The <u>maps</u> in $C_{\mathcal{O}}^{\mathrm{normal}}$ (resp: $C_{\mathcal{O}}$) from $(\underline{A}, C)$ into $(\underline{B}, D)$ are the pairs (F,f) where $F: \underline{A} \to \underline{B}$ is a homomorphism of rings (such that F maps the image of $\mathcal{O}$ in $\underline{A}$ into the image of $\mathcal{O}$ in $\underline{B}$) and where $f: (C \times_A B)_{red} \to D$ is a proper morphism of schemes over $\mathrm{Spec}(A_{red})$, where $A = \underline{A} \otimes_{\mathcal{O}} k$, $B = \underline{B} \otimes_{\mathcal{O}} k$. (We call such a map <u>a semi-linear proper map</u>). We also introduce the category $M_{\mathcal{O}}$, having for <u>objects</u> the pairs $(\underline{A}, M)$ where $\underline{A}$ is an $\mathcal{O}$-algebra and M is an $\underline{A}$-module, and such that the <u>maps</u> from $(\underline{A}, M)$ into $(\underline{B}, N)$ are the pairs (F,f) where $F: \underline{A} \to \underline{B}$ is a homomorphism of rings (such that F maps the image of $\mathcal{O}$ in $\underline{A}$ into the image of $\mathcal{O}$ in $\underline{B}$), and $f: M \otimes_{\underline{A}} \underline{B} \to N$ is a homomorphism of $\underline{B}$-modules (we call such a mapping (F,f) a semi-linear homomorphism). Then

Theorem 1. For each integer h there is induced a functor, the h'th lifted p-adic homology with compact supports, from the category $C_{\mathcal{O}}^{\text{normal}}$ (respectively: $C_{\mathcal{O}}$) into the category $M_{\mathcal{O}}$, such that the restriction of this functor to $C_{\mathcal{O},\underline{A}}$ is the functor H_h^c of Theorem 6 of chapter 1, all normal (respectively: all) $\mathcal{O}$-algebras $\underline{A}$.

Sketch of Proof: Let $(F,f): (\underline{A},C) \to (\underline{B},D)$ be a morphism in $C_{\mathcal{O}}^{\text{normal}}$. Then there exists $\underline{X}$ such that, in the notations of Chapter 1, $(C,\underline{X})$ is an object in $C'_{\mathcal{O},\underline{A}}$. Then $H_h^c(C,\underline{A}\dagger\otimes_{\mathbb{Z}}\mathbb{Q}) = H^{2N-h}(X,X-C,(\Gamma_{\underline{A}}^{*}(\underline{X})\dagger)\otimes_{\mathbb{Z}}\mathbb{Q})$ and $H_h^c((C\times_{\text{Spec}(A)}\text{Spec}(B))_{\text{red}}, (\underline{B}\dagger)\otimes_{\mathbb{Z}}\mathbb{Q}) = H^{2N-h}(X\times_A B, X\times_A B-C\times_A B, (\Gamma_{\underline{B}}^{*}(\underline{X}\times_{\underline{A}}\underline{B})\dagger)\otimes_{\mathbb{Z}}\mathbb{Q})$ where N, $\underline{X}$, X are as in Proposition 3 of chapter 1. Define $H_h^c(F,f)$ to be the composite:

$$H^{2N-h}(X,X-C,(\Gamma_{\underline{A}}^{*}(\underline{X})\dagger)\otimes_{\mathbb{Z}}\mathbb{Q}) \to H^{2N-h}(X\times_A B, X\times_A B-C\times_A B,(\Gamma_{\underline{B}}^{*}(\underline{X}\times_{\underline{A}}\underline{B})\dagger)\otimes_{\mathbb{Z}}\mathbb{Q})$$

$$\xrightarrow{H_h^c(f,\underline{B}\dagger\otimes_{\mathbb{Z}}\mathbb{Q})} H_h^c(D,(\underline{B}\dagger)\otimes_{\mathbb{Z}}\mathbb{Q}).$$

Example 1. Suppose that $\text{char}(k) = p \neq 0$, let $(\underline{A},C) \in C_{\mathcal{O}}^{\text{normal}}$ (or $C_{\mathcal{O}}$) and let $F: \underline{A} \to \underline{A}$ be a fixed endomorphism of the ring $\underline{A}$ such that $(F\otimes_{\mathcal{O}} k)_{\text{red}}: A_{\text{red}} \to A_{\text{red}}$ is the p'th power endomorphism of the ring in characteristic p A_{red} (that takes $x \in A_{\text{red}}$ into $x^p \in A_{\text{red}}$). (E.g., if $\mathcal{O} = \hat{\mathbb{Z}}_p$ and $\underline{A} = W^{-}(A)$, see [7], or $\underline{A} = W(A)$, then one can take F to be W^{-}(p'th power map), or W(p'th power map), respectively).

Let α_C be the p'th power endomorphism of the scheme C in characteristic p (that is set-theoretically the identity map, and that induces the p'th power endomorphism of the local ring $\mathcal{O}_{C,c}$, all $c \in C$). Then for each integer h, we have the endomorphism

$$H_h^c(F,\alpha_C) : H_h^c(C,(\underline{A}\dagger)\underset{\mathbb{Z}}{\otimes}\mathbb{Q}) \to H_h^c(C,(\underline{A}\dagger)\underset{\mathbb{Z}}{\otimes}\mathbb{Q})$$

of the abelian group $H_h^c(C,(\underline{A}\dagger)\underset{\mathbb{Z}}{\otimes}\mathbb{Q})$, which is semi-linear with respect to the ring endomorphism

$$(F\dagger)\underset{\mathbb{Z}}{\otimes}\mathbb{Q} : (\underline{A}\dagger)\underset{\mathbb{Z}}{\otimes}\mathbb{Q} \to (\underline{A}\dagger)\underset{\mathbb{Z}}{\otimes}\mathbb{Q}$$

of the $\hat{\mathbb{Q}}_p$-algebra $(\underline{A}\dagger) \underset{\mathbb{Z}}{\otimes} \mathbb{Q}$. We call this map the h'th <u>zeta endomorphism</u> of $H_h^c(C,(\underline{A}\dagger)\underset{\mathbb{Z}}{\otimes}\mathbb{Q})$, for each integer h.

<u>Example 2</u>. In Example 1, consider the special case in which $\underline{A} = \mathcal{O}'$, a discrete valuation ring containing $\mathcal{O}$ as a subring such that $M_{\mathcal{O}'} \cap \mathcal{O} = M_{\mathcal{O}}$. Let $K' = (\mathcal{O}'\dagger) \underset{\mathbb{Z}}{\otimes} \mathbb{Q} = \text{q.f.}(\mathcal{O}'\dagger)$ and let k' be the residue class field of $\mathcal{O}'$. Then by chapter 2 $H_h^c(C,K')$ is a finitely generated K' vector space, all integers h, $0 \leq h \leq 2 \dim C$ (and the groups vanish for h not in this range by the last Remark in Chapter 1). By Example 1, we have the <u>zeta endomorphism</u> $H_h^c(F,\alpha_C)$ of this K'-vector space, an endomorphism semi-linear with respect to the endomorphism $(F\dagger) \underset{\mathbb{Z}}{\otimes} \mathbb{Q}$ (call this endomorphism F') of the field K'. Therefore if we fix a basis for the finite dimensional vector space $H_h^c(C,K')$ over the field K', then the zeta endomorphism defines a $\beta_h \times \beta_h$ matrix $(\beta_h=\dim_{K'} H_h^c(C,K'))$, unique up to F'-<u>similarity</u>. (Two $\beta_h \times \beta_h$ matrices W, W_0 with coefficients in K' are F'-<u>similar</u> <u>iff</u> there exists an invertible $(\beta_h \times \beta_h)$-matrix B with coefficients in K' such that

$$B^{F'} \cdot WB^{-1} = W_0 \quad , \text{ where } B^{F'} \text{ is obtained by throwing}$$

the coefficients of B through F'). We denote this matrix, unique up to F'-similarity, by $W^h(C)$, and call it the <u>h'th zeta matrix of the algebraic variety</u> C <u>over the field</u> <u>k' of characteristic $p \neq 0$</u>, for $0 \leq h \leq 2\dim C$.

<u>Example 3.</u> Let k' be a <u>finite field</u> and for simplicity suppose that we take $\mathcal{O}' = W(k')$. Then there exists a unique endomorphism (in fact, automorphism) F of the ring $\mathcal{O}'$ such that F induces

the p'th power automorphism of the field k'. Then for every algebraic variety C over the field k' that is properly embeddable over $\mathcal{O}'$, we have that the hypotheses of Example 2 hold, where $K' = q.f.\mathcal{O}'$; a finite extension of $\hat{\mathbb{Q}}_p$ of degree r, where $r = [k':\mathbb{Z}/p\mathbb{Z}]$. Therefore by Example 2 we have the zeta matrices $W^h(C)$, square $\beta_h \times \beta_h$ matrices with coefficients in K', (each unique up to F'-similarity, where F' is the automorphism of K' induced by F) and where $\beta_h = \dim_{K'} H_h^c(C,K')$, $0 \leq h \leq 2 \dim C$. The composite of the p'th power endomorphism α_C of C as defined in Example 1 with itself r times is the standard Frobenius endomorphism f of the algebraic variety C over the finite field k', an (ordinary) map of algebraic varieties over k', where r is the degree of k' over $\mathbb{Z}/p\mathbb{Z}$: $\alpha_C^r = f$. It follows readily that

$$(1)\quad (W^h)^{(F')^{p^{r-1}}} \cdot (W^h)^{(F')^{p^{r-2}}} \cdots (W^h)^{(F')^{p^i}} \cdots (W^h)^{F'} \cdot W^h$$

= the matrix of the linear transformation $H_h^c(id_{\mathcal{O}'}, f)$ of the β_h-dimensional K'-vector space $H_h^c(C,K')$ into itself, $0 \leq h \leq 2\dim C$, where $W^h = W^h(C)$, $0 \leq h \leq 2\dim C$. (This latter linear transformation is an ordinary linear transformation, not merely semi-linear, since F^r = identity of $\mathcal{O}'$.) That is, the product matrix (1) is the matrix of the linear transformation of the h'th homology group with compact supports of C into itself induced by the Frobenius mapping, and is unique up to (ordinary) similarity.

<u>Remark</u>: As was shown in my Harvard seminar on "Zeta Matrices of an Algebraic Family", in the situation of Example 1, for every prime ideal $p \in \mathrm{Spec}(A)$ there is induced a quotient ring $\mathcal{O}'_p$ of $\underline{A}$ of mixed characteristic such that $(\mathcal{O}'_p \otimes_{\mathcal{O}} k)_{red} \approx \mathbb{k}(p)$ and such that F induces an endomorphism

F_p of $\mathcal{O}'_p$; and then the zeta endomorphisms of the algebraic family C over Spec(A) for all integers h determine, essentially, the h'th zeta endomorphism of the algebraic variety: $C_p = (C \times_A \mathbb{k}(p))_{red}$ over the field $\mathbb{k}(p)$, (an endomorphism of the h'th lifted p-adic homology group with compact supports : $H_h^c(C_p, (\mathcal{O}'_p\dagger) \otimes_{\mathbb{Z}} \mathbb{Q})$ of the algebraic variety C_p over $\mathbb{k}(p)$), for each integer h, all prime ideals $p \subset A$. (Also in that seminar it was shown that if, e.g., C is simple and proper over Spec(A_{red}) and liftable over $\underline{A}$, then $H_h^c(C, (\underline{A}\dagger) \otimes_{\mathbb{Z}} \mathbb{Q})$ is locally free of finite rank as $(\underline{A}\dagger) \otimes_{\mathbb{Z}} \mathbb{Q}$-module. Therefore in this case the h'th zeta endomorphism of $H_h^c(C, (\underline{A}\dagger) \otimes_{\mathbb{Z}} \mathbb{Q})$ can be expressed by a square matrix with coefficients in $(\underline{A}\dagger) \otimes_{\mathbb{Z}} \mathbb{Q}$, unique up to $(F\dagger \otimes_{\mathbb{Z}} \mathbb{Q})$-similarity, which we called the h'th zeta matrix of the algebraic family C over Spec(A_{red}) with coefficients in $(\underline{A}\dagger) \otimes_{\mathbb{Z}} \mathbb{Q}$, and denote by $W^h(C)$, all integers h. In this case, the zeta matrix $W^h(C)$ of the algebraic family C over Spec(A) determines the h'th zeta matrix $W^h(C_p)$ of the algebraic variety C_p over $\mathbb{k}(p)$ simply by throwing the coefficients of $W^h(C)$ through the epimorphism: $\underline{A} \to \mathcal{O}'_p$, all integers h, all prime ideals $p \subset A$. Thus, the zeta matrices of such an algebraic family determine the zeta matrices of each of the varieties in the family.)

The following completely proved theorem is a generalization of the first Weil Conjecutre [8], "Lefschetz Theorem."

Theorem 2. (Generalized Weil's Lefschetz Theorem Conjecture). Let k' be a finite field and let C be an algebraic variety over k' that is properly embeddable over $\mathcal{O}' = W(k)$. Then the zeta function $Z_C(T)$ can be written as an alternating product:

$$(2) \qquad Z_C(T) = \frac{P_1(T)\cdot P_3(T)\cdot P_5(T)\cdots P_{2d-1}(T)}{P_0(T)\cdot P_2(T)\cdot P_4(T)\cdots P_{2d}(T)}$$

when $d = \dim C$, and where

(3) $P_h(T) = \det(I_h - f_h \cdot T)$,

the reverse characteristic polynomial of f_h, where f_h is the map induced by the Frobenius endomorphism of C on the β_h dimensional K' vector space $H_h^c(C,K')$, the h'th lifted homology group with compact supports of C, and where I_h is the identity endomorphism of $H_h^c(C,K')$, $0 \leq h \leq 2d$. In particular, P_h is a polynomial of degree β_h with coefficients in K' and with constant term 1, $0 \leq h \leq 2d$. (The fact that P_h has degree β_h is equivalent to the fact that map f_h induced by the Frobenius on the h'th homology group with compact supports is <u>injective</u>, and this is proved by arguing using the perfection [7] $C^{p^{-\infty}}$ of C in a manner similar to the analogous assertion proved in [9].)

<u>Sketch of Proof</u>:

<u>Case 1</u>. C is a non-singular affine hypersurface, of the type $C = \mathrm{Spec}(k'[T_1,\ldots,T_{d+1},g^{-1}]/(H))$, $d = \dim C$, where g, $H \in k'[T_1,\ldots,T_{d+1}]$. Assume also that $T_1,\ldots,T_{d+1} | g$. Then let $\underline{g}$, $\underline{H} \in \mathcal{O}'[T_1,\ldots,T_{d+1}]$ be elements that map into g, H and let $\underline{C} = \mathrm{Spec}(\mathcal{O}'[T_1,\ldots,T_{d+1},\underline{g}^{-1}]/(\underline{H}))$. Then by definition (see chapter 1) we have that

(4) $H_h^c(C,K') = H^{2d-h}(C,(\Gamma^*_{\mathcal{O}'}(\underline{C})\dagger)\otimes_{\mathbb{Z}}\mathbb{Q})$, $0 \leq h \leq 2d$.

Since $\underline{C}$ is affine, by [3], II.3.1.1, pg. 190, we have that these latter groups are isomorphic to the cohomology of the global sections cochain complex, call it $C^*\dagger$, of $(\Gamma^*_{\mathcal{O}'}(\underline{C})\dagger) \otimes_{\mathbb{Z}} \mathbb{Q}$ over $\underline{C}$,

(4') $H_h^c(C,K') \approx H^{2d-h}(C^*\dagger)$, all integers h.

The group on the right side of equation (4) is the (2d-h)'th lifted p-adic cohomology group of C ([6]) and under the isomorphism (4) the endomorphism f_h induced by the Frobenius of C on the left side of (4) corresponds to $p^{rd}(f^{2d-h})^{-1}$, where f^{2d-h} is the endomorphism of the right side of equation (4), induced by the Frobenius mapping. (The key diagram is:

$$\begin{array}{ccccccc}
H^{2d-h}(C) & \xrightarrow{proj^{2d-h}} & H^{2d-h}(C^{\times}{}_k, A) & \xrightarrow[\text{with } u_{\Gamma_2}]{\text{cupping}} & H^{4d-h+2}(C^{\times}{}_k, A, C^{\times}{}_k, A-\Gamma_2) & \xleftarrow[\approx]{\text{excision}} & H^{4d-h+2}(\bar{X}^{\times}{}_k, A, \bar{X}^{\times}{}_k, A-\Gamma_2) \\
\Big\downarrow f_C^{2d-h} & & \Big\downarrow (f_{C^{\times}{}_k}, id_A)^{2d-h} & & \Big\downarrow (f_{C^{\times}{}_k}, id_A)^{4d-h+2} & & \Big\downarrow (f_{\bar{X}^{\times}{}_k}, A)^{4d-h+2} \\
H^{2d-h}(C) & \xrightarrow{proj^{2d-h}} & H^{2d-h}(C^{\times}{}_k, A) & \xrightarrow[\text{with } u_{\Gamma_{f_0}}]{\text{cupping}} & H^{4d-h+2}(C^{\times}{}_k, A, C^{\times}{}_k, A-\Gamma_{f_0}) & \xleftarrow[\approx]{\text{excision}} & H^{4d-h+2}(\bar{X}^{\times}{}_k, A, \bar{X}^{\times}{}_k, A-\Gamma_{f_0})
\end{array}$$

$$\begin{array}{ccccccc}
\xrightarrow{\text{restriction}} & H^{4d-h+2}(\bar{X}^{\times}{}_k, (A, A-C)) & \xrightarrow[\text{of Chap. 1}]{\text{maps 10-14}} & H^{4d-h+2}(P^{\times}{}_k, (A, A-C)) & \xrightarrow[\text{of Chap. 1}]{\text{Cor. 4.1}} & H^{2d-h}(A, A-C) \\
& \Big\downarrow (f_{\bar{X}^{\times}{}_k}, A)^{4d-h+2} & & \Big\downarrow (f_{P^{\times}{}_k}, idA)^{4d-h+2} & & \Big\downarrow (id_A)^{2d-h} \\
\xrightarrow{\text{restriction}} & H^{4d-h+2}(\bar{X}^{\times}{}_k, (A, A-C) & \xrightarrow[\text{of Chap. 1}]{\text{maps 10-14}} & H^{4d-h+2}(P^{\times}{}_k, (A, A-C)) & \xrightarrow[\text{of Chap. 1}]{\text{Cor. 4.1}} & H^{2d-h}(A, A-C)
\end{array}$$

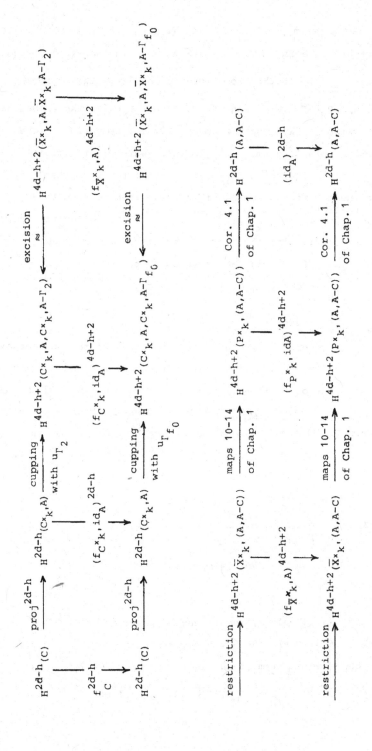

where $A=\mathrm{Spec}(k'[T_1,\ldots,T_{d+1},g^{-1}])$, $\bar{X}$=closure of C in $\mathbb{P}^{d+1}(k')$, Γ_ι is the graph of the inclusion $\iota: C \to A$, Γ_{f_0} is the graph of the composite, f_0, of the Frobenius f_C of C with the inclusion $\iota: C \to A$, $P = \mathbb{P}^d(k')$ and $f_{\bar{X}}$, resp: F_P , is the Frobenius endomorphism of $\bar{X}$, resp: P, over k'. The composite of the top row of this diagram is $H_h^c(id_C,K')$, which is the identity of $H_h^c(C,K')$. The composite of the bottom row is $f_h = H_h^c(f_C,K')$. The leftmost vertical map is f_C^{2d-h}, and the next-to-the-rightmost vertical map is $(f_P \underset{k'}{\times} id_A)^{4d-h+2}$. Since f_P^{2d} = (multiplication by p^{rd}), considering the maps of Corollary 4.1 of chapter 1, commutativity of the diagram tells us that

$$(f_C)_h \circ f_C^{2d-h} = (\text{multiplication by } p^{rd}),$$

as required.)

Therefore,

P_h = (reverse characteristic polynomial of the endomorphism $p^{rd}(f^{2d-h})^{-1}$ of $H^{2d-h}(C^*\dagger)$), all integers h. But, in his original proof of "rationality of the zeta function" [10], (we can choose $\underline{H} \in \mathcal{O}'[T_1,\ldots,T_{d+1}]$ so that its coefficients are Teichmuller representatives) Dwork proves that, if Q^{2d-h} is the reverse characteristic polynomial induced by $p^{rd}(f^{2d-h})^{-1}$ on C^{2d-h} (in the sense of characteristic polynomials of endomorphisms of p-adic Banach spaces), then

$$(5) \qquad Z_X(T) = \frac{Q^1 \cdot Q^3 \cdot \ldots \cdot Q^{2d-1}}{Q^0 \cdot Q^2 \cdot \ldots \cdot Q^{2d}} .$$

Since given a short exact sequence

$$0 \to B' \to B \to B'' \to 0$$

of p-adic Banach spaces, and an endomorphism of B that maps B' into itself, then the reverse characterstic polynomial P_B of the endomorphism of B is the product: $P_{B'} \cdot P_{B''}$ of the reverse characteristic polynomials $P_{B'}$, $P_{B''}$ of the endomorphisms induced on B' and on B", equations (4') and (5) prove equation (2) . (Since the alternating product of the reverse characteristic polynomials on the cohomology is the alternating product of the reverse characteristic polynomials on the cochains).

Case 2. General case. If U is a dense open subset of C, then we have (see Prop. 7 of chapter 1) the long exact sequence:

$$(6) \qquad \cdots \xrightarrow{\partial_{h+1}} H_h^c(C-U) \to H_h^c(C) \xrightarrow{\text{restriction}} H_h^c(U) \xrightarrow{\partial_h} H_{h-1}^c(C-U) \to \cdots$$

Since dimension of C-U < dim C, if we proceed by induction on d = dim C, then the assertion is true for C iff true for U. Therefore the inductive assumption implies that the verity of the assertion for C depends only on the birational equivalence class of C. Since every irreducible algebraic variety is birationally equivalent to a non-singular affine hypersurface as in Case 1, we are therefore through. Q.E.D.

(Remark: In the case that C is simple, proper and liftable, the above Theorem (which is then exactly the original Weil Lefschetz Theorem Conjecture [8] for p-adic cohomology) was first proved in [3] by a somewhat different method.)

Conjecture 3. (Generalized Weil's "Riemann Hypothesis" Conjecture). The hypotheses being as in Theorem 2, if we fix any complex embedding, $K' \subset \mathbb{C}$, then we can write

$$(7) \qquad P_h(T) = \prod_{i=1}^{\beta_h} (1-\alpha_{hi}T) \quad ,$$

where

$$(8) \qquad |\alpha_{hi}| = p^{rh'/2},\ 1 \leq i \leq \beta_h \ ,\ 0 \leq h \leq 2d, \text{ and where}$$

$h' = h'(i,h)$, dpending on h and i, is an integer between 0 and h, all integers i, h, $1 \leq i \leq \beta_h$, $0 \leq h \leq 2d$.

Remark 1. In the special case that the algebraic variety C is complete, non-singular and liftable, so that the functional equation ([3], III.2, pg. 254) holds, (for the non-liftable case, see [6]) so that the sequences: $(p^{dr}/\alpha_{h,1},\ldots,p^{dr}/\alpha_{h,\beta_h})$ and $(\alpha_{2d-h,1},\ldots,\alpha_{2d-h,\beta_h})$ are identical up to permutation, $0 \le h \le 2d$, then equations (7) and (8) would imply $|\alpha_{hi}| = p^{rh/2}$, $1 \le i \le \beta_h$, $0 \le h \le 2d$, the usual form of the original Weil "Riemann Hypothesis" Conjecture [8]. (Of course, if one deletes "liftable", then this Remark remains valid, see e.g. [6]).

2. Conversely, if Conjecture 3 is true in dimensions $\le d-1$, and if $d = \dim C$, then from the exact sequence (6) we see that Conjecture 3 is true for C <u>iff</u> Conjecture 3 is true for any algebraic variety birationally equivalent to C. Therefore, if Conjecture 3 were known, e.g., in the complete non-singular case, and if resolution of singularities in characteristic p were known, then Conjecture 3 would hold in general.

3. Therefore, Conjecture 3 is at the moment slightly stronger than the original Weil "Riemann Hypothesis" Conjecture, [8], (but would be essentially equivalent if one had some form of resolution of singularities in characteristic p).

CHAPTER 4

q-adic homology with compact supports and the zeta function

Much of the results of this paper, goes through to q-adic cohomology. The main difference is, of course, that q-adic cohomology doesn't give anything interesting for an algebraic family (Example 1 of chapter 3 doesn't go through) and there are no q-adic analogues of zeta matrices or of the zeta endomorphism. But the Frobenius map still acts on q-adic cohomology, and we obtain the following theorems.

Theorem 1. Let k be an algebraically closed field and let q be a rational prime $\neq$ char(k). Let C_k be the category having for objects all reduced algebraic varieties C over Spec(k) that are properly embeddable (see chapter 1) over k, and for maps all proper maps of varieties over k. Then for every integer $h \geq 0$, we have a covariant functor: $C \leadsto H_h^c(C,\hat{\mathbb{Z}}_q)$, the q-adic homology of C with compact supports, from the category C_k into the category of finitely generated $\hat{\mathbb{Z}}_q$-modules,[1] such that, whenever h is a positive integer and X is a non-singular, separated algebraic variety over k, of constant dimension n, containing C as a closed subvariety, then there is induced a canonical isomorphism

$$H^{2n-h}(X,X-C,\hat{\mathbb{Z}}_q) \approx H_h^c(C,\hat{\mathbb{Z}}_q)$$

all integers h, where the cohomology groups are the q-adic cohomology groups as defined in [9].

Sketch of Proof: One proceeds as in the proofs of Lemmae 1 and 2, and of Propositions 3 and 5 and Theorem 6 of chapter 1 to define the indicated functors. The proof that the groups are finitely

[1] (One first proves the corresponding theorem for coefficients in $\mathbb{Z}/q\mathbb{Z}$, and then uses Theorem 1, Chapter V of [5].)

generated over $\hat{\mathbb{Z}}_q$ proceeds exactly as in chapter 2.

Remark: The proof of Theorem 1 is slightly easier than its p-adic analogue (Theorem 6 of chapter 1). The reason is that certain maps can be constructed for q-adic cohomology, using the combinatorial definition given in [9], Chapter I, which cannot be constructed directly for p-adic, which allows a simplification in the proofs of e.g. the q-adic analogues of Lemmae 1 and 2 and Proposition 5 of chapter 1. (However, the proof of Theorem 4 above can be given exactly as in the p-adic case, if one wishes).

Theorem 2. (q-Adic Generalized Weil's Lefschetz Theorem). Let k' be a finite field and let C be an algebraic variety over k' that is properly embeddable over k'. Let $\overline{k}$ be the algebraic closure of k' and let $C_{\overline{k}} = C \times_{k'} \overline{k}$. Then the zeta function $Z_C(T)$ can be written as an alternating product, as in equation (2) of chapter 3, where $d = \dim C$, and where, as in equation (3) of chapter 3, $P_h(T)$ is the reverse characteristic polynomial of the endomorphism of the β_h-dimensional $\hat{\mathbb{Q}}_q$-vector space, the h'th homology group with compact supports $H_h^c(C_{\overline{k}}, \hat{\mathbb{Z}}_q) \otimes_{\hat{\mathbb{Z}}_q} \hat{\mathbb{Q}}_q$ of $C_{\overline{k}}$, induced by the Frobenius map of C over k'.

The proof is entirely similar to that of Theorem 2 of chapter 3. Finally,

Theorem 3. (q-Adic Generalized Weil's "Riemann Hypothesis" Theorem). The hypotheses being as in Theorem 2, if we fix any complex embedding: $\hat{\mathbb{Q}}_q \subset \mathbb{C}$, then, equations (7) and (8) of chapter 3 hold.

Remark: Notice that Theorem 6 is a theorem, rather than a conjecture, as is the case for the p-adic analogue (Conjecture 3 of chapter 3).

Proof: As in Remark 2 following Conjecture 3 of chapter 3, we see that to prove Theorem 6, if one proceeds by induction on $d = \dim C$, it suffices to prove the result for a variety birationally equivalent to C. Therefore one reduces to the case in which C is an affine non-singular hypersurface; and then the result follows from theorems in [11].

Remark: Analogues of the results of this paper can also be established for p-adic cohomolgy using the bounded Witt vectors ([7]).

CHAPTER 5

Lifted p-Adic Homology With Compact Supports on Affines. p-Adic "Riemann Hypothesis" in Special Cases.

In this chapter we study vanishing of the lower dimensional lifted p-adic homology groups with compact supports, when evaluated at an affine. Then we prove the p-adic "Riemann Hypothesis" for projective, non-singular, liftable varieties over finite fields.

Theorem 1. Let $\mathcal{O}$ be a complete discrete valuation ring with residue class field k and quotient field K, let $\underline{A}$ be a an $\mathcal{O}$-algebra, let $\underline{B}$ be a simple $\underline{A}$-algebra of finite presentation, and let $B = \underline{B} \otimes_{\mathcal{O}} k$. Suppose that all of the fibers of $\mathrm{Spec}(B)$ over $\mathrm{Spec}(A)$ are of the same fixed dimension N. Let C be a reduced closed subscheme of $\mathrm{Spec}(B_{red})$, such that we have an integer $d \geq 0$ such that the ideal of C on $\mathrm{Spec}(B_{red})$ is the radicle of an ideal that is generated by d elements.

Assume also the technical hypothesis, that either the ring $\underline{A}$ is normal, or else that $\underline{A}$ can be represented as a quotient of a polynomial ring $\underline{P} = \mathcal{O}[(T_i)_{i \in I}]$ over $\mathcal{O}$ such that there exists $\underline{B}_{\underline{P}}$ simple of finite presentation over $\underline{P}$ such that $(\underline{B}_{\underline{P}}) \otimes_{\underline{P}} \underline{A} \approx \underline{B}$ as $\underline{A}$-algebras.

Let $H_h^c(C, \underline{A}\dagger \otimes_{\mathbb{Z}} \mathbb{Q})$ be the lifted p-adic homology of C with compact supports, with coefficients in $(\underline{A}\dagger) \otimes_{\mathbb{Z}} \mathbb{Q}$, as defined in Theorem 6 of chapter 1. Then

$$H_h^c(C, \underline{A}\dagger \otimes_{\mathbb{Z}} \mathbb{Q}) = 0, \quad \text{all integers } h \leq N - d - 1.$$

Proof: First, notice that, since C is closed in $\mathrm{Spec}(B_{red})$, C is affine. Since also the ideal of C on B_{red} is the radicle of a finitely generated ideal, we have that C is properly embeddable over $\underline{A}$ (taking $\underline{X} = \mathrm{Spec}(\underline{B})$ proves this); and in the case $\underline{A}$ is not normal, C is polynomially properly embeddable over A (take $X = \mathrm{Spec}(\underline{B}_{\underline{P}})$). Therefore, in all cases, $H_h^c(C, \underline{A}\dagger \otimes_{\mathbb{Z}} \mathbb{Q})$, as defined in Theorem 6 of chapter 1, makes sense, all integers h. The proof reduces immediately to the case in which $\underline{A}$ is normal, which we assume.

The proof is by induction on d.

Case I. $d = 0$. Then $C = \mathrm{Spec}(B_{red})$. Then a simple, separated lifting of finite presentation of C over $\underline{A}$, having all fibers over Spec(A) of the same dimension, is $\underline{C} = \mathrm{Spec}(\underline{B})$. Therefore in this case if $\overline{\underline{X}}$ is any scheme proper over $\underline{A}$ containing $\underline{X} = \underline{C}$ as an open subscheme, then $(C, \overline{\underline{X}}) \in C_{Q\underline{A}}$, and

$$H_h^c(C, \underline{A}\dagger \otimes_{\mathbb{Z}} \mathbb{Q}) = H^{2N-h}(C, \Gamma_{\underline{A}}^*(\underline{C})\dagger \otimes_{\mathbb{Z}} \mathbb{Q}),$$

all integers h. But clearly $\Gamma_{\underline{A}}^p(\underline{C})\dagger = 0$ for $p \geq N + 1$ (since $\underline{B}$ is simple over $\underline{A}$, and the fibers of Spec(B) over Spec(A) are all of dimension N). Also, since $(C, \mathcal{O}_{\underline{C}}|C)$ is an affine $\mathcal{O}$-space, $H^q(C, \Gamma_{\underline{A}}^p(\underline{C})\dagger \otimes_{\mathbb{Z}} \mathbb{Q}) = 0$ for $q \geq 1$ all integers $p \geq 0$ (by [3], Chapter II, §3, Theorem 1, pg. 174). Therefore, in the first spectral sequence of hypercohomology ([3], Chapter I), which is the form

$$E_1^{p,q} = H^q(C, (\Gamma_{\underline{A}}^p(\underline{C})\dagger) \otimes_{\mathbb{Z}} \mathbb{Q}) \Rightarrow H^n(C, (\Gamma_{\underline{A}}^*(\underline{C})\dagger) \otimes_{\mathbb{Z}} \mathbb{Q}),$$

we have that $E_1^{p,q} = 0$ unless $q = 0$ and $0 \leq p \leq N$. Therefore, the groups in the abutment vanish for $n \geq N+1$. This proves Case I.

Case II. $d > 0$. We assume the assertion has been established for $d-1$; to establish it for d.

Let $f_1, \ldots, f_d$ be d elements of the ideal I_C of C on $\mathrm{Spec}(B_{red})$ such that the radicle of the ideal generated by $f_1, \ldots, f_d$ is all of I_C. Let C_{d-1} be the closed subset: $(f_1 = \ldots = f_{d-1} = 0)$ of $\mathrm{Spec}(B_{red})$, endowed with its induced reduced structure. Then C_{d-1} (respectively: $C_{d-1} - C_d$) is a reduced closed subscheme of $\mathrm{Spec}(B_{red})$ (respectively: of $\mathrm{Spec}((B_{f_d})_{red})$, and is such that the ideal of C_{d-1} (respectively: of $C_{d-1} - C_d$) on $\mathrm{Spec}(B_{red})$ (respectively: on $\mathrm{Spec}((B_{f_d})_{red})$) is the radicle of an ideal generated by $d-1$ functions, namely the functions $f_1, \ldots, f_{d-1}$. Let $\underline{f}_d$ be any element of $\underline{B}$ that maps into $f_d \in B_{red}$. Then by the inductive assumption, applied to the simple $\underline{A}$-algebra $\underline{B}$ (respectively: $\underline{B}_{\underline{f}_d}$) of finite presentation over $\underline{A}$, such that the fibers of $(\underline{B} \otimes_{\mathcal{O}} k)_{red}$ (respectively: of $(\underline{B}_{\underline{f}_d} \otimes_{\mathcal{O}} k)_{red}$) over A_{red} are all of the same dimension N, and to the reduced closed subset C_{d-1} (respectively: $C_{d-1} - C_d$) of $\mathrm{Spec}(B_{red})$ (respectively: of $\mathrm{Spec}((B_{f_d})_{red})$), we have, ($d-1$ replacing d), that

(1) $$H_h^c(C_{d-1}, (\underline{A}\dagger) \otimes_{\mathbb{Z}} \mathbb{Q}) = 0, \quad \text{all integers } h \leq N-d,$$

(respectively: that

(2) $$H_h^c(C_{d-1} - C_d, (\underline{A}\dagger) \otimes_{\mathbb{Z}} \mathbb{Q}) = 0, \quad \text{all integers } h \_ N-d).$$

We have the long exact sequence for lifted p-adic homology with compact supports:

$$(3) \quad \ldots \to H^c_{h+1}(C_{d-1} - C_d) \xrightarrow{\partial_{h+1}} H^c_h(C_d) \xrightarrow{H^c_h(\iota)} H^c_h(C_{d-1})$$
$$\xrightarrow{\text{restriction}} H^c_h(C_{d-1} - C_d) \xrightarrow{\partial_h} \ldots,$$

where the coefficients are in $\underline{A}^{\dagger} \otimes_{\mathbb{Z}} \mathbb{Q}$. (Chapter 1, Proposition 7, with $D = C_d$, $C = C_{d-1}$, $U = C_{d-1} - C_d$, and $\iota: C_d \to C_{d-1}$ the inclusion).

Substituting equations (1) and (2) into the long exact sequence (3) completes the induction. Q.E.D.

Example 1. Let $\mathcal{O}$ be a c.d.v.r. with residue class field k, let $\underline{A}$ be an $\mathcal{O}$-algebra, let $A = \underline{A} \otimes_{\mathcal{O}} k$, and let C be a reduced affine scheme over $\mathrm{Spec}(A_{red})$. Let n be the supremum of the dimensions of fibers of C over $\mathrm{Spec}(A_{red})$. Suppose that C is a (week) set - theoretic complete intersection over $\mathrm{Spec}(A_{red})$ - that is, suppose that there exists $\underline{B}$ simple of finite presentation over $\underline{A}$, such that the fibers of $\mathrm{Spec}(B_{red})$ over $\mathrm{Spec}(A_{red})$ are all of the same dimension N, where $B = \underline{B} \otimes_{\mathcal{O}} k$, and such that C is isomorphic to a reduced closed subscheme of $\mathrm{Spec}(B_{red})$, in such a way that the ideal of C on $\mathrm{Spec}(B_{red})$ is the radicle of an ideal generated by exactly $N - n$ functions.

(Notice that the definition of "(week) set - theoretic complete intersection" that we have given here is indeed less restrictive than the usual definition. In the usual definition of "(ordinary) set - theoretic complete intersection" over $\mathrm{Spec}(A_{red})$, one would insist, in addition, that $\underline{B} = \underline{A}[T_1, \ldots, T_N]$.) Then

Corollary 1.1. Let $\mathcal{O}$ be a c.d.v.r., let $\underline{A}$ be an $\mathcal{O}$-algebra and let C be a (week) set - theoretic complete intersection over $\mathrm{Spec}(A_{red})$. Then

(4) $$H^c_h(C, \underline{A}\dagger \otimes_{\mathbb{Z}} \mathbb{Q}) = 0, \quad \text{all integers} \quad h \leq n-1,$$

where n is the maximum dimension of all fibers of C over $\mathrm{Spec}(A_{red})$.

<u>Proof</u>: We can take $d = N - n$ in Theorem 1. Then by Theorem 1 the indicated groups vanish for $h \leq N - d - 1 = N - (N-n) - 1 = n - 1$. Q.E.D.

<u>Example 2</u>. Let $\underline{A} = \mathcal{O}'$, a discrete valuation ring such that $M_{\mathcal{O}'} \cap \mathcal{O} = M_{\mathcal{O}}$ and let C be an affine absolutely non-singular algebraic variety over the field $k' = \mathbb{k}(\mathcal{O}')$. Then <u>usually</u> C is a (week) set-theoretic complete intersection over k'. (E.g., this is the case if <u>either</u> C admits a flat lifting over $\mathcal{O}'$ <u>or</u> if C is an (ordinary) set-theoretic complete intersection over the field k').

<u>If this is the case</u>, then by Example 1, for the lifted p-adic homology with compact supports of C, we have

$$H^c_h(C, K') = 0 \quad \text{for} \quad h \leq n-1,$$

where $K' = (\mathcal{O}')\dagger \otimes_{\mathbb{Z}} \mathbb{Q}$, and n is the dimension of the affine algebraic variety C.

<u>Example 3</u>. Let $\underline{A}$ and D be as in the last Example of chapter 1. Then $C = D$ obeys the hypotheses of Theorem 1, with $\underline{B} = \underline{A}$, $N = 0$ and $d = n$. Therefore by Theorem 1 $H^c_h(C, (\underline{A}\dagger) \otimes_{\mathbb{Z}} \mathbb{Q}) = 0$ for $h \leq -n-1$, which is indeed easily seen directly. Note that as we have observed in the Example in chapter 1, $H^c_{-n}(C, (\underline{A}\dagger) \otimes_{\mathbb{Z}} \mathbb{Q}) \neq 0$, so this result is "best possible" in this case.

Note also that, in Example 3 above, the maximum dimension of fibers of C over $\mathrm{Spec}(A_{red})$ is zero, but that equation (4) of Example 1 of course fails very strongly if one takes $n = 0$ (since the (-n)'th group doesn't vanish). Therefore, in this example, C is indeed very far from being a "(week) set-theoretic complete intersection" over $\mathrm{Spec}(A_{red})$ in the sense of Example 1.

The next result in this chapter is somewhat special. Quite possibly a slightly longer study (than I have made at the moment) might yield more.

Proposition 2. Let X be a projective, non-singular liftable algebraic variety over a finite field k. Then the p-adic "Riemann hypothesis" (Conjecture 3 of chapter 3; see [8]) is true for X.

Proof: We prove the result using our p-adic cohomology for complete absolutely non-singular, liftable algebraic varieties over fields, as defined in [3], chapter III, §1, Theorem 8, pg. 250. This functor, evaluated on X as in the hypotheses of this Proposition, is the direct product of the functor evaluated on the connected components of X. Therefore we can assume that X is connected.

The proof is by induction on $n = \dim X$. The Proposition is trivial for $n = 0$. Suppose that the Proposition is known in dimension $< n$, where $n > 0$. To prove it in dimension n. Let X be connected, non-singular, complete and liftable of dimension n over k.

Let $\underline{X}$ be a simple, projective lifting of X over a complete discrete valuation ring $\mathcal{O}$ of mixed characteristic having k for residue class field. Let K be the quotient field of $\mathcal{O}$. Fix a complex embedding: $K \subset \mathbb{C}$.

Let $\underline{D}$ be a generic hyperplane section of $\underline{X}$ and let $D = \underline{D} \times_{\mathcal{O}} k$. Then D obeys all the hypotheses of the Proposition, and is of dimension $n-1$. Therefore, by the inductive assumption, we know that

(5) For every eigenvalue α, of the endomorphism $H^h(f_D, K)$ of $H^h(D,K)$ induced by the Frobenius endomorphism

f_D of the algebraic variety D over the finite field k, on the lifted p-adic cohomology $H^h(D,K)$ (as defined in [3], Chapter III, §1, Theorem 8, pg. 250), we have that $|\alpha| = p^{rh/2}$ (where $\#(k) = p^r$ and p is the characteristic of k). Let $D_K = \underline{D} \underset{0}{\times} K$, $X_K = \underline{X} \underset{0}{\times} K$. Then taking $\underline{D}$ for $\underline{X}$, $\underline{X}$ for $\underline{Y}$, the inclusion $\iota: D \to X$ for f and the inclusion $\underline{\iota}: \underline{D} \to \underline{X}$ for $\underline{f}$ in equation (2) of Theorem 8 of [3], Chapter III, §1, pg. 250, we have the commutative diagram:

$$(6) \qquad \begin{array}{ccc} H^*(X,K) & \xrightarrow{H^*(\iota,K)} & H^*(D,K) \\ \| & \iota_K^* & \| \\ H^*(X_K,\Gamma_K^*) & \longrightarrow & H^*(D_K,\Gamma_K^*), \end{array}$$

where $\iota_K = \underline{\iota} \underset{0}{\times} K$. Moreover, if $(X_{\mathbb{C}})_{top}$, resp: $(D_{\mathbb{C}})_{top}$, denotes the set of points rational over $\mathbb{C}$ of the projective, non-singular complex algebraic variety $X_{\mathbb{C}} = X_K \underset{K}{\times} \mathbb{C}$, respectively: $D_{\mathbb{C}} = D_K \underset{K}{\times} \mathbb{C}$, together with the classical topology, then by Theorem 1, conclusion (4), of [3], Chapter III, §1, pps. 238-239, we have that, for ι_K^* in the diagram (6) above, that, under the isomorphisms (4) of [3], Chapter III, §1, Theorem 1, pps. 238-239, we have that

(7) $\quad \iota_K^* \underset{K}{\otimes} \mathbb{C}$ is identified with $H^*((\iota_{\mathbb{C}})_{top},\mathbb{C})$, in which the rightmost map is the mapping induced on classical complex cohomology: $H^*((X_{\mathbb{C}})_{top},\mathbb{C}) \to H^*((D_{\mathbb{C}})_{top},\mathbb{C})$ by the set-theoretic inclusion

$$(\iota_{\mathbb{C}})_{top}\colon (D_{\mathbb{C}})_{top} \hookrightarrow (X_{\mathbb{C}})_{top}.$$

But $D_{\mathbb{C}}$ is a generic hyperplane section of the projective, non-singular complex algebraic variety $X_{\mathbb{C}}$. Therefore, by the well-known theorem of Solomon Lefschetz, about generic hyperplane sections of projective varieties over $\mathbb{C}$, we have that the mappings on the right side of equation (7) are

$$\begin{cases} \text{isomorphisms in dimension } \le n-2; \\ \text{a monomorphism in dimension } n-1. \end{cases}$$

From the commutative diagram (6), and the identification (7), if follows, likewise, that

(8) $$H^h(\iota,K) \text{ is } \begin{cases} \text{an isomorphism for } h \le n-2, \\ \text{a monomorphism for } h = n-1. \end{cases}$$

(For an alternative proof, and a more general such observation, see Remarks 1 and 2 below).

But then for $h \le n-1$, $H^h(X,K)$ is a vector subspace of the K-vector space $H^h(D,K)$, and is such that the K-endomorphism $H^h(f_X,K)$ of $H^h(X,K)$, induced by the Frobenius, f_X, of X over k, is induced by the endomorphism $H^h(f_D,K)$ of $H^h(D,K)$, where f_D is the Frobenius of D over k. By the inductive assumption we have equation (5) for D. Therefore,

(9) For every eigenvalue α of $H^h(f_X,K)$ we have that $|\alpha| = p^{rh/2}$, all integers $h \le n-1$.

The cohomology groups: $H^*(X,K)$ come equipped with cup products ([3], Chapter I, §7), which are preserved by maps over k ([3], Chapter III, §1, Theorem 6, pg. 247), and in particular by $H^*(f_X,K)$. And $H^*(X,K)$ obeys Poincaré duality ([3], Chapter III, §1, Theorem 1, pps. 238-239). We therefore have the functional equation ([3], Chapter III, §2, Proposition 2, pps. 253-254):

(10) If $(\alpha_{h,1},\ldots,\alpha_{h,\beta_h})$ are the eigenvalues with multiplicities of $H^h(f_X,K)$, where $\beta_h = \dim_K H^h(X,K)$, then for every integer h, $0 \le h \le n$, there exists a permutation π_h of $\{1,\ldots,\beta_h\}$ such that the sequences: $(p^{rn}/\alpha_{h,1},\ldots,p^{rn}/\alpha_{h,\beta_h})$ and $(\alpha_{h,\pi_h(1)},\ldots,\alpha_{h,\pi_h(\beta_h)})$ coincide. Equations (9) and (10) imply

(11) For every eigenvalue α of $H^h(f_X,K)$ we have that $|\alpha| = p^{rh/2}$, all integers $h \ge n+1$.

Therefore, by equations (9) and (11), for every integer $h \neq n$, we have that every eigenvalue of $H^h(f_X, K)$ has absolute value $p^{rh/2}$. It remains to prove the corresponding assertion for $h = n$.

But $\beta_h = \dim_K H^h(X,K)$ = the h'th Betti number, in the sense of combinatorial topology, of the complex algebraic variety $(X_{\mathbb{C}})_{top}$. (The latter equality by [3], Chapter III, §1, Theorem 1, pps. 238-239). If P_h is the reverse characteristic polynomial of $H^h(f_X,K)$, then P_h is of degree β_h, all integers h, and we have seen, in [3], Chapter III, §2, pg. 253, that the zeta function $Z_X(T)$ of the algebraic variety X over the finite field k can be written as the alternating product:

$$(12) \qquad Z_X(T) = \frac{P_1(T) \cdot P_3(T) \ldots P_{2n-1}(T)}{P_0(T) \cdot P_2(T) \ldots P_{2n}(T)} \quad .$$

Let q be any rational prime $\neq p$ and let Q_h be ther reverse characteristic polynomial induced by the Frobenius over the field k of the h'th q-adic cohomology group $H^h(X_{\bar{k}}, \hat{\mathbb{Q}}_q)$, of the algebraic variety over $\bar{k}$, $X_{\bar{k}} = X \times_k \bar{k}$ with coefficients in $\hat{\mathbb{Q}}_q$, where $\bar{k}$ is the algebraic closure of the field k, $0 \leq h \leq 2n$. Then we have seen in [9] that the dimension of the $\hat{\mathbb{Q}}_q$ - vector space $H^h(X_{\bar{k}}, \hat{\mathbb{Q}}_q)$ is β_h, the h'th Betti number in the sense of combinatorial topology of $(X_{\mathbb{C}})_{top}$, and that

$$(13) \qquad Z_X(T) = \frac{Q_1(T) \cdot Q_3(T) \cdots Q_{2n-1}(T)}{Q_0(T) \cdot Q_2(T) \cdots Q_{2n}(T)}$$

Therefore $\deg(Q_h) = \beta_h = \deg(P_h)$, all integers h. Fix a complex embedding, $\hat{\mathbb{Q}}_q \subset \mathbb{C}$. Then in [11], it is shown that the inverse roots of Q_h have absolute value $p^{rh/2}$, all integers h, $0 \leq h \leq 2n$. By equations (9) and (11), we have likewise that the inverse roots of P_h have absolute value $p^{rh/2}$, for all integers h, $0 \leq h \leq 2n$, $h \neq n$. Therefore, by equations (12) and (13) (and unique factorization in $\mathbb{C}[T]$) it follows that P_h must divide Q_h, for all integers h, $0 \leq h \leq 2n$, $h \neq n$. But since $\deg P_h = \deg Q_h$, and both P_h and Q_h have constant term 1 $(0 \leq h \leq 2n)$, this implies that $P_h = Q_h$, all integers $0 \leq h \leq 2n$, $h \neq n$. But then from equations (12) and (13) it follows that $P_n = Q_n$. Therefore the inverse roots of P_n have absolute value $p^{rn/2}$, completing the proof. Q.E.D.

Remarks 1. Another way of proving equation (8) of Proposition 2 is as follows. Let $U = X - D$. Then U obeys the hypotheses of Example 2 above, and therefore

$$H_h^c(U,K) = 0, \quad h \leq n-1.$$

From the long exact sequence of lifted p-adic homology with compact supports:

$$\cdots \to H_{h+1}^c(U,K) \xrightarrow{\partial_{h+1}} H_h^c(D,K) \xrightarrow{H_h^c(\iota,K)} H_h^c(X,K) \xrightarrow{\text{restriction}} H_h^c(U,K) \xrightarrow{\partial_h} \cdots,$$

see Chapter 1, Proposition 7, if follows that

$$(14) \quad H_h^c(\iota,K): \ H_h^c(D,K) \to H_h^c(X,K) \ \text{ is } \begin{cases} \text{an isomorphism}, & h \leq n-2, \\ \text{an epimorphism}, & h = n-1. \end{cases}$$

But $H_h^c(D,K) = H^{2n-h-2}(D,K)$, and $H_h^c(X,K) = H^{2n-h}(X,K)$, all integers h, and, with these identifications, $H_h^c(\iota,K)$ corresponds, under Poincaré duality, to the dual of the mapping $H^{2n-h-2}(\iota,K)$. Therefore equation (14) implies (and in fact is equivalent) to equation (8).

2. The proof given in Remark 1 above of equation (14) generalizes to the case in which X is an arbitrary projective, absolutely non-singular algebraic variety over a field k, <u>liftable</u> <u>or</u> <u>not</u>, and D is a generic hyperplane section of X, assuming that ` - D is a (week) set-theoretic complete inter-

section, as defined in Example 1. (And similarly, one proves the analogue of equation (8), for the lifted p-adic cohomology, (as defined in [6]), for such an X).

Bibliography

1. Lubkin S. and Kato, G. "Second Leray Spectral Sequence of Relative Hypercohomology", Proc. Natl. Acad. Sci. USA, in press.
2. Lubkin, S. "p-Adic Cohomology Theorems," Annals of Mathematics, in press.
3. Lubkin, S. "A p-Adic Proof of Weil's Conjectures" (1968), Annals of Math. 87, Nos. 1-2, 105-255.
4. Nagata, M. "A generalization of the imbedding problem of an abstract variety in a complete variety," J. Math. Kyoto University 3 (1963), 89-102.
5. Lubkin S. "Cohomology of Completions", Notas de Mat., N. Holland Pub. Co., in press.
6. Lubkin, S. "Lifted p-Adic Cohomology," Notas de Mat. N. Holland Pub. Co., to appear.
7. Lubkin, S. "Generalization of p-Adic Cohomology; Bounded Witt Vectors. A Canonical Lifting of a Variety in Characteristic $p \neq 0$ Back to Characteristic Zero", Compositio Math. 34, Fasc. 3, (1977), pps. 225-277, Noordhuff Intnl. Pub.
8. Weil, A. "Number of Solutions of Equations Over Finite Fields," Amer. Math. Soc. Bull., 55 (1949), 497-508.
9. Lubkin, S. "On a Conjecture of André Weil," Amer. J. Math., 89 (1967), 443-548.
10. Dwork, B. "On the rationality of the zeta function of an algebraic variety", Amer. J. Math. 83 (1960), 631-648.

11. Deligne, P. "La Conjecture de Weil. I", Inst. Hautes Ét. Sci. Publ. Math. No. 43 (1974), 273-307.

Problem. Let A be a commutative ring with identity that contains a field. Let X be a scheme simple and proper of finite presentation over A. Let $\Gamma_A^p(X)$ be the sheaf of germs of p-differential forms of X over A $(= \wedge_{\mathcal{O}_X}^p \Gamma_A^1(X))$, and let $\Gamma_A^*(X)$ be the corresponding cochain complex of sheaves of A-modules over X. Then the problem is: To prove that $H^h(X,\Gamma_A^*)$ is flat of finite presentation (= projective and finitely generated) as A-module, for each integer $h \geq 0$.

Remarks 1. The answer to the problem is "yes" if A contains a field of characteristic zero (see, e.g., my book "Lifted p-Adic Cohomology", in press).

2. Consider the special case of the Problem, in which $A = \mathcal{O}/M_{\mathcal{O}}^n$, where $\mathcal{O}$ is a complete discrete valuation ring that contains a field of characteristic p (p a fixed rational prime), $M_{\mathcal{O}}$ is the maximal ideal of $\mathcal{O}$, and n is any positive integer. I can show that the Problem is true for all rings A containing a field of characteristic p (for the fixed prime p) iff the Problem is true for all such rings A of the form $A = \mathcal{O}/M_{\mathcal{O}}^n$, $n \geq 1$.

3. If the ring A is a local ring that contains a field of characteristic $p > 0$, and if the maximal ideal I of A is such that $I^p = 0$, then the answer to the Problem is "yes". (This is not obvious. It is proved in my book, "Lifted p-Adic Cohomology", to appear).

4. In fact, if A and I are as in Remark 3 above, and if $B = A/I$, and if $\mathcal{C}$ is the category of all preschemes Y

simple of finite presentation over B, (notice that such a Y is liftable over A, and in fact admits a canonical lifting over A. To prove this, one can reduce to the case in which A is Noetherian. Then A contains its residue class field B, and $Y \times_B A$ is such a lifting), the maps in $\mathcal{C}$ being maps of preschemes over B, (whether liftable or not,) then in my book, "Lifted p-Adic Cohomology" (to appear), I define a functor: $Y \rightsquigarrow H^h(Y,A)$, $h \geq 0$, from the category $\mathcal{C}$ into the category of graded A-algebras such that, for every $Y \in \mathcal{C}$, and any X simple of finite presentation over A such that $X \times_A B$ is B-isomorphic to Y, then every such B-isomorphism induces an isomorphism of graded A-algebras: $H^h(Y,A) \approx H^n(X,\Gamma_A^*(X))$, $h \geq 0$, (where the groups on the right side are the hypercohomology groups of X with coefficients in the cochain complex of sheaves $\Gamma_A^*(X)$.)

(If the hypothesis that "$I^p = 0$" is dropped, then the assertion in Remark 4 becomes false, unless the characteristic is zero (in which case it becomes true again)). The result quoted in Remark 3 above follows easily from Remark 4.

5. The analogue of the Problem, if one deletes the hypothesis that "A contains a field", it easily shown to be false. Counterexamples are easily given even if A is the ring $\hat{\mathbb{Z}}_p$ of p-adic integers.

Department of Mathematics
University of Rochester
Rochester, NY 14627, U.S.A.

On a Problem of Grothendieck

Alexander Lubotzky

1. The group $cl_A(G)$.

Let A be a non-trivial commutative ring with a unit, G a discrete group. By $Rep_A(G)$ we shall denote the category of all the representations of G on finitely generated modules over A ("A,G-modules"). Mod(A) will be the category of all the finitely generated modules over A. $F:Rep_A(G) \to Mod(A)$ is the forgetful functor. Define $cl_A(G)$ as the group of all the automorphisms of F which commute with the tensor product operation. That is to say: $\alpha \in cl_A(G)$ means that for every $X \in Ob(Rep_A(G))$ there is an A-automorphism α_X of FX, such that if $\varphi: X \to Y$ is a morphism between two objects in $Rep_A(G)$, then the following diagram is commutative:

$$\begin{array}{ccc} FX & \xrightarrow{F\varphi} & FY \\ {\scriptstyle \alpha_X}\downarrow & & \downarrow{\scriptstyle \alpha_Y} \\ FX & \xrightarrow{F\varphi} & FY \end{array}$$

The condition "commute with the tensor product operation" means that $\alpha_{X\otimes Y} = \alpha_X \otimes \alpha_Y$ for every $X,Y \in Ob(Rep_A(G))$.

From the definition , it is clear that to every $g \in G$ there corresponds $t(g) \in cl_A(G)$ defined by $t(g)_X = \rho_X(g)$ where ρ is the representation of G on X. t is a homomorphism whose kernel is the intersection of the kernels of all the representations of G on finitely generated modules over A. If ρ is a representation of G on a finitely generated module V, it has a canonical continuation $\tilde{\rho}: cl_A(G) \to Aut\ V$.

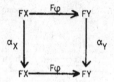

A group G is said to be <u>residually finite</u> if the intersection of all the subgroups of finite index in G is trivial. It is clear that if G is a residually finite group, then the homomorphism $t:G \to cl_A(G)$ is injective.

The group $cl_A(G)$ was introduced by Grothendieck in [2], in order to treat the following question: Let $\varphi:G' \to G$ be a homomorphism between two discrete groups, such that $\hat{\varphi}:\hat{G}' \to \hat{G}$ (the corresponding continuation between the pro-finite completions) is an isomorphism, When is φ itself an isomorphism? Grothendieck shows that if $\hat{\varphi}$ is an isomorphism then the categories $Rep_A(G')$ and $Rep_A(G)$ are canonically equivalent. Therefore, if there is a non-trivial ring A such that $cl_A(G') = G'$ and $cl_A(G) = G$, then $\varphi:G' \to G$ is an isomorphism.

Our interest in $cl_A(G)$ has an additional reason: Hochschild and Mostow, defined for a group G and a field K, another group $M_K(G)$ of the proper automorphisms of the algebra of the representative functions of G over K (see [3] for the definitions). Using $M_{\mathbb{C}}(G)$ they proved (following the results of Pontryagin, Tannaka, Harish-Chandra and others, for other classes of groups) a "duality theorem" for analytic Lie groups (i.e. a theorem about the possibility of reconstructing the group G from its category of representations and a method for its reconstruction). In [4], it is proven that $M_K(G)$ is canonically isomorphic to $cl_K(G)$. On the other hand, it is shown there that a straightforward application of $M_{\mathbb{C}}(G)$ for discrete groups does not yield satisfactory duality properties for these groups.

This leads us to consider the following question posed by Grothendieck ([2]) as a question about the "duality theory" of discrete groups: Is it true that $t:G \to cl_R(G)$ is an isomorphism for every finitely presented residually finite group, where R stands for the ring of rational integers or some other ring of algebraic integers?

Grothendieck showed that it is true in case G is an arithmetic group (e.g. SL(n,Z)) such that every subgroup of finite index in G is a congruence subgroup (e.g. $n \geq 3$, see below). In the following sections we shall give counter-examples to this question, using also the congruence subgroup problem.

For more information about $cl_A(G)$ and duality-theory the reader is referred to [4].

2. $cl_A(G)$ for a finite group.

We shall begin with a proposition stated without a proof in [2]. The method of proof gives the motivation for the definitions afterwards.

Proposition 1: Let A be a commutative ring with a unit 1, s.t. spec(A) is connected (i.e. $x \in A$ and $x^2 = x$ implies $x = 0$ or 1) and G a finite group. Then $t: G \to cl_A(G)$ is an isomorphism, i.e. $cl_A(G) \simeq G$.

Proof: Consider $B = A[G]$ the group-ring, as an A,G-module, by multiplication from the left by elements of G. Let $\alpha \in cl_A(G)$, and e be the identity element of G.

Write $\alpha_B(e) = \Sigma\, a_g g$ $(a_g \in A,\ g \in G)$. If V is any A,G-module and $v \in V$, then there is a unique A,G-homomorphism φ from B into V, determined by $\varphi(e) = v$. By the definition of $cl_A(G)$:

$$\alpha_V \circ \varphi = \varphi \circ \alpha_B.$$

In particular:

$$\alpha_V \circ \varphi(e) = \varphi \circ \alpha_B(e)$$

$$\alpha_V(v) = \varphi(\Sigma\, a_g(g \cdot e)) = \Sigma\, a_g g \cdot \varphi(e) = \Sigma\, a_g(g \cdot v).$$

This means that the action of α_B on $e \in B$ determines α.

On the other hand, the assumption that α commutes with the tensor product operation implies:

$$\sum_{i=1}^{n} a_{g_i} g_i \otimes \sum_{j=1}^{n} a_{g_j} g_j = \sum_{i=1}^{n} a_{g_i} (g_i \otimes g_i)$$

where $\{g_1, \ldots, g_n\}$ are the elements of G; that is

$$\sum a_{g_i} a_{g_j} (g_i \otimes g_j) = \sum a_{g_i} (g_i \otimes g_i).$$

As $\{g_i \otimes g_j\}_{i,j=1}^{n}$ is an independent set, we have: $a_{g_i} \cdot a_{g_j} = \delta_{ij} a_{g_i}$ $(1 \leq i,j \leq n)$, which implies that there exists an i_0 such that $a_{g_{i_0}}^2 = a_{g_{i_0}}$, and $a_{g_{i_0}} = 0$ for $i \neq i_0$. The only possibilities for such $a_{g_{i_0}}$ are 0 or 1. 0 is impossible, so $a_{g_{i_0}} = 1$. Therefore $\alpha_B(e) = g_{i_0}$, and $\alpha_V(v) = g_{i_0} \cdot v$ and so $\alpha = t(g_{i_0})$ and the proposition is proven.

3. $cl_A(G)$ and the pro-finite completion of G.

In this section we shall assume A is a Noetherian ring such that spec(A) is connected and A modulo every maximal ideal is finite. Let $\hat{G}$ be the pro-finite completion of G. As every continuous representation of $\hat{G}$ over A factors through a finite quotient of G, $Rep_A(\hat{G})$ (when we consider only continuous representations of $\hat{G}$) is a full sub-category of $Rep_A(G)$, so there is a map from $cl_A(G)$ into $cl_A(\hat{G}) = \hat{G}$.

It was noted by Grothendieck that this map is injective, since every representation-module V in $Rep_A(G)$ is determined by the action of G on the finite quotient modules $V/J \cdot V$ (J is a non-zero ideal in A) and the action of G on these finite modules factors through a finite quotient of G.

So one may identify $cl_A(G)$ as a subgroup of $\hat{G}$.

Let V be a finitely generated module over A, then $\hat{V} = \varprojlim V/J \cdot V$ (where J runs over the non-zero ideals of A) is a pro-finite module, whose group of automorphisms is a pro-finite group [8]. As Aut $V \subset$ Aut $\hat{V}$, a representation ρ of G on V has a unique continuation to a homomorphism $\hat{\rho}:\hat{G} \to$ Aut $\hat{V}$.

The following simple lemma gives a useful characterization of $cl_A(G)$ as a subgroup of $\hat{G}$.

Lemma 1. Let A be as above. Then

$$cl_A(G) = \{x \in \hat{G} \mid \hat{\rho}(x) \in \text{Aut } V \subset \text{Aut } \hat{V}, \text{ for every rep. } \rho \text{ on f.g. module } V\}.$$

Proof: If ρ is a representation of G on a module V, then $\tilde{\rho}:cl_A(G) \to$ Aut V satisfies $\tilde{\rho} \circ t = \rho$. On the other hand, $cl_A(G)$ is a subgroup of $\hat{G}$ and it is clear that $\hat{\rho}|_{cl_A(G)} = \tilde{\rho}$, so we get that the image of $cl_A(G)$ under $\hat{\rho}$ is in Aut V.

For the opposite inclusion: every element $x \in \hat{G}$, such that $\hat{\rho}(x) \in$ Aut V for every representation ρ of G, defines an automorphism of the functor $F:Rep_A(G) \to Mod(A)$. For topological reasons (x can be approximated arbitrarily close by elements of G), x also commutes with the tensor product operation, so $x \in cl_A(G)$. Q.E.D.

4. The Congruence Subgroup Problem (C.S.P.)

Let k be a number field, H an algebraic subgroup of GL_n, defined over k, S be a finite set of primes of k containing all Archimedean primes, and θ_S is the ring of S-integers, namely: $\theta_S = \{x \in k \mid v(x) \geq 0 \ \ \forall v \notin S\}$. Such a ring will be called a <u>number ring</u>. Denote $G = H(\theta_S)$. For every ideal J in θ_S, let

$$G_J = \{A \in G \mid A \equiv 1 (\text{mod } J)\}.$$

A subgroup of G which contains G_J for some non-zero ideal J will be called a congruence subgroup. Congruence subgroups are of finite index, and the congruence subgroup problem asks whether the converse is true, i.e. is every subgroup of finite index in G a congruence subgroup.

Define two topologies on G: the pro-finite topology (resp. the congruence topology) by taking the family of the subgroups of finite index (resp. the congruence subgroups) as a fundamental system of neighborhoods of the identity element of G. C.S.P. asks whether these two topologies are coincident.

Denote by $\hat{G}$ and $\tilde{G}$ the completions of G relative to the pro-finite topology and the congruence topology, respectively. As the first one is finer than the second one, there is an epimorphism $\pi:\hat{G} \to \tilde{G}$ and we have an exact sequence:

$$1 \to C(G) \to \hat{G} \to \tilde{G} \to 1.$$

(Note that G is a subgroup of $GL_n(\theta_S)$ and $\tilde{G}$ is a subgroup of $GL_n(\hat{\theta}_S)$.)

An affirmative solution to C.S.P. is equivalent to the vanishing of C(G).

C.S.P. was settled for many cases (although not for all!). See [6] for a complete description of the history of the problem. The main results are the following:

Let H be a simply-connected, absolutely-simple Chevalley group. Then whenever the k-rank of H is greater or equal to 2, C(G) is always finite; trivial if k has a real embedding; and isomorphic to μ_k (= the group of roots of unity in k) in the case k is purely imaginary. On the other hand, for SL_2: if k = Q or Q $(\sqrt{-d})$ $(1 \leq d \in Z)$ and $S = \{\infty\}$, then C(G) is an infinite group, while in the other cases C(G) is finite; isomorphic to μ_k if k is purely imaginary and trivial otherwise.

When C(G) is a finite group, we shall say that G has an <u>almost affirmative solution to C.S.P.</u>

5. <u>The stalk of the representations of</u> G.

Let A be a commutative ring with a unit. V is an object in $Rep_A(G)$. By $Rep_A^V(G)$ we shall denote the full subcategory generated by V, namely all the A,G-modules which are isomorphic to A,G-modules obtained from V by direct products, tensor products and taking quotient modules. Let (V_1,ρ_1), (V_2,ρ_2) be two representations of G over A. We shall say that ρ_1,ρ_2 are commensurable if there are A,G-submodules V_1', V_2' of finite index in V_1,V_2, respectively, and a subgroup of finite index L of G such that $(V_1',\rho_1|_L)$ is A,L-isomorphic with $(V_2',\rho_2|_L)$. The commensurability relation is an equivalence relation. The set of equivalence classes will be called the stalk of the representations of G over A. G will be said to have a cyclic stalk of representations over A if there is an A,G-module V such that every A,G-module is commensurable to an A,G-module in $\mathrm{Rep}_A^V(G)$. Such a V will be called a generator for the stalk of G over A.

Examples: 1) Let G be a finite group. Then every representation of G is commensurable to a trivial representation. So the stalk of G is cyclic, generated by the trivial representation of G on rank-one free module

2) Let G be a semi-simple, connected and simply-connected linear algebraic group over an algebraically-closed field K of characteristic zero. Since G has no subgroups of finite index, and modules over the field have no submodules of finite index, therefore, two representations are commensurable if and only if they are isomorphic. On the other hand, the stalk of the algebraic representations of G is cyclic, for it is well known that every faithful representation of G "generates" all the other representations.

Less trivial is the following example which we shall state as a proposition and outline the proof:

Proposition 2: Let H be a semi-simple connected and simply-connected linear algebraic group defined over a number field k; let θ_S be a number ring in k, G be $H(\theta_S)$ and assume G has an almost affirmative solution to the C.S.P. Then G has a cyclic stalk of representations over every number ring

Proof: By applying a method of Bass-Milnor-Serre in [1, §16], one can show that every representation of G is algebraic "up to a subgroup of finite index". A computation similar to that done by Serre ([7],§2.7) shows that every representation of G is commensurable to an algebraic representation. Using example 2 above we get our result.

Our interest in cyclic stalks comes from the following proposition:

Proposition 3: Let A be a Noetherian ring such that spec(A) is connected and A modulo every maximal ideal is finite. Let G be a discrete group whose stalk of representations over G is cyclic, generated by (V,ρ). Then

$$cl_A(G) = \{x\in\hat{G} \mid \hat{\rho}(x) \in \text{Aut } V \subset \text{Aut } \hat{V}\}.$$

The proof is technical and, therefore, we shall omit it here. The idea is, of course, that V generates $Rep_A(G)$ so every automorphism of $F:Rep_A(G) \to Mod(A)$ is determined by its action on V.

Now, we are in a position to prove the following theorems:

Theorem 1: Let $G = H(\theta_S)$ where H is a connected, simply-connected, semi-simple linear algebraic group defined over a number field k; where S is a finite set of primes in k, consisting of at least one prime $\nu(\neq\infty)$ such that H is isotropic over k_ν (= the ν-adic completion of k) and assume that G has an almost affirmative solution to C.S.P., then $cl_Z(G) = \hat{G}$.

Proof: It suffices to prove that every representation of G on a finitely generated module over Z factors through a finite quotient of G, for if so, then the trivial representation of G on a rank-one free module generates the stalk of the representations of G. By proposition 3 this implies $cl_Z(G) = \hat{G}$.

Assume the contrary that there is a representation of G with an infinite image. Then it is equal, up to a subgroup of finite index, to an algebraic representation of G. But it is impossible for G to have an algebraic representation over Z (see example 2 and proposition 2 above). Q.E.D.

The simplest examples of groups which satisfy all the assumptions of theorem 1, are $G_p = SL(n, Z(^1/_{p^\infty}))$, where $Z(^1/_{p^\infty})$ is the ring of all the rational numbers whose dominators are not divided by any prime other than p. So, $cl_Z(G_p) = \prod_{q \neq p} SL(n, \hat{Z}q)$, ($\hat{Z}q$ is the ring of q-adic integers).

Theorem 1 gives a counter-example to the problem of Grothendieck. On the other hand one may guess that when G is a "Lie group" over a number ring R (namely, G = H(R), where H is a linear algebraic group defined over k, the field of quotients of R) then $cl_R(G) = G$. The following theorem shows that this conjecture is also false:

Theorem 2: Let H be a connected, simply-connected, semi-simple algebraic group defined over a number field k. G = H(R) where $R = \theta_S$ is a number ring in k. Assume G has an almost affirmative solution to the congruence subgroup problem, i.e. C(G) is a finite group. Then $cl_R(G) = C(G)\cdot G$ (semi-direct product).

Moreover, if H has the property K-T ([6], p.111) then C(G) is an abelian group and $cl_R(G) = C(G)\times G$.

Proof: By proposition 1 we know that the stalk of the representations of G over R is cyclic. From proposition 2 we conclude that $cl_R(G)$ is the inverse limit of G = H(R) in Aut $\hat{V}$ (where V is a generator of the stalk). From the description of the C.S.P. in section 4, it follows that π^{-1} (Aut V) = $C(G)\cdot G$.

The second part of the theorem follows from considering the action of H(k) on C(G) and deducing that this action is trivial (see [6][5] for details). Q.E.D.

The simplest examples of groups which satisfy all the assumptions of theorem 2 but C(G) is a non-trivial finite group, are $G_n = SL(n, Z[\sqrt{-1}])$, $n \geq 3$. In this case C(G) is isomorphic to a cyclic group of order four, and by the theorem:

$$cl_{Z[\sqrt{-1}]}(G_n) = C(G_n) \times G_n$$

Note that if G = H(R) has a strictly positive solution to C.S.P. (i.e. C(G) = {1}), then $cl_R(G) = G$, a result which was proved by Grothendieck [2].

ACKNOWLEDGEMENT

The author wishes to express his gratitude to his thesis advisor, Professor H. Furstenberg for his encouragement and many helpful suggestions.

REFERENCES

1. H. Bass, J. Milnor, J.P. Serre, Solution of the congruence subgroup problem for SL(n) ($n \geq 3$) and SP(2n) ($n \geq 2$), Publ. Math. I.H.E.S., 33 (1967), 59-137.

2. A. Grothendieck, Representationes lineaires et compactification profinie des groupes discretes, Manuscripta Mathematica, Vol. 2 (1970), pp. 375-396.

3. G.P. Hochschild and G.D. Mostow, Representations and representative functions of Lie groups, Ann. of Math. Vol. 66 (1957), 495-542.

4. A. Lubotzky, Tannaka duality for discrete groups, (in preparation).

5. A. Lubotzky, Generalized congruence subgroup problem for discrete groups, (in preparation).

6. M.S. Raghunathan, On the congruence subgroup problem, Publ. Math. I.H.E.S., 46 (1976), 107-161.

7. J.P. Serre, Le problem de groupes de congruence pour SL_2, Ann. of Math Vol. 92 (1970), 489-527.

8. J. Smith, On products of profinite groups, Ill. J. Math. 13 (1969), 680-688.

Department of Mathematics and Computer Science
Bar-Ilan University
Ramat-Gan, Israel.

Faithfully representable analytic groups

by

Andy R. Magid
University of Oklahoma

In this paper, <u>analytic group</u> means a connected complex Lie group. If G is an analytic group and $p : G \to GL_n\mathbb{C}$, where $p(g) = [p_{ij}(g)]$, is an analytic representation, the functions $p_{ij} : G \to \mathbb{C}$ are called the coordinate functions of p. The set $R(G)$ of all coordinate functions of all analytic representations of G forms a complex algebra, under pointwise operations, called the <u>algebra of representative functions on G</u>. G acts on $R(G)$ in the following way: if $f \in R(G)$ and $x \in G$, $x \cdot f : G \to \mathbb{C}$ given by $(x \cdot f)(y) = f(yx)$ and $f \cdot x : G \to \mathbb{C}$ given by $(f \cdot x)(y) = f(xy)$ are in $R(G)$. For x in G, let $L_x : G \to G$ be given by $L_x(y) = xy$ and $R_x : G \to G$ be given by $R_x(y) = yx$. Then $x \cdot f = fR_x$ and $f \cdot x = fL_x$.

The algebra of representative functions on a group with a faithful representation has the following description, due to Hochschild and Mostow [3, Section 3]:

<u>Theorem</u>. Let G be an analytic group with a faithful representation. Then $R(G) = A[Q]$ where $Q = \exp(\mathrm{Hom}(G,\mathbb{C}))$ and

1) A is a finitely generated subalgebra of $R(G)$

2) $Ax = A$ for all x in G

3) $G \to \mathrm{alg}_{\mathbb{C}}(A,\mathbb{C})(= \mathrm{Max}(A))$ by $x \to$ (evaluation at x) is bijective.

Conversely, if A satisfies 1), 2), 3), then $R(G) = A[Q]$.

We note that if A satisfies 1), 2), 3) of the theorem, then (G,A) is an affine algebraic variety over $\mathbb{C}$ such that $L_x : G \to G$ is a morphism for all x in G ; we say that (G,A) is a <u>left algebraic group</u>. The above theorem shows that finding left algebraic group structures on G is equivalent to determing $R(G)$. If (G,A) is a left algebraic group structure, the <u>core</u> of the structure, $C(G)$, is the set of all x in G such that R_x is a morphism: $C(G) = \{x \mid xA = A\}$. It turns out that $C(G)$ is an algebraic group [4, Cor. 1.5, p. 1047] and that this algebraic group determines A [5, Thm. 2.3, p. 174].

Groups with a faithful representation have an intrinsic characterization [3, p. 113]; we will show here how, from this characterization, a representation can be constructed which yields a left algebraic group structure.

For our purposes, "algebraic group" will mean "affine complex algebraic group." Thus algebraic groups always possess faithful representations. Embedding an analytic group in an algebraic group, therefore, produces a faithful representation. We show that if G is an analytic group with a faithful representation, then G is a normal analytic subgroup of an algebraic group G' such that G' is a semi-direct product of G and an algebraic torus T ; then $\mathbb{C}[G']^T$ is the coordinate ring of a left algebraic group structure on G .

We also use the following notations and conventions: a <u>torus</u> always means a multiplicative algebraic torus (product of $GL_1\mathbb{C}$) . If G is a group and x in G , $I(x)$ is the inner antomorphism given by x . A <u>reductive</u> group is an analytic group with a faithful representation such that every representation is completely reducible. A reductive group is algebraic, and in fact the image of a reductive group under any representation is Zariski-closed.

Definition. An analytic group G is an FR group if G admits a faithful finite dimensional analytic representation; i.e. if there is an injective analytic homomorphism $G \to GL_n\mathbb{C}$ for some n .

FR groups have the following intrinsic characterization, due to Hochschild and Mostow:

Theorem. The analytic group G is FR if and only if G is a semi-direct product of a closed , solvable , simply connected normal subgroup K and a reductive subgroup P [3, p. 113].

Proofs of the "only if" assertion can be found in [2, Thm. 4.2, p. 96] (from a representation-theoretic point of view) and in [6, Thm. 10, p. 880] (from a group-theoretic point of view). This paper presents a new proof of the "if" assertion, which explains, among other things, why an FR group carries a left algebraic group structure.

It should be further mentioned that Hochschild and Mostow have a slightly stronger characterization of FR groups: a simply connected solvable normal subgroup L of an analytic group H is called a nucleus if H/L is reductive, and they show that there exists a reductive subgroup Q of H such that H is the semi-direct product of L and Q [2, Thm. 3.6, p. 95]. Their proof uses the fact that a reductive group is the complexification of a compact real Lie group. This paper also contains a new proof of this, which avoids the use of compact real forms.

We now fix the following notation: the analytic group G is a semi-direct product of the simply connected closed solvable normal subgroup K and the reductive subgroup P .

<u>Lemma</u>: Lie(K) is a sum (not necessarily direct) of Lie subalgebras N and C , where N is a nilpotent ideal of Lie(G) , C is a Cartan subalgebra of Lie(K) , and [Lie(P),C] = 0 [3, Lemma 2.1, p. 113].

Let $C = \exp_K(\mathcal{C})$ and $N = \exp_K(\mathcal{N})$. Then N and C are closed, simply connected analytic subgroups of K , with N normal in G and hence K , and the elements of C and P commute. In addition, we can (canonically) regard C and N as unipotant algebraic groups. Let $s : C \to \mathrm{Aut}(N)$ be given by $s(c) = I(c) \mid N$. Let K_1 be the (analytic) semi-direct product $N x_s C$; when convenient, we regard N and C as subgroups of K_1 . There is an analytic homomorphism $f : K_1 \to K$ given by $f(n,c) = nc$. Let L be its kernel and let L_c be the connected component of the identity in L . Then $K_1/L_c \to K$ has the discrete group L/L_c as kernel; since K is simply connected L/L_c is trivial so L is connected. Actually, $L = \{(x,x^{-1}) \mid x \in N \cap C\}$, so L is also nilpotent.

Next, we look at the image $s(C)$ of C under s . $\mathrm{Aut}(N) = \mathrm{Aut}(\mathrm{Lie}(N))$ is an algebraic group and hence the Zariski closure $\overline{s(C)}$ of $s(C)$ in Aut(N) is an algebraic group, which is nilpotent since C is nilpotent. Thus $\overline{s(C)} = U x T$ where U is the unipotent radical of $\overline{s(C)}$ and T is a torus. Let $p : \overline{s(C)} \to U$ and $q : \overline{s(C)} \to T$ be the projections. As noted above, we can regard as a unipotent algebraic group. Let D be the algebraic group $C x T$. Now $r = ps : C \to U$ is an analytic homomorphism of unipotent algebraic groups, and hence algebraic. Let $t : D \to \mathrm{Ant}(N)$ be given by $t(c,x) = (r(c)x)$. Since t is just $r x id_T$ followed by the inclusion of $U x T$ into Aut(N) , t is algebraic. We can thus form the semi-direct product $K_2 = N x_t D$ and K_2 is an algebraic group.

We will now embed K_1 in K_2. There is an analytic map $g : C \to D$ which sends c to $(c,qs(c))$. Let $h : K_1 \to K_2$ be given by $h(n,c) = (n,g(c))$. Then h is an injective analytic homomorphism by construction. Since $D = g(C)T$, $K_2 = h(K_1)T$. Moreover, it is clear that $T \cap h(K_1) = \{e\}$. We claim that $h(K_1)$ is Zariski-dense in K_2. It suffices to show that $g(C)$ is Zariski-dense in D. Let $\overline{g(C)}$ be the Zariski-closure of $g(C)$ in D. Since C is nilpotent, so is $\overline{g(C)}$, so $\overline{g(C)} = V \times S$ where V is the unipotent radical of $\overline{g(C)}$ and S is the unique maximal torus of $\overline{g(C)}$. The projection from $\overline{g(C)}$ to C is surjective and hence so is the projection from $\overline{g(C)}$ to C. Thus V projects onto C. The projection from $g(C)$ to T has $qs(C)$ as image, and hence the image of $g(C)$ in T is Zariski-dense. Thus $\overline{g(C)}$ projects onto T and hence S projects onto T. It follows that $\overline{g(C)} = D$.

Now $h(K_1)$ is Zariski-dense in K_2, so $(K_2,K_2) = (h(K_1), h(K_1)) \subseteq h(K_1)$ and $h(K_1)$ is normal in K_2. It follows that the algebraic group K_2 is an analytic semi-direct product of $h(K_1)$ and T, and $h : K_1 \to K_2$ is an analytic embedding of K_1 in an algebraic group with Zariski-dense image.

We now need to examine $h(L)$ (recall that L is the kernel of $K_1 \to K$). First, we look at $s(N \cap C)$. If we identify $\mathrm{Aut}(N)$ with $\mathrm{Aut}(\mathrm{Lie}(N))$, then $s(x) = \mathrm{Ad}(x)$ for x in C. Since N is nilpotent, if $x \in C \cap N$, $\mathrm{Ad}(x)$ is unipotent, and it follows that $s(N \cap C) \subseteq U$, so $qs(x) = e$ if $x \in C \cap N$. Thus $g(C \cap N) = (C \cap N) \times \{e\}$. Now $L = \{(x,x^{-1}) \mid x \in C \cap N\}$ so if $y = (x,x^{-1}) \in L$, $h(y) = (x,x^{-1},e)$. Hence $h(L)$ is contained in the unipotent radical of K_2 and it follows that $h(L)$ is Zariski-closed in K_2 (recall that L is simply connected). Since L is normal in K_1 and $h(K_1)$ is Zariski-dense in K_2, we also have that $h(L)$ is normal in K_2. We can thus form the algebraic group quotient $K' = K_2/h(L)$, and we have an analytic embedding $\ell : K \to K'$ induced from h. Moreover, $\ell(K)$ is Zariski-dense in the algebraic group K' and K' is the semi-direct product of $\ell(K)$ and a torus T' which is the image of T in K'.

Finally, we need to embed G in an algebraic group. This will be done by extending the action of P on K to an action on K' and then forming semi-direct products. We begin by looking at the action of P on N in G: we have a homomorphism $v : P \to \mathrm{Aut}(N)$ given by $v(p) = I(p) \mid N$. Since P is reductive, $v(P)$ is an algebraic subgroup of $\mathrm{Aut}(N)$ [6, Prop. 5, p. 878]. As noted above, P and C commute in G, so $v(P)$ and $s(C)$ commute in $\mathrm{Aut}(N)$, and hence $v(P)$ and the Zariski-closure $U \mathrm{x} T$ of $s(C)$ commute in $\mathrm{Aut}(N)$. In other words, the actions of P and D on N commute. As an algebraic variety, $K_2 = N \mathrm{x} D$. For p in P, $K_2 \to K_2$ by $(n,d) \to (pnp^{-1},d)$ is a morphism of varieties, and since P and D commute on N, this is an algebraic group automorphism of K_2. Thus we can form the semi-direct product algebraic group $K_2 \mathrm{x} P = G_2$. Since P commutes with $N \cap C$, L remains normal in G_2, and we have the algebraic group quotient $G' = G_2/L$; and G' is a semi-direct product of K' and P. Let $m : G \to G'$ be ℓ on K and the identity on P. Then m is an analytic embedding of G in an algebraic group. Moreover, since $\ell(K)$ is Zariski-dense in K', $m(G)$ is Zariski-dense in G'. Also, $G' = K'P = \ell(K)TP = m(G)T$, and the construction shows G' to be a semi-direct product of $m(G)$ and T, and that T and P commute in G'.

We have thus established the following theorem:

Theorem 1. Let G be an analytic group which is the semi-direct product of a solvable, normal, closed, simply connected subgroup K and a reductive subgroup P. Then G is a Zariski-dense analytic subgroup of an algebraic group G' such that G' is a semi-direct product of G and a torus T, and T and P commute.

We recall how embeddings as in Theorem 1 yield left algebraic group structures (see [6, Prop. 6, p. 878] for details): we have a bijection $G \to G'/T$, and G'/T is an affine algebraic variety. Since G'/T is right cosets of T, for any y in G', the map $xT \to yxT$ is a morphism of G'/T, and hence, regarding G as an algebraic variety via the above bijection, each L_x for x in G is a morphism. Moreover, if $x \in P$, R_x is a morphism of G since P and T commute. Thus P is contained in the core of the left algebraic group structure.

Theorem 1 has a converse:

Theorem: Let G be a Zariski-dense analytic subgroup of the algebraic group G'. Then there is a torus T in G' such that $G' = GT$ (not necessarily semi-direct product) and $\mathrm{Lie}(G') = \mathrm{Lie}(G) \oplus \mathrm{Lie}(T)$ (semi-direct product) [6, Theorem 3, p. 876].

Further information about these matters is contained in [6] and [7]. We now turn to the problem of showing that an analytic group with a nucleus is a semi-direct product of the nucleus and a reductive subgroup. We fix the following notations: G is an analytic group and K is a closed, solvable, simply-connected normal subgroup of G such that G/K is reductive.

We begin with the case that G is solvable, so G/K is a torus. We want to find a torus T in G such that $T \to G/K$ is an isomorphism. We use induction on the dimension of K. Let K_0 be the last non-vanishing turn in the derived series of K. K_0 is simply connected, closed, solvable, and normal in G, and $\overline{K} = K/K_0$ is a nucleus of $\overline{G} = G/K_0$. By induction, there is a torus $\overline{T}$ in $\overline{G}$ such that $\overline{T} \to \overline{G}/\overline{K} = G/K$ is an isomorphism. Let G_0 be the inverse image of $\overline{T}$ in G. Then $G_0/K_0 = \overline{T}$ so K_0 is a nucleus of G_0. We note that K_0 is abelian. If T is a torus in G_0 such that $T \to G_0/K_0$ is an isomorphism, then $T \to G/K$ is an isomorphism, so we may replace G by G_0 and K by K_0 and assume K is abelian, and hence a vector group. The action of G on K by conjunction factors through G/K, and since G/K is a torus this means there is a one-dimensional subgroup K_0 of K normal in G. By induction, using arguments similar to the above, we may assume $K_0 = K$, i.e. that K is one-dimensional. Then $\mathrm{Aut}(K) = \mathbb{C}^*$, and we have a homomorphism s : $G \to \mathrm{Aut}(K)$ where $s(g) = I(g) \mid K$. Let G_1 be the kernel of s and G_c be the connected component of the identity in G_1. Then $K \subsetneq G_c$ and we have a surjection $G/K \to G/G_c$. Since G/K is a torus, G/G_c is a direct product of a torus and a compact group A. Since $G/G_1 \subseteq \mathbb{C}^*$, A is containted in the kernel of $G/G_2 \to G/G_1$. But this kernel is the discrete group G_1/G_c, so A is trivial. Thus G/G_c is either trivial or is a one-dimensional torus. Thus the surjection $G/K \to G/G_c$ is split and $G/K = G/G_c x\overline{T}$ where $\overline{T}$ is a subtorus of G/K. Thus G_c/K is a torus (isomorphic to $\overline{T}$). Since K is central in G_c, G_c is actually abelian, so $G_c = VxT_1$ where V is a vector group and T is a torus. Let $\tilde{K}$ be the connected component of $\exp^{-1}(K)$ in $\mathrm{Lie}(G_c)$. $\tilde{K}$ is isomorphic to K and $\mathrm{Lie}(G_c)/\tilde{K}$ is the universal covering of G_c/K. It follows that G_c/K and G_c have the same fundamental group, and hence that

$\dim(T_1) = \dim(G_c/K) = \dim(G_c) - 1$, so V is one-dimensional. The projection $K \to V$ is either then trivial or an isomorphism. If it's trivial, $G_c/K = V\times(T_1/K)$, which is impossible since G_c/K is a torus. Thus $K \to V$ is an isomorphism. If we identify T_1 and $(\mathbb{C}^*)^{(n)}$ then $K = \{(v,\exp(\alpha_1,v),\ldots,\exp(\alpha_n v))\} \mid v \in V\}$ for appropriate $\alpha_1,\ldots,\alpha_n$ in $\mathbb{C}$. Thus $G_c = K\times T_1$ also and $T_1 \to \overline{T}$ is an isomorphism. T is the kernel of the additive characters of G_c and hence characteristic in G , and it follows that T_1 is normal, hence central, in G . Finally, let H be the inverse image of G/G_c regarded as a subgroup of G/K in G . Then $K \subseteq H$ and H/K is trivial or a one-dimensional torus. If $H = K$, $G = G_c$ and we are done. If not, H is an extension of $\mathbb{C}^*$ by K . If $H = \mathbb{C}^*\times K$, let $T_2 = \mathbb{C}^* \subsetneq H$; then $T_2 \to H/K$ is an isomorphism. Otherwise $\mathrm{Lie}(H)$ is the solvable non-abelian two-dimensional algebra, so the universal cover $\widetilde{H}$ of H is $\mathbb{C}\times_{\exp}\mathbb{C}$ (semi-direct product). The center of $\widetilde{H}$ is $\{0\}\times 2\pi i\mathbb{Z}$ and this contains the fundamental group of H . H is not simply connected (since H/K isn't) so $\pi_1(H) = \{0\}\times 2\pi i m\mathbb{Z}$ for some $m \neq 0$. Thus $H = \widetilde{H}/\pi_1(H)$ contains the torus $T_2 = 0\times\mathbb{C}/\pi_1(H)$. $T_2 \cap K$ is the kernel of $T_2 \to H/K$. This latter is a map of one-dimensional tori so its kernel is either all of T_2 or finite. Thus $T_2 \cap K = T_2$ or $T_2 \cap K$ is finite. The former implies that $T_2 \subseteq K$, which is impossible, while the latter implies that $T_2 \cap K$ is trivial since K has no finite subgroup. Thus $T_2 \to H/K$ is an isomorphism. Now $T_1 \cap T_2 \subseteq K \cap T_1 = \{e\}$, and T_1 is normal in G . It follows that $T = T_1T_2$ is a torus in G and $T \to G/K$ is an isomorphism.

So far, we have been assuming G solvable. We now drop that assumption. Let R be the radical of G . Then $K \subseteq R$ and $R/K = \overline{T}$ is the radical of $G/K = \overline{G}$. Since $\overline{G}$ is reductive, $\overline{T}$ is a torus. Thus K

is a nucleus of the solvable group R, and by the above argument for the solvable case there is a torus T in R such that $T \to \overline{T}$ is an isomorphism. Let $L = \mathrm{Lie}(G)$ and $\overline{L} = \mathrm{Lie}(\overline{G})$. T acts on L via the adjoint representation, and the projection $L \to \overline{L}$ is a T-morphism. Since T is a torus, the induced map $L^T \to \overline{L}^T$ is surjective, and since $\overline{T}$ is central in $\overline{G}$, $\overline{L}^T = \overline{L}$. Since $\overline{G}$ is reductive, there is a projection $\overline{L} \to [\overline{L},\overline{L}]$ and $[\overline{L},\overline{L}]$ is semi-simple. Combining, we have a surjection $L^T \to [\overline{L},\overline{L}]$, and hence there is a subalgebra $\mathcal{S}$ of L^T such that $\mathcal{S} \to [\overline{L},\overline{L}]$ is an isomorphism. Let S be the analytic subgroup of G with $\mathrm{Lie}(S) = \mathcal{S}$. Then S is semi-simple and since $\mathcal{S} \subseteq L^T$, S and T commute. Under the projection $G \to \overline{G}$, S maps onto $(\overline{G},\overline{G})$ inducing an isomorphism on Lie algebras. The kernel of $S \to (\overline{G},\overline{G})$ is thus central and hence finite. Since the kernel is $S \cap K$ and K has no finite subgroups, the map is an isomorphism. Let $P = TS$. P is a subgroup of G since S normalizes, in fact centralizes, T. Since $G/R = \overline{G}/\overline{T} = (\overline{G},\overline{G})$, $S \cap R$ is finite. Let $x = ts$ be in $K \cap P$ where $t \in T$ and $s \in S$. Since $T \subseteq R$ and $K \subseteq R$, $s \in S \cap R$, so $s^n = e$ for some n. Then $x^n = t^n$ is in $K \cap T = \{e\}$, so $x^n = e$. Since K has no elements of finite order, this means $x = e$. Thus $K \cap P = \{e\}$. Since $P \to \overline{G}$ is by construction surjective, $P \to \overline{G}$ is an isomorphism.

We have thus shown:

Theorem 2. Let G be an analytic group and K a nucleus of G. Then there is a reductive subgroup P of G such that G is the semi-direct product of K and P.

The above proof relies on the fact that the center of a semi-simple analytic group is finite. This can be established using complex groups via the classification theorem of semi-simple complex Lie algebras and the fact that all of these are algebraic, or by using compact real forms [1, Thm. 2.1, p. 198]. Since the point of the above proof is to avoid compact real forms, it would be nice to have an elementary direct proof of the fact.

References

1. G. Hochschild, The Structure of Lie Groups, Holden Day, San Francisco, 1965.

2. G. Hochschild and G. D. Mostow, "Representations and representative functions of Lie groups, III," Ann. Math. 70(1959), 85-100.

3. ________, "On the algebra of representative functions of an analytic group," Amer. J. Math. 83(1961), 111-136.

4. A. Magid, "Analytic left algebraic groups," Amer. J. Math. 99(1977), 1045-1059.

5. ________, "Analytic left algebraic groups, II," Trans. Amer. Math. Soc. 238(1978), 165-177.

6. ________, "Analytic subgroups of affine algebraic groups," Duke J. 44(1977), 875-882.

7. ________, "Analytic subgroups of affine algebraic groups," Pacific J. (to appear).

Andy R. Magid
Mathematics Department
University of Oklahoma
Norman, OK 73019
USA

THE POINCARE - SERRE - VERDIER DUALITY

by Zoghman MEBKHOUT

CONTENTS

§1 - INTRODUCTION

This lecture is mainly based on [23]. We shall see how the Poincaré, Serre and Verdier dualities are closely related as expected after Grothendieck's work on duality [8]. One of the oldest problem in algebraic or analytic geometry is to compute topological invariants of a variety having an extra structure by means of this extra structure. The Poincaré lemma computes the cohomology of a smooth complex manifold. The De Rham's theorem computes the cohomology of a differential manifold. The Grothendieck's theorem [7] computes the cohomology of a smoother algebraic variety over the complex number $\mathbb{C}$. Another example is the Hilbert-Riemann problem asking "whether any finite-dimensional complex representation of the fundamental group of a complex smooth quasi-projective variety can be obtained as the monodromy representation of a differential equation in this variety with regular singular points". This problem is solved by P. Deligne in [4]. There are many other examples of such kind. Our main new thing is that the "ALGEBRAIC ANALYSIS", sometimes called the "MICRO-LOCAL-ANALYSIS" which starts with Sato's hyperfunction - microfunction theory, is a powerful tool to understand the connections between the different structures of a variety. The algebraic analysis is the analysis on the cotangent bundle $T^{*}X$ of a complex smooth manifold X, which is much bigger than the base space X and has a symplectic structure. We shall see that the "Poincaré-Verdier" duality is closely related to the Hilbert-Riemann problem for the category of constructible sheaves of complex vector spaces which are the natural generalization of local systems of vector spaces. We use freely the Verdier's derived categories which is of the highest importance to understand the "duality theorem" in his whole generality. We refer the

Reader to the original paper of Verdier [27] which is the best as an introduction to the derived categories. I must now thank Professor LØNSTED, K. for inviting me. I must also thank Brigitte Saintonge for typing the manuscript.

§ 2 - CONNEXIONS

Let $(X, \mathcal{O}_X)$ be a smooth complex manifold of dimension n and $\mathcal{D}_X$ the sheaf of differential operators of finite order. The sheaf $\mathcal{D}_X$ is a coherent sheaf of non commutative rings. The sheaf $\mathcal{O}_X$ of the holomorphic functions in X is a subring of $\mathcal{D}_X$. We call a left $\mathcal{D}_X$-module $\mathcal{M}$ which is localy free as finite type as a $\mathcal{O}_X$-module a connexion. This is not the traditional definition of a connexion but it is convenient for your purpose. The sheaf $\mathcal{O}_X$ is a left $\mathcal{D}_X$-module and we can consider the sheaf

$$\underline{\text{hom}}_{\mathcal{D}_X}(\mathcal{M}, \mathcal{O}_X)$$

of the holomorphic solutions of the connexion $\mathcal{M}$. It plays the rôle of the horizontal section of a classical connexion. The "Cauchy existence theorem" says that $\underline{\text{hom}}_{\mathcal{D}_X}(\mathcal{M}, \mathcal{O}_X)$ is a local system of complex vector spaces. Now, the "Frobenius existence theorem" says that the functor

$$\mathcal{M} \to \underline{\text{hom}}_{\mathcal{D}_X}(\mathcal{M}, \mathcal{O}_X)$$

is an equivalence of categories between the category of the connexions and the category of the local systems of complex vector spaces. Its inverse associates to a local system $\mathcal{E}$ the connexion

$$\underline{\text{hom}}_{\mathbb{C}_X}(\mathcal{E}, \mathcal{O}_X) \quad .$$

Because of the Poincaré lemma, when we are interested only by the connexions, there are no higher solutions sheaves and the situation is very well understood since a long time.

But the local systems of complex vector spaces are not enough for our purpose. For example, if f is a proper morphism between two smooth manifolds X and Y, the cohomology sheaves of the complex $\mathbb{R}f_* \mathbb{C}_X$ are not any more local systems of complex vectors spaces if f is not smooth (See Deligne [4], p.106). They are only constructible sheaves of vector spaces. So they cannot be solution of a connexion. We must generalize the notion of a connexion.

3 - SYSTEM OF LINEAR DIFFERENTIEL EQUATIONS

Definition 3.1 - A linear system of differential equations is a left coherent $\mathcal{D}_X$-module $\mathcal{M}$.

Example 3.2 - Let

$$P = P\left(x, \frac{d}{dx}\right) = \sum_{|\alpha|=m} a_\alpha(x) \frac{d}{dx^\alpha} + \sum_{|\alpha|=m-1} a_\alpha(x) \frac{d}{dx^\alpha} + \dots \quad a_0(x)$$

be a differential operator of order m on X where $x = (x_1, \dots, x_n)$ is a local coordinate and $\frac{d}{dx^\alpha} = \frac{d}{dx_1^{\alpha_1}} \cdots \frac{d}{dx_n^{\alpha_n}}$. Then the left $\mathcal{D}_X$-module $\mathcal{M} = \mathcal{D}_X / \mathcal{D}_X P$ is a system of linear differential equations if $\mathcal{D}_X P$ is the left ideal generated by P. More generally to any matrice $(P)_{pq}$ of differential operators, we associate a left coherent $\mathcal{D}_X$-module, which is the cokernel of the homomorphism $\mathcal{D}^q \to \mathcal{D}^p$.

Example 3.3 - Any connexion is a system of linear differential equations.

Classically to a differential operator $P\left(x, \frac{d}{dx}\right)$ of order m, we associate its characteristic variety $\check{SS}(P)$ in the cotangent bundle. It is the hypersurface of T^*X defined by

$$\check{SS}(P) = \{(x, \xi) \in T^*X \;;\; \sum_{|\alpha|=m} a_\alpha(x)\xi^\alpha = 0\} .$$

To a system of differential equations $\mathcal{M}$, we associate its characteristic variety $\check{SS}(\mathcal{M})$ in T^*X. To definite it, we must introduce the sheaf $\check{\mathcal{E}}_X$ of the micro-differential operators of finite order. The sheaf $\check{\mathcal{E}}_X$ is a coherent sheaf of non-commutative rings on the cotangent bundle T^*X. By choosing a local coordinate system (x, ξ) for any open subset U in T^*X, we have

$$\Gamma(U, \check{\mathcal{E}}_X) = \left\{P_j(x,\xi)_{j\in\mathbb{Z}} \quad ; \quad P_j(x,\xi) \in \Gamma(U, \mathcal{O}_{T^*X})\right\}$$

such that

i) $P_j(x,\xi)$ is homogeneous of degree j with respect to ξ.

ii) $\operatorname{Sup}|P_j(x,\xi)| < (-j)!\, R_K^{-j}$ for any $K \ll U$ and $j < 0$.

iii) $P_j(x,\xi) = 0$ for $j \gg 0$.

(see [12] or [25]). $\check{\mathcal{E}}_X$ contains $\Pi^{-1}\mathcal{D}_X$ if Π is the projection from T^*X on X and is <u>flat</u> over it.

<u>Definition 3.4</u> - <u>A system of micro-differential operators is a left coherent $\check{\mathcal{E}}_X$-module $\check{\mathcal{M}}$</u>. We have the basic theorem proved for the <u>first</u> time in [25] :

<u>Theorem 3.5</u> - <u>The support $\check{SS}(\check{\mathcal{M}})$ of a micro-differential system is an involuntary analytic subspace of T^*X if $\check{\mathcal{M}}$ is not zero.</u>

We recall that an analytic subset of T^*X is involuntary if for any two functions f,g vanishing on it, then <u>Poisson brackets</u> $\{f,g\} = \sum_{j=1} \left(\frac{\partial f}{\partial \xi_j}\frac{\partial g}{\partial x_j} - \frac{\partial f}{\partial x_j}\frac{\partial g}{\partial \xi_j}\right)$ vanishes on it. An involuntary subset has always co-dimension equal or less than $n = \dim X$.

Now, if $\mathcal{M}$ is a $\mathcal{D}_X$-module coherent, we associate to it the $\check{\mathcal{E}}_X$-module

$$\check{\mathcal{M}} = \check{\mathcal{E}}_X \underset{\Pi^{-1}\mathcal{D}_X}{\otimes} \Pi^{-1}\mathcal{M} \quad .$$

Because of the flatness of $\check{\mathcal{E}}_X$ over $\Pi^{-1}\mathcal{D}$, the $\check{\mathcal{E}}$-module $\check{\mathcal{M}}$ is coherent and the caracteristic variety of $\mathcal{M}$ which is $\check{SS}(\mathcal{M}) \underset{\text{déf}}{=} \check{SS}(\check{\mathcal{M}})$ is an involuntary analytic subset of T^*X by theorem 3.5. The theorem 3.5 can be considered as a <u>precise</u> form of the <u>Hilbert's Nullstellensatz</u> in this case because it gives a <u>lower</u> bound for $\dim \check{SS}(\mathcal{M})$ which is $n = \dim X$. This theorem leads to make the definition :

<u>Definition 3.6</u> - An <u>holonomic system</u> (<u>maximally overdetermined system in the previous terminology) is a coherent $\mathcal{D}_X$-module $\mathcal{M}$ such that $\dim \check{SS}(\mathcal{M}) = n$.</u>

An involuntary subset of T^*X of dimension $n = \dim X$ is called Lagrangien or holonomic.

Example 3.7 - The de Rham system $\mathcal{O}_X$ is holonomic because $\check{SS}(\mathcal{O}_X) = X$ = zero section of T^*X. More generally, any connection $\mathcal{M}$ is holonomic because $\check{SS}(\mathcal{M}) = X$. So the holonomic systems are natural generalizations of the connections.

Here is a decisive example of an holonomic system which is not a connection.

Let Y be an hypersurface of X with any kind of singularities defined by an ideal $\mathcal{J}$. Following Grothendieck ([7], note n°5), the sheaf of meromorphic functions on X having poles on Y is defined by

$$\mathcal{O}[*Y] \underset{\text{déf}}{=} \varinjlim_{k} \underline{hom}_{\mathcal{O}_X}(\mathcal{J}^k, \mathcal{O}_X)$$

It is a left $\mathcal{D}_X$-module ; in fact, we have :

Theorem 3.8 - The $\mathcal{D}_X$-module $\mathcal{O}[*Y]$ is holonomic. In particular, it is $\mathcal{D}_X$-coherent.

As an $\mathcal{O}_X$-module, $\mathcal{O}[*Y]$ is not coherent. So, it cannot be a connection. The theorem 3.8 is not a trivial one and depends on works of I.N. Bernstein [2], J.E. Björk [3] and M. Kashiwara [12]. In fact, Kashiwara computes even $\check{SS}(\mathcal{O}[*Y])$ in [12]. Because of the theorem 3.8, we must consider the holonomic systems as a very important notion. In fact, the "Cauchy existence" theorem for holonomic systems goes like that :

Theorem 3.9 - The complex $\mathbb{R}\,\underline{hom}_{\mathcal{D}_X}(\mathcal{M}, \mathcal{O}_X)$ of holomorphic solutions of an holonomic system is a constructible complex of vector spaces.

This theorem is proved by M. Kashiwara in [11] using Whitney's theorems on stratifications. We recall that a constructible sheaf on X is a sheaf $\mathcal{F}$ of finite complex vector spaces such there exists a stratification $\cup_i X_i$ of X such as the restriction $\mathcal{F}_{|X_i}$ on each strata is locally constant. Of course, any local system is constructible. In fact, any $\mathbb{C}$-constructible sheaf is a local system outside a nowhere dense analytic subset of X. A complex is constructible if it is bounded and has constructible cohomology.

Example 3.10 - We have $\mathbb{R}\,\underline{hom}_{\mathcal{D}_X}(\mathcal{O}_X, \mathcal{O}_X) = \underline{\mathbb{C}}_X$. It is just as the usual Poincaré lemma. More generally, for any connection $\mathcal{M}$, $\mathbb{R}\,\underline{\hom}_{\mathcal{D}_X}(\mathcal{M}, \mathcal{O}_X)$ is a local system. It is the usual "Cauchy existence theorem".

§ 4 - DUALITY THEOREM FOR LINEAR SYSTEMS OF DIFFERENTIAL EQUATIONS.

To state the duality theorem, first some notations. If $\mathcal{A}$ is any sheaf of rings on X, we note $D(\mathcal{A})$ the derived category of the category of the left $\mathcal{A}$-modules. We note $D(\mathcal{D}_X)_c$ (resp. $D(\mathcal{O}_X)_c$, resp. $D(\mathbb{C}_X)_c$) the sub-category of $D(\mathcal{D}_X)$ (resp. $D(\mathcal{O}_X)$, resp. $D(\mathbb{C}_X)_c$) of bounded complex having $\mathcal{D}_X$-coherent (resp. $\mathcal{O}_X$-coherent, resp. $\mathbb{C}_X$-constructible) cohomology. Those categories are trianguled. We note finally $D(\mathcal{D}_X)_h$ the sub-category of $D(\mathcal{D}_X)_c$ of complex having $\mathcal{D}_X$-holonomic cohomology. The category $D(\mathcal{D}_X)_h$ is also trianguled.

Let $\mathcal{M}$ be a complex of $D(\mathcal{D}_X)_c$. Its De Rham complex $DR(\mathcal{M})$ (see for example [20]) is the complex $\mathbb{R}\,\underline{\hom}_{\mathcal{D}_X}(\mathcal{O}_X, \mathcal{M})$. We note

$$\mathbb{E}^i(\mathcal{M}) = \mathbb{E}\mathrm{xt}^i_{\mathcal{D}_X}(X\,;\mathcal{M}, \mathcal{O}_X)$$

the i-st cohomology space of the complex

$$\mathbb{R}\Gamma\Big(X\,;\mathbb{R}\,\underline{\hom}_{\mathcal{D}_X}(\mathcal{M}, \mathcal{O}_X)\Big) = \mathbb{R}\hom_{\mathcal{D}_X}(\mathcal{M}, \mathcal{O}_X)\ .$$

The complex spaces $\mathbb{E}^i(\mathcal{M})$ are the global holomorphic solutions of $\mathcal{M}$.

We note also $\mathbb{E}^i(\mathcal{M}) = \mathbb{E}\mathrm{xt}^{2n-i}_{\mathcal{D}_X,c}(X\,;\mathcal{O}_X, \mathcal{M})$ the (2n-i)st cohomology space of complex

$$\mathbb{R}\Gamma_c\Big(X\,;DR(\mathcal{M})\Big) = \mathbb{R}\hom_{\mathcal{D}_X,c}(\mathcal{O}_X, \mathcal{M})\ .$$

The complex spaces $\mathbb{E}^c_i(\mathcal{M})$ are the global De Rham of $\mathcal{M}$. The family of the compact of X is designed by c.

Then we have the Yoneda pairing :

$$\mathbb{E}^c_i(\mathcal{M}) \times \mathbb{E}^i(\mathcal{M}) \to \mathbb{E}^i_0(\mathcal{O}_X) = H^{2n}_c(X\,;\mathbb{C})\ .$$

By following this pairing by the trace map

$$H_c^{2n}(X\,;\mathbb{C}) \to \mathbb{C} \quad ,$$

we get a pairing

$$A(\mathcal{M}) : \mathbb{E}_i^c(\mathcal{M}) \times \mathbb{E}^i(\mathcal{M}) \to \mathbb{C} \quad .$$

Theorem 4.1 - (Duality theorem for L.S.D.E. [22]).

Let $\mathcal{M}$ be a complex of $D(\mathcal{D})_c$. Then there is an unique couple of topologies Q.F.S. - Q.D.F.S. on $\mathbb{E}^i(\mathcal{M})$ and $\mathbb{E}_i^c(\mathcal{M})$ such that $A(\mathcal{M})$ induces a perfect pairing between the separated spaces associated to them. If $\mathcal{M}$ is a complex of $D(\mathcal{D})_h$, then $A(\mathcal{M})$ induces an isomorphism between $\mathbb{E}^i(\mathcal{M})$ and the algebraic dual of $\mathbb{E}_i^c(\mathcal{M})$.

We recall that Q.F.S. means Quotient of Fréchet-Schwartz, and Q.D.F.S. means Quotient of Dual of Fréchet-Schwartz.

4.1 - The Poincaré duality.

Let $\mathcal{M}$ be the de Rham system $\mathcal{O}_X$ which is in $D(\mathcal{D}_X)_h$. By the Poincaré lemma, we have

$$\mathbb{E}^i(\mathcal{O}_X) = H^i(X\,;\mathbb{C}) \qquad \text{and}$$

$$\mathbb{E}_i^c(\mathcal{O}_X) = H_c^{2n-i}(X\,;\mathbb{C}) = H_i^c(X\,;\mathbb{C}) \ .$$

If we apply the "duality theorem" to $\mathcal{O}_X$, we get an analytic proof of the Poincaré duality for X which is not necessary compact.

More generally, if we apply the duality theorem to any connection, we find, by the aid of the "Cauchy existence theorem", the Poincaré duality for any local system of $\mathbb{C}$ - vector spaces.

4.2 - The Serre duality.

Let $\mathcal{M}$ be the Cauchy-Rieman system $\mathcal{D}_X$ which is in $D(\mathcal{D}_X)_c$. It is easy to see that [20]

$$\mathbb{E}^i(\mathcal{D}_X) = H^i(X\,;\,\mathcal{O}_X)$$

$$\mathbb{E}_i^c(\mathcal{D}_X) = H_c^{n-i}(X\,;\,\Omega_X)$$

where Ω_X is the sheaf of the holomorphic n forms. If we apply the "duality theorem" to $\mathcal{D}_X$, we find the Serre duality.

More generally, if we take for $\mathcal{M}$ any locally free $\mathcal{O}_X$-module, we find the Serre duality for locally free $\mathcal{O}_X$-modules.

4.3 - Serre duality for analytic coherent modules.

Let $\mathcal{F}$ be a complex of $D(\mathcal{O}_X)_c$. Then the complex

$$\mathcal{M}_{\mathcal{F}} = \mathcal{D}_X \underset{\mathcal{O}_X}{\otimes} \mathbb{R}\, \underline{hom}_{\mathcal{O}_X}(\mathcal{F}, \mathcal{O}_X)$$

is in $D(\mathcal{D})_c$ because $\mathcal{D}_X$ is flat over $\mathcal{O}_X$. It is easy to see [22] that

$$\mathbb{E}^i(\mathcal{M}_{\mathcal{F}}) = \mathbb{H}^i(X\,;\mathcal{F})$$

$$\mathbb{E}_i^c(\mathcal{M}_{\mathcal{F}}) = \mathbb{E}\mathrm{xt}^{n-i}_{\mathcal{O}_X,c}(X\,;\,\mathcal{F}, \Omega_X) \quad .$$

If we apply the duality theorem to $\mathcal{M}_{\mathcal{F}}$, we find the Serre duality for analytic coherent modules due to Malgrange B. ([16] preparation theorem), Snominen K. ([26] homology) and Ramis J.P., Ruget G. ([24] Cousin-Grothendieck's complex).

Next example is the most important and shows how powerfull is the duality theorem for $\mathcal{D}_X$-modules.

4.4 - Poincaré duality for analytic spaces.

Let Y be a complex analytic subspace of X defined by an ideal $\mathcal{J}$. Let us consider the "algebraic local cohomology" of Y

$$\underline{\mathbb{R}\Gamma}_{[Y]}(\mathcal{M}) \overset{\text{déf}}{=} \mathbb{R} \varinjlim_k \underline{hom}_{\mathcal{O}_X}(\mathcal{O}_X / \mathcal{J}^k, \mathcal{M})$$

with coefficient in a complex of $D(\mathcal{D}_X)$. This definition is due to Grothendieck for complex of $D(\mathcal{O}_X)$ [7]. In fact, if $\mathcal{M}$ is a complex of $D(\mathcal{D}_X)$ then $\underline{\mathbb{R}\Gamma}_{[Y]}(\mathcal{M})$ is a complex of $D(\mathcal{D}_X)$ because it is a sub-complex of $\mathbb{R}\Gamma_Y(\mathcal{M})$. Such a definition for $\mathcal{D}_X$-modules appears for the first time in [21]. We have the decisive theorem :

Theorem 4.4.1 - (Singular Poincaré lemma). The complex $\mathbb{R}\Gamma_{[Y]}(\mathcal{O}_X)$ is a complex of

$D(\mathcal{D})_h$ and we have the isomorphism in $D(\mathbb{C}_X)_c$:

$$\mathbb{R}\,\underline{\hom}_{\mathcal{D}_X}\left(\mathbb{R}\underline{\Gamma}_{[Y]}(\mathcal{O}_X)\,,\,\mathcal{O}_X\right) = \underline{\underline{\mathbb{C}}}_Y \quad .$$

This theorem is proved for the first time in ([21] theorem 1.1 ; see [23] chap.II for details). It is the "singular Poincaré lemma" and gives a realization of Grothendieck's idea about local cohomology who said in the introduction of S.G.A2 [5] : "Ce formalisme peut dans de nombreuses questions jouer un rôle de "localisation" analogue à celui joué par la considération de voisinages tubulaires de Y en géométrie différentielle". So the theorem 4.4.1 computes the cohomology of Y. In [21], we realize $\mathbb{R}\underline{\Gamma}_{[Y]}(\mathcal{O}_X)$ as a "relative Cousin" complex $L^{\cdot}_{[Y]}(\mathcal{O}_X)$ on X. This can be used to study the filtration of the cohomology $H^i(Y\,;C)$. To compute the homology of Y, we have the following theorem.

Theorem 4.4.2 - (Grothendieck's comparison theorem). The natural morphism $DR\left(\mathbb{R}\underline{\Gamma}_{[Y]}(\mathcal{O}_X)\right) \to DR\left(\mathbb{R}\underline{\Gamma}_Y(\mathcal{O}_X)\right)$ is an isomorphism in $D(\mathbb{C}_X)_c$.

This theorem is proved in ([21] theorem 3.1) by reducing it to the Grothendieck's theorem [7] by an "algebraic Mayer Vietoris sequences" as in M.Herrera - D. Libermann [10]. If X is a compactification of a smooth algebraic variety and Y is the infinite divisor, P. Deligne interpretes in [4] this theorem as the regularity of the De Rham connection $\mathcal{O}_X$. We see more later on about that. We have by theorem 4.4.1 that

$$\mathbb{E}^i\left(\mathbb{R}\underline{\Gamma}_{[Y]}(\mathcal{O}_X)\right) = H^i(Y\,;\mathbb{C})$$

and by theorem 4.4.2 that

$$\mathbb{E}^c_i\left(\mathbb{R}\underline{\Gamma}_{[Y]}(\mathcal{O}_X)\right) = H^c_i(Y\,;\mathbb{C}) \quad .$$

If we apply the "duality theorem" to $\mathbb{R}\underline{\Gamma}_{[Y]}(\mathcal{O}_X)$ which lives in $D(\mathcal{D})_h$, we get an analytic proof of the Poincaré duality for Y. In [10], M. Herrera - D. Libermann proved the analytic compact case (X = compact) by using a duality theorem for "first differential operators" and R.Hartshorne proved the algebraic proper case by using the formal "Serre duality". So the "duality theorem" says that it is not necessary to restrict ourselves to first order operators but we can take any "order operators".

Before going further, let us notice that the systems which have caracteristics variety of dimension $X = \dim X$ give the Poincaré duality and the systems with $2n = \dim T^*X$ caracteristic variety dimension give the Serre duality. It is a geometric interpretation of the spectral sequence which relies them when it exists. Now we have a new land in T*X, namely what is the meaning of the "duality theorem" for the systems $\mathcal{M}$ such that $n+1 \leq \dim \check{SS}(\mathcal{M}) \leq 2n-1$? For example, what is the meaning of duality theorem for sub-holonomic systems which have n+1 caracteristic variety dimension ?

4.5 - Verdier duality for constructible sheaves.

The example 4.4 concerns the caracteristic sheaf of a subspace Y of X. But we can prove the Poincaré duality by the aid of the local algebraic cohomology $\underline{\mathbb{R}\Gamma}_{[Y]}(\mathcal{O}_X)$ which is a very nice system of finite order. If we want to prove the Verdier duality for more general coefficients, we must introduce the sheaf $\mathcal{D}_X^{\infty}$ of differential operators of infinite order. Such a differential operator is just a power serie in the derivation with some conditions on the coefficients, such that $\mathcal{O}_X$ is a left $\mathcal{D}^{\infty}$-module (see [20] for a cohomological construction).

The sheaf $\mathcal{D}_X^{\infty}$ is not coherent, but it is faintfully flat over its subsheaf $\mathcal{D}_X$. The general $\mathcal{D}_X^{\infty}$-modules are difficult to handle. We must restrict ourselves to a smaller class of $\mathcal{D}_X^{\infty}$-module, namely the admissible $\mathcal{D}_X^{\infty}$-module of [25].

Definition 4.5.1 - A left $\mathcal{D}^{\infty}$-module $\mathcal{M}^{\infty}$ is admissible if locally on X, there exists a left coherent $\mathcal{D}_X$-module $\mathcal{M}$ and an isomorphism :

$$\mathcal{M}^{\infty} \simeq \mathcal{D}_X^{\infty} \underset{\mathcal{D}_X}{\otimes} \mathcal{M}.$$

The caracteristic variety $\check{SS}(\mathcal{M}^{\infty})$ for $\mathcal{D}^{\infty}$-admissible module is defined just as for $\mathcal{D}_X$-module and it is an involuntary analytic subspace of T*X. We note by $D(\mathcal{D}^{\infty})_c$ the subcategory of $D(\mathcal{D}_X^{\infty})$ of bounded complex with $\mathcal{D}^{\infty}$-admissible cohomology. A $\mathcal{D}^{\infty}$-

holonomic system is an admissible system $\mathcal{M}^\infty$ such that dim $\check{SS}(\mathcal{M}^\infty) = n$. We shall note $D(\mathcal{D}^\infty)_h$ the subcategory of $D(\mathcal{D}^\infty)_c$ of complex with $\mathcal{D}^\infty$-holonomic cohomology. The categories $D(\mathcal{D}^\infty)_c$ and $D(\mathcal{D}^\infty)_h$ are both triangulated categories. The problem is to decide whether a $\mathcal{D}^\infty$-module is or not admissible. Of course, the $\mathcal{D}^\infty$-module $\mathcal{O}_X$ is admissible

$$\mathcal{O}_X \simeq \mathcal{D}_X^\infty \underset{\mathcal{D}_X}{\otimes} \mathcal{O}_X .$$

Here is a decisive example. Let consider the analytic local cohomology of Y $\underline{\mathbb{R}\Gamma}_Y(\mathcal{O}_X)$.

We have a natural morphism

$$\underline{\mathbb{R}\Gamma}_{[Y]}(\mathcal{O}_X) \to \underline{\mathbb{R}\Gamma}_Y(\mathcal{O}_X)$$

which rises the morphism

$$\mathcal{D}_X^\infty \underset{\mathcal{D}_X}{\otimes} \underline{\mathbb{R}\Gamma}_{[Y]}(\mathcal{O}_X) \to \underline{\mathbb{R}\Gamma}_Y(\mathcal{O}_X) .$$

Theorem 4.5.2 - (Local analytic-algebraic comparison theorem). The morphism

$$\mathcal{D}_X^\infty \underset{\mathcal{D}_X}{\otimes} \underline{\mathbb{R}\Gamma}_{[Y]}(\mathcal{O}_X) \to \underline{\mathbb{R}\Gamma}_Y(\mathcal{O}_X)$$

is an isomorphism in $D(\mathcal{D}_X^\infty)$ and $\mathbb{R}\Gamma_Y(\mathcal{O}_X)$ belongs to $D(\mathcal{D}_X^\infty)_h$.

This theorem is proved for the first time in ([21], theorem 1.2 ; see [23] for details). For the special case when Y is normal crossing divisor, it is proved by a direct calculation in ([18],[19]). It was our starting point in studying function with essential singularities in the context of $\mathcal{D}_X^\infty$-modules. It dates from 1972 when we studied "Cauchy principal" value and residue of forms having essential singularities along an hypersurface Y [19]. The theorem 4.5.2 has many consequences because it solves many technical difficulties in analytic geometry due to the existence of the essential singularities which do not exist in algebraic geometry. For example, if X is a smooth noetherian scheme and Y is a sub-scheme, then $\mathcal{D}_X^\infty = \mathcal{D}_X$ and theorem 4.5.2 reduces to Grothendieck's theorem of S.G.A2 [5] which is of course much simpler. We must notice that in analytic geometry that we cannot take any $\mathcal{O}_X$-module $\mathcal{M}$ as in algebraic geometry, but the module $\mathcal{M}$ must have regular singularities along Y always because of the growth conditions.

The theorem 4.5.2 allows us to define the Leray-Herrera residue for forms having essential singularities in any co-dimension [20]. Let us give another application of theorem 4.5.2. Let x be a point of the smooth complex manifold X and consider the local ring $\mathcal{O}_x$. Then the affine scheme Spec $\mathcal{O}_x$ is noetherian and regular and it has a Grothendieck-Cousin dualizing complex

$$L^{\cdot}(\tilde{\mathcal{O}}_x) = \underline{H}^{\cdot}_{Z^{\cdot}|Z^{\cdot+1}}(\tilde{\mathcal{O}}_x) \quad .$$

If we take the global section, we get an $\mathcal{O}_x$-injective resolution $L^{\cdot}_x$ of $\mathcal{O}_x$. In fact, $L^{\cdot}_x$ is a resolution of left $\mathcal{D}_x$-modules.

Now, in analytic geometry, there is an $\mathcal{O}_x$-injective resolution $\widehat{L^{\cdot}_x}$ (see for example [1]). In fact, $\widehat{L^{\cdot}_x}$ is a resolution of left $\mathcal{D}^{\infty}_x$-modules. Thanks to theorem 4.5.2, we have the relation

$$\widehat{L}^{\cdot}_x = \mathcal{D}^{\infty}_x \underset{\mathcal{D}_x}{\otimes} L^{\cdot}_x \quad .$$

This relation can be used as an algebraic construction of $\widehat{L}^{\cdot}_x$! and was conjectured in [19].

The proof of theorem 4.5.2 relies on the theorem 4.4.1 and the "Frobenius principe" for $\mathcal{D}^{\infty}$-modules :

Theorem 4.5.3 - (Frobenius principe). The natural morphism $\mathcal{M}^{\infty} \to \mathbb{R}\,\mathrm{hom}_{\mathbb{C}_X}\left(\mathbb{R}\,\mathrm{hom}_{\mathcal{D}^{\infty}_X}(\mathcal{M}^{\infty}, \mathcal{O}_X), \mathcal{O}_X\right)$ of $D(\mathcal{D}^{\infty})_h$ is an isomorphism.

This theorem is proved in ([23] chap.III, theorem 2.1). It uses Verdier duality and nuclear topological vector spaces as in [14]. It says that the $\mathcal{D}^{\infty}$-module $\mathcal{O}_X$ is dualizing in $D(\mathcal{D}^{\infty})_h$.

Now let L be a complex of $D(\mathbb{C}_X)_c$. Then the complex $\mathbb{R}\,\mathrm{hom}_{\mathbb{C}_X}(L, \mathcal{O}_X)$ is a complex of $D(\mathcal{D}_X^\infty)$ because $\mathcal{O}_X$ is a $\mathcal{D}_X^\infty$-module. In fact, we have

Theorem 4.5.4 - The complex $\mathbb{R}hom_{\mathbb{C}_X}(L, \mathcal{O}_X)$ is in $D(\mathcal{D}_X^\infty)_h$ if L is in $D(\mathbb{C}_X)_c$ and we have the isomorphism in $D(\mathbb{C}_X)_c$

$$L \simeq \mathbb{R}hom_{\mathcal{D}_X^\infty}\left(\mathbb{R}hom_{\mathbb{C}_X}(L, \mathcal{O}_X), \mathcal{O}_X\right)$$

This theorem is proved in ([23] chap.III, prop.3.1). It reduces by "dévissage" of constructible sheaf to theorem 4.5.2.

So theorems 4.5.3 and 4.5.4 say that the functors as $\mathcal{M}^\infty \to \mathbb{R}hom_{\mathcal{D}_X^\infty}(\mathcal{M}^\infty, \mathcal{O}_X)$ and $L \to \mathbb{R}hom_{\mathbb{C}_X}(L, \mathcal{O}_X)$ are inverse of each other and provide an equivalence of categories between $D(\mathcal{D}^\infty)_h$ and $D(\mathbb{C}_X)_c$. This is the "Frobenius existence theorem" for $\mathcal{D}^\infty$-holonomic systems and show that the $\mathcal{D}$-holonomic are not enough.

Let us return to the "duality theorem". For a complex of $D(\mathcal{D}^\infty)_c$, the spaces $\mathbb{E}^i(\mathcal{M}^\infty)$, $\mathbb{E}_i^c(\mathcal{M}^\infty)$ and the pairing $A(\mathcal{M}^\infty)$ are defined and the "duality theorem" hold in $D(\mathcal{D}^\infty)_c$ because of the faintfull flatness of $\mathcal{D}_X^\infty$ over $\mathcal{D}_X$.

Let L be a complex of $D(\mathbb{C}_X)_c$; then we have

$$\mathbb{H}^i(X, L) = \mathbb{E}^i\left(\mathbb{R}hom_{\mathbb{C}_X}(L, \mathcal{O}_X)\right)$$

$$\mathbb{E}xt^{2n-i}_{\mathbb{C}_X, c}(X ; L, \mathbb{C}_X) = \mathbb{E}_i^c\left(\mathbb{R}hom_{\mathbb{C}_X}(L, \mathcal{O}_X)\right) .$$

(See [23], chap.IV).

If we apply the "duality theorem" to $\mathcal{M}^\infty = \mathbb{R}hom_{\mathbb{C}_X}(L, \mathcal{O}_X)$, we get the "Verdier duality" [28] for constructible coefficients in a purely analytic way. So far we are concerned with smooth complex manifold, the apparently topological notion of Poincaré-Verdier duality with constructible sheaves as coefficients, is in fact the purely analytic notion of duality for linear systems of differential equations. We can even ask more. Theorem 4.5.4 tells us that any complex of $D(\mathbb{C}_X)_c$ is a solution of $\mathcal{D}_X^\infty$-holonomic complex. But in [4], Deligne proves that any local system of vector spaces on smooth quasi-projective variety X is a solution of a connection on X having a regular singularity along the divisor Y at infinite of any compactification $\bar{X}$ of X.

This is the "Hilbert-Riemann" problem because any representation of the fundamental group comes from a local system. As constructible sheaves are natural generalizations of local systems, it is natural to ask the same problem for this category. So the first thing is to know what is a system on X having regular singularity along Y.

§ 5 - SYSTEMS WITH REGULAR SINGULARITIES AND THE HILBERT-RIEMANN PROBLEM.

Let us recall the definition of an ordinary linear equation on a Riemannsurface X. Let $P\left(x, \frac{d}{dx}\right) = a_m(x) \frac{d}{dx^m} + a_{m-1}(x) \frac{d}{dx^{m-1}} + \dots + a_0(x)$ be a linear differential equation. Classicaly, we say $P\left(x, \frac{d}{dx}\right)$ has a regular singularity at x_0 if and only if the functions $(x-x_0)^{m-i} \frac{a_i(x)}{a_m(x)}$ $i = m-1, \dots, 0$ are holomorphic at x_0. We can consider the operators

$$P : \mathcal{O}_{x_0} \to \mathcal{O}_{x_0}$$

$$P : \widehat{\mathcal{O}}_{x_0} \to \widehat{\mathcal{O}}_{x_0}$$

where $\widehat{\mathcal{O}}_{x_0}$ is the completion of $\mathcal{O}_{x_0}$ for the natural Krull topology. We have, if $\mathcal{M}$ is the system $\mathcal{D}/\mathcal{D}P$:

$$\mathrm{Ker}(P, \mathcal{O}_{x_0}) \simeq \underline{\mathrm{hom}}_{\mathcal{D}_X}(\mathcal{M}, \mathcal{O}_{x_0})$$

$$\mathrm{Coker}(P, \mathcal{O}_{x_0}) \simeq \mathcal{E}xt^1_{\mathcal{D}_X}(\mathcal{M}, \mathcal{O}_{x_0})$$

$$\mathrm{Ker}(P, \widehat{\mathcal{O}}_{x_0}) \simeq \underline{\mathrm{hom}}_{\mathcal{D}_X}(\mathcal{M}, \widehat{\mathcal{O}}_{x_0})$$

$$\mathrm{Coker}(P, \widehat{\mathcal{O}}_{x_0}) \simeq \mathcal{E}xt^1_{\mathcal{D}_X}(\mathcal{M}, \widehat{\mathcal{O}}_{x_0})$$

B.Malgrange proved in ([17], theorem 1.4) that $P\left(x, \frac{d}{dx}\right)$ has a regular singularity at x_0 if and only if we have the isomorphism

$$\mathrm{Ker}(P, \mathcal{O}_{x_0}) \simeq \mathrm{Ker}(P, \widehat{\mathcal{O}}_{x_0})$$

$$\mathrm{Coker}(P, \mathcal{O}_{x_0}) \simeq \mathrm{Coker}(P, \widehat{\mathcal{O}}_{x_0})$$

In another language, P has a regular singularity if and only if the natural isomorphism

$$\mathbb{R}\underline{hom}_{\mathcal{D}_X}(\mathcal{M}, \mathcal{O}_{x_0}) \to \mathbb{R}\underline{hom}_{\mathcal{D}_X}(\mathcal{M}, \hat{\mathcal{O}}_{x_0})$$

is an isomorphism in $D(\mathbb{C}_X)$.

Let us write it in a different way.

Because of the coherence, we have

$$\mathbb{R}\underline{hom}_{\mathcal{D}_X}(\mathcal{M}, \mathcal{O}_{x_0}) \simeq \mathbb{R}\underline{hom}_{\mathcal{D}_X}(\mathcal{M}, \mathcal{O}_X) \underset{\mathbb{C}}{\otimes} \mathbb{C}_{x_0} .$$

But $\hat{\mathcal{O}}_{x_0}$ is the formal completion

$$\mathcal{O}_{\widehat{X|x_0}} = \varprojlim_k \mathcal{O}_X / \mathcal{J}^k$$

of X along x_0 if $\mathcal{J}$ is the ideal defining x_0. If we put $x_0 = Y$, then the isomorphism become

$$\mathbb{R}\underline{hom}_{\mathcal{D}_X}(\mathcal{M}, \mathcal{O}_X) \underset{\mathbb{C}}{\otimes} \mathbb{C}_Y \xrightarrow{\sim} \mathbb{R}\underline{hom}_{\mathcal{D}_X}(\mathcal{M}, \mathcal{O}_{\widehat{X|Y}}) .$$

In this way, everything has a meaning if X is a complex smooth manifold of <u>any dimension</u> and Y is an analytic subspace of X defined by $\mathcal{J}$. So, if we defined a system $\mathcal{M}$ having <u>regular singularities</u> along Y by the previous isomorphism, it agrees with the case of Riemann surface by Malgrange's theorem.

But it is a <u>very difficult</u> problem to verify this isomorphism. In fact, if $\mathcal{M}$ is the De Rham system $\mathcal{O}_X$, we have the following theorem :

<u>Theorem 5.1 - (P. Deligne, M. Herrera, D. Libermann [10]). We have the following isomorphism in</u> $D(\mathbb{C}_X)_c$

$$\mathbb{C}_Y = \mathbb{R}\underline{hom}_{\mathcal{D}_X}(\mathcal{O}_X, \mathcal{O}_X) \underset{\mathbb{C}}{\otimes} \mathbb{C}_Y \simeq \mathbb{R}\underline{hom}_{\mathcal{D}_X}(\mathcal{O}_X, \mathcal{O}_{\widehat{X|Y}})$$

$$= DR(\mathcal{O}_{\widehat{X|Y}}) = \Omega^{\bullet}_{\widehat{X|Y}} .$$

The proof of <u>Deligne</u> (1969) is not published. <u>M. Herrera</u>, D. <u>Libermann</u> proved it in the case where X is compact [10] by using "duality". There is another proof due to <u>R.Hartshorne</u> [9]. In fact, we can derive theorem 5.1. from theorem 4.4.1 because for any connection $\mathcal{M}$ we have the proposition ([23], Chap.II, prop.6.1)

Proposition 5.2 - Let $\mathcal{M}$ be a connection. Then we have the isomorphism in $D(\mathbb{C}_X)_c$

$$\mathbb{R}\underline{hom}_{\mathcal{D}}\left(\mathbb{R}\underline{\Gamma}_{[Y]}(\mathcal{M}),\ \mathcal{O}_X\right) \simeq \mathbb{R}\underline{hom}_{\mathcal{D}_X}(\mathcal{M}, \mathcal{O}_{\widehat{X|Y}}) \quad .$$

If we apply this proposition to $\mathcal{M} = \mathcal{O}_X$, we get theorem 5.1 from theorem 4.4.1.

In [4], Deligne proved that a connection $\mathcal{M}$ is regular if and only if the natural morphism $DR\left(\mathbb{R}\underline{\Gamma}_{[Y]}(\mathcal{M})\right) \to DR\left(\mathbb{R}\underline{\Gamma}_Y(\mathcal{M})\right)$ is an isomorphism ([4], proposition 6.20).

In fact, from the "local duality theorem", for $\mathcal{D}_X$-holonomic system ([23], chap. III, theorem 1.1) and the "Frobenius principe", for $\mathcal{D}_X^\infty$-holonomic systems ([23], chap. III, theorem 2.1), we get very easily the following proposition.

Proposition 5.2 - Let $\mathcal{M}$ be a complex of $D(\mathcal{D}_X)_h$. Then the following conditions are equivalent.

(a) The natural morphism $DR\left(\mathbb{R}\underline{\Gamma}_{[Y]}(\mathcal{M})\right) \to DR\left(\mathbb{R}\underline{\Gamma}_Y(\mathcal{M})\right)$ is an isomorphism in $D(\mathbb{C}_X)_c$.

(b) We have the canonical isomorphism in $D(\mathbb{C}_X)_c$

$$\mathbb{R}\underline{hom}_{\mathcal{D}}(\mathcal{M}, \mathcal{O}_X) \underset{\mathbb{C}}{\otimes} \mathbb{C}_Y \simeq \mathbb{R}\underline{hom}_{\mathcal{D}}\left(\mathbb{R}\underline{\Gamma}_{[Y]}(\mathcal{M}),\ \mathcal{O}_X\right)$$
$$= \mathbb{R}\underline{hom}_{\mathcal{D}_X}(\mathcal{M}, \mathcal{O}_{\widehat{X|Y}}) \quad .$$

(c) We have the canonical isomorphism in $D(\mathcal{D}_X^\infty)_h$

$$\mathcal{D}_X^\infty \underset{\mathcal{D}_X}{\otimes} \mathbb{R}\underline{\Gamma}_{[Y]}(\mathcal{M}) \simeq \mathbb{R}\underline{\Gamma}_Y(\mathcal{D}_X^\infty \underset{\mathcal{D}_X}{\otimes} \mathcal{M}) \quad .$$

This proposition is proved in ([23], chap.III, proposition 3.3). For $\mathcal{M} = \mathcal{O}_X$, the condition (a) is the theorem 3.1 of [21] which is essentially the Grothendieck's theorem [7]. Condition (b) is the "singular Poincaré lemma" of ([21], theorem 1.1). Condition (c) is the comparison theorem of ([21], theorem 1.2). All those theorems depend on the "resolution of the singularities" and show how difficult is to verify such conditions on a system. Of course, we call a system $\mathcal{M}$ fulfilling the conditions (a), (b) and (c) a "regular singular system along Y".

See [13] for micro-local version.

Let L be a complex of $D(\mathbb{C}_X)_c$ and sing(L) the analytic subspace of X where its cohomology is not locally free as sheaves of $\mathbb{C}$-vector spaces.

Proposition 5.3 - Locally on X, any complex of $D(\mathbb{C}_X)_c$, L is the solution of a complex of $D(\mathcal{D}_X)_h$ having regular singular cohomology along sing(L).

By "dévissage", this proposition reduces to theorem 4.5.2.The global situation is the most difficult part and it is the analogue of the "Hilbert-Riemann" problem treated by Deligne for local systems in [4]. This is not done yet and it is the remark 5.1 of [22].

Let us say a word about the proof of the "duality theorem". It uses standard Grothendieck's techniques. The only difference is that the complex spaces $\mathbb{E}^i(\mathcal{M})$ and $\mathbb{E}^c_i(\mathcal{M})$ are not cohomology but hypercohomology even for a single $\mathcal{D}_X$-module. So they are cohomology of an object of a derived category which is definitively not a local object for the usual topology of X. We must use something more to compute the spaces $\mathbb{E}^i(\mathcal{M})$ and $\mathbb{E}^c_i(\mathcal{M})$. This thing more is the Deligne-Verdier "Méthode de la descente cohomologique" [29] which comes from the Grothendieck's topologies (see [23], chap.IV).

We hope we have showed that the $\mathcal{D}_X$-modules are a powerful tool with cotangent bundle T*X in geometry.

Let us give another example. Let $f : X \to Y$ be a projective morphism between smooth complex manifold, say of the same dimension. We have seen that the complex $\mathbb{R} f_* \underline{\mathbb{C}}_X$ cannot be a solution of a connection. But we have

$$\mathbb{R} f_* \underline{\mathbb{C}}_X = \mathbb{R} f_* \mathbb{R}\underline{hom}_{\mathcal{D}_X}(\mathcal{O}_X, \mathcal{O}_X) \sim \mathbb{R}\underline{hom}_{\mathcal{D}_Y}\left(\int_f \mathcal{O}_X, \mathcal{O}_Y\right) .$$

$\int_f \mathcal{O}_X$ is the complex $\mathbb{R} f_* \mathcal{D}_{Y\leftarrow X} \overset{L}{\underset{\mathcal{D}_X}{\otimes}} \mathcal{O}_X$ where $\mathcal{D}_{Y\leftarrow X}$ is the right $\mathcal{D}_X$-module defined in [12] for example. The complex $\int_f \mathcal{O}_X$ is the "integration along the fiber" of the De Rham system and lives in $D(\mathcal{D}_Y)_h$, [12].

So $\mathbb{R} f_* \underline{\mathbb{C}}_X$ is a solution of a system of $D(\mathcal{D}_Y)_h$ which has no connection as cohomology. This is the relative version of the "duality theorem for the $\mathcal{D}_X$-modules". It will be the theme of our forecoming paper and solves the Deligne's difficulty ([4], page 106).

REFERENCES

[1] Banica, C. and Stanasila, O., *Algebraic Methods in Global Theory of Complex Spaces*, John Wiley, New York (1976).

[2] Bernstein, I.N., *The Analytic Continuation of Generalized Functions with Respect to a Parameter*, Functional Anal. Appl. 6, 26-40 (1972).

[3] Björk, I.E., *Dimensions over Algebras of Differential Operators*, To appear in North-Holland, Amsterdam (197?).

[4] Deligne, P., *Equations Différentielles à Points Singuliers Réguliers*, Lecture Notes in Math. n°163, Berlin Heidelberg - New York Springer (1970).

[5] Grothendieck, A., *Cohomologie Locale des Faisceaux Cohérents et Théorèmes de Lefschetz Locaux et Globaux* (S.G.A.2), North-Holland, Amsterdam (1968).

[6] Grothendieck, A., *Théorème de Dualité pour les Faisceaux Algébriques Cohérents*, Séminaire Bourbaki n°149, Paris (1957).

[7] Grothendieck, A., *On the de Rham Cohomology of Algebraic Varieties*, Publ. Math. I.H.E.S. n°29, 95-103 (1966).

[8] Hartshorne, R., *Residues and Duality*, Lecture Notes in Math. n°20, Berlin Heidelberg - New York Springer (1966).

[9] Hartshorne, R., *On the de Rham Cohomology of Algebraic Varieties*, Publ. Math. I.H.E.S. n°45, 6-99 (1975).

[10] Herrera, M. and Libermann, D., *Duality and the de Rham Cohomology of Infinitesimal Neighbourhoods*, Invent. Math. 13, 97-126 (1971).

[11] Kashiwara, M., *On the Maximally Overdetermined Systems of Linear Differential Equations I*, Publ. R.I.M.S. Kyoto University 10, 655-579 (1975).

[12] Kashiwara, M., *B-Functions and Holonomic Systems*, Invent. Math. 38, 33-53 (1976).

[13] Kashiwara, M. and Oshima, T., *Systems of Differential Equations with Regular Singularities and their Boundary Value Problems*, Ann. of Math. 106, 145-200 (1977).

[14] Kiehl, R., Verdier, J.L., *Ein Einfacher Beweis des Kohäreuzsatzes von Grauert*, Math. Ann. 195, 24-50 (1971).

[15] Libermann, D., *Generalization of the de Rham Complex with Applications to Duality Theory and the Cohomology of Singular Varieties*, Proceedings of the Conf. of Complex Analysis, Rice (1972).

[16] Malgrange, B., *Systèmes Différentiels à Coefficients Constants*, Séminaire Bourbaki n°246, 1-15 (1962-63).

[17] Malgrange, B., *Sur les Points Singuliers des Equations Différentielles*, Enseignement Math. 20, 147-176 (1974).

[18] Mebkhout, Z., *La Valeur Principale des Fonctions à Singularités Essentielles*, C.R. Acad. Sc. Paris 280, 205-207 (1975).

[19] Mebkhout, Z., *La Valeur Principale et le Résidu Simple des Formes à Singularités Essentielles. In Fonctions de Plusieurs Variables Complexes II*, Lecture Notes in Math. n°482, 190-215, Berlin Heidelberg - New York Springer (1975).

[20] Mebkhout, Z., *La Cohomologie locale d'une Hypersurface. In fonctions de Plusieurs Variables Complexes III*, Lecture Notes in Math. n°670, 89-119, Berlin Heidelberg New York Springer (1977).

[21] Mebkhout, Z., *Local Cohomology of Analytic Spaces*, Publ. R.I.M.S. Kyoto Univers. 12, 247-256, Suppl. (1977).

[22] Mebkhout, Z., *Théorèmes de Dualité pour les $\mathcal{D}_X$-Modules Cohérents*, C.R. Acad. Sc. Paris 285, 785-787 (1977).

[23] Mebkhout, Z., *Cohomologie Locale des Espaces Analytiques Complexes et Théorèmes de Dualité Locale et Globale pour les $\mathcal{D}_X$-Modules Cohérents*, Thèse de Doctorat d'Etat, Université Paris VII (1978), To appear in Lecture Notes in Math.

[24] Ramis, J.P., Ruget, G., *Complexe Dualisant et Théorème de Dualité en Géométrie Analytique Complexe*, Publ. Math. I.H.E.S. n°38, 71-91 (1971).

[25] Sato, M., Kawai , T., Kashiwara, M., *Micro-Functions and Pseudo-Differential Equations*, Lecture Notes in Math. n°287, 265-529, Berlin Heidelberg - New York Springer (1973).

[26] Suominam, K., *Duality for Coherent Sheaves in Analytic Manifolds*, Ann. Acad. Sc. Fermicae, Helsinski 424, 1-19 (1968).

[27] Verdier, J.L., *Catégories Dérivées. Etat 0. In S.G.A. 4.1/2* , Lecture Notes in Math. n°569, 262-311, Berlin Heidelberg - New York Springer (1977).

28 Verdier, J.L., *Dualité dans la Cohomologie des Espaces Localement Compacts*, Séminaire Bourbaki n°300 (1965-66).

29 Verdier, J.L., *Topologies sur les Espaces de Cohomologies d'un Complexe de Faisceaux Analytiques à Cohomologie Cohérente*, Bull. Soc. Math. France 99 337-342 (1971).

Université d'Orléans
F-45045 Orléans
France

MUMFORD'S NUMERICAL FUNCTION AND STABLE PROJECTIVE HYPERSURFACES

by

Linda Ness

Let G be a connected algebraic group defined over an algebraically closed field k. Let $G \times V \to V$ be a representation of G on a k vector variety V. In this paper we study a numerical function $M: V \to \mathbb{R}$ defined, but little exploited, by Mumford on p. 65 of G.I.T. The numerical function M is invariant on G orbits, and the sign of $M(v)$ determines whether v is stable, unstable, or semistable. The function M is defined by

$$M(v) = \sup_{\lambda \in \Gamma^*(G)} \frac{m(v,\lambda)}{\|\lambda\|}$$

where $m(v,\lambda)$ is the numerical function Mumford uses in Chapter 2 of G.I.T. to give a numerical criterion for stability, $\Gamma^*(G)$ is the set of nontrivial one parameter subgroups of G, and $\| \ \|$ is a length function on $\Gamma^*(G)$. Mumford pointed out that the supremum was actually attained. He also showed that for each $v \in V$, the function

$$\mu(v,\lambda) : \Gamma^*(G) \to \mathbb{R} : \mu(v,\lambda) = \frac{m(v,\lambda)}{\|\lambda\|}$$

descends to a function on the Tits building of parabolic conjugacy classes of one parameter subgroups (the flag complex). Kempf proved in [K], that if v is unstable, M is attained on a unique parabolic conjugacy class of one parameter subgroups. He, then, used this to solve a rationality question posed by Mumford on p. 64 of G.I.T.

In this paper I prove that $|M_T(v)|$, the absolute value of the restriction of M to a torus T, is the distance from 0 to the boundary of the convex hull, in $\Gamma(T) \times \mathbb{R}$, of the duals of the T-weights for v. I can characterize the boundary of the convex hull in the case that v is stable. Since $M(v) = \max\{M_T(v) : T \subset G$ is a maximal torus , this gives a geometric characterization of M. (Theorem 3.7) The last three sections of the paper are devoted to projective hypersurfaces, i.e. the representation $SL(n,k) \times Sym^d(V_n^*) \to Sym^d(V_n^*)$. I prove, using the numerical criterion for stability, that nonsingular hypersurfaces are stable when $d > n \geq 3$. Mumford proves this, using invariant theory, in G.I.T. for $d \geq 3$, $n \geq 3$. For plane curves, I make precise the notion of the worst point on a curve. For nonsingular plane curve of degree $d \geq 3$, for stable and unstable plane cubic curves, and for stable and unstable plane quartic curves, I prove that M is determined by a "worst" point on the curve, and that M is attained on a finite number of parabolic conjugacy classes of one parameter subgroups. Furthermore, I can prove in these cases, that the associated parabolics fix the natural flag at the worst points.

The organization of the paper is as follows: In section one, we introduce the numerical function $m(v,\lambda)$ and give the numerical criterion for stability. Section two is entirely devoted to developing some linear algebra and convexity theory. In section three, we introduce the length function, the function $\mu(v,\lambda)$, and the final numerical function M. The last three sections are devoted to projective hypersurfaces, plane curves, and cubic and quartic curves, respectively.

SECTION 1

Let G be a connected algebraic group defined over a fixed algebraically closed field k. Throughout this paper we will be working within the category of k-varieties unless otherwise stated.

Let $G_m = \{t \neq 0\} \subset k$ be the multiplicative subgroup. A one-parameter subgroup of G is a homomorphism $\lambda : G_m \to G$. Let $\Gamma(G)$ denote the set of one-parameter subgroups of G; let $\Gamma^*(G)$ denote the set of non-trivial one-parameter subgroups of G. Note that G acts on $\Gamma(G)$ by conjugation, where $g*\lambda = g\cdot\lambda\cdot g^{-1}$. For any integer n there is an n-multiple of λ, $n\cdot\lambda$, defined by $n\cdot\lambda(t) = \lambda(t^n)$. A one-parameter subgroup λ is indivisible if $\lambda \neq n\,\lambda'$, where n is a positive integer, and λ' is some other element of $\Gamma(G)$.

A representation of G on a vector space V is a morphism $G\times V \to V$ sending (g,v) to $g\cdot v$, such that G acts on V by linear automorphisms.

In this paper, we will use the following definitions for stability of vectors in V in terms of the one-parameter subgroups of G.

Definition: A nonzero vector $v \in V$ is

(a) unstable if there exists $\lambda \in \Gamma^*(G)$ such that $\lim\limits_{t\to 0} \lambda(t)\cdot v = 0$.

(b) semistable if for all $\lambda \in \Gamma^*(G)$, $\lim\limits_{t\to 0} \lambda(t)\cdot v \neq 0$

(c) stable if for all $\lambda \in \Gamma^*(B)$, $\lim\limits_{t\to 0} \lambda(t)\cdot v$ does not exist.

Next we define Mumford's numerical function $m : V\times\Gamma(G) \to \mathbb{Z}$ and give the numerical criterion for stability in terms of this function. We begin by recalling the well-known

Lemma 1.1 Every representation of $\mathbb{G}_m$ is completely reducible; the irreducible invariant subspaces have dimension one.

The composition of the representation of G on V with a one-parameter subgroup $\lambda : \mathbb{G}_m \to G$ gives a representation of $\mathbb{G}_m$ on V. Thus for each $\lambda \in \Gamma(G)$, there is an eigen-decomposition of V, i.e.

(1) $$V = \bigoplus_{n\in\mathbb{Z}} V_n \qquad \text{where} \quad V_n = \{v \mid \lambda(t)\cdot v = t^n v\}$$

If $v \in V$, then $v = \Sigma v_n$, where $v_n \in V_n$. Define $m(v,\lambda)$ by

(2) $$m(v,\lambda) = \min\{n : v_n \neq 0\}$$

Three obvious properties of the numerical function m are:

(3) $$m(v,n\cdot\lambda) = n\cdot m(v,\lambda) \quad \text{for any positive integer } n,$$

(4) $$m(gv,g{*}\lambda) = m(v,\lambda) \quad \text{for all } g \in G, \text{ and}$$

(5) $$m(\alpha v,\lambda) = m(v,\lambda) \quad \text{for any nonzero } \alpha \in k.$$

Understanding the values that the function $m(v, \;) : \Gamma^*(T) \to \mathbb{Z}$ may take for a particular nonzero $v \in V$ is equivalent to understanding the degree of stability of v, because of the following numerical criterion:

Lemma 1.2 Suppose $v \in V$ is a nonzero vector. Then

(a) v is unstable $\Leftrightarrow m(v,\lambda) > 0$ for some $\lambda \in \Gamma^*(G)$

(b) v is semistable $\Leftrightarrow m(v,\lambda) \leq 0$ for some $\lambda \in \Gamma^*(G)$

(c) v is stable $\Leftrightarrow m(v,\lambda) < 0$ for all $\lambda \in \Gamma^*(G)$.

Proof: The lemma follows immediately from the definitions of stable, semi-stable, and unstable, and from the definition of the numerical function. Q.E.D.

We now want to introduce the parabolic subgroup $P(\lambda)$ associated to a one-parameter subgroup $\lambda \in \Gamma^*(G)$. Define $P(\lambda) \subset G$ by

(6) $$P(\lambda) = \{g \in G : \lim_{t\to 0} \lambda(t)g\lambda(t)^{-1} \text{ exists in } G\}.$$

Mumford proved in Proposition 2.6 of [G.I.T.], that this defines a parabolic subgroup of G. This is easy to see if G is a subgroup of the group of linear automorphisms of some vector space.

We can relate $P(\lambda)$ to the notions we discussed earlier in the section. Let $V = \bigoplus_{n\in\mathbb{Z}} V_n$ be the eigen-decomposition of V determined by λ. This decomposition determines a natural weight filtration of V. Namely, let $V^i = \bigoplus_{n\geq i} V_n$, so $\ldots V^{i+1} \subset V^i \subset \ldots$

Lemma 1.3 Let $\lambda \in \Gamma^*(G)$ and let $V^{i+1} \subset V^i \subset \ldots$ be the λ-weight filtration of V. Then

i) $m(v,\lambda) = \max\{i : v \subset V^i\}$,

ii) $P(\lambda)$ preserves the λ-weight filtration, and

iii) $m(v,\lambda) = m(p\cdot v,\lambda) = m(v,p^{-1}*\lambda)$ for $\lambda \in P(\lambda)$.

Proof: The lemma follows easily from the definition of m in (2), of $P(\lambda)$ in (6), and of V^2. The second inequality in iii) is a consequence of property (4) of m. Q.E.D.

We next want to study how $m(v, \ )$ varies when λ is contained in a torus of G. Recall that a torus of G is an algebraic subgroup of G, which is isomorphic to an ℓ-fold product $G_m \times \ldots \times G_m$, for some integer $\ell \geq 1$. Clearly the image, $\mathrm{Im}\lambda$, of any one-parameter subgroup $\lambda \in \Gamma^*(G)$ is a torus. The image of any one-parameter group and, in fact, any torus is contained in a maximal torus. Furthermore any two maximal tori are conjugate [B]. The essential fact concerning representations of a torus is

Lemma 1.4 Every representation of a torus is completely reducible.

Thus if $T \subset G$ is a torus, the representation of G on V induces a representation of T on V. Hence there is an eigen decomposition of V with respect to T, or a T-weight decomposition,

$$(7) \qquad V = \bigoplus_{\chi \in \chi(T)} V_\chi$$

where T acts on V_χ by the character χ, and $\chi(T)$ is the set of characters of T.

The set of T-weights of V, $S(T)$, is the finite set of characters of T,

$$(8) \qquad S(T) = \{\chi \in \chi(T) : V_\chi \neq 0\}.$$

Every $v \in V$ can be written $v = \Sigma v$, $v_\chi \in V_\chi$. The T-state of v, $S_v(T)$, is the subset (T) defined by

(9) $$S_v(T) \cdot \{\chi \in \chi(T) : v_\chi \neq 0\}$$

Recall that if T has dimension n, $\chi(T)$ and $\Gamma(T)$ are free abelian groups of rank n. In fact, for the torus $\mathbb{G}_m^n$, the characters are the homomorphisms $(t_1,\ldots,t_n) \to \prod_{j=1}^{n} t_j^{i_j}$, and the one-parameter subgroups are the homomorphisms $t \to (t^{a_1},\ldots t^{a_n})$, where both $i = i_1,\ldots,i_n)$ and $a = (a_1,\ldots,a_n)$ are n-tuples of integers.

There is a natural perfect $\mathbb{Z}$-bilinear pairing

(10) $$\chi(T) \times \Gamma(T) \to \mathbb{Z} : (\chi,\lambda) \to \langle \chi,\lambda \rangle \qquad \text{where} \quad \chi \circ \lambda(t) = t^{\langle \chi,\lambda \rangle}$$

Thus each character $\chi \in \Gamma(T)$ determines a $\mathbb{Z}$-linear functional $\langle \chi, \ \rangle : \Gamma(T) \to \mathbb{Z}$; dually each one-parameter subgroup λ determines a $\mathbb{Z}$-linear functional $\langle \ ,\lambda \rangle : \chi(T) \to \mathbb{Z}$. Hence the action of a one-parameter subgroup λ on $v = \Sigma v_\chi$ is given by

(11) $$\lambda(t) \cdot v = \Sigma t^{\langle \chi,\lambda \rangle} v_\chi \quad ,$$

so the T-weight decomposition of V is compatible with the λ-decomposition of V, as in (1). This, together with the definition (2) of $m(v,\lambda)$ shows that

(12) $$m(v,\lambda) = \min\{ \langle \chi,\lambda \rangle : \chi \in S_v(T) \}.$$

The previous discussion shows that it is fairly easy to study the restriction of the function $m(v, \ )$ to tori. We summarize the properties that we shall use most in

Lemma 1.5 For each $v \in V$, and for each torus $T \subset G$, the function $m(v, \ )$ on $\Gamma(T)$ takes integral values, and is the minimum of a finite set of $\mathbb{Z}$-linear functionals on $\Gamma(T)$. Furthermore, only a finite number of functions $m(v, \ )$ on $\Gamma(T)$ arise as v varies throughout V.

Proof: The first statement is implied by (10) and (12). The last statement is true because $m(v, \ )$ depends only on the T-state of v, and only a finite number of such states are possible. Q.E.D.

We conclude this section by giving a geometric criterion for stability. For each torus T, the T-weights form a finite subset $S(T)$ of the lattice $\chi(T)$ in the vector space $\chi(T) \otimes \mathbb{R}$. For each $v \in V$, the T-state of v, $S_v(T)$, is a subset of $S(T)$. Each one-parameter subgroup $\lambda \in \Gamma(T)$ determines a hyperplane $\langle \ ,\lambda \rangle = 0 \subset \chi(T) \otimes \mathbb{R}$. Note that we are now visualizing everything in the "weight space" $\chi(T) \otimes \mathbb{R}$, which is the dual, via $\langle \ , \ \rangle$, of the "one parameter subgroup space" where we have been working before. The geometric stability criterion is

Lemma 1.6 Suppose $v \in V$ is nonzero. Then

i) v unstable if and only if for some maximal torus $T \subset G$, and some $\lambda \in \Gamma^*(T)$, the weights in $S_v(T)$ lie in the positive half-space $\langle \ ,\lambda \rangle > 0 \subset \chi(T) \otimes \mathbb{R}$.

ii) v is semistable but not stable if and only if some maximal torus $T \subset G$ and some $\lambda \in \Gamma^*(T)$, the weights in $S_v(T)$ lie in the closed half-space $\langle \ ,\lambda \rangle \geq 0$, and some weight is on the hyperplane $\langle \ ,\lambda \rangle = 0 \subset \chi(T) \otimes \mathbb{R}$.

iii) v is stable if and only if for all maximal tori T, and all $\lambda \in \Gamma^*(T)$, the weights in $S_v(T)$ lie on both sides of the hyperplane $\langle \ ,\lambda \rangle = 0 \subset \chi(T) \otimes \mathbb{R}$.

Proof: The lemma follows from the numerical criterion Lemma 1.2, the expression for $m(v,\lambda)$ given in (12), and the following observations. For $\lambda \in \Gamma(T)$, $m(v,\lambda) > 0$ if and only if all the weights $\chi \in S_v(T)$ are in the positive half-space determined by $\langle \ ,\lambda \rangle > 0$. Thus i) holds. Part ii) holds for analogous reasons. To prove iii) we just remark that if $m(v,\lambda) \leq 0$, then $m(v,\lambda^{-1}) \ \ 0$, since $\langle \chi,\lambda^{-1} \rangle = -\langle \chi,\lambda \rangle$ for all $\lambda \in \Gamma(T)$. Q.E.D.

§2 Linear Algebra and Convexity

Throughout this section we will let W denote a real vector space of finite dimension n with positive definite inner product $\langle\ ,\ \rangle$. As usual we will let $\|w\|^2 = \langle w,w \rangle$. We will also assume that

*There is a fixed lattice $L \subset W$ of rank n, such that the inner product $\langle\ ,\ \rangle$ is integral on L.

We fix a finite nonempty subset A of L. Let L_A denote the corresponding set of linear functional on W, so $L_A = \{\langle a,\ \rangle : a \in A\}$. We define two functions $h : W \to \mathbb{R}$ and $g : W - \{0\} \to \mathbb{R}$ by

$$h(w) = \min \{\langle a,w \rangle : a \in A\} \text{ and} \tag{13}$$

$$g(w) = \frac{h(w)}{\|w\|} = \min \{ \langle a,\frac{w}{\|w\|} \rangle : a \in A\} \quad \text{for } w \neq 0. \tag{14}$$

Clearly g is constant on rays eminating from the origin and attains its supremum along some ray.

In this section, we will develop techniques using linear algebra and the notion of convexity, for finding and visualizing the supremum of g and the directions along which it is attained.

The first three lemmas will show that the supremum of g is attained in a rational direction. To show this, it suffices to prove that if the restriction of h to $S : \|w\| = 1$, the unit sphere, attains a relative maximum at w_0, then $\mathbb{R}\cdot w_0 \cap L$ is nonempty.

<u>Lemma 2.1</u> Suppose h attains a negative relative maximum on S at w_0. Then

i) n linearly independent functionals in L_A are equal to $h(w_0)$ at w_0, and

ii) the intersection $\mathbb{R}\cdot w_0 \cap L$ is nonempty.

<u>Proof</u>: Part ii) follows from part i) easily since the inner product is integral on the lattice. Part i) is trivial if $n = 1$. To prove part i) if $n > 1$, we first make the obvious, but useful,

Remark: Suppose ℓ is a nonconstant linear functional on W, where $\dim W \geq 2$. The restriction of ℓ to S has only two critical points, which are antipodal. In fact ℓ has negative minimum value and positive maximum value.

We can now prove the lemma. Suppose that $\ell_1, \ldots, \ell_m$ are the functionals in L_A with value $h(w_0)$ at w_0. Let $V \subset W$ be the subspace defined by $\ell_1 = \ldots = \ell_m$. Then in a neighborhood of w_0 in $V \cap S$, $h = \ell_1$. As h has a negative relative maximum on $V \cap S$, the remark implies that $\dim V = 1$. Hence, there are n linearly independent ℓ_i's. Q.E.D.

Lemma 2.2 Suppose h attains a zero relative maximum on S at w_0. Then in any neighborhood of w_0, there are points in $\mathbb{Q} \cdot L \cap \{v \in S : h(v) = 0\}$.

Proof: Let $\ell_1, \ldots \ell_m$ be the elements of L_A which have value zero at w_0. Let $V \subset W$ be the subspace defined by $\ell_1 = \ldots = \ell_m = 0$. Then h has value zero in a neighborhood of w_0 in $S \cap V$. The lemma follows because $\mathbb{Q} \cdot L \cap V \cap S$ is dense in $V \cap S$. Q.E.D.

Lemma 2.3 Suppose $h(w) > 0$ for some $w \in S$. Let $U \subset S$ be the set $U = \{w \in S : h(w) > 0\}$. Then the only relative maximum for h on U is an absolute maximum, say at w_o, and one of the following three conditions holds at w_0.

i) n linearly independent functionals in L_A are equal to $h(w_0)$ at w_0, or

ii) only one linear functional $\langle a, \ \rangle$ is equal to h at w_0, and the ray $\mathbb{R}^+ \cdot w_0$ is perpendicular to the hyperplane $\langle a, \ \rangle = 0$, or

iii) m linearly independent functionals $\langle a_i, \ \rangle$, $i = 1, \ldots,$ where $1 < m < n$, are equal to h at w_0, and in the linear subspace $V : \langle a_1, \ \rangle = \ldots = \langle a_m, \ \rangle$, the ray $\mathbb{R}^+ \cdot w_0$ is perpendicular to the hyperplane $\langle a_i, \ \rangle = 0$.

In all cases, the intersection $\mathbb{R} \cdot w_0 \cap L$ is nonempty.

Proof: The lemma is easily true if $n = 1$, so assume $n > 1$. We first note that the restriction of h to U is convex, so there is only one relative maximum, the absolute maximum. We next make the obvious

Remark: The restriction to S of a nonconstant linear functional $\langle a, \ \rangle$ on W, attains a relative maximum at w_0 only if the ray $\mathbb{R}^+ \cdot w_0$ is perpendicular to the hyperplane $\langle a, \ \rangle = 0$.

Thus, either condition i) holds or by the remark ii) or iii) holds. We still must prove the integrality statement. Clearly if i) holds, it is true. If ii) holds a is on the line $\mathbb{R} \cdot w_0$. If iii) holds, the common projection of the a_i's on V is on the line $\mathbb{R} \cdot w_0$. Q.E.D.

Note that $\langle a, \frac{w}{\|w\|} \rangle$ is the signed length of the projection of a in the direction of w. Thus a geometric restatement of the definition of g is

(15) $g(w)$ = the minimum of the set of signed lengths of the projections of a in the direction of w. where a ranges over the set A.

Thus A is a subset of the closed half-space $\langle \ , \frac{w}{\|w\|} \rangle \geq g(w)$ and the intersection of A with the bounding hyperplane $\langle \ , \frac{w}{\|w\|} \rangle = g(w)$ is nonempty. Note that $|g(w)|$ is the unsigned distance from 0 to the hyperplane $\langle \ , \frac{w}{\|w\|} \rangle = g(w)$ and $g(w) < 0$ if and only if 0 is the open halfspace $\langle \ , \frac{w}{\|w\|} \rangle > g(w)$. A geometric restatement of Lemma 2.1, thus, is

Lemma 2.4 Suppose $g(w) < 0$ for all $w \in W$, and suppose g attains a relative maximum at w_0. Then the hyperplane $\langle \ , \frac{w_0}{\|w_0\|} \rangle = g(w_0)$ contains n linearly independent vectors in A. Furthermore, the ray $\mathbb{R}^+ \cdot w_0$ has a nonempty intersection with the lattice L. Q.E.D. One can also restate Lemmas 2.2 and 2.3.

By considering the convex hull of A, which we will denote by $\hat{A}$, we can obtain geometric interpretations of both the sign of g and the maximum value of g. We first make the

Remark: $\hat{A}$ is the intersection of the closed half-spaces $\left\langle \ , \frac{w}{\|w\|} \right\rangle \geq g(w)$, where $w \in L$.

Lemma 2.5

i) The zero vector is not in $\hat{A}$ if and only if $g(w) > 0$ for some $w \in L$.

ii) The zero vector is on the boundary of $\hat{A}$ if and only if $g(w) \leq 0$ for all $w \in L$. and $g(w) = 0$ for some $w \in L$.

iii) The zero vector is in the interior of $\hat{A}$ if and only if $g(w) < 0$ for all $w \in L$.

Proof: It suffices to prove parts i) and iii). The remark implies part i) and the only if implication of part iii). Assume, then, that $g(w) < 0$ for all $w \in L$. The definition of g and Lemma 2.2 imply that $M = \max_{w \in W} g(w) < 0$, so each of the closed half-spaces $\left\langle \ , \frac{w}{\|w\|} \right\rangle \geq g(w)$ contains the open ball of radius $|M|$ centered at 0. Thus, the remark implies that 0 is in the interior of $\hat{A}$. Q.E.D.

An easy consequence of the previous lemma is

Lemma 2.6

i) The zero vector is not in $\hat{A}$ if and only if A is contained in the positive half-space $\langle \ ,w \rangle > 0$ for some $w \in L$.

ii) The zero vector is on the boundary of $\hat{A}$ if and only if A is contained in the closed half-space $\langle \ ,w \rangle \geq 0$ for some $w \in L$, but for no $w \in L$ is A contained in the open half-space $\langle \ ,w \rangle > 0$.

iii) The zero vector is in the interior of $\hat{A}$ if and only if for all $w \in L$, A intersects both the positive and negative half-spaces $\langle \ ,w \rangle > 0$ and $\langle \ ,w \rangle < 0$.

---

*We refer the reader to [E] for a discussion of convexity.

<u>Proof</u>: The lemma is implied by Lemma 2.5 together with the fact that for any $w \in W$, A is contained in the closed half-plane $\langle\ ,w\rangle \geq \|w\| \cdot g(w)$. Q.E.D.

This last lemma could also be proved independently without any mention of g.

A <u>face</u> of a convex set $C \subset W$ is a maximal convex subset of the boundary of C.

We can now give a geometric characterization of the maximum value attained by g in all cases.

<u>Theorem 2.7</u> Let M denote the maximum value attained by g in W. Then $|M|$ is the distance d from 0 to the boundary of the convex hull $\hat{A}$.

<u>Proof</u>: If 0 is on the boundary of $\hat{A}$, $d = 0$, and $M = 0$ also by Lemma 2.5, part ii. Suppose next that 0 is in the interior of $\hat{A}$. We observed in the proof of Lemma 2.5 that the largest closed ball in $\hat{A}$ centered at 0 has radius $|M|$. By Lemma 2.1, this ball meets the boundary of $\hat{A}$ in a direction in $Q \cdot L$ where there are n-linearly independent elements of A which span the face at this point of contact. Finally we assume that 0 is not in $\hat{A}$. In this case, d is the distance from 0 to $\hat{A}$. For each w, A is contained in the closed half-space $\langle \cdot, \frac{w}{\|w\|} \rangle \geq g(w)$, so $g(w) \leq d$ for all w and hence $M \leq d$. Lemma 2.3 implies that M is the distance from 0 to some point w_0 of $\hat{A}$. Thus $d \leq M$. Combining the inequalities gives $d = m$. Q.E.D.

We close this section by giving a "convex" analogue of Lemma 2.3.

<u>Lemma 2.8</u> Assume 0 is not in the convex hull $\hat{A}$ of A. The distance d from 0 to $\hat{A}$ is attained at a unique point w_0. Either

i) $w_0 \in \hat{A}$

ii) w_0 is perpendicular to and contained in a face of $\hat{A}$ of dimension n-1.

iii) w_0 is perpendicular to and contained in a face of the boundary of dimension m, $0 < m < n-1$.

In any of these cases, the ray $\mathbb{R}^+ \cdot w_0$ has a nonempty intersection with the lattice.

<u>Proof</u>: The lemma is an immediate consequence of the previous theorem and Lemma 2.3. Q.E.D.

§3 The G-invariant numerical function M

We return now to the context and notation of the first section, so let V be a representation of G. We will define, for each $v \in V$, a numerical function $\mu(v, \ ) : \Gamma^*(G) \to \mathbb{R}$. On tori, this function will be related to the numerical function $m(v, \ )$ of section 1 as $g(14)$ was to $h(13)$ in section two. Later we will study the supremum $M(v)$ of $\mu(v, \ )$, which we will show is invariant on G orbits in V.

A <u>length function</u> on $\Gamma(G)$ is a function $\| \ \| : \Gamma(G) \to \mathbb{R}$ with the properties

L_1: $\|\lambda\| \geq 0$ and $\|\lambda\| = 0$ if and only if λ is the trivial one parameter subgroup.

L_2: $\|\ell \cdot \lambda\| = |\ell| \cdot \|\lambda\|$ for $\ell \in \mathbb{Z}$

L_3: $\|g * \lambda\| = \|\lambda\|$ for all $g \in G$

L_4: If $T \subset G$ is a torus, there is an inner product $\langle\langle \ , \ \rangle\rangle$ on $\Gamma(T) \otimes \mathbb{R}$ which is integral on $\Gamma(T)$ and which has the property that $\langle\langle \lambda, \lambda \rangle\rangle \ \|\lambda\|^2$, for $\lambda \in \Gamma(T)$.

Next we will show how a representation of G determines a length function. Define $\| \ \| : \Gamma(G) \to \mathbb{R}$ by

$$(16) \qquad \|\lambda\|^2 = \sum_{n \in \mathbb{Z}} \dim V_n \cdot n^2$$

where $V = \bigoplus_{n \in \mathbb{Z}} V_n$ is the λ-decomposition of V.

<u>Lemma 3.1</u> The function $\| \ \| : \Gamma(G) \to \mathbb{R}$ defined in (16) is a length function on $\Gamma(G)$ if and only if the trivial one parameter subgroup of G acts trivially on V.

<u>Proof</u>: The assumption of the lemma is equivalent to the second part of L_1, i.e. $\|\lambda\|^2 \neq 0$ for $\lambda \in \Gamma^*(G)$. Assume now that only the trivial one parameter subgroup of G acts trivially on V. We must verify properties L_2, L_3, and L_4, and the first part of L_1. The definition (16) of $\| \ \|$ implies that the first part of L_1 does

hold. Let $V = \bigoplus_{n\in\mathbb{Z}} V_n$ be the λ-decomposition of V; then the $\ell\cdot\lambda$ decomposition of V is $V \bigoplus_{m\in\mathbb{Z}} \tilde{V}_m$ where $V_n = \tilde{V}_{\ell n}$ for all $n\in\mathbb{Z}$, so L_2 holds. Property L_3 holds since the $g*\lambda$ decomposition of V is the same as the λ-decomposition of V. To see that L_4 holds, define $\langle\!\langle\ ,\ \rangle\!\rangle : \Gamma(T)\otimes\mathbb{R} \times \Gamma(T)\otimes\mathbb{R} \to \mathbb{R}$ to be the unique bilinear form such that

$$(17)\qquad \langle\!\langle\lambda,\mu\rangle\!\rangle = \sum_{\chi\in\chi(T)} \dim V_\chi \cdot \langle\chi,\lambda\rangle\,\langle\chi,\mu\rangle$$

for λ and μ in $\Gamma(T)$, where $V = \bigoplus_{\chi\in\chi(T)} V_\chi$ is the T-weight decomposition of V. From (11) it is clear that $\langle\!\langle\lambda,\lambda\rangle\!\rangle = \|\lambda\|^2$. and $\langle\!\langle\ ,\ \rangle\!\rangle$ is integral valued on $\Gamma(T)$.

Fix a length function $\|\ \|$ on $\Gamma(G)$. We use this length function to define a new numerical function $\mu : V \times \Gamma^*(G) \to \mathbb{R}$, which normalizes the numerical function m. Namely, let

$$(18)\qquad \mu(v,\lambda) = \frac{m(v,\lambda)}{\|\lambda\|}$$

<u>Lemma 3.2</u> The numerical function $\mu : V \times \Gamma^*(G) \to \mathbb{R}$ has the following properties:

i) $\mu(v, \ell\cdot\lambda) = \mu(v,\lambda)$ for any positive $\ell\in\mathbb{Z}$

ii) $\mu(\alpha v,\lambda) = \mu(v,\lambda)$ for any $\alpha\in k$, $\alpha\neq 0$

iii) $\mu(g\cdot v, g*\lambda) = \mu(v,\lambda)$ for $g\in G$

iv) $\mu(v,\lambda) = \mu(p\cdot v,\lambda) = \mu(v, p^{-1}*\lambda)$ for $p\in P(\lambda)$

v) Lemma 1.2 remains true when $m(v,\lambda)$ is replaced by $\mu(v,\lambda)$.

<u>Proof:</u> The lemma follows immediately from the fact that $\|\ \|$ is a length function, properties (3), (4), (5) of m, and 1.3, part iii. Q.E.D.

Suppose now that $T\subset G$ is a torus. There is a canonical homomorphism $\chi(T)\otimes\mathbb{R} \to \Gamma(T)\otimes\mathbb{R} : \chi \to \lambda_\chi$ determined by $\langle\chi,\lambda\rangle = \langle\!\langle\lambda_\chi,\lambda\rangle\!\rangle$ for all $\lambda\in\Gamma(T)\otimes\mathbb{R}$. The image of the lattice $\chi(T)$ in $\chi(T)\otimes\mathbb{R}$ is a lattice which is contained in the lattice $\Gamma(T)$ for some integer m. For each $v \in V$, the function $\mu(v,\ ): \Gamma^*(T) \to \mathbb{R}$ extends to a continuous function on $\Gamma^*(T) \times \mathbb{R}$ since both $m(v,\ )$ and extend. Precisely (12) implies

(19) $\quad m(v,\lambda) = \min\{\langle\!\langle \lambda_\chi, \lambda \rangle\!\rangle : \chi \in S_v(T)\}$ for all $\lambda \in \Gamma(T) \otimes \mathbb{R}$ so

(20) $\quad \mu(v,\beta) = \min\{\langle\!\langle \lambda_\chi, \frac{\lambda}{\|\lambda\|} \rangle\!\rangle : \chi \in S_v(T)\}$ for all $\lambda \in \Gamma(T) \otimes \mathbb{R}$.

For each torus $T \subset G$ we also define a function $M_T : V \to \mathbb{R}$ by

(21) $$M_T(v) = \sup_{\lambda \in \Gamma^*(T)} \mu(v,\lambda)$$

Note that the theory developed in section two applies to the function $\mu(v,\ ) : \Gamma(T) \otimes \mathbb{R} \to \mathbb{R}$, for each $v \in V$. Denote the image of $\chi(T)$, $S(T)$, and $S_v(T)$ in $\Gamma(T) \otimes \mathbb{R}$ by $\chi(T)^*$, $S(T)^*$, and $S_v(T)^*$ respectively, and let m be an integer such that $\chi(T)^* \subset \frac{1}{m}$ (T). To apply the results of section two, simply let $L = \chi(T)^*$, $\langle\ ,\ \rangle = m^2 \langle\!\langle\ ,\ \rangle\!\rangle$, $A = S_v(T)^*$, $h = m(v,\ )$, and $g = \mu(v,\ )$. The theory implies, for example:

<u>Lemma 3.3</u> Suppose $v \in V$ and $T \subset G$ is a torus.

i) There exists $\lambda \in \Gamma^*(T)$ such that $M_T(v) = \mu(v,\lambda)$.

ii) If $v \neq 0$, $|M_T(v)|$ is the distance from 0 to the boundary of the convex hull $\widehat{S_v(T)^*}$ of the image of the T-state of v, $S_v(T)$, in $\Gamma(T) \otimes \mathbb{R}$.

<u>Proof</u>: If $v = 0$, then $S_v(T)$ is empty, so $\mu(v,\lambda) = \infty$ for all $\lambda \in \Gamma^*(T)$. If $v \neq 0$, then $S_v(T)$ is nonempty so Lemmas 2.1, 2.2, and 2.3 imply that there exists $\lambda \in \chi(T)^* \subset \frac{1}{m}\Gamma(T)$ such that $M_T(v) = \mu(v,\lambda)$. Since $m\cdot\lambda \in \Gamma(T)$, and $\mu(v,\lambda) = \mu(v, m\lambda)$ we have proved i). Part ii) follows from the definition of $M_T(v)$ and Theorem 2.9. Q.E.D.

An algorithm for computing $M_T(v)$ and $\Lambda_{v,T} = \{\lambda \in \Gamma^*(T) : M_T(v) = \mu(v,\lambda)\}$ is contained in Lemmas 2.1, 2.3, 2.8, and 2.10 and Theorem 2.9, in the cases $M_T(v) < 0$ and $M_T(v) > 0$. The algorithm is the nicest in the case that $M_T(v) < 0$.

We define a final numerical function $M : V \to \mathbb{R}$ by

(22) $$M(v) = \sup_{\lambda \in \Gamma^*(G)} \mu(v,\lambda)$$

Lemma 3.4 Suppose $T \subset G$ is a maximal torus. Then for each $v \in V$, the set of values $\{M_T(gv) : g \in G\}$ is finite and $M(v) = \max_{g \in G} M_T(gv)$.

Proof: If $\lambda \in \Gamma^*(G)$, $\lambda = g^{-1} * \lambda_0$ for some $\lambda_0 \in \Gamma^*(T)$ and some $g \in G$, so $\mu(v,\lambda) = \mu(gv,\lambda_0)$. Hence $M(v) = \sup_{g \in G} M_T(g \cdot v)$. By Lemma 1.5, as g ranges throughout G, only a finite number of functions $\mu(g \cdot v, \ ) : \Gamma^*(T) \to \mathbb{R}$ arise. Hence the set $\{M_T(gv) : g \in G\}$ is finite.

Lemma 3.5 For each $v \in V$, there exists $\lambda \in \Gamma^*(G)$ such that $M(v) = \mu(v,\lambda)$

Proof: The lemma is an immediate consequence of the two previous Lemmas 3.3 and 3.4. Q.E.D.

The usefulness of this new function $M : V \to \mathbb{R}$ is revealed in

Lemma 3.6

i) For each $v \in V$, the function M is constant on the G orbit of v, $G \cdot v$.

If v is a nonzero vector in V, then

ii) v is unstable $\Leftrightarrow M(v) > 0$,

iii) v is semistable $\Leftrightarrow M(v) \leq 0$,

iv) v is stable $\Leftrightarrow M(v) < 0$.

Proof: For part i) we note that

$$M(g \cdot v) = \sup_{\lambda \in \Gamma^*(G)} \mu(g \cdot v,\lambda) = \sup_{\lambda \in \Gamma^*(G)} \mu(v, g^{-1} * \lambda) = M(v).$$

Since $\mu(v, \ )$ actually attains the value $M(v)$ on some $\lambda \in \Gamma(G)$, and since Lemma 1.2 is true for $\mu(v, \ )$ as well as $m(v, \ )$, parts ii, iii, and iv are true. Q.E.D.

The theory of section two implies that the sign of M and the actual value of M have the following geometric interpretation.

Theorem 3.7 Suppose v is a nonzero vector in V and suppose $T \subset G$ is a fixed maximal torus. For any nonzero w in V let $\widehat{S_w(T)^*}$ denote the convex hull of the image of $S_w(T)$ in $\Gamma(T) \otimes \mathbb{R}$. Let $d_{w,T}$ denote the distance from 0 to the boundary of $\widehat{S_w(T)^*}$.

i) $M(v) > 0$ if and only if 0 is not in $S_{gv}(T)^*$ for some $g \in G$.

ii) $M(v) = 0$ if and only if 0 is on the boundary of $\widehat{S_{gv}(T)^*}$ for some $g \in G$, and for all $g \in G$, 0 is in $\widehat{S_{gv}(T)^*}$.

iii) $M(v) < 0$ if and only if for all $g \in G$, 0 is in the interior of $\widehat{S_{gv}(T)^*}$.

iv) If $M(v) > 0$, then $M(v) = \max \{d_{gv,T} : 0 \notin \widehat{S_{gv}(T)^*}\}$

v) If $M(v) < 0$, then $-M(v) = \min \{d_{gv,T} : g \in G\}$.

Proof: The theorem follows immediately from Lemma 2.5, Lemma 3.3, and Lemma 3.4. Q.E.D.

In order to compute $M(v)$ effectively, we would have to have a method for computing M_T on an orbit of G. We would like a method for computing all the convex hulls of weight that arise as G pushes v around in V. We would, of course, like to be able to pick out convex hulls, which have faces closest to 0.

A simplification of this problem is given in

Lemma 3.8 Let $T \subset G$ be a fixed maximal torus of G; let $B \supset T$ be a fixed Borel subgroup containing T; and let U be the unipotent radical of B. Then for $v \in V$, $v \neq 0$,

$$M(v) = \max_{u \in U} M_T(u \cdot v)$$

Proof: Clearly $M(v) \geq \max_{u \in U} M_T(u \cdot v)$. We must show that equality holds. By Lemma 3.5, there exists $\lambda_0 \in \Gamma^*(G)$ such that $M(v) = \mu(v, \lambda_0)$. Let $P(\lambda_0)$ be the associated parabolic subgroup.

The intersection $B \cap P(\lambda_0)$ contains a maximal torus T'. Since any two maximal tori in $P(\lambda_0)$ are conjugate, there exists $p \in P(\lambda_0)$ such that $p*\lambda_0 \subset T'$. By Lemma 1.3, $\mu(v,\lambda_0) = \mu(v,p*\lambda_0) = M(v)$. Hence we may assume $\mathrm{Im}\lambda_0 \in T' \subset B$. The fact that any two maximal tori in B are conjugate via an element of U implies $\lambda_0 = u^{-1}*\lambda$ for some $u \in U$, and some $\lambda \in \Gamma^*(T)$. Applying Lemma 3.2, gives $M(v) = \mu(v,\lambda_0) = \mu(v,u^{-1}*\lambda) = \mu(u\cdot v,\lambda)$. Q.E.D.

There is also a nice uniqueness result in the case that v is unstable, which is the case where our algorithm for $M_T(v)$ is poorest. Clearly if $M(v) = \mu(v,\lambda)$, then $M(v) = \mu(v,p*\lambda)$ for any $p \in P(\lambda)$. Let Λ'_v be the set of indivisible one parameter subgroups such that $M(v) = \mu(v,\lambda)$

Theorem (Kempf, Theorem 1.2 in [K]): Suppose $v \in V$ is unstable. Then there exists $\lambda \in \Lambda'_v$ such that $\Lambda'_v = \{p*\lambda : p \in P(\lambda)\}$.

To obtain more complete information about stability and instability via numerical functions, it would seem necessary to quantify other information about the convex hulls of the T states $S_{g\cdot v}(T)$, in addition to the distance from the origin.

§4 Projective Hypersurfaces*

In this section we want to restrict our attention to the representation

$$SL(n,k) \times \mathrm{Sym}^d(V_n^*) \to \mathrm{Sym}^d(V_n^*)$$

of $SL(n,k)$ on the space of homogeneous forms of degree d in the fixed variables $X_1,\dots,X_n$. The representation for linear forms $(d = 1)$ is obvious. The representation for forms of higher degree $d > 1$ is uniquely determined by

*The natural multilinear mapping from the d-fold product $V_n^* \times \dots \times V_n^* \to \mathrm{Sym}^d(V_n^*)$ given by multiplication is $SL(n,k)$ equivariant.

We will first make explicit some of the notions of the previous section for this representation. Let $T \subset SL(n,k)$ denote a maximal torus; let $Y_1,\dots,Y_n$ be a basis of V_n^* with respect to which T acts diagonally via the characters $t_1,\dots,t_n$. Then T acts diagonally with respect to the monomials of degree d in the Y_i's. In other words, the monomials of degree d in the Y_i's form a basis for the T-weight decomposition of $\mathrm{Sym}^d(V_n^*)$.

The characters $t_1,\dots,t_n$ induce an isomorphism of T with the algebraic subvariety of $(k^*)^n$ defined by $\prod_{i=1}^n t_i = 1$. Identify T with $\prod_{i=1}^n t_i = 1 \subset (k^*)^n$.

The free abelian group $\Gamma(T)$ of one-parameter subgroups is isomorphic to the intersection of the lattice $\mathbb{Z}^n$ with the hyperplane $\Sigma u_i = 0$ in $\mathbb{R}^n$, since every $\lambda \in \Gamma(T)$ can be uniquely described by the n-tuple of integers a such that $\lambda(t) = (t^{a_1},\dots,t^{a_n})$. Denote this λ by λ_a. Hence we may identify $\Gamma(T) \otimes \mathbb{R}$ with $\Sigma u_i = 0 \subset \mathbb{R}^n$.

The reader is also referred to G.I.T. Chapter 4, section 2 and to [M]

Clearly the Euclidean length function on $\Sigma u_i = 0$ determines a length function on $\Gamma(T)$. This is the length function that we shall use. One can show that the Euclidean length function differs from the length function induced by the representation (17) by a constant depending on d.

Every character of T has the form $\chi^i = \prod_{j=1}^{n} t_j^{i_j}$, $i \in \mathbb{Z}^n$. The pairing $\langle \chi^i, \lambda_a \rangle = \sum_{j=1}^{n} i_j a_j$, so is the pairing induced by the Euclidean pairing in $\mathbb{R}^n$. Hence, the free group of characters of T, $\chi(T)$, is isomorphic to $\mathbb{Z}^n/\mathbb{Z}^n \cap (\Sigma u_i = 0)$. For two such characters $\chi^i, \chi^{i'}$ are equal if and only if they agree on all one-parameter subgroups. The canonical homomorphism $\chi(T) \to \Gamma(T)$ is thus realized by perpendicular projection of $\mathbb{Z}^n \subset \mathbb{R}^n$ onto $\Sigma u_i = 0 \subset \mathbb{R}^n$.

Let $S^d(T)$ be the set of T weights for the representation of $SL(n,k)$ in the space of degree d forms. If $Y^i = Y_1^{i_1} \cdot \ldots \cdot Y_n^{i_n}$ is a monomial of degree d, T acts on Y^i by the character χ^i. Thus $S^d(T)$ may be identified with set of monomials of degree d in the Y_i's. $S^d(T)$ may also be identified the n-tuples of non-negative integers in the hyperplane $\Sigma u_i = d$. The convex hull of $S^d(T)$ has the monomials Y_i^d, $i = 1,\ldots,n$ as its extreme points, and has $d/n(1,\ldots,1)$ as its center. The image of $S^d(T)$ under the canonical homomorphism $\chi(T) \to \Gamma(T)$, is obtained by parallel translation of the plane $\Sigma u_i = d$ to $\Sigma u_i = 0$. This carries the center $d/n(1,\ldots,1)$ to the origin. Hence we may, equally well, work in the weight hyperplane $u_i = d$, by viewing one-parameter groups as vectors in the weight hyperplane emanating from $d/n(1,\ldots,1)$, the center of the convex hull of the set of T weights $S^d(T)$.

Every form F in $\mathrm{Sym}^d(V_n{}^*)$ defines a hypersurface of degree d, $H_F : F = 0$ in $\mathbb{P}(V_n)$. The hypersurface H_F determines a form of degree d, uniquely up to constant multiple. Lemma 3.2 implies that the degree of stability is a projective notion, so we can speak about the stability of projective hypersurfaces, of a fixed degree, and a fixed number of variables.

Two easy, general results which relate the geometry to stability are:

Lemma 4.1 Suppose $F \in \mathrm{Sym}^d(V_n^*)$, where $d \geq 3$ and $n \geq 3$. Let $T \subset SL(n,k)$ be a maximal torus, which acts diagonally with respect to the basis $Y,\ldots,Y_n$ of V_n^*. The hypersurface $H_F : F = 0 \subset \mathbb{P}(V_n)$ has a singular point at $[Y_1,\ldots,Y_n] = [1,0,\ldots,0]$ if

i) $d > n$ and $m(F,\lambda_a) \geq 0$ where $a = (a_1,\ldots,a_n)$ with $a_1 \leq a_2 \leq \ldots \leq a_n$.

ii) $d = n$ and $m(F,\lambda_a) > 0$ where $a = (a_1,\ldots,a_n)$ with $a_1 \leq a_2 \leq \ldots \leq a_n$.

Proof: It suffices to show that in the T-eigen decomposition $F = \Sigma c_i Y^i$, the coefficients of $Y^d_.$ and $Y_1^{d-1} Y_k$ for $K = 2,\ldots,n$ are zero. To show this, we just check that the λ_a weights on these T-eigenspaces are negative for i) and non-positive for ii). The λ_a weights are da_1 and $(d-1)a_1 + a_K$, respectively. Since $a_1 \leq \ldots \leq a_n$, we need only show that $(d-1)a_1 + a_n < 0$ if $d > n$ and $(d-1)a_1 + a_n \leq 0$ if $d = n$. Note that $\lambda_a \in \Gamma^*(T)$, and $T \subset SL(n,\mathbb{C})$ implies that $\Sigma a_i = 0$ and $a_1 \neq 0$, so $a_1 < 0$ and $a_n > 0$. The largest value of $a_n/|a_1|$ occurs in multiples of the one-parameter subgraoup λ_a, where $a_1 = \ldots = a_{n-1} = -1$ and $a_n = n-1$. There $(d-1)a_1 + a_n = n-d$. Q.E.D.

Lemma 4.2 Suppose $F \in \mathrm{Sym}^d(V_n^*)$ determines a nonsingular hypersurface $H_F : F = 0$ in $\mathbb{P}(V_n)$, where $d \geq 3$ and $n \geq 3$.

i) If $d > n$, then F is stable.

ii) If $d = n$; then F is semistable.

Proof: The lemma is an immediate consequence of the previous lemma if one uses the numerical criterion in Lemma 1.2. Q.E.D.

Mumford, using the discriminant and the invariant theory definition of stability, proves in Proposition 4.2 of [G.I.T.] a more general result than Lemma 4.2; namely: If $F \in \mathrm{Sym}^d(V_n^*)$, $d \geq 3$, $n \geq 3$, determines a nonsingular hypersurface in $\mathbb{P}(V_n)$, then F is stable. We have proven, directly, a stronger result in Lemma 4.1, in the case $d > n$.

Our Lemma 4.1 fails to extend to the case where $d < n$. To see this consider the

Example: Let $F = Y_1^{d-1}Y_n$. Let $a = (-1,-1,\dots,n-1)$. Then $m(F,\lambda_a) = n-d > 0$ if $n > d$, but the point $[Y_1,\dots,Y_n] = [1,0,\dots,0]$ is nonsingular.

§5 Plane Curves

We consider here the representations

$$SL(3,k) \times Sym^d(V_3^*) \to Sym^d(V_3^*)$$

Note that, if T is a maximal torus which acts diagonally with respect to the basis X, Y, and Z, the convex hull of the T weights, $\widehat{S^d(T)}$ is a triangle with vertices at the highest weights X^d, Y^d, Z^d

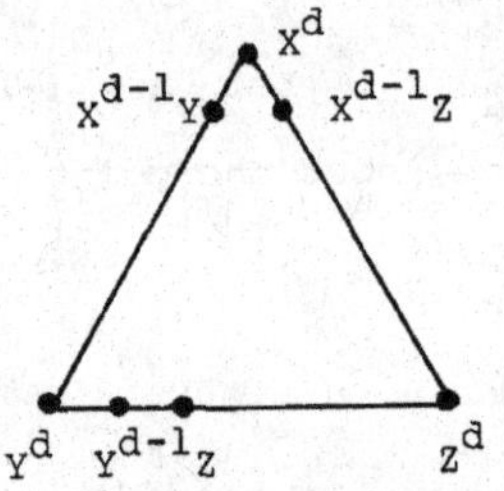

We begin by defining an order on the points of a plane curve. Intuitively, the maximal points will be the worst points. Let C_F denote the plane curve defined by $F = 0$ in $\mathbb{P}(V_3)$. Suppose $P \in C_F$. Let T_p denote the tangent cone to C_F at P. We define a triple

$$\sigma(P) = (m_T(P), m(P), m_I(P)$$

for each $P \in C_F$ as follows:

$$m_T(P) = \max\{\text{multiplicity of } L : L \in T_p\}$$
$$m(P) = \text{the multiplicity of } P \text{ on } C_F$$
$$m_I(P) = \max\{I(L \cap C_F; P) : L \in T_p\}$$

Note that $m_T(P)$ and $m(P)$ are always positive integers, while $m_I(P)$ may be a positive integer or ∞.

Let $<$ denote the following partial ordering on triples of extended real numbers: $i = (i_1, i_2, i_3) < j = (j_1, j_2, j_3)$ if and only if

1) $i_1 < j_1$ or

2) $i_1 = j_1$ and $i_2 < j_2$ or

3) $i_1 = j_1$, $i_2 = j_2$, and $i_3 \leq j_3$.

The ordering $<$ on triples of extended real numbers induces an ordering on the triples $\sigma(P)$.

A point $P \in C_F$ is maximal if and only if $\sigma(P) \geq \sigma(Q)$ for all $Q \in C_F$.

For each $P \in C_F$, there is a natural flag F_p of V_3 associated to C_F. Let $T'_p \subset T_p$ be the subset of the tangent cone having the highest order contact with C, so

$$T'_p = \{L \in T_p : I(L \cap C_F, P) = m_I\}$$

Let Span T'_p denote the smallest projective subspace of $\mathbb{P}(V_3)$ containing T'_p. Let $\Pi : V_3 \to \mathbb{P}(V_3)$ denote the canonical map. The flag at P is defined by

(23) $$F_p : \Pi^{-1}(P) \subset \Pi^{-1}(\text{Span } T'_p) \subset V_3$$

(24) Let $\Lambda'_{F,P} = \{\text{indivisible } \lambda \in \Lambda_F' : P(\lambda) \text{ fixes } F_p\}$.

Note that, for each flag F_p, there is a unique maximal parabolic subgroup fixing F_p. If Π^{-1} (Span T'_p) $\neq V_3$, there is a unique parabolic subgroup, a Borel subgroup, which fixes F_p. However, if Π^{-1} (Span T'_p) $= V_3$, there are two kinds of parabolics which fix F_p.

Lemma 5.1 Suppose for each maximal torus T, $\Lambda'_{F,P} \cap \Gamma^*(T)$ contains at most one one-parameter subgroup. If $\lambda_0 \in \Lambda'_{F,P}$, and if $P(\lambda_0) \supseteq P(\lambda)$ for all other $\lambda \in \Lambda'_{F,P}$, then

$$\Lambda'_{F,P} = \{p*\lambda_0 : p \in P(\lambda_0)\}.$$

Proof: If $p \in P(\lambda_0)$, the definition (6) of $P(\lambda_0)$ implies $P(p*\lambda_0) = P(\lambda_0)$, and Lemma 1.3 implies $\mu(F,\lambda_0) = \mu(F,p*\lambda_0)$, so $p*\lambda_0 \in \Lambda'_{F,P}$. Conversely, assume $\lambda \in \Lambda'_{F,P}$. Since $P(\lambda_0) \supseteq P(\lambda)$, λ and λ_0 are both in $P(\lambda_0)$. Thus there are maximal tori T and T_0 of $SL(3,k)$ contained in $P(\lambda_0)$, such that $\lambda \in \Gamma^*(T)$ and

$\lambda_0 \in \Gamma^*(T_0)$. Choose $p \in P(\lambda_0)$ such that $p*T_0 = T$. Then $p*\lambda_0$ and λ are both in $\Lambda'_{F,P} \cap \Gamma^*(T)$. By the uniqueness hypothesis, $p*\lambda_0 = \lambda$. Q.E.D.

By Lemma 4.2 and Mumford's result, nonsingular plane curves of degree $d \geq 3$ are stable. We can compute $M(F)$ for such curves in terms of the order of the worst flex on the curve.

Recall that $P \in C_F$ is a <u>flex</u> if and only if $m_I(P) = I(C_F \cap L_p, P) \geq 3$, where L_p is the tangent line at P. If P is a flex, we say that the <u>order of the flex</u> at P is $m_I(P) - 2$.

If C_F is a nonsingular plane curve of degree $d \geq 3$, then, the maximal points of C_F are precisely the flexes of highest order.

Let $T \subset SL(3,k)$ be a maximal torus which acts diagonally with respect to the basis Y_1, Y_2, Y_3 of V_3^*. Let P_i denote the point $Y_j = Y_k = 0$ for $j \neq i$, $k \neq i$. Let L_j denote the line $Y_j = 0$.

<u>Lemma 5.2</u> Suppose $F \in \mathrm{Sym}^d(V_3^*)$ $d \geq 3$ defines a nonsingular curve $C_F \subset \mathbb{P}(V_3)$. Assuming the above notation, let

$$e = \max\{I(C_F \cap L_j, P_i) : i,j \in \{1,2,3\},\ i \neq j\}$$

i) If the curve does not contain any of the points P_i, $i = 1,2,3$, or if $e \leq \frac{d}{2}\left[\frac{2d-4}{2d-3}\right]$, then $M_T(F) = \frac{d}{\sqrt{6}}$

ii) If $e > \frac{d}{2}\left[\frac{2d-4}{2d-3}\right]$ then $M_T(F) = \frac{d+(d-3)e}{\sqrt{6(e^2-e+1)}}$

iii) If $e = I(L_1 \cap C_F, P_3) > \frac{d}{2}\left[\frac{2d-4}{2d-3}\right]$ then there exists a unique indivisible one-parameter subgroup λ_a, $a_1 \geq a_2 \geq a_3$, such that $M_T(F) = |\mu(F,\lambda_a)|$. Here $a = (2e-1,\ 2-e,\ -(1+e))$ and $P(\lambda_a)$ acting on V_3 via the adjoint action preserves the flag $P_3 \subset L_1 \subset V_3$.

Before giving the proof we remark that for $d = 3$ and $d = 4$ $2 > \frac{d}{2}\left[\frac{2d-4}{2d-3}\right] \geq 1$, so the hypothesis of part ii) means that C_F is tangent to L_j at P_i, for some i and j. For $d > 4$, the hypothesis means that C_F has a flex of sufficiently high order at some P_i, and is tangent to some L_j there.

Proof: Since $d \geq 3$, F is stable. Thus 0 is in the interior of $\widehat{S_F^d(T)^*}$, the convex hull of the image of the T-state of F in $\Gamma(T) \otimes \mathbb{R}$, and $|M_F(T)|$ is the distance from 0 to the boundary of $\widehat{S_F^d(T)^*}$. One can check that $\frac{d}{\sqrt{6}}$ is the distance from 0 to the boundary of $\widehat{S_F^d(T)^*}$. Thus, if none of the points P_i are on C_F, $\widehat{S_F^d(T)^*} = \widehat{S^d(T)^*}$ since $\widehat{S_F^d(T)^*}$ contains all the extreme points of $\widehat{S^d(T)^*}$, so $M_T(F) = d/\sqrt{6}$

Next suppose ℓ is a line containing a face of the convex hull $\widehat{S_F^d(T)}$. Then ℓ must either be a part of the boundary of $\widehat{S^d(T)}$, or up to permutation of coordinates must be a line joining $Y_1Y_3^{d-1}$ and $Y_2^mY_3^{d-m}$. Suppose not. Then if ℓ were parallel to a face of $\widehat{S^d(T)}$, F would be reducible, hence singular, which contradicts the assumption on F. Otherwise, up to permutation of coordinates ℓ would be a line containing $Y_1^pY_3^{d-1}$ and $Y_2^mY_3^{d-m}$ for some m and p both greater than 1. This would imply that some P_i was a singular point of F, which again is a contradiction.

A line joining $Y_1Y_3^{d-1}$ and $Y_2^mY_3^{d-m}$ is on the boundary of the convex hull if and only if $I(L_1 \cap C_F, P_3) = m$. The if implication is clear. The only if implication is true because by Lemma 2.8 a face must contain two linearly independent weights. The only weights on ℓ are $Y_1Y_3^{d-1}$ and $Y_2^mY_3^{d-m}$. Clearly the analagous statement is true for any line obtained from ℓ by permutation of coordinates.

One can check that the distance from 0 to the line joining $Y_1Y_3^{d-1}$ and $Y_2^mY_3^{d-m}$ is

$$|\mu(F,\lambda_a)| = \frac{d + (d-3)\,m}{\sqrt{6}\,(m^2-m+1)}$$

If a is indivisible, $a = (2m-1,\ 2-m,\ -(1+m))$. One can also check that

$$\frac{d + (d-3)m}{\sqrt{6}\,(m^2-m+1)} < \frac{d}{\sqrt{6}} \Leftrightarrow m > \frac{d}{2}\left[\frac{2d-4}{2d-3}\right]$$

and that the expression on the left decreases as m increases. The uniqueness asserted in iii) holds since only one face of $\widehat{S^d_F(T)^*}$ cuts off the corner containing P_3. The last statement in iii) follows from the definition (6) of $P(\lambda)$. Q.E.D.

Theorem 5.3 Suppose $F \in \mathrm{Sym}^d(V_3^*)$ defines a nonsingular curve in $\mathbb{P}(V_3)$. Let $e-2$ denote the maximum of the order of the flexes on C_F. Then

i) if $e > \frac{d}{2}\left[\frac{2d-4}{2d-3}\right]$, $M(F) = \frac{-(d+(d-3)e)}{\sqrt{6(c^2-e+1)}}$ and $\Lambda'_F = \bigsqcup_{P \in C_F} \Lambda'_{F,P}$ and $\Lambda'_{F,P}$ is nonempty if and only if P is a maximal point, i.e. a flex of maximal order on C_F.

ii) If $e \le \frac{d}{2}\left[\frac{2d-4}{2d-3}\right]$, $M(F) = \frac{-d}{\sqrt{6}}$ and $\Lambda'_F = \{g * \lambda_0$, where λ_0 is indivisible and is perpendicular to the boundary of $\widehat{S^d(T)^*}$. Here T is the maximal torus which acts diagonally with respect to the λ-decomposition of V_3^*.

Proof: The first part of part i) follows from Lemma 5.2, part ii, and the fact that $M(F) = \max\{M_T(F) : T \subset SL(3,k)$ is a maximal torus$\}$. The second part of part i) follows from Lemma 5.2, part iii, and Lemma 5.1. The first part of part ii follows from Lemma 5.3, part i. The last statement in part ii holds because $\frac{d}{\sqrt{6}}$ is the distance from 0 to the boundary of $\widehat{S^d(T)^*}$ and because conjugation by g is an isometry. Q.E.D.

§6 Cubic and Quartic Curves

We will close this paper by computing the function M and the subset Λ'_F for stable and unstable, cubic and quartic plane curves. The summary of our results is

Theorem 6.1 Fix $d = 3$. Suppose F, F_1 and F_2 in $\mathrm{Sym}^d(V_3{}^*)$ are stable or unstable.

i) $M(F)$ depends only on the maximal points $P \in C_F$.

ii) If P_i is a maximal point for F_i, $i = 1, 2$ and if $\sigma(P_1) < \sigma(P_2)$, then $M(F_1) < M(F_2)$.

iii) $\Lambda'_F = \bigcup_{p \in C_F} \Lambda'_{F,P}$ and $\Lambda'_{F,P}$ is nonempty if and only if P is a maximal point of C_F.

Theorem 6.1': Theorem 6.1 is true for stable quartics, i.e. stable $F \in \mathrm{Sym}^4(V_3{}^*)$.

First we give the geometric meaning of the various notions of stability in these two cases.

Lemma 6.2 A form $F \in \mathrm{Sym}^3(V_3{}^*)$, $F \neq 0$ is

i) stable if and only if C_F is nonsingular

ii) unstable if and only if C_F has a cusp or a triple point

iii) semistable but not stable if and only if C_F has an ordinary double point.

Proof: Below we have drawn the configuration of T weights $S^3(T)$ for a maximal torus acting diagonally with respect to Y_1, Y_2, Y_3.

(25) 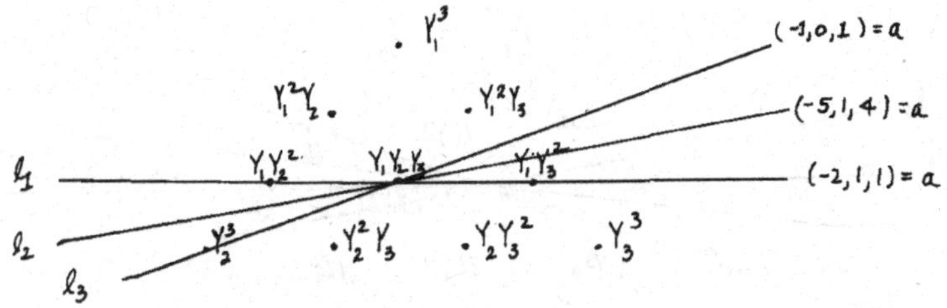

The lemma now follows easily from the diagram, using Lemmas 2.6, 3.6, and Theorem 3.7. For a form F is unstable if and only if for some maximal torus T, $S_T(F)$ is contained in an open half-plane determined by one of the lines. A form F is semistable but not stable if and only if for some maximal torus F lies in a closed, but not an open, half-plane determined by one of the three lines. Since a cubic curve can only have a double or triple point, the stable forms determine nonsingular curves. Q.E.D.

Recall that $P \in C_F$, degree $F \geq 4$, is a <u>tacnode</u> if and only if $\sigma(P) = (2,2,K)$ where $K > 3$. In other words P is a tacnode if and only if P is a cusp on C_F and the tangent line, $L = 0$, to P at C_F, agrees to higher than usual order.

<u>Lemma 6.3</u> Suppose $F \in \mathrm{Sym}^4(V_3{}^*)$.

i) F is unstable if and only if F has a triple point or F has a tacnode at P, and $F = L\cdot\tilde{F}$ where $L = 0$ is tangent to the tacnode.

ii) F is semistable but not stable if and only if F has a tacnode at P, and the linear form $L = 0$ which defines the tangent line at P is not a factor of F.

iii) F is stable if F is nonsingular or if F has a double point which is not a tacnode.

<u>Proof</u>: Consider the diagram below of T weights, $S^4(T)$, where T is a maximal torus acting diagonally with respect to Y_1, Y_2, Y_3.

(26) 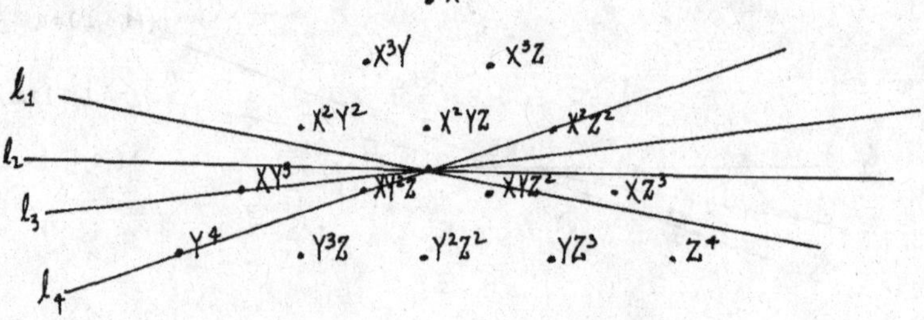

We have drawn the 4 determining lines. The proof now is the same as that of the previous lemma. Q.E.D.

We now begin to prove a sequence of three lemmas that will imply Theorem 6.1. We remark that Theorem 6.1 for stable cubics is implied by Theorem 5.3. Since a stable cubic is nonsingular its maximal points are precisely the 9 flexes of order 1. One can check that $M(F) = {}^{-3}/\sqrt{42}$ for stable cubics using Theorem 5.3. Hence M is constant on stable cubics.

Lemma 6.4 Suppose $F \in \mathrm{Sym}^3(V_3{}^*)$ is unstable. Let T be a fixed maximal torus which acts diagonally with respect to Y_1, Y_2, Y_3. The orbit of F under $SL(3,4)$ contains exactly one of the following forms, up to constant multiple.

$$F_1 = Y_1{}^2Y_3 + c_{03}Y_2{}^3, \quad c_{03} \neq 0$$
$$F_2 = Y_1{}^2Y_3 + c_{12}Y_1Y_2{}^2, \quad c_{12} \neq 0$$
$$F_3 = Y_1{}^3 + c_{21}Y_1{}^2Y_2 + c_{12}Y_1Y_2{}^2 + c_{03}Y_2{}^3, \quad c_{03} \neq 0$$
$$F_4 = Y_1{}^2Y_2$$
$$F_5 = Y_1{}^3$$

Proof: The lemma follows from Lemma 6.2, and the diagram (25). It is fairly easy to check that if F, with respect to some torus, lies in an open half-plane determined by one of the lines in (25), then the orbit of F contains one of the F_i's. Since the singularities of the F_i's are distinct, F can be brought into only one of these forms. Clearly each of these forms is unstable. One could also argue via maximal points. Q.E.D.

Remark: Each of the curves C_{F_i}, $i = 1,2,3,4$ have a unique maximal point j. For C_{F_i}, $i = 1,2,3$ it is the unique singular point. For C_{F_4} it is the triple point which is the intersection of the two components. Every point of C_{F_5} is maximal. If P_i denotes the unique maximal point, one can check that

$$\sigma(P_1) = (2,2,3) < \sigma(P_2) = (2,2,\infty) < \sigma(P_3) = (1,3,2) < \sigma(P_4) = (2,3,\infty) < \sigma(P_5) = (3,3,\infty)$$

Lemma 6.5 Suppose $F \in \mathrm{Sym}^3(V_3{*})$ is unstable. Let T be a torus as in Lemma 6.4.

a) For each of the unstable forms F_i, $M_T(F_i) = M(F_i)$. More precisely $M(F_5)=\sqrt{6}>M(F_4)=\sqrt{2}>M(F_3)=3/\sqrt{6}>M(F_2)=\frac{1}{\sqrt{2}}>M(F_1)=\frac{3}{\sqrt{42}}$

b) $\Lambda'_{F_i} = \{p*\lambda_{a^i} : p \in P(\lambda_{a^i})\}$ where $a^5 = (2,-1,-1)$, $a^4 = (1,0,-1)$, $a^3 = (1,1,-2)$, $a^2 = (1,0,-1)$, $a^1 = (4,1,-5)$

c) The parabolic subgroup $P(\lambda_a i)$ acting on V_3, via the adjoint action, fixes the flag at the maximal point of F_i.

Proof: Part c) follows easily from the second part of b) and the fact that the maximal points for the F_i are at $[0,0,1]$. Using the fact that $M_T(F_i)$ = the distance from the center $(1,1,1)$ of $S^3(T)$ to the boundary of $S_{F_i}(T)$ one can easily check that $M_T(F_i)$ has the value asserted in part a) and that the distance to $S_{F_i}(T)$ is attained in the direction a^i. Since Lemma 2.11 and Theorem 3.7, imply that the distance is attained in a unique direction, Lemma 5.1 and part c) imply the first part of b). It remains to show that $M(F_i) = M_T(F_i)$. To do this we use Lemma 3.8, which says that $M(F_i) = \sup_{u\in U} M_T(u\cdot F_i)$ where $u \in U$ and U is a unipotent radical of some Borel subgroup B containing T. Let U be the group of unipotent transformations of the form $Y_1 \quad Y_1$, $Y_2 \quad Y_2 + aY_1$, $Y_3 \quad Y_3 + bY_2 + cY_1$. It suffices to show that the distance from the center $(1,1,1)$ to $S_{u\cdot F_i}(T)$ equals the distance from the center $(1,1,1)$ to $S_{F_i}(T)$ for all $u \in U$, for each $i = 1,2,\ldots 5$. But one can check that $u \cdot F_5 = F_5$, $u\cdot F_4 = Y_1{}^2Y_2 + aY_1{}^3$, $u\cdot F_3$ is another F_3 from, $u\cdot F_2 = F_2 + eY_1{}^2Y_2 + fY_1{}^3$, and $u\cdot F_1 = F_1 + gY_1{}^2Y_2 + kY_1Y_a{}^2 + \ell Y_1{}^3$, for some e, f, g, h, and ℓ in k. If one next uses the diagram in the proof of Lemma 5.1 to compare the convex hull $S_{F_i}(T)$ with the convex hull $S_{u\cdot F_i}(T)$, one will see that the distance from the center to the boundary is independent of u, for each i. Q.E.D.

The proof of Theorem 6.1 follows immediately from the remarks preceding Lemma 6.4, the remark before Lemma 6.5 and Lemma 6.5.

Let $T \subset SL(3,k)$ be a maximal torus which acts diagonally with respect to the basis Y_1, Y_2, Y_3 of $V_3{}^*$. Let P_i denote the point $Y_j = Y_k = 0$, $i \neq j,k$. Let L_j denote the line $Y_j = 0$.

Lemma 6.6 Suppose $F \in Sym^4(V_3{}^*)$ is stable; let C_F denote the curve defined by $F = 0$ in $\mathbb{P}(V_3)$. Assume $\sigma(P_3) \geq \sigma(P_i)$, $i=1,2$.

a) If P_3 is not on C_F, or if neither of the lines, L_1 or L_2, through P_3 is tangent to C_F at P_3, then $M_T(F) = \frac{-4}{\sqrt{6}}$ i.e. $|M_T(F)|$ is the distance from the center to the boundary of $S^4(T)$.

b) Assume L_1 is tangent to C_F at P_3, $L_1 \in T_{P_3}$, and L_1 realizes $m_I(P)$.

i) $M_T(F) = -\sqrt{2}$ if $\sigma(P_3) = (1,1,2)$,

ii) $M_T(F) = -7/\sqrt{42}$ if $\sigma(P_3) = (1,1,3)$,

iii) $M_T(F) = -8/\sqrt{78}$ if $\sigma(P_3) = (1,1,4)$,

iv) $M_T(F) = -1/\sqrt{2}$ if $\sigma(P_3) = (1,2,2)$,

v) $M_T(F) = -4/\sqrt{42}$ if $\sigma(P_3) = (1,2,3)$ and F is not reducible

vi) $M_T(F) = -1/\sqrt{6}$ if $\sigma(P_3) = (1,2,\infty)$ or if $\sigma(P_3) = (1,1,\infty)$

vii) $M_T(F) = -2/\sqrt{42}$ if $\sigma(P_3) = (2,2,3)$

Furthermore if L is tangent to C_F at P_3, then $\sigma(P_3)$ is one of the above triples.

c) The possible values for $M_T(F)$ satisfy the inequalities
$-4/\sqrt{6} < -\sqrt{2} < -7/\sqrt{42} < -8/\sqrt{78} < -1/\sqrt{2} < -4/\sqrt{42} < -2/\sqrt{42}$

d) If L_1 is tangent to C_F at P_3, and if $L_1 \subset T'_{P_3}$, so L_1 is a tangent having the highest contact at P_3, then there exists a unique indivisible one parameter subgroup λ_a, $a_1 \geq a_2 \geq a_3$, such that $M_T(F) = \mu(F,\lambda_a)$.

Proof: One can easily check the inequalities in part c). Part a) and the first three parts of b) follow from Theorem 5.3. One can easily check the last statement in b) using the geometric condition for stability in Lemma 6.3. To prove the rest of part b, we consider the diagram (26) of $S^4(T)$, and argue as follows. Let d_i equal the minimum of the distances from the center $C = 4/3(1,1,1)$ of $S^4(T)$ to the lines, which contain the faces of $S^4_F(T)$, for which P_i is not in the same open half plane as C. Then $M_T(F) = \min\{d_i : i-1,2,3\}$. Now d_3 is the distance from C to the line ℓ containing $X^{m_T} Y^{m-m_T} Z^{d-m}$ and $Y^{m_I} Z^{d-m_I}$ where $\sigma(P_3) = (m_T, m, m_I)$. To justify this we first check that ℓ contains a face of $S^4_F(T)$. The weights which determine ℓ, $X^{m_T} Y^{m-m_T} Z^{d-m}$ and $Y^{m_I} Z^{d-m_I}$ must be in $S_F(T)$, for $\sigma(P_3)$ to equal (m_T, m, m_I). Since $m \le 2$ and $d = 4$, ℓ contains $S_F(T)$ in the closed half-plane not containing P_3. By Lemma 2.8, then ℓ contains a face of $S^4_F(T)$. Next we note that $d = 4$ implies that there are at most two lines containing faces for which P_3 is not in the same closed half plane as 0, (i.e. we have omitted the boundaries of $S^4_F(T)$ containing P_3). However L, by assumption was a tangent of highest order contact which realized m_I. Thus if there is another such face, it is no closer to C than ℓ. Hence d_3 is the distance from C to ℓ. One can check that the distance from C to ℓ is the proposed value for $M_T(F)$. Since the inequalities for part C are true, and since $\sigma(P_1) \le \sigma(P_3)$, $\sigma(P_2) \le \sigma(P_3)$ and since F is not reducible in cases iv) and v) we see that $d_3 \le d_1$ and $d_3 \le d_2$ so $M_T(F) = d_3$.

One can check that the distance from C to ℓ is attained at λ_a where $a_1 \ge a_2 \ge a_3$. Since there are no more than 2 faces which "cut off the corner containing P_3", the uniqueness statement in part d) holds. Q.E.D.

Proof of Theorem 6.1' Suppose F is stable quartic and $P \in C_F$ is a maximal point. Then $\sigma(P)$ can assume any of the values listed in Lemma 6.5, except $(1,1,\infty)$. For if F is reducible then $\sigma(P) \geq (1,2,\infty)$. We may assume that $P = (0,0,1)$ with respect to a fixed maximal torus T_0, and L_1 is a tangent of highest contact which realizes m_I. Then if T is any other maximal torus $M_T(F) \leq M_{T_0}(F)$ by Lemma 6.5. Since $M(F) = \sup\{M_T(F) : T \text{ is a maximal torus of } T\}$, $M_T = M_{T_0}(F)$. Thus part i) is proved. Lemma 6.5, part c, implies part ii). Part iii) follows from part d of Lemma 6.5 and Lemma 5.1. Q.E.D.

For unstable quartics, Theorem 1.6 is true, if we modify part a) and part b) a little. There is however, a much greater variety of unstable forms here.

Let $T \subset SL(3,k)$ denote a fixed maximal torus which acts diagonally with respect to the basis X,Y,Z of V_3^*.

<u>Lemma 6.6</u> If $F \in \mathrm{Sym}\, 4(V_3^*)$ is unstable, the orbit of F contains one and only one type of the following forms F_i up to constant multiple. For each i, $P_3 = [0,0,1]$ is a maximal point of F_i. Let σ_i equal $\sigma(P_3)$ for F_i. Then F_i, σ_i, and $M_T(F_i)$ are given by the following chart:

F_i	σ_i	$M_T(F_i)$
$F_1 = XY^3 + aX^2Z^2 + X^2(\mathbb{Q}(Y,Z))$, $a \neq 0$	$(2,2,\infty)$	$2/\sqrt{41}$
$F_2 = Y^3Z + aXY^2Z + bX^2YZ + cX^3Z + \mathbb{Q}(x,y)$, $a \neq 0$, $b \neq 0$	$(1,3,\ )$	$1/\sqrt{6}$
$F_3 = X^2YZ + aY^4 + bX^3Z + cXY^3 + dX^2Y^2 + eX^3Y + fX^4$, $a \neq 0$	$(2,3,4)$	$4/\sqrt{42}$
$F_4 = X^2YZ + aXY^3 + bX^2Y^2 + cX^3Y + dX^3Z + eX^4$, $a \neq 0$	$(2,3,\infty)$	$1/\sqrt{2}$
$F_5 = X^2YZ + bX^2Y^2 + cX^3Y + dX^3Z + eX^4$	$(2,3,\infty)$	$1/3\sqrt{6}$
$F_6 = Y^4 + aX^3Z + bY^3 + cX^2Y^2 + dX^3Y + eX^4$, $a \neq 0$	$(3,3,4)$	$8/\sqrt{78}$
$F_7 = XY^3 + aX^3Z + bX^2Y^2 + cX^3Y + dX^4$, $a \neq 0$	$(3,3,\infty)$	$7/\sqrt{42}$
$F_8 = X^2Y^2 + aX^3Z + bX^3Y + cX^4$, $a \neq 0$	$(3,3,\infty)$	$\sqrt{2}$
$F_9 = X^2Y^2 + aX^3Y + bX^4 + cXY^3 + dY^4$	$(2,4,\infty)$ or $(1,4,\infty)$	$2/3\sqrt{6}$
$F_{10} = X^3Y$	$(3,4,\infty)$	$1/3\sqrt{42}$
$F_{11} = X^4$ (Here $\mathbb{Q}$ denotes a quadratic polynomial)	$(4,4,\infty)$	$4/3\sqrt{6}$

Thus a) $M_T(F_i) < M_T(F_{i+1})$ for $i = 1,2,\dots,10$.

b) If $i = 11$, there is a unique maximal point of F_i.

c) The σ_i's are a nondecreasing function of i, for $i \in \{i \in \mathbb{Z} : 1 \le i \le 11$ and $i \ne 2,9\}$.

d) For each i, there is a unique indivisible one parameter subgroup λ_a, $a_1 \ge a_2 \ge a_3$, such that $\mu(F_i,\lambda_a) = M_T(F_i)$ and $P(\lambda_a)$ fixes the natural flag F_{P_3}.

Proof: The lemma is easy to check if one uses the geometric characterization of instability in Lemma 6.3, the diagram of $S^4(T)$ in (26), and Theorem 2.10. For it is easy to compute the distance d_i from the center $c = 4/3(1,1,1)$ of $S^4(T)$ to the boundary of $S^4{}_{F_i}(T)$, using the diagram 26) and $d_i = M_T(F_i)$, by Theorem 2.10.

The true analogue of Theorem 6.1 for quartics, thus is

Theorem 6.7 Suppose $F \in \mathrm{Sym}^4(V_3^*)$ is an unstable quartic. Then

a) $M_T(F)$ is uniquely determined by a maximal point of F, and the reducibility of F.

b) $\Lambda'_F = \Lambda'_{P,F}$ where P is a maximal point of F.

Proof: Follows immediately from Lemma 6.6 and Lemma 5.1. Q.E.D.

Bibliography

[B] Borel, A. Linear Algebraic Groups. Benjamin, New York, 1969.

[E] Eggleston, H.G. Convexity: Cambridge Tract in Mathematics and Mathematical Physics, No. 47. Cambridge University Press, 1958.

[K] Kempf, G. Instability in invariant theory, Annals of Mathematics. 108 (1978).

[G.I.T.] Mumford, D. Geometric Invariant Theory. Springer-Verlag, 1965.

[M] Mumford, D. Stability of Projective Varieties, L'Enseignement de mathématique, T. XXIII, fasc. 1-2, 1972

[W] Wolfe, P. Algorithm for a least-distance Programming Problem, Math. Programming Study I(1974) 190-205. North Holland.

Remark: Wolfe's paper contains a computer algorithm for obtaining $M_T(v)$. It would be interesting to exploit this.

University of Washington
Department of Mathematics
Seattle, Washington 98105
USA

THE TRACE OF FROBENIUS FOR ELLIPTIC CURVES WITH COMPLEX MULTIPLICATION

by

Loren D. Olson

Let E be an elliptic curve defined over $\mathbb{Q}$. Given a fixed integer $f_o \in \mathbb{Z}$, we may ask for a complete description of the set of all primes p such that the trace of Frobenius at p is f_o. In general, not much is known. In this article we shall provide an explicit description in the case of elliptic curves with complex multiplication. The case $f_o=0$ is classical. Our main result is the following.

<u>Theorem.</u> Let E be an elliptic curve defined over $\mathbb{Q}$ with complex multiplication in $\mathbb{Q}(\sqrt{m})$ with $m < 0$ and m square-free. Let N be 16(48 if m=-3) times the product of all odd primes $l \neq -m$ where E has bad reduction. Let $f_o \neq 0$ be a fixed integer. Then there exists a set T of residue classes modulo N such that $f_p = f_o$ (and E has good reduction at p) $\Leftrightarrow$ p is a member of the quadratic progression $(f_o^2-mh^2)/4$ and h represents an element of T (and E has good reduction at p).

The first three sections are devoted to the case of complex multiplication in $\mathbb{Q}(\sqrt{-3})$ as well as some applications of the main theorem in this case. § 4 discusses the asymptotic distribution of primes with a given trace of Frobenius. In the next section we treat the case $\mathbb{Q}(\sqrt{-1})$. § 6 treats the remaining cases uniformly and concludes the proof of the main theorem.

Contents.

§ 1.) Anomalous primes and quadratic progressions.

If E is an elliptic curve defined over $\mathbb{Q}$ with complex multiplication in $\mathbb{Q}(\sqrt{m})$ where $m < 0$ is a square-free integer with $m \equiv 1 \pmod 4$, then the anomalous primes for E must be members of the quadratic progression $[(-mf^2)t^2+1]/4$ where f is the conductor of End (E) in the ring of integers in $\mathbb{Q}(\sqrt{m})$.

Mazur [8] showed the importance of anomalous primes for the behavior of rational points on elliptic curves with values in towers of number fields. Anomalous primes for certain elliptic curves with complex multiplication were examined in Olson [11]. A corollary of the main theorem in this article is a concise description of the anomalous primes for any given elliptic curve defined over $\mathbb{Q}$ with complex multiplication.

It is of course an open question as to whether such quadratic progressions contain infinitely many primes or not. Certainly the existence of an infinite number of anomalous primes for E would imply this. However, the converse is not at all clear since it involves the distribution of certain roots of unity among these primes á la Kummer's conjecture (cf. Mazur [8, pp. 186-8]). We begin by providing an example of an elliptic curve where the two questions become equivalent; specifically we shall show that the anomalous primes for $Y^2 = X^3 + 2$ coincide precisely with the primes occurring in certain explicit quadratic progressions.

Let E be an elliptic curve defined over $\mathbb{Q}$ with j-invariant $j \neq 0, 2^6 3^3$. E may then be put in "standard form" as in Olson [10], i.e. there exist A, $B \in \mathbb{Z}$, $B > 0$, uniquely determined by j with the property that if E is an elliptic curve defined over $\mathbb{Q}$ with $j(E) = j$, then there exists a unique non-zero square-free $D \in \mathbb{Z}$ (the minimal D-factor of E) such that E is isomorphic over $\mathbb{Q}$ to the elliptic curve defined by the affine equation

$$Y^2 = X^3 + AD^2X + BD^3.$$

If $p \geq 5$ is a prime where E has good reduction, let f_p be the trace of Frobenius at p and N_p the number of $\mathbb{Z}/p\mathbb{Z}$-rational points on the reduced curve.

Let $H_p = \sum_{\substack{h\geq o,\ i\geq o \\ 2h+3i=P}} \frac{P!}{h!\,i!\,(P-h-i)!} A^h B^i$

with $P = (p-1)/2$. This is Deuring's formula for the Hasse invariant at p of the elliptic curve $Y^2 = X^3 + AX + B$. Thus $f_p \equiv D^P H_p \equiv (\frac{D}{p})\ H_p \pmod p$. Let T(E) denote the group of $\mathbb{Q}$-rational torsion points on E, and let t_E denote the order of T(E).

Now let E be an elliptic curve defined over $\mathbb{Q}$ with j-invariant $-2^{15}3^1 5^3$. E has complex multiplication in $\mathbb{Q}(\sqrt{-3})$ and the conductor of End(E) is f = 3. Let E_D denote the elliptic curve with j-invariant $j = -2^{15}3^1 5^3$ and minimal D-factor D, i.e. E_D is defined by the affine equation

$$Y^2 = X^3 - 2^3 3^1 5^1 D^2 X + 2^1 11^1 23^1 D^3.$$

Lemma 1.1. There exists an isogeny defined over $\mathbb{Q}$ of degree 3 between E_D and the elliptic curve defined by the affine equation $Y^2 = X^3 + 2D^3$.

Proof: $P = (-2D, \sqrt{-6D^3})$ is a point of order 3 on $Y^2 = X^3 + 2D^3$ rational over $\mathbb{Q}(\sqrt{-6D^3})$. $2P = -P = (-2D, -\sqrt{-6D^3})$. The group {e, P, 2P} is rational over $\mathbb{Q}$. Taking this to be the kernel of an isogeny, we obtain E_D as the image of this isogeny using the formulas of Vélu [12]. Conversely $Q = (6D, \sqrt{2D^3})$ is a point of order 3 on E_D and the group {e, Q, 2Q} is rational over $\mathbb{Q}$. The image of the isogeny with this group as its kernel is $Y^2 = X^3 + 3^6(2D^3)$ which is isomorphic over $\mathbb{Q}$ to $Y^2 = X^3 + 2D^3$.

Corollary 1.2. The elliptic curves E_D and $Y^2 = X^3 + 2D^3$ have the same Hasse invariant and trace of Frobenius at each prime p and their sets of anomalous primes coincide.

Consider the curves E_D. By the results of Olson [8], we know that $t_E = 3$ if the minimal D-factor of E is D = 2 and $t_E = 1$ otherwise. Let $p \equiv 1 \pmod 6$ where E has good reduction. Then $p = (f_p/2)^2 + 27t^2/4$ for a uniquely determined positive integer t. We have $4p = f_p^2 + 27t^2$. If $f_p \equiv 0 \pmod 3$, then $p \equiv 0 \pmod 3$, a contradiction. On the other hand, every prime $p \equiv 1 \pmod 6$ is such that 4p may be written as $4p = x^2 + 27y^2$ where y is a uniquely determined positive integer and x is an integer uniquely determined up to sign. Thus given such a p, we have exactly two choices for f_p. What is of interest here is to determine the sign of f_p. Since $N_p = 1 + p - f_p$, we see that $N_p \equiv 0 \pmod 3 \Leftrightarrow f_p \equiv 2 \pmod 3$.

Now let E_2 be the curve with 2 as its minimal D-factor. $f_p \equiv (\frac{2}{p}) H_p \pmod p$ for this curve. Since $3 | N_p$, $f_p \equiv 2 \pmod 3$ for E_2. If we are given a fixed integer $f_0 \equiv 2 \pmod 3$, then the primes p having f_0 as trace of Frobenius f_p for E_2 are those p in the quadratic progression $[f_0^2 + 27t^2]/4$. Notice that $-f_0$ cannot be f_p for E_2 since $-f_0 \not\equiv 2 \pmod 3$. Thus given any elliptic curve E defined over $\mathbb{Q}$ with $j = -2^{15}3^{1}5^{3}$ and any fixed integer $f_0 \not\equiv 0 \pmod 3$, it is easy to determine which primes p have f_0 as trace of Frobenius f_p for E. Suppose now that we take the curve E_1 with 1 as its minimal D-factor. The anomalous primes for E_1 will be those of the form $[1 + 27t^2]/4$ such that $(\frac{2}{p}) = -1$.

Recall that $(\frac{2}{p}) = -1 \Leftrightarrow p \equiv \pm 3 \pmod 8$. If $p = [1 + 27t^2]/4$, then t must be odd; let $t = 2u + 1$. Then $4p = 108u^2 + 108u + 28$, or $p = 27u^2 + 27u + 7$. An elementary calculation then shows that $p \equiv \pm 3 \pmod 8 \Leftrightarrow u \equiv 1, 3, 4,$ or $6 \pmod 8$. The latter values of u modulo 8 may be expressed as certain subprogressions of the original, and we summarize the discussion in the following theorem.

Theorem 1.3. Let E denote either the elliptic curve E_1 defined over $\mathbb{Q}$ by the affine equation $Y^2 = X^3 - 2^3 3^1 5^1 X + 2^1 11^1 23^1$ or the elliptic curve defined by $Y^2 = X^3 + 2$. A prime p is an anomalous prime for E $\Leftrightarrow$ p belongs to one of the quadratic progressions $27(8v + a)^2 + 27(8v + a) + 7$ for $a = 1, 3, 4,$ or 6 and $v \geqq 0$. E has infinitely many anomalous primes $\Leftrightarrow$ one of the above progressions contains infinitely many primes.

Remark. The quadratic progressions $27(8v + 1)^2 + 27(8v + 1) + 7$ and $27(8v + 6)^2 + 27(8v + 6) + 7$ yield the same primes for $v \in \mathbb{Z}$ [$v \Rightarrow -(v + 1)$ takes a prime in the first progression to the same prime in the second progression]. Similarly for the progressions $27(8v + 3)^2 + 27(8v + 3) + 7$ and $27(8v + 4)^2 + 27(8v + 4) + 7$.

Remark. A numerical computation using Theorem 1.3 reveals 199 anomalous primes less than 100,000,000 for the elliptic curves given there.

§ 2.) An explicit formula for $\binom{3n}{n}$ mod p, $p = 6n + 1$

We can now take advantage of the isogeny in Lemma 1.2 to determine the residue class of $\binom{3n}{n}$ modulo p for all primes p of the form $p = 6n + 1$. We may write $4p = L^2 + 27M^2$ with

$L, M \in \mathbb{Z}$. L and M are unique up to sign. We shall normalize L by requiring $L \equiv 1 \pmod 3$. The three cube roots of 1 modulo p are 1, $(L + 9M)/(L - 9M)$, and $(L - 9M)/(L + 9M)$. 2^{2n} is the cubic residue symbol of 2 mod p and $2^{2n} \equiv 1 \pmod p$ $\Leftrightarrow 2|M$. If $2\nmid M$ (and hence $2\nmid L$) and M is normalized by requiring $L \equiv M \pmod 4$, then $2^{2n} \equiv (L + 9M)/(L - 9M) \pmod p$. This formula for 2^{2n} may be found in Lehmer [7] or Williams [13].

<u>Theorem 2.1.</u> Let $p = 6n + 1$ be a prime and write $4p = L^2 + 27M^2$ with $L \equiv 1 \pmod 3$.

(1.) $\binom{3n}{n} \equiv -L \pmod p \Leftrightarrow 2|M$.

(2.) $\binom{3n}{n} \equiv -L(L + 9M)/(L - 9M) \equiv (L - 9M)/2$ $\pmod p \Leftrightarrow 2\nmid M$ and M is normalized by $M \equiv L \pmod 4$.

<u>First Proof:</u> Consider the elliptic curve defined by $Y^2 = X^3 + 2^4$. We have seen above that the trace of Frobenius f_p at p is given by $f_p = -L$. By Deuring's formula, f_p is also given modulo p by $f_p \equiv \binom{3n}{n}2^{4n}$. Thus $2|M \Rightarrow \binom{3n}{n} \equiv \binom{3n}{n}2^{4n} \equiv f_p \equiv -L \pmod p$. If $2\nmid M$, then $\binom{3n}{n}[(L + 9M)/(L - 9M)]^2 \equiv$ $\binom{3n}{n}2^{4n} \equiv f_p \equiv -L \pmod p$, i.e. $\binom{3n}{n} \equiv$ $-L(L + 9M)/(L - 9M) \equiv (L - 9M)/2 \pmod p$.

<u>Second Proof:</u> Gauss [1, Article 358] (see also Ireland and Rosen [3, p.97]) showed that the elliptic curve $X^3 + Y^3 = Z^3$ has $f_p = -L$ for $p \equiv 1 \pmod 6$. This curve is $\mathbb{Q}$-isomorphic to the elliptic curve given by $Y^2 = X^3 - 2^4 3^3$. We thus have $-L \equiv \binom{3n}{n}(-2^4 3^3)^n \equiv \binom{3n}{n}(-3/p)2^{4n} \pmod p$ or $\binom{3n}{n} \equiv 2^{2n}(-L)$ $\pmod p$. Using the known formula for 2^{2n} given above, we obtain again Theorem 2.1.

§ 3.) The trace of Frobenius f_p for elliptic curves with complex multiplication in $\mathbb{Q}(\sqrt{-3})$

Suppose first that E is an elliptic curve defined by $Y^2 = X^3 + a_6$ and that p is a prime of the form $p = 6n + 1$ where E has good reduction. There are exactly six possible values for the trace of Frobenius at p depending on which sixth root of unity a_6^n is. Theorem 2.1 and $f_p \equiv \binom{3n}{n} a_6^n \pmod p$ give us one means of computing these six possibilities. However the arithmetic of $\mathbb{Q}(\sqrt{-3})$ provides an alternative method of determining them. Write $4p = L^2 + 27M^2$ with $L \equiv 1 \pmod 3$ and $L, M \in \mathbb{Z}$. If $p = \pi_1\pi_2$ is a factorization into primes in $\mathbb{Q}(\sqrt{-3})$, then there are six possible choices for π_1 and π_2. One such is $\pi_1 = (L + 3\sqrt{-3}M)/2$ and $\pi_2 = (L - 3\sqrt{-3}M)/2$. All possible factorizations are then given by $p = (\zeta\pi_1)(\zeta^{-1}\pi_2)$ as ζ runs through the sixth roots of unity. The six possibilities for f_p are $f_p = \zeta\pi_1 + \zeta^{-1}\pi_2$, and these are $\pm L$, $\pm (L + 9M)/2$, $\pm (L - 9M)/2$.

Remark. We note here the connection with the Jacobsthal sum $\phi_3(D) = \Sigma_{x=1}^{p-1} (x/p)((x^3+D)/p)$. If $f_p(D)$ denotes the trace of Frobenius of $Y^2 = X^3 + D$, then $\phi_3(D) = -1 - f_p(D^2)$ and $f_p(D) = -(D/p)\,[\phi_3(D^{-1}) + 1]$.

Given any fixed integer $f_o \neq 0$, the primes ≥ 5 which have f_o as trace of Frobenius must be members of the quadratic progression $p = (f_o^2 + 3h^2)/4$. Clearly $f_o \equiv 0 \pmod 3$ is impossible, so we assume $f_o \not\equiv 0 \pmod 3$ from now on.
With $4p = L^2 + 27M^2$ as above, we also normalize M in the case $2 \nmid M$ by requiring $L \equiv M \pmod 4$. We have $f_o^2 + 3h^2 = 4p = L^2 + 27M^2$ or $(f_o + \sqrt{-3}\,h)(f_o - \sqrt{-3}\,h) = (L + 3\sqrt{-3}\,M)(L - 3\sqrt{-3}\,M)$.

Using unique factorization in $\mathbb{Q}(\sqrt{-3})$ and the fact that $L \equiv 1$ (mod 3), we obtain after an elementary calculation the following possibilities:

<u>Case A:</u> If $h \equiv 0 \pmod 3$, then $L = (f_o/3)f_o$ and $M = \pm h/3$. If $h \equiv 3 \pmod 6$ $[\Leftrightarrow 2 \nmid M]$, then $M = (-1)^a h/3$ where $a = (h - 3L)/6$.

<u>Case B:</u> If $h \not\equiv 0 \pmod 3$, then $L = [-(f_o/3)f_o - 3(h/3)h]/2$ and $M = (-1)^b[(h/3)h - (f_o/3)f_o]/6$ where $b = [(f_o/3)f_o + (h/3)h]/2$.

Since $f_p \equiv \binom{3n}{n} a_6^n \pmod p$ and since Theorem 2.1 gives us the value of $\binom{3n}{n}$ modulo p, we are in need of some information concerning a_6^n modulo p. Since a_6 has a factorization into primes, it suffices to assume $a_6 = q$ is a prime.

Assume $q \geq 5$. Cases A and B show that a knowledge of the residue class of h modulo 6q determines:

(i.) the residue class of h modulo 3 uniquely;

(ii.) the residue class of L modulo q uniquely;

(iii.) the residue class of M modulo q uniquely up to sign.

Choose once and for all a set U of $(q-1)/2$ non-zero residue classes modulo q such that $x \in U \Rightarrow -x \notin U$. The sign of M is also uniquely determined in all cases with the exception of $h \equiv 0 \pmod 6$, in which case we agree to take the residue class of M in $U \cup \{0\}$.

Let $s_1 = 1$, $s_2 = -1$, $s_3 = (L + 9M)/(L - 9M)$, $s_4 = -s_3$, $s_5 = (L - 9M)/(L + 9M)$, and $s_6 = -s_5$. Let $\mathcal{S} = \{s_1, \ldots, s_6\}$, which is the group of sixth roots of unity modulo p.

According to Williams [13] there exists a set $\mathcal{L}_1(q)$ of residue classes modulo q with the following properties:

1.) $q^{2n} \equiv 1 \pmod p \Leftrightarrow$ there does not exist a $k \in \mathcal{L}_1(q)$ such that $L^2 \equiv k^2M^2 \pmod q$.

2.) If $q^{2n} \not\equiv 1 \pmod p$, then M can be uniquely defined by requiring $L \equiv kM \pmod q$ for some $k \in \mathcal{L}_1(q)$ and with this normalization $q^{2n} \equiv (L + 9M)/(L - 9M) \pmod p$.

Theorem 3.1. Let f_o be a fixed integer, $f_o \not\equiv 0 \pmod 3$. Let $q \geqq 5$ be a prime and $d \in \mathbb{Z}$. There exist sets $\mathcal{T}_i$, $i = 1,\ldots, 6$, of residue classes modulo 48q (12 suffices if $q \equiv 1 \pmod 4$) such that if $h \in \mathcal{T}_i$, then $(q^d)^n \equiv s_i \pmod p$ for all primes p in the quadratic progression $p = (f_o^2 + 3h^2)/4$ and such that $\bigcup \mathcal{T}_i$ includes all residue classes modulo 48q. If q = 2 or 3, this result holds modulo 24q.

Proof: Clearly it suffices to prove the theorem for d = 1. $q^n \equiv q^{-2n}q^{3n} \equiv q^{-2n}(q/p) \pmod p$. Assume $q \geqq 5$. The residue class of h modulo 48q determines the residue class of h modulo 4 as well as the residue class of p modulo q. By quadratic reciprocity, (q/p) is therefore determined. Thus it suffices to show that each residue class modulo 24q gives rise to a unique i, $1 \leqq i \leqq 6$, such that $q^{2n} \equiv s_i \pmod p$. But we have seen above that a knowledge of h modulo 6q suffices to determine L and M uniquely modulo q. The existence of $\mathcal{L}_1(q)$ completes the proof for $q \geqq 5$. Suppose q = 2. (2/p) is determined by a knowledge of h modulo 16. To normalize L and M in accordance with the value of 2^{2n} discussed previously, we need only h modulo 6. Thus h modulo 48 suffices for q = 2. Suppose q = 3. Then $(3/p) = (-1/p)(-3/p) = (-1/p)$.

Thus a knowledge of h modulo 8 suffices to determine (3/p). A knowledge of h modulo 18 determines M modulo 3 and L and M uniquely modulo q, so 3^{2n} is uniquely determined by Theorem 1 of Williams [13]. Thus h modulo 72 suffices for q = 3. Let N be as described in the main theorem.

<u>Theorem 3.2.</u> Let f_o be a fixed integer, $f_o \not\equiv 0 \pmod 3$. Let E be an elliptic curve defined by $Y^2 = X^3 + a_6$. There exists a set $\mathcal{T}$ of residue classes modulo N such that the trace of Frobenius at a prime p = 6n + 1 where E has good reduction is $f_o \Leftrightarrow$ p is a member of one of the quadratic progressions $p = (f_o^2 + 3h^2)/4$ with $h \in \mathcal{T}$ and $p \nmid a_6$.

<u>Proof:</u> This now follows immediately from Theorems 2.1 and 3.1 since $f_p \equiv \binom{3n}{n} a_6^n \pmod p$.

<u>Corollary 3.3.</u> Let E_D denote the elliptic curve defined over $\mathbb{Q}$ with j-invariant $j = -2^{15}3^1 5^3$ and minimal D-factor D, i.e. E_D is defined by the affine equation $Y^2 = X^3 - 2^3 3^1 5^1 D^2 X + 2^1 11^1 23^1 D^3$. Let f_o be a fixed integer, $f_o \not\equiv 0 \pmod 3$. There exists a set $\mathcal{T}$ of residue classes modulo N such that the trace of Frobenius at a prime p = 6n + 1 where E has good reduction is $f_o \Leftrightarrow$ p is a member of one of the quadratic progressions $p = (f_o^2 + 3h^2)/4$ with $h \in \mathcal{T}$ and $p \nmid D$.

<u>Proof:</u> By Lemma 1.1 we are reduced to the case of the elliptic curve $Y^2 = X^3 + 2D^3$. Applying Theorem 3.2, we obtain the desired result.

There exists one additional class of elliptic curves defined over $\mathbb{Q}$ which have complex multiplication in $\mathbb{Q}(\sqrt{-3})$, namely those with j-invariant $j = 2^4 3^3 5^3$. These have the affine equations $Y^2 = X^3 - 3^1 5^1 D^2 X + 2^1 11^1 D^3$. $(2D,0)$ is a $\mathbb{Q}$-rational point of order 2 on this curve, and $(3D, 2D\sqrt{D})$ is a $\mathbb{Q}(\sqrt{D})$-rational point of order 3 on this curve. They generate the $\mathbb{Q}$-rational group $F = \{e, (2D,o), (3D, 2D\sqrt{D}), (3D, -2D\sqrt{D}), (-D, 6D\sqrt{D}), (-D, -6D\sqrt{D})\}$. Taking F to be the kernel of an isogeny, we obtain the following lemma.

Lemma 3.4. $Y^2 = X^3 - 3^1 5^1 D^2 X + 2^1 11^1 D^3$ is isogenous over $\mathbb{Q}$ to $Y^2 = X^3 - 3^3 D^3$.

Corollary 3.5. Let C_D denote the elliptic curve defined over $\mathbb{Q}$ with j-invariant $j = 2^4 3^3 5^3$ and minimal D-factor D, i.e. C_D is defined by the affine equation $Y^2 = X^3 - 3^1 5^1 D^2 X + 2^1 11^1 D^3$. Let f_o be a fixed integer, $f_o \not\equiv 0 \pmod 3$. There exists a set $\mathcal{T}$ of residue classes modulo N such that the trace of Frobenius at a prime $p = 6n + 1$ where E has good reduction is $f_o \Leftrightarrow p$ is a member of one of the quadratic progressions $p = (f_o^2 + 3h^2)/4$ with $h \in \mathcal{T}$ and $p \nmid D$.

Remark. Theorem 3.2 has the obvious interpretation for the Jacobsthal sums $\phi_3(D)$.

§ 4.) Asymptotic distributions of primes with fixed trace of Frobenius

Given a fixed f_o, it is natural to ask for the asymptotic distribution of those primes p which have f_o as trace of Frobenius. Lang and Trotter [6] have examined this problem at length for elliptic curves in general. As mentioned at the outset, for curves with complex multiplication, those primes p which have f_o as trace of Frobenius lie in a quadratic progression. In 1923 Hardy and Littlewood [2, p.48] made a conjecture (Conjecture F) concerning the distribution of primes in quadratic progressions. The results of § 1-3 here enable us to express the distribution of the primes p with f_o as trace of Frobenius for elliptic curves with complex multiplication in $\mathbb{Q}(\sqrt{-3})$ precisely in terms of the distribution of primes in certain quadratic progressions. Let $a,b,c \in \mathbb{Z}$ with $a > 0$, $(a,b,c) = 1$, not both a+b and c even, and with $D=b^2-4ac$ not a square. Then Conjecture F of Hardy and Littlewood predicts that the number P(n) of primes $\leq n$ in the quadratic progression $am^2 + bm + c$, $m \geq 0$, is given asymptotically by $P(n) \sim \frac{\varepsilon C}{\sqrt{a}} \frac{\sqrt{n}}{\log n} \prod (\frac{p}{p-1})$ where the product is taken over all odd common prime divisors of a and b. ε is 1 if a+b is odd and 2 otherwise, and $C = \prod (1 - \frac{1}{q-1}(\frac{D}{q}))$ where the product is taken over all primes q such that $q \geq 3$ and $q \nmid a$. Consider now an elliptic curve $Y^2 = X^3 + a_6$. Theorem 3.2 implies the existence of a set $\mathcal{T}$ of residue classes modulo N such that $f_p = f_o \Leftrightarrow p = (f_o^2 + 3h^2)/4$ for some $h \in \mathcal{T}$. After $\mathcal{T}$ has been calculated, the appropriate quadratic progressions may be written down and analyzed to obtain the fraction of primes with $f_p = f_o$ under the assumption of Conjecture F.

Mazur [8, p. 187] raised the question of what fraction of the primes in the quadratic progression $(1 + 3h^2)/4$ are anomalous for an elliptic curve of the form $Y^2 = X^3 + a_6$. Theorem 3.2 now enables us to answer this question for any given a_6 under the assumption of Conjecture F. As an example, we consider the case $a_6 = 2$.

Theorem 4.1. Assume that Conjecture F of Hardy-Littlewood is true. Then 1/6 of the primes in the quadratic progression $(1 + 3h^2)/4$ are anomalous for the elliptic curve defined by $Y^2 = X^3 + 2$.

Proof: If $p = (3h^2 + 1)/4$ is a prime, then $4p = 3h^2 + 1$ and $2 \nmid h$. We may assume $h \geqq 1$. Write $h = 2t + 1$. Then $4p = 12t^2 + 12t + 4$ or $p = 3t^2 + 3t + 1$. Call the progression P. The anomalous primes are given by the four quadratic progressions Q_a with $p = 27(8v+a)^2 + 27(8v+a) + 7$, $v \geqq 0$, for $a = 1, 3, 4$, and 6 according to Theorem 1.3.

The five quadratic progressions P, Q_1, Q_3, Q_4, and Q_6 all have the same ε, C, and $\Pi(\frac{p}{p-1})$. Thus the only difference in asymptotic distribution lies in the factor $1/\sqrt{a}$ if we assume Conjecture F. For P this is $1/\sqrt{3}$, and for each Q_a it is $1/\sqrt{(27)(64)}$. The ratio of these two is 1/24. Since Q_1, Q_3, Q_4, and Q_6 yield all the anomalous primes (and distinct primes), the fraction of anomalous primes is 1/6.

§ 5.) The trace of Frobenius f_p for elliptic curves with complex multiplication in $\mathbb{Q}(\sqrt{-1})$

Suppose first that E is an elliptic curve defined by $Y^2 = X^3 + a_4X$ and that $p = 4n + 1$ is a prime where E has good reduction. f_p is determined by $f_p \equiv \binom{2n}{n} a_4^n \pmod{p}$. It is a classical result of Gauss that if $p = a^2 + 4b^2$ with $a \equiv 1 \pmod{4}$, then $\binom{2n}{n} \equiv 2a \pmod{p}$. The possible values of a_4^n modulo p are the fourth roots of unity in $\mathbb{Z}/p\mathbb{Z}$ which are 1, -1, $a/2b = -2b/a$, and $-a/2b = 2b/a$. The possible values of f_p are thus $2a$, $-2a$, $4b$, and $-4b$. Clearly f_p is even.

$4 \nmid f_p \Leftrightarrow f_p = \pm 2a \Leftrightarrow (a_4/p) = 1$. If $a_4 \equiv m^2 \pmod{p}$, then $f_p = (m/p)2a$, a fact which was already known to Jacobsthal [4]. $f_p = -\phi_2(a_4)$ where $\phi_2(D) = \Sigma_{x=1}^{p-1} (x/p)((x^2+D)/p)$ is a Jacobsthal sum.

<u>Theorem 5.1.</u> Let E be given by $Y^2 = X^3 + a_4X$.

(1.) Let $f_o \in \mathbb{Z}$ be a fixed integer with $f_o \equiv 2 \pmod{4}$. Then there exists a set $\mathcal{T}$ of residue classes modulo N such that the trace of Frobenius $f_p = f_o \Leftrightarrow p$ is a member of the quadratic progression $(f_o/2)^2 + 4h^2$, $h \geqq 0$ with $h \in \mathcal{T}$ and $p \nmid a_4$.

(2.) Let $f_o \in \mathbb{Z}$ be a fixed integer with $f_o \equiv 0 \pmod{4}$, $f_o \neq 0$. Then there exists a set $\mathcal{T}$ of residue classes modulo N such that the trace of Frobenius $f_p = f_o \Leftrightarrow p$ is a member of the quadratic progression $(f_o/2)^2 + (2h+1)^2$, $h \geqq 0$ with $h \in \mathcal{T}$ and $p \nmid a_4$.

<u>Proof:</u> (1.) Assume $f_o \equiv 2 \pmod{4}$. Set $g = (f_o - 2)/4$. Then $\binom{2n}{n} \equiv (-1)^g f_o \pmod{p}$. We need the value of $(a_4)^n$ mod p. Clearly it suffices to know this for the case of $a_4 = q$ a prime and $a_4 = -1$. A knowledge of h modulo 2 determines $(-1)^n$. Put $a = (-1)^g f_o/2$ so that $a \equiv 1 \pmod{4}$. Define b as follows: $b = (-1)^{(h-1)/2}h$ if $h \equiv 1 \pmod{2}$ and $b = h$ if $h \equiv 0 \pmod{2}$. Lehmer [7] showed that $2^n \equiv (-1)^{b/2} \pmod{p}$ if $b \equiv 0 \pmod{2}$ and $2^n \equiv b/2a \pmod{p}$ if $b \equiv 1 \pmod{2}$. Thus a knowledge of h modulo 2 determines 2^n as well. We may assume that $a_4 = q$ is a prime ≥ 3. Put $\pi = a + 2bi$ and let $(\ /\ )_4$ denote the fourth-power residue symbol. By the generalized Euler criterion, $(q/\pi)_4 \equiv q^n \pmod{\pi}$.
However the law of biquadratic reciprocity implies that

$(q/\pi)_4$ depends only on the residue class of h modulo 2q. Thus the residue class of h modulo 2q determines whether q^n is congruent to 1, -1, a/2b, or -a/2b modulo p. Since $f_p \equiv \binom{2n}{n} a_4^n \equiv (-1)^g f_o a_4^n \pmod{p}$, we need $(-1)^g a_4^n \equiv 1 \pmod{p}$. Clearly the residue class of h modulo N determines this.

(2.) Assume now that $f_o \equiv 0 \pmod{4}$. Set $a = (-1)^h(2h + 1)$ so that $a \equiv 1 \pmod{4}$. Define b as follows: $b = (-1)^{(f_o-4)/8} f_o/4$ if $f_o \equiv 4 \pmod{8}$ and $b = f_o/4$ if $f_o \equiv 0 \pmod{8}$. Then we proceed as before with $f_p \equiv \binom{2n}{n} a_4^n \equiv (2a)\ a_4^n \pmod{p}$. For $f_p = f_o$, we need $a_4^n \equiv (-1)^{(f_o-4)/8}\ 2b/a \pmod{p}$ if $f_o \equiv 4 \pmod{8}$ and $a_4^n \equiv 2b/a \pmod{p}$ if $f_o \equiv 0 \pmod{8}$. As above, the residue class of h modulo N determines this.

Up to now in this §, we have considered elliptic curves with j-invariant $j = 2^6 3^3$. The other elliptic curves defined over $\mathbb{Q}$ which have complex multiplication in $\mathbb{Q}(\sqrt{-1})$ are characterized by having j-invariant $j = 2^3 3^3 11^3$. Such a curve E is isomorphic over $\mathbb{Q}$ to an elliptic curve given by an affine equation of the form

$$Y^2 = X^3 - 11^1 D^2 X + 2^1 7^1 D^3$$

with $D \in \mathbb{Z}$, $D \neq 0$. (2D,0) is a $\mathbb{Q}$-rational point of order 2 on this curve, and the group $F = \{e, (2D,0)\}$ is $\mathbb{Q}$-rational. Taking F to be the kernel of an isogeny, we obtain the following lemma.

Lemma 5.2. $Y^2 = X^3 - 11^1 D^2 X + 2^1 7^1 D^3$ is isogenous over $\mathbb{Q}$ to $Y^2 = X^3 - D^2 X$.

Corollary 5.3. If E is an elliptic curve defined over $\mathbb{Q}$ with j-invariant $j = 2^3 3^3 11^3$, then the conclusions of Theorem 5.1 hold for E.

§ 6.) The general case

We assume now that E is an elliptic curve defined over $\mathbb{Q}$ with complex multiplication in $\mathbb{Q}(\sqrt{m})$ with $m < 0$ and m square-free and $m \neq -1, -3$. We first notice one fact concerning elliptic curves with complex multiplication in $\mathbb{Q}(\sqrt{-7})$.

Lemma 6.1. Let E be an elliptic curve defined over $\mathbb{Q}$ with j-invariant $j=3^3 5^3 17^3$. Then E is isogenous over $\mathbb{Q}$ to an elliptic curve defined over $\mathbb{Q}$ with j-invariant $j=-3^3 5^3$.

Proof: E may be assumed to have the affine equation $Y^2=X^3-5^1 7^1 17^1 D^2 X + 2^1 3^1 7^2 19^1 D^3$. $Q = (2^1 7^1 D, 0)$ is a $\mathbb{Q}$-rational point of order 2 on E. Take $\{e,Q\}$ as the kernel of a $\mathbb{Q}$-rational isogeny. The formula of Vélu [12] give the desired result.

The lemma has as a consequence that we may assume that $\mathrm{End}(E)$ is the full ring of integers O in $\mathbb{Q}(\sqrt{m})$ in the situation under discussion here. We are interested in obtaining the proper normalization of a factorization of the prime p in $\mathbb{Q}(\sqrt{m})$ so that a factor induces Frobenius on the reduction of E modulo p and in showing that this normalization depends only on residue classes modulo a certain integer. This can be done by finding an abelian extension $K/\mathbb{Q}(\sqrt{m})$ such that:
(1). K gives us the necessary information concerning Frobenius, and (2) we can compute the conductor of $K/\mathbb{Q}(\sqrt{m})$.
We may regard $\sqrt{m}$ as an element of $\mathrm{End}(E)$.

Let $M=\ker(\sqrt{m})$ for $m \neq -2$ and $M=\ker(2\sqrt{m})$ for $m = -2$. Let $K=\mathbb{Q}(\sqrt{m})(M)$. Let $p=PP'$ be a prime which splits in $\mathbb{Q}(\sqrt{m})$ and is unramified in K. Let "~" denote reduction. Reduction is injective on M. The number of elements in M is always larger than $\deg(1-\zeta)$ where ζ is any root of unity in O. Let Π generate P so that Π induces the Frobenius. $\deg\Pi = p$ and $\deg(\Pi-\zeta\Pi)=p\deg(1-\zeta)$ and $(p, \text{card } M) = 1$ imply that $(\Pi-\zeta\Pi)\widetilde{M} = \widetilde{e} \Leftrightarrow \zeta = 1$. In other words we have shown that to achieve a proper normalization of Π it suffices to show that this choice of Π agrees with the Frobenius on the group $\widetilde{M}$, i.e. the desired condition (1.) on the field K is satisfied for the choice of K given here.

Let $G=\text{Gal}(K/\mathbb{Q}(\sqrt{m}))$ and let $(\;,K/\mathbb{Q}(\sqrt{m}))$ denote the Artin symbol. For each $\sigma \in G$ let $S_\sigma = \{$prime ideals $P \subseteq O$ such that card (O/PO) is a prime splitting in O and not ramified in K and such that $(P, K/\mathbb{Q}(\sqrt{m})) = \sigma\}$.

<u>Lemma 6.2.</u> If $p = P_1P_2$ and $q = Q_1Q_2$ are two primes splitting in O, not ramified in K, and such that $(P_1,K/\mathbb{Q}(\sqrt{m})) = (Q_1, K/\mathbb{Q}(\sqrt{m}))$ and $|f_p| = |f_q|$, then $\Pi_1 \equiv \gamma_1 \pmod{\sqrt{m}}$ where Π_1 (resp. γ_1) induces the Frobenius modulo P_1 (resp. Q_1) and $f_p = f_q$.

<u>Proof:</u> If $|f_p| = |f_q|$, then $p \equiv q \pmod m$ and this determines Π_1 and γ_1 modulo $\sqrt{m}$ up to sign. Assume $m \neq -2$. Suppose now that $\Pi_1 \equiv -\gamma_1 \pmod{\sqrt{m}}$ with $\Pi_1 = (a+b\sqrt{m})/2$ and $\gamma_1=(-a+d\sqrt{m})/2$ with $a,b,d \in \mathbb{Z}$. $(P_1,K/\mathbb{Q}(\sqrt{m}))=(Q_1,K/\mathbb{Q}(\sqrt{m})) \Rightarrow \Pi_1(x,y)=\gamma_1(x,y) \Rightarrow e=(2\Pi_1-2\gamma_1)(x,y) \Rightarrow (2a+(b-d)(\sqrt{m}))(x,y)=2a(x,y)$. Since $(2a,\text{card } M)$

$= 1$, $(x,y) = e$ and we have a contradiction. Assume $m = -2$. Assume $\Pi_1 \equiv -\gamma_1 \pmod{\sqrt{m}}$ and $\Pi_1 = a+b\sqrt{m}$, $\gamma_1 = -a+d\sqrt{m}$. As above, $e = 2a(x,y)$ for $(x,y) \in M$. Since $(a, \operatorname{card}M) = 1$, we have $e = 2(x,y)$. But M has 8 elements for $m = -2$ and $\ker(2)$ has only 4 elements. This gives the necessary contradiction.

<u>Corollary 6.3.</u> Let c be the conductor of $K/\mathbb{Q}(\sqrt{m})$. Let $f_o \neq 0$ be a fixed integer. If p and q are primes splitting in $\mathbb{Q}(\sqrt{m})$ and unramified in K having the same residue class in the ideal class factor group $I(c)/P_cN(c)$ and which are members of the quadratic progression $(f_o^2-mh^2)/4$, then $f_p=f_q$.

We must now calculate the conductor of $K/\mathbb{Q}(\sqrt{m})$. We do this by examining the higher ramification groups. If p is a prime such that E has good reduction at p, then p is unramified in K (Lang[5], pp. 113-114). Suppose that E has bad reduction at p with $P|p$. Assume first $p \neq 2,3,-m$. Then $K/\mathbb{Q}(\sqrt{m})$ is tamely ramified at P (cf. Ogg [9]). Hence P occurs with exponent 1 in the conductor. Assume bad reduction at $p=3$. Then the field F obtained from $\mathbb{Q}(\sqrt{m})$ by adjoining the x-coordinates of M is unramified at P (cf. Lang [5], p. 126). Thus $K/\mathbb{Q}(\sqrt{m})$ is tamely ramified at P and P appears with exponent 1. If $m \neq -2$ and $P= \sqrt{-m}$, then $K/\mathbb{Q}(\sqrt{m})$ is tamely ramified at P and P appears with exponent 1 by a similar argument. Assume $m \neq -2$ and $p=2$.

Then 2 is unramified in F, but ramified in K. Let Q be a prime ideal in F lying over 2, and let $Q=R^2$ with R a prime ideal in K. Then R appears with exponent 3 in the different of K/F and hence also in the different of $K/\mathbb{Q}(\sqrt{m})$. Let H be the Galois group of the completion K_R over the completion F_Q. H has 2 elements and "Hilbert's formula" implies that $H = H_0 = H_1 = H_2$ and $\{id\} = H_3 = H_4 = \ldots$ are the higher ramification groups. The conductor-ramification theorem gives us 3 for the exponent of P in the conductor. If $m = -2$ and $p=2$, then a similar examination reveals that 3 is also the correct exponent in this case. We now prove the main theorem.

Theorem 6.4. Let E be an elliptic curve defined over $\mathbb{Q}$ with complex multiplication in $\mathbb{Q}(\sqrt{m})$ with $m < 0$ and m square-free. Let N be 16 (48 if $m=-3$) times the product of all odd primes $l \neq -m$ where E has bad reduction. Let $f_o \neq 0$ be a fixed integer. Then there exists a set $\mathcal{T}$ of residue classes modulo N such that $f_p=f_o$ (and E has good reduction at p) $\Leftrightarrow$ p is a member of the quadratic progression $(f_o{}^2-mh^2)/4$ and h represents an element of $\mathcal{T}$ (and E has good reduction at p).

Proof: The cases $m=-1,-3$ have been dealt with previously. Lemma 6.1 allows us to assume that $\mathrm{End}(E)=\mathcal{O}$ and we may apply the remainder of the discussion in this §.

Write $c = \Pi L^{n(L)}$ where L are prime ideals in O. Let p and q be primes in the quadratic progression $(f_o^2-mh^2)/4$ where E has good reduction. Assume $m \neq -2$. Let $P = ((f_o+b\sqrt{m})/2)$ and $Q=((f_o+d\sqrt{m})/2)$ be prime ideals in O lying over p and q respectively. Let $l = \mathrm{char}(O/L)$. Assume $l \neq 2$ so that $n(L) = 1$. If $b \equiv d \pmod{l}$, then $(b-d)\sqrt{m} \equiv 0 \pmod{L}$ and $(f_o+b\sqrt{m})/2 \equiv (f_o+d\sqrt{m})/2 \pmod{L} \Leftrightarrow P \equiv Q \pmod{L}$. If $l=2$, then $b \equiv d \pmod{16} \Rightarrow P \equiv Q \pmod{L^3}$. Assume now that $m = -2$. Then $2 | f_p$ if E has good reduction at p, and we may therefore assume $2 | f_o$. Let $a = f_o/2$. Let $P = (a+b\sqrt{m})$ and $Q = (a+d\sqrt{m})$ be as above. If $l \neq 2$, then $b \equiv d \pmod{l} \Rightarrow P \equiv Q \pmod{L}$. If $l = 2$, then $b \equiv d \pmod{8} \Rightarrow P \equiv Q \pmod{L^3}$. In all cases, we have $b \equiv d \pmod{N} \Rightarrow P \equiv Q \pmod{L^{n(L)}}$ for all $L|c$; P and Q have the same residue class in $I(c)/P_cN(c)$. The preceding corollary concludes the proof.

BIBLIOGRAPHY

1.) C.F. Gauss, Disquisitiones Aritmeticae, (English translation), Yale University Press, New Haven, Conn., 1966.

2.) G.H. Hardy and J.E. Littlewood, Some problems of partitio numerorum III. Acta Math. 44 (1923), 1-70 Reprinted in: G.H. Hardy, Collected papers, Vol. 1. Oxford University Press, London, 1966, 561-630.

3.) K. Ireland and M.I. Rosen, Elements of Number Theory, Bogden and Quigley, Inc., Tarrytown-on-Hudson, New York, 1972.

4.) E. Jacobsthal, Anwendungen einer Formel aus der Theorie der quadratischen Reste, Dissertation, Berlin, 1906.

5.) S. Lang, Elliptic Functions, Addison-Wesley Publishing Company, Inc., Reading, Massachusetts, 1973.

6.) S. Lang and H. Trotter , Frobenius Distributions in GL_2-Extensions, Lecture Notes in Mathematics 504, Springer-Verlag, Berlin, 1976.

7.) E. Lehmner, On Euler's Criterion, J. Austral, Math. Soc. 1 (1959/61), part 1, 64-70.

8.) B. Mazur, Rational Points of Abelian Varieties with Values in Towers of Number Fields, Invent. Math. 18 (1972), 183-266.

9.) A. Ogg, Elliptic Curves and Wild Ramification, Amer. J. of Math. 89 (1967), 1 - 21.

10.) L.D. Olson, Points of Finite Order on Elliptic Curves with Complex Multiplication, manuscripta math. 14 (1974), 195 - 205.

11.) L.D. Olson, Hasse Invariants and Anomalous Primes for Elliptic Curves with Complex Multiplication, J. of Number Theory 8(1976), 397 - 414.

12.) J. Vélu, Isogénies entre courbes elliptiques, C.R. Acad. Sc. Paris 273 (1971), 238-241.

13.) K.S. Williams, On Euler's Criterion for Cubic Non-residues, Proc. Amer. Math. Soc. 49 (1975), 277 - 283.

Loren D. Olson
Department of Mathematics
University of Tromsø
9001 Tromsø
Norway

Abelian varieties: moduli and lifting properties

Frans OORT (Utrecht)

1. Introduction

Algebraic varieties are classified with the help of discrete invariants (dimension, genus, ...), and by continuous invariants (usually called "moduli"). In order to obtain good families one has to impose an extra structure, usually in the form of a polarization (and other additional structures). Hence we are interested in classifying isomorphism classes of pairs (X,λ); here

X is an abelian variety

(e.g. cf. [14], we write AV for abelian variety), and

$\lambda : X \to X^t$ is a polarization

(we write X^t for the dual of an abelian variety, in Mumford's notation it would be given by $\hat{X}$, and λ is an isogeny which is of the form

$$\lambda(x) = \mathrm{Cl}(D_x - D) \in \mathrm{Pic}^0(X) = X^t(k)$$

for some ample divisor D on X, cf [13], Definition 6.2; cf. [11], page 194). Discrete invariants of (X,λ) are

$$\dim X = g,$$

$$\deg \lambda = d^2$$

(and there may be more discrete invariants, cf. [15]). For fixed g and d the coarse moduli scheme $A_{g,d}$ exists cf. [13], page 139, Theorem 7.10) as a scheme over $\mathrm{Spec}(\mathbb{Z})$. We are interested in the structure of this scheme.

To some extent, the structure of the characteristic zero fibre

$A_{g,d} \otimes \mathbb{C}$

is fairly well understood (cf. [31]; cf. [10], Sections 14 γ, δ, pp. 408-426).

One of the first problems concerning the moduli space $A_{g,d}$ is the question whether every component of $A_{g,d} \otimes \mathbb{F}_p$ extends to characteristic zero:

Lifting problem. Suppose given an algebraically closed field k, with char(k) = p > 0, and a polarized abelian variety (X_0, λ_0) over k. Does there exist an integral domain R of characteristic zero, a polarized abelian scheme (X, λ) over R, and a reduction homomorphism $R \to k$ such that

$$(X,\lambda) \otimes_R k \cong (X_0, \lambda_0)?$$

For a discussion of the lifting problems, cf. [22]. In this survey-note we discuss a solution for the lifting problem just stated. Suppose the lifting problem has an affirmative answer (which in fact it does); then we have the following:

Application. Let k, and (X_0, λ_0) be as above, and $g = \dim X_0 \leq 3$. Suppose $\deg(\lambda_0) = 1$. Then there exists an algebraic curve C_0 over k (with genus$(C_0) = g$) such that its canonically polarized Jacobian variety equals

$$(\mathrm{Jac}(C_0), \Theta_{C_0}) \cong (X_0, \lambda_0).$$

In case $\dim X_0 = g = 2$, this was proved by Weil (cf. [34], Satz 2, p. 37). We sketch the proof for $g \leq 3$ which can be found in [23]. Let

$$j : M_g \to A_{g,1}$$

be the "Torelli-mapping". We know

	$\dim(M_g \otimes \mathbb{C})$	$\dim(A_{g,1} \otimes \mathbb{C})$
$g = 1$	1	1
$g = 2$	3	3
$g = 3$	6	6
$g \geq 2$	$3g-3$	$\frac{1}{2}g(g+1)$

Thus for $g \leq 3$ we conclude equality of the dimensions. The "Torelli theorem" tells us that j is an <u>injective</u> map (on geometric points). By topological (or analytical) arguments one shows that $A_{g,1} \otimes \mathbb{C}$ is <u>connected</u>, and by arguments from algebraic geometry it follows that it is irreducible (for n big, $A_{g,1,n}$ is smooth, etc.). Thus $j(M_g)$ is dense in $A_{g,1}$ for $g \leq 3$, its closure consists of Jacobians of stable curves, cf. [15], p. 462, hence the curve we are looking for exists in characteristic zero. By the lifting problem we can choose a polarized abelian scheme (X,λ) lifting (X_0,λ_0); by the previous arguments $(X,\lambda) \otimes \mathbb{C}$ is the polarized Jacobian variety of some curve, hence we can choose the ring R in such a way that it is a discrete valuation ring, that (X,λ) is defined over R, and that there exists $C \subset (X \otimes K)$, with K the field of fractions of R, such that

$$(X,\lambda) \otimes K \cong (\mathrm{Jac}(C), \Theta_C).$$

The flat extension of C inside X yields a curve $C \subset X$ over R, and it is straightforward to check that $C_0 = C \otimes k$ has the property

$$(X_0,\lambda_0) \cong (\mathrm{Jac}(C_0), \Theta_{C_0});$$

thus we have proved the result. Note that C_0 may be a reducible curve (a stable curve with smooth components, connected like a tree); our method of proof avoided the study of the different cases for this configuration.

<u>Corollary</u>. For $g \leq 3$ the moduli space

$$A_{g,1} \otimes \mathbb{F}_p$$

is irreducible.

This follows from the irreducibility of $M_g \otimes k$ (cf. [3]) for any field k.

For the lifting problem one can first observe that it suffices to find a formal abelian scheme $\mathcal{X}$ over some complete local domain R plus a polarization: in such a situation it follows from EGA, III^1.5.4.5 that $\mathcal{X}$ is algebraizable, i.e. that it is the formal completion of an abelian scheme over R. A result of Grothendieck (cf.[22], Theorem 2.2.1) states that the deformation problem for formal abelian schemes is smooth. Thus it suffies to lift the polarization also. In fact this method works if $\deg(\lambda) = 1$ (a principal polarization), and it follows that principally polarized abelian varieties can be lifted to characteristic zero (also cf. [2], Theorem 12). This direct approach, which uses only obstruction calculus, seems difficult once the deformation functor of (X_0,λ_0) is not smooth, and in fact, for the general lifting problem we have been unable to find a solution based on obstruction calculus only.

An important new method proposed by D. Mumford around 1967 is based on the following:

<u>Lemma</u>. Suppose (X_0,λ_0) is a specialization of (Y_0,μ_0), and suppose the latter one can be lifted to characteristic zero; then (X_0,λ_0) can be lifted to characteristic zero (we use the word "specialization" here to denote equi-characteristic families).

<u>Sketch</u> of the proof:

$$\begin{array}{ccccc} & & R & \hookrightarrow & W \\ & & \downarrow & & \downarrow \\ k & \longleftarrow & T & \hookrightarrow & K \end{array} \qquad\qquad \begin{array}{ccccc} & & \mathcal{X} & = & Z \\ & & \uparrow & & \uparrow \\ X_0 & \hookrightarrow & \mathcal{Y} & \hookleftarrow & Y \end{array}$$

Let T be a domain of characteristic p, let $(\mathcal{Y},\mu)$ be a polarized abelian scheme over T, with $(X_0,\lambda_0) = (\mathcal{Y},\mu) \otimes k$; let K be an algebraic closure of the field of fractions of T, and suppose that $(Y,\mu) = (\mathcal{Y},\mu) \otimes K$ can be lifted to $(\mathcal{Z},\mu)$ over $W \subset C$. Take for R the biggest subring of W which maps onto T. Observe that $(\mathcal{Z},\mu)$ in fact is defined over R; hence the lemma is proved. Observe that the ring R can be very big, e.g. let W be the ring of infinite Witt-vectors over K, then R is the ring of vectors with first component in T, and all other components in K (and K is an infinite extension of T if $k \neq T$).

Definitions: Let X be an abelian variety over an algebraically closed field k, with char(k) = p. We say that the p-rank of X equals $f = f(X)$ if

$$|{}_pX(\bar{k})| = p^f$$

(with ${}_pX$ = kernel of multiplication by p, i.e. p^f is the number of points on X of order dividing p). Note that

$$0 \leqslant f(X) = f \leqslant \dim(X);$$

we say that X is ordinary if $f = \dim X$.

By a theorem of Serre and Tate it follows that polarized ordinary abelian varieties can be lifted to characteristic zero. Thus, the lifting problem follows if we can give an affirmative answer to the

Question: Is every polarized abelian variety in characteristic $p > 0$ specialization of an ordinary AV? (or, equivalently: take the generic point of a component of $\mathcal{A}_{g,d} \otimes \mathbb{F}_p$, does it correspond to an ordinary AV?)

2. Stratifications of $A_{g,d} \otimes \mathbb{F}_p$

Instead of directly attacking the lifting problem (from char. p to char. zero) we are now interested in the deformation problem (staying in char. p). There are differences with the theory in characteristic zero.

Let N be a finite group scheme over a field k, let X be an abelian variety over k, and if char(k) = p > 0, moreover suppose that α_p cannot be embedded into N (here α_p = ker F : $G_a \to G_a$, cf. [21], p.I-11). Then

$$|\mathrm{Hom}(N,X)| < \infty,$$

i.e. there are no "continuous families" of subgroup schemes of X, each member being isomorphic to N. However, if $\alpha_p \subset X_1$ and $\alpha_p \subset X_2$, then there exists at least a 1-dimensional family of subgroupschemes

$$\phi_a : \alpha_p \to X_1 \times X_2, \ a \in \mathbb{P}^1.$$

(A consequence: an AV of CM type in characteristic zero is defined over an algebraic closure of $\mathbb{Q}$, but the analogous statement is false in characteristic p > 0, cf. [25]).

Furthermore, there is another difficulty in characteristic p > 0. Let $\phi : X \to Y$ be an isogeny. If we work in characteristic zero, then the differential of ϕ, maps

$$d\phi : TX \to TY,$$

the tangent space at $0 \in X$, isomorphically onto the tangent space at $0 \in Y$. However, in characteristic p > 0, it may happen that $d\phi = 0$, while ϕ is an isogeny (e.g. ϕ = Frobenius, or $\phi = p.\mathrm{id}_X$). Roughly speaking, the theory in characteristic p > 0 resembles the classical theory if we study group schemes N with $\alpha_p \not\subset N$ and isogenies (polarizations, etc.) with $d\phi$ an isomorphism. However, other cases cannot be avoided, and we indicate our strategy in those cases.

From now on, $A_{g,d}$ stands for the moduli space over $\mathbb{F}_p$. We try to construct a stratification of a component V of $A_{g,d}$ (i.e. V should be written as a disjunct union of locally closed subsets) such that the largest stratum consists of all points related to ordinary AV's (and that would solve the lifting problem). What is a good stratification?

<u>Example</u>, g = 1. Every component of the moduli space has dimension 1. An elliptic curve E has f = 0 iff E is supersingular. Thus the stratification is obvious: the supersingular curves correspond to points on the moduli space, and the ordinary elliptic curves make up a 1-dimensional subset in each component of the moduli space.

<u>Example</u>, g = 2. The structure of $A_{2,1}$ is quite well known by the work of Igusa, cf. [18]. Here the first interesting aspect of characteristic p > 0 shows up: let E_1 and E_2 be supersingular elliptic curves. Then $\alpha_p \subset E_1$ and $\alpha_p \subset E_2$, and we can construct for every $a \in \mathbb{P}^1$ an exact sequence ($a = (a_1 : a_2)$ and ϕ_a has components $\times\, a_i$):

$$0 \to \alpha_p \xrightarrow{\phi_a} E_1 \times E_2 \to Z_a \to 0.$$

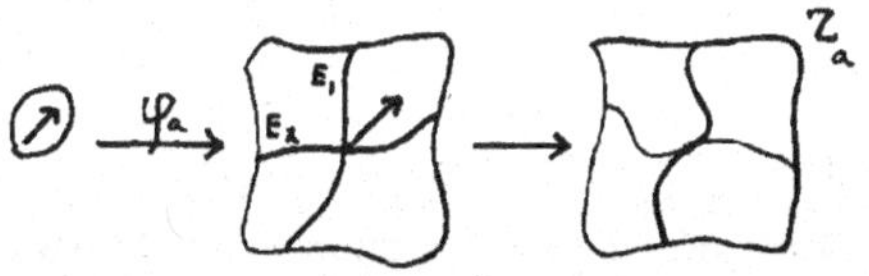

In this way be obtain a $\mathbb{P}^1$-family (Z_a) of AVs, which is not constant. Thus a polarization on this family produces a curve in the moduli space, the abelian surfaces are all isogenous to each other. More precisely, let us dualize the exact sequence above:

$$0 \to \alpha_p \to Y_a \xrightarrow{\psi_a} X \to 0$$

with $Y_a = (Z_a)^t$, and $X = (E_1 \times E_2)^t$. Take a principal polarization λ on X (e.g. the canonical principal polarization on $E_1^t \cong E_1$ and $E_2^t \cong E_2$), and take $\mu_a = \psi_a^*(\lambda)$.

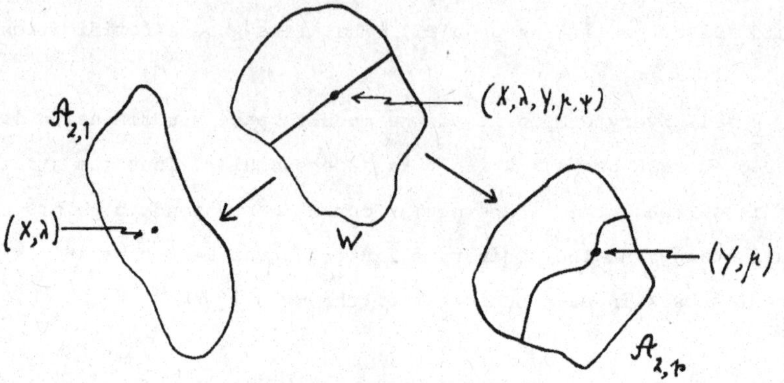

Let W be the isogeny correspondence, i.e. it is the coarse moduli scheme of the (X,λ,Y,μ,ψ) as above. Under this correspondence the point $(X,\lambda) \in A_{2,1}$ correspondes to a <u>curve</u> $\{(Y_a,\mu_a)\} \subset A_{2,p}$, i.e. at (X,λ) the correspondence is a blowing-up.

In general such an isogeny correspondence is <u>finite-to-one</u>: <u>at ordinary points</u>, <u>and at points which correspond to AVs with</u> p-rank = (dim X) - 1. However, if $f(X) < g - 1$ the correspondence might make up a whole component of some $A_{g,d}$. If $g = 2$ one can easily show (cf. [29]) that this is not the case, hence we obtain a cheap proof for liftability of abelian surfaces. However, for large g the situation is much more complicated. Although we know quite a lot about a stratification of $A_{g,1}$ (cf. [9], page 163, Theorem 7), how can we conclude things about $A_{g,d}$?

For p-divisible formal groups one can define its isogeny type (cf. [12]). This is a finer classification for AVs of dimension at least 3 than the p-rank. In order to study the stratification of the moduli space by the isogeny type (of the formal group obtained by completing the local ring at $0 \in X$) we recall (cf. [9], pp. 196/197):

Example, $g = 3$. We distinguish: type 1 = ($f = 3$) = ordinary case, type 2 = ($f = 2$), type 3 = ($f = 1$); in case $f = 0$, $g = 3$, the formal group G can have two distinct isogeny types:

type 4 = ($f = 0$ & $G \sim (G_{2,1} + G_{2,1})$),

type 5 = ($f = 0$ & $G \sim (G_{1,1})^3$)

(for the notation $G_{n,m}$, cf. [12]). The last case, $G \sim (G_{1,1})^g$ with $g = 3$, is called the supersingular case; it is known that in such a case the AV is isogenous to a product of (supersingular) elliptic curves (cf. [26], Theorem 4.2). Thus one can construct a family of dimension $(g(g-1))/2$ consisting of g-dimensional supersingular AVs (cf. [20], Theorem 2.4.(iii) and Corollary 2.6). For $g = 2$ it is known that such a 1-dimensional family can carry a principal polarization. For $g = 3$ Koblitz tried to answer the question, whether such a principal polarization exists, via a computation. He found that the numbers of hyperelliptic curves of genus three having a Jacobian of type 4, respectively of type 5, defined over $\mathbb{F}_p$ (with $3 \leqslant p \leqslant 13$) are "almost the same". However, it turns out that type 4 defines a 3-dimensional subset of $A_{3,1}$ and type 5 defines a 2-dimensional subset of $A_{3,1}$! An "explanation" of the descrepancy between the truth and these computer-results has not yet been given. Possibilities: (i) type 5 is not transversal to the hyperelliptic locus (cf. [9], p. 196, bottom; cf. Section 4 below, Question 5); (ii) curves of type 5 are defined over "small fields" (like in the case of elliptic curves); (iii) prime numbers $\leqslant 2g + 1$ are dangerous, so the results for $p = 3,5,7$ don't have much meaning (this last point was raised by K. Lønsted!) and the primes $p = 11,13$ do not give significant suggestions (cf. [9], page 197). Also cf. [35].

This last example suggests that the stratification by isogeny type is difficult to describe: inside $A_{3,1}$ a component of type 5 is contained in type 4, but for some d (e.g. $d = p^3$), there are components of type 5 of dimension 3 not contained in a component of type 4! So we restrict ourselves to another stratification:

$$A_g = \bigcup_d A_{g,d},$$

$$V_f = \{x \in A_g \mid f(X) \leq f\}$$

(here X denotes the AV defined over $\overline{k(x)}$ corresponding to x), and clearly:

$$V_0 \subset V_1 \subset \ldots \subset V_g = A_g.$$

Each V_f is closed, hence $V_f - V_{f-1}$ is locally closed (and the crucial question is: is $V_{g-1} \neq A_g$ in a very strong sense, i.e. in equality on every component of A_g).

3. $\dim V_f = \frac{1}{2}g(g+1) - g + f$

If V is a union of irreducible components, we allow ourselves only to write dim V if all components have the same dimension. The formula in the title in this section solves the lifting problem. The main difficulty in proving the formula lies in the fact that in general tangent space computations don't give good results: these closed sets may be singular; moreover, equations for V_f may be obtained, but the schemes thus defined may be not reduced. Hence our methods have to be more refined than merely computations in formal deformation theory Furthermore the explicit description of the deformations via Dieudonné-Cartier theory is difficult unless the Dieudonné-module associated to the formal group is generated by one element.

We now outline the program of the proof. The most important ideas underlying it are due to Mumford; details will be published by Norman and Oort (cf. [19]). We work entirely in characteristic $p > 0$.

If G is a groupscheme, or a formal group, and k is algebraically closed, we write

$$a(G) = \dim_k(\mathrm{Hom}(\alpha_p, G))$$

($\mathrm{End}_k(\alpha_p) \cong k$, hence $\mathrm{Hom}(\alpha_p, G)$ is a right-k-module). If Z is an irreducible closed subset of A_g, we write $a(-/Z \subset A_g)$ for the a-number of the AV corresponding to the generic point of Z; analogous notation $f(-/Z \subset A_g)$ for the p-rank. Steps in the proof are:

(A) Let G be a p-divisible formal group, suppose $a(G) > 0$. Consider the sequence of "generic quotients":

$$G = G_0 \to G_1 \to G_2 \to \dots$$

(i.e. embed $\alpha_p \to G_i$ in a generic way, and $G_{i+1} = G_i/\alpha_p$). Then there exists an integer $s \geqslant 0$ such that $a(G_s) = 1$ (cf. [27]).

(B) Let V' be a component of V_f. Then

$$\mathrm{codim}(V' \subset A_g) \leqslant g - f$$

(cf. [26], Lemma 1.6).

(C) Let $W \subset A_{g,d} \times A_{g,pd}$ be the isogeny correspondence, i.e. a coarse moduli scheme for quintuples $(X, \lambda, Y, \mu, \psi)$, cf. Section 2. Suppose $Z \subset A_{g,d}$ is an irreducible closed subset, and let $Z' \subset A_{g,pd}$ be a component of the closure of the projection of $W \cap (Z \times A_{g,pd})$. Then

$$\dim(Z) + a(-/Z \subset A_g) \leqslant \dim Z' + a(-/Z' \subset A_g).$$

(D) Any component of A_g has dimension at least $\frac{1}{2}g(g+1)$ (cf. [22], Theorem 2.3.3), and equality holds at points parametrizing ordinary AVs (cf. [2], Corollary 22).

(E) Suppose $Z \subset A_g$ is a component of V_f with $a(-/Z \subset A_g) = 1$. Then

$$\operatorname{codim}(Z \subset A_g) \geq g - f$$

(here we mean: there exists at least one component of A_g for which this codimension-inequality holds).

Some comments: (A) is proved with the help of Dieudonné-Manin theory, and it is crucial that any p-divisible formal group up to isogeny can be defined over a finite field. The inequality (B) is easy. The inequality (C) follows easily: the fibre

$$W \cap \{(X,\lambda)\} \times A_{g,pd}$$

is a finite-to-one image of $\mathbb{P}^{a-1}$, with $a = a(X) = a(X^t)$, etc.; inequality comes from the fact that λ produces $\psi^*(\lambda) = \mu$, a polarization on Y, but conversely μ need not to descend to X. The fact (D) follows from local deformation theory. The most difficult part is (E): by Dieudonné-Cartier theory Norman could write down deformations of formal groups (in case the module is monogenic); the Hasse-Witt matrix of the deformed module should give its p-rank: a combination of difficult methods, deep results on Dieudonné-modules, and tricks (cf. [17], and [19]).

From (A ~ E) we deduce the result:
(A,C) imply that a component Z of V_f (with $f < g$) of maximal dimension has the property $a(-/Z \subset A_g) = 1$. Thus (A,C,D,E) imply that each component of A_g has points corresponding with ordinary AVs, hence by (D):

$$\dim A_g = \tfrac{1}{2}g(g+1).$$

Take Z, an irreducible component of V_f, apply the method of (C), and repeat the process untill we obtain an irreducible closed set $Z' \subset A_g$ with $a(-/Z' \subset A_g) = 1$ (this is possibly by (A)). Let Z" be a component of V_f containing Z'. Then by (B), (C) and (E) we obtain:

$$\tfrac{1}{2}g(g+1) - g + f + a(-/Z \subset A_g) \leq$$
$$\leq \dim Z + a(-/Z \subset A_g) \leq$$
$$\leq \dim Z' + 1 \leq$$
$$\leq \dim Z'' + 1 \leq$$
$$\leq \tfrac{1}{2}g(g+1) - g + f + 1.$$

Thus these are all equalities, hence

$$\dim V_f = \tfrac{1}{2}g(g+1) - g + f, \text{ and}$$
$$a(-/Z \subset A_g) = 1.$$

This ends the proof, for details we refer to [19].

Corollary: Let U be a component of A_g, then

$$\tfrac{1}{2}g(g+1) = \dim U > \dim V_{g-1} = \tfrac{1}{2}g(g+1) - 1;$$

i.e. every polarized AV is a specialization of an ordinary polarized AV. Thus every polarized AV in characteristic $p > 0$ can be lifted to characteristic zero.

4. Some questions

In the topic we have discussed, there seem to be many open problems. We indicate some of these; we have no idea about the difficulty of some of these; also the list is far from complete; some of these problems were suggested by Mumford, Ueno, Norman, Grothendieck, etc.

Question 1. Is $A_{g,1} \otimes k$ irreducible?

We have seen that the answer is "yes" if $g = 1,2$, or 3. The answer is positive if $\mathrm{char}(k) > 2$ and $g = 4$ or $g = 5$ (cf.[1], 6.5). The answer is positive if $\mathrm{char}(k) = 0$. It might be that the question can be answered by constructing a good compactification of $A_{g,1}$; that seems one of the most difficult problems now, on which much work is going on.

Question 2. Let (X,λ) be a polarized AV (over a field of positive characteristic); find explicit equations for the local moduli space of (X,λ), or at least for the leading terms of such equations.

Norman (cf.[17], Section 3, and [18]) indicates a procedure, and gives examples.

Question 3. Components of $A_{g,1} \otimes \mathbb{F}_p$ may intersect (cf.[18]); characterize all components which intersect at a given point (X,λ).

Consider a sequence (δ) as in Mumford, [15]; suppose we know that $A^{(\delta)}$ is irreducible; which conditions for (δ) and (δ') are necessary and sufficient in order that $A^{(\delta)}$ and $A^{(\delta')}$ meet? In [18] Norman indicates two components in the case $g = 2$ belonging to $(\delta) = (p,p)$, respectively $(\delta') = (1,p^2)$ which meet (of course, in a point (X,λ) with $f(X) < g-1$, hence $f(X) = 0$, hence X supersingular in this case).

Question 4. Does there exist a polarized AV (X,λ) defined over k which does not lift to $W_\infty(k)$ (say, $\mathrm{char}(k) = p$, and k algebraically closed)?

We know (X,λ) lifts to a (ramified) extension of $W_\infty(k)$.

<u>Question 5.</u> Does there exist a 2-dimensional family (in M_3) of supersingular hyperelliptic curves of genus 3?

Compare the discussion in Section 2 above: example, $g = 3$. It is not so difficult to write down for a given prime p all hyperelliptic curves with $f(\mathrm{Jac}(C)) = 0$; but how can one decide which curves in such families are supersingular? Approaching from the other end: we can easily construct supersingular AVs, we can indicate on which sub-families there exists a principal polarization, but how can we recognize which of these Jacobians belong to hyperelliptic curves? Possibly both approaches could be sucessfull if we had more tools available (something like "higher Hasse-Witt caculus", respectively something like "hyperelliptic Schottky-relations").

<u>Question 6</u>. Let $\psi : (y,\mu) \to (X,\lambda)$ be an isogeny of polarized AVs in characteristic p; can it be lifted to characteristic zero?

It is known that an endomorphism in general cannot be lifted to an endomorphism (cf. [5], pp. 259-263; cf. [24]).

<u>Question 7.</u> Let M_g denote the moduli space of curves of genus g; the Torelli theorem tell us that the morphism

$$j : M_g \to A_{g,1}$$

is injective on geometric points. Is it true that j is an immersion?

The answer is affirmative for $g \leqslant 3$ (cf.[28], Proposition 8). The answer is also positive at non-hyperelliptic points (already a very classical result, easily proven by tangent-space considerations using the Max Noether theorem, cf.[30], Theorem 2.10). Added in proof: the answer to Question 7 is probably "yes" if $p \neq 2$ (Oort & J. Steenbrink).

Question 8. (cf. Grothendieck, [7], page 150). If a p-divisible group G specializes to G', then the Newton-polygon of G is below the Newton-polygon of G' (cf.[4], page 91, Theorem); is this condition sufficient? (in the following sense: given G', and a polygon below the Newton-polygon of G', does there exist a family of p-divisible groups whose generic fibre has the given Newton-polygon?)

In [33] we find a positive answer to the question if we are allowed to replace G' within its isogeny type.

Question 9. Let $s \in A_{2,1}$ be the generic point of $A_{2,1} \otimes \mathbb{C}$, and let X be the corresponding AV; can X be defined over $\mathbb{C}(s)$?

Shimura, [32], has shown that the answer is negative if we replace $\mathbb{C}$ by a field contained in $\mathbb{R}$.

Question 10. Can one develop a theory of Prym varieties in characteristic p = 2?

Cf. [16],[1].

RERERENCES

[1] A. Beauville, Prym varieties and the Schottky problem. Invent. Math. 41 (1977), 149-196.

[2] S. Crick, Local moduli of abelian varieties. Amer. J. Math. 97 (1975), 851-861.

[3] P. Deligne & D. Mumford, The irreducibility of the space of curves of given genus. Publ. Math. No. 36 (Volume dedicated to O. Zariski), IHÉS, 1969, 75-109.

[4] M. Demazure, Lectures on p-divisible groups. Lect. N. Math. 302, Springer-Verlag, 1972.

[5] M. Deuring, Die Typen der Multiplikatorenringe elliptischer Funktionenkörper. Abh. Math. Sem. Hamburg 14 (1941), 197-272.

[6] A. Grotendieck & J. Dieudonné, Éléments de géométrie algébrique. Chap. III (first part). Publ. Math. No. 11, IHÉS, 1967. EGA

[7] A. Grothendieck, Groupes de Barsotti-Tate et cristaux de Dieudonné. Sem. de Math. Sup., 1970. Press. Univ. Montreal, 1974.

[8] J.-I. Igusa, Arithmetic variety of moduli for genus two. Ann. Math. 72 (1960), 612-649.

[9] N. Koblitz, p-adic variation of the zeta-function over families of varieties defined over finite fields. Compos. Math. 31 (1975), 119-218.

[10] K. Kodaira & D.C. Spencer, On deformations of complex analytic structures I,II. Ann. Math. 67 (1958), 328-466.

[11] S. Lang, Abelian varieties. Intersc. Tracts N. 7. Intersc. Publ., New York, 1959.

[12] Yu. I. Manin, The theory of commutative formal groups over fields of finite characteristic. Russ. Math. Surveys 18 (1963), 1-80.

[13] D. Mumford, Geometric invariant theory. Ergebn. Math. Vol. 34, Springer-Verlag, 1965.

[14] D. Mumford, Abelian varieties. Tata Inst. F.R. Stud. in Math. Vol. 5, Bombay.Oxford Uni. Press 1974.

[15] D. Mumford, The structure of the moduli space of curves and abelian varieties. Actes, Congrès intern. math., 1970, Tome 1, 467-465. Gauthiers-Villars, Paris, 1971.

[16] D. Mumford, Prym varieties I. Contributions to analysis, 325-350. Acad. Press, New York, 1974.

[17] P. Norman, An algorithm for computing local moduli of abelian varieties. Ann. Math., 101 (1975), 499-509.

[18] P. Norman, Intersections of components of the moduli space of abelian varieties. To appear in Journ. Pure Appl. Algebra.

[19] P. Norman & F. Oort, Moduli of abelian varieties. To appear.

[20] T.Oda & F. Oort, Supersingular abelian varieties. To appear.

[21] F. Oort, Commutative group schemes. Lect. N. Math. 15, Springer-Verlag, 1966.

[22] F. Oort, Finite group schemes, local moduli for abelian varieties and lifting problems. In : Algebraic geometry, Oslo, 1970; Wolters-Noordhoff Publ. Cy, 1972 (also : Compos. Math. 23 (1971), 265-296).

[23] F. Oort & K. Ueno, Principally polarized abelian varieties of dimension two or three are Jacobian varieties. Journ. Fac. Sc. Univ. Tokyo 20 (1973), 377-381.

[24] F. Oort, Lifting an endomorphism of an elliptic curve to characteristic zero. Proc. Kon. Ned. Akad. Wetenschappen. 76 (1973) (Indag. Math. 35 (1973)), 466-470.

[25] F. Oort, The isogeny class of a CM-type abelian variety is defined over a finite extension of the prime field. Journ. Pure Appl. Algebra 3 (1973), 399-408.

[26] F. Oort, Subvarieties of moduli spaces. Invent. Math. 24 (1974), 95-119.

[27] F. Oort, Isogenies of formal groups, Proc. Kon. Ned. Akad. Wetensch. 78 (1975) (Indag. Math. 37 (1975)), 391-400.

[28] F. Oort, Fine and coarse moduli schemes are different. Dept. Math. Univ. Amsterdam, Report 74-10 (1974).

[29] F. Oort, Families of subgroup schemes of formal groups. Contributions to algebra. A collection of papers dedicated to Ellis Kolchin, Ed. H. Bass, P.J. Cassidy, J. Kovacic; Acad. Press, 1977; pp. 303-319.

[30] B. Saint-Donat, On Petri's analysis of the linear system of quadrics through a canonical curve. Math. Ann. 206 (1973), 157-175.

[31] Séminaire H. Gartan, Fonctions automorphes, 10e année, 1957-58. Secr. Math., Paris, 1958.

[32] G. Shimura, On the field of rationality for an abelian variety. Nagoya Math. J. 45 (1971), 167-178.

[33] G. Traverso. Families of Dieudonné modules and specialization of Barsotti-Tate groups.

[34] A. Weil, Zum Beweis des Torellischen Satzes. Nachr. Akad. Wiss. Göttingen, Math.-Phys. Kl., 1975, 33-53.

[35] N. Yui, On the Jacobian varieties of algebraic curves over fields of characteristic $p > 0$.

F. OORT
Mathematisch Instituut
Rijksuniversiteit Utrecht
De Uithof, Budapestlaan 6
3508 TA UTRECHT
The Netherlands

A Family of genus two Fibrations

by Ulf Persson *

Columbia University

This Note will be devoted to the following Theorem.

Theorem: Let x,y be strictly positive integers, satisfying

a) $2x-6 \leq y \leq 4x-4$ b) $y \neq 4x-5$.

Then there is a minimal surface of general type X,which in fact can be chosen to be a genus two fibration,with $\chi(X)=x$ $c_1^2(X)=y$.

Remark: The condition b) reflect a very annoying gap,which the author has as yet not been able to fill, hopefully in a forthcoming paper this will be settled.

The techniques used are completly elementary,namely that of double coverings and the structure of effective divisors on rational ruled surfaces.

To keep this note as brief as possible,they will be assumed to be known.

---

*This work has been partially supported by the National Science Foundation under Grant No. MCS77-07660

Notation, Terminology and Basic Facts

By F_N is meant a rational ruled surface (relative minimal), whose minimal section (section with lowest selfintersection) has self-intersection -N .

A standard way of constructing F_N is to add a section at infinity to the line bundle $O_{\mathbb{P}^1}(N)$.

The Neron-Severi group of F_N is generated by a section S , with $S^2=N$, and a fiber F ($F^2=0$). Furthermore SF=1 .

The minimal section is given by S-NF

A singularity of a (branch) curve is called in-essential or negligable iff its multiplicity is at most three, and after a blow-up at most two.

A triple point is called an infinitely close triple point, iff it remains a triple point after a blow-up .

Thus an infinitely close triple point is the simplest kind of an essential singularity, and can easily be visualized as a triple point with tangent branches.

The following is well-known , see e,g. [H] , and its formal consequence elementary to prove directly (see [P,remark after 2.5]

Proposition 0. The singularities of a double covering along a branch locus with no essential singularities are all rational double points .

Formal consequence: In the computation of the chern-numbers of the resolution of a double covering, the in-essential singularities can be ignored .

As is well-known there is a "canonical" way of resolving the singularities, resulting from the singularities of the branch locus, of a double covering (see [H] or [P,prop.2.3]).

Construction

Proposition 1: There exists on F_N a curve $C_{N,k,a}$ (a=0,1,2) equivalent to 6S+2aF with k infinitely close triple points and no other essential singularities.

Furthermore fixing N and a, the only restriction on k is $0 \leq k \leq 2N + 2[a/2]$.

Proof: There is a double covering $\pi : F_N \to F_{2N}$, branched at two disjoint sections S_o and S_∞ ,where $S_o = S$, $S_\infty = S - 2NF$.

Choose 2N+1 distinct points $q_1, \ldots q_{2N+1}$ on S_o and a point q_{2N+2} on S_∞ whose projection onto S_o is distinct from the previously chosen points.

Let $k \leq 2N$. It is then possible to find three distinct sections S_1, S_2, S_3 all equivalent to S but distinct from S_o, with $q_1, \ldots q_k$ as only common intersections. (You can think of the sections as given by polynomials in 2N variables, whose only common zeroes are $q_1, \ldots q_k$)

Similarly we can also find three distinct sections S_1', S_2', S_3'

all equivalent to S+F , with $q_1,\ldots q_k$ as the common intersections (now no further restriction on k is necessary).

Observe that the pullbacks of the above sections will all have vertical tangents at the points $\pi^*(q_i)$ $(i\leq k)$.

Define $C_{N,k,0} = \cup\, \pi^*(S_i)$; $C_{N,k,1} = \cup\, \pi^*(S_i) \cup 2F$ and $C_{N,k,2} = \cup\, \pi^*(S_i') \cup F$ (where i=1,2,3 and F are "generic" fibers). From the above observation it is clear that these curves will have infinitely close triple points at $\pi^*(q_i)$ $(i\leq k)$.

<u>Definition</u>: Let $X_{N,k,a}$ be the resolution of the double covering of F_N along $C_{N,k,a}$; with the exceptional divisors stemming from the resolutions of the infinitely close triple points blown down.

<u>Proposition 2</u>:

$$c_1^2(X_{N,k,a}) = 6N + 4a - 8 - k$$
$$\chi(X_{N,k,a}) = 3N + 2a - 1 - k$$

Proof: Clear from the standard formulas of double coverings. (see e.g. [P; 2.2. , 4.3.remarks following 4.4.])

<u>Proposition 3</u>: $X_{N,k,a}$ is minimal when either a=0 $N\geq 2$; a=1 $N\geq 1$; a=2 $k\leq 2N+1$ $N\geq 0$ or a=2, k=2N+2 $N\geq 1$.

This will follow as a straightforward application of the following two observations.

<u>Lemma i</u>: If $\pi: X \to Y$ is a double covering, X non-ruled, then the exceptional divisors of X fall into two types.

a) pullbacks of exceptional divisors of Y, disjoint from the branchlocus.

b) reduced pullbacks of rational components of the branchlocus with self-intersection -2 .

Proof:(cf.[P;2.6.] . Given an exceptional divisor E on X ; either E and $\pi(E)$ are disjoint then a) or they coincide (X is non-ruled) in the latter case if π did not fix E,it would be a pullback and hence $E^2=0(2)$, thus b) .

Lemma ii: Any exceptional divisor of a surface is a fixed component of a canonical curve (if such exists).
This is a standard fact in the theory of surfaces and hence its proof omitted, I am grateful to Moishezon who pointed out to me the pertinence of that observation.

Proof of proposition 3: Denote by E_i , E_i' the exceptional divisors corresponding to the blow-ups of the ith infinitely close triple point. (E_i' denotes the final blow up of the remaining triple point after the blow-up E_i).

The canonical divisor of $X_{N,k,a}$ is given by $\pi^*(K + E_i + E_i' - B - E_i - 2E_i')$ where K is the canonical divisor on F_N and $B = 3S + aF$ - modulo a slight modification indicated below. This can be simplified to $\pi^*(S - E_i + (N-2+a)F + E_i - E_i')$ $i \leq k$ .However the $\pi^*(E_i - E_i')$ are exceptional and hence blown down in our definition of $X_{N,k,a}$.

By slight abuse let S_o denote $\pi^*(S_o)$ (In the terminology of the proof of proposition 1), furthermore let us write
$\overline{S}_o = S_o - E_1 - \dots - E_k$ $(k \leq 2N+1)$; $\overline{S}_o = S_o - E_1 - \dots - E_{k-1}$ $(k=2N+2)$
Thus with the above restrictions on N,k and a , canonical curves are given by the following pullbacks. (Note that by our construction of $C_{N,k,a}$ $\overline{S}_o$ are effective !)

$\pi^*(\overline{S}_o + (N-2+a)F)$ $k \leq 2N+1$ or $\pi^*(\overline{S}_o + (F-E_{2N+2}) + (N-1)F)$ $k=2N+2$. By inspection these components are not part of the branchlocus $C_{N,k,a}$ nor disjoint from it, unless $a=0$ $k=2N$ or $a=2$ $k=2N+2$ when $\overline{S}_o$ becomes disjoint,but in neither case will it be exceptional.

The proof of the Theorem is now a straightforward application of Propositions 1,2 and 3.

Comments

The great flaw is the inability to fill the line $y=4x-5$.This could be done however if we could replace $2[a/2]$ with a in Prop 1 ! Such curves however cannot be pullbacks via $\pi:F_N \to F_{2N}$, so entirely new methods of construction have to be supplied.The following conjecture would fill the gap if true.(notation as in the proof of Proposition 1)

Conjecture:The linear subsystem of $|2S+F|$ on F_N ,defined by having vertical tangents at $\pi^*(q_1),\ldots\pi^*(q_{2N})$ is strictly bigger then the pullback of the linear subsystem of $|S+F|$ on F_{2N} consisting of sections passing through $q_1,\ldots q_{2N}$.

Note that the conjecture is false if $|2S+F|$ and $|S+F|$ are replaced by $|2S|$ and $|S|$ respectively,which is easy to see by intersection arguments.

There are of course many related (or merely subsidiary) results,which the author hopes ,along with the filling of the gap, to expound in a longer,fuller and less terse paper,hopefully to appear in the near future,

Bibliography

[H] Horikawa Algebraic Surfaces of general type with small c_1^2 I
Ann.of.Math II ser 104 (1976)

[P] Persson Double coverings and surfaces of general type.
In Algebraic Geometry, Proceedings, Tromsø, Norway 1977 (ed. L. D. Olson), Springer Lecture Notes vol 687, 1978.

Columbia University
New York, N.Y. 10027
U.S.A.

Ideals associated to a desingularization.

Ragni Piene*

1. Introduction.

Let X be a reduced algebraic curve with normalization $f : Z \to X$. Let J denote the jacobian ideal of X, I the ramification ideal of f, and C the conductor of Z in X. By comparing certain invertible sheaves on Z I showed in ([P], § 2) the equality

$$(*) \qquad J\mathcal{O}_Z = C \cdot I ,$$

when X is a local complete intersection.

As an example – where these three ideals can be easily computed – consider the plane curve

$$f(x, y) = x^b - y^a = 0 , \quad (a, b) = 1 , \quad a < b .$$

Its normalization k[t] is given by $x \mapsto t^a$, $y \mapsto t^b$. The jacobian ideal is $J = (bx^{b-1}, ay^{a-1})$, the ramification ideal $I = (bt^{b-1}, at^{a-1})$, and the conductor, as an ideal in k[t], is $C = (t^{(a-1)(b-1)})$.

When X is a variety of dimension ≥ 1, whose normalization Z is smooth, I also showed that (*) holds on the variety obtained by blowing up the ideal I in Z. However, examples indicated that (*) should hold already on Z. We shall prove this below, as a corollary of a more general result:

Consider a desingularization $f : Z \to X$ of a Cohen-Macaulay variety X, of dimension r. Define the ω-jacobian ideal J of X by

$$J = \mathrm{Ann}\ (\mathrm{Coker}\ (\varphi : \Omega_X^r \to \omega_X)) ,$$

where φ is the canonical map from r-differentials to dualizing sheaf. If X is a local complete intersection (l.c.i.) we show that J is equal

* Partially supported by the Norwegian Research Council for Science and the Humanities.

to the (usual) jacobian ideal

$$\tilde{J} = F^r(\Omega^1_X) ,$$

defined as the r-th Fitting ideal of the module of differentials.

Inspired by ([M - T]) we consider the adjunction conductor (or ideal of adjoint functions) $A \subseteq O_X$, defined by

$$A = \mathrm{Ann}\ (\mathrm{Coker}\ (\mathrm{Tr}_f : f_*\omega_Z \to \omega_X)) .$$

If f is finite (and X is Gorenstein), this ideal is equal to the usual conductor C.

We introduce an ideal $B \subseteq O_Z$ measuring where

$$\rho : f^*f_*\Omega^r_Z \to \Omega^r_Z$$

is not surjective,

$$B = \mathrm{Ann}\ (\mathrm{Coker}\ \rho) .$$

Note that $B = O_Z$ if f is finite, but that $B \neq O_Z$ for example if f factors through another desingularization of X.

Finally, let $I = F^o(\Omega^1_{Z/X})$ denote the ramification ideal of f. Then we prove (Thm. 1): Suppose X is a Gorenstein variety. There is an equality of ideals,

(**) $$B \cdot JO_Z = I \cdot AO_Z .$$

We also give some examples of the relations (*) and (**).

In the last section we look at Nash transformations and jacobian blow-ups: If X is a l.c.i., it is known ([N]) that the Nash transformation $\pi : \tilde{X} \to X$ is equal to the blow-up $\tilde{\pi}$ of the jacobian ideal $\mathfrak{J}$. Nobile ([N]) also showed that the Nash transformation is <u>locally</u> the blow-up of an ideal which is contained in the jacobian ideal, but whose associated subscheme might have support strictly bigger than the singular locus of X.

Here we show that, for a Gorenstein variety X, the Nash transformation π is (globally) equal to the blow-up of the ω-jacobian ideal, hence to the blow-up of a subscheme whose support is contained in the singular locus.

Moreover, π is equal to the blow-up $\tilde{\pi}$ of the jacobian ideal $\mathfrak{J}$ if and only if dim proj $\pi^*\Omega_X^1 \leq 1$ holds.

We know no explicit way of computing the ω-jacobian ideal J of a Gorenstein, non l.c.i. variety X. Therefore we can give no examples where $J \neq \mathfrak{J}$ (or $\pi \neq \tilde{\pi}$) holds, though such seem likely to exist.

2. Relations among the ideals.

Let k be an algebraically closed field. A variety is a reduced, equidimensional algebraic k-scheme. If X is a variety of dimension r, we let $\Omega_X^r = \Lambda^r\Omega_X^1$ denote its sheaf of r-differentials. If X is Cohen-Macaulay, it has a dualizing sheaf ω_X, and if $f : Z \to X$ is a proper map between Cohen-Macaulay varieties of the same dimension, there is a trace homomorphism ([H], VII, 3.4)

$$\mathrm{Tr}_f : f_*\omega_Z \to \omega_X .$$

Suppose X is Cohen-Macaulay, of dimension r, and let

$$n : X' \to X$$

denote its normalization. Let $i : U \hookrightarrow X'$ denote the inclusion of the smooth locus U of X'. Then $i^*\omega_{X'} = i^*\Omega_{X'}^r$. Since depth $\omega_{X'} \geq 2$ at all points of X' - U, we get

$$\omega_{X'} = i_*i^*\Omega_{X'}^r ,$$

and hence a natural homomorphism

$$\varphi_{X'} : \Omega_{X'}^r \to \omega_{X'} .$$

Define

$$\varphi_X : \Omega_X^r \to \omega_X$$

as the composition

$$\Omega_X^r \longrightarrow n_*\Omega_{X'}^r \xrightarrow{n_*\varphi_{X'}} n_*\omega_{X'} \xrightarrow{\mathrm{Tr}_n} \omega_X .$$

(The map φ_X can be defined in much greater generality, see [E].) For any proper birational map $f : Z \to X$ between Cohen-Macaulay varieties, there is a commutative diagram

$$(***)\qquad \begin{array}{ccc} \Omega_X^r & \xrightarrow{\varphi_X} & \omega_X \\ \downarrow & & \uparrow \mathrm{Tr}_f \\ f_*\Omega_Z^r & \xrightarrow{f_*\varphi_Z} & f_*\omega_Z \end{array} .$$

For the rest of this section we shall assume that X is a Gorenstein variety of dimension r, given with a desingularization

$$f : Z \to X ,$$

i.e., Z is smooth, f is proper and birational. As above we let

$$n : X' \to X$$

denote the normalization of X.

Note that ω_X is invertible, and that we have

$$\omega_Z = \Omega_Z^r .$$

The jacobian ideal $\tilde{J} \subseteq O_X$ of X is the rth Fitting ideal of Ω_X^1,

$$\tilde{J} = F^r(\Omega_X^1) .$$

We define the ω-jacobian ideal $J \subseteq O_X$ of X by

$$J = \mathrm{Ann}\ (\mathrm{Coker}\ \varphi_X) = \mathrm{Im}\ (\varphi_X \otimes \omega_X^{-1} : \Omega_X^r \otimes \omega_X^{-1} \to O_X) .$$

The conductor (of X' in X) is the ideal $C \subseteq O_X$,

$$C = \underline{\mathrm{Hom}}_X(n_*O_{X'}, O_X) = \mathrm{Ann}\ (\mathrm{Coker}\ O_X \to n_*O_{X'}) .$$

Note that the ideal $C^{\sim} \subseteq O_{X'}$ induced by C is equal to C.

Define the adjunction conductor $A \subseteq O_X$ by

$$A = \mathrm{Ann}\ (\mathrm{Coker}\ \mathrm{Tr}_f) = \mathrm{Im}\ (\mathrm{Tr}_f \otimes \omega_X^{-1} : f_*\omega_Z \otimes \omega_X^{-1} \to O_X) .$$

The ramification ideal $I \subseteq O_Z$ of f is the 0-th Fitting ideal of the sheet of relative differentials $\Omega^1_{Z/X}$,

$$I = F^o(\Omega^1_{Z/X}) .$$

Finally we define the ideal $B \subseteq O_Z$ by

$$B = \mathrm{Ann}\,(\mathrm{Coker}\ \rho) = \mathrm{Im}\,(\rho \otimes \omega_Z^{-1} : f^*f_*\omega_Z \otimes \omega_Z^{-1} \to O_Z) .$$

Remark that we always have inclusions

$$J \subseteq A \quad \text{and} \quad I \subseteq B .$$

The first is an immediate consequence of (***), while the second holds because of the equality

$$I = \mathrm{Im}\,(f^*\Omega^r_X \otimes \omega_Z^{-1} \to O_Z)$$

and the commutativity of

$$\begin{array}{ccc} f^*\Omega^r_X & \longrightarrow & f^*f_*\omega_Z \\ & \searrow & \downarrow \\ & & \omega_Z \end{array} .$$

Moreover, if X itself is smooth, φ_X and Tr_f are isomorphisms, so in that case

$$J = A = O_X \quad \text{and} \quad I = B$$

hold.

General duality theory for finite maps implies that Tr_n induces an isomorphism

$$n_*\omega_{X'} \xrightarrow{\sim} \underline{\mathrm{Hom}}_X(n_*O_{X'}, O_X) \otimes \omega_X = C \otimes \omega_X .$$

Since φ_X factors through Tr_n, we obtain the inclusion

$$A \subseteq C .$$

The conductor C is the ideal of <u>subadjoint</u> functions, while A is the ideal of <u>adjoint</u> functions on X (see [Z], [M - T]). Locally, this can be formulated as follows. Suppose α is a meromorphic differential which generates ω_X. Then $\psi \in C$ iff $\psi\alpha$ is regular at all points of codimension ≤ 1 on X', and $\psi \in A$ iff $\psi\alpha$ becomes regular on Z. One says that the singularities of X do not affect the adjunction conditions if A = C holds. This occurs for example when f is finite (i.e., Z = X'). The other classical example of such singularities are the rational double points (Du Val, [DV]). In that case, $f_*\omega_Z \xrightarrow{\sim} \omega_X$ holds, hence we get $A = C = O_X$.

Remark: Suppose $g : Y \to X$ is a finite map between Gorenstein schemes of the same dimension. In ([P], 2.9) we used the duality isomorphism

$$\omega_Y \simeq \underline{\mathrm{Hom}}_X(g_*O_Y, O_X)^\sim \otimes g^*\omega_X$$

to show that any such birational map is equal to the blow-up of the conductor

$$C_{Y/X} = \underline{\mathrm{Hom}}_X(g_*O_Y, O_X)$$

of Y in X. A special case – X is a hypersurface and g its normalization – has been proved independently by Wilson ([W], 2.7). ∎

When X is a l.c.i., we can give an explicit description of the map $\varphi_X : \Omega_X^r \to \omega_X$. This we shall use to prove the following.

Proposition 1: If X is a l.c.i., then

$$J = \mathfrak{J}$$

holds, i.e., the jacobian and ω-jacobian ideals are equal.

Proof: The question is local, so assume $X \subseteq Y$, Y smooth of dimension n, and X is defined by $f_1, \ldots, f_{n-r}$. Choose a system of coordinates $y_1, \ldots, y_n$ on Y around $x \in X$. We know

$$\omega_X = \underline{\mathrm{Ext}}_Y^{n-r}(O_X, \Omega_Y^n)$$

and, since X is a l.c.i. ([H], III, § 7),

$$\underline{\mathrm{Ext}}_Y^{n-r}(O_X, \Omega_Y^n) = \underline{\mathrm{Hom}}_X(\Lambda^{n-r}N, \Omega_Y^n|_X) ,$$

where $N = N_{X/Y}$ denotes the conormal bundle of X in Y,

$$N = (f_1, \ldots, f_{n-r})/(f_1, \ldots, f_{n-r})^2 .$$

The choices $(f_1, \ldots, f_{n-r})$ and $(y_1, \ldots, y_n)$ give a generator

$$\begin{bmatrix} dy_1 \wedge \ldots \wedge dy_n \\ f_1 \ldots f_{n-r} \end{bmatrix}$$

for ω_X at x, and the map

$$\Omega^r_Y|_X \to \Omega^r_X \overset{\varphi_X}{\to} \omega_X$$

is given by

$$dy_{i_1} \wedge \ldots \wedge dy_{i_r} \mapsto \begin{bmatrix} dy_{i_1} \wedge \ldots \wedge dy_{i_r} \wedge df_1 \wedge \ldots \wedge df_{n-r} \\ f_1 \ldots f_{n-r} \end{bmatrix}$$

$$= (-1)^\sigma \left| \frac{\partial(f_1, \ldots, f_{n-r})}{\partial(y_{j_1}, \ldots, y_{j_{n-r}})} \right| \cdot \begin{bmatrix} dy_1 \wedge \ldots \wedge dy_n \\ f_1 \ldots f_{n-r} \end{bmatrix}$$

where $\{j_1, \ldots, j_{n-r}\} = \{1, \ldots, n\} - \{i_1, \ldots, i_r\}$.

Hence J is the ideal generated by the (n-r)-minors of the jacobian matrix $\frac{\partial f}{\partial y}$, which is precisely the jacobian ideal

$$\mathfrak{J} = F^r(\Omega^1_X)$$

(since $N \overset{\frac{\partial f}{\partial y}}{\to} \Omega^1_Y|_X \to \Omega^1_X \to 0$ is a presentation). ∎

Theorem 1: There is an equality of ideals in O_Z

$$(**) \qquad B \cdot JO_Z = I \cdot AO_Z \, .$$

Moreover, AO_Z is invertible if and only if B is, and JO_Z is invertible if and only if I is.

Corollary 1: Suppose f is finite. Then C = A is invertible, and there is an equality

$$(*) \qquad JO_Z = I \cdot C \, .$$

Corollary 2: Suppose X is a l.c.i. and $\rho : f^*f_*\omega_Z \to \omega_Z$ is surjective. Then the jacobian ideal $\mathfrak{J} = F^r(\Omega^1_X)$ decomposes in O_Z as

$$\tilde{J}O_Z = I \cdot A\,O_Z\ ,$$

and $A\,O_Z$ is invertible.

Proof of the theorem: Consider the diagram

$$\begin{array}{ccccc} & & \xrightarrow{\quad f^*\varphi_X \quad} & & \\ & & & \overset{f^*Tr}{\nearrow} & f^*\omega_X \\ f^*\Omega^r_X & \longrightarrow & f^*f_*\omega_Z & & \\ & & & \underset{\rho}{\searrow} & \omega_Z \\ & & \xrightarrow[\quad \Lambda^r df \quad]{} & & \end{array}$$

Since torsionfree quotients of the same sheaf which are generically isomorphic are everywhere isomorphic, we deduce

$$B \otimes \omega_Z = \mathrm{Im}\,(\rho) = \mathrm{Im}\,(f^*Tr) = AO_Z \otimes f^*\omega_X$$

and

$$I \otimes \omega_Z = \mathrm{Im}\,(\Lambda^r df) = \mathrm{Im}\,(f^*\varphi_X) = JO_Z \otimes f^*\omega_X\ .$$

■

Observe that if $\rho : f^*f_*\omega_Z \to \omega_Z$ is surjective, then $A\,O_Z$ is invertible (in particular, f factors through the blow up of A). More generally, whenever f^*Tr_f factors through ρ, we get

$$\omega_Z \cong \tilde{A} \otimes f^*\,\omega_X$$

for some invertible ideal $\tilde{A}$, with $AO_Z \subseteq \tilde{A}$ and $B \cdot \tilde{A} = AO_Z$.

3. Examples.

Assume $f : Z \to X$ is as in Cor. 2. We shall explicit the homomorphisms occurring in the definitions of the ideals and thus obtain (locally) a generator for the invertible ideal AO_Z in terms of generators of $\tilde{J}$ and I.

Take $x \in X$ and $z \in f^{-1}(z)$. We assume $X \subseteq Y$ as in the proof of Prop. 1, and in addition we choose a system of coordinates $(z_1, .., z_r)$ on Z around z. Look at

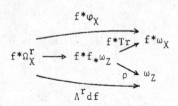

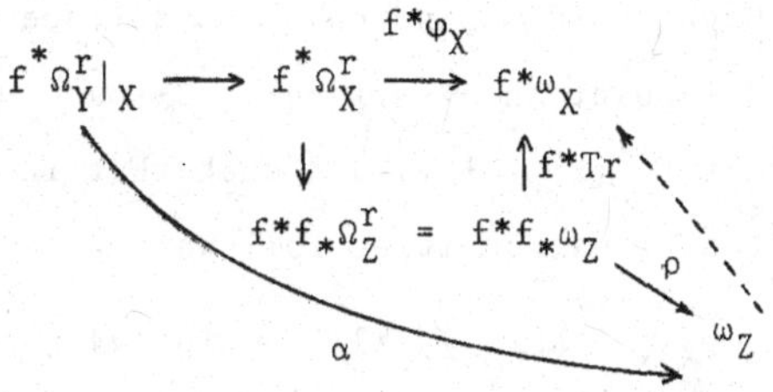

$$\begin{array}{ccccccc} f^*\Omega^r_Y|_X & \longrightarrow & f^*\Omega^r_X & \xrightarrow{f^*\varphi_X} & f^*\omega_X & & \\ & & \downarrow & & \uparrow f^*Tr & \nwarrow & \\ & & f^*f_*\Omega^r_Z & = & f^*f_*\omega_Z & \xrightarrow{\rho} & \omega_Z \end{array} \qquad \alpha : f^*\Omega^r_Y|_X \longrightarrow \omega_Z$$

The dotted arrow exists since ρ is surjective.

We have seen (proof of Prop. 1) that the element $dy_{i_1} \wedge \ldots \wedge dy_{i_r}$ goes to

$$(-1)^\sigma \left|\frac{\partial(f_1, \ldots, f_{n-r})}{\partial(y_{j_1}, \ldots, y_{j_{n-r}})}\right| \begin{bmatrix} dy_1 \wedge \ldots \wedge dy_n \\ f_1 \ldots f_{n-r} \end{bmatrix}$$

in ω_X. The same element goes, via α, to

$$\left|\frac{\partial(y_{i_1}, \ldots, y_{i_r})}{\partial(z_1, \ldots, z_r)}\right| dz_1 \wedge \ldots \wedge dz_r .$$

Hence $dz_1 \wedge \ldots \wedge dz_r$ goes to

$$(-1)^\sigma \frac{\left|\frac{\partial(f_1, \ldots, f_{n-r})}{\partial(y_{j_1}, \ldots, y_{j_{n-r}})}\right|}{\left|\frac{\partial(y_{i_1}, \ldots, y_{i_r})}{\partial(z_1, \ldots, z_r)}\right|} \begin{bmatrix} dy_1 \wedge \ldots \wedge dy_n \\ f_1 \ldots f_{n-r} \end{bmatrix}$$

in $f^*\omega_X$. So the element

$$a = \frac{(-1)^\sigma \left|\frac{\partial(f_1, \ldots, f_{n-r})}{\partial(y_{j_1}, \ldots, y_{j_{n-r}})}\right|}{\left|\frac{\partial(y_{i_1}, \ldots, y_{i_r})}{\partial(z_1, \ldots, z_r)}\right|}$$

is a generator for $A\mathcal{O}_Z$ at z. (The element a is independent of the particular rearrangement

$$\{i_1, \ldots, i_r, j_1, \ldots, j_{n-r}\} \text{ of } \{1, 2, \ldots, n\} .)$$

Example 1: Higher order pinchpoint on surface in 3-space.

Let $x \in X$ be a point on a surface in 3-space with smooth normalization Z such that $f^{-1}(x) = \{z\}$, and such that in the completion of the local rings the situation is as follows:

$$\hat{O}_{X,x} = k[\![x,\ y,\ z]\!]/f \to k[\![u,\ v]\!] = \hat{O}_{Z,z}\ ,$$

with $f(x,\ y,\ z) = y^{ab} - x^b z^a$, $x \mapsto u^a$, $y \mapsto uv$, $z \mapsto v^b$. ($a = 1$, $b = 2$ gives an ordinary pinch point (char $k \neq 2$).)

Then $AO_Z = C$ is generated in $O_{Z,z}$ by

$$- \frac{\frac{\partial f}{\partial x}}{\left|\frac{\partial(y,\ z)}{\partial(u,\ v)}\right|} = \frac{\frac{\partial f}{\partial y}}{\left|\frac{\partial(x,\ z)}{\partial(u,\ v)}\right|} = - \frac{\frac{\partial f}{\partial z}}{\left|\frac{\partial(x,\ y)}{\partial(u,\ v)}\right|} = u^{a(b-1)} v^{b(a-1)}\ .$$

(In the case $a = 1$, $b = 2$, one sees directly that C is generated in $\hat{O}_{X,x}$ by x and y, hence in $\hat{O}_{Z,z}$ by u.)

Example 2: A cone over a smooth plane curve.

Let X be the cone over a smooth plane curve of degree d, with vertex $x \in X$. Take $f : Z \to X$ to be the blow-up of the maximal ideal m_x. In this case, the map $\rho : f^*f_*\omega_Z \to \omega_Z$ is surjective.

An easy argument shows that, whenever ρ is surjective, the ideal A satisfies

$$A = f_*(AO_Z)\ .$$

Set $E = f^{-1}(x)$, $m_x O_Z = O_Z(-E)$. Clearly

$$\tilde{J}O_Z = O_Z(-(d-1)E)$$

and

$$I = O_Z(-E)\ .$$

Thus we obtain

$$AO_Z = O_Z(-(d-2)E)\ ,$$

$$A = f_*(AO_Z) = m_X^{d-2}\ .$$

Remark that the method of ([M - T], Th. 2.1.1) applies to compute A in this situation also.

Example 3: Different and discriminant.

Let X be a branch of a locally plane curve, with normalization $f : Z \to X$. Suppose $p : X \to Y$ is a projection onto a smooth curve satisfying

$$\Omega^1_{Z/X} = \Omega^1_{Z/Y} .$$

At a point $z \in Z$ we have the following situation ($x = f(z)$, $y = p(x)$)

$$\begin{array}{ccc} O_{Z,z} = k[t]_{(t)} & \leftarrow & (k[x,\ y]/f)_{(x,y)} = O_{X,x} \\ & & \uparrow \\ & & k[x]_{(x)} = O_{Y,y} \end{array} .$$

Assume $I = (\frac{\partial x}{\partial t}, \frac{\partial y}{\partial t})$ is generated by $\frac{\partial x}{\partial t}$. Then $\tilde{J}O_Z$ is generated by $\frac{\partial f}{\partial y}$. Since

$$I = F^o(\Omega^1_{Z/X}) = F^o(\Omega^1_{Z/Y}) ,$$

our equality (*) can be written (at z)

$$\frac{\partial f}{\partial y} \cdot O_{Z,z} = C \cdot F^o(\Omega^1_{Z/Y}) ,$$

where C is the conductor of Z in X.

In the present case, $\frac{\partial f}{\partial y} O_{X,x}$ is the <u>different</u> of $O_{X,x}$ over $O_{Y,y}$ ([Z], § 15), and $F^o(\Omega^1_{Z/Y}) = \text{Ann}\ (\Omega^1_{Z/Y})$ is the different of $O_{Z,z}$ over $O_{Y,y}$ ([S], III, Prop. 14). So here (*) turns out to be the formula relating differents and conductor ([S], III, Cor. 1, p. 65).

4. Nash transformation and jacobian blow-ups.

Let X be a r-dimensional variety. We let $\pi : \bar{X} \to X$ denote its Nash transformation, with Nash bundle

$$\pi^*\Omega^1_X \twoheadrightarrow \Omega .$$

Recall that $\bar{X}$ is the closure of the rational section

$$X \dashrightarrow \text{Grass}_r\ (\Omega^1_X)$$

defined on the smooth locus of X. Let $\tilde{\pi} : \tilde{X} \to X$ denote the blow-up of the jacobian ideal

$$\tilde{J} = F^r(\Omega_X^1)$$

of X. Since $\tilde{\pi}^*\Omega_X^1$ admits a r-quotient ([L], Lemma 1), $\tilde{\pi}$ factors through π, the latter being minimal with respect to that property.

If X is a l.c.i., then $\pi = \tilde{\pi}$ holds ([N]). We shall show that $\pi = \tilde{\pi}$ is equivalent to

$$\dim \operatorname{proj} \pi^*\Omega_X^1 \leq 1 .$$

(Note that $\dim \operatorname{proj} \tilde{\pi}^*\Omega_X^1 \leq 1$ always holds, by ([L], Lemma 1).)

Proposition 2: The Nash transformation π is equal to the jacobian blow-up $\tilde{\pi}$ if and only if

$$\dim \operatorname{proj} \pi^*\Omega_X^1 \leq 1$$

holds.

Proof: Suppose (locally) we have a presentation

$$0 \to E \to F \to \pi^*\Omega_X^1 \to 0 .$$

Set E' equal to the kernel of the composition

$$F \to \pi^*\Omega_X^1 \to \Omega .$$

Then E' is locally free, with the same rank as E. Hence

$$F^o(\operatorname{Coker}(E \to E'))$$

is invertible, therefore so is

$$\tilde{J}O_{\tilde{X}} = F^r(\pi^*\Omega_X^1) = F^o(\operatorname{Ker}(\pi^*\Omega_X^1 \to \Omega)) = F^o(\operatorname{Coker}(E \to E')) .$$

∎

The following proposition is a slight generalization of ([P], 1.1). The proof carries over – we need only replace the Grassmann variety of the bundle E by the Grassmann scheme representing m-quotients of the coherent sheaf E.

Proposition 3: Let Y be a scheme, E a coherent O_Y-module, F a locally free rank m O_Y-module, and $\alpha : E \to F$ a homomorphism.

The ideal $I = F^o(\operatorname{Coker} \alpha)$ is invertible if and only if $\operatorname{Im}(\alpha)$ is locally free with rank m.

We shall apply this to compare the Nash transformations and jacobian blow-ups of two varieties Z and X, with respect to a map $f : Z \to X$.

Proposition 4: Let $f : Z \to X$ be a proper, surjective, generically unramified map. Denote by $\pi_Z : \bar{Z} \to Z$ and $\pi_X : \bar{X} \to X$ (resp. $\tilde{\pi}_Z : \tilde{Z} \to Z$ and $\tilde{\pi}_X : \tilde{X} \to X$) the Nash transformation (resp. the blow-up of the jacobian ideal) of Z and of X. Set $\bar{Q}$ (resp. $\tilde{Q}$) equal to the cokernel of the composition

$$\bar{\alpha} : \pi_Z^* f^* \Omega_X^1 \to \pi_Z^* \Omega_Z^1 \to \bar{\Omega}_Z$$

$$(\text{resp. } \tilde{\alpha} : \tilde{\pi}_Z^* f^* \Omega_X^1 \to \tilde{\pi}_Z^* \Omega_Z^1 \to \tilde{\Omega}_Z)$$

where $\bar{\Omega}_Z$ (resp. $\tilde{\Omega}_Z$) denotes the canonical quotient on $\bar{Z}$ (resp. $\tilde{Z}$). Finally, let $\bar{\psi} : Z' \to \bar{Z}$ (resp. $\tilde{\psi} : Z'' \to \tilde{Z}$) denote the blow-up of $F^o(\bar{Q})$ (resp. $F^o(\tilde{Q})$).

Then there exist factorizations $\bar{g} : Z' \to \bar{X}$ and $\tilde{g} : Z'' \to \bar{X}$ of $f \circ \pi_Z \circ \bar{\psi}$ and $f \circ \tilde{\pi}_Z \circ \tilde{\psi}$ through π_X.

Proof: Apply Prop. 3 to the maps $\bar{\psi}^*(\bar{\alpha})$ and $\tilde{\psi}^*(\tilde{\alpha})$ to get the existence of quotients of Ω_X^1 on Z' and Z''. Then use the minimality property of π_X. ∎

Corollary 1: Suppose $\pi_X = \tilde{\pi}_X$ holds. Then $\bar{\psi}$ (resp. $\tilde{\psi}$) is equal to the blow-up of the ideal $\tilde{J}\mathcal{O}_{\bar{Z}}$ (resp. $\tilde{J}\mathcal{O}_{\tilde{Z}}$), where $\tilde{J} = F^r(\Omega_X^1)$ denotes the jacobian ideal of X. In other words, the ideals $\tilde{J}\mathcal{O}_{\bar{Z}}$ and $F^o(\bar{Q})$ (resp. $\tilde{J}\mathcal{O}_{\tilde{Z}}$ and $F^o(\tilde{Q})$) become invertible at the same time.

Proof: Consider

$$\pi_Z^* f^* \Omega_X^1 \xrightarrow{\bar{\alpha}} \bar{\Omega}_Z \to \bar{Q} \to 0 .$$

By Prop. 3, $F^o(\bar{Q})$ invertible $\iff$ Im $(\bar{\alpha})$ is locally free with rank $r = \dim X = \dim Z$.

This is equivalent to the existence of a r-quotient of $\pi_Z^* f^* \Omega_X^1$, therefore to $\mathfrak{J}\mathcal{O}_{\bar{Z}}$ being invertible, since $\pi_X = \tilde{\pi}_X$ holds.

∎

Corollary 2: Suppose $\pi_X = \tilde{\pi}_X$ holds, and that Z is smooth. Then the blow-up $\psi : Z' \to Z$ of $I = F^o(\Omega^1_{Z/X})$ is equal to the blow-up of $\tilde{J}O_Z$.

Earlier we showed (Thm. 1) that if X is Gorenstein, then I and JO_Z become invertible at the same time (here J is the ω-jacobian ideal). Hence, when X is a l.c.i., we have already proved Cor. 2 above.

Theorem 2: The Nash transformation $\pi : \bar{X} \to X$ of a Gorenstein variety X is equal to the blow-up of the ω-jacobian ideal J of X.

Proof: We want to show that, on any Y with $Y \to X$, J is invertible if and only if Ω^1_X admits a r-quotient.

Assume $\Omega_X \twoheadrightarrow \Omega$ is a r-quotient. Then $\Omega^r_X \to \Lambda^r\Omega$ is a 1-quotient which is generically isomorphic to the torsion-free quotient $\Omega^r_X \to J \otimes \omega_X$, hence everywhere isomorphic – therefore J is invertible.

If J is invertible there is a 1-quotient $\Omega^r_X \to \Omega^r = J \otimes \omega_X$. Consider (as in the proof of Prop. 3, see [P], 1.1)

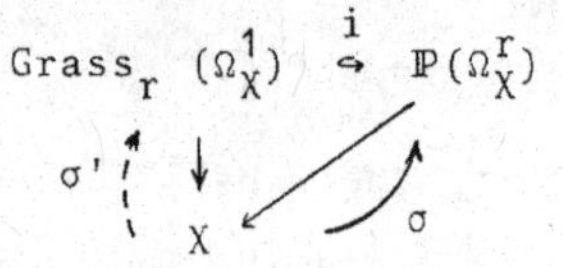

where σ is the section defined by $\Omega^r_X \to \Omega^r$, and σ' is the obvious rational section defined on the smooth locus of X. Clearly σ restricts to σ', so since the Plücker embedding i is closed, σ' extends to X. Hence there exists a r-quotient $\Omega^1_X \to \Omega$ (with $\Lambda^r\Omega = \Omega^r$). ∎

Remarks. 1) We are unable to give any examples of a Gorenstein, non l.c.i. variety X with $J \neq \tilde{J}$, because we do not know how to compute J in this case. Probably such varieties exist, however: Suppose $X \subseteq Y$, Y smooth of dimension n, and X defined by m equations. Then the jacobian ideal $\tilde{J}$ is generated by $\binom{n}{r}\binom{m}{n-r}$ elements (the (n-r)-minors of the jacobian matrix), while the ω-jacobian ideal J is generated by $\binom{n}{r}$ elements (it is the image of $\Omega^r_{Y|X} \otimes \omega_X^{-1} \to O_X$).

2) Since $\varphi_X : \Omega^r_X \to \omega_X$ is an isomorphism wherever X is smooth, there is an inclusion $\operatorname{supp}(O_X/J) \subseteq \operatorname{sing}(X) = \operatorname{supp}(O_X/\tilde{J})$. It seems

reasonable to believe $J \subseteq \tilde{J}$ holds – this would imply the other inclusion and also that φ_X surjective implies X smooth (and hence φ_X an isomorphism).

Suppose φ_X is surjective, so that $J = O_X$ (since $J \subseteq C$, this implies X normal). Then there exists (Thm. 2) a r-quotient of Ω_X^1. If char $k = 0$, this implies X is smooth ([L]). We do not know whether this is still true in positive characteristic. The usual positive characteristic counter-examples to "Ω_X^1 has a r-quotient $\Rightarrow$ X is smooth" do not apply as counter-examples to the above, because they do not have surjective φ_X (though the image of φ_X is invertible).

References

[DV] P. Du Val, "On isolated singularities of surfaces which do not affect the conditions of adjunction", Proc. Camb. Phil. Soc., 30 (1934).

[E] F. Elzein, Thèse, Univ. Paris VII, 1977.

[H] R. Hartshorne, Residues and duality, LNM Vol. 20, Springer-Verlag, 1966.

[L] J. Lipman, "On the jacobian ideal of the module of differentials", Proc. AMS, 21 (1969), 422-426.

[M-T] M. Merle, B. Teissier, "Conditions d'adjonction, d'après Du Val", Sem. sur les singularités des surfaces, Ecole Polytechnique 1976-77.

[N] A. Nobile, "Some properties of the Nash blowing-up", Pac. J. Math., 60 (1975), 297-305.

[P] R. Piene, "Polar classes of singular varieties", Ann. scient. Ec. Norm. Sup., 11 (1978).

[S] J.-P. Serre, Corps locaux, Paris 1962 (Hermann).

[W] P.M.H. Wilson, "On blowing up conductor ideals", Math. Proc. Camb. Phil. Soc., 83 (1978), 445-450.

[Z] O. Zariski, An introduction to the theory of algebraic surfaces, LNM Vol. 83, Springer-Verlag, 1969.

Institut Mittag-Leffler
Auravägen 17
S-182 62 Djursholm
Sweden

Schottky groups and Schottky curves.

by M. van der Put.

§o Introduction

This paper gives a summary of results obtained by L. Gerritzen and the author. The work is inspired by two articles of D. Mumford ([6].[7]). The purpose is to clarify especially the analytic part of Mumford's work on degenerating curves and abelian varieties.

In the papers [6],[7] analytic means a formal scheme over a (discrete) valuation ring. It is known that formal scheme's over a valuation ring V (not necessarily discrete) correpond to rigid-analytic spaces over the quotient field k of V (in the sense of J. Tate; H. Grauert; R. Remmert; R. Kiehl; L. Gerritzen; M. Raynaud; etal). In the sequel we will use the term k-holomorphic space for one of the possibilities of defining a rigid-analytic space over k. What we do is studying curves and abelian varieties over k as k-holomorphic spaces and applying the extensive literature on k-holomorphic spaces. In this way we are able to simplify and extend parts of [6],[7]. Another advantage of the use of k-holomorphic spaces is the possibility to compare with the complex-holomorphic spaces.

§1. k-holomorphic spaces and their reductions.

(1.1) For the convenience of the reader we recall some definitions and elementary properties. By $k < T_1,\dots,T_n >$ we denote the k-algebra of all power series $\Sigma\, a_\alpha T_1^{\alpha_1} \dots T_n^{\alpha_n}$ with $\lim |a_\alpha| = 0$. This k-algebra is noetherian and for every maximal ideal the residue-class field is a finite extension of k. (The k-algebra $k < T_1,\dots,T_n >$ behaves in fact very much like $k[T_1,\dots,T_n]$).

An affinoid algebra A over k (or Tate-algebra) is a ring of the type $k < T_1,\dots,T_n >/_I$ (some n, some ideal I). Further Sp(A) denotes the set of all maximal ideals of A provided with an obvious topology. The elements of A are seen as (continuous) functions on Sp(A). This topological space is given a Grothendieck-topology by prescribing a collection of "allowed" open subsets and a collection of "allowed" open coverings. For Sp A we make the following choice:

1) allowed open = rational subset = a subset of X = Sp(A) having the form $Y = \{x \in X \mid |f_o(x)| \geq |f_i(x)| \text{ all } i = 1,\dots,n\}$ where $f_o, f_1,\dots,f_n \in A$ generate the unit ideal.
2) allowed open covering = finite covering of a rational subset by rational subsets. Then X = Sp(A) provided with this Grothendieck-topology and the sheaf $\mathcal{O}_X$, defined by $\mathcal{O}_X(Y) = {A < T_1,\dots,T_n >}/_{(f_oT_i - f_i)_{i=1}^n}$, is called an affinoid space over k.

A k-holomorphic space is a topological space with a Grothendieck topology and a structure sheaf, which is locally an affinoid space over k.

(1.2) Examples

1. Any algebraic variety $/_k$ has a natural structure as k-holomorphic space.
2. Let C be a compact subset of $\mathbb{P}^1_k$, then $\mathbb{P}^1_k$-C has a natural structure as k-holomorphic

space.

3. Let Γ be a discrete subgroup of $(k^*)^n$ such that the subgroup of $\mathbb{R}$ generated by $\{\log|s| \mid s$ any coordinate of any $\gamma \in \Gamma\}$ is $\mathbb{Z}$. Then $(k^*)^n/_\Gamma$ has a natural structure as k-holomorphic space. The space $(k^*)^n/_\Gamma$ is called a k-holomorphic torus.

(1.3) The canonical reduction of an affinoid space $X = Sp(A)$ is given as follows: the V-algebra $A^o = \{f \in A \mid |f(x)| \leq 1$ for all $x \in X\}$ has the ideal $A^{oo} = \{f \in A \mid |f(x)| < 1$ for all $x \in X\}$. The residue-class ring $\bar{A} = A^o/_{A^{oo}}$ is a finitely generated, reduced, algebra over $\bar{k} = k^o/_{k^{oo}}$ = the residue field of k.

Then $\bar{X}$ = the maximal spectrum of $\bar{A}$ is the canonical reduction of X. There is an obvious map $p: X \to \bar{X}$. This map is surjective. A subset $Y \subset X$ is called formal if $Y = p^{-1}(Z)$ for some Zariski open affine subset of $\bar{X}$.

An allowed covering $\mathcal{U} = \{U_i\}$ of a k-holomorphic space X is called pure if all U_i are affinoid and $U_i \cap U_j$ is a formal subset of U_i (all $i \neq j$). With respect to a pure covering $\mathcal{U}$ of X, one defines a reduction $\bar{X}_{\mathcal{U}}$ of X as the gluing of $\{\bar{U}_i\}$ with respect to the open immersions $\overline{U_i \cap U_j} \to \bar{U}_i$; $\overline{U_i \cap U_j} \to \bar{U}_j$.

1.4. Examples.

1. $X = Sp(k < T_1,\ldots,T_n >)$" = "$\{(z_1,\ldots,z_n) \in k^n \mid$ all $|z_i| \leq 1\}$".
   Then $\bar{X} = \mathrm{Max}\ (\bar{k}[T_1,\ldots,T_n])$.
2. $X = Sp(k < T,S >/_{(TS-p)})$" = "$\{z \in k \mid |p| \leq |z| \leq 1\}$ where $p \in k$, $o < |p| < 1$.

   Then $\bar{X} = \max\ (\bar{k}[T,S]/_{TS})$. In picture 

3. $X = \mathbb{P}^1_k - C$ has (arbitrarily fine) pure coverings $\mathcal{U}$. The reduction $\bar{X}_{\mathcal{U}}$ consists of countably many $\mathbb{P}^1_{\bar{k}}$'s. Two of those lines intersect at most in one point. Every intersection is normal. The intersection graph of $\bar{X}_{\mathcal{U}}$ is a locally fine tree. If C is the set of limit points of a Schotkky group then this coincides with the tree constructed by Mumford in [6].

A simple case is: $C = \{o,\infty\} \subset \mathbb{P}^1_k$; let $\pi \in k$, $o < |\pi| < 1$, then $\mathcal{U} = \{\{z \in k \mid |\pi|^{n+1} \leq |z| \leq |\pi|^n\} \mid n \in \mathbb{Z}\}$ is a pure covering with reduction:

4. $X = k^*/_{<q>}$ where $o < |q| < 1$. Suppose for convenience that $\pi_2, \pi_1 \in k$ exists with $|q| < |\pi_2| < |\pi_1| < 1$. Define X_1, X_2 and X_3 to be the images in X of the affinoids:

$\{z \in k \mid |\pi_1| \leq |z| \leq 1\}$, $\{z \in k \mid |\pi_2| \leq |z| \leq |\pi_1|\}$ and $\{z \in k \mid |q| \leq |z| \leq |\pi_2|\}$.
Then $\{X_1, X_2, X_3\}$ is a pure covering of X and its reduction is

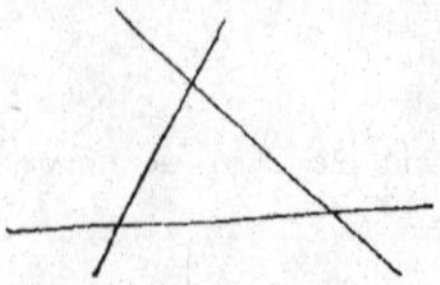 where each line is a $\mathbb{P}^1_{\bar{k}}$.

5. $X = {(k^*)^n}/_{\Gamma}$ is a <u>holomorphic torus</u>. After a change of coordinates Γ is generated by n elements of the form $a_1 = (a_{11}\pi^{k_1}, a_{12}, \ldots, a_{1n})$

$$a_2 = (a_{21}, a_{22}\pi^{k_2}, \ldots, a_{2n})$$
$$\vdots$$
$$a_n = (a_{n1}, \ldots, a_{n-1,1}, a_{nn}\pi^{k_n})$$

where $o < |\pi| < 1$, all $k_i > o$, all $|a_{ij}| = 1$.
For convenience we suppose that 3 divides all k_i. For a sequence $\alpha = (\alpha_1, \ldots, \alpha_n)$ with $\alpha_i \in \{1,2,3\}$ we define X_α = the image in X of the affinoid space $\{(z_1, \ldots, z_n) \in k^{*n} |\ |\pi^{k_i/3}|^{\alpha_i} \le |z_i| \le |\pi^{k_i/3}|^{\alpha_{i-1}}$ for all i}. The $\{X_\alpha\}$ form a pure covering of X. The corresponding reduction $\bar{X}$ is slightly more complicated then a product of n copies of the reduction in (1.4.4). Namely there is a sequence $\bar{X} = X_n \xrightarrow{\varphi_n} X_{n-1} \xrightarrow{\varphi_{n1}} \ldots \xrightarrow{\varphi_2} X_1$, where all maps are smooth, and all fibres and X_1 are isomorphic to three $\mathbb{P}^1_{\bar{k}}$'s intersecting like 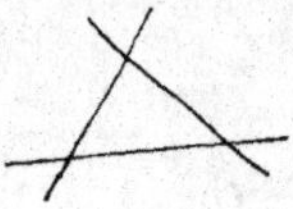

6. X <u>is a projective variety over</u> k. Let a projective model Y over V of X be given. Corresponding to Y one can find a pure covering $\mathcal{U} = \mathcal{U}_Y$ of X. The reduction $\bar{X}_{\mathcal{U}}$ turns out to be $(Y\times_V \bar{k})_{red}$. Moreover any finite covering of X by affinoids can be refined to a pure covering of the type $\mathcal{U}_Y$.

§ 2. Cohomology of constant sheaves .

For a constant sheaf F (for convenience the stalk F of F is a field) on a k-holomorphic space X we want to calculate $H^i(X, F)$ or dim $H^i(X, F)$.

First method. It can be shown that H^i coincides with the Čech-cohomology groups. For 1-dimensional spaces it is possible to make explicit calculations for the Čech-cohomology, because one "knows" the allowed coverings. One has the following result:

(2.1) Proposition

1) If X is an affinoid subset of $\mathbb{P}^1_k$ or $X = \mathbb{P}^1_k - C$ for some compact C, then $H^i(X, F) = 0$ for $i \neq 0$ and any constant sheaf F.

2) Let X be an elliptic curve over k. If X has good reduction then $H^1(X, F) = 0$. If X has bad reduction then $H^1(X, F) = F$.

For dim X > 1 one has no grip on the allowed covering of X, even for simple spaces like $X = \{(z_1, z_2) \in k^2 \mid |z_1| \leq 1 \text{ and } |z_2| \leq 1\}$. A partial remedy is a base change theorem. In this theorem we have to work with "constructible" sheaves. We will not explain this notion but just state that any constant sheaf is constructible and that the class of constructible sheaves is closed under all sorts of operations on sheaves.

Further, the ordinary points of X are insufficient to separate the sheaves on X. We have to work with geometric points instead. A closed geometric point p of X corresponds to a continuous k-homomorphism $\varphi : \mathcal{O}_X(U) \to K$, where U is some affinoid of X and K is a complete field extension of K and (im φ) generates K topologically. Then we can show the following:

(2.2) Proposition (Base change) Let $u : Y \to X$ be a morphism of k-holomorphic spaces such that for every affinoid $U \subset X$, $u^{-1}(U)$ has a finite covering by affinoids. Let p be a closed geometric point (with field K) of X. Then there exists K-holomorphic space $p \times_X Y$ and a morphism $v : p \times_X Y \to Y$ and for any sheaf S on Y canonical maps $\theta^i_S : (R^i u_* S)_p \to H^i(p \times_X Y, v^* S)$ $(1 \geq 0)$.

If S is constructible then all θ^i_S are isomorphisms.

(2.3) Consequences

1) $H^i(X, S) = 0$ if S is constructibel and i > dim X.

2) $H^i(X, F) = 0$ is $i \neq 0$, F constant and $X = \{(z_1, \ldots, z_n) \in k^n \mid \text{all } |z_i| \leq 1\}$

3) For a holomorphic torus $X = (K^*)^n/\Gamma$ and a constant sheaf F we have $H^i(X, F) \cong H^i(\Gamma, F)$ where F is considered as a trivial Γ-module. So dim $H^i(X, F) = \binom{n}{i}$.

4) Let a hyperelliptic curve $\varphi : X \to \mathbb{P}^1_k$ be given, ramified in a set A of rational points of $\mathbb{P}^1_k$. From (2.2) it follows that $H^1(X, F) = H^1(\mathbb{P}^1_k, \varphi_* F)$ and if the residue characteristic of k is $\neq 2$ then dim $H^1(\mathbb{P}^1_k, \varphi_* F)$ can be calculated in terms of the "geometry" of A.

There exists a (natural) reduction $\rho : \mathbb{P}^1_k \to Z$ corresponding to a pure covering of $\mathbb{P}^1_k$ such that:

1) Z has irreduible component $Z_1, \ldots, Z_s$; each Z_i is a projective line over $\bar{k}$; two lines meet at most in one point; every intersection is normal; the intersection graph is a finite tree.

2) Let $\{a_1, \ldots, a_t\}$ denote the points $\bigcup_{i\neq j} Z_i \cap Z_j$. Then $\rho(A) \cap \{a_1, \ldots a_t\} = \emptyset$ and $\rho(a) = \rho(a')$, with $a, a' \in A$ implies $a = a'$.

Then $\dim H^1(X, f) = \#\{i \mid Z-a_i$ has on both connected components an even number of elements of $\rho(A)\} - \#\{j \mid Z_j \cap \rho(A) = \emptyset$ and on each connected component of $Z - Z_j$ lies an even number of elements of $\rho(A)\}$.

Similar calculations can be made for "analytically tame" coverings $X \to \mathbb{P}^1_k$ of degree $n > 2$. Also in this case $\dim H^1(X, F)$ depends only on n and the "geometry" of the ramification points. However we have no closed formula and we do not know whether every non-singular curve X is an "analytically tame" covering of $\mathbb{P}^1_k$.

Second method

Let S be a sheaf on X and $\mathcal{U}$ a pure covering of X. Let $\rho : X \to \bar{X}_{\mathcal{U}}$ denote as before the reduction corresponding to U. It is easily seen that $\check{H}^i(\mathcal{U}, S) \cong \check{H}^i(\rho(\mathcal{U}), \rho_* S)$. Especially if S is a constant sheaf on X then $\rho_* S$ is a constant sheaf on $\bar{X}_{\mathcal{U}}$ with respect to the Zariski-Topology. This leads to the following results:

(2.4) Proposition Suppose that X has arbitrary fine, pure coverings. Then

1) $H^i(X, S) = 0$ for $i > \dim X$ and any sheaf S.

2) For any constant sheaf F one has $H^i(X, F) = \varinjlim_{\mathcal{U}} H^i(\bar{X}_{\mathcal{U}}, F)$

(2.5) Theorem Let X be a complete n.s. curve over k (the valuation of k is supposed to be discrete). Then

1) $\dim H^1(X, F) \leq g$ = genus of X and $H^2(X, F) = 0$.

2) $\dim H^1(X, F) = g$ if and only if X is totally split over k

3) Let $h = \dim H^1(X, F)$ and suppose that X has stable reduction over k. Then $(\bar{k}^*)^h = H^1(X, \bar{k}^*)$ is isomorphic to be toruspart of the closed fibre of the Néron minimal model of the Jacobian of X. In particular $h = 0$ if and only if the Jacobian of X has good reduction.

The complex-analytic analogue of (2.5) is: $\dim H^1(X, F) = 2g$, $\dim H^2(X, F) = 1$. We will sketch a proof of (2.5): Using the example (1.4,6) and (2.4) we have to calculate $\varinjlim H^1((Y \times_V \bar{k})_{red}, F)$ where Y in a projective model /V of X. Resolution of singularities in dimension 2 yields that we may suppose: 1) Y is non-singular.

2) $Y \times_V \bar{k} = \sum_{i=1}^{s} n_i C_i$, where each C_i is a non-singular curve of genus g_i over $\bar{k}$.

3) The C_i meet normally.

A cofinal set of reductions is now obtained by blowing up Y at points of $(Y \times_V \bar{k})_{red} \subset Y$. Those blowing ups do not change $H^1((Y \times_V \bar{k})_{red}, F)$. One easily sees

that $H^1((Y\times_V\bar{k})_{red}, F) = 1 - s + \frac{1}{2}\sum_{i\neq j}(C_i \cdot C_j)$.
Further g = genus of X = genus of $Y\times_V\bar{k} = 1 + \sum_{i=1}^{s} n_i(g_i - 1) + \frac{1}{2}\sum_{i\neq j} n_i(C_i \cdot C_j)$.
So 1) and 2) follow. The term "totally split over k" means "all $g_i = 0$ and all $n_i = 1$".
In order to prove 3) we consider the exact sequence $1 \to \mathcal{O}_Z^* \to \oplus\, \mathcal{O}_{C_i}^* \to \oplus_{i\,j}\, \mathcal{O}_{C_i \cap C_j}^* \to 0$
where $Z = Y\times_V\bar{k} = \Sigma\, C_i$.
It follows that $1 \to H^1(Z, \bar{k}^*) \to H^1(\mathcal{O}_Z^*) \to \oplus\, H^1(\mathcal{O}_{C_i}^*) \to 1$ is exact. Further $(\bar{k}^*)^h = H^1(Z, \bar{k}^*) = H^1(X, \bar{k}^*)$. Using [1] p 89, one finds that $H^1(\mathcal{O}_Z^*)^o$ = the closed fibre of the Néron minimal model of the Jacbi-variety of X. This proves 3).

Problem. Is dim $H^1(X, F)$ invariant under (finite) field-extensions?

Third method (Gerritzen [5]). For a k-holomorphic space X as usual, Pic $(X) = H^1(X, O_X^*)$ classifies the invertible sheaves on X. Let E and L denote the affinoid spaces $\{z \in k \mid |z| \leq 1\}$ and $\{z \in k \mid |z| = 1\}$. Then

(2.6) Theorem

1) Pic $(X \times E)$ = Pic $(X) \oplus P(X)$.

2) Pic $(X \times L)$ = Pic $(X) \oplus P_1(X) \oplus P_{-1}(X) \oplus H^1(X, \mathbb{Z})$.

3) $P(X) \simeq P_1(X) \simeq P_{-1}(X)$.

If k has characteristic 0, then P(X) can be computed. As a corollary one finds:

a) $H^1(X, \mathbb{Z}) = 0$ if $X = \{(z_1, \ldots, z_{n_1}) \in k^n \mid \text{all } |z_i| \leq 1\}$

b) If X is an elliptic curve, then $H^1(X, \mathbb{Z}) = 0$ or $\mathbb{Z}$ according to X has a good or a bad reduction.

§3. Discontinuous groups.

A subgroup Γ of PGl(2,k) acting on $\mathbb{P}^1_k$ is called discontinuous if

1) for every $z\in\mathbb{P}^1_k$, the closure of Γz is compact.

2) not every point of $\mathbb{P}^1_k$ is a limit point.

The group Γ is called a Schottky group if moreover Γ is finitely generated and contains no elements ($\neq 1$) of finite order. Let Γ be a discontinuous group with C as set of limit points. Then C is compact and nowhere dense. If C has more than 2 points then C is perfect. We will restrict our attention to that case. The k-holomorphic space $\Omega=\mathbb{P}^1_k - C$, of ordinairy points of Γ has a natural pure covering, invariant under Γ. The reduction $\overline{\Omega}$ and the corresponding tree T have a Γ-action. Using the action of Γ on T one can prove the following results.

(3.1) Theorem.

Let Γ be a finitely generated discontinuous group. Then

1) Γ contains a normal subgroup Γ_o of finite index, such that Γ_o is a Schottky group.

2) Any Schottky group is a free group.

3) The quotient Ω/Γ exists as a k-holomorphic space.

4) $\Omega/\Gamma = X$ is a non-singular complete algebraic curve of genus=rank $\Gamma/[\Gamma,\Gamma]$. Further X is totally split over k.

A converse of this theorem can be proved if the valuation of k is discrete. An

analytic proof alomg the lines of [6] and a combination with the results in §1,2 gives the following result.

(3.2) Theorem. Let X be a curve (complete and non-singular) over k. The following properties are equivalent:

(1) X is toally split over k.

(2) dim $H^1(X,F)$ = genus of X.

(3) X has a finite covering consisting of affinoid subsets of $\mathbb{P}^1_k$.

(4) $X=\Omega/\Gamma$ for a Schottky group Γ.

A curve satisfying the equivalent properties of (3.2) is called a Schottky curve.

(3.3) The theta-functions on a Schottky curve. Let Γ be a Schottky group on g generators with corresponding curve $X=\Omega/\Gamma$. A meromorphic function f on Ω (not identically zero) is called a theta-function if $f(\gamma z)=c(\gamma).f(z)$ holds for all $\gamma\in\Gamma$ and some $c(\gamma)\in k^*$. It follows that the automorphy-factor $\gamma \longmapsto c(\gamma)$ is a grouphomomorphism $\Gamma \longrightarrow k^*$. The basic example of a theta-function is $\theta(a,b;z)=\prod_{\gamma\in\Gamma}\frac{z-\gamma(a)}{z-\gamma(b)}$, where $a,b\in\Omega$. Using some function theory on Ω, one shows:

(3.3.1) Every theta-function on Ω has the form $c\prod_{i=1}^{r}\theta(a_i,b_i;z)$ where $c\in k^*$ and $a_i, b_i\in\Omega$. Using a fundamental domain for Γ one can show:

(3.3.2) The map ϕ: (the group of theta-functions) $\longrightarrow$ Hom(Γ,k*), which associates to every theta-function the automorphy factor, is surjective. The kernel of ϕ consists of the meromorphic functions on X.

On Ω we consider divisors with a discrete support. Let D^{Γ} denote the Γ-invariant divisors. Then D^{Γ} is isomorphic to the group of divisors on X. Let D_{finite} denote the divisors on Ω of finite support and let E be the subgroup of D_{finite} given by $E=\{\sum_{i=1}^{s}(\gamma_i(d_i)-d_i) \mid \gamma_i \in\Gamma;\ d_i \in D_{finite}\}$. Then there is an exact sequence:

(3.3.3) $o \longrightarrow E \longrightarrow D_{finite} \xrightarrow{\alpha} D^{\Gamma} \longrightarrow o$, where α is given by $\alpha(d)=\sum_{\gamma\in\Gamma}\gamma(d)$.

Consider the map ψ : (theta-functions on Ω) $\longrightarrow$ (divisors of degree o on X), defined by $\psi(\theta)$ is the Γ-invariant divisor of θ considered as a divisor on X. Then clearly ψ is surjective and according to (3.3.1) the kernel of ψ is generated by k* and $\{\theta(a,\alpha(a);z) \mid a\in\Omega, \alpha\in\Gamma\}$. Hence $\Lambda=\phi(\ker\psi) \subset (k^*)^g=\mathrm{Hom}(\Gamma,k^*)$ is also the image of the map $p:\Gamma \to \mathrm{Hom}(\Gamma,k^*)$ given by $p(\alpha)=$ the automorphy factor of $\theta(a,\alpha(a);z)$ (does not depend on a). In fact p can also be considered as a map $p:\Gamma_{ab}\times\Gamma_{ab} \longrightarrow k^*$ where Γ_{ab}= modulo its commutator subgroup = $\mathbb{Z}^g$. This p is easily seen to be symmetric. An explicite calculation on the fundamental domain of Γ shows:

(3.3.4) p is positive definite, i.e. the symmetric bilinear map $p^*:\Gamma_{ab}\times\Gamma_{ab} \longrightarrow \mathbb{R}$ given by $p^*(s,t)=-\log|p(s,t)|$ is positive definite.

As we will see in § 4 this implies that Λ is a discrete subgroup with rank g of $(k^*)^g$ and that the holomorphic torus $(k^*)^g/\Lambda$ is in fact an abelian variety. Finally from our construction above it follows that $(k^*)^g/\Lambda$ is isomorphic to $\frac{\text{(divisors on X of degree o)}}{\text{(principal divisors)}}$. So we have obtained an analytic construction of the Jacobi-variety J of X. The map p is the canonical polarization of J. As in the complex case one can show the Riemann-vanishing theorem: the theta-function on $(k^*)^g$ with respect to Λ corresponding with the polarization p vanishes on a translate of X in J.

§4. Abelian varieties.

The following data are given:

1) a split algebraic torus $(k^*)^n$ with character group H.
2) a subgroup Λ of $(k^*)^n$, isomorphic to Z^n.
3) a "polarization" $p:\Lambda \longrightarrow H$ i.e. a homomorphism p such that $\langle\lambda_1,\lambda_2\rangle = p(\lambda_2)(\lambda_1)\in k^*$ is symmetric and positive definite.

Then Mumford constructs in [7] a smooth commutative groupscheme G over Spec(V) (V is the discrete valuation ring of k) such that $G\times_V k$ is an abelian variety and $G\times_V\bar{k} = (\bar{k}^*)^n$ here $\bar{k}$ is the residue field of V. It is stated (without proof) that any such groupscheme G can be obtained from the data 1,2,3. The approach of Gerritzen [3] is different. First of all 3) implies that Λ is a discrete subgroup and that $(k^*)^n/\Lambda$ is a holomorphic torus. The polarization p gives rise to theta-functions on

$(k^*)^n$. One can show that there are enough theta-functions and that the function-field of $(k^*)^n/\Lambda$ has transcendence degree n over k. Using arguments of "GAGA" type it follows that $(k^*)^n/\Lambda$ is an abelian variety. And in fact $(k^*)^n/\Lambda = G \times_V k$. We conjecture the following:

a) G can be obtained from a pure covering of the holomorphic space $(k^*)^n/\Lambda$.

b) An abelian variety A/k of dimension n has a presentation $(k^*)^n/\Lambda$ if and only if $H^1(A,\mathbb{Z})=\mathbb{Z}^n$.

Finally, full details of statements and proofs will be published by L.Gerritzen and M.van der Put in the "Lecture notes in Mathematics".

References.

[1] P.Deligne- D.Mumford: The irreducibility of the space of curves of a given genus. Publ.I.H.E.S. 36,1969.

[2] V.Drinfeld-Yu.Manin: Periods of p-adic Schottky groups. J.r.angewan.Math. 262/263,239-247,1973.

[3] L.Gerritzen : On non-Archimedean representations of Abelian varieties. Math.Ann. 196,323-346, 1972.

[4] L.Gerritzen : Zur nichtarchimedischen Uniformisierung von Kurven. Math. Ann. 210,321-337,1974.

[5] L.Gerritzen : Zerlegung der Picard-Gruppe nichtarchimedischer holomorpher Räume. Compositio Math. 35, 23-38,1977.

[6] D.Mumford : An analytic construction of degenerating curves over complete local rings. Compositio Math. 24,129-174,1972.

[7] D.Mumford : An analytic construction of degenerating abclian varieties over complete rings. Compositio Math. 24,239-272,1972.

MODULI FOR PRINCIPAL BUNDLES

A. RAMANATHAN

After A. Weil's initial work, Narasimhan, Seshadri and Mumford took up the study of vector bundles over complete nosingular algebraic curves. For a survey of this field see (Ramanan).

Here we consider the analogous questions for principal fiber bundles with reductive group as structure group. We give definitions and statements of results witha sketch of methods of proof. Details will appear elsewhere.

Let X be a projective nonsingular algebraic curve over the field k. We assume in general the genus of X to be greater than 2. Let G be a connected reductive algebraic group over k. Typical examples for G to be kept in mind are GL(n), SO(n), SP(n).

A principal bundle E over a space X (or a G-bundle, for short) is a space E onwhich G operates (from the right) and a G-invariant morphism $p: E \longrightarrow X$ which is locally trivial in the etale topology, i.e. for every point x of X there is a neighbourhood U of x and an etale covering $f: U' \longrightarrow U$ such that there is a G-equivariant isomorphism of f(E) with U'x G over U' where G operates by right translations on the second factor of U' x G.

If G operates on a space F we can form the associated fiber space E(F) over X in the usual way : E(F) is the quotient of E x F for the action of G given by $g(e,f) = (eg, g^{-1} f)$, where $e \in E$, $f \in F$, $g \in G$. Then the fibers of $E(F) \longrightarrow X$ are (noncanonically) isomorphic to F. If F is a vector space and $G \longrightarrow GL(F)$ is a representation E(F) is a vector bundle

In particular if $G = GL(n)$, for the natural action of $GL(n)$ on k^n we get the vector bundle $E(k^n)$. We can recover the $GL(n)$-bundle E as the bundle of frames from $E(k^n)$. Thus the notion of a $GL(n)$-bundle is equivalent to a vector bundle.

If H is another group and $G \longrightarrow H$ is a homomorphism then E(H) is in a natural way a H-bundle. We then say (especially when G is a subgroup of H) that E(H) is got from E by extension of structure group and that E is got from E(H) by reduction of structure group.

By taking a representation $G \longrightarrow GL(n)$ one can think of a G-bundle E as the vector bundle $E(k^n)$ together with a G-structure. For example we can think of an $O(n)$-bundle as a vector bundle together with quadratic forms on the fibers.

We will be interested in constructing a moduli space for G-bundles. Roughly speaking we wish to give the set M of isomorphism classes of G-bundles on X the structure of an algebraic scheme in a natural way. Natural in particular would mean that if we have a family of G-bundles on X parametrised by an algebraic variety T, i.e. a G-bundle $\underline{E}$ on $T \times X$ the natural map from T to M sending t in T to the isomorphism class of $\underline{E}_t$, the restriction of $\underline{E}$ to $t \times X \approx X$, is a morphism. As is usual with moduli problems (Cf. (Mumford)) there can be bad G-bundles which we cannot hope to include in a global moduli space. For example there are G-bundles E such that wecan contruct a family parametrised by the affine line A^1 , $\underline{E} \longrightarrow A^1 \times X$,such that $\underline{E}_t$ is isomorphic to E for every t other than o and $\underline{E}_o$ is not isomorphic to E. This forces us to either avoid such a E or identify E and $\underline{E}_o$. For vector bundles Mumford introduced the notion of stability which characterises good bundles.

DEFINITION 1.: A vector bundle $V \longrightarrow X$ is stable (resp. semistable) if for every nontrivial proper subbundle W of V we have

$$\frac{\deg W}{\operatorname{rk} W} < \ (\leq) \ \frac{\deg V}{\operatorname{rk} V}$$

Here rk W = dimension of the fiber of W and deg W = degree of the line bundle det W = the rk W - th exterior power of W. (Degree of a line bundle is got, for example, by taking a meromorphic section of it and counting its zeroes and poles with proper multiplicities).

The motivation for this definition comes from Mumford's geometric invariant theory.

We notice that a subbundle W of V gives rise to a reduction of structure group of the associated principal bundle to the subgroup which leaves the subspace $k^{\operatorname{rk} W}$ of $k^{\operatorname{rk} V}$ invariant. the general structure theory of algebraic groups shows that this subgroup is an instance of the general notion of a maximal parabolic subgroup of reductive group. So we formally generalise the above definition.

DEFINITION 2.: A G-bundle $E \longrightarrow X$ is stable (resp. semistable) if for any reduction of structure group to any maximal parabolic subgroup P of G we have that the line bundle associated to the reduced P-bundle by the dominant character $P \longrightarrow k^{\times}$ has degree strictly less than (resp. less than or equal to) zero.

Note that the group of characters of P which are trivial on the center of G is isomorphic to $\mathbb{Z}$ and we can single out a generator by requiring that the line bundle associated by it to $G \longrightarrow G/P$ is ample.

When the base field k is $\mathbb{C}$ we again have that stable G-bundles on X correspond to the representations of the fundamental (or a kind of ramified) group, as it happens for vector bundles. We explain this.

Let Γ be the fundamental group of X. Let K be a maximal compact subgroup of G. A homomorphism $\rho : \Gamma \longrightarrow K$ is called a <u>unitary representation</u> of Γ. We call $\rho : \Gamma \longrightarrow K$ <u>irreducible</u> if for the action of Γ (through ρ and the adjoint representation of K) on the Lie algebra <u>k</u> of K the only fixed vectors are in the center of <u>k</u>. Let $\tilde{X} \longrightarrow X$ be the universal covering of X. This is a Γ-bundle. The associated K-bundle as well as the associated G-bundle is called a <u>unitary bundle</u>.

THEOREM 1.: For simplicity let us assume G to be semisimple. Then a G-bundle is stable if and only if it is associated to an irreducible unitary representation of Γ in K. Also a G-bundle associated to a unitary representation is semistable.

For a proof of this theorem and also a general version for reductive groups see (Ramanathan,1).

To get a complete moduli space we take the set of semistable G-bundles on X (not only stable ones) and introduce an equivalence relation between them. The equivalence amoung stable bundles turns out to be isomorphism. The equivalence is defined as follows. If E is a semistable bundle we prove that there is an admissible reduction of structure group to a parabolic subgroup P = M.U where M is a Levi-component (i.e. a maximal reductive subgroup of P) and U is the unipotent radical of P, such that the M-bundle obtained by the extension of structure group $P \longrightarrow M$ is stable. (Here admissible reduction to P means that for any character χ of P which is trivial on the center of G the associated line bundle has degree zero) Moreover the G-bundle obtained from this M-bundle by the extension of structure group $M \hookrightarrow G$ depends only on E and we denote it by gr E.

DEFINITION 3.: The semistable G-bundles E_1 and E_2 are equivalent iff gr E_1 is isomorphic to gr E_2.

Our result then is that there is a 'coarse moduli scheme' for equivalen-ce classes of semistable G-bundles. To make this precise let us define F: (Schemes) $\longrightarrow$ (Sets) to be the functor which associates to a scheme T the set of isomorphism classes of G-bundles on T x X. For any scheme M define h_M: (Schemes) $\longrightarrow$ (Sets) to be h_M = Morphisms (T, M).

THEOREM 2.: There is a unique scheme M (each of whose component is projective) such that

1) there is a morphism of functors $\varphi: F \longrightarrow h_M$

2) the set of equivalence classes of semistable G-bundles is identified by φ with the k-valued points of M.

3) if there is an M' satisfying 1) and 2) with $\varphi': F \longrightarrow h_{M'}$, there is a natural morphism $\chi: h_M \longrightarrow h_{M'}$ such that $\varphi' = \chi \cdot \varphi$

The method of proof is to reduce it to a quotient space problem and to use Mumford's geometric invariant theory. We take a suitable representation $\rho: G \longrightarrow GL(n)$ (usually the adjoint representation) and look upon a G-bundle E as a vector bundle together with a G-structure. By tensoring $E(\rho)$ with a very ample line bundle we can assume roughly that all $E(\rho)$ as Evaries over semistable G-bundles (algebraically equivalent to a fixed G-bundle) occur as quotients of a trivial vector bundle I^N. (This uses the fact that such G-bundles form a bounded family). Then for any point x in X the surjection $I^N \longrightarrow E(\rho) \rightarrow 0$ evaluated at x gives a point $E(\rho)_x$ of the Grassmanian $\mathrm{Grass}(I^N)$ of quotient spaces of I^N of rank = rank of $E(\rho)$. Moreover the G-structure of $E(\rho)$ gives rise to an element of some suitable tensor space $T(E(\rho)_x)$ For example if we take the adjoint representation we will have the Lie bracket $E(\rho) \otimes E(\rho) \longrightarrow E(\rho)$ which is an element of the tensor space $E(\rho)^* \otimes E(\rho)^* \otimes E(\rho)$. By evaluating at several points $x_1, \ldots, x_r$ in X each E gives a point of $Q = T(U) \times \ldots \times T(U)$, r factors, where $U \longrightarrow \mathrm{Grass}\,(I^N)$ is the universal quotient bundle and T(U) is the appropriate tensor space. The group GL(n) operates naturaly on Q and one can check that,for large r, two bundles E_1 and E_2 are isomorphic iff the corresponding points of Q lie in the same orbit under GL(n). After a careful analysis one can show that the existence of the quotient by GL(n) of (a subspace of) Q will provide the existence of a coarse moduli for equivalence classes of semistable G-bundles. In particular for proving this one has to check that the points of Q corresponding to semistable G-bundles are semistable for the action of GL(n) in the sense of Mumford(cf. (Mumford)). It is here that one uses the definition of stability strongly. Indeed the definition of a stable bundle is so made just to give this. One also makes use of the 'rgidity' of the tensor given by the G-structure.

For unstable,i.e. not semistable, G-bundles one can prove the following result on canonical reduction to a parabolic:

PROPOSITION 1: Let E be an unstable G-bundle. Then there is a unique reduction of structure group to a parabolic subgroup such that

1) the P/U-bundle (U being the unipotent radical of P) obtained by the extension of structure group $P \longrightarrow P/U$ from the reduced P-bundle is a semistable P/U-bundle.

2) for any nontrivial character of P which can be expressed as a positive combination of simple roots the associated line bundle has degree strictly greater than zero (simple roots taken w.r.t. any Borel subgroup contained in P).

For higher dimensional base spaces X (nonsingular, projective) it seems that the right objects to consider are G-bundles in codim 1, i.e.

G- bundles E over any open subsets U of X such that $\dim(X - U) \leq \dim X - 2$. This has been noticed already for vector bundles for which case the objects to be considered are torsionfree sheaves on X. Then one defines stability w.r.t. a polarisation given by an ample line bundle H of X.

DEFINITION 3.: A G-bundle $E \longrightarrow U \subset X$ is <u>stable</u> (resp. <u>semistable</u>) w.r.t. H iff for any reduction to any maximal parabolic subgroup P of the bundle E over any open subset of U the line bundle associated to the dominant character has degree $<$ (resp. $\leq$) 0.

Here the degree (w.r.t. H) of a line bundle is the intersection $c_1(L).H^{n-1}$ where $n = \dim X$.

For surfaces, i.e. dim X = 2, one can prove the existence of coarse moduli scheme for equivalence classes of semistable G-bundles using results from (Gieseker).

REFERENCES

1. Gieseker, D.: On the Moduli of Vector Bundles on an Algebraic Surface. Ann. Math. 106(1977) 45-60.

2. Mumford,D.:Geometric invariant theory. Berlin-Heidelberg-New York: Springer 1965

3. Ramanan, S.: Vector Bundles on algebraic curves. Talk given at the International Congress of Mathematicians in Helsinki (1978)

4. Ramanathan,A.: Stable principal bundles on a compact Riemann Surface. Math.Ann. 213, (1975) 129-152.

5. Ramanathan, A.: Thesis, Bombay 1976.

A. Ramanathan
Tata Institute of Fundamental Research, Bombay 400 005, India and
Sonderforschungsbereich, University of Bonn, Bonn.

Tata Institute of Fundamental Research
Homi Bhabha Road
Bombay 400 055 - INDIA

π_1 FOR SURFACES WITH SMALL K^2

Miles Reid

§0. Introduction.

This paper is an extract from a forthcoming more complete article [1]; I refer to [1] for some technical results. Varieties are defined over an algebraically closed field of characteristic 0. $\pi_1 X$ is the algebraic (profinite) fundamental group. (Analogous results to Theorem 1 have been obtained by E. Horikawa: Surfaces with small c_1^2, V.)

Theorem 1. Let X be a minimal surface of general type, and suppose that $K_X^2 < \frac{1}{3}c_2(X)$ (equivalently, $K_X^2 < 3\chi(\mathcal{O}_X)$); then either

(i) $\pi_1 X$ is finite

or (ii) there exists an etale Galois cover $Y_0 \to X$, Y_0 having a morphism $Y_0 \xrightarrow{f} C_0$ to a curve of genus $p > 0$ inducing an isomorphism $f_*\colon \pi_1 Y_0 \overset{\approx}{\to} \pi_1 C_0$.

Furthermore in (ii) the fibres of $f\colon Y_0 \to C_0$ are hyperelliptic curves of genus $g \leq 5$.

Corollary 2. (i) $\pi_1 X$ is an extension of $\pi_1 C_0$ by a finite group G_0; (ii) for every etale cover $Y \to X$ with $q(Y) \neq 0$ the Albanese map $\alpha\colon Y \to \mathrm{Alb}\, Y$ maps onto a curve; (iii) if $X \xrightarrow{f} C$ is a non-constant morphism of X with connected fibres to a curve C of genus $p > 0$ then $q(X) = p$.

Dividing $Y_0 \to C_0$ by the equivariant action of G_0 gives a restatement of (ii):

(ii)' $\pi_1 X$ is infinite, and contains a normal subgroup A of order ≤ 4; there exists a morphism $X \to B$ to a curve B; and every etale cover $Y \to X$ which corresponds to a finite quotient of $\pi_1 X/A$ is obtained by making a normalized pull-back diagram

$$\begin{array}{ccc} Y = \widetilde{Y \times_B C} & \longrightarrow & X \\ \downarrow & & \downarrow \\ C & \longrightarrow & B \end{array}$$

over a cover $C \to B$ ramified only at points of B corresponding to multiple fibres of $X \to B$.

Corollary 3. Let X be a minimal surface with $K_X^2 < 3p_g(X) - 7$, $K_X^2 < 3\chi(\mathcal{O}_X)$, and $p_g \geq 8$; then one of the following 4 cases hold:

(i) $|K_X|$ is composed of a rational pencil $X \dashrightarrow P^1$, and $q(X) = 0$;

(ii) $|K_X|$ is composed of an irrational pencil $X \to C$ with C a curve of genus $p > 0$, and $q(X) = p$;

(iii) $\varphi_{K_X} : X \dashrightarrow F \subset P^{p_g(X)-1}$ is generically 2-to-1 onto a rational surface F, and $q(X) = 0$;

or (iv) $\varphi_{K_X} : X \dashrightarrow F \subset P^{p_g(X)-1}$ is generically 2-to-1 onto a ruled surface of genus $p > 0$, and $q(X) = p$.

It is quite likely that (i) and (ii) cannot actually occur (see Problem R5); I also do not know if fibres of genus 4 or 5 can occur in Theorem 1 - if this happens then $|K_Y|$ must have a fixed part having large intersection number with the fibres.

There is an element of ineffectivity in Theorem 1, there being no bound for the finite $\pi_1 X$ which can occur; the problem is to bound the 2-torsion in Pic X (see Problem R2).

A positive answer to Problems R1 and R8 would be one step in the direction of the following:

Conjecture 4. The hypothesis in Theorem 1 can be weakened to $K_X^2 < \frac{1}{2}c_2(X)$ (equivalently $K_X^2 < 4\chi(\mathcal{O}_X)$).

The conjecture can be weakened to only ask that for surfaces with $K_X^2 < \frac{1}{2}c_2(X)$ (ii) and (iii) of Corollary 2 hold; perhaps the natural approach to this conjecture would be through differential-geometrical methods, which could also shed light on the problem of the topological fundamental group for surfaces in this range.

F. Sakai points out that the hypothesis of Theorem 1 cannot be weakened to $K_X^2 < c_2(X)$, since for every surface of general type S there exists a cyclic branched cover $T \to S$ with $K_T^2 < c_2(T)$. See however Problem R3.

§1. Proof of Theorem 1 and its corollaries.

Suppose that X is as in Theorem 1, and that $\pi_1 X$ is infinite. I will prove (ii).

Let $Y \to X$ be an etale Galois cover such that $G = \mathrm{Gal}\ Y/X$ has $|G| = n$; Y has the invariants

$$K_Y^2 = nK_X^2 \quad ; \quad \chi(\mathcal{O}_Y) = n\chi(\mathcal{O}_X).$$

Since $K_X^2 \leqq 3\chi(\mathcal{O}_X) - 1$ I have

$$K_Y^2 \leqq 3p_g(Y) + 3 - 3q(Y) - n .$$

Exactly as in [1], Step 5 of §1, if $n \geqq 11$ then $K_Y^2 < 3p_g - 7$, so that by [1], Theorem 8.1 (ii)(see also [2, Lemma (1.1) p. 122]) φ_{K_Y} cannot be birational; suppose that $|K_Y|$ is not composed of a pencil. Then φ_{K_Y} is generically m-to-1 onto a surface $\overline{Y}$, and the standard argument as in [1] §1, Step 2 gives

$$m.\deg \overline{Y} \leqq K_Y^2 .$$

Since $\overline{Y}$ is a surface spanning $P^{p_g(Y)-1}$, for $n \geqq 10$ I must have $n = 2$, and $\deg \overline{Y} < \frac{3}{2}(p_g(Y)-1)$. Thus using [1] Theorem 6.2 and Corollary 6.5 there are the following 3 possibilities for φ_{K_Y}:

(i) $|K_Y|$ is composed of a pencil ;

(ii) $\varphi_{K_Y} : Y \dashrightarrow \overline{Y}$ is generically 2-to-1 onto a rational surface $\overline{Y}$;

(iii) $\varphi_{K_Y} : Y \dashrightarrow \overline{Y}$ is generically 2-to-1 onto a surface $\overline{Y}$ having an irrational ruling by lines or conics.

In (ii) there is a biregular involution i of Y such that the quotient $F = Y/i$ is rational. As in [1] §1, Step 3 it follows from Lemma 4.2 (ii) of [1] that $G = (\mathbb{Z}/2)^a$. It follows that $(\mathbb{Z}/2)^a \subset \text{Pic } X$.

Proposition 1. There is a bound T (depending on X) such that $(\mathbb{Z}/2)^a \subset \text{Pic } X$ implies that $a \leqq T$.

Proof. T can be taken as $2q(X)$ plus the number of generators for the (finitely generated) Neron-Severi group.

Thus under the hypothesis that $\pi_1 X$ is infinite there exist etale Galois covers $Y \to X$ for which Y falls under (i) or (iii) above.

Proposition 3. Suppose that for some $Y \to X$ with $p_g(Y) \geqq 8$ $|K_Y|$ is composed of a pencil; then $\varphi_{K_Y} : Y \to C \subset P^{p_g(Y)-1}$ is a morphism with irreducible generic fibres of genus 2.

Proof. Write

$$|K_Y| = |E^{(r)}| + F ,$$

with F the fixed part and $E^{(r)} = E_1 + \cdots + E_r$, the E_i being irreducible fibres of a map $Y \dashrightarrow C$. Let E be the numerical class of E_i.

Then as in [1], Step 1 of §1 $K_Y^2 \geqq r^2E^2$, where furthermore $r \geqq p_g(Y) - 1$, so that $E^2 = 0$; also

$$K_YE \leqq \frac{1}{r}K_Y^2 \leqq \frac{3p_g(Y)+3}{p_g(Y)-1} < 4 ,$$

so that $K_YE = 2$.

Thus φ_{K_Y} is obtained by composing the morphism $Y \to C$ with the rational map $C \to P^{p_g(Y)-1}$ given by a linear system $|d|$ on C of degree r; if this map is not birational, then $r \geqq 2p_g(Y) - 2$, again contradicting the numerical conditions. The proposition is proved.

I write $\varphi_{K_Y} : Y \dashrightarrow \overline{Y} \subset P^{p_g(Y)-1}$ for the canonical rational map, regardless of the dimension of $\overline{Y}$.

Proposition 4. Let $Y_2 \xrightarrow{\psi_{2,1}} Y_1 \to X$ be an etale tower, with $p_g(Y_1) \geqq 8$; then there exists a rational map $\overline{Y}_2 \dashrightarrow \overline{Y}_1$ making the diagram

$$\begin{array}{ccc} Y_2 & \longrightarrow & \overline{Y}_2 \subset P^{p_g(Y_2)-1} \\ \downarrow & & \downarrow \\ Y_1 & \longrightarrow & \overline{Y}_1 \subset P^{p_g(Y)-1} \end{array}$$

commutative. Furthermore (i) if $\overline{Y}_2$ is a curve then so is $\overline{Y}_1$; (ii) if $\overline{Y}_2$ is a surface ruled by lines then either $\overline{Y}_1$ is a curve, or it is a surface ruled by lines.

Proof. The composite

$$Y_2 \longrightarrow Y_1 \dashrightarrow \overline{Y}_1 \subset P^{p_g(Y_1)-1}$$

is defined by the subspace $\psi_{2,1}^*H^0(K_Y) \subset H^0(K_{Y_2})$; hence the required map is defined by the linear projection from $P^{p_g(Y_2)-1}$ onto $P^{p_g(Y_1)-1}$. The final assertion is obvious.

Thus my surface X falls under one of the following 3 cases:

Case 0. There exists some etale Galois cover $Y_0 \to X$ such that for every etale cover $Y \to Y_0$ $|K_Y|$ is composed of a pencil of curves of genus 2;

Case i. (for $i = 1$ or 2). There exists an etale Galois cover $Y_0 \to X$ such that

(*) for every etale cover $Y \to Y_0$, $\varphi_{K_Y} : Y - \to \overline{Y} \subset P^{p_g(Y)-1}$ is a double cover of a surface $\overline{Y}$ having an irrational pencil of rational curves of degree i.

It seems quite likely that Case 0 cannot occur (see Problem R5); in any case it is easy to deal with:

Proposition 5. In Case 0 there exists a morphism $X \xrightarrow{f} B$ inducing an isomorphism $f_*: \pi_1 X \overset{\approx}{\to} \pi_1 B$.

Proof. Y_0 has a G-equivariant pencil $Y_0 \to C_0$ of curves of genus 2; by [1] Lemma 4.2 (i) G must act freely on C_0, so that the following diagram is a pull-back:

$$\begin{array}{ccc} Y_0 & \to & C_0 \\ \downarrow & & \downarrow \\ X = Y_0/G & \to & B = C_0/G . \end{array}$$

Now for every etale cover $Y_1 \to Y_0$ $\varphi_{K_{Y_1}}$ defined a pencil $Y_1 \to C_1$, and by Proposition 4 there is a morphism $C_1 \to C_0$ compatible with the canonical maps of Y_1 and Y_0. But these maps then form a pull-back diagram, and Y_1 can also be obtained as the pull-back $X \times_B C_1$; since every etale cover $Y \to X$ fits into a Galois tower under some such Y_1 the proposition is proved.

In Cases 1 and 2 the irrational pencil on $\overline{Y}$ defined a pencil $Y \to C$ on Y.

Lemma 6. In Case 1 the fibres of $Y \to C$ have genus ≤ 3; in Case 2 they have genus ≤ 5.

Proof. Since each fibre of $Y \to C$ goes into a rational curve of degree i each fibre imposes at most $i + 1$ conditions on a divisor in $|K_Y|$, and setting $r = \left[\frac{p_g-2}{i+1}\right]$ it follows that a divisor of $|K_Y|$ can be found containing r fibres $Y \to C$; thus

$$3p_g(Y) + 3 - 3q(Y) - n \geq K_Y{}^2 \geq rK_YE .$$

In Case 1 it then follows that $K_YE \leq 4$, and in Case 2 $K_YE \leq 8$, as required.

I can now make the further requirement on Y_0:

(**) as in (*), and for some g, for every etale $Y \to Y_0$ the irrational pencil on $\overline{Y}$ defined a pencil of curve of genus g on Y.

Now consider the multiple fibres of $Y_0 \to C_0$; by making a pull-back by a cover of C_0 ramified only in the points corresponding to the multiple fibres I arrive at a cover Y_0 which satisfies in addition

(***) as in (**), and $Y_0 \to C_0$ has no multiple fibres.

Now let $Y_1 \to Y_0$ be an etale Galois cover; by Proposition 4 the double covers $Y_i \to \overline{Y}_i$ fit into a commutative diagram with a map $\overline{Y}_1 \to \overline{Y}_0$; on the other hand, both $\overline{Y}_i$ are ruled by lines or conics, so that the map $\overline{Y}_1 \to \overline{Y}_0$, which is defined by linear projection, induces a map $C_1 \to C_0$ between the curves parametrising the lines or conics of $\overline{Y}_1$ and $\overline{Y}_0$. I thus get a diagram:

$$\begin{array}{ccccc} Y_1 & \dashrightarrow & \overline{Y}_1 & \dashrightarrow & C_1 \\ \downarrow & & \downarrow & & \downarrow \\ Y_0 & \dashrightarrow & \overline{Y}_0 & \dashrightarrow & C_0 . \end{array}$$

Now since both $Y_i \to C_i$ have fibres of the same genus it follows that no element of Gal Y_1/Y_0 acts trivially on C_1. Thus $C_1 \to C_0$ is also Galois, with Gal Y_1/Y_0 = Gal C_1/C_0; under these circumstances it follows that Y_1 is birational to the pull-back $Y_0 \times_{C_0} C_1$, and then since the pencil $Y_0 \to C_0$ is without multiple fibres it follows that $C_1 \to C_0$ is also etale. This proves Theorem 1, via the assertion

<u>Theorem 7</u>. X has a cover Y_0 satisfying (***) above. For any such Y_0 $f: Y_0 \to C_0$ induces an isomorphism $f_*: \pi_1 Y_0 \overset{\approx}{\to} \pi_1 C_0$.

<u>Proof of Corollary 0.2 (ii)</u>. Since $Y_0 \to C_0$ induces an isomorphism on π_1 it also induces an isomorphism f_*: Alb $Y_0 \overset{\approx}{\to} JC_0$, so that the Albanese map of Y_0 is just the composite of $Y_0 \to C_0$ with the embedding of C_0 in its Jacobian JC_0.

For X itself the Albanese map of X fits into a commutative diagram

$$\begin{array}{ccc} Y_0 \to C_0 \subset & & \text{Alb } Y_0 \\ \downarrow & & \downarrow \psi^* \\ \alpha: X & \longrightarrow & \text{Alb } X , \end{array}$$

so that $\alpha(X) \neq \psi_*(C_0)$; this proves (ii).

Proof of Corollary 0.2 (iii). Given $f: X \to C$ consider the diagram

$$\begin{array}{ccc} \alpha: X & \to & \mathrm{Alb}\, X \\ \downarrow & & \downarrow f_* \\ C & \to & JC \end{array} \quad ;$$

the image $\alpha(X)$ is a curve D which maps onto C under f_*; since f has irreducible fibres it follows that $D = C$, proving (iii).

Proof of Corollary 0.3. In view of Corollary 0.2 (iii) and [2, Lemma (1.1), p. 122] the only non-trivial assertion remaining to prove is that if $q(X) \neq 0$ then (i) and (iii) are impossible.

Let $Y_0 \to X$ be the cover as in Theorem 1; since $q(X) \neq 0$ the curve $C_0 \subset \mathrm{Alb}\, Y_0$ has a non-trivial map to the curve $\alpha(X) \subset \mathrm{Alb}\, X$, and this implies that C_0/G_0 has genus > 0.

The map $Y_0 \dashrightarrow \overline{X}$ provided by Proposition 4 fits into a diagram

$$\begin{array}{ccccc} Y_0 & \dashrightarrow & \overline{Y}_0 & \subset & \mathbb{P}^{p_g(Y_0)-1} \\ \downarrow & & \swarrow \ \ \downarrow & & \\ & \nearrow \overline{Y}_0/G_0 \searrow & & & \\ X & \dashrightarrow & \overline{X} & \subset & \mathbb{P}^{p_g(X)-1} \end{array} ,$$

together with the quotient map $X = Y_0/G_0 \to \overline{Y}_0/G_0$. Now split into cases: if $\overline{Y}_0$ is a curve then $\overline{Y}_0$ is birational to C_0; $\overline{Y}_0/G_0$ is a non-rational curve, birational to $\overline{X}$ according to Proposition 3.

On the other hand if $\overline{Y}_0$ is a surface then it is ruled over C_0, so that $\overline{Y}_0/C_0$ is ruled over C_0/G_0; if $X \dashrightarrow \overline{X}$ is generically 2-to-1 onto a surface then this surface is birational to $\overline{Y}_0/G_0$. If $\overline{X}$ is a curve then by Proposition 3 $X \dashrightarrow \overline{X}$ has irreducible fibres, so that the same holds for $\overline{Y}_0/G \dashrightarrow \overline{X}$. The ruled surface $\overline{Y}_0/G$ is birationally a product $\mathbb{P}^1 \times C$, with $C = C_0/G_0$ of genus $p > 0$. There are thus just two possibilites for the rational map $\overline{Y}_0/G \dashrightarrow \overline{X}$, the projections on the two factors. I am home if it is projection onto C. But the other case is impossible: for $X \overset{\varphi}{\dashrightarrow} Y_0/G$ is a double cover, and φ composed with the first projection is birationally the canonical map of X, and thus by Proposition 3 has fibres curves of genus 2. On the other hand φ composed with the second projection is the map $X \to C_0/G_0$ deduced from $Y_0 \to C_0$ by quotienting by C_0, and has thus fibres of genus $\leqq 5$. Thus the branch locus of the birational double cover $X \overset{\varphi}{\dashrightarrow} \mathbb{P}^1 \times C$ has degree $\leqq 12$ on the first factor, degree 2 on the second factor, and C has genus 1. It follows easily that then $p_g \leqq 7$.

References.

[1] M. Reid, Surfaces with $p_g = 0$, $K^2 = 2$, to appear.

[2] E. Horikawa, Surfaces with small c_1^2, II, Invent. Math., 37 (1976) 121 - 155.

Problems

R.1

A subscheme $X \subset \mathbb{P}^n$ imposes r conditions on quadrics if $h^0(I_X \cdot O_{\mathbb{P}^n}(2)) = \binom{n+2}{2} - r$.

Problem: Let $C \subset \mathbb{P}^{n+1}$ be an irreducible curve of degree d. Determine the function $f(n,r)$ such that C imposes $\leq r$ conditions on quadrics implies either

(i) $d \leq f(n,r)$

or

(ii) $C \subset F \subset \mathbb{P}^{n+1}$, with F a component of dimension ≥ 2 of the intersection of all quadrics through C.

If $r \leq 2n$ then $f = r$ and if $r = 2n + 1$ or $r = 2n + 2$ then $f = 2r - 2n$.

Conjecture: For $2n \leq r \leq 3n - 2$ one has $f(n,r) = 2r - 2n$, and for $r = 3n - 1$ one has $f(n,r) = 4n$.

This conjecture implies the following 3rd Castelnuovo Inequality

Conjecture: If X is a surface such that φ_{K_X} is birationel, then either

(i) $K_X^2 \geq 4p_g - 12$

or

(ii) $\varphi_{K_X}(X) = \overline{X}$ is contained in a 3-fold component of the intersection of all quadrics through $\overline{X}$.

References.

D.W. Babbage, A note on quadrics through a canonical curve, J. London Math. Soc. 14: 4 (1939), 310-314.

A. Tjurin, Prym Varieties, Izv. Akad. Nauk 39: 5 (1975), 1003-1043.

M. Reid, Surfaces with $p_g = 0$, $K^2 = 2$ (to appear).

R.2

Find an effective bound (or better still, the correct bound)

$T(p_g,q,K^2)$ such that for a minimal surface of general type X with given $p_g(X)$, $q(X)$, K_X^2 one has

$$(\mathbb{Z}/2)^a \subset \operatorname{Pic} X \Rightarrow a \leqq T.$$

<u>Example</u>: For $p_g = q = 0$, $K^2 = 2$, Miyaoka gives the bound $T = 3$.

<u>Conjecture</u>: For $p_g = q = 0$, $K^2 = 3$, $T = 4$ or 3.

<u>Reference</u>. Y. Miyaoka, Tricanonical maps of numerical Campedelli surfaces, in <u>Complex Analysis and Algebraic Geometry</u>, ed. Baily and Shioda, Iwanami Shoten, Tokyo, 1977.

<u>R.3</u>

<u>Conjecture</u>: If X is a surface of general type such that the Albanese map $\alpha: X \to \operatorname{Alb} X$ is birational, then $c_1^2(X) \geqq c_2(X)$.

<u>R.4</u>

For a surface X let $s \in H^0(S^m\Omega_X^1 \otimes \mathcal{O}_X(D))$ be a non-zero section; let X_s denote the divisor in the 3-fold $\mathbb{P}(\Omega_X^1)$:

$$\begin{array}{ccccc} Y_s & \to & X_s & \subset & \mathbb{P}(\Omega_X^1) \\ & \searrow^{\psi} & \downarrow^{\pi} & \swarrow & \\ & & X & & \end{array},$$

and let Y_s be a non-singular model of one component of X_s. Then ψ^*s has an expression as a product of m meromorphic sections of $\Omega_{Y_s}^1$, s_i.

<u>Problem</u>: Estimate the poles of s_i in terms of the poles of s and the ramification of ψ.

<u>R.5</u>

<u>Conjecture</u>: Let X be a minimal surface of general type for which $|K_X|$ is composed of a pencil of curves of genus 2.

Then $K_X^2 \geqq 4p_g - 6$.

(This may have been proved by E. Horikawa in case $q(X) = 0$.)

<u>Reference</u>: E. Horikawa, Surfaces with a pencil of curves of genus 2, in <u>Complex Analysis and Algebraic Geometry</u>, ed. Baily and Shioda, Iwanami Shoten, Tokyo, 1977.

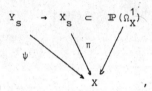

R.6

The problem is to define a birationally invariant value "K_X^3" for X a 3-fold of general type.

If the canonical ring $R(X,K_X)$ is finitely generated (which is totally unknown) then I define

$$"K_X^3" = \frac{1}{2} \cdot \lim_{\rightarrow} \frac{P_n}{n^3} .$$

This limit exists by the algebra of finitely generated graded rings. Otherwise the correct "K_X^3" could lie between $\frac{1}{2}\ \overline{\lim}\ \frac{P_n}{n^3}$ and $\frac{1}{2}\ \underline{\lim}\ \frac{P_n}{n^3}$.

R.7

Conjecture: For X a 3-fold of general type one has $P_2 \geqq 4p_g - 6$ and $P_3 \geqq 10p_g - 20$.

This is a very easy to check if $\varphi_{K_X}(X)$ is a 3-fold; and the limit cases correspond to double coverings of normal rational scrolls.

R.8

Conjecture: Let $F \subset \mathbb{P}^n$ be a 3-fold spanning $\mathbb{P}^n$, of degree $d \leq 2n - 4$. Then F is birational to $\mathbb{P}^3$ or to $\mathbb{P}^2 \times C$, with C a curve of genus $p > 0$.

R.9

Let $X \subset A^4$ be a 3-fold with an isolated singularity at $0 \in X$; let $f: X' \rightarrow X$ be a resolution, so that duality provides an injection $f_*\omega_{X'} \subset \omega_X$. Define the nth adjunction ideal

$$I_n = \mathrm{Hom}_{O_X}(\omega_X^{\otimes n}, f_*\omega_{X'}^{\otimes n}) \subset O_X ,$$

for $n \geqq 1$.

Problem: Is the sequence $\{I_n\}_{n \geqq 1}$ finitely generated?

This problem seems to be difficult, but is a rather weak form of the general problem of knowing if the canonical ring of 3-folds of general type is finitely generated.

For curves, $I_1 = G$ is the conductor of $f_*O_{X'}$ in O_X and $I_n = I_1^n$; for surfaces one can also prove (using a relatively minimal non-singular model for X') that for $p,q \geqq 3$: $I_p \cdot I_q = I_{p+q}$.

<u>Conjecture</u>. For X quasi-homogeneous $\{I_n\}_{n\geq 1}$ is finitely generated.

Mathematical Institute
University of Warwick
Coventry CV4 7AL
England

SYMMETRIC POWERS OF THE COTANGENT BUNDLE AND CLASSIFICATION OF ALGEBRAIC VARIETIES

Fumio SAKAI

Introduction. By an algebraic manifold we shall mean a non-singular complete algebraic variety defined over the complex number field $\mathbb{C}$. Let X be an algebraic manifold or more generally a compact complex manifold. We denote by Ω_X^1 the sheaf of germs of holomorphic 1-forms on X and by $S^m\Omega_X^1$ the m-th symmetric power of Ω_X^1. We define

$$Q_m(X)=\dim H^0(X,S^m\Omega_X^1).$$

Let us call $Q_m(X)$ the cotangent m-genus of X. If $\varphi'\in H^0(X,S^{m'}\Omega_X^1)$ and $\varphi''\in H^0(X,S^{m''}\Omega_X^1)$ are two symmetric 1-forms, then there is a product $\varphi'\varphi''$ $\in H^0(X,S^{m'+m''}\Omega_X^1)$. This multiplication makes $\Omega(X)=\oplus_{m=0}^{\infty}H^0(X,S^m\Omega_X^1)$ a graded ring. This ring $\Omega(X)$ is called the cotangent ring of X. We introduce the cotangent dimension $\lambda(X)$ of X, which is defined by

$$\lambda(X)=\text{tr.deg.}\,\Omega(X)-\dim X.$$

The cotangent dimension takes one of the values $-\dim X,\dots,0,\dots,\dim X$. Asymptotically $Q_m(X)\sim O(m^{n-1+\lambda})$. In particular $\lambda(X)=-\dim X$ if and only if $Q_m(X)=0$ for all $m>0$. If Ω_X^1 is a trivial bundle, then $Q_m(X)=\binom{m+n-1}{m}$, which implies $\lambda(X)=0$. If Ω_X^1 is ample, then $\lambda(X)=\dim X$. We shall see the formula $\lambda(X\times Y)=\lambda(X)+\lambda(Y)$.

In the case in which X is algebraic or X has a Kähler metric, the first cotangent genus $Q_1(X)$ coincides with the irregularity $q(X)$ of X ($\dim H^1(X,\mathcal{O})$), which is also equal to half of the first Betti number of X. Therefore $Q_1(X)$ is a topological invariant. For instance, if X is simply connected, then $Q_1(X)$ vanishes. But the higher terms $\{Q_m(X)\}$ behave differently. There are examples of simply connected algebraic manifolds with the property that $Q_m(X)\to\infty$ as $m\to\infty$.

As for the canonical bundle $K_X=\det(\Omega_X^1)$ of X, the geometric genus p_g

$=\dim H^0(X,K_X)$ was first studied and afterwards the importance of the plurigenus $P_m(X)=\dim H^0(X,K_X^m)$ was noticed. An algebraic surface with $q=p_2=0$ is a rational surface(Castelnuovo). An algebraic surface with $P_{12}=0$ is a ruled surface (Enriques), etc. The <u>canonical</u> <u>dimension</u> (the <u>Kodaira</u> <u>dimension</u>) $\kappa(X)$ of X is by definition

$$\kappa(X)=\text{tr.deg.}\oplus_{m=0}^{\infty}H^0(X,K_X^m)-1.$$

Sometimes the convention $\kappa(X)=-\infty$ instead of $\kappa(X)=-1$ is useful in some formulae. The notion $\kappa(X)$ has a fundamental importance in classification theory of algebraic varieties and compact complex manifolds (See Iitaka [7], Ueno [11]). It is natural to expect that the cotangent dimension $\lambda(X)$ added to the κ-classification will lead us to a finer classification theory. The purpose of this note is to give some figure of this λ-classification of algebraic varieties. Details will be discussed eleswhere.

In §1, we shall prepare some results concerning symmetric powers of vector bundles. In §2, we shall study basic properties of the cotangent dimension. We shall see the behavior of λ for a fiber space $f:X\to Y$. For instance, the inequality $\lambda(X)\leqq\lambda(X_y)+\dim Y$ holds for a general fiber X_y. Since the behavior of λ differs slightly from that of κ , we shall conjecture the converse inequality $\lambda(X)\geqq\lambda(X_y)+\lambda(Y)$, unless $\lambda(Y)=-\dim Y$. It turns out that λ is no longer a deformation invariant for a certain family of elliptic surfaces. In §3, we shall consider subvarieties in an abelian variety and in the projective space. Namely we shall show the followings:Let X be a submanifold in an abelian variety with ample normal bundle. Then $\lambda(X)=\min(\dim X, \operatorname{codim} X)$. Next let X be a complete intersection submanifold in the projective space $\mathbb{P}_N$. Then $Q_m(X)=0$ for $m<\dim X$ and if $\dim X>\operatorname{codim} X$, then $Q_m(X)=0$ for all m ($\lambda(X)=-\dim x$). In §4, we shall discuss the λ-classification of algebraic surfaces. Ruled surfaces ($\kappa=-\infty$) are divided into three classes (i) $\kappa=-2$, the base curve is $\mathbb{P}_1$, (ii)$\lambda=-1$, the base curve is an elliptic curve, (iii)$\lambda=0$, the base curve is a curve of genus$\geqq2$. Those surfaces with $\kappa=0$ are divided into two classes: (i) $\lambda=-2$, K3 surfaces, Enriques surfaces, (ii) $\lambda=0$, abelian surfaces, hyperelliptic surfaces. Those surfaces with $\kappa=1$ have four possibilities $\lambda=-2,-1,0,1$. Surfaces of general type ($\kappa=2$) have $\lambda=-2,-1,0,1,2$. Although the classification is incomplete in these cases, we shall give several exmples for every possible λ (TABLE III). It is very interesting to know the whole picture of λ-classification of algebraic surfaces.

Contents

It is my pleasure to thank my colleagues in SFB and in Bonn University for helpful conversations.

---

*) Sonderforschungsbereich Theoretische Mathematik 40, Universität Bonn

§1. Symmetric Powers of Vector Bundles.

By a vector bundle we shall mean a locally free coherent sheaf of finite rank. Let E be a vector bundle on a compact complex manifold X of dimension n. Let S^mE denote the m-th symmetric power of E. For two global sections $\varphi'\in H^0(X,S^{m'}E)$, $\varphi''\in H^0(X,S^{m''}E)$, there is a natural product $\varphi'\varphi''\in H^0(X,S^{m'+m''}E)$. Thus we obtain a graded ring $\Omega(X,E)=\oplus_{m=0}^{\infty}H^0(X,S^mE)$.

Definition. The E-dimension $\lambda(E,X)$ of X is defined by

$$\lambda(E,X)=\text{tr.deg.}\Omega(X,E)-\text{rank } E.$$

Let P(E) be the projective bundle $\text{Proj}(\oplus_{m=0}^{\infty}S^mE)$ with the projection $\pi:P(E)\longrightarrow X$ and let $O_{P(E)}(1)$ be the tautological line bundle on P(E). The isomorphism $\pi_*O_P(m)\simeq S^mE$ holds ($m\geqq 0$) and then $H^0(P(E),O_P(m))\simeq H^0(X,S^mE)$. Therefore the ring $\Omega(X,E)$ is isomorphic to the ring $\oplus_{m=0}^{\infty}H^0(P(E),O_P(m))$ $=\Omega(P(E),O_P(1))$. The dimension of P(E) is n+r-1, if we put r=rank E.

Hence the dimension $\lambda(E,X)$ can take one of the values $-r,\dots,0,\dots,n$. If there is a positive integer m_0 such that $\dim H^0(X,S^{m_0}E)>0$, then the following estimate holds for large m

$$\alpha m^{r-1+\lambda} \leq \dim H^0(X,S^{mm_0}E) \leq \beta m^{r-1+\lambda}, \tag{1.1}$$

where α,β are positive numbers and $\lambda=\lambda(E,X)$ (cf.[7], Theorem 2). We note that $\lambda(E,X)$ is equal to $\kappa(O_P(1),P(E))-(r-1)$ unless $\lambda(E,X)=-r$ (In this case $\kappa(O_P(1),P(E))=-\infty$ by definition). In what follows in this section, we study the basic property of the E-dimension.

<u>Definition</u>. The <u>inverse</u> <u>Chern</u> <u>classes</u> $\tilde{c}_i(E)\in H^{2i}(X,Z)$ are given by

$$\Sigma\tilde{c}_i(E)=(\Sigma(-1)^i\tilde{c}_i(E))^{-1}=(c(E^{\vee}))^{-1},$$

where $c(E)=\Sigma c_i(E)$ is the total Chern class of E and $E^{\vee}$ is the dual of E. For instance, $\tilde{c}_1(E)=c_1(E)$, $\tilde{c}_2(E)=c_1^2(E)-c_2(E)$.

If we write $\xi=c_1(O_P(1))$, then we have the formula (See [5])

$$\xi^{n+r-1}[P(E)]=\tilde{c}_n(E)[X]. \tag{1.2}$$

<u>Proposition</u> 1. Let E be a vector bundle on X. Suppose that S^mE is generated by global sections for some $m>0$. Then we have $\lambda(E,X)\geq 0$. Moreover $\lambda(E,X)=n$ if and only if $\tilde{c}_n(E)>0$.

<u>Proof</u>. By (1.2), it suffices to note that $O_P(m)$ is generated by global sections if and only if S^mE is generated by global sections.

<u>Proposition</u> 2. Let $0\to E'\to E\to E''\to 0$ be an exact sequence of vector bundles on X. Then

$$\dim H^0(X,S^mE)\leq\Sigma_{\substack{p+q=m\\ p,q\geq 0}}\dim H^0(X,S^pE'\otimes S^qE'').$$

If moreover the sequence splits (i.e., $E=E'\oplus E''$), then the equality holds.

<u>Proof</u>. This follows easily from the fact that there is a filtration $S^mE=F^0\supset F^1\supset\cdots\supset F^{m+1}=0$ such that $F_j/F_{j+1}\simeq S^jE'\otimes S^{m-j}E''$ for each j.

<u>Corollary</u>. If E'(resp.E'') is a trivial bundle, then

$$\lambda(E,X)\leq\lambda(E'',X)\quad(\text{resp. }\lambda(E,X)\leq\lambda(E',X)). \tag{1.3}$$

The equality holds, if the above sequence splits.

Example 1. (a) Let E be a rank 2 vector bundle on $\mathbb{P}_1$. It is known that $E \cong O_{\mathbb{P}_1}(p) \oplus O_{\mathbb{P}_1}(q)$ for some integers p,q. In this case, we have

$$\lambda(E,\mathbb{P}_1)=\begin{cases} 1 & \text{if } p>0 \text{ or } q>0 \\ 0 & \text{if } p=q=0 \\ -1 & \text{if } p=0,\ q<0 \text{ or } p<0,\ q=0 \\ -2 & \text{if } p<0,\ q<0. \end{cases}$$

(b)Let C be an elliptic curve. Let F_r denote the unique extension

$$0 \to O_C \to F_r \to F_{r-1} \to 0, \qquad r=2,3,\ldots.$$

where $F_1=O_C$. Atiyah [1] showed that $S^m(F_2)=F_{m+1}$ and that $\dim H^0(C,F_r)=1$ for every r. Hence $\lambda(F_2,C)=-1$.

Proposition 3. Let $0\to E'\to E\to E''\to 0$ be an exact sequence of vector bundles on X. Then there is a spectral sequence

$$E_1^{pq}=H^q(X,\textstyle\bigwedge^{-p} E'\otimes S^{m+p}E) \Rightarrow H^{p+q}(X,S^mE'') \qquad (-m\leq p\leq 0).$$

Proof. There is a Koszul type sequence

$$0\to \textstyle\bigwedge^m E'\to \bigwedge^{m-1} E'\otimes E\to\cdots\to E'\otimes S^{m-1}E\to S^mE\to S^mE''\to 0 \qquad \text{(exact)}$$

for each m. We define a complex $K_m^t=\bigwedge^{-t} E'\otimes S^{m+t}E$ for $-m\leq t\leq 0$, which induces the above spectral sequence.

Proposition 4. Let $f:X\longrightarrow Y$ be a surjective morphism of compact complex manifolds X,Y and E a vector bundle on Y. Then

$$\lambda(f^*E,X)=\lambda(E,Y).$$

Proof. There is a surjective morphism $\tilde{f}:P(f^*E)\longrightarrow P(E)$ such that $\tilde{f}^*O_{P(E)}(1)=O_{P(f^*E)}(1)$. By Theorem 4 in [7] , this gives the desired result.

Proposition 5. Let $f:X\longrightarrow Y$ be a fiber space between compact complex manifolds X,Y and E a vector bundle on X. Then

$$\lambda(E,X)\leq\lambda(E_y,X_y)+\dim Y,$$

for a general fiber $X_y=f^{-1}(y)$, where E_y is the restriction of E to X_y.

Proof. Let $\pi: P(E) \to X$ be the projection. Put $g = f \circ \pi$ and $P_y = g^{-1}(y)$. Clearly $P(E_y) = P_y$, $O_{P(E)}(1)_{|P_y} = O_{P(E_y)}(1)$. Applying the corresponding inequality of κ ([11] Theorem 5.11) to the fiber space $g: P(E) \to Y$ and $O_{P(E)}(1)$, we get the desired inequality.

Proposition 6. Let E, F be vector bundles on compact complex manifolds X, Y, respectively. Then

$$\lambda(\pi_X^* E \oplus \pi_Y^* F, X \times Y) = \lambda(E, X) + \lambda(F, Y),$$

where π_X, π_Y are the projections from the product $X \times Y$ to X, Y, respectively.

Proof. The Künneth formula gives $\Omega(X \times Y, \pi_X^* E \oplus \pi_Y^* F) \simeq \Omega(X, E) \otimes \Omega(Y, F)$. Hence $\mathrm{tr.deg.}\, \Omega(X \times Y, \pi_X^* E \oplus \pi_Y^* F) = \mathrm{tr.deg.}\, \Omega(X, E) + \mathrm{tr.deg.}\, \Omega(Y, F)$.

§2. The Cotangent Dimension and Fiber Spaces.

Let X be a compact complex manifold of dimension n. Let Ω_X^1 be the sheaf of germs of holomorphic 1-forms (the cotangent bundle).

Definition. $Q_m(X) = \dim H^0(X, S^m \Omega_X^1)$ (the cotangent m-genus)

$\lambda(X) = \lambda(\Omega_X^1, X)$ (the cotangent dimension)

Proposition 7. Let $f: X \to Y$ be a generically surjective meromorphic map of compact complex manifolds X, Y. Then for each integer $m > 0$, there is an injection

$$f^*: H^0(Y, S^m \Omega_Y^1) \hookrightarrow H^0(X, S^m \Omega_X^1)$$

and then

$$\lambda(X) + \dim X \geq \lambda(Y) + \dim Y.$$

If moreover $\dim X = \dim Y$ and f is bimeromorphic, then f^* is an isomorphism.

Corollary. $Q_m(X)$ and $\lambda(X)$ are bimeromorphic invariants of X.

Remark. If X is a singular complex space, we may define $Q_m(X)$,

$\lambda(X)$ to be $Q_m(X^*)$ and $\lambda(X^*)$, respectively, where X^* is a non-singular model of X.

Remark. We write $P_X=P(\Omega^1_X)$ and $L_X=O_{P_X}(1)$. Let $u_m:P_X \longrightarrow \mathbb{P}_N$ be the meromorphic map defined by a basis of the vector space $H^0(P_X,L^m_X)$, when $N+1=Q_m(X)>0$. Then $\max_{m>0}(\dim u_m(P_X))=\lambda(X)+n-1$.

Theorem 1. Let $f:X \longrightarrow Y$ be an unramified covering of compact complex manifolds X, Y. Then

$$\lambda(X)=\lambda(Y).$$

Proof. Since $\Omega^1_X=f^*\Omega^1_Y$, this follows from Proposition 4.

Theorem 2. $\lambda(X\times Y)=\lambda(X)+\lambda(Y)$ for compact complex manifolds X, Y.

Proof. Since $\Omega^1_{X\times Y}=\pi^*_X\Omega^1_X\oplus\pi^*_Y\Omega^1_Y$, this is immediate by Proposition 6.

Example 2. Let C be a projective curve. Then $\lambda(C)=-1$ ($\mathbb{P}_1$), 0 (elliptic curve), 1 (genus$\geqq$2).

Remark. Let $\overline{X}$ be a compact complex manifold and D a divisor with at most normal crossings on $\overline{X}$. We put $X=\overline{X}-D$. Then we define

$\overline{Q}_m(X)=\dim H^0(\overline{X},S^m(\Omega^1_{\overline{X}}(\log D)))$ (the logarithmic cotangent m-genus)

$\overline{\lambda}(X)=\lambda(\Omega^1_{\overline{X}}(\log D),\overline{X})$ (the logarithmic cotangent dimension)

By a fiber space we shall mean a surjective morphism $f:X \longrightarrow Y$ between compact complex manifolds X, Y such that the general fiber is connected.

Theorem 3. Let $f:X \longrightarrow Y$ be a fiber space. Then

$$\lambda(\Omega^1_{X,y},X_y)\leqq\lambda(X_y),$$

for a general fiber X_y, where $\Omega^1_{X,y}$ is the restriction of Ω^1_X to X_y. If moreover this fiber space is generically locally trivial, then the equality holds.

<u>Proof</u>. Let X_y be a general fiber of f. The conormal bundle $N^{\vee}_{X_y/X}$ is a trivial bundle I. Hence

$$0 \to I \to \Omega^1_{X,y} \to \Omega^1_{X_y} \to 0. \qquad \text{(exact)} \tag{2.1}$$

By Corollary (1.3), we get the desired inequality. The extension class of the sequence (2.1) corresponds to the Kodaira-Spencer class of this family at y. Hence it splits if and only if this fiber space is generically locally trivial.

<u>Theorem</u> 4. Let $f:X \longrightarrow Y$ be a fiber space. Then

$$\lambda(X) \leqq \lambda(X_y) + \dim Y,$$

for a general fiber X_y.

<u>Proof</u>. Immediate by Proposition 5 and Theorem 3.

<u>Theorem</u> 5. Let $f:X \longrightarrow Y$ be a fiber space. Suppose that $\lambda(Y) = \dim Y$. Then

$$\lambda(X) = \lambda(\Omega^1_{X,y}, X_y) + \lambda(Y),$$

for a general fiber X_y.

<u>Proof</u>. There is a generically injective homomorphism $\sigma: f^*\Omega^1_Y \longrightarrow \Omega^1_X$. Noting that $\lambda(\Omega^1_{X,y}, X_y) \neq -\dim X$, the following lemma completes the proof.

<u>Lemma</u> 1. Let $f:X \longrightarrow Y$ be a fiber space. Let E, F be vector bundles on X, Y, respectively. Assume that $\lambda(F,Y) = \dim Y$ and that there is a generically injective homomorphism $\sigma: f^*F \longrightarrow E$. Then

$$\lambda(E,X) = \lambda(E_y, X_y) + \dim Y, \qquad \text{unless } \lambda(E_y, X_y) = -\operatorname{rank} E,$$

for a general fiber X_y.

<u>Proof</u>. (If both E and F are line bundles, this has been proven by Fujita[4].) We consider the diagram:

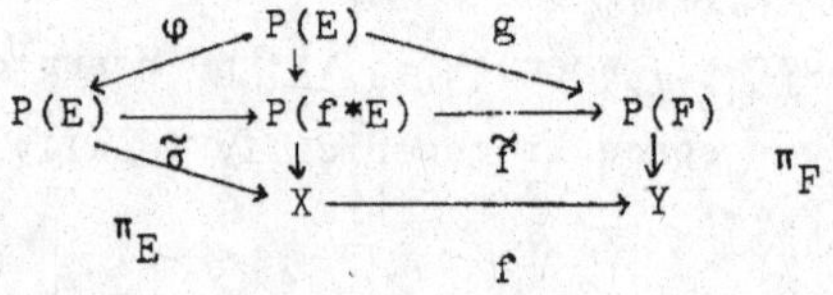

Here φ is a desingularization of the meromorphic map $\tilde{\sigma}$. We put $g = \tilde{f} \circ \tilde{\sigma} \circ \varphi$.

We have $\pi_{E*}\varphi_*Hom(g^*O_{P(F)}(1),\varphi^*O_{P(E)}(1))\simeq Hom(f^*F,E)$. Hence σ induces a non-trivial homomorphism $s:g^*O_{P(F)}(1)\longrightarrow\varphi^*O_{P(E)}(1)$. By applying the corresponding result for line bundles (Proposition 1, [4]) to our situation $g:\tilde{P}(E)\longrightarrow P(F)$, we can complete the proof.

Corollary. Let $f:X\longrightarrow Y$ be a fiber bundle with a fiber F. Suppose that $\lambda(Y)=\dim Y$, then

$$\lambda(X)=\lambda(F)+\lambda(Y).$$

Example 3. We consider an elliptic surface $f:S\longrightarrow B$ over a curve B with genus(B)≧2. Let C (an elliptic curve) be a general fiber of f. We can write the exact sequence (2.1) as follows.

$$0\to O_C\to\Omega^1_{S|C}\to O_C\to 0. \tag{2.2}$$

It occurs two cases (i) $\lambda(\Omega^1_{S|C},C)=0$, (2.2) splits, (ii) $\lambda(\Omega^1_{S|C},C)=-1$, (2.2) is a non-trivial extension (See Example 1 (b)). Combining these with Theorem 3, 5, we conclude

$$\lambda(S)=\begin{cases}1 & \text{constant moduli (constant j-invariant)}\\ 0 & \text{non-constant moduli (non-constant j-invariant).}\end{cases}$$

Remark. We take a family of elliptic surfaces $\Phi:\mathcal{S}\rightarrow\mathcal{T}$ such that each fiber S_t is an elliptic surface over a curve B_t of genus≧2. Suppose that $j(t_0)$=constant but $j(t)$ is not constant in general, where j is the j-invariant of S_t. Since $\lambda(S_{t_0})=1$, $\lambda(S_t)=0$, this gives an example that the cotangent dimension is not a deformation invariant.

Example 4. Next we consider an algebraic surface S with a fibration $f:S\longrightarrow B$ such that genus(B)≧2 and the genus of a general fiber C≧2. The sequence (2.1) has the form

$$0\to O_C\to\Omega^1_{S|C}\to K_C\to 0.$$

It follows that $\lambda(\Omega^1_{S|C},C)=1$. In fact, this is obvious if the above sequence splits (Theorem 3). If this is a non-trivial extension, by Theorem 2.2 in [5], $\Omega^1_{S|C}$ is ample, since K_C is ample, which implies that $\lambda(\Omega^1_{S|C},C)=1$. In view of Theorem 5, $\lambda(S)=2$ ($\lambda(S)=\lambda(C)+\lambda(B)$).

Theorem 6. Let X be a compact complex manifold. Then

$$\lambda(X) \leqq a(X),$$

where a(X) is the transcendental degree of the meromorphic function field of X.

Proof. Immediate by Theorem 3.8 in [11].

Conjecture Λ. Let X be an algebraic manifold (or a compact complex manifold). Then

$$\Lambda_+ : \quad \lambda(X) \leqq \kappa(X) \qquad \text{if } \kappa(X) \geqq 0,$$

$$\Lambda_{-\infty} : \quad \lambda(X) \leqq \dim X - 2 \qquad \text{if } \kappa(X) = -\infty.$$

The Cojecture Λ_+ is a consequence of the following special case.

Conjecture Λ_0. If $\kappa(X)=0$, then $\lambda(X) \leqq 0$.

Conjecture Λ_0 is true in the Kähler case, if $c_1(X)=0$ in $H^2(X,Q)$ (Kobayashi [8]). Cojecture Λ is true for algebraic surfaces.

Conjecture M. Let $f:X \longrightarrow Y$ be a fiber space. Then

$$\lambda(X) \leqq \lambda(X_y) + \lambda(Y) \qquad \text{unless } \lambda(Y) = -\dim Y,$$

for a general fiber X_y. (We have to add the condition that f admit no multiple fiber. See §4, (H), Remark.)

§3. Subvarieties in an Abelian Variety and in the Projective Spase.

In this section, we consider subvarieties. First we show

Proposition 8. Let X be a submanifold in a compact complex manifold Y. Then there is a spectral sequence

$$E_1^{pq} = H^q(X, \overset{-p}{\Lambda} N^\vee_{X/Y} \otimes S^{m+p}\Omega^1_{Y|X}) \Rightarrow H^{p+q}(X, S^m\Omega^1_X) \qquad (-m \leqq p \leqq 0),$$

where $N^\vee_{X/Y}$ is the conormal bundle of X in Y.

Proof. We can apply Proposition 3 to the exact sequence

$$0 \to N^\vee_{X/Y} \to \Omega^1_{Y|X} \to \Omega^1_X \to 0.$$

Let A be an abelian variety and X a subvariety of dimension n.

The argument of Ueno [11], p.116 shows that $Q_m(X) \geq \binom{m+n-1}{m}$. Hence $\lambda(X) \geq 0$.

Theorem 7. Let X be a submanifold in an abelian variety A. If the normal bundle $N_{X/A}$ is ample, then

$$\lambda(X) = \min(\dim X, \operatorname{codim} X).$$

Proof. Case (i). dim X>codim X. Since $\Omega^1_{A|X}$ is a trivial bundle I, we get a spectral sequence (by Proposition 8)

$$E_1^{pq} = H^q(X, \overset{-p}{\wedge} N^\vee_{X/A} \otimes S^{m+p} I) \Rightarrow H^{p+q}(X, S^m \Omega^1_X) \qquad (-m \leq p \leq 0).$$

Since $N_{X/A}$ is ample by hypothesis, we have by the vanishing theorem of Sommese ([10] Proposition (1.14)),

$$E_1^{pq} = 0 \qquad \text{for } p<0,\ p+q \leq (\dim X - \operatorname{codim} X) - 1.$$

Since dim X>codim X in this case, we obtain the following result.

$$E_\infty^{p,-p} = \begin{cases} 0 & \text{if } p<0, \\ H^0(X, S^m I) & \text{if } p=0. \end{cases}$$

Hence we get $Q_m(X) = \binom{m+\dim A-1}{m}$ and $\lambda(X) = \operatorname{codim} X$.

Case (ii). dim X$\leq$codim X. Clearly $\tilde{c}(\Omega^1_X) = \tilde{c}(\Omega^1_{A|X}) c(N_{X/A})$ (See §1, for the definition). Hence $\tilde{c}_n(\Omega^1_X) = c_n(N_{X/A})$, where n=dim X. Since $N_{X/A}$ is ample and dim X$\leq$codim X=rank $N_{X/A}$, we get $c_n(N_{X/A})>0$ (Bloch-Gieseker [2]). Noting that Ω^1_X is generated by global sections, by Proposition 1 we infer that $\lambda(X)=\dim X$.

Now we consider complete intersection submanifolds in the projective space. Let X be a complete intersection of r hypersurfaces of degree $d_1, \ldots, d_r$ in $\mathbb{P}_{n+r}$. The normal bundle can be written as a direct sum:

$$N = N_{X/\mathbb{P}_{n+r}} = H^{d_1} \oplus \cdots \oplus H^{d_r},$$

where H is the restriction of the hyperplane bunle of $\mathbb{P}_{n+r}$ to X. We want to use the spectral sequence in Proposition 8. For k$\leq$r, we have

$$\overset{k}{\wedge} N^\vee = \bigoplus_{1 \leq i_1 \leq \cdots \leq i_k \leq r} H^{-d_{i_1} \cdots d_{i_k}}$$

Theorem(cf.[9], p.521). $H^i(X, S^m \Omega^1_{\mathbb{P}|X} \otimes H^t) = 0$ for $i<n$, $m \geq t+2$.

As a consequence, $E_1^{pq}=H^q(X,\overset{-p}{\wedge}N^\vee\otimes S^{m+p}\Omega^1_{\mathbb{P}|X})=0$ for $p\neq n$. Hence if $m<n$, then $E_1^{p,-p}=0$ and also $E_\infty^{p,-p}=0$ for $-m\leq p\leq 0$, which implies $Q_m(X)=0$ for $m<n$. In a similar manner, one can show that if $n>r$, then $Q_m(X)=0$ for all m. Summarizing these results, we obtain

Theorem 8. Let X be a complete intersection submanifold in the projective space $\mathbb{P}_N$. Then $Q_m(X)=0$ for $m<\dim X$. If $\dim X>\operatorname{codim} X$. then $Q_m(X)=0$ for all m ($\lambda(X)=-\dim X$)).

Remark. In case $\dim X\leqq\operatorname{codim} X$, $Q_m(X)$ may not vanish. We examine this in the 2-dimensional case. Let S be a complete intersection surface in $\mathbb{P}_{2+r}$ of type $(d_1,\ldots,d_r)$. We have

$$\tilde{c}_2(S)=\tilde{c}_2(T_S)=\tilde{c}_2(\Omega^1_S)=c_1^2(S)-c_2(S)$$

$$=[\Sigma_{i<j}d_id_j-(\Sigma_jd_j)(r+3)+\frac{(r+3)(r+4)}{2}]\Pi_id_i$$

We shall see in §4 (G) that the condition $\tilde{c}_2(S)>0$ implies $\lambda(S)=2$. We present here the table of the sign of $\tilde{c}_2(S)$ for each type $(d_1,\ldots,d_r)$, which shows that in general $\tilde{c}_2(S)>0$ holds except some finite number of infinite series. It is known that for any complete intersection surface S, the fundamental group $\pi_1(S)$ is trivial. Thus we obtain many simply connected surfaces S with the property that $Q_m(S)\to\infty$ as $m\to\infty$.

TABLE I.

r	κ	$\tilde{c}_2$	type $(d_1,\ldots,d_r)$ $2\leq d_1\leq\cdots\leq d_r$	λ
2	$-\infty$	-	(2,2)	-2
	0	-	(2,3)	-2
	2	-	(2,d) $d\geq4$, (3,d) $d\geq3$, (4,d) $d\geq4$, (5,d) $d\geq5$, (6,d) $6\leq d\leq14$, (7,d) $7\leq d\leq9$, (8,8)	?
		0	(6,15), (7,10)	?
		+	the rest	2
3	0	-	(2,2,2)	-2
	2	-	(2,2,d) $d\geq2$, (2,3,d) $d\geq3$, (2,4,d) $d\geq4$, (2,5,d) $5\leq d\leq10$ (2,6,d) d=6,7, (3,3.d) $d\geq3$, (3,4,d) $4\leq d\leq8$, (3,5,5),	?

3		-	(4,4,d) $d=5,6$	?
	2	0	(2,5,11), (3,4,9), (3,5,6)	?
		+	the rest	2
		-	(2,2,2,d) $d\geq 2$, (2,2,3,d) $d\geq 3$, (2,2,4,d) $4\leq d\leq 7$, (2,2,5,5,), (2,3,3,d) $3\leq d\leq 6$, (2,3,4,4), (3,3,3,3)	?
4	2	0	(2,2,4,8), (2,3,3,7), (3,3,3,4)	?
		+	the rest	2
		-	(2,2,2,2,d) $d\geq 2$, (2,2,2,3,d) $3\leq d\leq 5$, (2,2,3,3,3)	?
5	2	0	(2,2,2,3,6), (2,2,2,4,4)	?
		+	the rest	2
		-	(2,2,2,2,2,d) $2\leq d\leq 4$	?
6	2	0	(2,2,2,2,2,5), (2,2,2,2,3,3)	?
		+	the rest	2
7	2	-	(2,2,2,2,2,2,2)	?
		+	the rest	2
≥ 8	2	+	all	2

§4. Classification of Algebraic Surfaces.

In this section, we shall discuss the classification of algebraic surfaces by the cotangent dimension λ. At the moment, the λ-classification of algebraic surfaces of general type ($\kappa=2$) is not yet established. For a brief outline of the κ-classification of surfaces, we refer Ueno [11], Appendix.

(A) $\kappa=-\infty$. Any surface S in this class is birational to a product $\mathbb{P}_1\times C$ with a curve C. $\lambda(S)=\lambda(\mathbb{P}_1)+\lambda(C)=\lambda(C)-1$ (Theorem 2). Hence $\lambda(S)=-2$ (if $C=\mathbb{P}_1$), -1 (if C is an elliptic curve), 0 (if genus$(C)\geqq 2$).

(B) $\kappa=0$. (i) Abelian surfaces and hyperelliptic surfaces. Clearly λ(abelian surface)=0. Since every hyperelliptic surface has an abelian surface as an unramified covering, we have λ(hyperelliptic surface)=0.

(ii) K3 surfaces and Enriques surfaces. In §3, we have seen $\lambda=-2$ for quartic surfaces in $\mathbb{P}_3$. We shall also verify this for Kummer surfaces in (D). The general case is answered by Kobayashi.

Theorem ([8]). Let S be a Kähler K3 surface. Then $\lambda(S)=-2$.

More generally, he shows that if X is a simply connected compact Kähler manifold with $K_X=0$ (trivial canonical bundle), then $Q_m(X)=0$ for all $m>0$. Each Enriques surface has a K3 surface as a 2-sheeted unramified covering. Therefore λ(Enriques surface)$=-2$.

(C) $\kappa=1$, $\lambda=0$, 1. We have seen examples in these classes in Example 3.

(D) $\kappa=1$, $\lambda=-2$, -1 (Special Quotient Surfaces). We shall construct
$\kappa=2$, $\lambda=-2$, -1, 0. surfaces belonging to these classes by quotients of products of two curves. Let C', C" be two curves which are double coverings of C_0', C_0'', respectively. We denote by ι', ι'' the involutions corresponding to the double coverings. We put $\iota=\iota'\times\iota''$. Let S be the minimal non-singular model of the quotient $C'\times C''/\langle\iota\rangle$. Let g', g_0', g'', g_0'' denote the genera of C', C_0', C", C_0'', respectively. In particular if $g'=g''=1$, $g_0'=g_0''=0$, S is nothing but a Kummer surface. We obatin the following table of those values κ, λ, Q_m of S for various genera.

TABLE II

g'	g_0'	g''	g_0''	κ	λ	Q_m
1	0	1	0	0	-2	0
1	0	$\geqq 2$	0	1	-2	0
1	0	$\geqq 2$	1	1	-1	1
$\geqq 2$	0	$\geqq 2$	0	2	-2	0
$\geqq 2$	1	$\geqq 2$	0	2	-1	1
$\geqq 2$	1	$\geqq 2$	1	2	0	m+1

Outline of the proof. We have $H^0(S,S^m\Omega_S^1)\hookrightarrow H^0(C'\times C'',S^m\Omega^1_{C'\times C''})^{\langle\iota\rangle}$ (invariant part). An element $\varphi\in H^0(S,S^m\Omega_S^1)$ can be written as $\varphi=\Sigma_{p+q=m}\alpha_p\otimes\beta_q$, with $\alpha_p\in H^0(C',K_{C'}^p)$, $\beta_q\in H^0(C'',K_{C''}^q)$. Let n' (resp. n") be the number of the fixed points of ι' (resp. ι''). Namely $n'=2g'-4g_0'+2$ ($n''=2g''-4g_0''+2$). Let x_i' (resp. x_j'') be fixed points of ι' (resp. ι''). The quotient surface $C'\times C''/\langle\iota\rangle$ has ordinary double points under $x_{ij}=(x_i',x_j'')$, which can be desingularized by a non-singular rational curve L_{ij}. We take a point x_{ij}. Let z, w be coordinates with center x_i', x_j'', respectively, such that $\iota':z\to -z$, $\iota'':w\to -w$. For a point $y\in L_{ij}$, we can choose a coordinate system (u,v) with center y satisfying ($z^2=u$, $w=zv$) (or ($z=wu$, $w^2=v$)). If we write $\alpha_p=a_p(z)(dz)^p$, $\beta_q=b_q(w)(dw)^q$, then

$$\varphi=\Sigma a_p(\sqrt{u})b_q(\sqrt{u}v)\frac{(du)^p\ (udv+vdu)^q}{(2\sqrt{u})^m} \qquad \text{near } y\in S.$$

Let p_i (resp. q_j) be the order of zero of α_p at x_i' (resp. β_q at x_j''). Then φ is holomorphic if and only if $p_i+q_j\geqq m$ for all combinations of i, j. Summing them for all i,j, we get $n''(\Sigma p_i)+n'(\Sigma q_j)\geqq n'n''m$. We note that $\deg(\alpha_p)=p(2g'-2)\geqq\Sigma p_i$, $\deg(\beta_q)=q(2g''-2)\geqq\Sigma q_j$. Calculating these together, we get $pg_0'+qg_0''\geqq m$. In case $g_0'=g_0''=0$, this is impossible. Therfore $Q_m(X)=0$. In case $g_0'=0$, $g_0''=1$, this is the case only if p=0, q=m and then $q_j=m$ for all j, from which follows $Q_m(X)=1$. In case $g_0'=g_0''=1$, we get $p_i=p$, for all i and $q_j=q$ for all j. Let γ_p be the unique non-vanishing form in $H^0(C_0',K_{C'}^p)$. The lifting $\tilde{\gamma}_p$ on C' satisfies the condition $(\tilde{\gamma}_p)=\Sigma p[x_i']$. $\deg(K_{C'}-\Sigma p[x_i'])=0$, so $(\tilde{\gamma}_p)$ is the unique divisor in $|K_{C'}-\Sigma p[x_i']|$. Similarly, there is a unique divisor $(\tilde{\gamma}_q)\in|K_{C''}-\Sigma q[x_j'']|$. As a consequence, we infer that $Q_m(X)=m+1$.

(E) $\kappa=2$, $\lambda=-2$. Hypersurfaces in $\mathbb{P}_3$ (Theorem 8)** . Any quotient (with $\kappa=2$) of a surface in this class belongs to the same class (Proposition 7). For instance, Godeaux surfaces, etc.

(F) $\kappa=2$, $\lambda=1$. Hypersurfaces (with $\kappa=2$) in an abelian 3-fold (Theorem 6).

**) I learned from Van de Ven that he also knew the fact $Q_m(S\subset\mathbb{P}_3)=0$.

(G) $\kappa=2$, $\lambda=2$. Surfaces with $\tilde{c}_2=c_1^2-c_2>0$ belong to this class (Bogomolov [3]). Here we write $\tilde{c}_2=\tilde{c}_2(S)=\tilde{c}_2(T_S)$ $(=\tilde{c}_2(\Omega_S^1)$, in this case, where T_S is the tangent bundle). In fact, by the Riemann-Roch formula, we have $\chi(S,S^m\Omega_S^1)=\frac{1}{6}m^3\tilde{c}_2+\{\text{lower terms}\}$. Since $K\otimes(\Omega_S^1)^\vee=\Omega_S^1$, we obtain by duality $H^2(S,S^m\Omega_S^1)=H^0(S,S^m\Omega_S^1\otimes K^{-(m-1)})$. If $P_{m-1}>0$, then $\dim H^2(S,S^m\Omega_S^1)\leqq Q_m(S)$. Hence $Q_m(S)\geqq\frac{1}{12}m^3\tilde{c}_2+\{\text{lower terms}\}$, for large m, which implies $\lambda(S)=2$.

__Examples of surfaces with__ $\tilde{c}_2>0$. (i) General complete intersections in $\mathbb{P}_{2+r}$, $r\geqq 2$ (See §3), (ii) Kodaira surfaces (See also Example 4), (iii) Quotients of symmetric bounded domains (ball, product of discs), in particular a product of two curves of genus$\geqq$2, (iv) some Hilbert modular surfaces ($\pi_1=0$).

Any covering of a surface in this class belongs to the same class. Let S be a surface with $\tilde{c}_2>0$. Let L be an ample line bundle on S and B_m a non-singular curve in the complete linear system $|2mL|$. We can construct a double covering $\pi:\tilde{S}\longrightarrow S$ which ramifies over B_m. Then

$$\tilde{c}_2(\tilde{S})=2\tilde{c}_2(S)+mKB_m-\tfrac{1}{2}B_m^2=2\tilde{c}_2(S)+2(mKL-m^2L^2).$$

Thus by letting m large enough, we obtain many surfaces with $\kappa=2$, $\lambda=2$ but $\tilde{c}_2<0$. Since $K_{\tilde{S}}=\pi^*(K_S+mL)$, we can even make $K_{\tilde{S}}$ ample for large m.

__Remark__. The recent beautiful result of Bogomolov can be stated as follows.

__Theorem__ ([3]). Let S be a surface of general type with $\lambda=2$. Then curves with bounded geometric genus form an algebraic family.

__Corollary__. There are only a finite number of rational curves and elliptic curves on a surface with $\kappa=2$, $\lambda=2$.

__Remark for Non-Algebraic Surfaces__. Conjecture Λ is true for compact complex surfaces up to non-Kähler K3 surfaces (whose existence is doubted). It is easy to see λ(Hopf surface)$=-2$. Non-algebraic surfaces with $\kappa=0$ are classified into three classes:(i) K3 surfaces,

$\lambda=-2$, if Kähler, (ii) Torus, $\lambda=0$, (iii) Certain elliptic bundles over an elliptic curve or their finnte unramified quotients, $\lambda=-1$. We know that $\lambda\leq 0$ for surfaces with no meromorphic function (Theorem 6).

Finally, we give the table of the λ-classification of algebraic surfaces.

TABLE III

κ	λ	structure (examples)
	-2	rational surfaces
$-\infty$	-1	ruled surfaces of genus 1
	0	ruled surfaces of genus$\geq$2
0	-2	K3 surfaces, Enriques surfaces
	0	abelian surfaces, hyperelliptic surfaces
	-2	special quotient surfaces
	-1	special quotient surfaces
1	0	elliptic surfaces over a curve of genus$\geq$2 (non-constant moduli)
	1	elliptic surfaces over a curve of genus$\geq$2 (constant moduli)
	-2	hypersurfaces in $\mathbb{P}_3$ and their quotients
	-1	special quotient surfaces
2	0	special quotient surfaces
	1	hypersurfaces in an abelian 3-fold
	2	surfaces with $\tilde{c}_2>0$ and their coverings

Added in November 1978. (i) By definition, we understand the transcendental degree of the ring $\Omega(X)$ as sup(tr.deg.R), where R moves over all finitely generated subrings of $\Omega(X)$. An alternative definition can be given through (1.1). (ii) There are two articles on topics in §3 (informed by Bogomolov). P.Brjukman: Tensor differential forms on algebraic varieties. Math. USSR. Izvestia 5 (1971), F.A.Bogomolov: Holomorphic symmetric tensors on projective surfaces. preprint in russian.

(H) <u>Elliptic</u> <u>Surfaces</u>.*** (i) (Multiple fibers) Let $f:S\longrightarrow B$ be an elliptic surface. Suppose f has multiple fibers over a_i with multiplicity ν_i, for $i=1,\dots,k$. We write $\lambda^f(B)=\lambda(K_B+\Sigma(1-\frac{1}{\nu_i})a_i,B)$.

<u>Claim</u>. If $\lambda^f(B)=1$, then $\lambda(S)=\begin{cases}1 & \text{constant moduli case}\\ 0 & \text{non-constant moduli case}\end{cases}$

<u>Proof</u>. There is a homomorphism from $f^*(K_B^m\otimes\Pi_i[a_i]^{[m(1-(1/\nu_i))]})$ to $S^m\Omega^1$. So we can use the argument of Theorem 5 and Example 3.

(ii) (non-constant moduli) Let $f:S\longrightarrow B$ be an elliptic surface with non-constant moduli. Let a_i, ν_i have the same meaning as above. Then

$$Q_m(S)=\dim H^0(B,\mathcal{O}(K_B^m\otimes\Pi_i[a_i]^{[m(1-\frac{1}{\nu_i})]})).$$

<u>Proof</u>. Take a small disc $\Delta\subset B$ such that $S_{|\Delta}$ contains no singular fiber. $S_{|\Delta}\simeq\Delta\times\mathbb{C}/G$, where G is the automorphisms $(u,\zeta)\longrightarrow(u,\zeta+n_1\omega(u)+n_2)$, $n_1,n_2\in\mathbb{Z}$ and $\{1,\omega(u)\}$ is the period of $f^{-1}(u)$. Then $\varphi\in H^0(S_{|\Delta},S^m\Omega^1)$ can be written as $\varphi=\Sigma_{p=0}^{m}a_p(u,\zeta)(du)^p(d\zeta)^{m-p}$, with G-invariant holomorphic functions $a_p(u,\zeta)$. By an induction, one can show $a_p=0$ $p<m$, $a_m=a_m(u)$. Consequently $\varphi=f^*\psi$, $\psi\in H^0(\Delta,K^m)$, from which follows the result.

(iii) (constant moduli) Let $f:S\longrightarrow B$ be an elliptic surface with at least one singular fiber(other than multiple non-singular fibers). Then $Q_m(S)=1$ if $\lambda^f(B)=0$ and $Q_m(S)=0$ if $\lambda^f(B)=-1$ (cf. the argument in (D))

The λ-classification of elliptic surfaces follows from (i),(ii),(iii)

TABLE IV

(constant moduli)

λ	$\lambda^f(B)$	g(B)	κ
-2	-1	0	$-\infty$, 0, 1
	-1	0	$-\infty$(ruled)
-1	0	0,1	0(non-Kähler)
	0	0,1	1
0	0	0,1	0(Kähler)
1	1	$g\geq 0$	1

(non-constant moduli)

λ	$\lambda^f(B)$	g(B)	κ
-2	-1	0	$-\infty$, 0, 1
-1	0	0	1
0	1	$g\geq 0$	1

***) Added on July 31,1978. I thank K.Ueno for a stimulative discussion.

Remark. Suppose $f:S \longrightarrow B$ is an elliptic surface with constant moduli over an elliptic curve B. If $\lambda^f(B)=1$, then $\lambda(S)=1(>\lambda(\text{fiber})+\lambda(B)=0)$. Therefore Conjecture M is not true in its full general form (This fact is first noticed by Ueno by an example). We may call $\lambda^f(B)$ the virtual cotangent dimension of B with respect to f.

REFERENCES

[1] Atiyah,M.: Vector bundles over an elliptic curve. Proc. London Math. Soc. (3) 7, 27 (1957) 414-452

[2] Bloch,S. & Gieseker,D.: The positivity of the Chern classes of an ample vector bundles. Invent. Math. 12 (1971) 112-117

[3] Bogomolov,F.A.: Families of curves on a surface of general type. Soviet Math. Dokl. 18 (1977) 1041-1044

[4] Fujita,T.:Some remarks on Kodaira dimensions of fiber spaces. Proc. Japan Acad. 53 Ser A (1977) 28-30

[5] Gieseker,D.: P-Ample bundles and their Chern classes. Nagoya Math. J. 43 (1971) 91-116

[6] Hartshorne,R.:Ample vector bundles. Publ. Math. IHES 29 (1966) 63-94

[7] Iitaka,S.:On D-dimension of algebraic varieties. J. Math. Soc. Japan 23 (1971) 356-373

[8] Kobayashi,S.: The first Chern class and holomorphic symmetric tensor fields. To appear

[9] ______. & Ochiai,T.: On complex manifolds with positive tangent bundles. J. Math. Soc. Japan 22 (1970) 499-525

[10] Sommese,A.J.: Complex subspaces of homogeneous complex manifolds I. Submanifolds of abelian varieties. Math. Ann. 233 (1978) 229-256

[11] Ueno,K.: Classification theory of algebraic varieties and compact complex spaces. Lecture Notes in Math. 439, Springer (1975)

SUPERSINGULAR K3 SURFACES

by

Tetsuji Shioda*)

Contents

---

*) Partially supported by SFB Theoretische Mathematik 40, Universität Bonn.

§0. Introduction

Let X be a K3 surface defined over an algebraically closed field k of characteristic $p > 0$. The Picard number of X, $\rho = \rho(X)$, is by definition the rank of the Néron-Severi group $NS(X)$. By Igusa's inequality, we have

(0.1) $$1 \leq \rho \leq b_2 = 22 ,$$

b_2 being the second Betti number of X. A K3 surface X is called supersingular if $\rho(X) = 22$.

In his paper [A], M. Artin introduced new invariants h, σ_0 and σ, and defined a beautiful filtration of the moduli space of K3 surfaces in terms of them. The invariant h is the height of the formal Brauer group $\hat{B}r(X)$, and is related to the Picard number ρ by Artin-Mazur inequality ([A], (0.1), and [2]):

(0.2) $$\rho \leq b_2 - 2h \quad \text{if} \quad h < \infty.$$

It follows from (0.1) and (0.2) that

(0.3) $$\rho = 1, 2, \ldots, 20 \quad \text{or} \quad 22$$

and

(0.4) $$h = 1, 2, \ldots, 10 \quad \text{or} \quad \infty .$$

A K3 surface with $h = \infty$ is called supersingular in the sense of Artin. Any supersingular surface is supersingular in the sense of Artin, while the converse is still conjectural in general but is true for elliptic K3 surfaces.

The invariants σ_0 and σ are defined only for supersingular surfaces and given as follows. The discriminant, $d(NS(X))$, of the intersection form on the Néron-Severi group $NS(X)$ of a supersingular K3 surface X has the form

(0.5) $$d(NS(X)) = -p^{2\sigma_0} \quad (1 \leq \sigma_0 \leq 10),$$

and one defines σ by $\sigma + \sigma_0 = 11$. Thus the possible values of σ are

(0.6) $$\sigma = 1, 2, \ldots, 10.$$

In the present paper, we shall study Kummer surfaces associated with abelian surfaces (in characteristic $\neq 2$) with the application to Artin's filtration in mind.

In §1, we prove that any infinitesimal deformation of a Kummer surface $Km(A)$ preserving the Néron-Severi group is trivial, if A is a product of two mutually isogenous elliptic curves with non-trivial endomorphisms (Theorem 1.4).

In §2, we look at Artin's filtration of the moduli space of K3 surfaces at $X_0 = Km(A_0)$ where A_0 is a product of two supersingular elliptic curves, and obtain Theorem 2.1. In particular, it follows that, in every characteristic $p > 2$,

(i) supersingular K3 surfaces depend on 9 moduli, and

(ii) the invariants h and σ take on all possible values (0.4) and (0.6);

these facts were previously known only in case $p \equiv 3 \pmod 4$. Also we have $\sigma_0(X_0) = 1$, i.e.,

(0.7) $$d(NS(X_0)) = -p^2.$$

In §3, we give a more elementary proof of (0.7) by considering the relation between the Néron-Severi group of an abelian surface and that of the associated Kummer surface. We include a proof, due to Serre, of an interesting theorem stating that two abelian varieties of the same dimension > 1, which are products of supersingular elliptic curves, are isomorphic to each other (Theorem 3.5).

In §4, we prove the uniqueness of the supersingular K3

surface with the property (0.7), following an idea of Ogus.

In §5, we compute the discriminant of the Néron-Severi group of the elliptic modular surface of level 4 and that of the Fermat quartic surface, in all characteristic $p \neq 2$.

In this paper, we consider K3 surfaces only in characteristic different from 2, because our main tool is the theory of Kummer surfaces. (For the supersingular K3 surfaces in characteristic 2, the reader is referred to the recent splendid work of Rudakov and Šafarevič [14].) Throughout this paper, k will denote an algebraically closed ground field of characteristic p, $p = 0$ or $p > 2$.

Finally the author wishes to thank M. Artin and A. Ogus for very stimulating conversations and J.-P. Serre for communicating him the proof of Theorem 3.5. The author would like to thank F. Hirzebruch and SFB, University of Bonn, for the warm hospitality during his stay in Bonn.

§1. Infinitesimal deformation of a Kummer surface preserving the Néron-Severi group

Let A be an abelian surface over k, and let $X = \mathrm{Km}(A)$ be the associated Kummer surface. By definition X is a K3 surface obtained from the quotient A/ι of A by the inversion automorphism ι by blowing up the 16 singular points corresponding to the points of order 2 on A. Equivalently, X is obtained in the following way. Let $\tilde{\beta}: \tilde{A} \longrightarrow A$ be the blowing-up of A at the 16 points of order 2. Then the inversion automorphism ι naturally extends to an involution $\tilde{\iota}$ of $\tilde{A}$ which fixes the 16 blown-up exceptional curves pointwise. The quotient $\tilde{A}/\tilde{\iota}$ is non-singular and it can be identified with $X = \mathrm{Km}(A)$. Thus we have a commutative diagram:

$$(1.1)\qquad \begin{array}{ccc} A & \xleftarrow{\tilde{\beta}} & \tilde{A} \\ {\scriptstyle\pi}\downarrow & & \downarrow{\scriptstyle\tilde{\pi}} \\ A/\iota & \xleftarrow{\beta} & \tilde{A}/\tilde{\iota} = X, \end{array}$$

where π and $\tilde{\pi}$ are quotient morphisms and β and $\tilde{\beta}$ are blowing-ups. Letting ${}_2A$ be the group of points of order 2 on A, we denote by E_a (resp. $\tilde{E}_a$) the non-singular rational curve on X (resp. on $\tilde{A}$) corresponding to $a \in {}_2A$:

$$(1.2)\qquad E_a = \beta^{-1}(\pi(a)), \qquad \tilde{E}_a = \tilde{\beta}^{-1}(a) \quad (a \in {}_2A).$$

It is known that

$$(1.3)\qquad \sum_{a\in{}_2A} E_a \quad \text{is divisible by } 2 \text{ in } \mathrm{Pic}(X).$$

If we replace the ground field k by a k-scheme S and an abelian surface A by an abelian scheme A_S over S with relative dimension 2, we can define the "Kummer surface over S", $\mathrm{Km}(A_S)$, in the same way as above; it is a smooth family of Kummer surfaces parametrized by S.

Proposition 1.1 Let $X = \mathrm{Km}(A)$ be a Kummer surface and consider any infinitesimal deformation X_S of X over $S = \mathrm{Spec}\, k[t]/(t^2)$, such that the Néron-Severi group of X extends to X_S. Then there exists an abelian scheme A_S over S with the special fibre A such that

$$X_S \simeq \mathrm{Km}(A_S).$$

Proof. Since $H^1(X, \mathcal{O}) = 0$ for a K3 surface X, the assumption implies that any invertible sheaf L on X extends uniquely to an invertible sheaf L_S on X_S ([6], (4.6.5)). Let Γ be an irreducible curve on X, and let $L = \mathcal{O}(\Gamma)$. Then it is easily seen that $H^1(X, L) = 0$. Hence Γ is induced by an effective Cartier divisor Γ_S on X_S. (This reasoning is due to Artin [A], p.564.)

In particular, the exceptional curves E_a $(a \in {}_2A)$ are induced by Cartier divisors E_{a_S} on X_S. By (1.3) and by the uniqueness of the extension, we have

(1.4) $\sum_{a \in {}_2A} E_{a_S}$ is divisible by 2 in $\mathrm{Pic}(X_S)$.

Let

(1.5) $$Y_S \rightarrow X_S$$

be the double covering of X_S branched exactly along $\bigcup E_{a_S}$; Y_S exists by (1.4) and is smooth over S. Let E'_{a_S} denote the curve in Y_S lying over E_{a_S}. Then, by (1.1), Y_S induces over $o \in S$ the surface $\tilde{A}$, and each E'_{a_S} induces the exceptional curve of the first kind $\tilde{E}_a$ on $\tilde{A}$. By Castelnuovo's theorem, there exists an S-morphism

(1.6) $$Y_S \rightarrow A_S,$$

contracting each E'_{a_S} to a single (S-valued) point $a_S \in A_S$. By construction, A_S is smooth and proper over S, and induces the abelian surface A we started from; the 16 sections a_S of A_S over S $(a \in {}_2A)$ induce on A the 16 points a of order 2. Therefore, A_S is an abelian scheme over S with the identity section o_S ([9], Proposition 6.15). Moreover the non-trivial automorphism (, say $\tilde{\tau}$,) of the double covering $Y_S \rightarrow X_S$ induces an involution, say τ, of A_S which fixes a_S $(a \in {}_2A)$ and which induces the inversion automorphism ι on A. Hence τ is the inversion automorphism of A_S over S ([9], Corollary 6.2). Thus we have proved

$$X_S \simeq Y_S/\tilde{\tau} = \mathrm{Km}(A_S).$$

q.e.d.

Now a Kummer surface $X = \mathrm{Km}(A)$ is called <u>special</u> if A is a reducible abelian variety (cf. [12], §4). In this case, there is a surjective homomorphism, say f, of A onto an elliptic curve C with connected fibres, and it induces an elliptic fibration on X (the so-called Kummer pencil):

(1.7)
$$\begin{array}{ccccc} A & \xrightarrow{\pi} & A/\iota & \xleftarrow{\beta} & X \\ f\downarrow & & \bar{f}\downarrow & & \downarrow\varphi \\ C & \xrightarrow{\pi_1} & C/\iota_1 & = & \mathbb{P}^1 \end{array}$$

Here ι_1 is the inversion automorphism on C and π_1 is the quotient morphism of degree 2. The singular fibres of φ lie over the 4 points $\bar{b}_1, \ldots, \bar{b}_4$ of $\mathbb{P}^1$, coming from the 4 points $b_1, \ldots, b_4$ of order 2 on C, and have the following form

(1.8)
$$\varphi^{-1}(\bar{b}_i) = 2F_i + \sum_{a \in {}_2A \cap f^{-1}(b_i)} E_a \quad (1 \leq i \leq 4),$$

in which F_i is the non-singular rational curve on X obtained as the image of $f^{-1}(b_i)$ under $\beta^{-1} \circ \pi$. (Note that (1.3) follows from (1.8) in this case, because all fibres of φ are linearly equivalent to each other.)

Proposition 1.2 With the notation of Proposition 1.1, assume further that $X = Km(A)$ is a special Kummer surface with the Kummer pencil φ associated with $f : A \to C$. Then there exist an elliptic fibration $\varphi_S : X_S \to \mathbb{P}^1_S$ and a surjective S-homomorphism $f_S : A_S \to C_S$, C_S being an elliptic curve over S, such that the diagram (1.7) extends to the similar one over S.

Proof. Applying the reasoning at the beginning of the proof of Proposition 1.1 to elliptic curves in the pencil φ and to the irreducible components of singular fibres of φ, we can find an elliptic fibration $\varphi_S : X_S \to \mathbb{P}^1_S$, inducing φ, which has also 4 singular fibres of type (1.8) over the 4 S-valued points $\bar{b}_{i_S} \in \mathbb{P}^1_S$ (inducing $\bar{b}_i \in \mathbb{P}^1$):

(1.9) $$\varphi_S^{-1}(\bar{b}_{i_S}) = 2F_{i_S} + \sum_{a \in {}_2A \cap f^{-1}(b_i)} E_{a_S} \qquad (1 \leq i \leq 4).$$

Let C_S denote the double covering of $\mathbb{P}^1_S$, branched at the 4 points $\bar{b}_{i_S}$ of $\mathbb{P}^1_S$; $C_S \to S$ is an elliptic curve over S (abelian scheme of relative dimension 1) inducing C. Obviously the formation of this double covering is compatible with the previous one (1.5) so that we have morphisms $g_S : Y_S \to C_S$ and $f_S : A_S \to C_S$ fitting in the commutative diagram:

(1.10) $$\begin{array}{ccccc} X_S & \xleftarrow{(1.5)} & Y_S & \xrightarrow{(1.6)} & A_S \\ \varphi_S \downarrow & & g_S \downarrow & & \downarrow f_S \\ \mathbb{P}^1_S & \longleftarrow & C_S & = & C_S \end{array}$$

Since f_S takes the identity section of A_S to that of C_S, it is a homomorphism of abelian schemes over S. q.e.d.

<u>Proposition</u> 1.3 With the notation of Proposition 1.1, assume further that $X = Km(A)$ with $A = C \times C'$, where C and C' are elliptic curves. Then there exist elliptic curves C_S and C'_S over S, inducing respectively C and C', such that

$$A_S \simeq C_S \underset{S}{\times} C'_S .$$

<u>Proof</u>. Let

$$f : A \to C \quad \text{and} \quad f' : A \to C'$$

be the projections of A to the first and the second factor. By Proposition 1.2, they extend to S-homomorphisms

$$f_S : A_S \to C_S \quad \text{and} \quad f'_S : A_S \to C'_S .$$

Thus we have a homomorphism

(1.11) $$(f_S, f'_S) : A_S \longrightarrow C_S \times_S C'_S .$$

On the other hand, the map

$$h : C \rightarrow A , \quad h(x) = (x, 0)$$

defines a section of f, and it is easily seen that h extends to a section h_S of f_S. Similarly a section h'_S of f'_S is defined. Consider the homomorphism

(1.12) $$h_S + h'_S : C_S \times_S C'_S \longrightarrow A_S .$$

Now the maps (1.11) and (1.12) are the inverse of each other, because they induce such on the special fibre A (cf. [9], Corollary 6.2).

q.e.d.

Theorem 1.4 Let $A = C \times C'$ be a product of two elliptic curves such that $\rho(A) \geq 4$, and let $X = Km(A)$. Then any infinitesimal deformation X_S of X over $S = \operatorname{Spec} k[t]/(t^2)$, such that the Néron-Severi group of X extends to X_S, is trivial.

Proof. By the above propositions, we have

(1.13) $$X_S \simeq Km(A_S), \quad A_S \simeq C_S \times_S C'_S ,$$

where C_S and C'_S are elliptic curves over S, inducing C and C'. Since $NS(X)$ extends to X_S by the assumption, the group $2NS(A)$, at least, also extends to A_S (cf. (3.4) below). Now $NS(A)$ is generated by the curves $C \times o$, $o \times C'$ and the graphs Γ_α ($\alpha \in Hom(C, C')$), and similarly for $NS(A_S)$. Hence it follows that the rank of $Hom_S(C_S, C'_S)$ must be the same as that of $Hom(C, C')$, which is ≥ 2 by the assumption on A (cf. [17], Appendix). Therefore both C_S and C'_S are families of elliptic curves over S with non-trivial endomorphisms. Hence they must be trivial, because the j-invariant of an elliptic curve with non-trivial multiplication is constant. It follows from (1.13) that A_S and X_S must also be trivial. q.e.d.

Remark 1.5 Artin proved the above result when X is the elliptic modular surface of level 4, by using the universal property of this surface ([A], Proposition (7.5)). Note that this is a special case of Theorem 1.4, because this surface is known to be isomorphic to a Kummer surface of the type under consideration (see [17], Theorem 1, and Example 5.1 below).

Proposition 1.6 A Kummer surface has no regular vector field.

Proof. Although this follows from the result of Rudakov and Šafarevič [13], we give a proof because it is elementary in the case of Kummer surfaces. Let θ be a regular vector field on a Kummer surface $\mathrm{Km}(A)$. Then θ can be viewed as a regular vector field on the quotient A/ι minus 16 singular points. Letting $\pi : A \longrightarrow A/\iota$ denote the quotient morphism, $\theta_1 = \pi^*(\theta)$ is a regular vector field on A minus 16 points of order 2, and hence it must be regular everywhere. Moreover θ_1 is invariant under the inversion automorphism ι. Since ι acts on the tangent space to A at o by multiplication by -1 and since $p \neq 2$, we conclude that $\theta_1 = 0$ and $\theta = 0$. q.e.d.

§2. Artin's filtration of the moduli space

Let us briefly recall its definition (for details, see [A], §7; compare [3] for the hypothesis (4.1) of [A]). Let X_0 be a K3 surface in characteristic p, and let $\tilde{M}$ denote the formal versal space for deformation of X_0 as unpolarized surface of characteristic p. Then the said filtration has the form:

$$(2.1) \qquad \tilde{M} \supset M = M_1 \supset \cdots \supset M_{10} \supset M_{11} \supseteq M_\infty \supseteq \Sigma_1 \supset \cdots \supset \Sigma_{10},$$

where M is the closed subset of $\tilde{M}$ defined by some polarization on

X_0, and where M_i and Σ_j are defined by the conditions:

(2.2) $$M_i : h \geq i \qquad (i = 1, \ldots, \infty)$$

(2.3) $$\Sigma_j : \sigma \geq j \qquad (j = 1, \ldots, 10),$$

h and σ being the invariants mentioned in the Introduction. Moreover each member of (2.1) is a closed subscheme of codimension ≤ 1 in its predecessor, and M_{11}, M_∞ and Σ_1 have the same reduced structure.

Now we choose X_0 as follows

(2.4) $$X_0 = \mathrm{Km}(A_0), \quad A_0 = C \times C' ,$$
$$C, C' : \text{supersingular elliptic curves.}$$

This is a supersingular K3 surface if $p > 2$ (cf. [17], Proposition 3, or § 3 below). By Proposition 1.6 and [A], (8.4), the space $\tilde{M}$ is smooth and of dimension 20. Hence we have

(2.5) $$\dim M_i \geq 20 - i \qquad (1 \leq i \leq 10)$$

(2.6) $$\dim \Sigma_j \geq 10 - j \qquad (1 \leq j \leq 10).$$

If we put $j_0 = \sigma(X_0)$, then Theorem 1.4 implies that Σ_{j_0} is the "origin" in $\tilde{M}$ scheme-theoretically. Hence it follows from (2.6) that $j_0 = 10$, and the equalities hold in (2.5) and (2.6). Thus we have proved:

<u>Theorem</u> 2.1 Let X_0 be the supersingular Kummer surface (2.4) in characteristic $p > 2$. Then Artin's filtration of the moduli space of K3 surfaces at X_0 has the following properties:

(2.7) $$\begin{cases} \text{(i)} & \dim M_i = 20 - i \quad (1 \leq i \leq 10) \\ \text{(ii)} & \dim M_\infty = 9 \\ \text{(iii)} & \dim \Sigma_j = 10 - j \quad (1 \leq j \leq 10) \\ \text{(iv)} & d(NS(X_0)) = -p^2 . \end{cases}$$

Corollary 2.2 In every characteristic $p > 2$, the height h of a K3 surface takes on all possible values (0.4) and the invariant σ takes on all possible values (0.6).

In fact, the K3 surface corresponding to the generic point of an irreducible component of M_i (resp. of Σ_j) has the height $h = i$ (resp. the invariant $\sigma = j$). Note that, as is clear from this argument, such a K3 surface is not necessarily defined over the ground field k, but over some finitely generated extension of k.

If we replace K3 surfaces by abelian surfaces in the above consideration, we obtain the analogous results by similar (and easier) methods:

For an abelian surface A, we denote by $h = h(A)$ the height of the formal Brauer group $\hat{B}r(A)$ which is again a 1-dimensional formal group. We have

(2.8) $$h(A) = 1, \quad 2 \quad \text{or} \quad \infty .$$

An abelian surface A is called supersingular if $\rho(A) = b_2(A) = 6$, or equivalently, if A is isogenous to a product of two supersingular elliptic curves ([17] Proposition 3). This condition implies $h = \infty$ by (0.2). If A is supersingular, then the discriminant of the Néron-Severi group is

(2.9) $$d(NS(A)) = -p^{2\sigma_0(A)},$$

where

(2.10) $$\sigma_0(A) = 1 \text{ or } 2.$$

Letting A_0 be as in (2.4), we consider the filtration of the local moduli space of abelian surfaces at A_0, which is similar to (2.1):

(2.11) $$\tilde{M}' \supset M' = M_1' \supset M_2' \supset M_3' \supseteq M_\infty' \supseteq \Sigma_1' \supset \Sigma_2'.$$

Here the formal versal space $\tilde{M}'$ of unpolarized abelian surfaces is smooth and of dimension 4, and M' is the closed set of $\tilde{M}'$ defined by some polarization. The subsets M'_i and Σ'_j are defined by the conditions:

$$M'_i : \ h \geq i \qquad (i = 1, 2, \infty)$$

$$\Sigma'_j : \ \sigma = 3 - \sigma_0 \geq j.$$

Arguing as in the proof of Theorem 2.1, we have

Theorem 2.3 With the above notation, the filtration (2.11) of the moduli space of abelian surfaces at A_0 has the properties:

$$(2.12) \quad \begin{cases} \text{(i)} & \dim M'_1 = 3, \quad \dim M'_2 = 2 \\ \text{(ii)} & \dim M'_\infty = 1 \\ \text{(iii)} & \dim \Sigma'_j = 2 - j \qquad (j = 1, 2) \\ \text{(iv)} & d(NS(A_0)) = -p^2. \end{cases}$$

In the next section, we shall see that the statements (2.7, iv) and (2.12, iv) are equivalent to each other, and give a more direct proof for them.

Remark 2.4 In our Copenhagen talk, we mentioned the following result as a consequence of Theorem 1.4.

(*) "In every characteristic $p > 2$, all possible values (0.3) for the Picard number of a K3 surface are actually taken on."

In writing up this paper, we have found a gap in our arguments. Since we could not fix it in time, let us raise a related problem.

First of all, we have

Proposition 2.5 Suppose that X_R is a smooth projective family of surfaces (or varieties) over a discrete valuation ring R, with the generic fibre X and the special fibre X_0. Assume that the Picard number of X_0 is the same as that of X. Then the natural map, induced by the specialization,

(2.13) $$\varphi : NS(X) \rightarrow NS(X_0)$$

is injective and the cokernel is a finite p-group, where p is the residue characteristic of R.

<u>Proof</u>. Let ξ_0 be an element of $NS(X_0)$ such that a multiple $\ell \cdot \xi_0$ is in the image of $NS(X)$ under φ. Assume that ℓ is a prime number $\neq p$. Let us consider the commutative diagram

(2.14) $$\begin{array}{ccc} NS(X) & \xrightarrow{\varphi} & NS(X_0) \\ \gamma \big\downarrow & & \big\downarrow \gamma_0 \\ H^2_{et}(X, \mathbf{Z}_\ell(1)) & \xrightarrow{\varphi'} & H^2_{et}(X_0, \mathbf{Z}_\ell(1)) \end{array},$$

where γ and γ' are natural injective maps coming from the Kummer theory and the map φ' is an isomorphism, because X_R is smooth and proper over R. The element

$$\eta = \varphi'^{-1}(\gamma_0(\xi_0)) \in H^2_{et}(X, \mathbf{Z}_\ell(1))$$

has the property that $\ell \cdot \eta$ belongs to the image of γ. However it is easily seen that the cokernel of γ is free from ℓ-torsion. Hence there exists an element $\xi \in NS(X)$ such that $\eta = \gamma(\xi)$, and hence $\xi_0 = \varphi(\xi)$. This shows that $\mathrm{Coker}(\varphi)$ is ℓ-torsion free for any prime $\ell \neq p$. The injectivity of φ is also clear from (2.14).

q.e.d.

Now the cokernel of φ can be non-trivial, but the only examples we know are in the case of supersingular surfaces.

<u>Problem</u> 2.6 In the situation of Proposition 2.5, assume further that X_0 is not supersingular, i.e. $\rho(X) = \rho(X_0) < b_2(X_0)$. Is the map φ an isomorphism?

If the answer is affirmative in the case of K3 surfaces, the statement (*) on the Picard numbers can be proved by applying Theorem 1.4 to the case where C and C' are elliptic curves with complex multiplication and by considering a suitable filtration of the moduli space $\tilde{M}$. At any rate, this method gives an <u>algebraic proof</u> of the well-known fact that the Picard number of a K3 surface in characteristic 0 takes on all possible values $\rho = 1, \dots, 20$.

§3. Néron-Severi group of a Kummer surface

In this section, we shall study the relation between the Néron-Severi group of an abelian surface and that of the associated Kummer surface, and give a direct proof for the statement (iv) of Theorem 2.1.

<u>Proposition</u> 3.1 Let A be an abelian surface in characteristic $\neq 2$ and let $X = \mathrm{Km}(A)$. Then the ranks and the discriminants of the Néron-Severi groups of X and A are related by the following:

$$(3.1) \qquad \rho(X) = \rho(A) + 16$$

$$(3.2) \qquad d(NS(X)) = 2^{\nu} d(NS(A)) \quad \text{for some } \nu \in \mathbb{Z}.$$

<u>Proof</u>. By the diagram (1.1), we have

$$(3.3) \qquad \begin{array}{ccc} NS(A) & \overset{\tilde{\beta}^*}{\hookrightarrow} & NS(\tilde{A}) \simeq NS(A) \oplus \sum_{a \in {}_2A} \mathbb{Z}\tilde{E}_a \\ & & \uparrow \tilde{\pi}^* \\ & & NS(X) \end{array}$$

Since $\tilde{E}_a$ $(a \in {}_2A)$ are exceptional curves of the first kind, the discriminants of $NS(\tilde{A})$ and $NS(A)$ are equal. Moreover we have the inclusion relation:

$$(3.4) \qquad NS(\tilde{A}) \supset \tilde{\pi}^* NS(X) \supset 2NS(\tilde{A}),$$

because $\tilde{\pi}^* E_a = 2\tilde{E}_a$ and, for any effective divisor D on A, one has

$$2D \approx D + \iota^*(D) \in \tilde{\pi}^* NS(X) \qquad \text{(cf. [17])}.$$

Hence it follows that $\rho(X) = \rho(\tilde{A})$ and that $d(NS(X))$ differs from $d(NS(\tilde{A}))$ only by a power of 2. q.e.d.

<u>Proposition</u> 3.2 Suppose that A is an abelian surface in characteristic $p \neq 2$ which lifts to an abelian surface A' in characteristic 0 such that $\rho(A') = \rho(A)$. Then the integer ν of (3.2) is given by

$$\nu = b_2(A) - \rho(A) = b_2(X) - \rho(X). \tag{3.5}$$

<u>Proof</u>. The second equality follows from (3.1). First consider the case $p = 0$. Then we may assume that $k = \mathbb{C}$ and $A' = A$, and use the transcendental theory. For a moment, let Y be one of the surfaces A, $\tilde{A}$ or X, and let T_Y denote the lattice of transcendental cycles on Y, i.e., the orthogonal complement of $NS(Y)$ in $H_2(Y, \mathbb{Z})$. Since $H_2(Y, \mathbb{Z})$ is unimodular by Poincaré duality, the discriminants of T_Y and $NS(Y)$ are equal up to sign, and the rank of T_Y is $b_2(Y) - \rho(Y)$. By (1.1), we have the natural map

$$\pi_* : T_A \simeq T_{\tilde{A}} \xrightarrow{\tilde{\pi}^*} T_X . \tag{3.6}$$

Hence the assertion (3.5) follows from the fact that (3.6) is a bijection of T_A onto T_X such that

$$(\pi_*\xi, \pi_*\eta) = 2(\xi, \eta) \qquad (\xi, \eta \in T_A); \tag{3.7}$$

for the latter, we refer to [12], §5, or [19].

Next let us consider the case $p > 2$. By assumption and by Proposition 2.5, we have

$$\begin{cases} d(NS(A)) = p^{\lambda} d(NS(A')) \\ d(NS(X)) = p^{\mu} d(NS(X')) \qquad (X' = Km(A')) \\ b_2(A) - \rho(A) = b_2(A') - \rho(A'). \end{cases} \tag{3.8}$$

Comparing these with (3.2), the assertion reduces to the characteristic 0 case. q.e.d

Remark 3.3 It would be interesting to know whether or not any non-supersingular abelian surface A has a lifting A' such that $\rho(A') = \rho(A)$. Also it would be nice if one could prove the above proposition without using the transcendental theory. In all cases, (3.5) is equivalent to the statement:

$$[NS(\tilde{A}) : \tilde{\pi}^* NS(X)] = 2^{11}. \tag{3.9}$$

The following proposition shows that (3.5) is also true in case A is supersingular. Namely we have

Proposition 3.4 Let A be a supersingular abelian surface in characteristic $p > 2$, and let $X = Km(A)$. Then X is a supersingular K3 surface with

$$d(NS(X)) = d(NS(A)), \tag{3.10}$$

which is a power of p.

Proof. Look at the map γ considered in (2.14):

$$\gamma : NS(A) \otimes \mathbf{Z}_\ell \longrightarrow H^2_{et}(A, \mathbf{Z}_\ell(1)) \quad (\ell \neq p). \tag{3.11}$$

This is an isomorphism in case A is supersingular, because both members have the same rank and the cokernel is free from ℓ-torsion. It follows that $d(NS(A))$ is a power of p, and similarly for $d(NS(X))$ (cf. [A], p.555). Since $p \neq 2$, this implies that $\nu = 0$ in (3.2), and hence (3.10) holds. q.e.d.

Theorem 3.5 (Deligne) Let $g \geq 2$, and let C_i and C_i' $(1 \leq i \leq g)$ be arbitrary supersingular elliptic curves. Then one has

$$C_1 \times \cdots \times C_g \simeq C_1' \times \cdots \times C_g'. \tag{3.12}$$

(This kind of phenomenon, especially the non-validity of cancellation for elliptic curves, was observed earlier by [20]. According to Ogus [11], Deligne's proof makes use of the cristalline (as well as etale) cohomology theory and the strong approximation theorem for a suitable algebraic group over $\mathbb{Q}$. The proof given below is due to Serre, but is reformulated in the way we understand it.)

Proof. Let C be a supersingular elliptic curve, and let $R = \mathrm{End}(C)$, which is a maximal order in a quaternion algebra over $\mathbb{Q}$. For each i, $P_i = \mathrm{Hom}(C, C_i)$ is a projective module of rank 1 over R. Now

$$\mathrm{Hom}(C, C_1 \times \cdots \times C_g) = P_1 \oplus \cdots \oplus P_g$$

is a projective module of rank $g \geq 2$ over R, and hence is free by Eichler's theorem [5]. Therefore we find an R-isomorphism

$$(3.13) \qquad \mathrm{Hom}(C, C_1 \times \cdots \times C_g) \xrightarrow{\sim} \mathrm{Hom}(C, C \times \cdots \times C).$$

Let $h_i \in \mathrm{Hom}(C, C_1 \times \cdots \times C_g)$ be the element corresponding to the embedding of C to the i-th factor of $C \times \cdots \times C$. If we define the homomorphism h of $C \times \cdots \times C$ to $C_1 \times \cdots \times C_g$ by

$$h(x_1, \cdots, x_g) = h_1(x_1) + \cdots + h_g(x_g),$$

it follows from (3.13) that every homomorphism of $C \times \cdots \times C$ to $C_1 \times \cdots \times C_g$ factors through this h. Identifying the products with their duals, we consider the dual homomorphism

$$f = {}^t h : C_1 \times \cdots \times C_g \longrightarrow C \times \cdots \times C.$$

We claim that f is an isomorphism.

Suppose that $a = (a_1, \ldots, a_g) \in \mathrm{Ker}(f)$. Then a is annihilated by all homomorphisms of $C_1 \times \cdots \times C_g$ to $C \times \cdots \times C$. In particular, a_1 is annihilated by all homomorphisms of C_1 to C. Hence it suffices to prove the following:

Lemma 3.6 Let C and C' be supersingular elliptic curves. If $a \in C$ is a common zero of all homomorphisms of C to C', then $a = 0$.

Proof. Since there is an isogeny of C to C', a must be of finite order, say n. We may assume that n is prime. Let ${}_nC$ (resp. ${}_nC'$) denote the subgroup (scheme) of C (resp. C') of points of order n. The restriction map

$$\mathrm{Hom}(C, C') \longrightarrow \mathrm{Hom}({}_nC, {}_nC')$$

induces an isomorphism

$$(3.14) \qquad \mathrm{Hom}(C, C') \otimes \mathbb{Z}/n\mathbb{Z} \xrightarrow{\sim} \mathrm{Hom}({}_nC, {}_nC'),$$

because both members are free $\mathbb{Z}/n\mathbb{Z}$-modules of rank 4. (cf. Lemma 2.1 of [20] for $n \neq p$. Note that this argument works even when $n = p$ — a remark due to N. Katz.) Hence a must be 0. q.e.d.

Proposition 3.7 Let A_0 be the unique supersingular abelian surface which is isomorphic to a product of elliptic curves. Then

$$(3.15) \qquad d(NS(A_0)) = -p^2.$$

Proof. By Theorem 3.5, we may assume that $A_0 = C \times C$ where C is a supersingular elliptic curve. Now it is easily seen that the Néron-Severi group of A is generated by 6 curves $C \times o$, $o \times C$ and Γ_{α_i} $(1 \leq i \leq 4)$, where $\{\alpha_i\}$ forms a basis of $R = \mathrm{End}(C)$ and where Γ_α denotes the graph of $\alpha : C \longrightarrow C$. We have

$$\begin{cases} \Gamma_\alpha(C \times o) = \deg(\alpha) \\ \Gamma_\alpha(o \times C) = 1 \\ \Gamma_\alpha \cdot \Gamma_\beta = \deg(\alpha - \beta) \qquad (\alpha, \beta \in R). \end{cases}$$

We choose the following basis $\{D_i\}$ of $NS(A)$:

$$\begin{cases} D_i = \Gamma_{\alpha_i} - (E \times o) - \deg(\alpha_i)(o \times E) \qquad (1 \leq i \leq 4) \\ D_5 = E \times o, \qquad D_6 = o \times E. \end{cases}$$

Note that, for $1 \leq i, j \leq 4$, we have

$$(D_iD_j) = \deg(\alpha_i - \alpha_j) - \deg(\alpha_i) - \deg(\alpha_j).$$

Hence the intersection matrix of NS(A) has the form

$$\left(\begin{array}{c|cc} -\Delta & \multicolumn{2}{c}{0} \\ \hline \multirow{2}{*}{0} & 0 & 1 \\ & 1 & 0 \end{array}\right), \tag{3.16}$$

where Δ is the intersection matrix of the quadratic form, defined by the norm form "deg" on $R = \mathrm{End}(C)$. Since R is a maximal order of a quaternion algebra over $\mathbb{Q}$ ramified only at p and ∞, we have $\det \Delta = p^2$ (cf. [4]), which proves our proposition. q.e.d.

<u>Corollary</u> 3.8 Let $X_0 = \mathrm{Km}(A_0)$, A_0 being as above. Then

$$d(NS(X_0)) = -p^2. \tag{3.17}$$

<u>Proof</u>. This follows immediately from (3.10) and (3.15). q.e.d.

Thus we have reproved the statement (iv) of Theorems 2.1 and 2.3, obtaining the additional information that the abelian surface A_0 and the Kummer surface $X_0 = \mathrm{Km}(A_0)$ are independent of the factorization: $A_0 \simeq C \times C'$.

<u>Remark</u> 3.9 Another proof of (3.17) can be deduced from Artin-Tate's formula, as proved by Milne [7]. Take a supersingular elliptic curve C defined over the prime field $\mathbb{F}_p$ $(p > 2)$, and consider the Kummer surface $X_0 = \mathrm{Km}(C \times C)$ over the field $k_0 = \mathbb{F}_{p^2}$. The Néron-Severi group $NS_{k_0}(X_0)$ of X_0 relative to k_0 has rank 22. Then the said formula reads:

$$1 = |\mathrm{Br}(X_0/k_0)| \cdot |d(NS_{k_0}(X_0))|/p^2,$$

where $\mathrm{Br}(X_0/k_0)$ is the Brauer group of X_0 over k_0, and its order is known to be a square. By the Hodge index theorem, it follows that

(3.18) $$d(NS_{k_0}(X_0)) = -p^2, \qquad Br(X_0/k_0) = \{0\} .$$

Hence the (absolute) Néron-Severi group $NS(X_0)$ has a basis consisting of k_0-rational divisors, and also has the discriminant $-p^2$.

§4. Uniqueness of supersingular K3 surface with the discriminant $-p^2$

Recently Ogus [11] has obtained some remarkable results on supersingular K3 surfaces, using the cristalline cohomology theory. In particular, he has proved*):

Theorem 4.1 Let X be a supersingular K3 surface in characteristic $p > 2$. Then the equivalence class of the quadratic form on the Néron-Severi group $NS(X)$ is uniquely determined by the invariant σ, or equivalently, by the discriminant.

Theorem 4.2 The supersingular K3 surface X with $d(NS(X)) = -p^2$ is unique up to isomorphisms.

Theorem 4.3 A supersingular K3 surface is a Kummer surface if and only if its invariant σ_0 is 1 or 2 (i.e. its discriminant is $-p^2$ or $-p^4$).

The proof of Theorem 4.1 is based on the cristalline theory and the strong approximation theory. Assuming it, we shall give a proof of Theorems 4.2 and 4.3, which is very close to the results in the preceding sections.

Proof of Theorem 4.2 Let X be any supersingular K3 surfaces with $d(NS(X)) = -p^2$ $(p > 2)$, and let X_0 be as in (2.4). By Theorem 4.1 and Corollary 3.8 (or Theorem 2.1 (iv)), there is an isometry $\varphi : NS(X) \to NS(X_0)$. We may assume without loss of generality

*) These results have independently been proved by Rudakov and Šafarevič.

that φ preserves the effective cycles (cf. [12], §6 or [21], §7).

We shall first show that X is a Kummer surface. Recall that $X_0 = \mathrm{Km}(A_0)$ contains 16 non-singular rational curves E_a $(a \in {}_2A_0)$, cf. (1.2). Each class $\varphi^{-1}(E_a)$ contains an effective divisor, say D_a. If D_a were reducible, then E_a would be linearly equivalent to a reducible effective divisor, and hence $h^0(E_a) \geq 2$, which contradicts the fact that $h^0(E_a) = (E_a^2)/2 + 2 = 1$. Thus D_a is irreducible, and is a non-singular rational curve, since $D_a^2 = E_a^2 = -2$. Moreover these 16 curves D_a are disjoint to each other (because $D_aD_b = E_aE_b = 0$ if $a \neq b$), and $\sum_a D_a$ is divisible by 2 in $\mathrm{Pic}(X)$ since $\sum_a E_a$ has the same property (1.3). By the same argument as in the proof of Proposition 1.1, we have a double covering Y of X ramified along $\bigcup D_a$, and obtain an abelian surface, say B, by blowing down the 16 exceptional curves on Y. Hence $X \simeq \mathrm{Km}(B)$ is a Kummer surface. Furthermore, considering the elliptic fibrations as in the proof of Propositions 1.2 and 1.3, we see that B is isomorphic to a product of elliptic curves $E \times E'$. Since $\rho(X) = \rho(X_0) = 22$, the elliptic curves E and E' are supersingular. Hence, by Theorem 3.5, $B = E \times E'$ is isomorphic to A_0. Therefore $X = \mathrm{Km}(B)$ is isomorphic to $X_0 = \mathrm{Km}(A_0)$. q.e.d.

<u>Proof of Theorem</u> 4.3 The only if part follows from Proposition 3.4 and (2.9). To prove the if part, it suffices to consider the case $\sigma_0 = 2$ because of Theorem 4.2. We take a supersingular Kummer surface X_0 with $\sigma_0 = 2$, whose existence is assured by Theorem 2.3. Let X be arbitrary supersingular K3 surface with $\sigma_0 = 2$. Then, by Theorem 4.1, we can find an isometry of $NS(X)$ to $NS(X_0)$. By the same reasoning as in the proof of Theorem 4.2, we conclude that X is a Kummer surface. q.e.d.

Theorem 4.1 holds true if one replaces K3 surface by abelian surface. Using it, we have

Theorem 4.4 Let A be a supersingular abelian surface. Then A is isomorphic to a product of elliptic curves if and only if $d(NS(A)) = -p^2$.

Proof. The only if part was proved in Proposition 3.7 or in Theorem 2.3. The if part follows from the above variant of Theorem 4.1 and the following:

Lemma 4.5 Let A be an abelian surface in any characteristic. Then A is isomorphic to a product of two elliptic curves if and only if there exist two divisors D_1 and D_2 on A such that

$$D_1^2 = D_2^2 = 0, \quad D_1 D_2 = 1. \tag{4.3}$$

Proof. The only if part is obvious. To prove the converse assume the existence of D_1 and D_2 as above. Letting T_x denote the translation on A by a point $x \in A$, we consider the subgroup scheme of A:

$$K(D_i) = \{x \in A \mid T_x^* D_i \sim D_i\} \quad (i = 1, 2), \tag{4.4}$$

where $\sim$ means linear equivalence. Since $D_1 D_2 \neq 0$, each D_i is not numerically equivalent to 0. Hence $K(D_i) \neq A$, so that $\dim K(D_i) \leq 1$. If $\dim K(D_1) = 0$, then, by the vanishing theorem of Mumford ([10], p.150), there exists a unique q_0 such that

$$H^q(A, \mathcal{O}(D_1)) \neq 0 \Longleftrightarrow q = q_0.$$

But then Riemann-Roch theorem reads

$$\pm \dim H^{q_0}(A, \mathcal{O}(D_1)) = (D_1^2)/2,$$

which contradicts the assumption $D_1^2 = 0$. Thus we have

$$\dim K(D_1) = \dim K(D_2) = 1.$$

Let E_i denote the connected component of the identity of $K(D_i)$; E_i is an elliptic curve on A. On the other hand, the divisor $D = D_1 + D_2$ has the self-intersection number $D^2 = 2$ by (4.3). Therefore the group scheme $K(D)$, defined in the same way as (4.4), is $\{0\}$ scheme-theoretically (cf. [10], loc.cit.). Obviously we have

$$E_1 \cap E_2 \subset K(D_1) \cap K(D_2) \subset K(D).$$

It follows that the natural homomorphism

$$E_1 \times E_2 \to A, \qquad (x_1, x_2) \mapsto x_1 - x_2$$

is an isomorphism. q.e.d.

§5. Examples. Néron-Severi group of the Fermat quartic surface.

Example 5.1 Let $X(p)$ denote the elliptic modular surface of level 4 in characteristic $p \neq 2$. Then

(i) if $p = 0$ or $p \equiv 1 \pmod 4$, then

$$\rho(X(p)) = 20, \qquad d(NS(X(p))) = -4^2;$$

(ii) if $p \equiv -1 \pmod 4$, then

$$\rho(X(p)) = 22, \qquad d(NS(X(p))) = -p^2.$$

Proof. By Theorem 1 of [17], $X(p)$ is isomorphic to the Kummer surface associated with $A(p) = C \times C$, where C is the elliptic curve

$$y^2 = x^4 - 1. \tag{5.1}$$

In case (ii), C is a supersingular elliptic curve, and the assertion follows from Theorem 2.1 (iv) or Corollary 3.8 (or Remark 3.9). In case (i), C is a non-supersingular elliptic curve with complex multiplications and hence $\rho = 20$. Moreover, if $p = 0$ (and the ground field k

is the field of complex numbers), then $A = A(0)$ is a "singular" abelian surface with $d(NS(A)) = -2^2$, because the intersection matrix on the lattice of transcendental cycles $T_A \subset H_2(A, \mathbf{Z})$ is given by $\begin{pmatrix} 2 & 0 \\ 0 & 2 \end{pmatrix}$ (cf. [18]). Hence we have $d(NS(X(0))) = -4^2$ by (3.2). If $p \equiv 1 \pmod 4$, consider the specialization map

$$NS(X(0)) \hookrightarrow NS(X(p)). \tag{5.2}$$

Letting ν be the order of the cokernel, we have

$$d(NS(X(p))) = d(NS(X(0)))/\nu^2 .$$

But ν is a power of p by Proposition 2.5, and hence $\nu = 1$, proving the assertion. (For a different proof, see [15], [16].) q.e.d.

Example 5.2 Let $Y(p)$ denote the Fermat quartic surface

$$x_0^4 + x_1^4 + x_2^4 + x_3^4 = 0 \tag{5.3}$$

in characteristic $p \neq 2$. Then

(i) if $p = 0$ or $p \equiv 1 \pmod 4$, then

$$\rho(Y(p)) = 20, \quad d(NS(Y(p))) = -8^2;$$

(ii) if $p \equiv -1 \pmod 4$, then

$$\rho(Y(p)) = 22, \quad d(NS(Y(p))) = -p^2.$$

Proof. Let C be the elliptic curve (5.1) with the origin $(x, y) = (1, 0)$. By a result of Mizukami [8], $Y(p)$ is the Kummer surface associated with $B(p) = C \times C'$, where C' is the elliptic curve obtained as the quotient of C by the point of order 2 $(-1, 0)$.

In case (ii), we have $A(p) \simeq B(p)$ by Theorem 3.5, or, more directly, by [20], § 3. Hence $Y(p) \simeq X(p)$, and the assertion follows from Example 5.1 (ii).

In case $p = 0$ $(k = \mathbf{C})$, we have

$$d(NS(B(0))) = - \det \begin{pmatrix} 4 & 0 \\ 0 & 4 \end{pmatrix} = - 4^2 ,$$

because

$$C \simeq \mathbb{C}/\mathbb{Z} + \mathbb{Z}\sqrt{-1} \qquad \text{and} \qquad C' \simeq \mathbb{C}/\mathbb{Z} + \mathbb{Z}\cdot 2\sqrt{-1}$$

(cf. [18]). The rest of the proof for (i) is similar to the case (i) of Example 5.1. q.e.d.

Finally we mention (without proof) the following result due to Mizukami and Swinnerton-Dyer:

Proposition 5.3 The Néron-Severi group of the Fermat quartic surface $Y(p)$ is spanned by the classes of lines (= 1-dim. linear subspaces) lying on $Y(p)$ if and only if

$$p = 0, \quad p \equiv 1 \pmod 4 \quad \text{or} \quad p = 3.$$

References

[1] Artin, M.: Supersingular K3 surfaces, Ann. scient. Éc. Norm. Sup. 7 (1974), 543-568 (cited as [A]).

[2] Artin, M. and Mazur, B.: Formal groups arising from algebraic varieties, Ann. scient. Éc. Norm. Sup. 10 (1977), 87-132.

[3] Artin, M. and Milne, J. S.: Duality in the flat cohomology of curves, Inventiones math. 35 (1976), 111-129.

[4] Deuring, M.: Algebren, Springer-Verlag, Berlin-Heidelberg-New York, 1968.

[5] Eichler, M.: Über die Idealklassenzahl hyperkomplexer Systeme, Math. Z. 43 (1938), 481-494.

[6] Grothendieck, A.: Éléments de Géométrie Algébrique III, Publ. Math. 11, 1961.

[7] Milne, J. S.: On a conjecture of Artin and Tate, Ann. of Math. 102 (1975), 517-533.

[8] Mizukami, M.: Birational morphisms from certain quartic surfaces to Kummer surfaces, Master Thesis, Univ. of Tokyo, 1976.

[9] Mumford, D.: Geometric Invariant Theory, Springer-Verlag, Berlin-Heidelberg-New York, 1965.

[10] Mumford, D.: Abelian Varieties, Bombay, Oxford Univ. Press, 1970.

[11] Ogus, A.: Cristaux K3 supersinguliers, Journées de Géométrie Algébrique de Rennes, 3-7 Juillet, 1978.

[12] Pjateckii-Šapiro, I. I. and Šafarevič, I. R.: A Torelli theorem for algebraic surfaces of type K3, Izv. Akad. Nauk SSSR 35 (1971), 530-572.

[13] Rudakov, A. N. and Šafarevič, I. R.: Inseparable morphisms of algebraic surfaces, Izv. Akad. Nauk SSSR 40 (1976), 1269-1307.

[14] Rudakov, A. N. and Šafarevič, I. R.: Supersingular K3 surfaces over a field of characteristic 2, Ibid. 42 (1978), 848-869.

[15] Shioda, T.: On elliptic modular surfaces, J. Math. Soc. Japan 24 (1972), 20-59.

[16] Shioda, T.: On rational points of the generic elliptic curve with level N structure over the field of modular functions of level N, J. Math. Soc. Japan 25 (1973), 144-157.

[17] Shioda, T.: Algebraic cycles on certain K3 surfaces in characteristic p, in "Proc. Int. Conf. on Manifolds (Tokyo, 1973)", Univ. Tokyo Press, 1975.

[18] Shioda, T. and Mitani, N.: Singular abelian surfaces and binary quadratic forms, in "Classification of algebraic varieties and compact complex manifolds", Springer Lecture Notes 412, 1974.

[19] Shioda, T. and Inose, H.: On singular K3 surfaces, in "Complex analysis and algebraic geometry", Iwanami Shoten, Tokyo, and Cambridge Univ. Press, 1977.

[20] Shioda, T.: Some remarks on abelian varieties, J. Fac. Sci. Univ. Tokyo, Sec. IA, 24 (1977), 11-21.

[21] Horikawa, E.: On the periods of Enriques surfaces I, Math. Ann. 234 (1978), 73-88.

Department of Mathematics
Faculty of Science
University of Tokyo
Hongo, Tokyo, JAPAN

Rational singularities in dimension ≥ 2.

by

Robert Treger

Introduction. Rational singularities in dimension greater than two were considered by Kempf [4] and Burns [2].

Definition. Let x be a (normal) point in X = Spec (R), a noetherian affine scheme in characteristic zero. We call x a rational singular point if there is a resolution of singularities $f: X' \to X$ such that $(R^i f_* \mathcal{O}_{X'})_x = 0$, for $i > 0$.

Burns gave a few examples of rational singularities.

1. Simple singularities (over $\mathbb{C}$) [2]: $A_n(m)$: $X_1^2 + \ldots + X_m^2 + X_{m+1}^{n+1} = 0$ $(n \geq 1, m \geq 2)$; $D_n(m)$: $X_1^2 + \ldots + X_{m-1}^2 + X_m(X_m^{n-2} + X_{m+1}^2) = 0$ $(n \geq 4, m \geq 2)$; $E_6(m)$: $X_1^2 + \ldots + X_{m-1}^2 + X_m^3 + X_{m+1}^4 = 0$ $(m \geq 2)$; $E_7(m)$: $X_1^2 + \ldots + X_{m-1}^2 + X_m(X_m^2 + X_{m+1}^3) = 0$ $(m \geq 2)$; $E_8(m)$: $X_1^2 + \ldots + X_{m-1}^2 + X_m^3 + X_{m+1}^5 = 0$ $(m \geq 2)$.

2. A cone over a cubic threefold $V \subset \mathbb{P}^4_{\mathbb{C}}$ has a rational singularity at the origin [2].

3. Let M be a complex manifold, and G a properly discontinuous group of automorphisms of M. Then X = M/G has only rational singularities [2].

In the first two sections of this paper we are studying absolutely isolated Cohen - Macaulay singularities (in any characteristic) satisfying the following numerical condition: (embedding dim.) = (dim.) + (multiplicity) - 1. Presumably such singularities are rational ([10, 5.9.] and § 4). Thus, we exclude cones over cubic threefolds as well as many quotient rational singularities in dimension > 2. We hope the theory of such singularities will find applications in the classification of n-dimensional varieties.

In the first section we give an abstract characterization of the simple singularities. Actually, Theorem 1 was essentially proved by Herszberg in 1957 in the unnoticed paper [3]. Brieskorn informed me about Herszberg's result. In the second section we prove a theorem about tangent cones to the singularities satisfying our numerical condition.

In the last two sections we consider tangent cones to more general singularities. In the section three we construct a scheme which parametrize tangent cones to the singularities satisfying more general numerical conditions and give an application to the deformation theory. In the last section we generalize some results of Wahl [10] as well as results of section two.

This paper was influenced by papers of Lipman [6] and Wahl [10].

I would like to thank The Mittag-Leffler Institut for its support during the final stage of the preparation of this paper.

§ 1. Our first result is a generalization of an old result of D. Kirby [5]. As we mentioned in the introduction, Theorem 1 was essentially proved by Herszberg.

A singularity is called <u>absolutely isolated</u> iff it can be resolved by successive blowing up of closed points.

<u>Theorem 1</u>. Let (R, M) be a complete local Cohen - Macaulay k-algebra of dimension $\nu \geq 2$, where k is an algebraically closed field of characteristic zero, $k \cong R/M$. Suppose the multiplicity of R is e = 2. Then R is a simple singularity iff it is absolutely isolated.

<u>Proof</u>. Exactly as in the case $\nu = 2$, one can check that simple singularities are absolutely isolated.

To prove the converse assertion, we can suppose that $R = k[[X_1, \ldots, X_u]]/I$ where u is the embedding dimension of R. By Abhyankar's lemma (see, also, Lemma 2), $u \leq e + \dim. (R) - 1 = \nu + 1$, hence $u = \nu + 1$. Obviously, X = Spec (R) has an isolated singularity.

Furthermore, X satisfies Serre's condition S_2 and R_1. Hence, it is a normal domain and I = (f) where f is a power series with the leading form of order 2.

We proceed by induction on ν. For $\nu = 2$, we can apply the result of Kirby [5]. Suppose $\nu > 2$. After changing variables and applying the Weierstrass Preparation Theorem we can suppose $f = y_1^2 + y_1 p_1 + p_o$ where $p_o, p_1 \in k[[y_2, \ldots, y_{\nu+1}]]$. By setting $x_1 = y_1 + p_1/2$, $x_i = y_i$ for $i > 0$, we can make f of the form: $f = x_1^2 + g(x_2, \ldots, x_{\nu+1})$. We claim that the leading form of g is of order two. Indeed, assuming converse, we blow up the maximal ideal of R. If $g = ax_2^3 + \ldots$, $a \neq 0$, we consider the blowing up via x_3. We obtain a new hypersurface $f' = z_1^2 + z_3(az_2^3 + \ldots)$ where $z_3 = x_3$, $z_i x_3 = x_i$ for $i \neq 3$. Singularities of this hypersurface contain a set of dimension ≥ 1, a contradiction. If the coefficient $a = 0$ we consider the blowing up via x_2. As before, we get a contradiction.

Consider, now, $Y = \mathrm{Spec}\,(k[[x_2, \ldots, x_{\nu+1}]]/g) \subset X$. It is easy to see that singularities of Y can be resolved by successive blowing up of closed points. Therefore, induction hypothesis imply the theorem.

Remark 1. An easy induction shows that simple singularities are rational (see, also, § 4). On the other hand, there are many rational hypersurface singularities in dimension > 2 with multiplicity 2 which are not the simple singularities [9].

§ 2. In this section we suppose the ground field k to be infinite.

Lemma 1. Let $V^d \subset \mathbb{P}_k^\nu = \mathbb{P}$ be a connected purely d-dimensional ($d > 0$) arithmetically Cohen - Macaulay projective subscheme. Suppose V^d satisfies the following condition:

(*) there exist $d - 1$ hyperplane sections $H_1, \ldots, H_{d-1}$ such that $H^1(V \cap H_1 \cap \ldots H_{d-1}, O) = 0$ and $V \cap H_1 \cap \ldots H_i$ are arithmetically Cohen - Macaulay of pure dimension $d - i$, for $1 \leq i \leq d - 1$.

Then $H^d(V, O_V(1 - d)) = 0$. Conversely, if $H^d(V, O_V(1 - d)) = 0$ then for $d - 1$ generic sections $H_1, \ldots, H_{d-1}$, $H^1(V \cap H_1 \cap \ldots H_{d-1}, O) = 0$.

Proof. The lemma is trivial for d = 1. Suppose d > 1. Consider an exact sequence

$$0 \to O_V(-1) \to O_V \to O_{V\cap H_1} \to 0$$

where H_1 is a generic hyperplane. For any $p \geq 2$ this sequence yields the cohomology sequence

$$H^{d-1}(O_{V\cap H_1}(p-d)) \to H^d(O_V(p-1-d)) \to H^d(O_V(p-d)) \to H^d(O_{V\cap H_1}(p-d)) .$$

We claim that $H^{d-1}(O_{V\cap H_1}(p-d)) = 0$, for $p \geq 2$, so $H^d(O_V(1-d)) = H^d(O_V(p-d))$, hence $H^d(O_V(1-d)) = 0$. Let $I = I_{V\cap H_1} \subset O_{\mathbb{P}^{\nu-1}}$ be the ideal of the subscheme $V \cap H_1 \subset \mathbb{P}^{\nu-1}$. The exact sequence

$$0 \to I(m-i) \to O_{\mathbb{P}^{\nu-1}}(m-i) \to O_{V\cap H_1}(m-i) \to 0 , \quad i, m > 0 ,$$

yields the exact sequence

$$H^i(O_{\mathbb{P}^{\nu-1}}(m-i)) \to H^i(O_{V\cap H_1}(m-i)) \to H^{i+1}(I(m-i)) \to H^{i+1}(O_{\mathbb{P}^{\nu-1}}(m-i))$$

Induction hypothesis and Serre's theorem on cohomology of $O_{\mathbb{P}^n}$ imply that I is 2-regular in the sence of Castelnuovo [7]. Hence $H^i(O_{V\cap H_1}(m-i)) = H^{i+1}(I(m-i)) = 0$, for $i > 0$, $m \geq 1$. The lemma is proved.

The next lemma is an easy generalization of one Lipman's result in [6].

Lemma 2. Let (R, M) be a local Cohen - Macaulay ring with infinite residue field k. Let dim (R) = d + 1, emb dim (R) = ν + 1, mult (R) = e. Let $\{x_1, \ldots, x_{d+1}, \ldots, x_{\nu+1}\}$ be a minimal bases of M such that $\{x_1, \ldots, x_{d+1}\}$ is a minimal reduction of M. The following conditions are equivalent:

(i) $M^2 = M(x_1, \ldots, x_{d+1})$,

(ii) $\nu + 1 = e + d$,

(iii) Ker $(k[x_1, \ldots, x_{\nu+1}] \overset{\Theta}{\to} \oplus_{n\geq 0} M^n/M^{n+1})$ is generated by $(\nu - d)(\nu - d + 1)/2$ linearly independent quadrics of the form:

$$Q_{ij} = x_i x_j + \Sigma_{l=d+2}^{\nu+1} \Sigma_{s=1}^{d+1} d_{ijls} x_l x_s + q_{ij}(x_1, \ldots, x_{d+1}) ,$$

$$d + 2 \leq i \leq j \leq \nu + 1 .$$

Proof (i) $\Longleftrightarrow$ (ii) (cf. [6] or [8]). Since e = length $(R/(x_1, \ldots, x_{d+1}))$, $d + 1$ = length $((x_1, \ldots, x_{d+1})/M(x_1, \ldots, x_{d+1}))$, $\nu + 1$ = length (M/M^2), we have $\nu + 1 \leq d + e$ and the equality holds iff $M^2 = M(x_1, \ldots, x_{d+1})$.

(i), (ii) $\Longleftrightarrow$ (iii). The kernel of Θ contains $(\nu - d)(\nu - d - 1)/2$ linearly independent forms Q_{ij}. Let us show that the ideal $I = \{Q_{ij}\}$ = Ker (Θ). Let $S_2 = \oplus_{n \geq o} S_{2,n} \subset S = k[x_1, \ldots, x_{\nu+1}]$ be the graded $k[x_1, \ldots, x_{d+1}]$-submodule generated by $\nu - d + 1$ element $1, x_{d+2}, \ldots, x_{\nu+1}$. The canonical map $S \to S/I$ restricts to a surjection $\Theta_2 \colon S_2 \to S/I$. We need only check that $\bar{\Theta} \cdot \Theta_2 \colon S_2 \to \oplus_{n \geq o} M^n/M^{n+1}$ is injective where $\bar{\Theta}$ is induced by the map Θ.

We have $\dim S_{2,n} = \binom{n+d}{d} + (\nu - d)\binom{n-1+d}{d} = (\nu - d + 1)n^d/d! + \ldots$ If we had a non-zero element in the Ker $(\bar{\Theta} \cdot \Theta_2)$, then $\dim (M^n/M^{n+1}) \leq (\nu - d)n^d/d! + \ldots$, for $n >> 0$, contradicting (ii). We omit the proof of (iii) $\Rightarrow$ (i).

Corollary 1. (i) The Hilbert function of R is

$$H_R(n) = \binom{n+d}{d} + (\nu - d)\binom{n-1+d}{d} = \dim (M^n/M^{n+1}) , \quad \text{for all } n \geq 0 .$$

(ii) (J. Sally, [8]). The ring $\oplus_{n \geq o} M^n/M^{n+1}$ is Cohen - Macaulay.

The Corollary follows immediately from the proof of Lemma 2.

Theorem 2. Let (R, M) be a local ring of dimension $d + 1 \geq 2$ and emb dim. $(R) = \nu + 1$ with infinite residue field. Let V = Proj $(\bar{R})$ where $\bar{R} = \oplus_{n \geq o} M^n/M^{n+1}$.

Suppose R, $\bar{R}$ are Cohen - Macaulay rings. If condition (*) (see Lemma 1) is satisfied then it is satisfied with $H_i = \{x_i = 0\}$ where $x_1, \ldots, x_{d-1}$ is a part of a regular system of parameters for R. Furthermore, the equivalent conditions (i) - (iii) of Lemma 2 hold.

Conversely, if R is Cohen - Macaulay then the conditions (i) - (iii) imply (iv) $H^d(V, O_V(1 - d)) = 0$, (V)$H^1(V \cap H_1 \cap \dots H_{d-1}, O) = 0$, and $V \cap H_1 \cap \dots H_i$ are arithmetically Cohen - Macaulay schemes of pure dimension d - i where $H_i = \{x_i = 0\}$ $(1 \leq i \leq d - 1)$.

Thus, if R, $\bar{R}$ are Cohen - Macaulay rings then the conditions (i) - (v) are equivalent.

Proof. If $H_i = \{x_i = 0\}$ then $V \cap H_1 \cap \dots H_i$ are arithmetically Cohen - Macaulay of pure dimension d - i. Hence, Lemma 1 implies the first assertion. We, now, prove that $M^2 = (x_1', \dots, x_{d+1}')M$. By Lemma 1, $H^1(V \cap H_1 \cap \dots H_{d-1}, O) = 0$. By induction, it is enough to prove the assertion for the ring $R' = R/(x_1, \dots, x_{d-1})$. In this situation we can apply Lipman's result ([6, the proof of (1.8)]; his proof is valid for non-normal rings), which gives the assertion in the two-dimensional case.

Now, we shall prove the converse. Consider the hyperplanes $H_i = \{x_i = 0\}$. By the theorem of J. Sally (see Corollary 1 (ii)), $\bar{R}$ is Cohen - Macaulay, and $V \cap H_1 \cap \dots H_i$ are arithmetically Cohen - Macaulay of pure dimension d - i. So, we left to prove $H^1(V \cap H_1 \cap \dots H_{d-1}) = 0$. We can suppose dim (R) = 2. In this case we can apply Lipman's remark [6, p. 163]. By Corollary 1 (i), $\dim (M^n/M^{n+1}) = \nu n + 1$, for $n \geq 0$. Hence, if $W \to \operatorname{Spec}(R)$ is the blowing up of $M \subset R$ then $H^1(W, O) = 0$. The Theorem is proved.

Corollary 2. With notation as in Theorem 2, suppose R is Cohen - Macaulay and satisfies the numerical condition: emb dim = mult. + dim. - 1. Assume R = P/I, where P is a regular local k-algebra of dimension $\nu + 1$; $\bar{R} = \bar{P}/\bar{I}$ where $\bar{P} = k[X_1, \dots X_{\nu+1}]$. Then there exist minimal projective resolutions:

$$(1) \quad 0 \to P^{b_{\nu-d}} \to \dots \to P^{b_2} \xrightarrow{\varphi_2} P^{b_1} \xrightarrow{\varphi_1} P \to P/I \to 0 ,$$

$$(2) \quad 0 \to \bar{P}^{b_{\nu-d}} \to \dots \quad \dots \to \bar{P}^{b_1} \xrightarrow{\bar{\varphi}_1} \bar{P} \to \bar{P}/\bar{I} \to 0 ,$$

such that (a) (2) is the associated graded complex attached to (1);

(b) $\bar{\varphi}_i$ is homogeneous of degree 1 ($i > 1$) or 2 ($i = 1$); (c) the b_i's depend only on multiplicity and are inductively computable from the equality: $(1-t)^{\nu+1} \Sigma_{n=o}^{\infty} [\binom{n+d}{d} + (\nu - d)\binom{n-1+d}{d}]t^n = 1 + \Sigma_{i=1}^{\nu-d} (-1)^i b_i t^{i+1}$.

Proof. It follows immediately from Wahl's theorem [10, Theorem 1.7], Theorem 2 and Corollary 1.

§ 3. We shall now consider a more general situation. Fix a number $m \geq 2$. Let $S = k[x_1, \ldots, x_{d+1}, \ldots, x_{\nu+1}]$, $\nu > d > 0$, be a polynomial ring over an infinite field k. We denote by S_m a graded $k[x_1, \ldots, x_{d+1}]$-submodule of S generated by 1; $x_{d+2}, \ldots, x_{\nu+1}$; $x_{i_1 i_2} = x_{i_1} x_{i_2}$ $(d+2 \leq i_1 \leq i_2 \leq \nu+1)$; ...; $x_{i_1 \ldots i_{m-1}} = x_{i_1} \ldots x_{i_{m-1}}$ $(d+2 \leq i_1 \leq \ldots \leq i_{m-1} \leq \nu+1)$. The Hilbert function of S_m is

$$H_m(n) = H_{S_m}(n) = \binom{n+d}{d} + \Sigma_{t=1}^{m-1} \binom{t+\nu-d-1}{\nu-d-1}\binom{n+d-t}{d} .$$

So, $H_m(n) = e \cdot n^d/d! +$ lower terms. We shall call $e = e(d, \nu, m) = \Sigma_{t=o}^{m-1}\binom{t+\nu-d-1}{\nu-d-1}$ the multiplicity of S_m.

Lemma 3. With (R, M) as in Lemma 2, suppose $M^m = M^{m-1}(x_1, \ldots, x_{d+1}) (\mathrm{mod}\ M^{m+1})$ and $e = e(d, \nu, m)$. Then the natural homomorphism $\Theta: S \to \bar{R} = \oplus M^n/M^{n+1}$ induces the isomorphism of graded $k[x_1, \ldots, x_{d+1}]$-modules $\Theta_m: S_m \rightleftarrows \bar{R}$.

Proof. Since $M^n = M^{m-1}(x_1, \ldots, x_{d+1})^{n+1-m} (\mathrm{mod}\ M^{n+1})$, the map Θ_m is surjective. If we had an element $f \in \mathrm{Ker}\,(\Theta_m)$, $f \neq 0$, $\deg(f) = q$, then, as in the proof of Lemma 2, $e < e(d, \nu, m)$, a contradiction.

Remark 2. If $e = e(d, \nu, 2)$ then $e = 1 + \nu - d$. Hence, by Lemma 2, $M^2 = M(x_1 \ldots, x_{d+1})$. Thus, Θ_2 is an isomorphism.

Lemma 4. Let $S_B = B[x_1, \ldots, x_{d+1}, \ldots x_{\nu+1}]$ be a polynomial ring over a k-algebra B. Fix a number $m \geq 2$. Consider an ideal $I_B = \{Q_{(i_1 \ldots i_m)}\} \subset S_B$ generated by

$$Q_{(i_1 \ldots i_m)} = x_{i_1} \ldots x_{i_m} + \varphi_{(i_1 \ldots i_m)}, \quad d+2 \leq i_1 \leq \ldots \leq i_m \leq \nu+1,$$

where $\varphi_{(i_1 \dots i_m)}$ are elements of degree m in the $B[x_1 \dots x_{d+1}]$-module $S_{B,m} = B \otimes_k S_m$. The following conditions are equivalent:

(i) the coefficients $\{a_\omega\}_{\omega \in \Omega_m}$ involved in all $\varphi_{(i_1 \dots i_m)}$ satisfy relations R_m arising from the following equalities in S_B:

$$\varphi_{(i_1 \dots \hat{i}_1 \dots i_{m+1})} \circ x_{i_1} = \varphi_{(i_1 \dots \hat{i}_s \dots i_{m+1})} \circ x_{i_s} ,$$

$$d + 2 \leq i_1 \leq \dots \leq i_{m+1} \leq \nu + 1 ,$$

where (by definition) $x_{i_1} \dots x_{i_{m-1}} \circ x_j = - \varphi_{(i_1 \dots i_k j i_{k+1} \dots i_{m-1})}$ if $i_k \leq j \leq i_{k+1}$, and the other operations are as in S_B.

(ii) the natural map Θ: $S_B \to S_B/I_B$ induces an isomorphism Θ_m: $S_{B,m} \cong S_B/I_B$, which provides $S_{B,m}$ with a ring structure $\circ$.

The conditions (i) and (ii) imply that all closed fibres of the map: Spec $(S_B/I_B) \to$ Spec (B) are Cohen - Macaulay of dimension d + 1 and embedding dimension $\nu + 1$, moreover, $I \cap S_{B,m} = \{0\}$.

<u>Proof</u> (ii) $\Rightarrow$ (i). The elements $(Q_{(i_1 \dots \hat{i}_1 \dots i_{m+1})} - x_{i_1 \dots \hat{i}_1 \dots i_{m+1}})x_{i_1}$

$$- (Q_{(i_1 \dots \hat{i}_s \dots i_{m+1})} - x_{i_1 \dots \hat{i}_s \dots i_{m+1}})x_{i_s},$$

$d + 2 \leq i_1 \leq \dots \leq i_{m+1} \leq \nu + 1$, lie in I_B. Therefore $\varphi_{(i_1 \dots \hat{i}_1 \dots i_{m+1})} \circ x_{i_1} = \varphi_{(i_1 \dots \hat{i}_s \dots i_{m+1})} \circ x_{i_s}$ in the ring $(S_{B,m}, \circ)$. By expending these equalities we obtain the relations R_m.

(i) $\Rightarrow$ (ii). We can introduce in $S_{B,m}$ a multiplication $\circ$ as follows:

$$(x_1 \dots x_{m-1}) \circ x_m = - \varphi_{(i_1 \dots i_m)} .$$

An easy induction shows that $\circ$ can be uniquely extended. Obviously, the map Θ_m is surjective. We can define a B-algebra homomorphism f: $S \to S_m$ such that $f \cdot g = 1$, where g: $S_m \to S$, by giving it, first, on the generators of the free commutative B-algebra S. Since $I \subseteq$ Ker (f), these ideals coincide. The lemma is proved.

<u>Proposition 1</u>. With notation as in Lemma 4, the functor $\$_m$: <u>Aff/k</u> $\to$ <u>Sets</u>, Spec (B) $\mapsto$ {the set of graded ring structures on $S_{B,m}$ preserving the given $k[x_1 \dots x_{d+1}]$-module structure} is representable by

the affine scheme Spec (A_m) where $A_m = k[a_\omega;\ \omega \in \Omega_m]/R_m$.

<u>Proof</u>. We must show $\mathcal{S}_m(B) = \mathrm{Hom}_k(A_m, B)$. To define a graded ring structure on $S_{B,m} = B \otimes_k S_m$ is the same as to choose an isomorphism Θ_m, as in Lemma 4 (ii), for some ideal $I_B \subset S_B$. This defines a B-point of Spec (A_m). On the other hand, by Lemma 4, each such point defines a ring structure. Now, it is obvious that $S_m(\ldots)$ and $\mathrm{Hom}_k(A_m, \ldots)$ are isomorphic as functors.

We shall denote Spec (A_m) by $\mathcal{S}_m$.

<u>Notation</u> (cf. [1]): $m \geq 2$; (B, N) is a local ring; $C = B[[x_1 \ldots \ldots x_{d+1} \ldots x_{\nu+1}]]$; $\mathcal{J} = \{Q_{(i_1 \ldots i_m)} + \psi_{(i_1 \ldots i_m)}\} \subset (x)C$ where $\psi_{(i_1 \ldots i_m)}$ are power series of order $> m$ in $x_1, \ldots, x_{\nu+1}$ and $Q_{(i_1 \ldots i_m)}$ are polynomials as in Lemma 4; $R_B = C/\mathcal{J}$.

We assume the natural map p: Spec $(R_B) \to$ Spec (B) is a Cohen - Macaulay morphism.

<u>Proposition 2</u>. With the above notation and assumption, if the multiplicities of all closed fibres of p are equal to $e(d, \nu, m)$, for fixed d, ν, m, then there is a natural isomorphism

$$S_{B,m} \cong \mathrm{Gr}_{(x)}(R_B) .$$

In particulary, $\mathrm{Gr}_{(x)}(R_B)$ is flat over B.

<u>Proof</u>. Apply Lemma 3.

Wahl conjectured [10, Conjecture 5.2] that every two-dimensional rational singularity R is a normally flat specialization of a cone (see also [10, 5.3. - 5.7.4.]).

<u>Corollary 3</u>. If $C = \mathrm{Proj}\ (\bar{R})$ is smoothable then the conjecture is valid.

<u>Proof</u>. If C is smoothable, it is smoothable in $\mathcal{S}_2$. Therefore, Proposition 2 implies the conjecture for a smoothable C.

§ 4. We shall generalize results of [10]. Let $V \subset \mathbb{P}_k^\nu = \mathbb{P}$ be a connected purely d-dimensional projective subscheme, with $\nu > d > 0$. Let $I \subset O_{\mathbb{P}}$ is the ideal sheaf of V. Assume $V \subset \mathbb{P}$ is arithmetically

Cohen - Macaulay. For simplicity we shall assume that d = 1. Our results can be generalized (see, also, Lemma 1).

Lemma 5. With the above notation and assumptions, the following conditions are equivalent for any $m \geq 2$:

(i) $H^2(I(m-2)) = 0$; (i') $H^1(O_V(m-2)) = 0$; (ii) I is m-regular in the sense of Castelnuovo, i.e., $H^i(I(m-i)) = 0$, all $i > 0$.

Proof. Consider an exact sequence

$$0 \to I(m-i) \to O_{\mathbf{P}}(m-i) \to O_V(m-i) \to 0\ , \qquad i > 0\ .$$

This sequence yields an exact cohomology sequence

$$\ldots \to H^{i-1}(O_V(m-i)) \to H^i(I(m-i)) \to H^i(O_{\mathbf{P}}(m-i)) \to H^i(O_V(m-i)) \to \ldots$$

Hence, (i) ~ (i'). Obviously, (ii) ⇒ (i). Further, since $V \subset \mathbf{P}$ is arithmetically Cohen - Macaulay, $H^1(I(n)) = 0$, for $n \geq 0$, and $H^i(O_V(n)) = 0$, $i > 1$, all n. Suppose $i \geq 3$. Then $H^i(I(m-i)) = H^i(O_{\mathbf{P}}(m-i)) = 0$, by Serre's theorem.

Proposition 3. With the above notation and assumptions, if $H^1(O_V(m-2)) = 0$ for the fixed number $m \geq 2$ and no k-form, $k < m$, vanishes on O_V then the affine cone over V, $\bar{R}$ = Cone (V), has a minimal graded projective resolution

$$0 \to \bar{P}^{b_{\nu-d}} \to \ldots \to \bar{P}^{b_1} \xrightarrow{\varphi_1} \bar{P} \to \bar{R} \to 0$$

where $\bar{P} = k[x_1, \ldots, x_{\nu+1}]$ and φ_i is homogeneous of degree $1 (i > 1)$ or $m (i = 1)$. Furthermore, the Betti numbers b_i are inductively computable from $H(n) = \dim H^0(O_V(n))$, viz,

$$(1-t)^{\nu+1}\ \Sigma_{n=0}^{\infty}\ H(n)t^n = 1 + \Sigma_{i=1}^{\nu-d}\ (-1)^i b_i t^{i+1}\ .$$

The proof is analogues to the Wahl's proof of [10, Proposition 1.4.]. We omit details.

Lemma 6. With the above notation and assumption, if $\chi(O_V(m-2)) = H_m(m-2)$ and $H^0(I(m-2)) = 0$, $m \geq 2$, then V is m-regular in the sense of Castelnuovo [7, § 14].

Proof. By definition, $\chi(O_V(m-2)) = h^o(O_V(m-2)) - h^1(O_V(m-2))$. Since $H^k(I(m-2)) = 0$, for $k = 0, 1$, $h^o(O_V(m-2)) = h^o(O_{\mathbf{P}^\nu}(m-2)) = \binom{m-2+\nu}{\nu}$. Further, $H_m(m-2) = \binom{m-2+\nu}{\nu}$ (see the beginning of § 3; $d = 1$). Hence $H^1(O_V(m-2)) = 0$, so, Lemma 6 follows from Lemma 5.

Proposition 4. If V, as above, is corresponding to some $(S_m, \circ)$, $m \geq 2$, then $\chi(O_V(m-2)) = H_m(m-2)$. Moreover, $(S_m, \circ)$, for any multiplication $\circ$, $m \geq 2$, $d = 1$, are m-regular in the sense of Castelnuovo [7, § 14].

Proof. It is well-known that $H_m(n)$ and $\chi(O_V(n))$ are polynomial functions for $n \geq m-2$. Since they coinside for $n >> 0$, they coinside for $n = m-2$, and we can apply Lemma 6.

Elsewhere, we shall construct an explicit minimal projective resolution of the universal ideal $I_{A_m} \subset S_{A_m} = A_m[x_1, \ldots, x_{\nu+1}]$ (see Lemma 4).

Remark 3. We can generalize Wahl's results [10, Propositions 2.3 and 3.2]. Instead of rational singularities one should consider (absolutely isolated) singularities with tangent cones isomorphic to some $(S_m, \circ)$, $m \geq 2$. We omit details.

The following conjecture is a generalization of Wahl's conjecture ([10, Conjecture 5.9]; see also [10, 5.13]).

Conjecture. Any absolutely isolated singularity (R, M), with the tangent cone isomorphic to some $(S_2, \circ)$, is rational.

Reduction. Using Theorem 2, Leray spectral sequence (see also [10, p. 258]) and induction, we can easily reduce the Conjecture to the following equivalent

Conjecture. Let X = Spec (R) be a two-dimensional absolutely isolated Cohen - Macaulay singularity with multiplicity e(R) = emb dim (R) - 1. Then c(R') = emb dim (R') - 1, where (R', M) is a local ring of a singular point x' in the first blowing up f: X' → X.

Proposition 5. Suppose $x' \in X'$ is a non-smooth point. Denote by B the one-dimensional Cohen - Macaulay ring R'/MR'. If emb dim (B) = e(B)

then emb dim (R') = e(R') + 1.

Proof. Case 1: $MR' \not\subset M'^2$. Then emb dim (R') = emb dim (B) + 1. Since emb dim (R') - 1 $\leq$ e(R') $\leq$ e(B) = emb dim (B) = emb dim (R') - 1, emb dim (R') = e(R') + 1.

Case 2: $MR' = aR' \subset M'^2$ ($a \in R'$). Then emb dim (R') = emb dim (B) = e(B). We claim that e(B) $\geq$ e(R') + 1, hence they are equal.

Let $w \in R'$ be an element such that its image $\bar{w} \in B$ is a superficial element. Then, the image of a, $\bar{a} \in R'/wR'$, is not a zero divisor in the ring $A = R'/wR'$. Since A is a Cohen - Macaulay ring, $e(\bar{a}A)$ = length $(A/\bar{a}A)$ where $e(\bar{a}A)$ is the multiplicity of the ideal $\bar{a}A \subset A$ [11, Macaulay rings, Theorem 3]. If $e(R'/MR') \leq e(R')$ then they are equal. Hence, $e(\bar{a}A)$ = length $(A/\bar{a}A)$ = length $(R'/(a, w))$ = $e(R'/MR')$ = $e(R') \leq e(A)$. By a known result, $e(\bar{a}A) = e(A)$ = length $(A/\bar{a}A)$ and $\bar{a}$ is a superficial element of A. Since $\bar{a} \in \bar{M}'^2 \subset A$, a contradiction.

Corollary 4. If the fibre $f^{-1}(M)$ is reduced then emb dim (R') = e(R') + 1.

Elsewhere, we shall return to the conjecture as well as its generalization to (S_m, o), for $m \geq 2$.

References

[1] B. Bennett, Normally Flat Deformation. Trans. A.M.S., vol. 225, 1977 (1-58).

[2] D. Burns, On Rational Singularities in Dimension > 2. Math. Ann. 211, 1974 (237-244).

[3] J. Herszberg, Classification of isolated double points of rank zero on primals in S_n. J. of the London Math. Soc., vol. 32, 1957 (198-203).

[4] G. Kempf, F. Knudsen, D. Mumford and B. Saint-Donat. Toroidal Embeddings I. Springer Lecture Notes in Math. 339, Berlin-Heidelberg -New York, 1973.

[5] D. Kirby, The Structure of Isolated Multiple Points of a Surface, II. Proc. London Math. Soc., vol. 7, 1958 (1-28).

[6] J. Lipman, Desingularization of two-dimensional schemes. Ann. of Math., 107, 1978 (151-207).

[7] D. Mumford, Lectures on Curves on an Algebraic Surface, Annals of Math. Studies, vol. 59, Princeton, 1966.

[8] J. Sally, On the Associated Graded Ring of a Local Cohen - Macaulay Ring, J. Math. Kyoto Univ., 17, No. 1, 1977 (19-22).

[9] E.Viehweg, Rational singularities of higher dimensional schemes, Proc. A.M.S., 63, 1977 (6-8).

[10] J. Wahl, Equations Defining Rational Singularities. Ann. Sc. Éc. Norm. Sup., 4e série, t. 10, 1977 (231-264).

[11] O. Zariski and P. Samuel, Commutative Algebra, vol. II, Princeton, N.J, 1960.

Institut Mittag-Leffler
Auravägen 17
S-18262 Djursholm
Sverige

DEFORMATIONS AND LOCAL TORELLI THEOREM
FOR CERTAIN SURFACES OF GENERAL TYPE

by Sampei USUI

Table of contents.

Introduction.

After the systematic investigation by Griffiths on the period maps [3], several researches have been made on the problem of their injectivity (Torelli-type problem) and of their surjectivity (for K3 surfaces, cf. [17], [10]; for Enriques surfaces, cf. [11]; for surfaces of general type, cf. II in [3], [15], [16], [20], [21], [12] etc.).

The purpose of this paper is to show the following: Let X' be a smooth projective surface of general type obtained by the normalization of a hypersurface X in a projective 3-space P only with ordinary singularities. Let D be the singular locus of X with reduced structure and let n be the degree of X in P. Then the period map is unramified at the origin of the parameter space of the Kuranishi family of the deformations of X' in one of the following cases:

(1) D is a complete intersection in P. (In this case we have few exceptions. For detail, see the theorem (3.5).)

(2) n is sufficiently large enough comparing to D. (See the theorem (5.8).)

The result in case (1) contains some examples of minimal surfaces of general type with non-ample canonical divisor, for which the local Torelli theorem holds. The result in case (2) gives some evidence that if there would be sufficiently many 2-forms on a given surface, their periods of integrals should determine the surface itself (cf. the remark (5.9)).

We recall here the definition of a surface X in a projective 3-space P only with ordinary singularities: Taking a suitable local coordinate system (x, y, z) at each point in P, the local equation of X in P is one of the following forms:

(i) 1,

(ii) z,

(iii) yz,

(iv) xyz,

(v) $xy^2 - z^2$.

These surfaces are attractive because every smooth projective surface can be obtained as the normalization of such a surface X. More precisely, in characteristic 0, via generic projection, every smooth projective surface can be projected onto such a surface X. These surfaces, especially their deformations, are intensively studed by Kodaira in [13] and, when their singular loci are smooth curves of complete intersections in the ambiant space, by Horikawa in [9] and by Tsuboi in [19].

We also recall that, given a smooth projective surface Y, the morphism

$$H^1(T_Y) \longrightarrow H^1(\Omega_Y^1) \otimes H^0(\Omega_Y^2)^{\vee}$$

induced from the contraction $T_Y \otimes \Omega_Y^2 \xrightarrow{\sim} \Omega_Y^1$ is called the infinitesimal period map at Y in the second cohomology (for the background, cf. [3], [20]).

This work was started on the joint research with Professor S. Tsuboi at the Research Institute for Mathematical Sciences in Kyoto previous year. The author expresses his hearty thanks to Professor S. Tsuboi and Professor K. Miyajima and the other professors at Kagoshima University who received him warmly in the previous summer.

March 10, 1978.

Notatios and conventions.

The category, which we treat, is schemes over the field $\mathbb{C}$ of complex numbers.

$h^i(F) = \dim_{\mathbb{C}} H^i(F)$ for a coherent $\underline{O}_X$-module F.

$\check{F} = \underline{Hom}_{\underline{O}_X}(F, \underline{O}_X)$ for a coherent $\underline{O}_X$-module F.

$\check{V} = Hom_{\mathbb{C}}(V, C)$ for a $\mathbb{C}$-vector space V.

$|L|$ denotes the complete linear system associated to an invertible $\underline{O}_X$-module L.

$Bs|L|$ denotes the set of the base points of $|L|$.

S_a and M_a denote the set of homogeneous elements of degree a of a graded ring S and that of a graded module M respectively $(a \in \mathbb{Z})$.

Ω_f denotes the sheaf of relative Kähler differentials for a morphism f of schemes.

$\Omega_X = \Omega_f$, where $f : X \longrightarrow \operatorname{Spec} \mathbb{C}$ is the structure morphism.

ω_X denotes the dualizing $\underline{O}_X$-module of a scheme X.

$$C(a) = \begin{cases} \dfrac{(a+3)(a+2)(a+1)}{6} & \text{if } a \text{ is a non-negative integer,} \\ 0 & \text{if } a \text{ is a negative integer.} \end{cases}$$

1. preliminaries.

In this section we summarize the preparatory results for the later use.

(1.1) Let P denote the projective 3-space. Let X be a hypersurface in P only with ordinary singularities, D be its double curve, that is, $D=\mathrm{Sing}(X)$ and T be the triple points of X, namely, $T=\mathrm{Sing}(D)$. Let $f: X'\longrightarrow X$ be the normalization. We set $D'=f^{-1}(D)$, $g=\mathrm{res}(f): D'\longrightarrow D$, $T'=f^{-1}(T)$ and n, d and t being the degrees of X, of D and of T respectively.

(1.2) We also use the following notations.

ω_X, ω_D, $\omega_{X'}$: the dualizing sheaves of X, of D and of X' respectively. Note that $\omega_X=\underline{\mathrm{Ext}}^1_{\underline{O}_P}(\underline{O}_X,\omega_P)\simeq\underline{O}_X(n-4)$.

$\underline{O}_X(a-bD)=\mathrm{Im}\{\underline{O}_X(a)\otimes\mathcal{J}_D^b\longrightarrow\underline{O}_X(a)\}$ $(a, b\in\mathbf{Z}$ and $b>0)$, where $\mathcal{J}_D$ denotes the $\underline{O}_P$-ideal of D.

$\underline{O}_{X'}(a)=f^*\underline{O}_X(a)$ $(a\in\mathbf{Z})$.

$\underline{O}_{D'}(a)=g^*\underline{O}_D(a)$ $(a\in\mathbf{Z})$.

$\underline{O}_{X'}(a-bD')=\underline{O}_{X'}(a)\otimes\underline{O}_{X'}(-D)^{\otimes b}$ $(a, b\in\mathbf{Z})$.

The following lemma can be found in [18].

Lemma (1.3) (Roberts).

(1.3.1) D is locally Cohen-Macauley and of pure codimension 1 in X.

(1.3.2) $0\longrightarrow\underline{O}_X\longrightarrow f_*\underline{O}_{X'}\longrightarrow\omega_D\otimes\check{\omega}_X\longrightarrow 0$ exact.

Lemma (1.4) (Kodaira).

(1.4.1) $\omega_{X'}\simeq\underline{O}_{X'}((n-4)-D')$.

(1.4.2) $f_*\underline{O}_{X'}(a-bD')\simeq\underline{O}_X(a-bD)$ $(a, b\in\mathbf{Z}$ and $b>0)$.

Proof. (1.4.1) is just the adjunction formula. (1.4.2) is obtained by a direct computation by using the local coordinates mentioned in the introduction. More precisely see Kodaira [13]. QED.

(1.5) Let $\mathcal{N}_{X/P}$ be the coherent $\underline{O}_{X'}$-module introduced by Horikawa (in [9], his notation is $\mathcal{T}_{X/P}$), which is defined for making the following sequence exact:

(1.5.1) $0\longrightarrow T_{X'}\longrightarrow f^*(T_P\otimes\underline{O}_X)\longrightarrow\mathcal{N}_{X/P}\longrightarrow 0.$

Let $\mathcal{J}_X$ be the $\underline{O}_P$-ideal of X. Since $\mathcal{J}_X/\mathcal{J}_X^2\simeq\underline{O}_X(-n)$, we have the exact sequence

(1.5.2) $$0 \longrightarrow \underline{O}_X(-n) \longrightarrow \Omega_P \otimes \underline{O}_X \longrightarrow \Omega_X \longrightarrow 0.$$

Dualizing (1.5.2), we get the exact sequence

$$0 \longrightarrow T_X \longrightarrow T_P \otimes \underline{O}_X \longrightarrow \underline{O}_X(n) \xrightarrow{\alpha} \underline{\mathrm{Ext}}^1_{\underline{O}_X}(\Omega_X, \underline{O}_X) \longrightarrow 0,$$

where we denote $\underline{\mathrm{Hom}}_{\underline{O}_X}(\Omega_X, \underline{O}_X)$ by T_X. We define $\mathcal{N}_{X/P} = \mathrm{Im}\{T_P \otimes \underline{O}_X \xrightarrow{\alpha} \underline{O}_X(n)\}$, which is nothing but the coherent $\underline{O}_X$-module introduced by Kodaira (in [13], he uses the notation Ψ). Similarly we denote $\underline{\mathrm{Hom}}_{\underline{O}_D}(\Omega_D, \underline{O}_D)$ by T_D and $\mathrm{Coker}\{T_D \longrightarrow T_P \otimes \underline{O}_D\}$ by $\mathcal{N}_{D/P}$. $\mathcal{N}_{D/P}$ is just the sheaf N in Kodaira [13]. In case that D is smooth, we also use the notation $N_{D/P}$ for $\mathcal{N}_{D/P}$.

The following results can be found in [9] and in [13].

<u>Lemma</u> (1.6).

(1.6.1) $f_*\mathcal{N}_{X/P} \cong \mathcal{N}_{X/P}$ (Horikawa).

(1.6.2) $0 \longrightarrow \underline{O}_X(n-2D) \longrightarrow \mathcal{N}_{X/P} \longrightarrow \mathcal{N}_{D/P} \longrightarrow 0$ exact (Kodaira).

The following formulae are calculated in [22] (cf. also [23]):

<u>Lemma</u> (1.7). X' has the following numerical characters:

(1.7.1) $c_1^2 = n(n-4)^2-(5n-24)d-4\chi(\underline{O}_D)+t.$

(1.7.2) $c_2 = n(n^2-4n-6)-(7n-24)d-8\chi(\underline{O}_D)-t.$

(1.8) We summarize here the results concerning the spectral sequence of the Koszul complex introduced by Lieberman-Wilsker-Peters in [14].

Let Y be a complete smooth scheme. Let M be an invertible $\underline{O}_X$-module, V be a subspace of $H^0(M)$ and E be a locally free $\underline{O}_Y$-module. Choose a basis f_1, ..., f_m of V and let e_1, ..., e_m be the dual basis. For the triple (M, V, E) we denote by $K^{\cdot}(M, V, E)$ the Koszul complex consisting of the $\underline{O}_Y$-modules

$$K^p(M, V, E) = (E \otimes_{\underline{O}_Y} M^{\otimes p}) \otimes_{\mathbb{C}} \overset{p}{\wedge} V$$

together with the coboundary maps defined by

$$d(x) = \sum_{i_1<\dots<i_{p+1}} (\sum_j (-1)^j f_{i_j} x_{i_1\dots\hat{i}_j\dots i_{p+1}}) e_{i_1} \wedge \dots \wedge e_{i_{p+1}}$$

for $x = \sum_{i_1<\dots<i_p} x_{i_1\dots i_p} e_{i_1} \wedge \dots \wedge e_{i_p} \in K^p(M, V, E)$.

The E_2 terms of the spectral sequence of the hypercohomology of this complex

$K^{\cdot}(M, V, E)$ with respect to the first filtlation are given by

$$'E_2^{p,q}(M, V, E) = \frac{\operatorname{Ker}\{H^q(E\otimes M^{\otimes p}) \otimes \bigwedge^{p} \check{V} \longrightarrow H^q(E\otimes M^{\otimes(p-1)}) \otimes \bigwedge^{p+1}\check{V}\}}{\operatorname{Im}\{H^q(E\otimes M^{\otimes(p-1)}) \otimes \bigwedge^{p-1} \check{V} \longrightarrow H^q(E\otimes M^{\otimes p}) \otimes \bigwedge^{p}\check{V}\}}.$$

Since $'E_2^{p,q}(M, V, E) = 0$ $(p<0$ or $q<0)$ by definition, we have the following well-known exact sequence:

$$(1.8.1) \qquad 0 \longrightarrow E_2^{1,0} \longrightarrow \mathbb{H}^1 \longrightarrow E_2^{0,1} \longrightarrow E_2^{2,0} \longrightarrow \mathbb{H}^2.$$

<u>Lemma</u> (1.9) (Lieberman-Wilsker-Peters). If $\operatorname{codim}_Y \operatorname{Bs}|M| \geqslant 2$, then we have $\mathbb{H}^1(M, V, E) = 0$.

<u>Proof</u>. This lemma is proved easily by observing the E_2 terms of the spectral sequence with respect to the second filtration (for detail, cf. [14]). QED.

We conclude this section by adding one more lemma which is easy to prove and is rather useful.

<u>Lemma</u> (1.10). Let p be an integer with $p \neq m$. If $h^0(E\otimes M^{\otimes p}) \leqslant 1$, then we Have $'E_2^{p,0}(M, V, E) = 0$.

<u>Proof</u>. In case $h^0(E\otimes M^{\otimes p}) = 0$, it is trivial. We assume $h^0(E\otimes M^{\otimes p}) = 1$. Let t be a basis of $H^0(E\otimes M^{\otimes p})$ and let ρ be the map making the following diagram commutative:

$$\begin{array}{ccc} H^0(E\otimes M^{\otimes p}) \otimes \bigwedge^{p}\check{V} & \xrightarrow{\ d\ } & H^0(E\otimes M^{\otimes(p+1)}) \otimes \bigwedge^{p+1}\check{V}. \\ {\scriptstyle t\otimes}\ \uparrow\ \wr & \nearrow_{\rho} & \\ \bigwedge^{p}\check{V} & & \end{array}$$

Explicitly ρ is given by

$$\rho(y) = \sum_{i_1<\dots<i_{p+1}} \Big(\sum_j (-1)^j y_{i_1\dots\hat{i}_j\dots i_{p+1}}\, t\cdot f_{i_j}\Big) e_{i_1}\wedge\dots\wedge e_{i_{p+1}} \qquad (y_{i_1\dots i_p} \in \mathbb{C}).$$

Since $M \xrightarrow{t\otimes} E\otimes M^{\otimes(p+1)}$ is injective, $H^0(M) \xrightarrow{t\cdot} H^0(E\otimes M^{\otimes(p+1)})$ is injective, and hence $t\cdot f_i$ $(1 \leqslant i \leqslant m)$ are linearly independent in $H^0(E\otimes M^{\otimes(p+1)})$. Therefore $\rho(y) = 0$ implies $y = 0$. This completes the proof. QED.

2. <u>deformations</u>.

In this section we study the small deformations of the surface X' in (1.1) when D is a smooth curve of complete intersection in P.[1)]

(2.1) If D is a curve of complete intersection in P, then the homogeneous equation of the hypersurface X in P is the following form:

$$AF^2 + 2BFG + CG^2,$$

where F, G, A, B and C are also homogeneous polynomials in $\mathbb{C}[X_0, \ldots, X_3]$. In this case D is given by $F = G = 0$.

Let I be the homogeneous $\mathbb{C}[X_0, \ldots, X_3]$- ideal generated by F and G. We set $n_1 = \deg F$ and $n_2 = \deg G$. We may assume $n_1 \geq n_2$ because of symmetry.

Note that $N_{D/P} \simeq \underline{O}_D(n_1) \oplus \underline{O}_D(n_2)$ and that $\omega_D \simeq \underline{O}_D(n_1+n_2-4)$.

From now on, except explicitly indicating the contrary, we assume that the surface X' is the normalization of a surface X just mentioned above.

(2.2) Let $P' \longrightarrow P$ be the blowing-up of P along D and let E' be the exceptional divisor. Then we have the following diagram:

$$\begin{array}{ccccccc} P' & \supset & X' & \supset & D' & \subset & E' \\ \downarrow & & \downarrow f & & \downarrow g & \swarrow \tilde{g} & \\ P & \supset & X & \supset & D. & & \end{array}$$

Since D is smooth, $\tilde{g} : E' = \mathrm{Proj}(\check{N}_{D/P}) \longrightarrow D$ is a $\mathbb{P}^1$-bundle. We denote by $L = \underline{O}_{P'}(-E') \otimes \underline{O}_{E'}$ the tautological invertible sheaf of the $\mathbb{P}^1$-bundle $\tilde{g}$. Note that $\tilde{g}_* L^{\otimes a} \simeq S^a(\check{N}_{D/P})$ $(a \in \mathbb{Z})$, where $S^a(\ )$ denotes the a-th symmetric tensor product.

<u>Lemma</u> (2.3). The surface X' has the following numerical characters.

$$q = 0.$$

$$p_g = C(n-n_1-4) + C(n-n_2-4) - C(n-n_1-n_2-4).$$

$$c_1^2 = n(n-4)^2 + n_1 n_2 \{2(n_1+n_2) + 16 - 5n\}.$$

We obtain the following table:

n	3	4	4	5	5	5	6	6	7	
n_1	1	1	2	1	2	2	2	3	3	otherwise
n_2	1	1	1	1	1	2	2	2	3	
p_g		0		2	1	0	2	1	2	$4 \leq$
c_1^2	+	0	+	0	-1	+		0		+
X'				not general type						general type

.

Proof. From (1.3.2), we have an exact sequence

$$H^1(\underline{O}_X) \longrightarrow H^1(f_*\underline{O}_{X'}) \longrightarrow H^1(\omega_D \otimes \check{\omega}_X) \xrightarrow{\beta} H^2(\underline{O}_X).$$

Since $H^1(\underline{O}_X) = 0$ and the dual map of β is surjective, we see that $H^1(f_*\underline{O}_{X'}) = 0$, that is, $q = 0$. By (1.4.1) and (1.4.2), we obtain that

$$H^0(\omega_{X'}) \cong H^0(\underline{O}_X((n-4)-D)) \cong I_{n-4}.$$

From this, we get the formula for p_g. The formula for c_1^2 is the direct consequence of (1.7.1). As for the last line of the table we will check them case by case. In case $n = 2n_1$, it is easy to see that $\omega_{X'} \cong \underline{O}_{X'}(n-4-n_2)$ and hence X' is of general type if and only if $n-4-n_2 > 0$. In case $(n, n_1, n_2) = (3, 1, 1)$, $(4, 1, 1)$ or $(5, 2, 2)$, the direct calculation shows that the 2-ple genus of X' is zero and hence X' is not of general type by the criterion of Kodaira. In case $(n, n_1, n_2) = (5, 1, 1)$, $(6, 2, 2)$ or $(7, 3, 3)$, we have, by (1.4.1), $\omega_{X'} \cong \underline{O}_{X'}(n_1 - D')$ and hence $D_1' = f^{-1}(F=0) - D'$ and $D_2' = f^{-1}(G=0) - D'$ form a basis of the complete linear system $|\omega_{X'}|$. If there would exist an exceptional curve of the first kind on X', say C', $(C' \cdot D_1') = (C' \cdot D_2') = -1$ and hence $C' \subset D_1' \cap D_2'$. This is a contradiction, since it is easy to see that $D_1' \cap D_2' \neq \emptyset$ imposes the existence of a point x in P satisfying $A(x) = B(x) = C(x) = F(x) = G(x) = 0$ and hence the existence of a point on X with its multiplicity ≥ 3. Thus, in the considering case, X' is relatively minimal and its $c_1^2 = 0$ whence X' is not of general type by the criterion of Kodaira. In case $(n, n_1, n_2) = (5, 2, 1)$, $|\omega_{X'}|$ has only one element, namely, $D_2' = f^{-1}(G=0) - D'$, which is isomorphic to the line $A = G = 0$ in P via res(f) and hence D_2' is the exceptional curve of the first kind. Contracting D_2', we get the relatively minimal model whose canonical invertible sheaf is trivial. The last assertion in the table, corresponding to "otherwise", can be easily verified. QED.

Lemma (2.4).

(2.4.1) $H^1(\underline{O}_{X'}(a-bD')) = 0$ $(a \in \mathbb{Z},\ b = 1 \text{ or } 2)$.

(2.4.2) $H^2(\underline{O}_{X'}(a-2D')) = 0$ if and only if $a > \max\{n-4,\ 2n_1+n_2-4\}$.

Proof. We use the following exact sequences:

$$0 \longrightarrow \underline{O}_X(a-2D) \longrightarrow \underline{O}_X(a-D) \longrightarrow \check{N}_{D/P} \otimes \underline{O}_D(a) \longrightarrow 0.$$

$$0 \longrightarrow \underline{O}_X(a-D) \longrightarrow \underline{O}_X(a) \longrightarrow \underline{O}_D(a) \longrightarrow 0.$$

Since $H^0(\underline{O}_X(a-D)) \longrightarrow H^0(\check{N}_{D/P} \otimes \underline{O}_D(a))$ and $H^0(\underline{O}_X(a)) \longrightarrow H^0(\underline{O}_D(a))$ are surjective and since $H^1(\underline{O}_X(a)) = 0$, we obtain (2.4.1). By (2.4.1), we get that

$$\begin{aligned} h^2(\underline{O}_X(a-2D)) &= h^2(\underline{O}_X(a)) + h^1(\underline{O}_D(a)) + h^1(\check{N}_{D/P} \otimes \underline{O}_D(a)) \\ &= h^0(\underline{O}_X(n-4-a)) + h^0(\underline{O}_D(n_1+n_2-4-a)) + h^0(\underline{O}_D(2n_1+n_2-4-a)) \\ &\quad + h^0(\underline{O}_X(n_1+2n_2-4-a)). \end{aligned}$$

This proves (2.4.2). QED.

Lemma (2.5). Let δ be the connecting homomorphism

$$\delta : H^0(\mathcal{N}_{X/P}) \longrightarrow H^1(T_{X'})$$

obtained from the exact sequence (1.5.1). We have the following table:

(2.5.1)

<table>
<tr><td>n</td><td>5</td><td>6</td><td>7</td><td>5</td><td>6</td><td>6</td><td>7</td><td>7</td><td>8</td><td>8</td><td>8</td><td rowspan="3">otherwise</td></tr>
<tr><td>n_1</td><td>1</td><td>2</td><td>3</td><td>2</td><td>3</td><td>2</td><td>3</td><td>3</td><td>4</td><td>4</td><td>4</td></tr>
<tr><td>n_2</td><td>1</td><td>2</td><td>3</td><td>1</td><td>2</td><td>1</td><td>1</td><td>2</td><td>1</td><td>2</td><td>3</td></tr>
<tr><td>$h^1(f^*(T_P \otimes \underline{O}_X))$</td><td colspan="3">2</td><td colspan="8">1</td><td>0</td></tr>
<tr><td>$h^1(\mathcal{N}_{X/P})$</td><td colspan="8">0</td><td colspan="3">1</td><td></td></tr>
<tr><td>δ</td><td colspan="8">not surjective</td><td colspan="4">surjective</td></tr>
</table>

.

Proof. By the duality theorem and by (1.4), we get that

$$h^1(f^*(T_P \otimes \underline{O}_X)) = h^1(f^*(\Omega_P \otimes \underline{O}_X) \otimes \omega_{X'}) = h^1(\Omega_P \otimes \underline{O}_X((n-4)-D)).$$

Now we use the following exact sequence:

$$0 \longrightarrow \Omega_P \otimes \underline{O}_X((n-4)-D) \longrightarrow \underline{O}_X((n-5)-D)^{\oplus 4} \longrightarrow \underline{O}_X((n-4)-D) \longrightarrow 0.$$

By using (2.4.1), we get that

$$H^1(\Omega_P \otimes \underline{O}_X((n-4)-D)) \cong \mathrm{Coker}\{I_{n-5}^{\oplus 4} \longrightarrow I_{n-4}\}.$$

From this, we can fill up the table concerning $h^1(f^*(T_P \otimes \underline{O}_X))$, by an elementary calculation. By (1.6) and by (2.4), we have, in the cases in the table (2.5.1) that

$$H^1(\mathcal{N}_{X/P}) \cong H^1(\mathcal{N}_{X/P}) \cong H^1(N_{D/P}).$$

From this we can complete the table concerning $h^1(\mathcal{N}_{X/P})$. As for the surjectivity of δ, Horikawa proved it in case $n = 2n_1$ and $n-n_2-4 > 0$ in [9]. QED.

(2.6) Let R denote the localization of $\mathbf{C}[T_1, \ldots, T_m]$ by the maximal ideal $\mathfrak{m}=(T_1, \dots, T_m)$, where

$$m = C(n_1)+C(n_2)+C(n-2n_1)+C(n-n_1-n_2)+C(n-2n_2).$$

Put $S = \operatorname{Spec} R$. Let $\tilde{F}$, $\tilde{G}$, $\tilde{A}$, $\tilde{B}$ and $\tilde{C}$ be the first order perturbations of F, of G, of A, of B and of C respectively, namely, $\tilde{F}=F+F_1$, where $F_1 = \sum_{1\le i\le C(n_1)} M_i T_i$ ($M_1, \dots, M_{C(n_1)}$ are the monomials of degree n_1 in $\mathbf{C}[X_0, \dots, X_3]$) etc. Let $\mathfrak{X}$ and $\mathfrak{D}$ be the subschemes in $P\times S$ defined by $\tilde{A}\tilde{F}^2+2\tilde{B}\tilde{F}\tilde{G}+\tilde{C}\tilde{G}^2=0$ and by $\tilde{F}=\tilde{G}=0$ respectively and let $\mathfrak{X}'$ be the blowing-up of $\mathfrak{X}$ along $\mathfrak{D}$ (or equivalently the normalization of $\mathfrak{X}$). Then we have the natural morphism:

$$(2.6.1) \qquad \mathfrak{X}' \longrightarrow S.$$

<u>Theorem</u> (2.7). Let X' be a surface of general type which is the normalization of such a hypersurface X as in (2.1). Except the cases

(2.7.1)

n	6	7	7
n_1	2	3	3
n_2	1	1	2

,

(2.6.1) gives a complete family of the deformations of X'. In particular, except the cases in the table (2.7.1), the parameter space of the Kuranishi family of the deformations of X' is smooth at the origin.[2)]

<u>Proof</u>. The map $\tau : T_S\otimes k(\mathfrak{m}) \longrightarrow H^0(\mathcal{N}_{X/P})$ is given by

$$\tilde{A}\tilde{F}^2+2\tilde{B}\tilde{F}\tilde{G}+\tilde{C}\tilde{G}^2 \quad \operatorname{mod} \mathfrak{m}^2,$$

that is, for $s\in T_S\otimes k(\mathfrak{m})\simeq \mathbf{A}^m$,

$$\tau(s)=(A_1F^2+2B_1FG+C_1G^2+2AFF_1+2BF_1G+2BFG_1+2CGG_1)\otimes k(\mathfrak{m}) \quad \operatorname{mod} (AF^2+2BFG+CG^2),$$

and it is easily verified that τ is surjective. Tensoring non-zero $\omega\in H^0(\omega_{X'})$ gives an injective morphism $T_{X'} \xrightarrow{\otimes\omega} T_{X'}\otimes\omega_{X'}\simeq\Omega_{X'}$ and hence an injection $H^0(T_{X'}) \longrightarrow H^0(\Omega_{X'})$. Since $q=0$ by (2.3), we see $H^0(T_{X'})= H^0(\Omega_{X'})=0$. The other assertions in the theorem are the consequences of (2.5) and of $H^0(T_{X'})=0$ by the general theory of deformations. QED.

<u>Remark</u> (2.8). The argument in this section also holds in the case that D is globally Cohen-Macauley, that is,

$$H^0(\underline{O}_D(a)) \simeq (\mathbb{C}[X_0, \ldots, X_3]/I)_a \qquad (a \in \mathbb{Z}),$$

where $I = \bigoplus_{a \in \mathbb{Z}} H^0(\mathcal{J}_D(a))$. Hence, in this case, the last statement of (2.7) is valid (several cases occur according to the degrees of the generators of I).

3. local Torelli theorem.

In this section we assume that X' is a surface of general type which is the normalization of such a surface X as in (2.1). We give a proof of the local Torelli theorem for such a surface X'.

Lemma (3.1).

(3.1.1) $|\underline{O}_{X'}(1)|$ is fixed points free.

(3.1.2) $|\underline{O}_{X'}(n_1 - D')|$ is fixed components free. [3)]

Proof. (3.1.1) is obvious. By (1.4.2), we have

$$H^0(\underline{O}_{X'}(n_1 - D')) \cong H^0(\underline{O}_X(n_1 - D)) \simeq I_{n_1}.$$

Put $D_1' = f^*(F=0) - D'$ and $D_2' = f^*(G=0) - D'$. Then

$$|\underline{O}_{X'}(n_1 - D')| = \{C' + D_2' \mid C' = f^*(C),\ C \in |\underline{O}_X(n_1 - n_2)|\} + \{D_1'\}.$$

Since $\bigcap_{C'} (C' + D_2') = D_2'$ and since $f^*(F=0) \cap f^*(G=0) = D'$, we get the assertion (3.1.2). QED.

Remark (3.2). In case $n = 2n_1$, it is easy to see that

$$\omega_{X'} \simeq \underline{O}_{X'}(n - n_2 - 4).$$

On the other hand, in case $n \neq 2n_1$, $\omega_{X'}$ is not a non-trivial power of an invertible sheaf.

The next lemma is the essential part in the proof of the local Torelli theorem. The proof of the lemma will be found in the next section.

Lemma (3.3).

(3.3.1) $$h^0(\Omega_{X'} \otimes \underline{O}_{X'}(1)) = \begin{cases} \leq 1 & \text{if } n = n_1 + n_2 + 1, \\ 0 & \text{otherwise.} \end{cases}$$

(3.3.2) In case $n \neq 2n_1$, except the case $n = 2n_1 + 1$ and $n_2 = 1$, we have

$$H^0(T_{X'} \otimes \underline{O}_{X'}(2n_1 - 2D')) = 0.$$

Lemma (3.4). In case $n_1 > n_2 + 1$, the map

$$\mathrm{Im}(\delta) \longrightarrow H^1(T_{X'} \otimes \underline{O}_{X'}(n_2 - D'))$$

is injective, where δ is the map in (2.5).

Proof. By (1.6.1), it is easy to see that

$$(3.4.1) \qquad f_*(\mathcal{N}_{X/P} \otimes \underline{O}_{X'}(n_2-D')) \cong \mathcal{N}_{X/P}\cdot \underline{O}_X(n_2-D),$$

where $\mathcal{N}_{X/P}\cdot \underline{O}_X(n_2-D) = \mathrm{Im}\{\mathcal{N}_{X/P} \otimes \underline{O}_X(n_2-D) \longrightarrow \mathcal{N}_{X/P} \otimes \underline{O}_X(n_2)\}$. By (1.5.1), (1.6.1) and (3.4.1), we have a commutative diagram:

$$\begin{array}{ccc}
H^1(T_{X'}) & \longrightarrow & H^1(T_{X'} \otimes \underline{O}_{X'}(n_2-D')) \\
{\scriptstyle \text{induced from } \delta}\uparrow & & \uparrow{\scriptstyle \gamma} \\
\dfrac{H^0(\mathcal{N}_{X/P})}{\mathrm{Im}\{H^0(T_P \otimes \underline{O}_X) \longrightarrow H^0(\mathcal{N}_{X/P})\}} & \xrightarrow{\ \psi\ } & \dfrac{H^0(\mathcal{N}_{X/P}\cdot \underline{O}_X(n_2-D))}{\mathrm{Im}\{H^0(T_P \otimes \underline{O}_X(n_2-D)) \longrightarrow H^0(\mathcal{N}_{X/P}\cdot \underline{O}_X(n_2-D))\}}.
\end{array}$$

Since γ is injective, it is enough to show that ψ is injective. The injectivity of ψ follows from the following assertion:

(3.4.2) Put $\Phi = AF^2 + 2BFG + CG^2$. If we assume that, for a given Ψ in $\mathbb{C}[X_0, \dots, X_3]_n$, there exist P_i in I_{n_2+1} $(0 \leq i \leq 3)$ satisfying $G\Psi \equiv \sum_{0\leq i\leq 3} P_i \frac{\partial \Phi}{\partial X_i} \bmod \Phi$, then there exist Q_i in $\mathbb{C}[X_0, \dots, X_3]_1$ $(0 \leq i \leq 3)$ such that $\Psi \equiv \sum_{0\leq i\leq 3} Q_i \frac{\partial \Phi}{\partial X_i} \bmod \Phi$.

The assertion (3.4.2) is trivially valid, because $I_{n_2+1} = G\mathbb{C}[X_0, \dots, X_3]_1$ by the assumption $n_1 > n_2 + 1$ and because G is a regular element modulo Φ. QED.

Theorem (3.5). If X' is a surface of general type which is the normalization of a hypersurface X defined in (2.1), then the local Torelli theorem holds for X' except the case $(n, n_1, n_2) = (7, 3, 1)$. As for this exception, still the map

$$\mathrm{Im}(\delta) \longrightarrow H^1(\Omega_{X'}) \otimes H^0(\omega_{X'})^{\vee}$$

is injective, where δ is the map in (2.5).

Proof. We use the notations in (1.8). By (3.3.1) and (1.10), we get

$${}'E_2^{2,0}(\underline{O}_{X'}(1), H^0(\underline{O}_{X'}(1)), T_{X'} \otimes \underline{O}_{X'}(i-D')) = 0 \qquad (i < n-4),$$

and by (3.1.1) and (1.9), we have

$$\mathbb{H}^1(\underline{O}_{X'}(1), H^0(\underline{O}_{X'}(1)), T_{X'} \otimes \underline{O}_{X'}(i-D')) = 0 \qquad (\forall i),$$

and hence by (1.8.1) we see that

$${}'E_1^{1,0}(\underline{O}_{X'}(1), H^0(\underline{O}_{X'}(1)), T_{X'} \otimes \underline{O}_{X'}(i-D')) = 0 \qquad (i < n-4).$$

Namely

(3.5.1) $\quad H^1(T_{X'}\otimes \underline{O}_{X'}(i-D')) \longrightarrow H^1(T_{X'}\otimes \underline{O}_{X'}(i+1-D'))\otimes H^0(\underline{O}_{X'}(1))^\vee$

is injective $(i < n-4)$. In case $n \neq 2n_1$, except $n = 2n_1+1$ and $n_2 = 1$, a similar argument applied for $(M, V, E) = (\underline{O}_{X'}(n_1-D'), H^0(\underline{O}_{X'}(n_1-D')), T_{X'})$ deduces, by using (3.3.2), (1.10), (3.1.2) and (1.9), that

(3.5.2) $\quad H^1(T_{X'}) \longrightarrow H^1(T_{X'}\otimes \underline{O}_{X'}(n_1-D'))\otimes H^0(\underline{O}_{X'}(n_1-D'))^\vee$

is injective. In case $n = n_2+1$, the map

(3.5.3) $\quad \operatorname{Im}(\delta) \longrightarrow H^1(T_{X'}\otimes \underline{O}_{X'}(n_2-D'))\otimes H^0(\underline{O}_{X'}(n_2-D'))^\vee$

is injective, which is an immediate consequence of (3.4), and, by (2.5), we see that $\operatorname{Im}(\delta) = H^1(T_{X'})$ except the case $(n, n_1, n_2) = (7, 3, 1)$.

Combining the above results, we conclude the proof of the theorem as follows. In case $n = 2n_1$, by a successive use of (3.5.1) and by the remark (3.2), we have the following commutative diagram:

$$\begin{array}{ccc} H^1(T_{X'}) & \hookrightarrow & H^1(T_{X'}\otimes \underline{O}_{X'}(n-n_2-4)) \otimes \{H^0(\underline{O}_{X'}(1))^\vee\}^{\otimes(n-n_2-4)} \\ & \searrow^{\varphi} & \uparrow \\ & & H^1(T_{X'}\otimes \underline{O}_{X'}(n-n_2-4))\otimes H^0(\underline{O}_{X'}(n-n_2-4))^\vee, \end{array}$$

where φ is the infinitesimal period map. In case $n \neq 2n_1$, except $n = 2n_1+1$ and $n_2 = 1$, by a successive use of (3.5.1) and by (3.5.2) we get the following diagram:

$$\begin{array}{ccc} H^1(T_{X'}) & \hookrightarrow & H^1(T_{X'}\otimes \underline{O}_{X'}(n-4-D'))\otimes H^0(\underline{O}_{X'}(n_1-D'))^\vee \otimes \{H^0(\underline{O}_{X'}(1))^\vee\}^{\otimes(n-n_1-4)} \\ & \searrow^{\varphi} & \uparrow \\ & & H^1(T_{X'}\otimes \underline{O}_{X'}(n-4-D'))\otimes H^0(\underline{O}_{X'}(n-4-D'))^\vee. \end{array}$$

In case $n_1 = n_2+1$, by (3.5.3) and (3.5.1), we obtain

$$\begin{array}{ccc} \operatorname{Im}(\delta) & \hookrightarrow & H^1(T_{X'}\otimes \underline{O}_{X'}(n-4-D'))\otimes H^0(\underline{O}_{X'}(n_2-D'))^\vee \otimes \{H^0(\underline{O}_{X'}(1))^\vee\}^{\otimes(n-n_2-4)} \\ & \searrow^{\varphi} & \uparrow \\ & & H^1(T_{X'}\otimes \underline{O}_{X'}(n-4-D'))\otimes H^0(\underline{O}_{X'}(n-4-D'))^\vee. \end{array}$$

Hence we get the injectivity of the infinitesimal period map φ in every case.

QED.

4. proof of the lemma (3.3).

We use the following well-known facts:

(4.1) $0 \longrightarrow \check{N}_{X'/P'} \longrightarrow \Omega_{P'} \otimes \underline{O}_{X'} \longrightarrow \Omega_{X'} \longrightarrow 0$ exact.

(4.2) $0 \longrightarrow f^*(\Omega_P \otimes \underline{O}_X) \longrightarrow \Omega_{P'} \otimes \underline{O}_{X'} \longrightarrow \Omega_{\tilde{g}} \otimes \underline{O}_{D'} \longrightarrow 0$ exact.

(4.3) $\Omega_{\tilde{g}} \simeq L^{\otimes(-2)} \otimes \tilde{g}^*(\det \check{N}_{D/P})$.

By (4.1) and (4.2), we have the following diagram:

$$\begin{array}{ccccc}
H^0(\Omega_{\tilde{g}} \otimes \underline{O}_{D'}(1)) & & & & \\
\uparrow & & & & \\
H^0(\Omega_{P'} \otimes \underline{O}_{X'}(1)) & \longrightarrow & H^0(\Omega_{X'} \otimes \underline{O}_{X'}(1)) & \longrightarrow & H^1(\check{N}_{X'/P'} \otimes \underline{O}_{X'}(1)). \\
\uparrow & & & & \\
H^0(f^*(\Omega_P \otimes \underline{O}_X) \otimes \underline{O}_{X'}(1)) & & & &
\end{array}$$

In order to prove (3.3.1), it is enough to show the following:

(4.4) $H^0(f^*(\Omega_P \otimes \underline{O}_X) \otimes \underline{O}_{X'}(1)) = 0.$

(4.5) $H^0(\Omega_{\tilde{g}} \otimes \underline{O}_{D'}(1)) = \begin{cases} \text{1-dimensional} & \text{if } n = 2n_1 \text{ and } n_1 = n_2 + 1, \\ 0 & \text{otherwise.} \end{cases}$

(4.6) $H^1(\check{N}_{X'/P'} \otimes \underline{O}_{X'}(1)) = \begin{cases} \text{1-dimensional} & \text{if } n = 2n_1 + 1 \text{ and } n_1 = n_2, \\ 0 & \text{otherwise.} \end{cases}$

By (1.3.2), we have an exact sequence

$$0 \longrightarrow \Omega_P \otimes \underline{O}_X(1) \longrightarrow \Omega_P \otimes \underline{O}_X(1) \otimes f_*\underline{O}_{X'} \longrightarrow \Omega_P \otimes \underline{O}_D(1-n+n_1+n_2) \longrightarrow 0.$$

From this, (4.4) can be verified by an easy calculation.

Tensoring $\Omega_{\tilde{g}} \otimes \tilde{g}^*\underline{O}_D(1)$ to the exact sequence

(4.7) $0 \longrightarrow \underline{O}_{E'}(-D') \longrightarrow \underline{O}_{E'} \longrightarrow \underline{O}_{D'} \longrightarrow 0$

and taking the direct image, we get, by using the relative duality theorem, the exact sequence

$$0 \longrightarrow \tilde{g}_*(\Omega_{\tilde{g}} \otimes \underline{O}_{D'}(1)) \longrightarrow S^2(N_{D/P}) \otimes \underline{O}_D(1-n) \xrightarrow{\ \varepsilon\ } \underline{O}_D(1) \longrightarrow 0,$$

and hence we obtain

$$H^0(\Omega_{\tilde{g}} \otimes \underline{O}_{D'}(1)) \simeq \operatorname{Ker}\{H^0(\underline{O}_D(2n_1+1-n) \oplus \underline{O}_D(n_1+n_2+1-n) \oplus \underline{O}_D(2n_2+1-n) \xrightarrow{\ \varepsilon\ } H^0(\underline{O}_D(1))\}.$$

Note that the map ε, considered as a homomorphism of graded modules, is as follows:

$$\varepsilon : (Q_1, Q_2, Q_3) \longrightarrow AQ_1 + 2BQ_2 + CQ_3,$$

where A, B and C are the polynomials in (2.1). In case $n > 2n_1+1$, obviously $H^0(\Omega_{\tilde{g}} \otimes \underline{O}_{D'}(1)) = 0$. In case $n = 2n_1+1$, the map ε is injective and hence $H^0(\Omega_{\tilde{g}} \otimes \underline{O}_{D'}(1)) = 0$. In case $n = 2n_1$ and $n_1 > n_2+1$, the same assertion holds. In case $n=2n_1$ and $n_1=n_2+1$, the map ε has the 1-dimensional kernel. Hence the assertion (4.5) is verified.

In order to prove (4.6), we may consider $H^1(\mathcal{O}_{X'}(2n-5-3D'))$ by the duality theorem. By the exact sequence

$$0 \longrightarrow \mathcal{O}_{X'}(2n-5-3D') \longrightarrow \mathcal{O}_{X'}(2n-5-2D') \longrightarrow \mathcal{O}_{X'}(2n-5-2D')\otimes\mathcal{O}_{D'} \longrightarrow 0,$$

and by (2.4.1), we get

$$H^1(\mathcal{O}_{X'}(2n-5-3D')) \cong \mathrm{Coker}\{H^0(\mathcal{O}_{X'}(2n-5-2D')) \longrightarrow H^0(\mathcal{O}_{X'}(2n-5-2D')\otimes\mathcal{O}_{D'})\}.$$

Note that $\mathcal{O}_{X'}(2n-5-2D')\otimes\mathcal{O}_{D'} \cong L^{\otimes 2}\otimes\mathcal{O}_{D'}(2n-5)$. Tensoring $L^{\otimes 2}\otimes\mathcal{O}_{D'}(2n-5)$ to the exact sequence (4.7), and taking the direct image, we get the exact sequence

$$0 \longrightarrow \mathcal{O}_D(n-5) \longrightarrow S^2(\check{N}_{D/P})\otimes\mathcal{O}_D(2n-5) \longrightarrow g_*(L^{\otimes 2}\otimes\mathcal{O}_{D'}(2n-5)) \longrightarrow 0.$$

Since $H^0(\mathcal{O}_X(2n-5-2D)) \longrightarrow H^0(S^2(\check{N}_{D/P})\otimes\mathcal{O}_D(2n-5))$ is surjective, we have

$$H^1(\mathcal{O}_{X'}(2n-5-3D)) \cong \mathrm{Coker}\{H^0(S^2(\check{N}_{D/P})\otimes\mathcal{O}_D(2n-5)) \longrightarrow H^0(g_*(L^{\otimes 2}\otimes\mathcal{O}_{D'}(2n-5)))\}$$

$$\cong \mathrm{Ker}\{H^1(\mathcal{O}_D(n-5)) \longrightarrow H^1(S^2(\check{N}_{D/P})\otimes\mathcal{O}_D(2n-5))\},$$

and hence, by the duality theorem, $H^1(\mathcal{O}_{X'}(2n-5-3D'))$ is dual to

$$\mathrm{Coker}\{H^0(\mathcal{O}_D(3n_1+n_2+1-2n)\oplus\mathcal{O}_D(2n_1+2n_2+1-2n)\oplus\mathcal{O}_D(n_1+3n_2+1-2n)) \xrightarrow{\ \varepsilon\ } H^0(\mathcal{O}_D(n_1+n_2+1-n))\}.$$

In case $n>n_1+n_2+1$, it is obvious that $H^1(\mathcal{O}_{X'}(2n-5-3D'))=0$. In case $n=2n_1$ and $n_1=n_2+1$, ε is surjective. In case $n=2n_1+1$ and $n_1=n_2$, ε has the 1-dimensional cokernel. This proves the assertion (4.6), and hence completes the proof of (3.3.1).

The proof of (3.3.2) is similar to that of (3.3.1). First note that $T_{X'}\otimes\mathcal{O}_{X'}(2n_1-2D') \cong \Omega_{X'}\otimes\mathcal{O}_{X'}(2n_1+4-n-D')$. We use the following exact diagram obtained from (4.1) and (4.2):

$$\begin{array}{ccccc}
H^0(\Omega_g\otimes\mathcal{O}_{X'}(2n_1+4-n-D')) & & & & \\
\uparrow & & & & \\
H^0(\Omega_{P'}\otimes\mathcal{O}_{X'}(2n_1+4-n-D')) & \longrightarrow & H^0(\Omega_{X'}\otimes\mathcal{O}_{X'}(2n_1+4-n-D')) & \longrightarrow & H^1(\check{N}_{X'/P'}\otimes\mathcal{O}_{X'}(2n_1+4-n-D')). \\
\uparrow & & & & \\
H^0(f^*(\Omega_P\otimes\mathcal{O}_X)\otimes\mathcal{O}_{X'}(2n_1+4-n-D')) & & & &
\end{array}$$

We will show the following (assuming X' of general type with $n\neq 2n_1$):

(4.8) $$H^0(f^*(\Omega_P\otimes\mathcal{O}_X)\otimes\mathcal{O}_{X'}(2n_1+4-n-D')) = \begin{cases} \text{6-dimensional} & \text{if } n=2n_1+1 \ \& \ n_2=1, \\ 0 & \text{otherwise.} \end{cases}$$

(4.9) $H^0(\Omega_{\tilde{g}} \otimes \underline{O}_{X'}(2n_1+4-n-D')) = 0.$

(4.10) $H^1(\check{N}_{X/P} \otimes \underline{O}_{X'}(2n_1+4-n-D')) = 0.$

From the exact sequence

$$0 \longrightarrow \Omega_P \otimes \underline{O}_X(2n_1+4-n-D) \longrightarrow \underline{O}_X(2n_1+3-n-D)^{\oplus 4} \longrightarrow \underline{O}_X(2n_1+4-n-D) \longrightarrow 0,$$

we have

$$0 \longrightarrow H^0(\Omega_P \otimes \underline{O}_X(2n_1+4-n-D)) \longrightarrow I^{\oplus 4}_{2n_1+3-n} \xrightarrow{\sigma} I_{2n_1+4-n}.$$

Since we assume that X' is of general type and $n \neq 2n_1$, the case $2n_1+3-n \geq n_1$ is decluded. Hence we have

$$I_{2n_1+3-n} = \mathbb{C}[X_0, \ldots, X_3]_{2n_1+3-n-n_2}.$$

On the other hand, from the inequality $2n_1+3-n-n_2 \geq 0$, one of the following three cases occurs:

(4.11) $n = 2n_1+1$ and $n_2 = 1$.

(4.12) $n = 2n_1+1$ and $n_2 = 2$.

(4.13) $n = 2n_1+2$ and $n_2 = 1$.

In cases (4.12) and (4.13), the map σ is injective. In case (4.11), the map σ has 6-dimensional kernel consisting of the Koszul relations. Hence (4.8) is verified.

Tensoring $\Omega_{\tilde{g}} \otimes g^*\underline{O}_D(2n_1+4-n) \otimes L$ to the exact sequence (4.7) and taking the direct image, we obtain

$$g_*(\Omega_{\tilde{g}} \otimes \underline{O}_{X'}(2n_1+4-n-D')) \cong R^1g_*(\Omega_{\tilde{g}} \otimes L^{-1} \otimes g^*\underline{O}_D(2n_1+4-2n)) \simeq N_{D/P} \otimes \underline{O}_D(2n_1+4-2n).$$

Since X' is of general type with $n \neq 2n_1$, we can declude the case $3n_1+4-2n \geq 0$ and hence $H^0(N_{D/P} \otimes \underline{O}_D(2n_1+4-2n)) = 0$. This proves the assertion (4.9).

(4.10) follows from (2.4.1) by taking its dual. This completes the proof of (3.3.2).

5. appendix.

In this appendix, we prove that, in the situation in (1.1), the period map is unramified at the origin of the parameter space of the Kuranishi family of the deformations of X' provided that n is sufficiently large enough comparing to D. We use the notations at the end of the introduction and in the section 1.

(5.1) Let $q_1 : P_1 \longrightarrow P$ be the blowing-up of P with the center T, let E_1 be the exceptional divisor and set $\tilde{h}_1 : E_1 \longrightarrow T$. X_1 and D_1 denote the proper transforms of X and of D respectively and set $f_1 = \mathrm{res}(q_1) : X_1 \longrightarrow X$ and $g_1 = \mathrm{res}(q_1) : D_1 \longrightarrow D$. Set $T_1 = f^{-1}(T)$ and $h_1 = \mathrm{res}(f_1) : T_1 \longrightarrow T$.

Let $q_2 : P_2 \longrightarrow P_1$ be the blowing-up of P_1 along D_1, let E_2 be the exceptional divisor of q_2 and set $\tilde{g}_2 : E_2 \longrightarrow D_1$. X_2, T_2 and E_1' denote the proper transformations of X_1, of T_1 and of E_1 respectively and set $f_2 = \mathrm{res}(q_2) : X_2 \longrightarrow X_1$, $h_2 = \mathrm{res}(q_2) : T_2 \longrightarrow T_1$ and $p_2 = \mathrm{res}(q_2) : E_1' \longrightarrow E_1$. Set $D_2 = f_2^{-1}(D_1)$ and $g_2 = \mathrm{res}(f_2) : D_2 \longrightarrow D_1$.

It is easy to see that D_1, X_2 and D_2 are smooth and that T_2 consists of the exceptional curves of the first kind on X_2. The surface contracted T_2 on X_2 coincides with the normalization X' of X by virtue of the Zariski's Main Theorem. Set $f_3 : X_2 \longrightarrow X'$, $g_3 : D_2 \longrightarrow D'$.

The above things form the following commutative diagram:

(5.1.1)

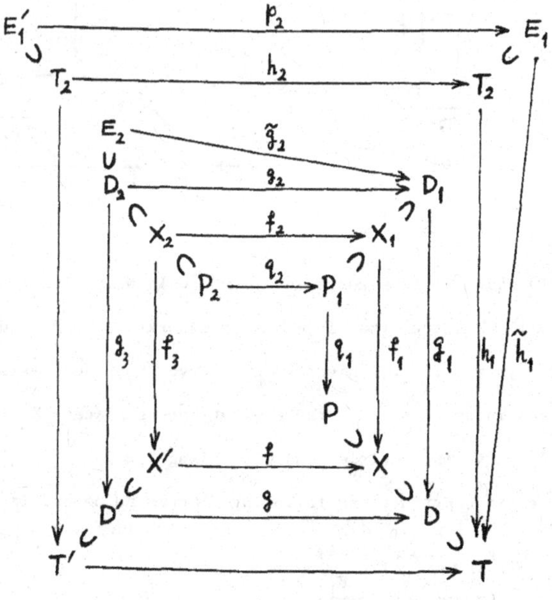

Note that $g_1 : D_1 \longrightarrow D$ is the normalization, $\tilde{h}_1 : E_1 \longrightarrow T$ is a $\mathbb{P}^2$-bundle, namely, a disjoint union of projective 2-spaces, $f_2 : X_2 \longrightarrow X_1$ is the normalization, $g_2 : D_2 \longrightarrow D_1$ is a ramified double covering, $\tilde{g}_2 : E_2 \longrightarrow D_1$ is a $\mathbb{P}^1$-bundle, $p_2 : E_1' \longrightarrow E_1$ is the blowing-up of the three points on $\mathbb{P}^2$ the component of E_1, $f_3 : X_2 \longrightarrow X'$ is the blowing-up of X' with the center T' and T_2 is its exceptional divisor, $g_3 : D_2 \longrightarrow D'$ is the normalization and that T' are the nodes on D' and by g each three nodes go down to a triple point of X.

In the neighborhood of a triple point of X, the figure of the above construction is as follows:

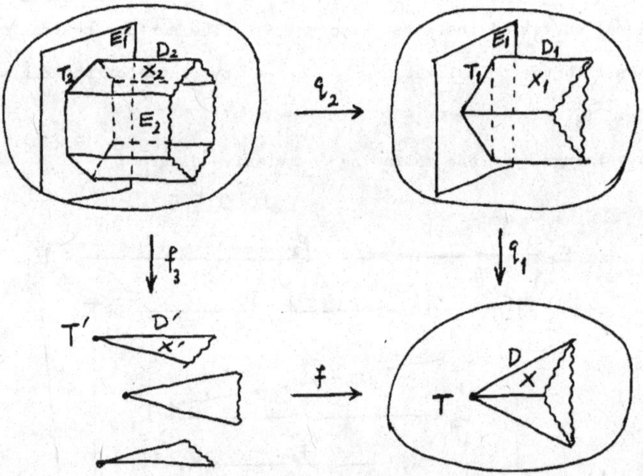

<u>Remark</u> (5.2) Let D be a curve in $P = \mathbb{P}^3$ only with singularities like coordinate axes in 3-space, that is, $D : yz = zx = xy = 0$ for a suitable local coordinates (x, y, z) in P. Given such a curve D, there exists an integer n_0 so that, for each integer $n \geq n_0$, there exist hypersurfaces X in P of degree n only with ordinary singularities and with $\mathrm{Sing}(X) = D$.

Actually, we can construct the following diagram as before in (5.1):

$$\begin{array}{ccccc}
E_1' & \xrightarrow{p_2} & E_1 & \xrightarrow{\tilde{h}_1} & T \\
 & & & & \cap \\
E_2 & \xrightarrow{\tilde{g}_2} & D_1 & \xrightarrow{g_1} & D \\
\cap & & \cap & & \cap \\
P_2 & \xrightarrow{q_2} & P_1 & \xrightarrow{q_1} & P
\end{array}$$

The composite morphism $q_1 \circ q_2$ can be also obtained by once blowing-up (cf. (2.3.7) in [7]) and its exceptional divisor can be easily calculated as $E_2 + 2E_1'$. Thus we see that $\underline{O}_{P_2}(-E_2) \otimes q_2^* \underline{O}_{P_1}(-2E_1)$ is $(q_1 \circ q_2)$-very ample and hence there exists an integer n_0 such that, for $n \geqslant n_0$, $\underline{O}_{P_2}(-2E_2) \otimes q_2^* \underline{O}_{P_1}(-4E_1) \otimes (q_1 \circ q_2)^* \underline{O}_P(n)$ is very ample. Set

$$M_n = \underline{O}_{P_2}(-2E_2) \otimes q_2^* \underline{O}_{P_1}(-4E_1) \otimes (q_1 \circ q_2)^* \underline{O}_P(n) \quad \text{and}$$

$$N_n = \underline{O}_{P_2}(-2E_2) \otimes q_2^* \underline{O}_{P_1}(-3E_1) \otimes (q_1 \circ q_2)^* \underline{O}_P(n).$$

We will compute $Bs|N_n|$. Tensoring N_n to the exact sequence

$$0 \longrightarrow \underline{O}_{P_2}(-E_1') \longrightarrow \underline{O}_{P_2} \longrightarrow \underline{O}_{E_1'} \longrightarrow 0$$

and using the Kodaira's vanishing theorem, we get the exact sequence

$$0 \longrightarrow H^0(M_n) \longrightarrow H^0(N_n) \longrightarrow H^0(N_n \otimes \underline{O}_{E_1'}) \longrightarrow 0.$$

Since M_n is very ample, the maximal fixed component of $|\mathrm{Im}(M_n \longrightarrow N_n)|$ is E_1'. Recall that E_1 is the disjoint union of $\mathbb{P}^2$ and set $E_1 = \coprod_{1 \leqslant i \leqslant t} E_1^{(i)}$ $(E_1^{(i)} = \mathbb{P}^2)$, $E_1' = \coprod_{1 \leqslant i \leqslant t} E_1'^{(i)}$ and $p_2^{(i)} = \mathrm{res}(p_2) : E_1'^{(i)} \longrightarrow E_1^{(i)}$. Note that

$$p_{2*}^{(i)}(N_n \otimes \underline{O}_{E_1'^{(i)}}) \simeq p_{2*}^{(i)}(\underline{O}_{E_1'^{(i)}}(-2E_2 \cdot E_1'^{(i)}) \otimes p_2^{(i)*}\underline{O}_{E_1^{(i)}}(3)) \simeq \mathcal{J}^2_{D \cdot E_1^{(i)}} \otimes \underline{O}_{E_1^{(i)}}(3),$$

where $\underline{O}_{E_1^{(i)}}(3) = \underline{O}_{\mathbb{P}^2}(3)$. $|\mathcal{J}^2_{D_1 \cdot E_1^{(i)}} \otimes \underline{O}_{E_1^{(i)}}(3)|$ contains only one member, say $T_1^{(i)}$, which is the uniquely determined three lines on $E_1^{(i)}$, and hence $|N_n \otimes \underline{O}_{E_1'^{(i)}}|$ consists of only one member, say $T_2^{(i)}$, which is the proper transform of $T_1^{(i)}$. Setting $T_2 = \coprod_{1 \leqslant i \leqslant t} T_2^{(i)}$, the above reasoning shows that $Bs|N_n| = T_2$. Hence $|N_n|$ defines a birational morphism $P_2 - T_2 \longrightarrow \mathbb{P}^r$. Let X_2 be the closure in P_2 of the pull-back of a generic hyperplane by this morphism, then, since T_2 is of codimension 2 in P_2, we see that $X_2 \in |N_n|$ and that X_2 is smooth outside E_1'. It is easily seen that the image X of X_2 in P is just what we want.

(5.3) From now on we assume the following conditions:

(5.3.1) $\quad H^1(\mathcal{J}_D \otimes \underline{O}_P(n-4)) = 0.$

(5.3.2) $\quad H^1(\Omega_P \otimes \mathcal{J}_D \otimes \underline{O}_P(n-4)) = 0.$

(5.3.3) $\quad H^0(\Omega_P \otimes \omega_D \otimes \underline{O}_D(5-n)) = 0.$

(5.3.4) $\quad H^0(S^2(N_{D_1/P_1}) \otimes \underline{O}_{D_1}(4E_1 \cdot D_1) \otimes g_1^* \underline{O}_D(1-n)) = 0.$

(5.3.5) $H^0(\Omega_{P_1} \otimes \omega_{D_1} \otimes \underline{O}_{D_1}(2E_1 \cdot D_1) \otimes g_1^* \underline{O}_D(5-n)) = 0$.

(5.3.6) $\underline{O}_{P_2}(-2E_2) \otimes q_2^* \underline{O}_{P_1}(-4E_1) \otimes (q_1\ q_2)^* \underline{O}_P(n-1)$ is ample.

(5.3.7) There exists an integer m satisfying the following conditions:

(5.3.7.1) $m \leqslant n-4$.

(5.3.7.2) $|\underline{O}_{X'}(m-D')|$ is fixed components free.

(5.3.7.3) $H^0(\underline{O}_{X'}(n-2m-3+D')) = 0$.

Note that these conditions are fulfilled by the Serre's theorem provided that the degree n of X is sufficiently large enough comparing to D.

Proposition (5.4). Under the conditions in (5.3), the parameter space of the Kuranishi family of the deformations of X' is smooth at the origin.

Proof. The conditions (5.3.1) and (5.3.2) imply $H^1(\mathcal{J}_D \otimes \underline{O}_P(n-5)) = 0$. From this and from (5.3.2), it follows $H^1(f^*(T_P \otimes \underline{O}_X)) = 0$ by the same argument in the proof of (2.7) ((5.3.7.1) and (5.3.7.2) assure the existence of non-zero element in $H^0(\omega_{X'})$). The rest is the consequence of the general theorey of deformations. QED.

The following lemma can be proved in the same way as (1.3.2):

Lemma (5.5). $0 \longrightarrow \underline{O}_{X_1} \longrightarrow f_{2*}\underline{O}_{X_2} \longrightarrow \omega_{D_1} \otimes \check{\omega}_{X_1} \longrightarrow 0$ exact.

Since $f_3 : X_2 \longrightarrow X'$ is the blowing-up of the smooth scheme X' with the smooth center T', and T_2 is its exceptional divisor, we have the following lemma:

Lemma (5.6).

(5.6.1) $f_{3*}\underline{O}_{X_2}(aT_2) \simeq \underline{O}_{X'}$, $(a \geqslant 0)$.

(5.6.2) $R^1 f_{3*}\underline{O}_{X_2}(aT_2) = 0$ $(a \leqslant 1)$.

Lemma (5.7). Under the conditions in (5.3), we have

(5.7.1) $H^0(\Omega_{X'} \otimes \underline{O}_{X'}(1)) = 0$ and

(5.7.2) $H^0(T_{X'} \otimes \underline{O}_{X'}(2m-2D')) = 0$.

Proof. Taking the direct image of the exact sequence

$$0 \longrightarrow f_3^*(\Omega_{X'} \otimes \underline{O}_{X'}(1)) \otimes \underline{O}_{X_2}(T_2) \longrightarrow \Omega_{X_2} \otimes (f_1 \circ f_2)^* \underline{O}_X(1) \otimes \underline{O}_{X_2}(T_2)$$

$$\longrightarrow \Omega_{f_3} \otimes (f_1 \circ f_2)^* \underline{O}_X(1) \otimes \underline{O}_{X_2}(T_2) \longrightarrow 0,$$

we get that, by (5.6.1),

$$\Omega_{X'} \otimes \underline{O}_{X'}(1) \simeq f_{3*}(\Omega_{X_2} \otimes (f_1 \circ f_2)^* \underline{O}_X(1) \otimes \underline{O}_{X_2}(T_2)).$$

Hence to prove (5.7.1) is equivalent to prove

(5.7.3) $\quad H^0(\Omega_{X_2} \otimes (f_1 \circ f_2)^* \underline{O}_X(1) \otimes \underline{O}_{X_2}(T_2)) = 0.$

By the exact sequence

$$0 \longrightarrow \check{N}_{X_2/P_2} \otimes (f_1 \circ f_2)^* \underline{O}_X(1) \otimes \underline{O}_{X_2}(T_2) \longrightarrow \Omega_{P_2} \otimes (f_1 \circ f_2)^* \underline{O}_X(1) \otimes \underline{O}_{X_2}(T_2)$$
$$\longrightarrow \Omega_{X_2} \otimes (f_1 \circ f_2)^* \underline{O}_X(1) \otimes \underline{O}_{X_2}(T_2) \longrightarrow 0,$$

to prove (5.7.3) it is enough to show the following:

(5.7.4) $\quad H^0(\Omega_{P_2} \otimes \underline{O}_{X_2}(T_2) \otimes (f_1 \circ f_2)^* \underline{O}_X(1)) = 0.$

(5.7.5) $\quad H^1(\check{N}_{X_2/P_2} \otimes \underline{O}_{X_2}(T_2) \otimes (f_1 \circ f_2)^* \underline{O}_X(1)) = 0.$

Since $\check{N}_{X_2/P_2} \otimes \underline{O}_{X_2}(T_2) \otimes (f_1 \circ f_2)^* \underline{O}_X(1) \simeq \underline{O}_{X_2}(D_2 + 2T_2)^{\otimes 2} \otimes (f_1 \circ f_2)^* \underline{O}_X(1-n)$, (5.7.5) follows the condition (5.3.6) by virtue of the Kodaira vanishing theorem.

Next we will prove (5.7.4). By the exact sequence

$$0 \longrightarrow f_2^*(\Omega_{P_1} \otimes f_1^* \underline{O}_X(1)) \otimes \underline{O}_{X_2}(T_2) \longrightarrow \Omega_{P_2} \otimes \underline{O}_{X_2}(T_2) \otimes (f_1 \circ f_2)^* \underline{O}_X(1)$$
$$\longrightarrow \Omega_{q_2} \otimes \underline{O}_{X_2}(T_2) \otimes (f_1 \circ f_2)^* \underline{O}_X(1) \longrightarrow 0,$$

it is enough to show the following:

(5.7.6) $\quad H^0(f_2^*(\Omega_{P_1} \otimes f_1^* \underline{O}_X(1)) \otimes \underline{O}_{X_2}(T_2)) = 0.$

(5.7.7) $\quad H^0(\Omega_{q_2} \otimes \underline{O}_{X_2}(T_2) \otimes (f_1 \circ f_2)^* \underline{O}_X(1)) = 0.$

We first prove (5.7.7). Tensoring $\Omega_{q_2} \otimes \underline{O}_{X_2}(T_2) \otimes (f_1 \circ f_2)^* \underline{O}_X(1)$ $\simeq \Omega_{\tilde{g}_2} \otimes \tilde{g}_2^* \underline{O}_{D_1}(E_1 \cdot D_1) \otimes (g_1 \circ \tilde{g}_2)^* \underline{O}_D(1)$ to the exact sequence

$$0 \longrightarrow \underline{O}_{E_2}(-D_2) \longrightarrow \underline{O}_{E_2} \longrightarrow \underline{O}_{D_2} \longrightarrow 0$$

and taking the direct image, we have the exact sequence

$$0 \longrightarrow \tilde{g}_{2*}\Omega_{\tilde{g}_2} \otimes \underline{O}_{D_1}(E_1 \cdot D_1) \otimes g_1^* \underline{O}_D(1) \longrightarrow S^2(N_{D_1/P_1}) \otimes \underline{O}_{D_1}(4E_1 \cdot D_1) \otimes g_1^* \underline{O}_D(1-n)$$
$$\longrightarrow \underline{O}_{D_1}(E_1 \cdot D_1) \otimes g_1^* \underline{O}_D(1) \longrightarrow 0$$

by the same argument as in the proof of (4.5). Hence (5.7.7) follows the condition (5.3.4), since $g_2 : D_2 \longrightarrow D_1$ is a finite morphism.

To prove (5.7.6), by using the exact sequence

$$0 \longrightarrow \Omega_{P_1} \otimes \underline{O}_{X_1}(T_1) \otimes f_1^*\underline{O}_X(1) \longrightarrow f_{2*}\underline{O}_{X_2} \otimes \Omega_{P_1} \otimes \underline{O}_{X_1}(T_1) \otimes f_1^*\underline{O}_X(1)$$
$$\longrightarrow \omega_{D_1} \otimes \check{\omega}_{X_1} \otimes \Omega_{P_1} \otimes \underline{O}_{X_1}(T_1) \otimes f_1^*\underline{O}_X(1) \longrightarrow 0$$

obtained from (5.5), it suffices to show the following:

(5.7.8) $H^0(\Omega_{P_1} \otimes \underline{O}_{X_1}(T_1) \otimes f_1^*\underline{O}_X(1)) = 0.$

(5.7.9) $H^0(\omega_{D_1} \otimes \check{\omega}_{X_1} \otimes \Omega_{P_1} \otimes \underline{O}_{X_1}(T_1) \otimes f_1^*\underline{O}_X(1)) = 0.$

To prove (5.7.8), we use the exact sequence

$$0 \longrightarrow f_1^*(\Omega_P \otimes \underline{O}_X(1)) \otimes \underline{O}_{X_1}(T_1) \longrightarrow \Omega_{P_1} \otimes \underline{O}_{X_1}(T_1) \otimes f_1^*\underline{O}_X(1)$$
$$\longrightarrow \Omega_{q_1} \otimes \underline{O}_{X_1}(T_1) \otimes f_1^*\underline{O}_X(1) \longrightarrow 0.$$

Since $f_{3*}f_2^*(f_1^*(\Omega_P \otimes \underline{O}_X(1)) \otimes \underline{O}_{X_1}(T_1)) \simeq f_{3*}(f_3^*f^*(\Omega_P \otimes \underline{O}_X(1)) \otimes \underline{O}_{X_2}(T_2)) \simeq f^*(\Omega_P \otimes \underline{O}_X(1))$ by (5.6.1), $H^0(f_1^*(\Omega_P \otimes \underline{O}_X(1)) \otimes \underline{O}_{X_1}(T_1)) = 0$ follows $H^0(f^*(\Omega_P \otimes \underline{O}_X(1))) = 0$, and the latter follows the exact sequence

$$0 \longrightarrow \Omega_P \otimes \underline{O}_X(1) \longrightarrow f_*\underline{O}_{X'} \otimes \Omega_P \otimes \underline{O}_X(1) \longrightarrow \omega_D \otimes \check{\omega}_X \otimes \Omega_P \otimes \underline{O}_X(1) \longrightarrow 0$$

obtained from (1.3.2) and the condition (5.3.3). On the other hand, E_1 is a disjoint union of $\mathbb{P}^2$ and since T_1 appears as three lines on each $\mathbb{P}^2$, setting $T_1 = \coprod_{1 \leq i \leq t} T_1^{(i)}$, we see that $\Omega_{q_1} \otimes \underline{O}_{X_1}(T_1) \otimes f_1^*\underline{O}_X(1) \simeq \Omega_{\tilde{h}_1} \otimes \underline{O}_{P_1}(E_1) \otimes \underline{O}_{E_1} \otimes \underline{O}_{T_1}$ is the disjoint union of $\Omega_{\mathbb{P}^2} \otimes \underline{O}_{\mathbb{P}^2}(-1) \otimes \underline{O}_{T_1^{(i)}}$ $(1 \leq i \leq t)$ and hence we can get $H^0(\Omega_{q_1} \otimes \underline{O}_{X_1}(T_1) \otimes f_1^*\underline{O}_X(1)) = 0$. Thus we have proven (5.7.8).

(5.7.9) follows the condition (5.3.5), since an easy computation shows that $\omega_{X_1} \simeq f_1^*\underline{O}_X(n-4) \otimes \underline{O}_{X_1}(-T_1)$. This completes the proof of (5.7.1).

Tensoring a non-zero element in $H^0(\underline{O}_{X'}(n-2m-3+D'))$ (such an element exists by the condition (5.3.7.3)) gives an injection

$$T_{X'} \otimes \underline{O}_{X'}(2m-2D') \longrightarrow T_{X'} \otimes \underline{O}_{X'}(n-3-D') \simeq \Omega_{X'} \otimes \underline{O}_{X'}(1)$$

and hence (5.7.2) follows (5.7.1). QED.

<u>Theorem</u> (5.8). In the case that the degree n of X in P is sufficiently large enough comparing to the singular locus D of X in the sense that the conditions in (5.3) are fulfilled, the local Torelli theorem holds for the

normalization X' of X.

Proof. We can derive this theorem from (5.3.7.1), (5.3.7.2) and (5.7) just in the same way as in proving (3.5). QED.

Remark (5.9). The moduli space of Gieseker ([6]) is divided by the Hilbert polynomial of $\omega_{X'}$, that is,

$$(\omega_{X'}^{\otimes s}) = \frac{1}{2}c_1^2 s - \frac{1}{2}c_1^2 s + \frac{1}{12}(c_1^2 + c_2).$$

(1.7) says that, fixing D and increasing n, c_1^2 is increasing and (1.4.1) says that $\omega_{X'}$ is getting "ampler and ampler". Hence (5.8) gives some evidevce to the naive feeling that if X' would have sufficiently many 2-forms, their periods of integrals should determine X' itself. (Note that the Kŷnef's example [12] has $p_g = c_1^2 = 1$.)

Notes

1) If X has only ordinary singularities and its singular locus D is a complete intersection in P, D becomes automatically smooth. Actually, blowing-up P along D, the fact that D is a complete intersection imposes that the exceptional divisor becomes a $\mathbb{P}^1$-bundle. On the other hand, if X would have triple points, the fibres over such points are 2-dimensional.

2) By using the result (3.3.2) below, in cases $(n, n_1, n_2) = (6, 2, 1)$, $(7, 3, 2)$ we see that $H^2(T_{X'}) = 0$ by duality and also $H^0(T_{X'}) = 0$ and hence the parameter space of the Kuranishi family of deformations of X' is smooth at the origin. On the contrary, in case $(n, n_1, n_2) = (7, 3, 1)$ we see that $\dim H^2(T_{X'}) = 6$ by (3.3.2) and so the smoothness of the parameter space is still unknown.

3) Actually $|\mathcal{O}_{X'}(n_1 - D')|$ is fixed points free, but (3.1.2) is enough for our later use. Note also that, by (3.1.2), $|\omega_{X'}|$ is fixed components free and hence X' is minimal.

References

[1] Altman, A. & Kleiman, K., Introduction to Grothendieck duality theory, Lecture Notes in Math. No 146, Springer-Verlag.

[2] Deligne, P., Travaux de Griffiths, Sem. Bourbaki 376 (1969/70) 213-237.

[3] Griffiths, P. A., Periods of integrals on algebraic manifolds I, II, III, Amer. J. Math. 90 (1968) 568-626; 805-865; Publ. Math. I. H. E. S. No 38 (1970), Paris.

[4] Griffiths, P. A., Periods of integrals on algebraic manifolds: Summary of main results and discussion of open probrems, Bull. Amer. Math. Soc. 75 (1970) 228-296.

[5] Griffths, P. A. & Schmid, W., Recent developments in Hodge theory, Proc. Symp. Bombey 1973, Oxford Univ. Press (1975).

[6] Gieseker, D., Global moduli for surfaces of general type, to appear.

[7] Grothendieck, A. & Diedonné, J., Éléments de Géométrie Algébrique III, Publ. Math. I. H. E. S. No 11, Paris.

[8] Hartshorne, R., Residues and duality, Lecture Notes in Math. No 20, Springer-Verlag.

[9] Horikawa, E., On the number of moduli of certain algebraic surfaces of general type, J. Fac. Sci. Univ. Tokyo (1974) 67-78.

[10] Horikawa, E., Surjectivity of the period map of K3 surfaces of degree 2, Math. Ann. 228 (1977) 113-146.

[11] Horikawa, E., On the periods of Enriques surfaces I, II, Proc. Japan. acad. 53-3 (1977) 124-127; 53-A-2 (1977) 53-55.

[12] Kĭnef, F., I., A simply connected surface of general type for which the local Torelli theorem does not hold, Cont. Ren. Acad. Bulgare des Sci. 30-3 (1977) 323-325. (Russian).

[13] Kodaira, K., On the characteristic systems of families of surfaces with ordinary singularities in a projective space, Amer. J. Math. 87 (1965) 227-255.

[14] Lieberman, D. & Wilsker, R. & Peters, C., A theorem of local Torelli-type, Math. Ann. 231 (1977) 39-45.

[15] Peters, C., The local Torelli theorem I, complete intersections, Math. Ann. 217 (1975) 1-16; Erratum, Math. Ann. 223 (1976) 191-192.

[16] Peters, C., The local Torelli theorem II, cyclic branched coverings, Ann. Sc. Normale di Pisa IV 3 (1976) 321-340.

[17] Pjateckiĭ-Šapiro, I. I. & Šafarevič, I. R., A Torelli theorem for algebraic surfaces of type K3, Izv. Acad. Nauk. 35 (1971) 530-572.

[18] Roberts, J., Hypersurfaces with non-singular normalization and their double loci, to appear.

[19] Tsuboi, S., On the sheaves of holomorphic vector fields on surfaces with ordinary singularities in a projective space I, II, Sci. Rep. Kagoshima Univ. 25 (1976) 1-26; II is to appear.

[20] Usui, S., Local Torelli theorem for non-singular complete intersections, Japan. J. Math. 2-2 (1976) 411-418.

[21] Usui, S., Local Torelli theorem for some non-singular weighted complete intersections, Proc. Internat. Symp. Algebraic Geometry, Kyoto 1977.

[22] Enriques, F., Le Superficie Algebaiche, Nicola Zanichelli, Bolona, 1949.

[23] Kodaira, K., Algebraic surfaces, Todai Seminary Notes No 20, Tokyo, 1968. (Japanese).

Faculty of Science
Kochi University
Asakura 1000, Kochi-shi
Japan

Formal groups and some arithmetic properties of elliptic curves

Noriko Yui

1. Introduction.

Let E be an elliptic curve, i.e, an abelian variety of dimension 1, defined over a field K. E has a plane cubic model given by a Weierstrass equation of the form

(1.1) $$y^2 + a_1xy + a_3y = x^3 + a_2x^2 + a_4x + a_6$$

where $a_i \in K$ for all i and x, y are affine coordinates.

Let $\mathfrak{D}_1$ denote the K-vector space of all differential 1-forms of the first kind on E. $\mathfrak{D}_1$ has dimension 1 over K and its canonical basis is given by

(1.2) $$\omega_0 = \frac{dx}{2y+a_1x+a_3}.$$

We call ω_0 the canonical invariant differential on E.

Let $u = -\frac{x}{y}$ be a local parameter of E near the point at infinity $(0,1,0)$ of E. Put $w = -\frac{1}{y}$. Then we can express the Weierstrass equation for E in (u,w)-coordinate as

(1.3) $$w = u^3 + a_1uw + a_2u^2w + a_3w^2 + a_4uw^2 + a_6w^3.$$

Substitute w recursively in the right hand side of (1.3). We obtain the formal power series expansion for E in u. In the same fashion, ω_0 can be expressed as a formal power series in u as

(1.4) $$\begin{aligned}\omega_0 &= du\{1 + a_1u + (a_1^2+a_2)u^2 + (a_1^3+2a_1a_2+2a_3)u^3 \\ &\quad + (a_1^4+3a_1^2a_2+6a_1a_3+a_2^2+2a_4)u^4 + \cdots\} \\ &= \sum_{n=1}^{\infty} a(n)u^{n-1}du\end{aligned}$$

where $a(1) = 1$ and $a(n) \in \mathbb{Z}[a_1,a_2,a_3,a_4,a_6]$ for all n.

When the characteristic of K is different from 2 or 3, E

can be defined by the equation of the form

(1.1') $$y^2 = x^3 + ax + b$$

where $a, b \in K$ and the cubic $x^3 + ax + b$ has distinct roots. In this case, the canonical invariant differential ω_0 on E is given by

(1.2') $$\omega_0 = \frac{dx}{2y}.$$

E is defined in (u,w)-plane by the equation

(1.3') $$w = u^3 + auw + bw^3,$$

and ω_0 has the formal power series expansion in u as follows.

(1.4') $$\omega_0 = \frac{dx}{2y} = \frac{dx/du}{2y}du = (-\frac{1}{2} + \frac{u}{2w}\frac{dw}{du})$$
$$= du\{1 + 2au^4 + 3bu^6 + 6a^2u^8 + 20abu^{10} + \cdots\}$$
$$= \sum_{n=1}^{\infty} a(n)u^{n-1}du. \quad \text{(See Yui [15].)}$$

The objects of our discussion in this talk are the coefficients $a(n)$ of the canonical invariant differential ω_0 on E. These coefficients provide us with subtle arithmetic information on elliptic curves, e.g, the Hasse invariant, the liftability of the Frobenius morphism, the Atkin and Swinnerton-Dyer congruences and so on. In the present lecture, we shall review the papers [15]-[16].

The theory of (commutative 1-dimensional) formal groups is extensively used as a tool in the investigation of arithmetic properties of elliptic curves. So here is the appropriate place and time to recall the definitions and some basic properties of (commutative 1-dimensional) formal groups, which we need in the succeeding discussions. For a thorough discussion of (commutative 1-dimensional) formal groups, see Fröhlich [5].

Let R be a commutative ring with the identity 1. We denote by $R[[x_1,\ldots,x_n]]$ the ring of formal power series in the variables $x_1,\ldots,x_n$. For f and g in $R[[x_1,\ldots,x_n]]$, we write $f \equiv g \pmod{\deg r}$ if $f-g$ contains no monomials of total degree less than r.

<u>(1.5) Definition.</u> <u>A (commutative</u> 1-<u>dimensional) formal group over</u> R is a formal power series $\Phi(x,y)$ over R in two variables x, y satisfying the following axioms:

(1) $\Phi(x,y) \equiv x + y \pmod{\deg 2}$,

(2) $\Phi(\Phi(x,y),z) = \Phi(x,\Phi(y,z))$,

(3) $\Phi(x,y) = \Phi(y,x)$.

(In this paper, we discuss only commutative 1-dimensional formal groups, so we simply use the terminology <u>formal groups</u> to mean those commutative 1-dimensional ones.)

If Φ and Ψ are formal groups over R, an R-<u>homomorphism of</u> Φ to Ψ is a formal power series $\lambda(x) \in R[[x]]$ without constant term satisfying $\lambda(\Phi(x,y)) = \Psi(\lambda(x), \lambda(y))$. Such a λ is called an R-<u>isomorphism</u> if $\lambda(x) \equiv x \pmod{\deg 2}$. If there is an R-isomorphism of Φ to Ψ, Φ and Ψ are said to be <u>isomorphic over</u> R, or simply R-<u>isomorphic</u>. The set $\mathrm{Hom}_R(\Phi,\Psi)$ of all R-homomorphisms of Φ to Ψ forms an abelian group under the addition $(\lambda_1+\lambda_2)(x) = \Psi(\lambda_1(x), \lambda_2(x))$ for $\lambda_1, \lambda_2 \in \mathrm{Hom}_R(\Phi,\Psi)$. In particular, $\mathrm{End}_R(\Phi) = \mathrm{Hom}_R(\Phi,\Phi)$ is a ring. We denote by $[n]_\Phi \in \mathrm{End}_R(\Phi)$ the image of $n \in \mathbb{Z}$ under the natural embedding $\mathbb{Z} \to \mathrm{End}_R(\Phi)$. In characteristic $p > 0$, $[p]_\Phi$ has the form

$$[p]_\Phi(x) = c_1 x^{p^h} + c_2 x^{p^{2h}} + \cdots$$

.(See Lazard [10] and Lubin [11].)

If $c_1 \neq 0$, <u>the height of</u> Φ is defined to be the integer h in this expression. If $[p]_\Phi = 0$, Φ is said to have <u>infinite height</u>. We denote by h or $\mathrm{ht}(\Phi)$ the height of Φ.

2. Formal groups of elliptic curves.

Let E be an elliptic curve over a field K defined by the Weierstrass equation (1.1). Let $E(K)$ be the set of all K-rational points on E and of the point at infinity $(0,1,0)$. It is a well known result that $E(K)$ forms an abelian group under the group law :

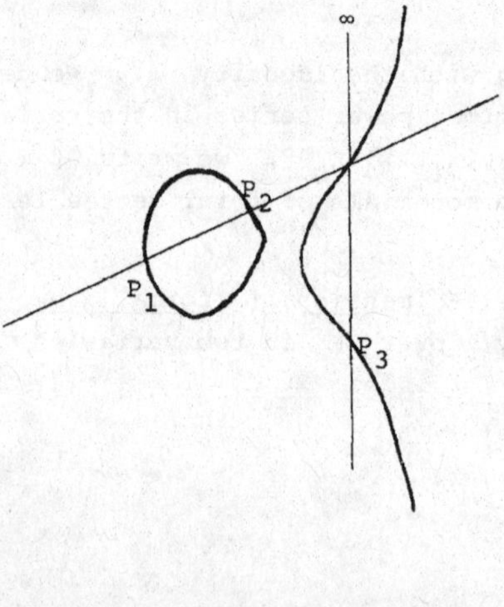

$P_1 + P_2 =: P_3$

with the point at infinity as its identity (zero). As we get formal power series expansions for E and ω_0 in the local parameter u, we can expand the group law of E into a formal power series in u. Let $P_i = (u_i, w_i)$, $i = 1,2,3$ be K-rational points on E such that $P_3 = P_1 + P_2$. Then we have

$$\begin{aligned} u_3 &= \Gamma(u_1,u_2) \\ &= u_1 + u_2 - a_1u_1u_2 - a_2(u_1^2u_2+u_1u_2^2) - 2a_3(u_1^3u_2+u_1u_2^3) \\ &\quad + (a_1a_2-3a_3)u_1^2u_2^2 + \dots \in \mathbb{Z}[a_1,a_2,a_3,a_4,a_6][[u_1,u_2]]. \end{aligned}$$

This Γ is <u>the formal group (law on one parameter) of</u> E. (See Tate [14].)

If the characteristic of K is different from 2 or 3, E can be defined by an equation of the form (1.1') and the formal group Γ of E is explicitly given by the formal power series as

$$\begin{aligned} \Gamma(u_1,u_2) &= u_1 + u_2 - 2au_1u_2 - 4a(u_1^3u_2^2+u_1^2u_2^3) \\ &\quad - 16b(u_1^3u_2^4+u_1^4u_2^3) - 9b(u_1^5u_2^2+u_1^2u_2^5) + \dots \\ &\quad \in \mathbb{Z}[a,b][[u_1,u_2]]. \quad \text{(See Yui [15].)} \end{aligned}$$

Let $m > 1$ be a rational integer. Then the rational map "multiplication by m ": $x \to mx$ (x a generic point) of E into itself is an isogeny of degree m^2 (see Cassels [3] or Lang [9]) and the endomorphism $[m]_\Gamma$:multiplication by m on Γ, is given recursively by

$$[1]_\Gamma(x) = x, \quad [m]_\Gamma(x) = \Gamma(x,[m-1]_\Gamma(x)).$$

<u>(2.1) Proposition.</u> (Honda [7].) <u>Let</u> E <u>be an elliptic curve over a field</u> K <u>and</u> $\omega_0 = \sum_{n=1}^{\infty} a(n)u^{n-1}du$ <u>the canonical invariant differential on</u> E. <u>Put</u> $f(u) = \sum_{n=1}^{\infty} \frac{a(n)}{n}u^n$ <u>and define</u> $\Gamma(u,v)$ <u>by</u> $\Gamma(u,v) = f^{-1}(f(u)+f(v))$. <u>Then</u> Γ <u>is the formal group of</u> E <u>and</u> $[p]_\Gamma(u) = f^{-1}(pf(u))$. (In this case, f(u) is called <u>the logarithm of</u> Γ.) ∎

3. Elliptic curves and formal groups over fields of finite characteristic.

Let K be a field of characteristic $p > 0$ and let $\overline{K}$ be the algebraic closure of K. Let E, u, ω_0 and Γ be as above. We know that $p : E \to E$ is an isogeny of degree p^2 and $[p]_\Gamma : \Gamma \to \Gamma$ has the form

$$(3.1) \qquad [p]_\Gamma(u) = c_1 u^{p^h} + c_2 u^{p^{2h}} + \cdots \quad \text{with } c_1 \neq 0.$$

(3.2) Theorem. *The formal group Γ of E has height $h = 1$ or 2. Moreover, we have the following assertions.*

(1) $h = 1 \Longleftrightarrow E$ *has p points of order p in $\overline{K}$.*

(2) $h = 2 \Longleftrightarrow E$ *has no points of order p in $\overline{K}$.*

Proof. We know that the isogeny $p : E \to E$, which has degree p^2, is not separable (as $p = \mathrm{char.}(K)$). p^h in the expression (3.1) gives the inseparable degree of this isogeny. So p^h must divide p^2 and hence we get $h = 1$ or 2. For the second assertions, let ${}_pE(\overline{K})$ denote the group of rational points on E of order p defined over $\overline{K}$. Then we know that the order of ${}_pE(\overline{K})$ is equal to the separable degree of the isogeny $p : E \to E$. Hence the order of ${}_pE(\overline{K})$ is given by p^{2-h}, from which the assertions (1) and (2) follow immediately. QED

(3.3) Proposition. *Let $\omega_0 = \sum_{n=1}^{\infty} a(n) u^{n-1} du$ be the canonical invariant differential on E. Then we have*

(1) $a(p) \neq 0$ *in* $K \Longleftrightarrow h = 1 \overset{\text{defn}}{\Longleftrightarrow} E$ *is ordinary.*

(2) $a(p) = 0$ *in* $K \Longleftrightarrow h = 2 \overset{\text{defn}}{\Longleftrightarrow} E$ *is supersingular.*

Proof. By Proposition (2.1), we know that $[p]_\Gamma(u) = f^{-1}(pf(u))$ with $f(u) = \sum_{n=1}^{\infty} \frac{a(n)}{n} u^n$. On the other hand, we also have the expression (3.1). So by comparing the coefficients of u^p of both equations, we get

$$a(p) = c_1 \quad \text{in } K.$$

Hence the assertions (1) and (2) follow. QED

Now we consider the differential 1-forms on E. We denote by $K(E) = K(x,y)$ the function field of E over K and by $\Omega^1(K(E))$

the K-vector space of all differential 1-forms on E over k. Then every element ω of $\Omega^1(K(E))$ can be expressed uniquely in the form

(3.4) $$\omega = d\phi + \eta^p x^{p-1}\, dx, \qquad \phi, \eta \in K(E)$$

(once the p-variable x is fixed).

<u>(3.5) Definition.</u> (Cf. Cartier [1].) <u>The Cartier operator</u>

$$\mathcal{C} : \Omega^1(K(E)) \longrightarrow \Omega^1(K(E))$$

is defined for ω of (3.4) by letting

$$\mathcal{C}(\omega) = \eta\, dx.$$

$\mathcal{C}$ is well defined independently of the choice of p-variable x. $\mathcal{C}$ is a p^{-1}-linear operator, that is,

$$\mathcal{C}(\phi^p \omega_1 + \psi^p \omega_2) = \phi\mathcal{C}(\omega_1) + \psi\mathcal{C}(\omega_2)$$

for $\phi, \psi \in K(E)$ and $\omega_1, \omega_2 \in \Omega^1(K(E))$.
For an arbitrary $\phi \in K(E)$, we have

$$\mathcal{C}(\phi^{n-1} d\phi) = \begin{cases} d\phi & \text{if } n = p, \\ 0 & \text{if } (n,p) = 1. \end{cases}$$

$\omega \in \Omega^1(K(E))$ is said to be <u>logarithmic</u> $\Longleftrightarrow \mathcal{C}(\omega) = \omega \Longleftrightarrow \omega = \frac{d\phi}{\phi}$ with some $\phi \in K(E)$. $\omega \in \Omega^1(K(E))$ is said to be <u>exact</u> $\Longleftrightarrow \mathcal{C}(\omega) = 0$ $\Longleftrightarrow \omega = d\phi$ with some $\phi \in K(E)$.

Now we apply the Cartier operator $\mathcal{C}$ to the canonical invariant differential ω_0 on the elliptic curve E of the form (1.1). The result is the following

<u>(3.6) Theorem.</u> <u>The image of</u> ω_0 <u>under the Cartier operator</u> $\mathcal{C}$ <u>is given by</u>

$$\mathcal{C}(\omega_0) = A^{1/p}\, \omega_0$$

<u>where</u> $\mathcal{C}$ <u>is represented by the element</u> $A^{1/p}$ <u>in</u> K <u>and</u> A <u>is explicitly given by the following value</u> :

$$A = \begin{cases} a_1 & \text{if } p = 2, \\ a_1^2 + a_2 & \text{if } p = 3, \\ \displaystyle\sum_{2i+3j=\frac{p-1}{2}} \frac{(\frac{p-1}{2})!}{i!j!(\frac{p-1}{2} - i - j)!} a^i b^j \, 4^{\frac{p-1}{2} - i - j} & \text{if } p \geq 5 \\ \text{where } a = \dfrac{-(a_1^2+4a_2)^2}{12} + 4a_4 + 2a_1a_3 & \\ \text{and } b = \dfrac{(a_1^2+4a_2)^3}{216} - \dfrac{(a_1^2+4a_2)(a_1a_3+2a_4)}{6} + a_3^2 + 4a_6. & \end{cases}$$

Proof. To apply the Cartier operator $\mathcal{C}$ on ω_0, we put ω_0 into the following form .

$$\omega_0 = \frac{dx}{2y+a_1x+a_3} = \frac{1}{(2y+a_1x+a_3)^p}(2y+a_1x+a_3)^{p-1}dx.$$

Then it suffices to compute the coefficient A of x^{p-1} in $(2y+a_1x+a_3)^{p-1}$, because all other terms give exact differentials. We have immediately that $A = a_1$ for $p = 2$, $a_1^2+a_2$ for $p = 3$. For $p \geq 5$, we replace x and y by

$$X = x + \frac{a_1^2+4a_2}{12}, \qquad Y = 2y+a_1x+a_3.$$

Then E can be defined by the equation of the form

$$Y^2 = 4X^3 + aX + b$$

where a and b are as in the statement of the theorem. Then the coefficient A of $x^{p-1}(= X^{p-1})$ in $Y^{p-1} = (4X^3+aX+b)^{\frac{p-1}{2}}$ is given by the Deuring formula as above. (Cf. Deuring [4].) QED

(3.7) Definition. The value A obtained in Theorem (3.6) is called the Hasse invariant of E.

(3.8) Theorem. Let A be the Hasse invariant of E. Put $H = \{ \alpha \in K \mid A\alpha^p = 0 \}$ and $G = \{ \alpha \in K \mid A\alpha^p = \alpha \}$. Then H is a K-vector space and G generates a K-vector space $\langle G \rangle$. Moreover, we have the following assertions:

(1) $\mathfrak{D}_1 \simeq H\,\omega_0 \Longleftrightarrow A = 0 \Longleftrightarrow$ <u>every</u> $\omega \in \mathfrak{D}_1$ <u>is exact</u>.

(2) $\mathfrak{D}_1 \simeq \langle G \rangle\,\omega_0 \Longleftrightarrow A \neq 0 \Longleftrightarrow$ <u>every</u> $\omega \in \mathfrak{D}_1$ <u>is logarithmic</u>.

Proof. The first assertions are clear. To prove the second assertions, we recall that $\mathfrak{D}_1$ is a K-vector space of dimension 1 with the canonical basis ω_0. So every element $\omega \in \mathfrak{D}_1$ can be written as $\omega = \alpha\,\omega_0$ with $\alpha \in K$. Now $H\,\omega_0$ and $\langle G \rangle\,\omega_0$ are K-vector subspaces of $\mathfrak{D}_1$, so it follows that they are either $\{0\}$ or $\mathfrak{D}_1$ itself. Hence we get what we claimed. QED

4. Elliptic curves and formal groups over $\mathfrak{p}$-adic integer rings.

In this section, let us assume that K is a field complete with respect to a rank-one valuation ν (additively written) which is the extension of the p-adic valuation ord_p of $\mathbb{Q}_p$ normalized so that $\nu(p) = 1$. Let R be the ring of intgers in K with maximal ideal $\mathfrak{p}$ and with the residue field k of characteristic $p > 0$.

Let E be an elliptic curve defined over K. Then there exists the Weierstrass minimal model of the form (1.1) for E with $a_i \in R$ for every i and with the discriminant of minimal order. So we can define $E^* =: E \bmod \mathfrak{p}$ by the equation obtained from (1.1) by replacing a_i by $a_i^* =: a_i \bmod \mathfrak{p}$ for every i. If E^* is also an elliptic curve over k, we say that E has <u>good reduction at</u> $\mathfrak{p}$. If E^* no longer defines an elliptic curve over k, E is said to have <u>bad reduction at</u> $\mathfrak{p}$. $\omega_0^* = \sum_{n=1}^{\infty} a(n)^* u^{n-1} du$ where $a(n)^* =: a(n) \bmod \mathfrak{p}$ is the canonical invariant differential on the elliptic curve E^* over k. Let Γ be the formal group of E. We denote by Γ^* the formal group of E^*, which is defined by $\Gamma^* =: \Gamma \bmod \mathfrak{p}$ over k. If E has good reduction at $\mathfrak{p}$, Γ^* has height $h = 1$ or 2 by (3.2). If E has bad reduction at $\mathfrak{p}$, i.e, E^* has a singularity, we have the following possibilities. If the singularity is a cusp, the group law of E^* is given by a usual addition of point coordinates. Hence $\Gamma^*(u,v) = u + v$, the additive (formal) group, and $h = \infty$. If the singularity is an ordinary double point with tangent rational over k (resp. with tangent not defined over k), the group law of E^* is given by multiplication of point coordinates. So $\Gamma^*(u,v) = u + v - uv$ (resp. $u + v + uv$, the multiplicative (formal) group). Hence in both cases, $h = 1$ because

$$[p]_{\Gamma^*}(u) = (1 \mp u)^p - 1 \equiv \mp u^p \pmod{\mathfrak{p}}.$$

(4.1) Proposition. With E, Γ, u and $\omega_0 = \sum_{n=1}^{\infty} a(n)u^{n-1}du$ as above, let

$$[p]_\Gamma(u) = pug_0(u) + \sum_{i=1}^{h-1} b(p^i)u^{p^i} g_i(u) + b(p^h)u^{p^h} g_h(u)$$

where $\nu(b(p^i)) > 0$ for each $1 \le i \le h-1$, $\nu(b(p^h)) = 0$ and $g_0(u)$, $g_i(u)$, $1 \le i \le h-1$ are units in $R[[u]]$ and $g_h(u) \in R[[u]]$, be the endomorphism multiplication by p on Γ. Assume that E has good reduction at $\mathfrak{p}$. Let $\omega_0^* = \sum_{n=1}^{\infty} a(n)^*u^{n-1}du$ be the canonical invariant differential on $E^* = E \bmod \mathfrak{p}$ and let $A^* = :A \bmod \mathfrak{p}$ be the Hasse invariant of E^*, where A is the value given in Theorem (3.6) with $a_i \in R$ for all i. Then we have the congruence:

$$a(p) \equiv b(p) \equiv A \pmod{\mathfrak{p}}.$$

Proof. We know that

$$[p]_\Gamma(u) = f^{-1}(pf(u)) \in R[[u]] \text{ with } f(u) = \sum_{n=1}^{\infty} \frac{a(n)}{n}u^n.$$

So by looking at the coefficients of u^p of this equation in characteristic $p > 0$, i.e, in $k = R/\mathfrak{p}$, we get the congruence

$$a(p) \equiv b(p) \pmod{\mathfrak{p}}.$$

To show the second congruence, we apply the Cartier operator $\mathcal{C}$ to ω_0^*. We get

$$\mathcal{C}(\omega_0^*) = A^{*1/p}\omega_0^* = A^{*1/p}\, du + \cdots .$$

On the other hand, we also have the equation

$$\mathcal{C}(\omega_0^*) = \mathcal{C}\left(\sum_{n=1}^{\infty} a(n)^*u^{n-1}du\right) = \sum_{n=1}^{\infty} a(np)^{*1/p}\, u^{n-1}\, du$$

$$= a(p)^{*1/p}\, du + \cdots .$$

Hence we get the required congrence

$$A \equiv a(p) \pmod{\mathfrak{p}}.$$ QED

Denote by $\bar{K}$ the algebraic closure of K, by $\bar{R}$ the integral closure of R in $\bar{K}$ and by $\bar{\mathfrak{p}}$ the maximal ideal of $\bar{R}$. The unique extension to $\bar{K}$ of the valuation ν will be also denoted by ν.

Let Γ be the formal group of E defined over R. Then $\bar{\wp}$ forms an abelian group $\Gamma(\bar{R})$ under Γ by defining the operation as follows: $\alpha \underset{\Gamma}{*} \beta = \Gamma(\alpha,\beta)$ for $\alpha, \beta \in \bar{\wp}$. The elements of $\Gamma(\bar{R})$ of finite order form a torsion subgroup. In particular, $\mathrm{Ker}\,[p]_\Gamma$ is a p-torsion subgroup of $\Gamma(\bar{R})$ (as one sees easily that

$$[p]_\Gamma(\alpha \underset{\Gamma}{*} \beta) = \Gamma([p]_\Gamma(\alpha),[p]_\Gamma(\beta)) = 0 \quad \text{for any } \alpha, \beta \in \mathrm{Ker}\,[p]_\Gamma).$$

For any positive real number $r \in \mathbb{R}^+$, we define (after Lubin [12]),

$$\Gamma(\bar{R})_r = \{ \alpha \in \Gamma(\bar{R}) \mid \nu(\alpha) \geq r \}.$$

Then $\Gamma(\bar{R})_r$ is a subgroup of $\Gamma(\bar{R})$.

(4.2) Definition. (See Lubin [12].) A subgroup S of $\Gamma(\bar{R})$ is called a *congruence torsion subgroup of* Γ, if there is a positive real number $r \in \mathbb{R}^+$ for which

$$S = \left\{ \alpha \in \Gamma(\bar{R})_r \;;\; \begin{array}{l} \text{there is } n \in \mathbb{N} \text{ such that} \\ \alpha \in \mathrm{Ker}\,[p^n]_\Gamma \end{array} \right\}.$$

A canonical subgroup $\mathrm{can}(\Gamma)$ of Γ is a congruence torsion subgroup of order p in $\mathrm{Ker}\,[p]_\Gamma$.

A natural question one can ask is "When does Γ have a canonical subgroup $\mathrm{can}(\Gamma)$?"

(4.3) Theorem. (Cf. Lubin [12].) *With* E, Γ, Γ^* *and* $[p]_\Gamma(u)$ *as above, we assume that* E *has good reduction at* $\wp$. *Then we have the following assertions.*

(1) *If* $h = 1$, *then a canonical subgroup* $\mathrm{can}(\Gamma)$ *of* Γ *always exists and it is explicitly given by*

$$\mathrm{can}(\Gamma) = \{ 0 \} \cup \{ \alpha \in \Gamma(\bar{R})_{\frac{1}{p-1}} \mid \nu(\alpha) = \tfrac{1}{p-1} \}.$$

(2) *If* $h = 2$, *then a canonical subgroup* $\mathrm{can}(\Gamma)$ *of* Γ *exists if and only if* $\nu(b(p)) < \frac{p}{p+1}$. *When* $\mathrm{can}(\Gamma)$ *exists, it is explicitly given by*

$$\mathrm{can}(\Gamma) = \{\,0\,\} \cup \{\,\alpha \in \Gamma(\bar{R})_{\frac{1-\nu(b(p))}{p-1}} \;\Big|\; \nu(\alpha) = \frac{1-\nu(b(p))}{p-1} \text{ with } \nu(b(p)) < \frac{p}{p+1}\,\}.$$

Proof. First we note that $[p]_\Gamma(u) = 0$ has p^h distinct roots in $\bar{\mathfrak{p}}$. (In fact, by differentiating the equation

$$[p]_\Gamma(\Gamma(u,v)) = \Gamma([p]_\Gamma(u),[p]_\Gamma(v))$$

with respect to v, we get

$$[p]'_\Gamma(\Gamma(u,v))\cdot\Gamma_2(u,v) = \Gamma_2([p]_\Gamma(u),[p]_\Gamma(v))\cdot[p]'_\Gamma(v).$$

Let $\alpha \in \bar{\mathfrak{p}}$ be a root of $[p]_\Gamma(u) = 0$. Put $u = \alpha$ and $v = 0$ in the above equation. Then we get

$$[p]'_\Gamma(\alpha)\cdot\Gamma_2(\alpha,0) = \Gamma_2(0,0)\cdot[p]'_\Gamma(0) = [p]'_\Gamma(0) \neq 0$$

and hence we have $[p]'_\Gamma(\alpha) \neq 0$.)

(1) If $h = 1$, then $\nu(b(p)) = 0$ and $\mathrm{Ker}\,[p]_\Gamma$ has order p. Hence by Definition (4.2), $\mathrm{can}(\Gamma) = \mathrm{Ker}\,[p]_\Gamma$. Now look at the Newton polygon $\mathcal{N}([p]_\Gamma)$ of $[p]_\Gamma(u)$. It has the shape as

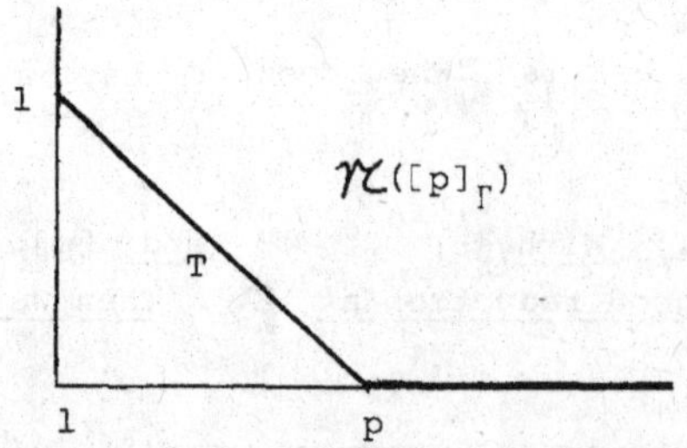

$$T : y = -\frac{1}{p-1}x + \frac{p}{p-1}$$

Hence every element of $\mathrm{can}(\Gamma)$ has order $\nu(\alpha) = \frac{1}{p-1}$ and we can take $r = \frac{1}{p-1}$.

(2) If $h = 2$, then $\mathrm{Ker}\,[p]_\Gamma$ has order p^2 and the Newton polygon $\mathcal{N}([p]_\Gamma)$ of $[p]_\Gamma(u)$ has the shape as below with slope of T in the interval $\left(-\frac{1}{p-1}, -\frac{1}{p^2-1}\right]$.

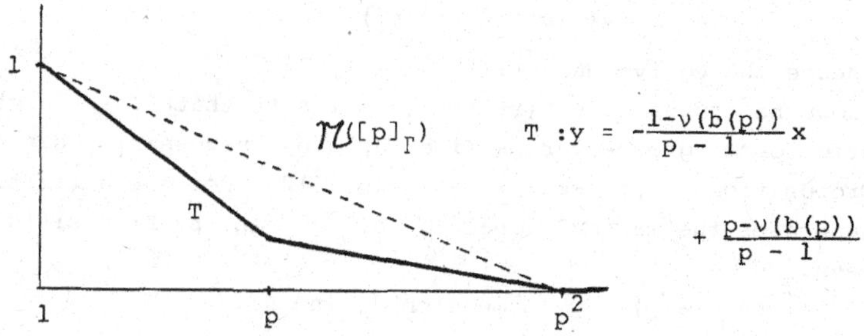

Now suppose that a canonical subgroup $\mathrm{can}(\Gamma)$ of Γ exists. Then $[p]_\Gamma(u)$ must have the factor $g(u)$ of degree p. This means that the Newton polygon $\mathcal{N}([p]_\Gamma)$ must have a vertex at $(p, \nu(b(p)))$. This is possible only if $\nu(b(p)) < \frac{p}{p+1}$. Conversely, if this inequality holds true, then the segment T of the Newton polygon $\mathcal{N}([p]_\Gamma)$ gives rise $p-1$ roots of $[p]_\Gamma(u) = 0$ with order $\frac{1-\nu(b(p))}{p-1}$. These $p-1$ roots together with 0 form a canonical subgroup $\mathrm{can}(\Gamma)$ with $r = \frac{1-\nu(b(p))}{p-1}$. QED

<u>(4.4) Theorem.</u> <u>With</u> $E, \Gamma, u, E^*, A, \omega_0 = \sum_{n=1}^{\infty} a(n)u^{n-1}du$ <u>and</u> $[p]_\Gamma(u) = pug_0(u) + b(p)u^p g_1(u) + \cdots$ <u>as above, suppose that</u> $h = 2$. <u>Then the following conditions are equivalent.</u>

(i) Γ <u>possesses a canonical subgroup</u> $\mathrm{can}(\Gamma)$.

(ii) $0 < \nu(b(p)) < \frac{p}{p+1}$.

(iii) $0 < \nu(a(p)) < \frac{p}{p+1}$.

(iv) $0 < \nu(A) < \frac{p}{p+1}$.

Proof. We have only to show the equivalences (ii) $\Leftrightarrow$ (iii) $\Leftrightarrow$ (iv). Let $f(u) = \sum_{n=1}^{\infty} \frac{a(n)}{n}u^n$ be the logarithm of Γ. An examination of the coefficients of the u^p-term of the equation $[p]_\Gamma(u) = f^{-1}(pf(u))$ gives the following identity:

$$a(p) = b(p)g_1(0) + p\cdot(\text{some element in } R).$$

Compare the ν-order of both sides. Then we obtain (by noting that $g_1(u)$ is a unit in $R[[u]]$),

$$\nu(b(p)) = \nu(a(p))$$

and hence the equivalence (ii) $\Leftrightarrow$ (iii).
To show the equivalence (iii)<=>(iv), note that Γ is a standard generic formal group over R (i.e, $[p]_\Gamma(u)$ necessarily has the form of Proposition (4.1), see also Lubin [12] for the definition). So p generates the maximal ideal $\wp$ of R, and by Proposition (4.1), we have

$$A = a(p) + p\cdot(\text{some element in } R).$$

Hence A and a(p) have the same ν-order, from which the equivalence (iii)<=>(iv) follows immediately. QED

(4.6) Examples. (1) Suppose that the residue characteristic $p = 2$. Let E be dfined by the equation

$$y^2 + xy = x^3 + a_2x + a_6 \quad \text{with} \quad \nu(a_6) = 0.$$

Then E* is ordinary (since $\nu(a_1) = \nu(1) = 0$) and hence Γ always has a canonical subgroup $\mathrm{can}(\Gamma)$.

If E is given by the equation

$$y^2 + a_3y = x^3 + a_4x + a_6 \quad \text{with} \quad \nu(a_3) = 0,$$

then E* is supersingular and Γ does not have a canonical subgroup (because $\nu(a_1) = \nu(0) = \infty$).

If E is given by the equation

$$y^2 + a_1xy + a_3y = x^3 \quad \text{with} \quad \nu(a_3) = 0\ ,$$

then E* is ordinary $\Leftrightarrow$ $\nu(a_1) = 0$, respectively E* is supersingular and Γ has a canonical subgroup $\mathrm{can}(\Gamma)$ $\Leftrightarrow$ $0 < \nu(a_1) < 2/3$.

(2) Suppose that the residue characteristic $p = 3$.

If E is given by the equation

$$y^2 = x^3 + a_2x^2 + a_6 \quad \text{with} \quad \nu(a_6) = 0,$$

then E* is ordinary $\Leftrightarrow$ $\nu(a_2) = 0$, respectively E* is supersingular and Γ has a canonical subgroup $\mathrm{can}(\Gamma)$ $\Leftrightarrow$ $0 < \nu(a_2) < 3/4$.

If E is given by the equation

$$y^2 = x^3 + a_4x + a_6 \quad \text{with} \quad \nu(a_4) = 0,$$

then E* is supersingular and Γ does not possess a canonical subgroup $\mathrm{can}(\Gamma)$ (because $\nu(A) = \nu(0) = \infty$).

(3) Suppose that the residue characteristic $p \geq 5$.
Let E be given by the equation

$$y^2 = x^3 + a_4x + a_6 \quad \text{with} \quad \nu(a_6) = 0.$$

Then some results are summarized in the following table.

p	i	j	E* and Γ
5	1	0	E* ordinary $\Leftrightarrow$ $\nu(a_4) = 0$
			E* supersingular & $\exists$ can(Γ) $\Leftrightarrow 0 < \nu(a_4) < \frac{5}{6}$
7	0	3	E* ordinary
11	1	1	E* ordinary
13	3	0	E* ordinary $\Leftrightarrow \nu(a_4) = 0$
			E* supersingular & $\exists$ can(Γ) $\Leftrightarrow 0 < \nu(a_4) < \frac{13}{42}$
	0	2	E* ordinary
17	4	0	E* ordinary $\Leftrightarrow \nu(a_4) = 0$
			E* supersingular & $\exists$ can(Γ) $\Leftrightarrow 0 < \nu(a_4) < \frac{17}{72}$
	1	2	E* ordinary
19	3	1	E* ordinary
	0	3	E* ordinary

We see tnat the condition $p \equiv 3 \pmod 4$ is sufficient for E* to be ordinary (because there always exists an integer $j > 0$ satisfying $2i + 3j = 2n + 1$, $n \in \mathbb{N}$). If $p \equiv 1 \pmod 4$, then there is the possibility that E* becomes supersingular (since there always exists an integer $i > 0$ satisfying $2i + 3j = 2n$, $n \in \mathbb{N}$).

(4.7) Theorem. (Cf. Lubin [12].) With E and Γ as above, assume that E has good reduction at $\mathfrak{p}$. Let F be the Frobenius morphism of E* and Γ* induced by the p-th power map $x \to x^p$ of k. Suppose that Γ has a canonical subgroup can(Γ), then the Frobenius morphism F can be lifted back to cahracteristic 0, i.e, to R[[u]].

Proof. Put

$$\tilde{F}(x) = \prod_{\alpha \in \operatorname{can}(\Gamma)} (x - \alpha).$$

Then $\tilde{F}(x)$ is a monic polynomial over R of degree p and moreover by the construction,

$$\tilde{F}(x) \equiv x^p \pmod{\wp}.$$

Here we claim that $\tilde{F}(x)$ is indeed the lifting of the Frobenius morphism F to $R[[u]]$. For this, we observe that $\tilde{F}(\Gamma(u,v))$ is the ideal $(\tilde{F}(u),\tilde{F}(v))$ which is the set of all formal power series $\Phi(u,v) \in R[[u,v]]$ satisfying $\Phi(\alpha,\beta) = 0$ for any $\alpha,\ \beta \in \mathrm{can}(\Gamma)$. (For any $\alpha,\ \beta \in \mathrm{can}(\Gamma)$, $\Gamma(\alpha,\beta) = \alpha \underset{\Gamma}{*} \beta \in \mathrm{can}(\Gamma)$. This implies that $\tilde{F}(\Gamma(u,v)) \in (\tilde{F}(u),\tilde{F}(v))$. The other implication is clear.)
We have the following commutative diagram:

$$\begin{array}{ccc}
\Gamma(u,v) & \xrightarrow{\ \tilde{F}\ } & \tilde{F}(\Gamma(u,v)) \\
\Big\downarrow \text{mod } \wp & & \Big\downarrow \text{mod } \wp \\
\Gamma^*(u,v) & \xrightarrow{\ F\ } & \Gamma^{*(p)}(u^p,v^p)
\end{array}$$

where $\Gamma^{*(p)}(u^p,v^p)$ is the formal power series in u^p, v^p with coefficients of the p-th power of those of Γ^*.
Hence $\tilde{F}(x)$ is in fact the lifting back of the Frobenius morphism F to characteristic 0, i.e, to $R[[u]]$. QED

5. Elliptic curves and formal groups over $\mathbb{Q}$.

Let E be an elliptic curve over $\mathbb{Q}$. Then there exists an (essentially unique) global Weierstrass minimal equation for E of the form (1.1) with $a_i \in \mathbb{Z}$ for every i and with the discriminant as small as possible. So $E^* = E \bmod p$ is defined over $\mathbb{F}_p$ for any rational prime p. As before, let $u = -\frac{x}{y}$ be a local parameter of E at the point at infinity of E and let

$$\omega_0 = \frac{dx}{2y+a_1x+a_3} = \sum_{n=1}^{\infty} a(n)u^{n-1}du, \quad a(1) = 1$$

be the canonical invariant differential on E. Then $a(n) \in \mathbb{Z}$ for all n.

The following two theorems (5.1) and (5.2) are very classical and well known, but we include them here for the sake of completeness.

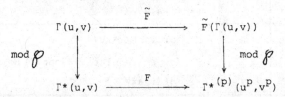

(5.1) Theorem. (Cf. Tate [14].) Let E be an elliptic curve over $\mathbb{Q}$ given by the global Weierstrass minimal equation. For a rational prime p, let $E^* = E \bmod p$. Let N_p denote the number of rational points on E^* defined over $\mathbb{F}_p$. Put $f_p = p+1-N_p$. If E has good reduction at p, then $f_p = \mathrm{Tr}(\pi_{E^*/\mathbb{F}_p})$ (the trace of the Frobenius endomorphism $\pi_{E^*/\mathbb{F}_p}$ of E^* relative to $\mathbb{F}_p$ in its ℓ-adic representation) and it satisfies the Riemann hypothesis $|f_p| \leq 2\sqrt{p}$. If E^* has a node with tangent defined over $\mathbb{F}_p$ (resp. not defined over $\mathbb{F}_p$), then $f_p = +1$ (resp. -1). If E^* has a cusp, then $f_p = 0$. ∎

(5.2) Theorem. (Cf. Olson [13].) Let E be an elliptic curve over $\mathbb{Q}$ defined by the equation of the form (1.1') with $a, b \in \mathbb{Z}$. Let p be a rational prime $\neq 2,3$. Assume that E has good reduction at p. Let $\omega_0 = \sum_{n=1}^{\infty} a(n)u^{n-1}du$, $a(1) = 1$ and $a(n) \in \mathbb{Z}$ for all n be the canonical invariant differential on E given by (1.4'). Then

$$a(p) \equiv -\sum_{x \in CSR(p)} \left(\frac{x^3+ax+b}{p}\right) \pmod{p}$$

where $CSR(p)$ stands for the complete system of residues mod p and $\left(\frac{\cdot}{p}\right)$ denotes the Legendre symbol.

In particular, if E is given by the equation $y^2 = x^3 + ax$ (resp. $y^2 = x^3 + b$), then

$$a(p) \equiv \begin{pmatrix} \frac{p-1}{2} \\ \frac{p-1}{4} \end{pmatrix} a^{\frac{p-1}{4}} \pmod{p} \quad \left(\text{resp. } \begin{pmatrix} \frac{p-1}{2} \\ \frac{p-1}{6} \end{pmatrix} b^{\frac{p-1}{6}} \pmod{p}\right). \ ∎$$

Now we are about to state one of the main theorems of this paper.

(5.3) Theorem. Let E be an elliptic curve over $\mathbb{Q}$ given by the global Weierstrass minimal equation (1.1) with $a_i \in \mathbb{Z}$ for all i and let $\omega_0 = \sum_{n=1}^{\infty} a(n)u^{n-1}du$ be the canonical invariant differential on E given by (1.4). Assume that E has good reduction at a *rational prime* p *and put* $E^* = E$ *mod* p. *Then we have the following assertions.*

(1) $a(p) \equiv \mathrm{Tr}(\pi_{E^*/\mathbb{F}_p}) \pmod{p}$

where $\pi_{E^*/\mathbb{F}_p}$ *is the Frobenius endomorphism of* E^* *relative to* $\mathbb{F}_p$ *and* $\mathrm{Tr}(\pi_{E^*/\mathbb{F}_p})$ *is the trace of* $\pi_{E^*/\mathbb{F}_p}$ *in its* ℓ*-adic representation.*

(2) $a(np) \equiv a(n)a(p) \pmod{p}$ *for* $(n,p) = 1$.

(3) $a(np) - \mathrm{Tr}(\pi_{E^*/\mathbb{F}_p})a(n) + pa(n/p) \equiv 0 \pmod{p^\alpha}$

for $n \equiv 0 \pmod{p^{\alpha-1}}$, $\alpha \geq 1$.

Historical remark on Theorem (5.3). The congruence relation (2) was discovered by Atkin and Swinnerton-Dyer, but their proof was never published. At the same time they conjectured the higher congruence (3). Cartier announced a proof of (3) in [1]. A complete proof of Theorem (5.3) was given by the author in [15].

In the rest of this paper, we shall survey the proofs of the congruences in Theorem (5.3). Our proof is based on deep study of formal groups associated to formal Dirichlet series with Euler products. So first of all, we shall look into the formal groups associated to certain formal Dirichlet series over integral domains of characteristic 0.

Let R be an integral domain of characteristic 0 with the field K of quotients. Let Φ be a formal group defined over R. Then over a $\mathbb{Q}$-algebra K, Φ is isomorphic to the additive (formal) group $G_a(x,y) = x + y$ and there exists a unique formal power series $\phi \in \mathrm{Hom}_K(\Phi, G_a)$, i.e, the logarithm of Φ, such that

$$\Phi(x,y) = \phi^{-1}(\phi(x) + \phi(y)).$$

We can write

$$\phi(x) = \sum_{n=1}^{\infty} \frac{\alpha(n)}{n} x^n, \qquad \alpha(n) \in R \quad \text{for all } n.$$

Associated to the logarithm ϕ of Φ, we put

$$\omega = d\,\phi(x) = \phi'(x)dx = \sum_{n=1}^{\infty} \alpha(n) x^{n-1} dx.$$

This ω is called the invariant differential of Φ. We now define a formal Dirichlet series $D_\Phi(s)$ associated to Φ by letting

$$D_\Phi(s) = \sum_{n=1}^{\infty} \alpha(n)n^{-s}.$$

Then Honda [7] has shown that the following assertion holds ture.

(5.4) Proposition. Over a $\mathbb{Q}$-algebra K, we have the following correspondences :

$$\Phi \longleftrightarrow \phi \longleftrightarrow \omega \longleftrightarrow D_\Phi(s)$$

that is, if one of Φ, ϕ, ω and $D_\Phi(s)$ is given, then the rest are uniquely determined from the given one. ∎

Proposition (5.4) asserts, among other things, that a formal Dirichlet series over R yields a formal group over K. However, in most of the cases, formal groups obtained by this construction are not defined over R. So a basic number theoretic problem left to be settled is formulated as follows : What kinds of formal Dirichlet series over R give rise to formal groups defined over R?

For the sake of applications and simplicity, we confine our discussion to formal Dirichlet series with Euler products defined over $\mathbb{Z}$ or rather $\mathbb{Z}_p$ for all rational prime p. Progress in this direction has been made by Honda [6] and Hill [7].

(5.5) Theorem. Let

$$D(s) = (1-b_1p^{-s}- \cdots - b_ip^{-is} - \cdots - b_np^{-ns})^{-1} \sum_{m=1}^{\infty} u_m m^{-s}$$

be a formal Dirichlet series with b_i, $u_m \in \mathbb{Z}_p$, $u_1 = 1$ and $u_m \in m\mathbb{Z}_p$ for every m. If $\mathrm{ord}_p(b_n) = n-1$ and $\mathrm{ord}_p(b_i) \geq i-1$ for $1 \leq i \leq n-1$, then the formal group Φ_D associated to $D(s)$ by the construction of Proposition (5.4) is defined over $\mathbb{Z}_p$. Moreover the characteristic polynomial of $\Phi_D^* =: \Phi_D \bmod p$, which is Eisenstein over $\mathbb{Z}_p$ of degree h = height of Φ_D^*, divides the polynomial

$$S(x) = -\frac{p^n}{b_n} g(x/p) \quad \text{where} \quad g(x) = 1 - b_1x - \cdots - b_nx^n. \quad ∎$$

(Theorem (5.5) needs some comments. Let $\bar{k}$ be an algebraically closed field of characteristic $p > 0$ and let Φ be a formal group over $\bar{k}$ of height $h < \infty$. Then the structure theorem of Dieudonné and Lubin asserts that $\mathrm{End}_{\bar{k}}(\Phi)$ is the maximal order in the central division algebra of rank h^2 over $\mathbb{Q}_p$ with invariant $1/h$ (see Honda [7], Theorem 1). Thus it follows that if k is an arbitrary field of characteritic $p > 0$, $\mathrm{End}_k(\Phi)$ is an order in a division algebra of rank dividing h^2 over $\mathbb{Q}_p$ (Lubin [11]) and, moreover, any $\lambda \in \mathrm{End}_k(\Phi)$ sits in a commutative field extension of $\mathbb{Q}_p$ of degree h and it is integral over $\mathbb{Z}_p$ and hence it satisfies a monic irreducible polynomial of degree dividing h over $\mathbb{Z}_p$: <u>the characteristic polynomial of</u> λ. If Φ is defined over a finite field with q elements, we mean by <u>the characteristic polynomial of</u> Φ, that of the Frobenius endomorphism x^q of Φ. If $k = \mathbb{F}_p$, the characteristic polynomial of Φ is an Eisenstein polynomial of degree h over $\mathbb{Z}_p$ (Cartier [1], Hill [6]).)

<u>(5.6) Observation.</u> <u>With</u> $D(s)$ <u>and</u> Φ_D <u>as in Theorem (5.5), put</u>

$$D(s) = \sum_{n=1}^{\infty} \alpha(n) n^{-s} \quad \underline{\text{with}} \quad \alpha(n) \in \mathbb{Z}_p \quad \underline{\text{for all}} \quad n.$$

<u>Then the coefficients</u> $\alpha(n)$ <u>satisfy the following equivalent conditions.</u>

(i) $\alpha(m) \equiv b_1\alpha(m/p) + b_2\alpha(m/p^2) + \cdots + b_n\alpha(m/p^n) \pmod{m}$

<u>where</u> $\alpha(r) = 0$ <u>if</u> $r \notin \mathbb{Z}$.

(ii) <u>Let</u> $\phi(x) = \sum_{n=1}^{\infty} \frac{\alpha(n)}{n} x^n$ <u>be the logarithm of</u> Φ_D.

<u>Put</u> $\Theta(x) = \phi(x) - \sum_{i=1}^{n} \frac{b_i}{p^i} \phi(x^{p^i})$. <u>Then</u> $\Theta(x) \in \mathbb{Z}_p[[x]]$.

Proof. (i) follows immediately from the definition of the coefficients $\alpha(n)$ and from the hypothesis on $D(s)$. To show the equivalence (i) $\Longleftrightarrow$ (ii), look at the coefficients of the term $x^{p^\ell m}$ with $(p,m) = 1$ of $\Theta(x)$. It is given by

$$\frac{a(p^\ell m)}{p^\ell m} - \sum_{i=1}^{n} \frac{b_i}{p^i} \frac{a(p^{\ell-i} m)}{p^{\ell-i} m} = \frac{1}{p^\ell m} \left\{ a(p^\ell m) - \sum_{i=1}^{n} b_i a(p^{\ell-i} m) \right\}.$$

Hence this is in $\mathbb{Z}_p$ if and only if the condition (i) holds true.

QED

(5.7) Lemma. Let $D_1(\mathbb{Z}_p)$ be the set of all formal power series over $\mathbb{Z}_p$ of the form $\sum_{m=1}^{\infty} u_m m^{-s}$ with $u_1 = 1$ and $u_m \in m\mathbb{Z}_p$ for all m. (We call such a formal Dirichlet series a "1-Dirichlet series".) Then $D_1(\mathbb{Z}_p)$ forms a ring.

Proof. It is easy to see that formal Dirichlet series over any ring R form a ring $D(R)$ and $R \to D(R)$ is a functor. In particular, if R is a field of characteristic 0, then for any $r \in \mathbb{Z}$, the formal substitution $s \to s+r$ defines an automorphism of $D(R)$ as follows.

$$\sum_{m=1}^{\infty} u_m m^{-s} \longrightarrow \sum_{m=1}^{\infty} u_m m^{-s-r} = \sum_{m=1}^{\infty} (u_m/m^r) m^{-s}.$$

We see that

$$D_1(\mathbb{Z}_p) = \{ \sum_{m=1}^{\infty} u_m m^{-s} \mid u_1 = 1,\ u_m \in m\mathbb{Z}_p \}$$

$$= \{ D(s) \in D(\mathbb{Z}_p) \mid D(s+1) \in D(\mathbb{Z}_p) \}$$

is the inverse image of $D(\mathbb{Z}_p)$ under the automorphism of $D(\mathbb{Q}_p)$ induced by the formal substitution $s \to s+1$. Hence it is a ring.

QED

(5.8) Theorem. Let $\Phi = \Phi_D$ be the formal group over $\mathbb{Z}_p$ constructed in Theorem (5.5) and let $\Phi^* =: \Phi \bmod p$ be the formal group defined over $\mathbb{F}_p$. Let $P(x) \in \mathbb{Z}_p[x]$ be the characteristic polynomial of Φ^*. Then the formal Dirichlet series $D_\Phi(s)$ ($= D(s)$ in Theorem (5.5)) associated to Φ has the "canonical factorization"

$$D_\Phi(s) = P(0)P(p^{1-s})^{-1}U(s) \quad \text{with} \quad U(s) \in D_1(\mathbb{Z}_p).$$

Proof. Let $S(x)$ be as in Theorem (5.5), i.e,

$$S(x) = -\frac{p^n}{b_n} g(x/p)$$

$$= x^n + \left(\frac{b_{n-1}}{b_n} p\right)x^{n-1} + \cdots + \left(\frac{b_i}{b_n}p^{n-i}\right)x^i + \cdots - \left(\frac{1}{b_n}p^n\right)$$

$$= \sum_{i=0}^{n} c_i x^i.$$

Then

$$\mathrm{ord}_p(c_0) = \mathrm{ord}_p(-\frac{p^n}{b_n}) = 1,$$

$$\mathrm{ord}_p(c_i) = \mathrm{ord}_p(\frac{p^n}{b_n}\cdot\frac{b_i}{p^i}) \geq n-(n-1)+i-1-i \geq 0.$$

As the characteristic polynomial $P(x)$ of Φ_D^* is an Eisenstein factor of $S(x)$ of degree h = height of Φ_D^* (so $1 \leq h \leq n$), we can factor $S(x)$ into the product:

$$S(x) = P(x)\,Q(x).$$

Now we have

$$g(x) = -\frac{b_n}{p^n} P(px)\,Q(px) = \overline{P}(px)\overline{Q}(px)$$

where

$$\overline{P}(x) = : P(x)/P(0), \qquad \overline{Q}(x) = : (-\frac{b_n}{p^n})\,P(0)Q(x).$$

So we can write

$$D(s) = g(p^{-s})^{-1} \sum_{m=1}^{\infty} u_m m^{-s} = \overline{P}(p^{1-s})^{-1}\,\overline{Q}(p^{1-s})^{-1} \sum_{m=1}^{\infty} u_m m^{-s}$$

$$= P(0)P(p^{1-s})^{-1}\,\overline{Q}(p^{1-s})^{-1} \sum_{m=1}^{\infty} u_m m^{-s}.$$

It remains to show that $\overline{Q}(p^{1-s})^{-1} \sum_{m=1}^{\infty} u_m m^{-s} \in D_1(\mathbb{Z}_p)$. Note that $\overline{Q}(x)$ has the constant term 1. So it follows immediately that $\overline{Q}(x)^{-1} \in \mathbb{Z}_p[[x]]$ and $\overline{Q}(p^{-s})^{-1} \in D(\mathbb{Z}_p)$. Hence $\overline{Q}(p^{1-s})^{-1} \in D_1(\mathbb{Z}_p)$. Thus we complete the proof by Lemma (5.7). QED

Hill [6] has shown that formal groups over $\mathbb{Z}_p$ are isomorphic over $\mathbb{Z}_p$ if and only if their reductions over $\mathbb{F}_p$ have the same characteristic polynomial. Here we shall investigate what happens to the formal Dirichlet series of formal groups over $\mathbb{Z}_p$, furnished with the canonical factorization, under $\mathbb{Z}_p$-isomorphisms of formal groups.

(5.9) Theorem (The isotypic theorem of formal Dirichlet series).

With Φ, $D_\Phi(s)$ and $P(x)$ as in Theorem (5.8), let Ψ be a formal group over $\mathbb{Z}_p$ isomorphic over $\mathbb{Z}_p$ to Φ. Then the formal Dirichlet series $D_\Psi(s)$ associated to Ψ has the same type of factorization as $D_\Phi(s)$, that is,

$$D_\Psi(s) = P(0)P(p^{1-s})^{-1}\,\tilde{U}(s) \quad \text{with} \quad \tilde{U}(s) \in D_1(\mathbb{Z}_p).$$

Proof. Put $\Phi^* = \Phi \bmod p$ and $\Psi^* = \Psi \bmod p$. Let $P(x) = \sum_{i=0}^{h} c_i x^i$, $c_h = 1$, $c_i \equiv 0 \pmod p$ for $i \le i \le h-1$, $c_0 \not\equiv 0 \pmod{p^2}$ be the characteristic polynomial of Φ^* and Ψ^*. Then by the hypothesis, $D_\Phi(s)$ has the form:

$$D_\Phi(s) = \left[1+(\frac{c_1}{c_0}p)p^{-s} + \cdots + (\frac{c_i}{c_0}p^i)p^{-is} + \cdots + (\frac{1}{c_0}p^h)p^{-hs}\right]^{-1}U(s)$$

$$= \sum_{n=1}^{\infty} \alpha(n)n^{-s}, \quad \text{with} \quad U(s) \in D_1(\mathbb{Z}_p).$$

Let $\phi(x) = \sum_{n=1}^{\infty} \frac{\alpha(n)}{n} x^n$ be the logarithm of Φ. Then by applying the same argument as Observation (5.6), we get

$$\phi(x) + \sum_{i=1}^{h-1} (\frac{c_i}{c_0})\phi(x^{p^i}) + (\frac{1}{c_0})\phi(x^{p^h}) \in \mathbb{Z}_p[[x]] .$$

Now let $\lambda(x) \in \mathbb{Z}_p[[x]]$ be a $\mathbb{Z}_p$-isomorphism of Φ to Ψ. Then it is easy to see that $\lambda(\phi(x)) =: \psi(x)$ is the logarithm of Ψ, i.e, $\Psi(x,y) = \psi^{-1}(\psi(x)+\psi(y))$. Here we claim that the following relation (*) holds true.

$$(*) \qquad \psi(x) + \sum_{i=1}^{h-1} (\frac{c_i}{c_0})\psi(x^{p^i}) + (\frac{1}{c_0})\psi(x^{p^h}) \in \mathbb{Z}_p[[x]].$$

For this, first of all, note that we have the congruence

$$\Psi\sum_{i=0}^{h} [c_i]_\Psi(x^{p^i}) \equiv 0 \pmod{p\,\mathbb{Z}_p[[x]]}$$

where the sum is taken with respect to Ψ. (This is because $\sum_{i=0}^{h} c_i x^i$ is the characteristic polynomial of Ψ^*.) Then we get

$$\psi\Big(\Psi\big(\sum_{i=0}^{h} [c_i]_\Psi (x^{p^i}) \big)\Big) \equiv 0 \pmod{p\,\mathbb{Z}_p[[x]]}.$$

But by the fact that ψ is the logarithm of Ψ, this congruence is read as

$$\sum_{i=0}^{h} \psi([c_i]_\Psi (x^{p^i})) \equiv 0 \pmod{p\,\mathbb{Z}_p[[x]]}$$

where the sum is the ordinary one.
Thus we obtain, by noting that $[c_i]_\Psi \in \mathrm{End}_{\mathbb{Z}_p}(\Psi)$,

$$\sum_{i=0}^{h} c_i \psi(x^{p^i}) \equiv 0 \pmod{p\,\mathbb{Z}_p[[x]]}.$$

As c_0 is a p-adic prime, by dividing the above congruence by c_0, we get the required relation (*).

Now put $\psi(x) = \sum_{n=1}^{\infty} \frac{\beta(n)}{n} x^n$ and $D_\Psi(s) = \sum_{n=1}^{\infty} \beta(n) n^{-s}$. Then an examination of the coefficients of x^n-term in the relation (*) gives

$$\beta(n) + \sum_{i=1}^{h-1} \left(\frac{c_i}{c_0} p^i\right) \beta(n/p^i) + \left(\frac{1}{c_0} p^h\right) \beta(n/p^h) \in n\,\mathbb{Z}_p.$$

Therefore $D_\Psi(s)$ has the factorization as

$$D_\Psi(s) = P(0)P(p^{1-s})^{-1}\, \tilde{U}(s) \quad \text{with} \quad \tilde{U}(s) \in D_1(\mathbb{Z}_p).$$

QED

Now we shall discuss formal groups of L-series of elliptic curves (cf. Tate [14]). Let E be an elliptic curve defined over $\mathbb{F}_p$. Then the L-series $L(E:s)$ of E is defined by

$$L(E:s) = (1 - \mathrm{Tr}(\pi_{E/\mathbb{F}_p})p^{-s} + p^{1-2s})^{-1}.$$

One can define the L-series even if E is not an elliptic curve, but a curve of the form (1.1) with a singularity. If E has an ordinary double point with tangent rational (resp. not rational) over F_p, then

$$L(E:s) = (1-p^{-s})^{-1} \quad (\text{resp. } (1+p^{-s})^{-1}).$$

If E has a cusp, then

$$L(E:s) = 1.$$

If E is an elliptic curve over $\mathbb{Q}$ defined by the global Weierstrass minimal equation, we define the local L-series $L_p(E:s)$ of E at each rational prime p by $L(E^*:s)$ with $E^* = E \bmod p$. The global L-series $L(E:s)$ of E is then given by putting together all local $L_p(E:s)$.

$$L(E:s) = 1\cdot \prod_{p|\Delta} (1 \pm p^{-s})^{-1} \prod_{p\nmid\Delta} (1-\mathrm{Tr}(\pi_{E^*/\mathbb{F}_p})p^{-s}+p^{1-2s})^{-1}$$

where Δ = discriminant of E.
Observe that when E has good reduction at p, $L_p(E:s)$ satisfies the conditions of Theorem (5.5). Moreover, we have

<u>(5.10) Theorem.</u> (Honda [7, 8].) <u>Let</u> E <u>be an elliptic curve over</u> $\mathbb{Q}$ <u>given by the global Weierstrass minimal equation (1.1). For each rational prime</u> p, <u>let</u> $L_p(E:s)$ <u>be the local</u> L-<u>series of</u> E <u>at</u> p. <u>Then the formal group</u> Φ_p <u>associated to</u> $L_p(E:s)$ <u>is defined over</u> $\mathbb{Z}_p$ <u>and moreover, it is</u> $\mathbb{Z}_p$<u>-isomorphic to the formal group (law)</u> Γ <u>of</u> E. <u>If</u> p <u>is a ratioanl prime such that</u> E^* <u>has a node with tangent rational over</u> $\mathbb{F}_p$ <u>(resp. tangent not defined over</u> $\mathbb{F}_p$), <u>then</u> Φ_p <u>is isomorphic over</u> $\mathbb{Z}_p$ <u>to the formal group</u> $x + y - xy$ (<u>resp.</u> $x + y + xy$ <u>the multiplicative (formal) group). If</u> p <u>is a rational prime such that</u> E^* <u>has a cusp, then</u> Φ_p <u>is isomorphic over</u> $\mathbb{Z}_p$ <u>to the additive (formal) group</u> $x + y$. ∎

<u>(5.11) Proof of Theorem (5.3).</u>

<u>Step</u> 1. Let p be a rational prime at which E has good reduction. Let Φ_p be the formal group over $\mathbb{Z}_p$ associated to the local L-series

$$L_p(E:s) = (1 - \mathrm{Tr}(\pi_{E^*/\mathbb{F}_p})p^{-s} + p^{1-2s})^{-1}.$$

Put $\Phi_p^* = \Phi_p \bmod p$. Then by Theorem (5.5), the characteristic polynomial $P_{\Phi_p}(x)$ of Φ_p^* divides the polynomial $x^2 - \mathrm{Tr}(\pi_{E^*/\mathbb{F}_p})x + p$. Suppose that $h = \mathrm{ht}(\Phi_p^*) = 1$, then $P_{\Phi_p}(x)$ must be an Eisenstein factor of the polynomial $x^2 - \mathrm{Tr}(\pi_{E^*/\mathbb{F}_p})x + p$. The existence of such a factor is assured by the Riemann hypothesis. We put $P_{\Phi_p}(x) = x + \xi$ with a p-adic prime ξ. Then

$$x^2 - \mathrm{Tr}(\pi_{E/\mathbb{F}_p})x + p = P_{\Phi_p}(x)(x + p/\xi) \quad \text{in } \mathbb{Z}_p[x].$$

Suppose now that $h = ht(\Phi_p^*) = 2$. Then if $p \geq 5$, $P_{\Phi_p}(x) = x^2 + p$. (Because $P_{\Phi_p}(x)$ is Eisenstein over $\mathbb{Z}_p$ of degree 2 and it must divide the polynomial $x^2 - Tr(\pi_{E^*/\mathbb{F}_p})x + p$. So $Tr(\pi_{E^*/\mathbb{F}_p}) \equiv 0 \pmod{p} \Longleftrightarrow Tr(\pi_{E^*/\mathbb{F}_p}) = 0$ or p. But by the Riemann hypothesis, $Tr(\pi_{E^*/\mathbb{F}_p}) = 0$ only can occur.) If $p = 2$ or 3, we have $P_{\Phi_p}(x) = x^2 \pm px + p$ or $x^2 + p$ as we can have $Tr(\pi_{E^*/\mathbb{F}_p}) = \pm p$ or 0.

Step 2. Apply Theorem (5.8) to $L_p(E:s)$. We can put $L_p(E:s)$ into the canonical form. If $h = ht(\Phi_p^*) = 1$,

$$L_p(E:s) = \left(1 + \frac{p}{\xi}p^{-s}\right)^{-1} \sum_{m=1}^{\infty} u_m m^{-s} \quad \text{with} \quad u_m = \begin{cases} (-\xi)^{\nu} & \text{if } m = p^{\nu}, \\ 0 & \text{otherwise.} \end{cases}$$

If $h = ht(\Phi_p^*) = 2$,

$$L_p(E:s) = (1 + p^{1-2s})^{-1} \quad \text{or} \quad (1 \pm p \cdot p^{-s} + p^{1-2s})^{-1}.$$

Step 3. Now we consider the formal group (law) Γ of E. The formal Dirichlet series $D_\Gamma(s)$ associated to Γ is given by

$$D_\Gamma(s) = \sum_{n=1}^{\infty} a(n) n^{-s}$$

where $a(n)$ are the coefficients of the canonical invariant differential ω_0 on E given by (1.4). Since Γ is isomorphic over $\mathbb{Z}_p$ to Φ_p, we can apply Theorem (5.9) to $D_\Gamma(s)$ and we get the factorization :

$$D_\Gamma(s) = \begin{cases} \left(1 + \frac{p}{\xi}p^{-s}\right)^{-1} \cdot U_\Gamma(s) & \text{with } U_\Gamma(s) \in D_1(\mathbb{Z}_p) \text{ if } h = 1, \\ (1 + p^{1-2s})^{-1}\, U_\Gamma(s) & \text{with } U_\Gamma(s) \in D_1(\mathbb{Z}_p) \\ \quad \text{or} & \\ (1 \pm p \cdot p^{-s} + p^{1-2s})^{-1}\, U_\Gamma(s) & \text{with } U_\Gamma(s) \in D_1(\mathbb{Z}_p) \text{ if } h = 2. \end{cases}$$

Step 4. If $h = 1$, the canonical factorization of $D_\Gamma(s)$ gives the following congruences :

$$a(np) + \frac{p}{\xi} a(n) \equiv 0 \pmod{np\,\mathbb{Z}_p},$$

$$a(n) + \frac{p}{\xi} a(n/p) \equiv 0 \pmod{n\,\mathbb{Z}_p}.$$

Multiplying the p-adic prime ξ to the second congruence and adding it to the first one, we get

$$a(np) + (\xi + \frac{p}{\xi}) a(n) + pa(n/p) \equiv 0 \pmod{np\,\mathbb{Z}_p}.$$

Since we have the equality

$$\xi + \frac{p}{\xi} = Tr(\pi_{E^*/\mathbb{F}_p}),$$

we finally obtain the congruence (*) :

$$(*) \qquad a(np) - Tr(\pi_{E^*/\mathbb{F}_p}) a(n) + pa(n/p) \equiv 0 \pmod{np\,\mathbb{Z}_p}.$$

If $h = 2$, we have immediately from the canonical factorization of $D_\Gamma(s)$, the congruence (*) :

$$a(np) + pa(n/p) \equiv 0 \pmod{np\,\mathbb{Z}_p},$$

(*) or

$$a(np) \pm pa(n/p) + pa(n/p) \equiv 0 \pmod{np\,\mathbb{Z}_p}.$$

In particular, the Atkin and Swinnerton-Dyer congruence (3) follows immediately from (*) if we take $n \equiv 0 \pmod{p^{\alpha-1}}$, $\alpha \geq 1$.

Step 5. Now let $\lambda(x) = \sum_{i=1}^{\infty} r_i x^i \in \mathbb{Z}_p[[x]]$, $r_1 = 1$ be the isomorphism of Φ_p to Γ over $\mathbb{Z}_p$. The logarithm $\phi_p(x)$ of Φ_p is the formal power series (refer the construction)

$$\phi_p(x) = x - \frac{\xi + p/\xi}{p} x^p + \cdots \in \mathbb{Q}_p[[x]],$$

and the logarithm of Γ is given by

$$(**) \qquad \sum_{n=1}^{\infty} \frac{a(n)}{n} x^n = \lambda(\phi_p(x)).$$

So by comparing the coefficients of x^p-term of (**) modulo p, we get the congruence (1) :

$$a(p) = -(\xi + p/\xi) + pr_p \equiv Tr(\pi_{E^*/\mathbb{F}_p}) \pmod{p}.$$

Step 6. Take n so that $(n,p) = 1$. Then the congruence $(*)$ is read

$$a(np) \equiv Tr(\pi_{E^*/\mathbb{F}_p})a(n) \pmod{p\mathbb{Z}_p} \quad \text{if } h = 1,$$

$$a(np) \equiv 0 \pmod{p\mathbb{Z}_p} \quad \text{if } h = 2.$$

This congruence, together with the congruence (1) then gives the congruence (2) :

$$a(np) \equiv a(n)a(p) \pmod{p} \quad \text{for } (n,p) = 1.$$

This concludes the proof of Theorem (5.3).

6. Appendix : Elliptic curves and formal groups over algebraic number fields.

After I had finished writing this paper, Professor I. Barsotti of University of Padova, Italy kindly communicated to me that the Atkin and Swinnerton-Dyer congruence (3) of Theorem (5.3) holds true for slightly more general rings than $\mathbb{Z}$.

(6.1) Theorem. Let R be a Dedekind ring of characteristic 0 in which a rational prime p decomposes as $p = \mathfrak{p}\mathfrak{p}'$, $(\mathfrak{p}, \mathfrak{p}') = 1$. Let $k = R/\mathfrak{p}$ denote the finite field with $q = p^e$ elements. Let E be an elliptic curve over R defined by the equation (1.1) with $a_i \in R$ for every i, and with good reduction at $\mathfrak{p}$. Put $E^* = E \bmod \mathfrak{p}$. Let $u = -\frac{x}{y}$ be a local parameter of E at the point at infinity $(0,1,0)$ and let

$$\omega_0 = \sum_{n=1}^{\infty} a(n)u^{n-1}du, \quad a(1) = 1 \text{ and } a(n) \in R \text{ for all } n$$

be the canonical invariant differential on E given by (1.4). Then the coefficients $a(n)$ satisfy the generalized Atkin and Swinnerton-Dyer congruence :

$$a(nq) - Tr(\pi_{E^*/k})a(n) + qa(n/q) \equiv 0 \pmod{p^r\mathfrak{p}}$$

for $n \equiv 0 \pmod{p^r}$ with $r \geq e$.
If $r < e$, the above congruence holds true with $a(n/q) = 0$.
(Here $Tr(\pi_{E^*/k})$ denotes the trace of the Frobenius endomorphism $\pi_{E^*/k}$ of E^* relative to k in its ℓ-adic representation.)

Proof. Let F be the Frobenius morphism and $V = p/F$ the Verschiebung morphism of E^*. Then $\pi_{E^*/k} = F^e$ and $\pi'_{E^*/k} = :V^e$ are endomorphisms of E^* with $\pi_{E^*/k}\pi'_{E^*/k} = p^e = q$ and they are conjugates of each other over $\mathbb{Q}$. So we have

$$\mathrm{Tr}(\pi_{E^*/k}) = \pi_{E^*/k} + \pi'_{E^*/k}.$$

Now let $f(u) = \sum_{n=1}^{\infty} \frac{a(n)}{n} u^n$ be the logarithm of the formal group Γ of E, i.e, $f'(u)du = \omega_0$. Then we have the identity (A*) :

$$\text{(A*)} \qquad \mathrm{Tr}(\pi_{E^*/k}) f(u) = \pi_{E^*/k} f(u) + \pi'_{E^*/k} f(u).$$

By examining the actions of the Frobenius and Verschiebung morphisms on $f(u)$ modulo $\mathfrak{p}$, we find

$$F f(u) \equiv \sum_{p|n} \frac{pa(n/p)^{(p)}}{n} u^n \pmod{\mathfrak{p}R[[u]]},$$

$$V f(u) \equiv \sum_{n=1}^{\infty} \frac{a(np)^{(1/p)}}{n} u^n \pmod{\mathfrak{p}R[[u]]}.$$

Hence we obtain

$$\text{(A**)} \qquad \pi_{E^*/k} f(u) \equiv \sum_{q|n} \frac{qa(n/q)}{n} u^n \pmod{\mathfrak{p}R[[u]]},$$

$$\text{(A***)} \qquad \pi'_{E^*/k} f(u) \equiv \sum_{n=1}^{\infty} \frac{a(nq)}{n} u^n \pmod{\mathfrak{p}R[[u]]}.$$

Thus, if $n \equiv 0 \pmod{p^r}$ with $r \geq e$, we obtain by putting together the above relations (A*), (A**) and (A***), the following :

$$a(nq) + qa(n/q) \equiv \mathrm{Tr}(\pi_{E^*/k})a(n) \pmod{p^r\mathfrak{p}}.$$

If $n \equiv 0 \pmod{p^r}$ with $r < e$, this congruence is read with $a(n/q) = 0$ and hence we get

$$a(nq) \equiv \mathrm{Tr}(\pi_{E^*/k})a(n) \pmod{p^r\mathfrak{p}}.$$

QED

Acknowledgement. I would like to express my heartfelt thanks to Professor I. Barsotti for his kind advice and to Professor K. Lønsted for encouragement.

References.

[1] Cartier,P., Une nouvelle opération sur les formes différentielles, C.R.Acad. Sci. Paris 244 (1957) 429-428.

[2] Cartier,P., Groupes formels, fonctions automorphes et fonctions zeta des courbes elliptiques, Actes Congrès intern. Math. Nice (1970) T2. 291-299.

[3] Cassels, J.W.S., Diophantine equations with special reference to elliptic curves, survey article, J. London Math. Soc. 41 (1962), 193-291.

[4] Deuring,M., Die Typen der Multiplikatorenringe elliptischer Funktionenkörper, Abh. Math. Sem. Hamburg 41 (1941), 197-272.

[5] Fröhlich,A., Formal Groups, Springer Lecture Notes in Mathematics No. 74 (1968).

[6] Hill,W., Formal groups and zeta functions of elliptic curves, Inventiones Math. 12 (1971), 337-345.

[7] Honda,T., Formal groups and zeta fucntions, Osaka J. Math. 5 (1968), 199-213.

[8] Honda,T., On the theory of commutative formal groups, J. Math. Soc. Japan 22 (1970), 213-246.

[9] Lang,S., Elliptic Fucntions, Addison-Wesley (1973), New York.

[10] Lazard,M., Sur les groupes de Lie formels à un paramètre, Bull. Soc. Math. France (1955), 251-274.

[11] Lubin,J., One-parameter formal Lie groups over $\mathfrak{p}$-adic integer rings, Ann. of Math. 80 (1964), 464-484.

[12] Lubin,J., Canonical subgroups of formal groups, Københavns Universitets Matematiske Institut Preprint Series No. 8 (1975).

[13] Olson,L., Hasse invariant and anomalous primes for elliptic curves with complex multiplication, J. Number Theory 8, No. 4 (1976), 397-414.

[14] Tate,J., The arithmetic of elliptic curves, Inventiones Math. 23 (1974), 179-206.

[15] Yui,N., Formal groups and p-adic properties of elliptic curves, (1974) Preprint.

[16] Yui,N., Elliptic curves and canonical subgroups of formal groups, (1977), to appear in J.reine u. angewandte Math. (Crelles Jour.).

Noriko Yui
Matematisk Institut
Københavns Universitet
Universitetsparken 5
2100 København Ø
Danmark

(Current address :
Department of Mathematics
University of Ottawa
Ottawa, Ontario
Canada, K1N 6N5)